FACTORS

To convert		to	multiply by
6. FORCE (continued)	newton (N)	pound	0.2248
	newton (N)	kg force	0.102
	kilo newton (kN)	kip k	0.225
7. FORCE/ LENGTH	pound/ft	kg force/m	1.488
	kip/ft	kg force/m	1488
	kip/ft	kN/m	14.59
	kg force/meter	lb/ft	0.672
	kN/meter	lb/ft	68.52
	kN/meter	kip/ft	0.06852
8. FORCE/AREA (STRESS)	pound/in.^{2} (psi)	kg force/cm^2	0.07031
	pound/in.^{2} (psi)	N/cm^2	0.6895
	pound/in.^{2} (psi)	N/mm^2 (MPa)	0.0069
	kip/in.^{2} (ksi)	MN/m^2	6.895
	kip/in.^{2} (ksi)	N/mm^2	6.895
	pound/ft^2	kg force/m^2	4.88
	pound/ft^2	kN/m^2	0.04788
	pound/ft^2	N/m^2	47.88
	kip/ft^2	kN/m^2	47.88
	kg force/cm^2	lb/in.^{2} (psi)	14.22
	kg force/m^2	lb/ft^2(psf)	0.205
	kg force/cm^2	N/cm^2	9.8
	newton/mm^2	ksi	0.145
	kN/m^2	k/ft^2	0.0208
	kN/m^2	lb/ft^2	20.8
9. MOMENTS	ft·kip	kg force·m	138.2
	ft·kip	kN·m	1.356
	in.·kip	kN·m	0.113
	in.·kip	kg force·m	11.52
	kg force·m	ft·lb	7.233
	ton force·m	ft·kip	7.233
	kN·m	ft·kip	0.7375

Design of Reinforced Concrete Structures

DESIGN
OF
REINFORCED
CONCRETE
STRUCTURES

M. NADIM HASSOUN
Professor of Civil Engineering
South Dakota State University

PWS Engineering
Boston

PWS PUBLISHERS

Prindle, Weber & Schmidt • 🐝 • Duxbury Press • ♠ • PWS Engineering • △
Statler Office Building • 20 Park Plaza • Boston, Massachusetts 02116

PWS Publishers is a division of Wadsworth, Inc.

Library of Congress Cataloging in Publication Data

Hassoun, M. Nadim.
  Design of reinforced concrete structures.

  Includes index.
  1. Reinforced concrete construction. 2. Structural
design. I. Title.
TA683.2.H36 1985   624.1′8341   84-25372
ISBN 0-534-03759-3

ISBN 0-534-03759-3

Printed in the United States of America

10 9 8 7 6 5 4 3 2 —85 86 87 88 89

PRODUCTION: Hal Lockwood, Bookman Productions
MANUSCRIPT EDITOR: Andrew Alden
INTERIOR DESIGN: Rodelinde Albrecht
COVER DESIGN: Rodelinde Albrecht
ILLUSTRATIONS: John Foster
COMPOSITION: Bi-Comp
PRINTING AND BINDING: Halliday Lithograph

CONTENTS

* Asterisks in the Table of Contents designate optional sections, as distinct from those that form the basic requirements of the first course in reinforced concrete design (see the Preface).

PREFACE

The main objective of a course in reinforced concrete design is to develop, in the engineering student, the ability to analyze and design a reinforced concrete member subjected to different types of forces in a simple and logical manner using the basic principles of statics and some empirical formulas based on experimental results. Once the analysis and design procedure is fully understood, its application to different types of structures becomes simple and direct, provided that the student has a good background in structural analysis.

The material presented in this book is based on the requirements of the American Concrete Institute (ACI) 1983 building code (318-83). Also, information has been presented on material properties and specifications adopted by some European codes. The purpose of this presentation is to make the engineering student aware of the requirements of other national codes for the design of reinforced concrete structures. The student may need this information if he or she becomes involved in the design or construction of projects outside of the United States.

Metrication has been introduced in this book using the international system of units in addition to the primary use of the U.S. customary units. Sample examples and problems in S.I. units are given throughout the text. All problems at the end of each chapter include numerical data in S.I. units; teachers may assign some problems and ask their students to use S.I. units in their solutions. It is believed that this approach will serve the needs of the students during the transition period from one system of units to the other. The 1977 ACI Code provided S.I. and Standard Metric (MKS) equivalent tables in its Appendices D and E, respectively. In 1983, a separate version of the ACI Code in metric equivalents was published by the American Concrete Institute.

A large number of examples are presented in the text explaining the general procedure for the analysis and design of reinforced concrete structural members. Two concepts of analysis and design are presented, namely, the strength design and working stress design (alternate design) methods. The first is based on ultimate strength and failure conditions, while the second concept is based on elastic analysis. The current ACI Code gives emphasis to the strength design method, which has been adopted throughout the text except in Chapter 5, where the analysis of reinforced concrete sections by the working stress design method is explained in detail.

This book is written to cover two courses in reinforced concrete design. Depending on the proficiency required, the first course may cover Chapters 1 through 11 and part of Chapter 13 (sections 1 through 6). The second course may start with the design of long columns in Chapter 12, then followed by the remaining sections of Chapter 13 on combined and eccentrically loaded footings and the final five chapters.

A number of optional sections have been included in various chapters. These sections are indicated by an asterisk (*) in the Table of Contents, and may easily be distinguished from those which form the basic requirements of the first course. The optional sections may be covered in the second course or relagated to a reading assignment.

Special features of the textbook include comprehensive information on properties of reinforced concrete, behavior of flexural members at failure, tests performed on some beams during the course to compare actual ultimate strength capacity with the calculated strength, shear resistance and stirrup spacing diagrams, shear design of members with variable depth, introducing a simple approach to the design of columns under eccentric loading, and frame hinges and footings under eccentric loading. This is in addition to an adequate number of examples, explained in steps, and the structural aid tables in Appendix C.

My sincere thanks to the reviewers of the manuscript for their constructive comments and valuable suggestions. Special thanks are due the civil engineering students at South Dakota State University for their feedback while studying from the manuscript form of this book.

Finally, I hope that this book provides the basic material on the analysis and design of reinforced concrete in a simple approach. Any comments, identification of errors, or suggestions will be highly appreciated.

M. Nadim Hassoun
Brookings, S. D.

NOTATION

a depth of the equivalent rectangular concrete stress block

a_b value of a for a balanced condition

A effective tension area of concrete surrounding one bar. This value is used for control of cracking

A_b area of an individual bar

A_c area of core of spirally reinforced column, area of concrete at the cross-section considered

ACI   American Concrete Institute

A_g gross (total) area of cross-section

A_l area of longitudinal reinforcement to resist torsion

A_{ps} area of prestressed reinforcement in the tension zone

A_s area of nonprestressed tension steel

A_s' area of compression steel

A_{sb} area of balanced steel, area of one steel bar

A_{st} total area of longitudinal steel

A_{sf} area of steel reinforcement to develop compressive strength in the overhanging flanges in T- or L-sections

A_t area of one leg of closed stirrups to resist torsion

A_{tc} transformed concrete area

A_v total area of shear reinforcement within a spacing s

A_1 loaded area

A_2 maximum area of supporting surface geometrically similar and concentric with the loaded area

b width of compression zone at extreme fibers

b_e effective width of flange

b_o perimeter of critical section for punching shear

b_r reduced width of flanged sections

b_w width of beam web

c distance from extreme compression fibers to neutral axis; cohesion

c_b c for a balanced section

c_1 side of a rectangular column measured in the direction of span

c_2 side of a rectangular column measured transverse to the span

C cross-sectional constant =

$\Sigma(1 - 0.63x/y)x^3y/3$; compression force

C_m — a factor relating actual moment diagram to an equivalent uniform moment diagram

C_r — creep coefficient = creep strain per unit stress per unit length

C_s — force in the compression steel

C_t — factor relating shear and torsional stress properties = $b_w d/\Sigma x^2 y$

C_w — compression force in web

d — distance from the extreme compression fibers to the centroid of the tension steel

d' — distance from the extreme compression fibers to the centroid of the compression steel

d_b — nominal diameter of a reinforcing bar

d_c — distance from tension extreme fiber to center of closest bar (used in crack control)

D — dead load; diameter of a circular section; bar diameter

e — eccentricity of load

e' — eccentricity of load with respect to the centroid of tension steel

E — load effects of earthquake

E_c — modulus of elasticity of concrete = $33w^{1.5}\sqrt{f'_c}$

E_{cb} — modulus of elasticity of beam concrete

E_{cc} — modulus of elasticity of column concrete

E_{cs} — modulus of elasticity of slab concrete

EI — flexural stiffness of compression member

E_s — modulus of elasticity of steel = 29,000 ksi

f — flexural stress

f_c — flexural compressive stress in concrete due to service loads

f_{ca} — allowable compressive stress in concrete (alternate design method)

f'_c — 28-day compressive strength of concrete (standard cylinder strength)

f_{ci} — compressive strength of concrete at transfer (initial prestress)

f_{pc} — compressive stress in concrete due to prestress after all losses

f_{pe} — compressive stress in concrete at extreme fiber due to the effective prestressing force after all losses

f_{ps} — stress in prestress steel at nominal strength

f_{pu} — tensile strength of prestressing tendons

f_{py} — yield strength of prestressing tendons

f_r — modulus of rupture of concrete = $7.5\sqrt{f'_c}$

f_s — stress in the tension steel due to service load

f'_s — stress in the compression steel due to service load

f_{se} — effective stress in prestressing steel after all losses

f_t — tensile strength of concrete

f_y — yield strength of nonprestressed steel

F — lateral pressure of liquids

G_c — shear modulus of concrete = $0.45E_e$

h — total depth of member

h_f — depth of flange in flanged sections

h_v — total depth of shearhead cross-section

H — lateral earth pressure

I	moment of inertia; impact	l_{hb}	basic development length of a standard hook
I_b	moment of inertia of gross section of beam about its centroidal axis	l_n	clear span
		l_u	unsupported length of compression member
I_c	moment of inertia of gross section of column	l_v	length of shearhead arm
I_{cr}	moment of inertia of a cracked transformed section	l_1	span length in the direction of moment
I_e	effective moment of inertia to compute deflection	l_2	span length in direction transverse to span l_1
I_g	moment of inertia of gross section of concrete neglecting steel	L	live load; span length
		M	design bending moment
I_s	moment of inertia of gross section of slab	M_a	moment in member at stage deflection is computed
I_{se}	moment of inertia of steel reinforcement about centroidal axis of section	M_b	balanced moment in columns
		M_{cr}	cracking moment
		M_m	modified moment
j	a factor relating the internal couple arm to d in working stress design	M_n	nominal ultimate moment = M_u/ϕ
J	polar moment of inertia	M_o	total factored moment
k	factor to compute the effective column length; wobble friction coefficient in prestressed concrete	M_p	plastic moment
		M_u	factored moment
		M_{u1}	part of M_u computed as singly reinforced
K	a factor relating the position of the neutral axis to d in working stress design	M_{u2}	part of M_u due to compression reinforcement
		M'_u	factored moment using an eccentricity e'
K_b	flexural stiffness of beam	M_v	shearhead moment resistance
K_c	flexural stiffness of column	M_{1b}	smaller factored end moment at the end of column due to loads that do not cause sidesway
K_{ec}	flexural stiffness of equivalent column		
K_s	flexural stiffness of slab	M_{2b}	larger factored end moment at the end of column due to loads that do not cause sidesway
K_t	torsional stiffness of torsional member		
l	length of prestressing tendon; span length	M_{2s}	Same as M_{2b} except due to loads that cause sidesway
l_c	vertical distance between supports	n	modular ratio = E_s/E_c
		N	normal load
l_d	development length	N_u	factored normal load
l_{dh}	l_{hb} times the applicable modification factor	N_1	normal force in bearing at base of column

P_b	balanced load in column at balanced strain conditions		T_u	factored torsional moment
P_c	Euler buckling load		u	bond stress
P_n	nominal axial strength of column for a given e		U	design strength to resist factored loads
P_{nw}	nominal axial strength of wall		v	design shear stress (working stress design method)
P_o	axial strength of loaded column at zero eccentricity		v_c	shear stress carried by concrete
P_s	prestressing force in the tendon at the jacking end		v_{cr}	shear stress at which diagonal cracks develop
P_u	factored axial load at a given eccentricity		v_h	horizontal shear stress
			V	unfactored shear force
P_x	prestressing force in the tendon at any point x		V_c	nominal shear strength provided by concrete
q	soil-bearing capacity, $\rho f_y/f'_c$		V_{ci}	nominal shear strength of concrete when diagonal cracking results from combined shear and moment
q_a	allowable bearing capacity of soil			
q_u	bearing capacity of soil using factored loads		V_{cw}	nominal shear strength of concrete when diagonal cracking results from excessive principal tensile stress in web
Q	static moment about the neutral axis of a portion of section through a line parallel to the neutral axis			
			V_d	shear force at section due to unfactored dead load
r	radius of gyration; radius of a circle		V_n	nominal shear strength
R	resultant of force system; reduction factor for long columns		V_p	vertical component of effective prestress force at section
			V_s	nominal shear strength resisted by shear reinforcement
R_u	a factor that depends on the steel ratio ρ, $R_u = M_u/bd^2$		V_u	factored shear force
s	spacing between bars, stirrups, or ties		w	width of crack; unit weight of concrete
t	thickness of slab		w_c	weight of concrete, pcf
T	torque; tension force; time-dependent factor for sustained load		w_d	factored dead load per unit area
			w_l	factored live load per unit area
			w_u	factored load per unit length of beam or per unit area of slab
T_c	nominal torsional moment strength provided by concrete		W	wind load
T_n	nominal torsional moment strength		x	shorter dimension of rectangular part of cross-section
T_s	nominal torsional moment strength provided by torsion reinforcement		x_1	shorter center-to-center dimension of closed rectangular stirrup

y	longer dimension of rectangular part of cross-section	β_a	ratio of unfactored dead load per unit area to unfactored live load per unit area
y_t	distance from centroidal axis of gross section, neglecting reinforcement, to extreme top fiber	β_c	ratio of long side to short side of column or loaded area
y_b	same as y_t, except to extreme bottom fibers	β_d	ratio of maximum factored dead load moment to maximum factored total load moment (always positive)
y_1	longer center-to-center dimension of closed rectangular stirrup	β_s	ratio of length of continuous edges to total perimeter of a slab panel
z	factor related to width of crack $= f_s^3\sqrt{Ad_c}$	β_t	ratio of torsional stiffness of edge beam section to flexural stiffness of slab: $E_{cb}C/2E_{cs}I_s$
α	angle between inclined stirrups and longitudinal axis of member; total angular change of prestressing tendon profile, in radians, from tendon jacking end to any point x; ratio of flexural stiffness of beam section to flexural stiffness of slab at a joint: $E_{cb}I_b/E_{cs}I_s$	β_1	ratio a/c. This factor $= 0.85$ for $f'_c \le 4$ ksi and decreases by 0.05 for each 1 ksi increase in f'_c
		γ_p	factor for type of prestressing tendon (0.4 or 0.28)
		γ_f	fraction of unbalanced moment transferred by flexure at slab-column connections
α_c	ratio of flexural stiffness of columns to combined flexural stiffness of the slabs and beams at a joint: $(\Sigma K_c)/\Sigma(K_s + K_b)$	γ_v	fraction of unbalanced moment transferred by eccentricity of shear at slab-column connections
α_{ec}	ratio of flexural stiffness of equivalent column to combined flexural stiffness of the slabs and beams at a joint: $(K_{ec})/\Sigma(K_s + K_b)$	δ	moment magnification factor
		δ_b	moment magnification factor for frames braced against sidesway
α_f	angle between shear friction steel and shear plane	δ_s	moment magnification factor for frames not braced against sidesway
α_m	average value of α for all beams on edges of a panel	Δ	deflection
α_t	coefficient as a function of y_1/x_1	ϵ	strain
α_v	ratio of stiffness of shearhead arm to surrounding composite slab section	ϵ_c	strain in concrete
		ϵ_s	strain in tension steel
α_1	α in direction of l_1	ϵ'_s	strain in compression steel
α_2	α in direction of l_2	ϵ_y	yield strain of steel $= f_y/E_s$
β	ratio of clear spans in long to short direction of two-way slabs; ratio of long side to short side of footing	θ	slope angle
		λ	multiplier for additional long-time deflection
		μ	Poisson's ratio; coefficient of

friction; curvature friction coefficient (prestress)

ρ ratio of nonprestressed tension steel $= A_s/bd$

ρ' ratio of compression steel $= A_s'/bd$

ρ_1 $(\rho - \rho')$

ρ_b balanced steel ratio

ρ_g ratio of total steel area to total concrete area

ρ_h ratio of vertical shear reinforcement area to gross concrete area of horizontal section

ρ_p ratio of prestressed reinforcement A_{ps}/bd

ρ_s ratio of volume of spiral reinforcement to total volume of core of a spirally reinforced compression member

ρ_v $(A_s + A_h)/bd$

ρ_w $A_s/b_w d$

ϕ strength reduction factor

ω $\rho f_y/f_c'$

ω' $\rho' f_y/f_c'$

ω_p $\rho_p f_{ps}/f_c'$

$\omega_w,$ $\omega_{pw},$ ω_w' reinforcement indices for flanged sections computed as for ω, ω_p, and ω' except that b shall be the web width, and reinforcement area shall be that required to develop compressive strength of web only

CHAPTER

1

Introduction

1.1   General

Reinforced concrete is one of the structural materials that is commonly used all over the world. Its two component materials, concrete and steel, work together to form structural members that can resist many types of loadings. The key to its performance lies in strengths that are complementary: concrete resists compression and steel reinforcement resists tension forces.

Although reinforced concrete is a heterogeneous complicated material, theoretical and experimental work makes the analysis and design of structural members, to some extent, straightforward and easy. It finds wide application in buildings and other structures because of its durability, high resistance to static and dynamic loads, resistance to fire and weathering, the availability of raw materials, and low maintenance costs. Structures such as bridges, water tanks, factories, tunnels, dams, viaducts, roads, pavings, retaining walls, and foundations can be constructed from reinforced concrete. It also has a long service life because the strength of concrete actually increases with time, provided that the steel bars are protected from corrosion.

The joint behavior of steel and concrete is based on the following properties:

- A bond is maintained between steel and concrete after the concrete hardens. The use of deformed steel bars greatly improves the bonding, and both materials deform together under load.
- The coefficients of thermal expansion of concrete (10 to $15 \times 10^{-6}/°C$) and steel ($12 \times 10^{-6}/°C$) are very close. Under changes of temperature not exceeding 80°C, differential strains are not observed and slipping of the steel bars is not expected.
- Concrete protects the steel reinforcement against corrosion and improves the fire resistance of the whole structural member.

Concrete is made of cement, aggregate, and water, proportioned in such a way as to produce the required designed strengths. Figure 1.1 shows some structures constructed of reinforced concrete.

1

(a)

(b)

(c)

(d)

FIGURE 1.1. Structures built with reinforced concrete: (a) Water Tower Place (building at right) in Chicago, the world's tallest reinforced concrete building (859 ft); (b) Marina City Towers, Chicago; (c) an office building in Amman, Jordan; and (d) CN Tower, Toronto, the world's tallest free-standing concrete structure (1465 ft).

1.2   Historical Background

Concrete was used by the ancient Romans in the construction of walls and roofs, but concrete has its first modern record as early as 1760, when in Britain John Smeaton used it in the first lock on the river Calder.[1] The walls of the lock were made of stones filled in with concrete. In 1796, J. Parker rediscovered the Roman natural cement, and 15 years later Vicat burned a mixture of clay and lime to produce cement. In 1824, Joseph Aspdin manufactured portland cement in Wakefield, Britain. It was called portland cement because when hardened it resembled stone from the quarries of the Isle of Portland.

In Britain, reinforced concrete was used in 1832 by Sir Marc Isambard Brunel in an experimental arch. In France, François Marten Le Brun built a concrete house in 1832 in Moissac, in which he used concrete arches of 18-foot spans. He used concrete to build a school in St. Aignan in 1834 and a church in Corbarièce in 1835. Joseph Louis Lambot[2] exhibited a small rowboat made of reinforced concrete at the Paris Exposition of 1854. In the same year, W. B. Wilkinson of England obtained a patent for a concrete floor reinforced by twisted cables. The Frenchman François Coignet obtained his first patent in 1855 for the system he used of iron bars, embedded in concrete floors, that extended to the supports. One year later, he added nuts at the screw ends of the bars, and in 1869 he published a book describing the applications of reinforced concrete.

Many investigators give credit for the invention of reinforced concrete to the gardener Joseph Monier, who obtained his patent in Paris on July 16, 1867.[3] He made garden tubs and pots of concrete reinforced with iron mesh, which he exhibited in Paris in 1867. In 1873, he registered a patent to use reinforced concrete in tanks and bridges, and four years later, he registered another patent to use it in beams and columns.[1]

Thaddeus Hyatt, a lawyer, was a pioneer in experimenting with reinforced concrete in the United States. He conducted flexural tests on 50 beams that contained iron bars as tension reinforcement and published the results in 1877. In other experiments, he exposed reinforced concrete slabs to high temperature to explore their fire resistance. Investigating the shrinkage of concrete and coefficients of expansion of concrete and steel, he found that both concrete and steel can be assumed to behave in a homogeneous manner for all practical purposes. This assumption was important for the design of reinforced concrete members using elastic theory. He used prefabricated slabs in his experiments and considered that prefabricated units were best cast of T-sections placed side by side to form a floor slab. Hyatt is generally credited with developing the principles upon which the analysis and design of reinforced concrete is based.

The first reinforced concrete house in the United States was built by W. E. Ward near Port Chester, New York. It used reinforced concrete for walls, beams, slabs, and staircases. The use of cavity walls was suggested by Ward to protect the house from storms, fire, and humidity. P. B. Write wrote in the *American Architect and Building News* in 1877, describing the applications of reinforced concrete in Ward's house as a new method in building construction. Ward gave a lecture on concrete and steel as structural materials, suggesting the use of reinforced concrete on a large scale.

In San Francisco, E. L. Ransome, head of the Concrete-Steel Company, used reinforced concrete in 1870 and deformed bars for the first time in 1884. During 1889–1891, he built the two-story Leland Stanford Museum in San Francisco using reinforced concrete. He also built a reinforced concrete bridge in San Francisco in addition to some industrial buildings for the Pacific Coast Borax Company. In 1900, after Ransome introduced the reinforced concrete skeleton, the thick wall system started to disappear in construction. He registered the skeleton type of structure in 1902 and used it in the Kelly and Jones four-story factory in Greensburg, Pennsylvania. He used spiral reinforcement in the columns as was suggested by Armand Considère of France.

A. N. Talbot, at the University of Illinois, and F. E. Turneaure and M. O. Withey, of the University of Wisconsin, conducted extensive tests on concrete to determine its behavior, compressive strength, and modulus of elasticity.

In Germany, G. A. Wayss bought the French Monier patent in 1879 and published his book on Monier methods of construction in 1887. Rudolph Schuster bought the patent rights in Austria, and the name of Monier spread in Europe. This is the main reason for crediting Monier as the inventor of reinforced concrete.

France played a major role in the development of reinforced concrete thanks to François Hennebique, who established an engineering office and employed many engineers to design thousands of reinforced concrete structures and develop reinforced concrete as a practical structural material. Hennebique conducted many experiments on reinforced concrete beams and used stirrups and bent bars for the first time to resist the shearing forces in a structural member. He issued a monthly magazine on reinforced concrete in Paris in 1898. He used 3-meter cantilever beams in a theatre in Morges, Switzerland, in 1899 and designed reinforced concrete helical staircases for the Grand and Petit Palais in Paris in 1898, using spans of 11 meters and cantilevers of 3.6 meters.

In 1900, the Ministry of Public Works in France called for a committee headed by Armand Considère, Chief Engineer of roads and bridges, to establish specifications for reinforced concrete, which were published in 1906.

Reinforced concrete was further refined by introducing some precompression in the tension zone, to decrease the excessive cracks. This was the preliminary introduction of partial and full prestressing. In 1928, Eugene Freyssinet established the practical technique of using prestressed concrete.[4]

From 1915 to 1935, research was conducted on axially loaded columns and creep effects on concrete; in 1940, eccentrically loaded columns were investigated. Ultimate-strength design started to receive special attention, in addition to diagonal tension and prestressed concrete. The American Concrete Institute Code specified the use of ultimate-strength design in 1963 and included this method in the 1971, 1977, and 1983 codes. Building codes and specifications for the design of reinforced concrete structures are established in most countries, and research continues on developing new applications and more economical designs.

1.3   Advantages and Disadvantages of Reinforced Concrete

Reinforced concrete, as a structural material, is widely used in many types of structures. It is competitive with steel if economically designed and executed.

The advantages of reinforced concrete can be summarized as follows:

1. It has a relatively high compressive strength.
2. It has better resistance to fire than steel.
3. It has a long service life with low maintenance cost.

4. In some types of structures, such as dams, piers, and footings, it is the most economical structural material.
5. It can be cast to take the shape required, making it widely used in precast structural components.
6. It yields rigid members with minimum apparent deflection.

The disadvantages of reinforced concrete can be summarized as follows:

1. It has a low tensile strength of about one-tenth of its compressive strength.
2. It needs mixing, casting, and curing, all of which affect the final strength of concrete.
3. The cost of the forms used to cast concrete is high. The cost of form material and workmanship may equal the cost of concrete placed in the forms.
4. It has a low compressive strength as compared to steel (the ratio is about $1:10$, depending on the materials), which leads to large sections in columns of multi-story buildings.
5. Cracks develop in concrete due to shrinkage and the application of live loads.

1.4   Units of Measurement

The United States is the last major country to use the traditional system of measurement, based on the yard and the pound, for engineering purposes. The advantages of the U.S. customary system (lying mostly in its human scale and its ingenious use of simple numerical proportions) are generally superseded by the precision and international uniformity made possible by S.I. (Système Internationale) or "metric" measurements. Even the customary units are now legally defined in terms of the meter and kilogram. Nevertheless, this book will make liberal use of both systems of measurement as the customary system is not likely to be abolished outright in this country or this generation.

The metric system is planned to be in universal use within the coming few years. The United States is committed to change to S.I. units, but there is no definite date for complete conversion. Great Britain, Canada, and Australia have been using S.I. units for several years.

The base units in the S.I. system are the units of length, mass, and time, which are the meter (m), the kilogram (kg), and the second (s), respectively. The unit of force, a derived unit called the newton (N), is defined as the force that gives an acceleration of 1 meter per second per second (1 m/s^2) to a mass of 1 kg:

$$1 \text{ N} = 1 \text{ kg} \cdot \text{m/s}^2$$

An earlier version of the metric system used the kilogram-force (kgf) for force and kgf/cm^2 for stress, where $1 \text{ kgf} = 9.80665 \text{ N}$.

Examples of some derived S.I. units used in structural analysis are shown in Table 1.1.

The weight of a body W, which is equal to the mass m multiplied by the local gravitational acceleration g (9.81 m/s^2), is expressed in newtons (N). The weight of a body of 1 kg mass is

$$W = mg = 1 \text{ kg} \times 9.81 \text{ m/s}^2 = 9.81 \text{ N}$$

TABLE 1.1. Examples of derived S.I. units

Quantity	S.I. units	Name of unit
Area	m²	(square meter)
Volume	m³	(cubic meter)
Volume (liquids)	$L = 0.001 \text{ m}^3$	(liter)
Density	kg/m³	(kilogram/cubic meter)
Speed, velocity	m/s	(meter/second)
Acceleration	m/s²	(meter/second squared)
Force	$N = kg \cdot m/s^2$	(newton)
Pressure, stress	$Pa = N/m^2 = kg/(m \cdot s^2)$	(pascal)
Energy, work, or quantity of heat	$J = N \cdot m = kg \cdot m^2/s^2$	(joule)
Power, radiant flux	$W = J/s = kg \cdot m^2/s^3$	(watt)
Moment of a force	$N \cdot m = kg \cdot m^2/s^2$	(newton meter)
Surface tension	$N/m = kg/s^2$	
Angle	rad	(radian: 2π rad = 360°)
Angular velocity	rad/s	(radian per second)
Angular acceleration	rad/s²	(radian per second squared)
Frequency	$Hz = 1/s$	(hertz)
Impulse	$N \cdot s = kg \cdot m/s$	(newton second)
Mass	$t = 1000 \text{ kg}$	(ton or metric ton)
Force	$dyn = 10^{-5} \text{ N}$	(dyne)

Multiples and submultiples of the base S.I. units can be expressed through the use of prefixes. The prefixes most frequently used in structural calculations are kilo (k), mega (M), milli (m), and micro (μ). For example,

$$1 \text{ km} = 1000 \text{ m}, \quad 1 \text{ mm} = 0.001 \text{ m}, \quad 1 \text{ } \mu m = 10^{-6} \text{ m}$$
$$1 \text{ kN} = 1000 \text{ N}, \quad 1 \text{ Mg} = 1000 \text{ kg} = 10^6 \text{ g}$$

1.5   Codes of Practice

In most countries, the design engineer is guided by specifications called the codes of practice. Engineering specifications are set up by various organizations to represent the minimum requirements necessary for the safety of the public, although they are not necessarily for the purpose of restricting engineers.

The codes specify design loads, allowable stresses, material quality, construction types, and other requirements related to building construction. The most significant code for reinforced concrete design in the United States is the Building Code Requirements for Reinforced Concrete, **ACI-318-83,** or the ACI Code. Most of the design examples of this book are based on this code. Other codes of practice and material specifications in the United States include the Uniform Building Code, Standard Building Code, National Building Code, Basic Building Code, South Florida Building Code, American Association of State Highway and Transportation Officials (AASHTO) specifications, and specifications issued by the American Society for Testing and Materials (ASTM), American Railway Engineering Association (AREA), and Bureau of Reclamation, Department of the Interior.

Recommendations for an international code of practice were issued in 1963 by the European Committee for Concrete (Comité Européen du Béton), or CEB. The members of the various subcommittees that established the recommendations represent different countries in Europe and the United States.

Different codes other than those of the United States include the British Standard (B.S.) Code of Practice for Reinforced Concrete, CP-110; the National Building Code of Canada; the German Code of Practice for Reinforced Concrete, DIN 1045; Specifications for the Design of Reinforced Concrete Structures and GOST Specifications for Steel Reinforcement (U.S.S.R.); and Technical Specifications for the Theory and Design of Reinforced Concrete Structures, CC-BA-78 (France).

Some of the codes of practice issued by European countries are translated into English by the Cement and Concrete Association, London.

1.6   Design Philosophy and Concepts

The design of a structure may be regarded as the process of selecting the proper materials and proportioning the different elements of the structure according to state-of-the-art engineering science and technology. In order to fulfill its purpose, the structure must meet the conditions of safety, serviceability, economy, and functionality. Two design concepts of reinforced concrete members are discussed in this book: the strength design and alternate design methods.

The strength design method is based on the ultimate strength of structural members assuming a failure condition, whether due to the crushing of the concrete or to the yield of the reinforcing steel bars. Although there is some additional strength in the bars after yielding (due to strain hardening), this additional strength is not considered in the analysis of reinforced concrete members. In the strength design method, the actual loads, or working loads, are multiplied by load factors to obtain the ultimate design loads. The load factors represent a high percentage of the factor of safety required in the design. Details of this method are presented in Chapters 3 and 4. The ACI Code emphasizes this method of design.

The alternate design method is also called the working stress design or the elastic design method. The design concept is based on elastic theory, assuming a straight-line stress variation along the depth of the concrete section. The actual loads or working loads acting on the structure are estimated and members are proportioned on the basis of certain allowable stresses in concrete and steel. The allowable stresses are fractions of the crushing strength of concrete f_c' and the yield strength of steel f_y. For example, the allowable concrete strength in flexure is equal to 0.45 of the ultimate strength of concrete f_c'. The allowable stress for steel with a yield strength of 40 ksi (thousand pounds per square inch) is equal to 20 ksi. Therefore, an adequate factor of safety is maintained. This method is presented in Chapter 5.

1.7   Loads

Loads are those forces for which a given structure should be proportioned. In general, loads may be classified as dead or live.

Dead loads include the weight of the structure (its self-weight) and any permanent material placed on the structure, such as tiles, roofing materials, and walls. Dead loads can be determined with a high degree of accuracy from the dimensions of the elements and the unit weight of materials.

Live loads are all other than dead loads. They may be steady or unsteady; movable or moving; applied slowly, suddenly, vertically, or laterally; and their magnitudes may fluctuate with time. In general, live loads include

- Occupancy loads caused by the weight of the people, furniture, and goods.
- The weight of snow if accumulation is probable.
- Forces resulting from wind action and temperature changes.
- The pressure of liquids or earth on retaining structures.
- The weight of traffic on a bridge.
- Dynamic forces resulting from moving loads (impact) or from earthquakes.
- Forces resulting from blast loading.

The ACI Code does not specify loads on structures; however, occupancy loads on different types of buildings are prescribed by the American National Standards Institute (ANSI).[5] Some typical values are shown in Table 1.2. Table 1.3 shows weights and specific gravity of various materials.

TABLE 1.2. Typical uniformly distributed design loads[5]

Occupancy	Contents	Design live load	
		lb/ft²	*kN/m²*
Assembly hall	Fixed seats	60	2.9
	Movable seats	100	4.8
Hospital	Operating rooms	60	2.9
	Private rooms	40	1.9
Hotel	Guest rooms	40	1.9
	Public rooms	100	4.8
	Balconies	100	4.8
Housing	Private houses and apartments	40	1.9
	Public rooms	100	4.8
Institution	Classrooms	40	1.9
	Corridors	100	4.8
Library	Reading rooms	60	2.9
	Stack rooms	150	7.2
Office building	Offices	50	2.4
	Lobbies	100	4.8
Stairs (or balconies)		100	4.8
Storage warehouses	Light	100	4.8
	Heavy	250	12.0
Yards and terraces		100	4.8

TABLE 1.3. Density and specific gravity of various materials

Material	Density lb/ft³	Density kg/m³	Specific gravity
Building materials			
Bricks	120	1924	1.8–2.0
Cement, portland, loose	90	1443	—
Cement, portland, set	183	2933	2.7–3.2
Earth, dry, packed	95	1523	—
Earth, moist, packed	96	1539	—
Glass	156	2500	2.4–2.6
Lime mortar, set	103	1651	1.4–1.9
Sand or gravel, dry, packed	100–120	1600–1924	—
Sand or gravel, wet	118–120	1892–1924	—
Timber, pine	30–44	480–705	0.48–0.70
Timber, oak	41–59	657–946	0.65–0.95
Liquids			
Oils	58	930	0.9–0.94
Water (at 4°C)	62.4	1000	1.0
Ice	56	898	0.88–0.92
Seawater	64	1026	1.02–1.03
Metals and minerals			
Aluminum	165	2645	2.55–2.75
Copper	556	8913	9.0
Iron	450	7214	7.2
Lead	710	11380	11.37
Steel, rolled	490	7855	7.85
Gypsum	159	2549	2.3–2.8
Limestone or marble	165	2645	2.5–2.8
Sandstone	147	2356	2.2–2.5
Shale or slate	175	2805	2.7–2.9
Normal weight concrete			
Plain	145	2324	2.2–2.4
Reinforced	150	2405	2.3–2.5
Prestressed	150	2405	2.3–2.5

AASHTO and AREA specifications prescribe vehicle loadings on highway and railway bridges, respectively. These loads are given in references 6 and 7.

Snow loads on structures may vary between 10 and 40 lb/ft² (0.5 and 2 kN/m²) depending on the local climate.

Wind loads may vary between 15 and 30 lb/ft² depending on the velocity of wind. The wind pressure on a structure F can be estimated from the following equation:

$$F = \frac{1}{2} C_s \cdot \rho \cdot v^2 \tag{1.1}$$

where

v = velocity of air (feet per second)

ρ = mass density of air at sea level (0.0765 lb/ft³ at 59°F, 1.224 kg/m³ at 15°C)

C_s = shape factor of the structure

Substituting the mass density in the above equation, then

$$F = \left(\frac{1}{2}\right)\left(\frac{0.0765}{32.2}\right)\left(\frac{5280}{3600}\right)^2 C_s V^2 = 0.00256 C_s V^2 \tag{1.2}$$

where F is the dynamic wind pressure in pounds per square foot and V is the wind speed in miles per hour.

For a wind of 100 mph with $C_s = 1$, the wind pressure is equal to 25.6 lb/ft². It is sometimes necessary to consider the effect of gusts in computing the wind pressure by multiplying the wind velocity in equation 1.2 by a gust factor, which generally varies between 1.1 and 1.3.

The shape factor C_s varies with the horizontal angle of incidence of the wind. On vertical surfaces of rectangular buildings, C_s may vary between 1.2 and 1.3. Detailed information on wind loads can be found in reference 5.

1.8   Safety Provisions

Structural members must always be proportioned to resist loads greater than the service or actual loads in order to provide proper safety against failure. In ultimate-strength design, the member is designed to resist factored loads, which are obtained by multiplying the service loads by load factors. Different factors are used for different loadings. Because dead loads can be estimated quite accurately, their load factors are smaller than those of live loads, which have a high degree of uncertainty. Several load combinations must be considered in the design to compute the maximum and minimum design forces. Reduction factors are used for some combinations of loads to reflect the low probability of their simultaneous occurrences. The ACI Code presents specific values of load factors to be used in the design of concrete structures (see Chapter 3, section 3.5).

In addition to load factors, the ACI Code specifies another factor to allow an additional reserve in the capacity of the structural member. The nominal strength is generally calculated using accepted analytic procedure based on statics and equilibrium; however, in order to account for the degree of accuracy within which the nominal strength can be calculated, and for adverse variations in materials and dimensions, a strength reduction factor ϕ should be used in the strength design method. Values of the strength reduction factors are given in Chapter 3, section 3.6.

To summarize the above discussion, the ACI Code has separated the safety provision into an overload or load factor and to an undercapacity (or strength reduction) factor ϕ. A safe design is achieved when the structure's strength, obtained by multiplying the nominal strength by the reduction factor ϕ, exceeds or equals the strength needed to withstand the factored loadings (service loads times their load factors). For example:

$$M_u \leq \phi M_n \quad \text{and} \quad V_u \leq \phi V_n \tag{1.3}$$

where

M_u and V_u = factored moment and shear forces and

M_n and V_n = nominal ultimate moment and shear capacity of the member, respectively

Given a load factor of 1.4 for dead load and a load factor of 1.7 for live load, the overall safety factor for a structure loaded by a dead load D and a live load L is

$$\text{Factor of safety} = \frac{1.4D + 1.7L}{D + L} \, (1/\phi) = \frac{1.4 + 1.7(L/D)}{1 + (L/D)} \, (1/\phi) \tag{1.4}$$

The factor of safety for various values of ϕ and L/D ratios is shown below.

ϕ	0.9				0.8				0.7			
L/D	0	1	2	3	0	1	2	3	0	1	2	3
Safety factor	1.56	1.72	1.78	1.81	1.75	1.94	2.00	2.03	2.00	2.21	2.29	2.32

For members subjected to flexure (beams), $\phi = 0.9$, and the factor of safety ranges between 1.56 for $L/D = 0$ and 1.81 for $L/D = 3$. The higher value applies to the member that is subjected to the greater live load.

For members subjected to axial forces (columns), $\phi = 0.7$, and the factor of safety ranges between 2.00 for $L/D = 0$ and 2.32 for $L/D = 3$. The increase in the factor of safety in columns reflects the greater overall safety requirement of these critical building elements.

A general format of equation 1.3 can be written as follows:[8]

$$\phi R_n \geq \nu_0 \Sigma(\nu_i Q_i) \tag{1.5}$$

where

R_n = nominal strength of the structural number

ϕ = undercapacity factor (<1.0)

ΣQ_i = sum of load effects

ν_i = overload factor

ν_0 = analysis factor (>1.0)

The subscript i indicates the load type, such as dead load, live load, and wind load. The analysis factor ν_0 is greater than 1.0 and is introduced to account for uncertainties in structural analysis.

The overload factor ν_i is introduced to account for several factors such as an increase in live load due to a change in the use of the structure and variations in erection procedures.

Equation 1.5 is proposed by Pinkham and Hansell[8] to introduce a general ultimate-strength approach for the design of steel buildings. The design concept is referred to as load and resistance factor design (LRFD).[8,9]

The National Building Code of Canada[10] uses the following probability factor format to specify the basic loading cases:

$$\text{Factored load effects (F.L.E.)} = u[\nu_D D_n + \alpha(\nu_L L_n + \nu_W W_n)] \tag{1.6}$$

and

$$\phi R_n \geq \text{F.L.E.}$$

where u refers to the load effects (dead, live, wind, etc.),

D_n, L_n, W_n = nominal or specified dead, live, and wind loads

ν_D, ν_L, ν_W = load factors

The factor α is a load combination probability factor and equal to 1.0, 0.7, and 0.6 if one, two, or three loads, respectively, are included in equation 1.6. It accounts for the reduced probability that maximum dead, live, wind, and other loads act simultaneously.

Another format similar to that of equation 1.6 is presented by the European Committee on Concrete (CEB).[11] Studies to develop a universal set of load factors for use in the design of buildings were begun in 1979 by the Center for Building Technology at the National Bureau of Standards. Results of these studies have been published in references 12 through 14. In 1983, new load and resistance factors were proposed by MacGregor[15] to be incorporated in the ACI Code.

1.9   High Concrete Buildings

High-rise buildings are becoming the dominant feature of many U.S. downtowns; a great number of these buildings are designed and constructed in reinforced concrete.

TABLE 1.4. Some reinforced concrete skyscrapers

Year	Structure	Location	Height, feet (m)
1959	Executive House Hotel	Chicago	371 (113)
1960	Bank of Georgia Building	Atlanta	391 (119)
1962	Marina City Towers	Chicago	601 (180)
1965	1000 Lake Shore Plaza Apartments	Chicago	640 (195)
1967	Lake Point Tower	Chicago	645 (197)
1969	One Shell Plaza	Houston	714 (218)
1975	Peachtree Center Plaza Hotel	Atlanta	723 (220)
1976	Water Tower Place (World's tallest reinforced concrete building)	Chicago	859 (262)
1976	CN Tower (World's tallest free-standing structure)	Toronto	1465 (447)
1977	Renaissance Center Westin Hotel	Detroit	740 (226)
1978	MLC Tower	Sydney	808 (246)
1980	Trump Tower	New York	664 (203)
1983	City Center	Minneapolis	528 (158)

Although at the beginning of the century the properties of concrete and the joint behavior of steel and concrete were not fully understood, a 16-story building, the Ingalls Building, was constructed in Cincinnati in 1902 with a total height of 210 feet (64 m). In 1922, the Medical Arts Building with a height of 230 feet (70 m) was constructed in Dallas, Texas. The design of concrete buildings was based on elastic theory concepts and a high factor of safety, resulting in large concrete sections in beams and columns. Not until 1956, when the ACI Code permitted the use of ultimate-strength design, did concrete buildings rise higher. After extensive research, high-strength concrete and high-strength steel were allowed in the design of reinforced concrete members. Consequently, small concrete sections as well as savings in materials were achieved, and new concepts of structural design were possible.

To visualize how high concrete buildings can be built, some reinforced concrete skyscrapers are listed in Table 1.4.[16]

REFERENCES

1. Ali Ra'afat: *The Art of Architecture and Reinforced Concrete,* Halabi, Cairo, 1970.
2. R. S. Kirby, S. Withington, A. B. Darling and F. G. Kilgour: *Engineering in History,* McGraw-Hill, New York, 1956.
3. Hans Straub: *A History of Civil Engineering,* Leonard Hill, London, 1952.
4. E. Freyssinet: "The Birth of Prestressing," Cement and Concrete Association Translation No. 29, London, 1956.
5. American National Standards Institute, *ANSI A58.1,* 1982.
6. American Association of State Highway and Transportation Officials (AASHTO): *Standard Specifications for Highway Bridges,* 12th edition, Washington, D.C., 1977.
7. American Railway Engineering Association (AREA): *Specifications for Steel Railway Bridges,* Chicago, 1965.
8. C. W. Pinkham and W. C. Hansell: "An Introduction to Load and Resistance Factor Design for Steel Buildings," *Engineering Journal AISC,* 15, 1978 (first quarter).
9. M. K. Ravindra and T. V. Galambos: "Load and Resistance Factor Design for Steel," *Journal of Structural Division, ASCE,* 104, September 1978.
10. National Research Council of Canada: *National Building Code of Canada,* Ottawa, 1980.
11. Comité Euro-International du Béton: *CEB-FIP Model Code for Concrete Structures,* 3rd edition, Paris, 1978.
12. B. Ellingwood, T. V. Galambos, J. G. MacGregor and C. A. Cornell: "Development of a Probability Based Load Criterion for American National Standards, A58," National Bureau of Standards Special Publication 577, 1980.
13. T. V. Galambos, B. Ellingwood, J. G. MacGregor and C. A. Cornell: "Probability-Based Load Criteria: Assessment of Current Design Practice," *ASCE Proceedings,* 108, May 1982.
14. B. Ellingwood, J. G. MacGregor, T. V. Galambos and C. A. Cornell: "Probability-Based Load Criteria: Load Factors and Load Combinations," *ASCE Proceedings,* 108, May 1982.
15. J. G. MacGregor: "Load and Resistance Factors for Concrete Design," *ACI Journal,* No. 4, July–August 1983.
16. *Concrete Construction,* 28, no. 2, February 1983.

Properties of
Reinforced Concrete

2.1   Strength of Concrete

Normal concrete consists of coarse and fine aggregate, cement, and water. The materials are mixed together until a cement paste is developed, filling most of the voids in the aggregates and producing a uniform dense concrete. The plastic concrete is then placed in a mold and left to set, harden, and develop adequate strength.

The strength of concrete depends upon many factors and may vary within wide limits with the same production method. The main factors that affect the strength of concrete will be described below.

2.1.1   Water-Cement Ratio

The water-cement ratio is the most important factor affecting the strength of concrete. For complete hydration of a given amount of cement, a water-cement ratio (by weight) equal to 0.25 is needed. A water-cement ratio more than 0.1 higher is needed for the concrete to be reasonably workable. This means that a water-cement ratio of more than 0.35 by weight must be chosen. This ratio corresponds to 4 gallons of water per sack of cement (94 lb) (or 17.5 L per 50 kg of cement). Good workability is attained when the ratio exceeds 0.5. The relation between water-cement ratio and compressive and flexural strength of normal weight concrete is shown in Table 2.1.

2.1.2   The Properties and Proportions of Concrete Constituents

Concrete is a mixture of cement, aggregate, and water. An increase in the cement content in the mix and the use of well-graded aggregate both increase the strength of concrete.

TABLE 2.1. Typical relation between water-cement ratio and compressive and flexural strength of normal weight concrete[1]

| By weight | Water-cement ratio | | Probable strength of concrete at 28 days | | | |
| | Gallons per sack (94 lb) | Liters per sack (50 kg) | Compressive | | Flexural | |
			psi	N/mm²	psi	N/mm²
0.35	4.0	17.5	6300	41	650	4.5
0.40	4.5	20.0	5800	40	610	4.2
0.44	5.0	22.0	5400	37	590	4.1
0.49	5.5	24.5	4800	33	560	3.9
0.53	6.0	26.5	4500	31	540	3.7
0.58	6.5	29.0	3900	27	500	3.5
0.62	7.0	31.0	3700	25	490	3.4
0.67	7.5	33.5	3200	22	450	3.1
0.71	8.0	35.5	2900	20	430	3.0

2.1.3 The Method of Mixing, Placing, Degree of Compaction, and Curing

The use of mechanical concrete mixers and the proper time of mixing both have favorable effects on strength of concrete. Also the use of vibrators produces dense concrete with minimum percentage of voids. A void ratio of 5 percent may reduce the concrete strength about 30 percent.

The curing conditions exercise an important influence on the strength of concrete. Both moisture and temperature have a direct effect on the hydration of cement. The longer the period of moist storage, the greater the strength. If the curing temperature is higher than the initial temperature of casting, the resulting 28-day strength of concrete is reached earlier than 28 days.

2.1.4 The Age of Concrete

The strength of concrete increases appreciably with age, and hydration of cement continues for months. In practice, the strength of concrete is determined from cylinders or cubes tested at the age of 7 days and 28 days. As a practical assumption, concrete at 28 days is 1.5 times as strong as at 7 days; the range varies between 1.3 and 1.7. The British code of practice[2] accepts concrete if the strength at 7 days is not less than two-thirds of the required 28-day strength. The strength of concrete at 28 days can be calculated by an empirical formula[3] relating this strength with that at 7 days as follows

$$f'_{c(28)} = f'_{c(7)} + 30\sqrt{f'_{c(7)}} \text{ (psi)} \tag{2.1}$$

where $f'_{c(28)}$ and $f'_{c(7)}$ are ultimate strengths at 28 and 7 days, respectively, in pounds per square inch. In S.I. units,

$$f'_{c(28)} = f'_{c(7)} + 2.5\sqrt{f'_{c(7)}} \text{ N/mm}^2$$
$$f'_{c(28)} = f'_{c(7)} + 8\sqrt{f'_{c(7)}} \text{ kgf/cm}^2$$

For a normal portland cement, the increase of strength with time, relative to 28-day strength, is as follows:

Age	7 days	14 days	28 days	3 months	6 months	1 year	2 years	5 years
Strength ratio	0.67	0.86	1.0	1.17	1.23	1.27	1.31	1.35

2.1.5 Loading Conditions

The compressive strength of concrete is estimated by testing a cylinder or cube to failure in a few minutes. Under sustained loads for years, the ultimate strength of concrete is reduced by about 30 percent. Under 1-day sustained loading, concrete may lose about 10 percent of its compressive strength. Sustained loads, as well as dynamic and impact effects if they occur on the structure, should be considered in the design of reinforced concrete members.

2.1.6 The Shape and Dimensions of the Tested Specimen

The common sizes of concrete specimens used to predict the compressive strength are either 6 by 12 in. (150 by 300 mm) cylinders or 6-in. (150-mm) cubes. When a given concrete is tested in compression by means of cylinders of like shape but of different sizes, the larger specimens give lower strength indexes. Table 2.2[4] gives the relative strength, for various sizes of cylinders, as a percentage of the strength of the standard cylinder; the heights of all cylinders are twice the diameters.

Sometimes concrete cylinders of nonstandard shape are tested. The greater the ratio of specimen height to diameter, the lower the strength indicated by the compression test. To compute the equivalent strength of the standard shape, the results must be multiplied by a correction factor. Approximate values of the correction factor are given in Table 2.3, extracted from ASTM C42-57. The relative strengths of a cylinder and a cube for different compressive strengths are shown in Table 2.4.

TABLE 2.2. Effect of size of compression specimen on strength of concrete

Size of cylinder (inch)	(mm)	Relative compressive strength
2 × 4	50 × 100	1.09
3 × 6	75 × 150	1.06
6 × 12	150 × 300	1.00
8 × 16	200 × 400	0.96
12 × 24	300 × 600	0.91
18 × 36	450 × 900	0.86
24 × 48	600 × 1200	0.84
36 × 72	900 × 1800	0.82

TABLE 2.3. Strength correction factor for cylinders of different height-diameter ratios

Ratio	2.0	1.75	1.50	1.25	1.10	1.00	0.75	0.5
Strength correction factor	1.00	0.98	0.96	0.94	0.90	0.85	0.70	0.50
Strength relative to standard cylinder	1.00	1.02	1.04	1.06	1.11	1.18	1.43	2.00

2.2   Required Strengths of Concrete

The different strengths of a particular concrete may be required in the design of reinforced concrete structures; these include the compressive, tensile, flexural, and shear strengths. Although it is sometimes difficult to perform tests to predict these strengths, it is possible to relate them to the compressive strength.

2.2.1   Compressive Strength

In designing structural members, it is assumed that the concrete resists compressive stresses and not tensile stresses; therefore, compressive strength is the criterion of quality of concrete. The other concrete stresses can be taken as a percentage of the compressive strength, which can be easily and accurately determined from tests. Specimens used to determine compressive strength may be cylindrical, cubical, or prismatic.

Test specimens in the form of a cylinder 6 in. (150 mm) across and 12 in. (300 mm) high are usually used to determine the strength of concrete. Details of the procedure are prescribed in ASTM Standard C39-72.

Cube specimens with sides of 6 in. (150 mm) or 8 in. (200 mm) are used in Great Britain, Germany, and other parts of Europe.

Prism specimens are used in France, the USSR, and other countries and are usually 70 by 70 by 350 mm or 100 by 100 by 500 mm. They are cast with their longer sides horizontal and tested, like cubes, in a position normal to the position of cast.

Before testing, the specimens are moist-cured and then tested at the age of 28 days by gradually applying a static load until rupture occurs. The rupture of the concrete specimen may be caused by: the applied tensile stress (failure in cohesion), the applied shearing stress (sliding failure), the compressive stress (crushing failure), or combinations of the above stresses.

The failure of the concrete specimen can be in one of three modes,[5] as shown in Figure 2.1. First, under axial compression, the concrete specimen may fail in shear

TABLE 2.4. Relative strengths of cylinder vs. cube[7]

Compressive strength	(psi)	1000	2200	2900	3500	3800	4900	5300	5900	6400	7300
	(N/mm²)	7.0	15.5	20.0	24.5	27.0	24.5	37.0	41.5	45.0	51.5
Strength ratio of cylinder to cube		0.77	0.76	0.81	0.87	0.91	0.93	0.94	0.95	0.96	0.96

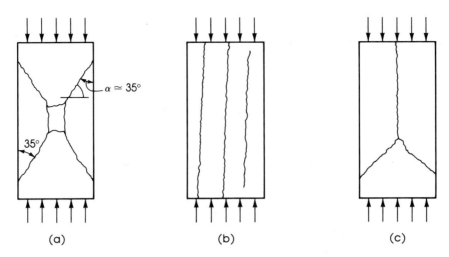

FIGURE 2.1. Modes of failure of standard concrete cylinders.

(Figure 2.1(a)). Resistance to failure is due to both cohesion and internal friction. The shear resistance may be represented by Coulomb's equation and Mohr's circle, shown in Figure 2.2:

$$v = c + P \tan \phi \tag{2.2}$$

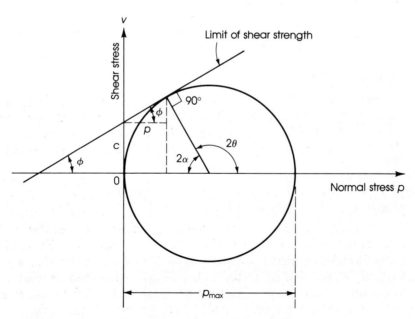

FIGURE 2.2. Mohr's circle showing the relation between angle of internal friction ϕ and angle of rupture α.

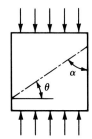

FIGURE 2.3. Mode of failure of standard concrete cube.

where

v = shear resistance

c = cohesion

ϕ = angle of internal friction

P = pressure between granules across the shear plane

If cohesion did not exist, then the angle α would be 45°, but due to the cohesion in the concrete particles, the angle α is $(45 - \phi/2)°$. The angle of internal friction of concrete can be assumed equal to 20°, in which case $\alpha = (45 - 20/2)° = 35°$.

The second type of failure (Figure 2.1(b)) results in the separation of the specimen into columnar pieces by what is known as splitting or columnar fracture. This failure occurs when the strength of concrete is high, and lateral expansion at the end bearing surfaces is relatively unrestrained.

The third type of failure (Figure 2.1(c)) is seen when a combination of shear and splitting failure occurs. Figure 2.3 shows the mode of failure of a standard cube in shear.

2.2.2   Stress-Strain Curves

The performance of a reinforced concrete member under load depends, to a great extent, on the stress-strain relationship of concrete and steel and on the type of stress applied to the member. Stress-strain curves for concrete are obtained by testing a concrete cylinder to rupture at the age of 28 days, and recording the strains at different load increments.

Figure 2.4 shows typical stress-strain curves for concretes of different strengths. All curves consist of an initial relatively straight elastic portion, reaching maximum stress at a strain of about 0.002; then rupture occurs at a strain of about 0.003.

Concrete having a compressive strength between 3000 and 6000 psi (21 and 42 N/mm²) is usually specified for reinforced concrete structures. For prestressed concrete, higher strengths, between 4000 and 8000 psi (28 and 56 N/mm²), may be adopted. Concrete of lower strength is less brittle. This means that concrete of low strength fails at a strain greater than that of high-strength concrete. Figure 2.5 shows strain values in concrete of different strengths plotted versus the stress/strength ratio.

At working loads, the concrete stress may reach half the maximum strength. If this ratio of 0.5 is considered, it can be seen that the stronger concrete has a higher deforma-

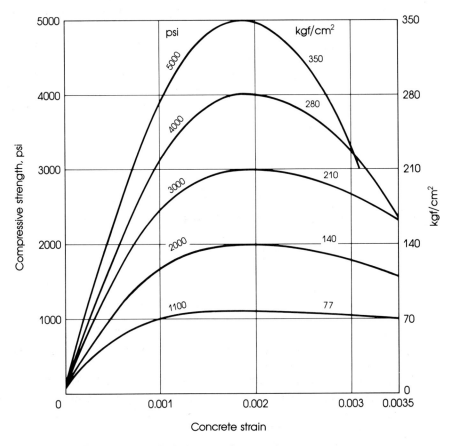

FIGURE 2.4. Typical stress-strain curves of concrete.

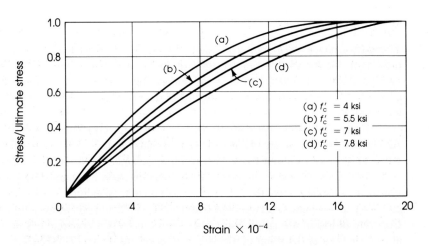

FIGURE 2.5. Stress/ultimate stress ratio versus strain for concrete strengths f'_c of (a) 4 ksi, (b) 5.5 ksi, (c) 7 ksi, (d) 7.8 ksi.

tion. But it can also be noticed that for any two concretes, the ratio of deformation is smaller than the ratio of strengths.

If a concrete test specimen is loaded over a part of its area, the strength is increased owing to the retaining influence of the surrounding concrete of the unloaded part. The ultimate strength in local compression or crushing may be found from the following expression:[6]

$$f'_{cl} = (A_2/A_1)^{1/3} f'_c \qquad (2.3)$$

where

f'_{cl} = ultimate strength of specimen partially loaded
f'_c = ultimate strength of cylinder or prism fully loaded
A_2 = total area of specimen
A_1 = loaded area

The areal ratio

$(A_2/A_1)^{1/3} \leq 1.5$ for the action of a local load only
$\qquad\qquad \leq 2.0$ for the action of local and other loads (2.4)

Given a nonsymmetrical application of the local load, A_2 is taken as symmetrical with respect to the center of gravity of the loaded area A_1.

For the above-mentioned condition, the **ACI Code, section 10.16,** specifies that the design bearing strength on the loaded area may be multiplied by the ratio $\sqrt{A_2/A_1}$, but not more than 2.

2.2.3   Tensile Strength

Concrete is a brittle material, and it cannot resist the high tensile stresses that are important when considering cracking, shear, and torsional problems. The low tensile capacity can be attributed to the high stress concentrations in concrete under load, so that a very high stress is reached in some portions of the specimen, causing microscopic cracks, while the other parts of the specimen are subjected to low stress.

Direct tension tests are not reliable for predicting the tensile strength of concrete, due to minor misalignment and stress concentrations in the gripping devices. An indirect tension test in the form of splitting a 6 by 12 in. (150 by 300 mm) cylinder was suggested by the Brazilian Fernando Carneiro. The test is usually called the splitting test, but it was also named the Brazilian test (although it was developed independently in Japan). In this test, the concrete cylinder is placed with its axis horizontal in a compression testing machine. The load is applied uniformly along two opposite lines on the surface of the cylinder through two plywood pads, as shown in Figure 2.6. Considering an element on the vertical diameter and at a distance y from the top fibers, the element is subjected to a compressive stress

$$f_c = \frac{2P}{\pi LD} \left(\frac{D^2}{y(D-y)} - 1 \right) \qquad (2.5)$$

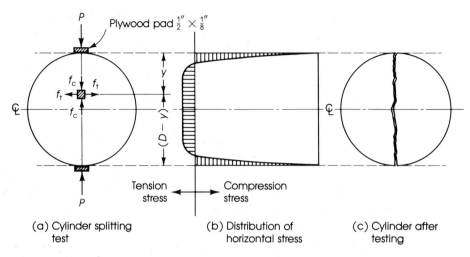

(a) Cylinder splitting   (b) Distribution of   (c) Cylinder after
test   horizontal stress   testing

FIGURE 2.6. Cylinder splitting test:[7] (a) configuration of test, (b) distribution of horizontal stress, and (c) cylinder after testing.

and a tensile stress

$$f'_{sp} = \frac{2P}{\pi L D} \tag{2.6}$$

where P = the compressive load on the cylinder and D and L = the diameter and length of the cylinder. For a 6 by 12 in. (150 by 300 mm) cylinder and at a distance $y = D/2$, the compression strength is

$$f_c = 0.0265P \tag{2.7}$$

and the tensile strength

$$f'_{sp} = 0.0088P = f_c/3 \tag{2.8}$$

Cubes can also be subjected to a splitting test. The load is applied through semicylindrical steel rollers resting against the cube.

The splitting strength f'_{sp} can be related to the compressive strength of concrete in that it varies between 6 and 7 times $\sqrt{f'_c}$ for normal concrete and between 4 and 5 times $\sqrt{f'_c}$ for lightweight concrete.* The direct tensile stress f'_t can also be estimated from the split test; its value varies between $0.5\,f'_{sp}$ and $0.7\,f'_{sp}$. The smaller of the above values applies to higher strength concrete. The splitting strength f'_{sp} can be estimated as 10 percent of the compressive strength up to $f'_c = 6000$ psi (42 N/mm²). For higher values of compressive strength, f'_{sp} can be taken as 9 percent of f'_c.

In general, the tensile strength of concrete ranges from 7 to 11 percent of its compressive strength with an average of 10 percent. The lower the compressive strength, the higher the relative tensile strength.

2.2.4   Flexural Strength (Modulus of Rupture)

Experiments on concrete beams have shown that ultimate tensile strength in bending is greater than the tensile stresses obtained by direct or splitting tests.

---

* $\sqrt{f'_c}$ psi = 0.083 $\sqrt{f'_c}$ MPa = 0.265 $\sqrt{f'_c}$ kgf/cm².

Flexural strength is expressed in terms of the modulus of rupture of concrete (f_r), which is the maximum tensile stress in concrete in bending. The modulus of rupture can be calculated from the flexural formula used for elastic materials, $f_r = (M \cdot c)/I$, by testing a plain concrete beam. The beam of 6 by 6 by 28 in. (150 by 150 by 700 mm) is supported on a 24-in. (600-mm) span and loaded to rupture by two loads, 4 in. (100 mm) on either side of the center. A smaller beam of 4 by 4 by 20 in. (100 by 100 by 500 mm) on a 16-in. (400-mm) span may also be used.

The modulus of rupture of concrete ranges between 11 and 23 percent of the compressive strength. A value of 15 percent can be assumed for strengths of about 4000 psi (28 N/mm²).

A number of empirical formulas relate the modulus of rupture of concrete to its compressive strength. The CEB gives the value of the modulus of rupture f_r as

$$f_r = 9.5\sqrt{f'_c} \text{ (psi)}$$
$$\qquad = 0.79\sqrt{f'_c} \text{ (N/mm}^2) \tag{2.9}$$

The **ACI Code (318-83)** prescribes the value

$$f_r = 7.5\sqrt{f'_c} \text{ (psi)}$$
$$\qquad = 0.62\sqrt{f'_c} \text{ (N/mm}^2) \tag{2.10}$$

The modulus of rupture as related to the strength obtained from the split test on cylinders may be taken as

$$f_r = (1.25 \text{ to } 1.50)f'_{sp} \tag{2.11}$$

2.2.5   Shear Strength

Pure shear is seldom encountered in reinforced concrete members, as it is usually accompanied by the action of normal forces. An element subjected to pure shear breaks transversely into two parts. Therefore, the concrete element must be strong enough to resist the applied shear forces.

Shear strength may be considered as 20 to 30 percent greater than the tensile strength of concrete, or about 12 percent of its compressive strength. The ACI Code allows an ultimate shear strength of $2\sqrt{f'_c}$ psi ($0.17\sqrt{f'_c}$ N/mm²) on plain concrete sections.

2.3   Modulus of Elasticity of Concrete

One of the most important elastic properties of concrete is its modulus of elasticity. This can be obtained from compressive tests on concrete cylinders. The modulus of elasticity E_c can be defined as the change of stress with respect to strain in the elastic range:

$$E_c = \frac{\text{unit stress}}{\text{unit strain}} \tag{2.12}$$

2.3.1   Modulus of Elasticity of Concrete in Compression

The modulus of elasticity is a measure of stiffness, or the resistance of the material to deformation. In concrete, as in any elastoplastic material, the stress is not proportional to the strain, and the stress-strain relationship is a curved line. A general form of expressing this relationship is

$$\text{Strain} = K(\text{stress})^m \qquad (2.13)$$

or

$$\varepsilon = K f^m$$

where ε = strain, f = stress, and K and m are constants. The actual stress-strain curve of concrete can be obtained by measuring the strains under increments of loading on a standard cylinder.

The "initial tangent modulus" (Figure 2.7) is represented by the slope of the tangent to the curve at the origin under elastic deformation. This modulus is of limited value and cannot be determined with accuracy. Geometrically, the tangent modulus of elasticity of concrete E_c is the slope of the tangent to the stress-strain curve at a given stress. Under long-time action of load, and due to the development of plastic deformation, the stress-to-total-strain ratio becomes a variable quantity; hence, the variable modulus of elasticity of concrete E_c is a derivative of the stress with respect to strain:

$$E_c = \frac{df_c}{d\varepsilon_c} \qquad (2.14)$$

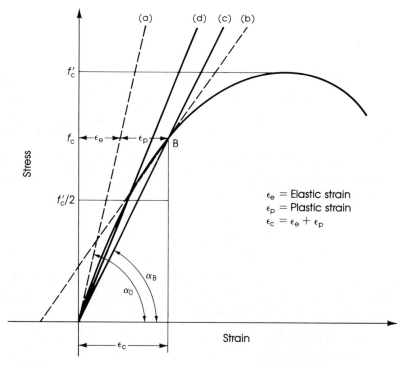

FIGURE 2.7. Stress-strain curve and modulus of elasticity of concrete. Lines a–d represent (a) initial tangent modulus, (b) tangent modulus at a stress f_c, (c) secant modulus at a stress f_c, and (d) secant modulus at a stress $f_c'/2$.

Since it is difficult to determine E_c this way, the "secant modulus" can be used. The secant modulus is represented by the slope of a line drawn from the origin to a specific point of stress (B) on the stress-strain curve. Referring to Figure 2.7, and letting ε_e and ε_p be the elastic and plastic strains at a level of stress f_c, then the total concrete strain $\varepsilon_c = \varepsilon_e + \varepsilon_p$. If

$$\beta = \frac{\varepsilon_p}{\varepsilon_c} = \text{coefficient of plastic strain}$$

and

$$\gamma = \frac{\varepsilon_e}{\varepsilon_c} = \text{coefficient of elastic strain}$$

$$= \frac{\varepsilon_c - \varepsilon_p}{\varepsilon_c} = (1 - \beta)$$

and

$$E_e = \tan \alpha_0$$

then

$$f_c = E_c \varepsilon_c = E_e \varepsilon_e$$

or

$$E_c = E_e \frac{\varepsilon_e}{\varepsilon_c} = \gamma E_e = (1 - \beta) E_e \tag{2.15}$$

For an ideally elastic material, $\varepsilon_p \to 0$, $\beta \to 0$, and $E_c = E_e$. For an ideally plastic material, $\varepsilon_p \to \infty$ and $\gamma \to 0$. Since concrete is an elastoplastic material, the value of γ depends on the magnitude of stress and the duration of the load that causes creep action on concrete. The value of β may vary from zero to a maximum (under long-time loading) of 0.8.

The **ACI Code (318-83)** gives a simple formula for calculating the modulus of elasticity of normal and lightweight concrete considering the secant modulus at a level of stress f_c equal to half the ultimate concrete strength f'_c:

$$E_c = 33 w^{1.5} \sqrt{f'_c} \text{ psi } (w \text{ in pcf})$$
$$= 0.043 w^{1.5} \sqrt{f'_c} \text{ N/mm}^2 \tag{2.16}$$
$$= 0.01368 w^{1.5} \sqrt{f'_c} \text{ kgf/cm}^2 \ (w \text{ in kg/m}^3)$$

where w = unit weight of concrete (between 90 and 155 pounds per cubic foot or 1400–2500 kg/m³), and f'_c = ultimate strength of a standard concrete cylinder. For normal weight concrete, w is approximately 145 pcf (2320 kg/m³); thus

$$E_c = 57,400 \sqrt{f'_c} \text{ psi}$$
$$= 4730 \sqrt{f'_c} \text{ MPa} \tag{2.17}$$
$$= 15,200 \sqrt{f'_c} \text{ kgf/cm}^2$$

The modulus of elasticity E_c for different values of f'_c are shown in Table A.10 of Appendix A.

2.3.2   Shear Modulus

The modulus of elasticity of concrete in shear ranges from about 0.4 to 0.6 of the corresponding modulus in compression. From the theory of elasticity, the shear modulus is taken as follows:

$$G_c = \frac{E_c}{2(1 + \mu)} \tag{2.18}$$

where μ = Poisson's ratio of concrete. If μ is taken equal to $\frac{1}{6}$, then $G_c = 0.43 E_c$.

2.4   Poisson's Ratio

Poisson's ratio μ is the ratio of the transverse to the longitudinal strains under axial stress within the elastic range. This ratio varies between 0.15 and 0.20 for both normal and lightweight concrete. Poisson's ratio is used in structural analysis of flat slabs, tunnels, tanks, arch dams, and other statically indeterminate structures. For isotropic elastic materials, Poisson's ratio is equal to 0.25. An average value of 0.18 can be used for concrete.

2.5   Modular Ratio

Modular ratio n is the ratio of the modulus of elasticity of steel to the modulus of elasticity of concrete:

$$n = \frac{E_s}{E_c}$$

Since the modulus of elasticity of steel is considered constant and is equal to 29×10^6 psi **(ACI Code 318-83)**, and $E_c = 33 w^{1.5} \sqrt{f_c'}$, therefore

$$n = \frac{29 \times 10^6}{33 w^{1.5} \sqrt{f_c'}} \tag{2.19}$$

For normal weight concrete, $E_c = 57,400 \sqrt{f_c'}$. Hence n can be taken as

$$n = 500/\sqrt{f_c'} (f_c' \text{ in psi})$$
$$= 42/\sqrt{f_c'} (f_c' \text{ in N/mm}^2) \tag{2.20}$$
$$= 130/\sqrt{f_c'} (f_c' \text{ in kgf/cm}^2)$$

The significance and the use of the modular ratio are explained in Chapter 5.

2.6   Volume Changes of Concrete

Concrete undergoes volume changes during hardening. If it loses moisture by evaporation, it shrinks, but if the concrete hardens in water, it expands. The causes of the volume changes in concrete can be attributed to changes in moisture content, chemical reaction of the cement with water, variation in temperature, and applied loads.

2.6.1   Shrinkage

The change in the volume of drying concrete is not equal to the volume of water removed.[7] The evaporation of free water causes little or no shrinkage. As concrete continues to dry, water evaporates and the volume of the restrained cement paste changes, causing concrete to shrink, probably due to the capillary tension that develops in the water remaining in concrete. Emptying of the capillaries causes a loss of water without shrinkage. But once the absorbed water is removed, shrinkage occurs.

Many factors influence the shrinkage of concrete caused by the variations in moisture conditions.[5]

1. *Cement and water content.* The more cement or water content in the concrete mix, the greater the shrinkage.
2. *Composition and fineness of cement.* High-early-strength and low-heat cements show more shrinkage than normal portland cement. The finer the cement, the greater the expansion under moist conditions.
3. *Type, amount, and gradation of aggregate.* The smaller the size of aggregate particles, the greater the shrinkage. The greater the aggregate content, the smaller the shrinkage.[14]
4. *Ambient conditions, moisture and temperature.* Concrete specimens subjected to moist conditions undergo an expansion of 200 to 300×10^{-6}, but if they are left to dry in air, they shrink. High temperature speeds the evaporation of water and consequently increases shrinkage.
5. *Admixtures.* Admixtures that increase the water requirement of concrete increase the shrinkage value.
6. *Size and shape of specimen.* Under drying conditions, shrinkage of large masses of concrete varies from the surface to the interior, causing different shrinkage values. This will cause internal tensile stresses, and cracks may develop.
7. *Amount and distribution of reinforcement.* As shrinkage takes place in a reinforced concrete member, tension stresses develop in the concrete and equal compressive stresses develop in the steel. These stresses are added to those developed by the loading action. Therefore, cracks may develop in concrete when a high percentage of steel is used. Proper distribution of reinforcement, by producing better distribution of tensile stresses in concrete, can reduce differential internal stresses.

The values of final shrinkage for ordinary concrete vary between 200 and 700×10^{-6}. For normal weight concrete, a value of 300×10^{-6} may be used. The British Code CP110[12] gives a value of 500×10^{-6}, which represents an unrestrained shrinkage of 1.5 mm in 3 m length in thin plain concrete sections. If the member is restrained, a tensile stress of about 10 N/mm^2 (1400 psi) arises. If concrete is kept moist for a certain period after setting, shrinkage is reduced; therefore it is important to cure the concrete for a period of no fewer than 7 days.

Exposure of concrete to wind increases the shrinkage rate on the upwind side. Shrinkage causes an increase in the deflection of structural members, which in turn increases with time. Symmetrical reinforcement in the concrete section may prevent curvature and deflection due to shrinkage.

Generally, concrete shrinks at a high rate during the initial period of hardening, but at later stages the rate diminishes gradually. It can be said that 15 to 30 percent of the shrinkage value occurs in 2 weeks, 40 to 80 percent in 1 month, and 70 to 85 percent in 1 year.

2.6.2   Expansion Due to Rise in Temperature

Concrete expands with increasing temperature and contracts with decreasing temperature. The coefficient of thermal expansion of concrete varies between 4 and 7×10^{-6} per degree Fahrenheit. An average value of 5.5×10^{-6} per degree Fahrenheit (12×10^{-6} per degree Celsius) can be used for ordinary concrete. The B.S. Code[12] suggests a value of 10^{-5} per degree Celsius. This value represents a change of length of 10 mm in a 30-m member subjected to a change in temperature of 33°C. If the member is restrained and unreinforced, a stress of about 7 N/mm^2 (1000 psi) may develop.

In long reinforced concrete structures, expansion joints must be provided at lengths of 100 to 200 ft (20–60 m). The width of the expansion joint is about 1 in. (25 mm). Concrete is not a good conductor of heat, while steel is a good one. The ability of concrete to carry load is not much affected by temperature.

2.7   Creep

Concrete is an elastoplastic material, and beginning with small stresses, plastic strains develop in addition to elastic ones. Under sustained load as shown in Figure 2.8, plastic deformation continues to develop over a period that may last for years. Such deformation increases at a high rate during the first 4 months after application of the load. This slow plastic deformation under constant stress is called creep.

Figure 2.9 shows a concrete cylinder that is loaded. The instantaneous deformation is ε_1, which is equal to the stress divided by the modulus of elasticity. If the same stress is kept for a period of time, an additional strain ε_2, due to creep effect, can be recorded. If load is then released, the elastic strain ε_1 in addition to some creep strain will be recovered. The final permanent plastic strain ε_3 will be left, as shown in Figure 2.9. In this case $\varepsilon_3 = (1 - \alpha)\varepsilon_2$, where α is the ratio of the recovered creep strain to the total creep strain. The ratio α ranges between 0.1 and 0.2. The magnitude of creep recovery varies with the previous creep and depends appreciably upon the period of the sustained load. Creep recovery rate will be less if the loading period is increased, probably due to the hardening of concrete while in a deformed condition.

The ultimate magnitude of creep varies between 0.2×10^{-6} and 2×10^{-6} per unit stress (lb/in.^{2}) per unit length. A value of 1×10^{-6} can be used in practice. The ratio of creep strain to elastic strain may be as high as 4.

Creep takes place in the hardened cement matrix around the strong aggregate. It may be attributed to slippage along planes within the crystal lattice, internal stresses caused by changes in the crystal lattice, and gradual loss of water from the cement gel in the concrete.

The different factors that affect the creep of concrete can be summarized as follows.[15]

1. *The level of stress.* Creep increases with an increase of stress in specimens made from concrete of the same strength and with the same duration of load.

FIGURE 2.8. Creep test on concrete cylinders.

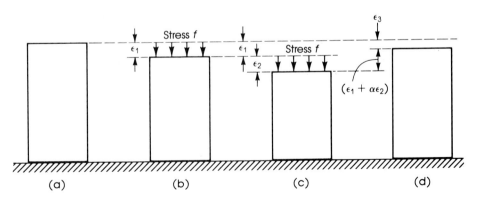

FIGURE 2.9. Deformation in a loaded concrete cylinder: (a) specimen unloaded, (b) elastic deformation, (c) elastic plus creep deformation, (d) permanent deformation after release of load.

2. *Duration of loading.* Creep increases with the loading period. About 80 percent of the creep occurs within the first 4 months, 90 percent after about 2 years.

3. *Strength and age of concrete.* Creep tends to be smaller if concrete is loaded at a late age. Also, creep of 2000 psi (14 N/mm^2) strength concrete is about 1.41×10^{-6} while that of 4000 psi (28 N/mm^2) strength concrete is about 0.8×10^{-6} per unit stress and length of time.

4. *The ambient conditions.* Creep is reduced with an increase in the humidity of the ambient air.

5. *Rate of loading.* Creep increases with an increase in the rate of loading when followed by prolonged loading.

6. *The percentage and distribution of steel reinforcement in a reinforced concrete member.* Creep tends to be smaller for higher proportion or better distribution of steel.

7. *Size of the concrete mass.* Creep decreases with an increase in the size of the tested specimen.

8. *The type, fineness, and content of cement.* The amount of cement greatly affects the final creep of concrete as cement creeps about 15 times as much as concrete.

9. *The water-cement ratio.* Creep increases with an increase in the water-cement ratio.

10. *Type and grading of aggregate.* Well-graded aggregate will produce dense concrete and consequently a reduction in creep.

11. *Type of curing.* High-temperature steam curing of concrete as well as the proper use of a plasticizer will reduce the amount of creep. Figure 2.10 shows the variation of creep deformation with time.[5]

Creep develops not only in compression, but also in tension, bending, and torsion.

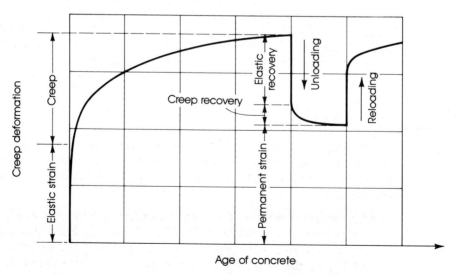

FIGURE 2.10. Elastic and creep deformation of concrete under loading, release of load, and reloading conditions.

The ratio of the rate of creep in tension to that in compression will be greater than 1 in the first two weeks, but this ratio decreases over longer periods.[5]

Creep in concrete under compression has been tested by many investigators. Troxell, Davis, and Raphael[16] measured creep strains periodically for up to 20 years and estimated that of the total creep after 20 years, 18–35 percent occurred in 2 weeks, 30–70 percent occurred in 3 months, and 64–83 percent occurred in 1 year.

Various expressions relating creep with time have been suggested by different investigators. One simple expression is

$$C_r = x \sqrt[y]{t} \qquad\qquad (2.21)$$

where

C_r = creep strain per unit stress per unit length

x, y = constants to be determined by experiments

t = time in days

For normal concrete loaded after 28 days, $C_r = 0.13\sqrt[3]{t}$. The stress range limit due to creep as suggested by **ACI Committee 215** for a minimum stress f_{min} is

$$f_{cr} = 0.4f'_c - \tfrac{1}{2}f_{min} \qquad\qquad (2.22)$$

Creep augments the deflection of reinforced concrete beams appreciably with time. In the design of reinforced concrete members, long-term deflection may be critical and has to be considered in proper design. Extensive deformation may influence the stability of the structure.

Sustained loads affect the strength as well as the deformation of concrete. A reduction of up to 30 percent of the strength of unreinforced concrete may be expected when concrete is subjected to a concentric sustained load for 1 year.

The fatigue strength of concrete is much smaller than its static strength. Repeated loading and unloading cycles in compression lead to a gradual accumulation of plastic deformations. If concrete in compression is subjected to about 2 million cycles, its fatigue limit is about 50 to 60 percent of the static compression strength. In beams, the fatigue limit of concrete is about 55 percent of its static strength.[17]

2.8   Unit Weight of Concrete

The unit weight w of hardened normal concrete ordinarily used in buildings and similar structures depends on the concrete mix, maximum size and grading of aggregates, water-cement ratio, and strength of concrete. The following values of the unit weight of concrete may be used:

- Unit weight of plain concrete using maximum aggregate size of $\tfrac{3}{4}$ in. (20 mm) varies between 145 and 150 lb/ft³ (2320–2400 kg/m³). For concrete of strength less than 4000 psi (280 kg/cm²) a value of 145 lb/ft³ (2320 kg/m³) can be used, while for higher strength concretes, w can be assumed equal to 150 lb/ft³ (2400 kg/m³).
- Unit weight of plain mass concrete of maximum aggregate size of 4 to 6 in.

(100–150 mm) varies between 150 and 160 lb/ft³ (2400–2560 kg/m³). An average value of 155 lb/ft³ (2500 kg/m³) may be used.

- Unit weight of reinforced concrete, using about 0.7 to 1.5 percent of steel in the concrete section, may be taken as 150 lb/ft³ (2400 kg/m³). For higher percentages of steel, the unit weight w can be assumed to be 155 lb/ft³ (2500 kg/m³).
- Unit weight of structural lightweight concrete varies between 90 and 120 lb/ft³ (1440–1920 kg/m³).
- Unit weight of lightweight concrete used for fireproofing, masonry, or insulation purposes varies between 20 and 90 lb/ft³ (320–1440 kg/m³). Concretes of upper values may be used for load-bearing concrete members.
- Unit weight of heavy concrete varies between 200 and 270 lb/ft³ (3200–4300 kg/m³). Heavy concrete made with natural barite aggregate of $1\frac{1}{2}$ in. maximum size (38 mm) weighs about 225 lb/ft³ (3600 kg/m³). Iron ore sand and steel-punchings aggregate produce a unit weight of 270 lb/ft³ (4320 kg/m³).[19]

2.9   Fire Resistance

Fire resistance of a material is its ability to resist fire for a certain time without serious loss of strength, distortion, or collapse.[20] In the case of concrete, fire resistance depends on the thickness, type of construction, type and size of aggregates, and cement content. It is important to consider the effect of fire on tall buildings more than low or single-story buildings, because occupants need more time to escape.

Reinforced concrete is a much better fire-resistant material than steel. Steelwork heats rapidly and its strength drops appreciably in a short time. Concrete itself has low thermal conductivity. The effect of temperatures below 250°C is small on concrete, but definite loss is expected at higher temperatures. Flexural strength more than compressive strength seems to be affected by fire.[7]

The use of 1-in. insulation materials such as vermiculite on top of concrete slabs will increase the fire resistance by about 2 hours.[21]

2.10   Steel Reinforcement

Reinforcement, usually in the form of steel bars, is placed in the concrete member, mainly in the tension zone, to resist the tensile forces resulting from external load on the member. Reinforcement is also used to increase the member's compression resistance. Steel costs more than concrete, but it has a yield strength about 10 times the compressive strength of concrete. The function and behavior of both steel and concrete in a reinforced concrete member are discussed in Chapter 3.

Longitudinal bars taking either tensile or compression forces in a concrete member are called main reinforcement. Additional reinforcement in slabs, in a direction perpendicular to the main reinforcement, is called secondary or distribution reinforcement. In reinforced concrete beams, another type of steel reinforcement is used, transverse to the direction of the main steel, and bent in a box or U shape. These are called stirrups. Similar reinforcements are used in columns, where they are called ties (Figure 2.11).

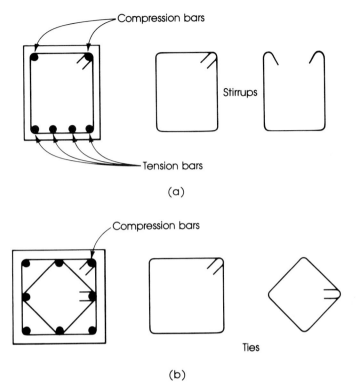

FIGURE 2.11. Reinforced concrete reinforcement: (a) beam and (b) column.

2.10.1   Types of Steel Reinforcement

Different types of steel reinforcement are used in various reinforced concrete members. These types can be classified as follows:

Round bars.   These are used most widely for reinforced concrete. Round bars are available in a large range of diameters from $\frac{1}{4}$ in. (6 mm) to $1\frac{3}{8}$ in. (36 mm), plus two special types of $1\frac{3}{4}$ in. (45 mm) and $2\frac{1}{4}$ in. (57 mm). Round bars, depending on their surfaces, are either plain or deformed bars. Plain bars are used mainly for secondary reinforcement or in stirrups and ties. Deformed bars have projections or deformations on the surface for the purpose of improving the bond with concrete and reducing the width of cracks opening in the tension zone.

The diameter of a plain bar can be measured easily, but for a deformed bar a nominal diameter is used that is the diameter of a circular surface with the same area as the section of the deformed bar. Requirements of surface projections on bars are specified by ASTM Specification A305, or (A615-75). The bar sizes are designated by numbers 2 through 11, corresponding to the diameter in eighths of an inch. For instance, No. 7 bar has a nominal diameter $\frac{7}{8}$ in. and No. 4 bar has a nominal diameter of $\frac{1}{2}$ in. The two largest sizes are designated No. 14 and No. 18, respectively. American standard bar marks are shown on the steel reinforcement to indicate the initial of the producing mill, the bar size, and the type of steel (Figure 2.12). The grade of the reinforcement is indicated on the bars by either the continuous line system or the number system. In the first system, one longitudinal line is added to the bar, in addition to the main ribs, to indicate the high-strength grade of 60 ksi (420 N/mm²) according to ASTM Specification A617. If only the main ribs are shown on the bar without any additional lines, the steel is of the ordinary grade according to ASTM A615 for the structural grade (f_y = 40 ksi, or 280 N/mm²). In the number system, the yield strength of the high-strength grades is marked clearly on every bar. For ordinary grades, no strength marks are indicated. The two types are shown in Figure 2.12.

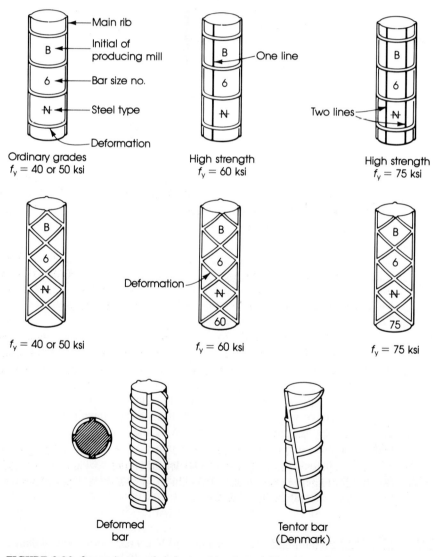

FIGURE 2.12. Some types of deformed bars and American standard bar marks.

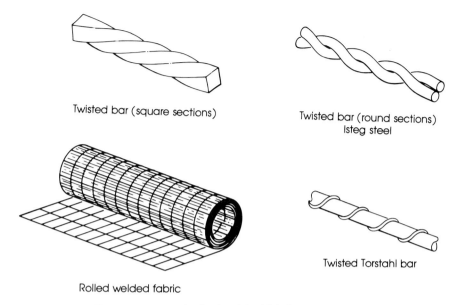

Twisted bar (square sections)

Twisted bar (round sections)
Isteg steel

Twisted Torstahl bar

Rolled welded fabric

FIGURE 2.13. Twisted bars and rolled welded fabrics.

Twisted bars.   These are used in reinforced concrete members in Europe, specifically in Austria. One type is I-Steg steel bars, made by cold-twisting two round mild steel bars in special machines. Torstahl steel bars are made by cold-twisting a deformed round mild steel bar (Figure 2.13).

Welded fabrics, mats, expanded metal, and cages.   These consist of a series of longitudinal and transverse cold-drawn steel wires, generally at right angles, and welded together at all points of intersections. Steel reinforcement may be built up into three-dimensional cages before being placed in the forms.

Prestressed concrete wires and strands.   These use special high-strength steel (see Chapter 18). High-tensile steel wires of diameters 0.192 in. (5 mm) and 0.276 in. (7 mm) are used to form the prestressing cables by winding six steel wires around a seventh wire of slightly larger diameter.

2.10.2   Grades and Strength

Different grades of steel are used in reinforced concrete. Limitations on the minimum yield strength, ultimate strength, and elongation are explained in ASTM Specifications for reinforcing steel bars (Table 2.5). The grades of some types of reinforcing steel used in Europe are shown in Table 2.6.

2.10.3   Stress-Strain Curves

The most important factor affecting the mechanical properties and stress-strain curve of the steel is its chemical composition. The introduction of carbon and alloying additives

TABLE 2.5. Grades of ASTM reinforcing steel bars

Steel	Minimum yield strength f_y			Ultimate strength f_{su}		
	ksi	kgf/cm²	MPa	ksi	kgf/cm²	MPa
Billet steel						
Grade 40	40	2800	276	70	4900	483
60	60	4200	414	90	6300	621
75	75	5200	518	100	7000	690
Rail steel						
Grade 50	50	3500	345	80	5600	551
60	60	4200	414	90	6300	621
Deformed wire						
Reinforcing	75	5200	518	85	6000	586
Fabric	70	4900	483	80	5600	551
Cold-drawn wire						
Reinforcing	70	4900	483	80	5600	551
Fabric	65	4500	448	75	5200	518
Fabric	56	3900	386	70	4900	483

in steel increases its strength but reduces its ductility. Commercial steel rarely contains more than 1.2 percent carbon; the proportion of carbon used in structural steels varies between 0.2 and 0.3 percent. Figure 2.14 shows a typical stress-strain curve of a mild steel reinforcing bar.

Two other properties are of interest in the design of reinforced concrete structures; first is the modulus of elasticity E_s. It has been shown that the modulus of elasticity is constant for all types of steel. The **ACI Code (318-83)** has adopted a value of E_s of 29 × 10⁶ psi (2.1 × 10⁶ kgf/cm², or 2.0 × 10⁵ MPa). The modulus of elasticity is the slope of the stress-strain curve in the elastic range up to the proportional limit:

$$E_s = \frac{\text{stress}}{\text{strain}}$$

Second is the yield strength f_y. A typical stress-strain curve for a bar of low-carbon steel with a yield strength of 50 ksi (350 MPa) or less usually shows an elastic portion followed by a definite yield point at the yield stress level (see Figure 2.15). Immediately after the

TABLE 2.6. Grades of some European mild steel

Grade	Elastic limit			Yield strength			Ultimate strength		
	kgf/cm²	ksi	MPa	kgf/cm²	ksi	MPa	kgf/cm²	ksi	MPa
Steel 33	2200	32	216	2400	34	235	3300	47	323
37	2400	34	235	2800	40	276	3700	53	363
44	2800	40	276	3300	47	323	4400	63	431
52	3800	54	372	4200	60	412	5200	75	518
Torstahl	3500	50	343	3800	54	372	4800	68	470
I-Steg	3500	50	343	3800	54	372	4800	68	470

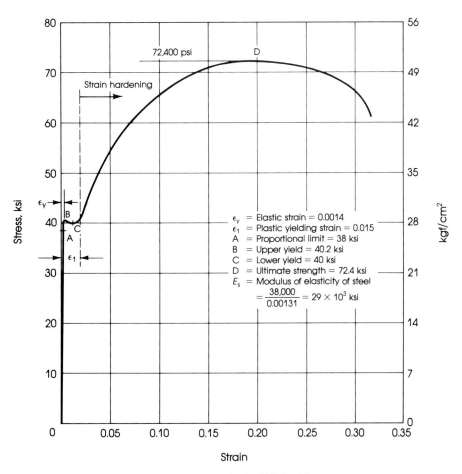

FIGURE 2.14. Stress-strain curve for a tested mild steel bar.

yield stress is reached, the strain increases by about 8 to 15 times the initial yield strain ε_y.

In high-tensile steel, a definite yield point may not show on the stress-strain curve. In this case, ultimate strength is reached gradually under an increase of stress (Figure 2.15). The yield strength or proof stress is considered as that stress which leaves a residual strain of 0.2 percent on the release of load, or a total strain of 0.5 to 0.6 percent under load.

SUMMARY

SECTION 2.1

The main factors that affect the strength of concrete are the water-cement ratio, properties and proportions of materials, age of concrete, loading conditions, and shape of tested specimen.

$$f'_c \text{ (cylinder)} = 0.85 f'_c \text{ (cube)} = 1.10 f'_c \text{ (prism)}$$

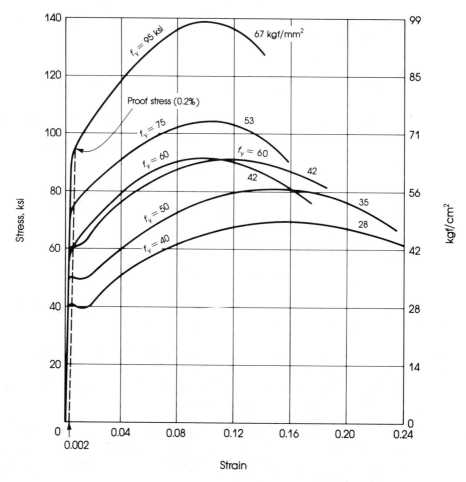

FIGURE 2.15. Typical stress-strain curves for some reinforcing steel bars of different grades. Note that 60-ksi steel may or may not show a definite yield point.

1. The usual specimen used to determine the compressive strength of concrete at 28 days is a 6 by 12 in. (150 by 300 mm) cylinder. Compressive strength between 3000 and 6000 psi is usually specified for reinforced concrete structures. Maximum stress f'_c is reached at an estimated strain of 0.002, while rupture occurs at a strain of about 0.003.

2. Tensile strength of concrete is measured indirectly by a split-test performed on a standard cylinder using the formula

$$f'_{sp} = \frac{2P}{\pi LD} \tag{2.6}$$

Tensile strength of concrete is approximately $0.1f'_c$.

3. Flexural strength (modulus of rupture f_r) of concrete is calculated by testing a 6 by 6 by 28 in. plain concrete beam.

$$f_r = 7.5\sqrt{f'_c} \tag{2.10}$$

4. Ultimate shear strength $= 2\sqrt{f'_c}$

SECTION 2.3

The modulus of elasticity of concrete E_c, for unit weight w between 90 and 155 pcf, is

$$E_c = 33w^{1.5}\sqrt{f'_c} \text{ (psi)} \tag{2.16}$$

For normal weight concrete ($w = 145$ pcf),

$$E_c = 57,400\sqrt{f'_c} \tag{2.17}$$

The shear modulus of concrete G_c is

$$G_c = \frac{E_c}{2(1 + \mu)} \tag{2.18}$$
$$= 0.43E_c \text{ for a Poisson's ratio } \mu = 1/6$$

SECTIONS 2.4 and 2.5

1. Poisson's ratio μ varies between 0.15 and 0.20 with an average value of 0.18.
2. Modular ratio $n = E_s/E_c = 500/\sqrt{f'_c}$ where f'_c is in psi.

SECTION 2.6

1. Values of shrinkage for normal concrete fall between 200×10^{-6} and 700×10^{-6}. An average value of 300×10^{-6} may be used.
2. The coefficient of expansion of concrete falls between 4×10^{-6} and $7 \times 10^{-6}/°F$. The B.S. Code suggests a value of $10^{-5}/°C$.

SECTION 2.7

The ultimate magnitude of creep varies between 0.2×10^{-6} and 2×10^{-6} per unit stress per unit length. An average value of 1×10^{-6} may be adopted in practical problems. Of the ultimate (20-year) creep, 18–35 percent occurs in 2 weeks, 30–70 percent in 3 months, and 64–83 percent in 1 year.

SECTION 2.8

The unit weight of normal concrete is 145 pcf for plain concrete and 150 pcf for reinforced concrete.

SECTION 2.9

Reinforced concrete is a much better fire-resistant material than steel. Concrete itself has low thermal conductivity. An increase in concrete cover in structural members such as walls, columns, beams, and floor slabs will increase the fire resistance of these members.

SECTION 2.10

The grades of steel mainly used are grade 40 ($f_y = 40$ ksi) and grade 60 ($f_y = 60$ ksi). The modulus of elasticity of steel $E_s = 29 \times 10^6$ psi.

REFERENCES

1. Portland Cement Association: *Design and Control of Concrete Mixtures,* Chicago, 1979.
2. British Standard Institution: *B. S. Code of Practice For Reinforced Concrete,* CP 114, 2nd edition, 1957.

3. H. E. Davis, G. E. Troxell and C. T. Wiskocil: *The Testing and Inspection of Engineering Materials,* McGraw-Hill, New York, 1964.

4. United States Bureau of Reclamation: *Concrete Manual,* 7th edition, 1963.

5. G. E. Troxell and H. E. Davis: *Composition and Properties of Concrete,* McGraw-Hill, New York, 1956.

6. V. Murashev and V. B. Sigalov: *The Design of Reinforced Concrete Structures,* Mir Publications, Moscow, 1968.

7. A. M. Neville: *Properties of Concrete,* Pitman and Sons, London, 1965.

8. H. Gonnerman and E. C. Shuman: "Compression, Flexural and Tension Tests of Plain Concrete," *ASTM Proceedings,* 28, Part II, 1928.

9. S. Musa: "The Effect of Steel Fibers on the Behavior of High Strength Reinforced Concrete Beams," M. S. Thesis, South Dakota State University, August 1982.

10. American Concrete Institute and the Cement and Concrete Association: *Recommendations for an International Code of Practice for Reinforced Concrete,* London, 1963.

11. T. C. Powers: "Measuring Young's Modulus of Elasticity by Means of Sonic Vibrations," *ASTM Proceedings,* 40, 1940.

12. British Standard Institution: *B. S. Code of Practice for Reinforced Concrete,* CP 110, London, 1973.

13. L. M. Legatski: *Cellular Concrete,* Elastizell Corp. of America, Ann Arbor, Michigan, 1980.

14. G. Pickett: "Effect of Aggregate on Shrinkage of Concrete and Hypothesis Concerning Shrinkage," *ACI Journal,* 52, Jan. 1956.

15. "Symposium on Shrinkage and Creep of Concrete," *ACI Journal,* 53, Dec. 1957.

16. G. E. Troxell, J. M. Raphael and R. E. Davis: "Long Time Creep and Shrinkage Tests of Plain and Reinforced Concrete," *ASTM Proceedings,* 58, 1958.

17. "Fatigue of Concrete—Reviews of Research," *ACI Journal,* 55, 1958.

18. A. Ruettgers, E. N. Vidal and S. P. Wing: "An Investigation of the Permeability of Mass Concrete with Particular Reference to Boulder Dam," *ACI Journal,* 31, 1935.

19. E. J. Callan: "Concrete for Radiation Shielding," *ACI Journal,* 50, 1954.

20. J. Faber and F. Mead: *Reinforced Concrete,* Spon Ltd., London, 1967.

21. F. Walley and S. C. Bate: *A Guide to the B. S. Code of Practice for Prestressed Concrete,* Concrete Publications, London, 1959.

PROBLEMS

2.1. Discuss the effect of the following factors on the compressive strength of concrete: the water-cement ratio, proportions of concrete materials, mixing, placing, vibration, and the age, shape, and dimensions of the concrete specimen.

2.2. Estimate the ratio of the compressive strength of concrete to its tensile, flexural, and shear strength.

2.3. What are the possible types of failure of a concrete specimen?

2.4. How is the tensile strength of concrete determined? Explain what is meant by the splitting test of a concrete cylinder.

2.5. Define the modulus of rupture and modulus of elasticity of concrete.

2.6. What is meant by the initial tangent and the secant moduli?

2.7. Explain the modulus of elasticity of concrete in compression and the shear modulus.

2.8. Determine the modulus of elasticity of concrete by the ACI formula for a concrete cylinder that has a unit weight of 120 pcf (1920 kg/m^3) and a compressive strength of 3000 psi (21 MPa).

2.9. Estimate the modulus of elasticity and the shear modulus of a concrete specimen with a dry density of 150 pcf (2400 kg/m^3) and compressive strength of 4500 psi (31 MPa) using Poisson's ratio, $\mu = 0.18$.

2.10. What is meant by the modular ratio and Poisson's ratio? Give approximate values for concrete.

2.11. What factors influence the shrinkage of concrete?

2.12. What factors influence the creep of concrete?

2.13. What are the types and grades of the steel reinforcement used in reinforced concrete?

2.14. On the stress-strain diagram of a steel bar, show and explain the following: proportional limit, yield stress, ultimate stress, yield strain, and modulus of elasticity.

Strength Design Method: Flexural Analysis of Reinforced Concrete Beams

3.1   Introduction

Ultimate-strength design is a method of determining the dimensions of a structural member based on ultimate loads and ultimate strengths of sections. Ultimate loads are found by multiplying the working loads, the dead load and the assumed live loads, by load factors. Ultimate strength of sections is reached by the yielding of steel, followed by the crushing of concrete. Ultimate loads cause external ultimate forces such as bending moments, shear, or thrust, depending on how these loads are applied to the structure. The section of the member is designed in such a way that its internal ultimate capacity is equal to or greater than the external ultimate forces acting on the member.

In proportioning reinforced concrete structural members, it will be necessary to investigate three main items:

1. *The safety of the structure.* This is maintained by providing adequate internal ultimate strength capacity.
2. *Deflection of the structural member under working loads.* The maximum value of this must be limited, usually specified as a factor of the span, to preserve the appearance of the structure.
3. *Width of cracks under working loads.* Visible cracks spoil the appearance of the structure and also permit humidity to penetrate into the concrete, causing corrosion of steel and weakening the reinforced concrete member. The ACI Code of 1983 limits crack widths for interior and exterior exposure to 0.016 in. (0.40 mm) and 0.013 in. (0.33 mm), respectively.

It is worth mentioning that the ultimate-strength design method was first permitted in Britain in 1957, in the United States in 1956, and in the USSR in 1935. The ACI Code of 1963 put equal emphasis on both ultimate-strength design and working-stress design methods, while the ACI Codes of 1971, 1977, and 1983 emphasized ultimate-strength design.

3.2   Assumptions

Reinforced concrete sections are heterogeneous (nonhomogeneous) as they are made of two different materials, concrete and steel. Therefore proportioning structural members by ultimate-strength design is based on the following assumptions:

1. Strain in concrete is the same as in reinforcing bars at the same level, provided that the bond between the steel and concrete is adequate.
2. Strain in concrete is linearly proportional to the distance from the neutral axis.
3. The modulus of elasticity of all grades of steel is taken as

$$E_s = 29 \times 10^6 \ \text{lb/in.}^2 \qquad (200{,}000 \ \text{MPa or N/mm}^2 \ \text{or} \ 2.039 \times 10^6 \ \text{kgf/cm}^2)$$

   The stress in the elastic range is equal to the strain multiplied by E_s. In the plastic range, the steel stress to be considered is the yield stress f_y.
4. Plane cross sections continue to be plane after bending.
5. Tensile strength of concrete is neglected because (1) concrete's tensile strength is about 10 percent of its compressive strength, (2) cracked concrete is assumed to be not effective, and (3) before cracking, the entire concrete section is effective in resisting the external moment.
6. The method of elastic analysis, assuming an ideal behavior at all levels of stress, is not valid. At high stresses, nonelastic behavior is assumed, which is in close agreement with the actual behavior of concrete and steel.
7. At ultimate strength, the maximum strain at the extreme compression fibers is assumed equal to 0.003, by the ACI Code provision.
8. At ultimate strength, the shape of the compressive concrete stress distribution may be assumed to be rectangular, parabolic, or trapezoidal. In this text, a rectangular shape will be assumed.

3.3   Behavior of a Simply Supported Reinforced Concrete Beam Loaded to Failure

Concrete being weakest in tension, a concrete beam under an assumed working load will definitely crack at the tension side, and the beam will collapse if tensile reinforcement is not provided. Concrete cracks occur at a loading stage when its maximum tensile stress reaches the modulus of rupture of concrete. Therefore, steel bars are used to increase the moment capacity of the beam; the steel bars resist the tensile force while the concrete resists the compressive force.

To study the behavior of a reinforced concrete beam under increasing load, let us examine how two beams were tested to failure. Details of the beams are shown in Figure 3.1. Both beams had a section of 4.5 by 8 in. (110 by 200 mm), reinforced only on the tension side by two No. 5 bars. They were made of the same concrete mix. Beam 1 had no stirrups while beam 2 was provided with No. 3 stirrups spaced at 3 inches. The loading system and testing procedure were the same for both beams. To determine the compressive strength of the concrete and its modulus of elasticity E_c, a standard concrete cylinder was tested and strain measured at different load increments. Results of tests are shown in Tables 3.1 and 3.2 and in Figure 3.2. The following observations were

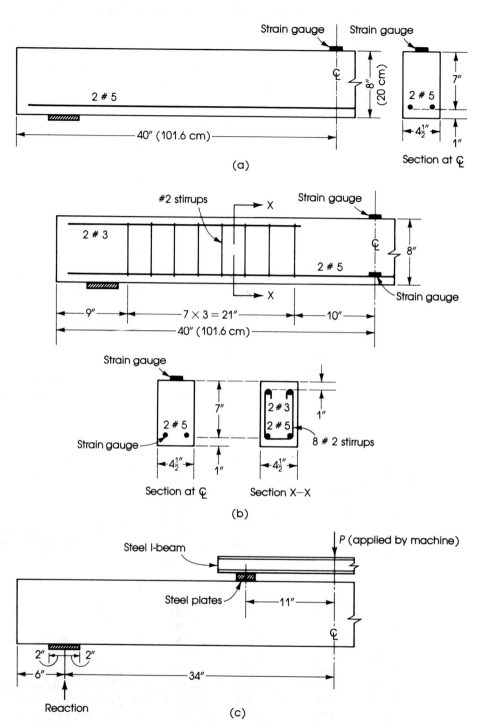

FIGURE 3.1. Details of tested beams: (a) beam 1, (b) beam 2, and (c) loading system. All beams are symmetrical about the centerline.

TABLE 3.1. Main readings of testing beam 1 (rectangular section, tension steel only, no stirrups)

| Load | | Deflection | Concrete strain | Observations |
kN	lb	mm	$\times 10^{-6}$	
0	0	0	0	
4.50	1000	0.329	60	
9.00	2000	0.633	130	
13.50	3000	0.990	220	
18.00	4000	1.396	315	
22.50	5000	1.878	429	
27.00	6000	2.350	529	Vertical crack about midspan
31.50	7000	2.810	625	
36.00	8000	3.260	723	
40.50	9000	3.750	829	
42.75	9500	3.995	889	Diagonal crack appeared at
45.00	10000	4.225	935	one end of beam
49.50	11000	4.760	1060	
54.00	12000	5.295	1170	
58.50	13000	5.850	1300	
60.75	13500	6.200	1365	
61.20	13600	—	—	Concrete failure and collapse of beam

TABLE 3.2. Main readings of testing beam 2 (rectangular section, tension steel only, and stirrups)

| Load | | Deflection | Concrete strain | Steel strain | Observations |
kN	lb	mm	$\times 10^{-6}$	$\times 10^{-6}$	
0	0	0	0	0	
4.50	1000	0.305	70	50	
9.00	2000	0.610	140	120	
13.50	3000	0.965	235	270	
18.00	4000	1.438	345	430	
22.50	5000	1.880	440	550	
27.00	6000	2.362	550	685	Vertical crack at midspan
36.00	8000	3.175	720	940	
45.00	10000	4.064	915	1180	
49.50	11000	4.520	1030	1310	Diagonal crack
54.00	12000	4.980	1120	1430	
63.00	14000	5.944	1370	1690	
66.38	14750	9.804	1500	—	Steel yields
70.88	15750	14.620	1790	—	
72.90	16200	—	—	—	

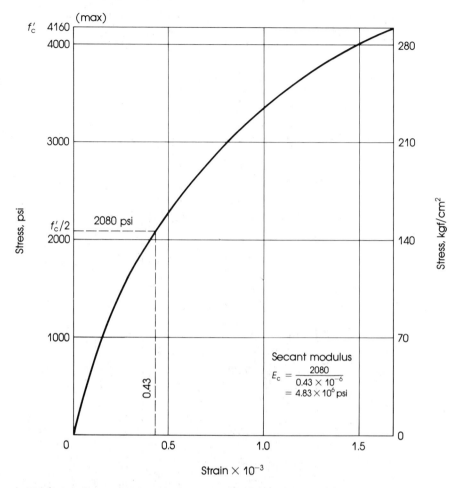

FIGURE 3.2. Stress-strain curve for a standard 6 by 12 in. cylinder.

noted at different distinguishable stages of loading. Each stage is depicted in a stress-strain diagram in Figure 3.3.

Stage 1: At zero external load, each beam carried its own weight in addition to that of the loading system, which consisted of an I-beam and some plates. Weights of beam 1, beam 2, and the loading system were 270, 271, and 88.5 lb (1200, 1205, and 394 N), respectively. Both beams behaved similarly at this stage. At any section, the entire concrete section in addition to the steel reinforcement resisted the bending moment and shearing forces. Maximum stresses occurred at the section of maximum bending moment, that is, at midspan. Maximum tension stress at the bottom fibers was much less than the modulus of rupture of concrete. Compressive stress at the top fibers was much less than the ultimate concrete compressive stress f'_c. No cracks were seen at this stage.

Stage 2: This stage was reached when the external load P was increased from zero to P_1, which produced tensile stresses at the bottom fibers equal to the modulus of rupture of concrete. At this stage the entire concrete section was effective, with the steel bars at the tension side sustaining a strain equal to that of the surrounding concrete.

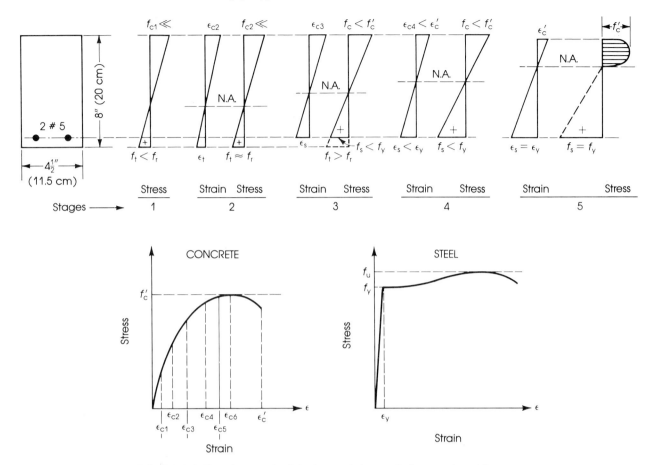

FIGURE 3.3. Development of strains and stresses in beams 1 and 2 under load (midspan sections).

Stress in the steel bars was equal to the stress in the adjacent concrete multiplied by the modular ratio n, the ratio of the modulus of elasticity of steel to that of concrete. The compressive stress of concrete at the top fibers was still very small as compared with the compressive strength f'_c. The behavior of the beams was elastic within this stage of loading.

At a load of 6000 lb (26.7 kN), vertical cracks were seen in both beams. The strains in beams 1 and 2 were read to be 529×10^{-6} and 550×10^{-6}, which corresponds to stresses of 2500 psi (17.24 N/mm²) and 2600 psi (17.90 N/mm²), respectively, on the graph of Figure 3.2.

The strain in the steel bar of beam 2 was 685×10^{-6}, which corresponds to a steel stress of 20,000 psi (138 N/mm²) when the strain is multiplied by the modulus of elasticity of steel.

Stage 3: When the load was increased beyond P_1 (greater than 6000 lb, 26.7 kN), tensile stresses in concrete at the tension zone increased until they were greater than the modulus of rupture f_r, and cracks developed. The neutral axis shifted upward, and cracks extended close to the level of the shifted neutral axis. Concrete in the tension zone lost its tensile strength, and the steel bars started to work effectively and to resist the entire tensile force. Between cracks, the concrete bottom fibers had tensile stresses, but of negligible value. It can be assumed that concrete below the neutral axis did not participate in resisting external moments.

In general, the development of cracks and the spacing and maximum width of cracks depend on many factors, such as the level of stress in the steel bars, distribution of steel bars in the section, concrete cover, and grade of steel used.

At this stage, the deflection of the beams increased clearly as the moment of inertia of the cracked section was less than that of the uncracked section. Cracks started about the midspan of the beam, while other parts along the length of the beam did not crack. When load was again increased, new cracks developed, extending toward the supports. The spacing of these cracks depends on the concrete cover and the level of steel stress. The width of cracks also increased. One or two of the central cracks were most affected by the load, and their crack widths increased appreciably while the other crack widths increased much less. It is more important to investigate those wide cracks than to consider the larger number of small cracks.

If the load was released within this stage of loading, it would be observed that permanent fine cracks, of no significant magnitude, were left. On reloading, cracks would open quickly, as the tensile strength of concrete had already been lost. Therefore, it can be stated that the second stage, once passed, does not happen again in the life of the beam. When cracks develop under working loads, the resistance of the entire concrete section and the gross moment of inertia are no longer valid.

At high compressive stresses, the strain of the concrete increased rapidly, and the stress of concrete at any strain level was estimated from a stress-strain graph obtained by testing a standard cylinder to failure for the same concrete (Figure 3.2). As for the steel, the stresses were still below the yield stress, and the stress at any level of strain was obtained by multiplying the strain of steel ε_s by the modulus of elasticity of steel E_s.

Stage 4: In beam 1, at a load value of 9500 lb (42.75 kN), shear stresses at a distance of about the depth of the beam from the support increased and caused diagonal cracks at approximately 45° from horizontal in the direction of principal stresses resulting from the combined action of bending moment and shearing force. The diagonal crack extended downward to the level of the steel bars, then extended horizontally at that level toward the support. When the crack, which had been widening gradually, reached the end of the beam, a concrete piece broke off and failure occurred suddenly (Figure 3.4). The failure load was 13,600 lb (61.2 kN). Stresses in concrete and steel at the midspan section did not reach their failure stresses. (The shear behavior of beams will be discussed in Chapter 8.)

In beam 2, at a load of 11,000 lb (49.5 kN), a diagonal crack developed similar to that of beam 1, then other parallel diagonal cracks appeared and the stirrups started to take effective part in resisting the principal stresses. Cracks did not extend along the horizontal main steel bars as in beam 1. On increasing the load, diagonal cracks on the other end of the beam developed at a load of 13,250 lb (59.6 kN). Failure did not occur at this stage because of the presence of stirrups.

Stage 5: When the load on beam 2 was further increased, strains increased rapidly until the maximum carrying capacity of the beam was reached at ultimate load $P_u = 16,200$ lb (72.9 kN).

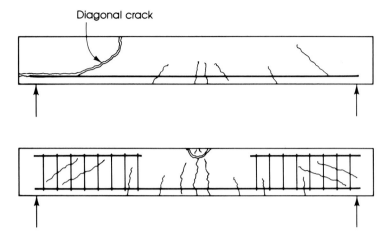

FIGURE 3.4. Shape of beam 1 at shear failure (top) and beam 2 at bending moment failure (bottom).

In beam 2, the amount of steel reinforcement used was relatively small. When the strain in the steel reached the yield strain, which can be considered equal to yield stress divided by the modulus of elasticity of steel $\varepsilon_y = f_y/E_s$, the strain in the concrete ε_c was less than the strain at maximum compressive stress f'_c. The steel bars yielded and the strain in steel increased to about 12 times that of the yield strain without increase in load. Cracks widened sharply, deflection of the beam increased greatly, and the compressive strain on the concrete increased. After another very small increase of load, steel strain hardening occurred and concrete reached its maximum strain ε'_c, and it started to crush under load; then the beam collapsed. Figure 3.4 shows the failure shapes of the two beams.

3.4 Types of Flexural Failure

Three types of flexural failure of a structural member can be expected, depending on the percentage of steel used in the section.

- Steel may reach its yield strength before the concrete reaches its maximum strength (Figure 3.5a). In this case failure is due to yielding of steel. The section contains a relatively small amount of steel and is called an *under-reinforced section*.
- Steel may reach its yield strength at the same time as concrete reaches its ultimate strength (Figure 3.5b). The section is called a *balanced section*. Steel and concrete fail simultaneously.
- Concrete may fail before the yield of steel (Figure 3.5c), due to the presence of a high percentage of steel in the section. In this case the concrete strength f'_c and maximum strain ε'_c are reached while the steel stress is less than the yield strength: $f_s < f_y$. The section is called an *over-reinforced section*.

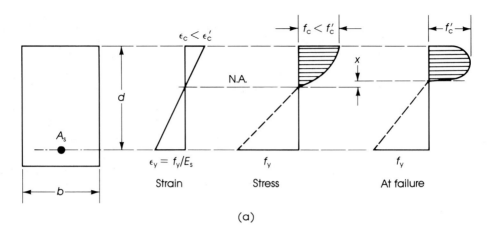

(a)

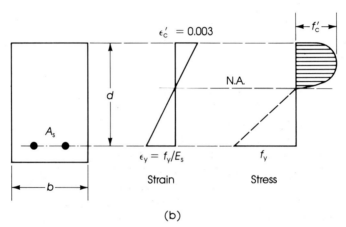

(b)

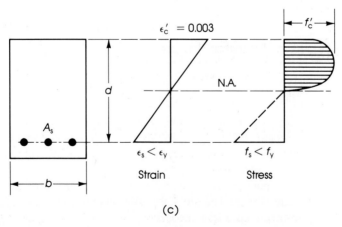

(c)

FIGURE 3.5. Stress and strain diagrams for (a) under-reinforced, (b) balanced, and (c) over-reinforced sections.

It can be assumed that concrete fails in compression when the concrete strain reaches 0.003. A range of 0.0025 to 0.004 has been obtained from tests; therefore, ε'_c can be taken as 0.003.

In a structure based on under-reinforced sections, steel yields before the crushing of concrete. Cracks widen extensively, giving a warning before the concrete crushes and the structure collapses. This type of design is adopted by the ACI Code. In a structure designed with balanced or over-reinforced conditions, the concrete fails suddenly and collapse occurs immediately with no warning. This type of design is not allowed by the ACI Code. There is no doubt that to utilize steel and concrete most efficiently, a balanced section is most suitable, but due to safety needs the ACI Code limits the maximum steel percentage to 75 percent of the balanced steel ratio.

3.5   Load Factors

The types of loads and the safety provisions were explained in Chapter 1, sections 1.7 and 1.8.

In ultimate-strength design, the factored design load is obtained by multiplying the dead load by a load factor, and multiplying the specified live load by another load factor. The magnitude of the load factor must be adequate to limit the probability of failure and to permit an economical structure. The choice of a proper load factor, or in general a proper factor of safety, depends mainly on the importance of the structure (whether a courthouse or a warehouse), the degree of warning needed prior to collapse, the importance of each structural member (whether a beam or a column), the expectation of overload, the accuracy of workmanship, and the accuracy of calculations.

Based on historical studies of various structures, experience, and the principles of probability, the ACI Code adopts a load factor of 1.4 for dead loads and 1.7 for live loads. The dead load factor is smaller because the dead load can be computed with a greater degree of certainty than the live load. Moreover, the choice of the factors reflects the degree of economical design as well as the degree of safety and serviceability of the structure. It is also based on the fact that the performance of the structure under actual loads must be satisfactory within specific limits.

If the ultimate load is denoted by U and those due to wind load and earthquake are W and E, respectively, then according to the ACI Code, the ultimate required strength U shall be the most critical of the following:

1. In the case of dead, live, and wind loads,

$$U = 1.4D + 1.7L \qquad (3.1a)$$
$$U = 0.75[1.4D + 1.7L + 1.7W] \qquad (3.1b)$$
$$U = 0.9D + 1.3W \qquad (3.1c)$$

2. In cases when earthquake forces E must be included in the design, $1.1E$ shall be substituted for W in the above equations.
3. In cases when earth pressure load H must be included in the design,

$$U = 1.4D + 1.7L + 1.7H \qquad (3.2a)$$

But where dead load D and live load L reduce the effect of H, U shall be checked for

$$U = 0.9D + 1.7H \tag{3.2b}$$

But for any combination of D, L, or H,

$$U = 1.4D + 1.7L$$

4. If weight and pressure loads from liquids F must be included in the design,

$$U = 1.4D + 1.7L + 1.4F \tag{3.3a}$$

But, where dead load D and live load L reduce the effect of F,

$$U = 0.9D + 1.4F \tag{3.3b}$$

For any combination of D, L, or F,

$$U = 1.4D + 1.7L$$

The vertical pressure of liquids shall be considered as dead load with due regard to variation in liquid depth.

5. When impact effects are taken into account, it shall be included in the live load. The ACI Code of 1983 does not specify a value, but AASHTO specifications give the impact effect I as a percentage of the live load L as follows:

$$I = \frac{50}{125 + S} \le 30\% \tag{3.4}$$

where

I = percentage of impact (with a maximum of 30 percent)

S = part of the span loaded in ft

When a better estimation is known from experiments or experience, the actual value shall be used:

Live load including impact to be used = $L(1 + I)$

6. Where the structural effects of differential settlement, creep shrinkage, or temperature change may be significant, they shall be included with the dead load D:

$$U = 0.75(1.4D + 1.7L) \tag{3.1d}$$

Equation 3.1a is most generally used. The dead load factor is equal to 1.4 while the live load factor is equal to 1.7. These values are less than those specified by the ACI Code of 1963 of 1.5 for the dead load and 1.8 for the live load. The decrease was suggested because of the more comprehensive code provisions, additional research and experience, and improved concrete and steel control.

For applied concentrated dead and live loads, P_D and P_L, the ultimate concentrated load, $P_U = 1.4P_D + 1.7P_L$; and also $M_U = 1.4M_D + 1.7M_L$, where M_D and M_L are the actual dead load and live load moments, respectively.

3.6   Capacity Reduction Factor

The nominal strength of sections is reduced by a factor ϕ to account for small adverse variations in material strengths, workmanship, dimensions, control, and degree of supervision. The factor ϕ constitutes a portion of the factor of safety as was discussed in section 1.8.

The **ACI Code, section 9.3.2**, specifies the following values to be used:

- Bending, axial tension, bending and axial tension $\phi = 0.90$
- Shear and torsion $\phi = 0.85$
- Bearing on concrete $\phi = 0.70$
- Bending in plain concrete or in concrete with minimum
  reinforcement of $200/f_y$ $\phi = 0.65$
- Axial compression, with or without bending:
     Members with spiral reinforcement $\phi = 0.75$
     Other reinforced members $\phi = 0.70$

Limitations of the axial-compression case are explained in detail in Chapter 10.

3.7   Significance of Analysis and Design Expressions

Two approaches for the investigations of a reinforced concrete member will be used in this book:

Analysis of a section. This implies that the dimensions and steel used in the section (in addition to concrete and steel yield strengths) are given, and it is required to calculate the internal ultimate moment capacity of the section so that it can be compared with the applied external ultimate moment. It may also be necessary to check maximum stresses under external loads.

Design of a section. This implies that the external ultimate moment is known from structural analysis, and it is required to compute the dimensions of an adequate concrete section and the amount of steel reinforcement. Concrete strength and yield strength of steel used are given.

3.8   Equivalent Compressive Stress Distribution

The distribution of compressive concrete stresses at failure may be assumed to be a rectangle, trapezoid, parabola, or any other shape that is in good agreement with test results.

When a beam is about to fail, the steel will yield first if the section is under-reinforced, and in this case the stress in steel is equal to the yield stress. If the section is over-reinforced, concrete crushes first and the strain is assumed equal to 0.003, which agrees with many tests of beams and columns. A compressive force C develops in the compression zone and a tension force T develops in the tension zone at the level of the steel bars. The position of the force T is known, as its line of application coincides with the center of gravity of the steel bars. The position of the compressive force C is not known unless the compressive volume is known and its center of gravity is located. If that is done, the *moment arm*, which is the vertical distance between C and T, will consequently be known.

In Figure 3.6, if concrete fails, $\varepsilon_c' = 0.003$, and if steel yields, as in the case of a balanced section, $f_s = f_y$.

The compression force C is represented by the volume of the stress block, which has the non-uniform shape of stress over the rectangular hatched area of bc. This volume may be considered equal to $C = bc(\alpha_1 f_c')$, where $(\alpha_1 f_c')$ is an assumed average stress of the non-uniform stress block.

The position of the compression force C is at a distance z from the top fibers, which can be considered as a fraction of the distance c (the distance from the top fibers to the neutral axis), and z can be assumed equal to $\alpha_2 c$, where $\alpha_2 < 1$. The values of α_1 and α_2 have been estimated from many tests and their values, as suggested by Mattock, Kriz, and Hognestad,[3] are as follows:

$\alpha_1 = 0.72$ for $f_c' \leq 4000$ psi (27.6 MPa) and decreases linearly by 0.04 for every 1000 psi (6.9 MPa) greater than 4000 psi

$\alpha_2 = 0.425$ for $f_c' \leq 4000$ psi (27.6 MPa) and decreases linearly by 0.025 for every 1000 psi (6.9 MPa) greater than 4000 psi

The decrease in the values of α_1 and α_2 is related to the fact that high-strength concretes show more brittleness than low-strength concretes.[2]

To derive a simple rational approach for calculations of the internal forces of a section, the 1983 ACI Code adopted an equivalent rectangular concrete stress distribution, which was first proposed by C. S. Whitney and checked by Mattock and others.[3] A concrete stress of $0.85f_c'$ is assumed uniformly distributed over an equivalent compression zone bounded by the edges of the cross section and a line parallel to the neutral axis at a distance $a = \beta_1 c$ from the fiber of maximum compressive strain, where c is the distance between the top of the compressive section and the neutral axis (Figure 3.7).

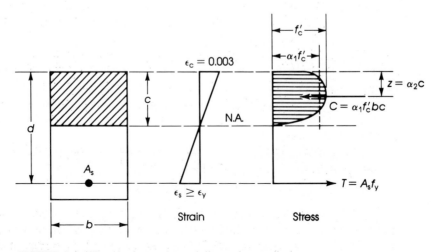

FIGURE 3.6. Ultimate forces in a rectangular section.

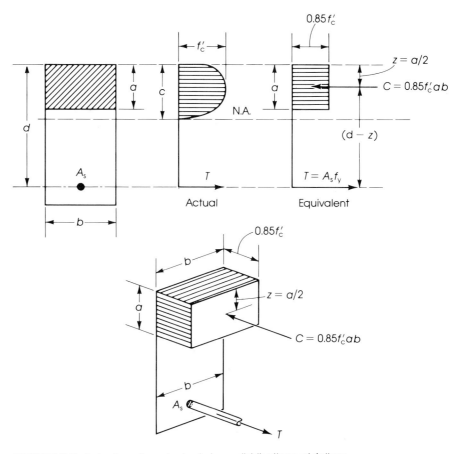

FIGURE 3.7. Actual and equivalent stress distributions at failure.

The fraction $\beta_1 = 0.85$ for concrete strengths $f'_c \leq 4000$ psi (27.6 MPa) and is reduced linearly at a rate of 0.05 for each 1000 psi (6.9 MPa) of stress greater than 4000 psi (Figure 3.8), but not less than 0.65.

The above discussion applies in general to any section, and it is not confined to a rectangular shape. In rectangular sections, the area of the compressive zone is equal to ba, and every unit area is acted on by a uniform stress equal to $0.85f'_c$, giving a total stress volume equal to $0.85f'_c ab$ that corresponds to the compressive force C. For any other shape, the force C is equal to the area of the compressive zone multiplied by a constant stress equal to $0.85f'_c$.

For example, in the section shown in Figure 3.9, the force C is equal to the hatched area of the cross section multiplied by $0.85f'_c$:

$$C = 0.85f'_c(6 \times 3 + 10 \times 2) = 32.3f'_c \text{ lb}$$

The position of the force C is at a distance z from the top fibers, at the position of the resultant force of all small-element forces of the section. As in this case when the stress is uniform and equals $0.85f'_c$, the resultant force C is located at the center of gravity of the compressive zone, which has a depth of a.

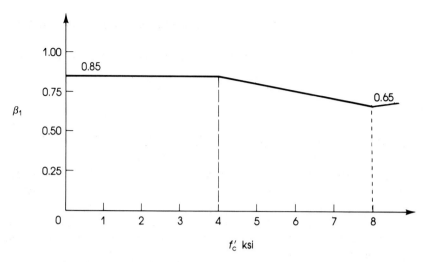

FIGURE 3.8. Values of β_1 for different compressive strengths of concrete f_c'.

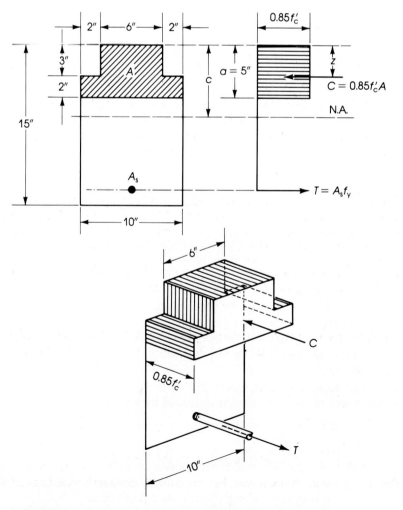

FIGURE 3.9. Ultimate forces in a non-rectangular section.

In this example z is calculated by taking moments about the top fibers:

$$z = \frac{(6 \times 3 \times \frac{3}{2}) + 10 \times 2(1 + 3)}{6 \times 3 + 10 \times 2} = \frac{107}{38} = 2.82 \text{ in.}$$

3.9 Singly Reinforced Rectangular Section in Bending

It was explained previously that a balanced condition is achieved when steel yields at the same time as the concrete fails, and that failure usually happens suddenly. This implies that the yield strain in the steel is reached ($\varepsilon_y = f_y/E_s$), and that the concrete has reached its maximum strain of 0.003. The percentage of reinforcement used to produce a balanced condition is called the balanced steel ratio ρ_b. This is equal to the area of steel A_s divided by the effective cross-section bd:

$$\rho_b = \frac{A_s(\text{balanced})}{bd} \tag{3.5}$$

where

b = width of the compression face of the number

d = distance from the extreme compression fiber to the centroid of the tension reinforcement

If a smaller percentage of reinforcement is used, the section is under-reinforced, and steel bars yield before the crushing of concrete, which avoids sudden failure. To ensure this, the ACI Code limits the maximum tension reinforcement to 0.75 of that which produces a balanced condition at ultimate loading:

$$\rho_{\max} = 0.75\rho_b \tag{3.6}$$

If a greater percentage of reinforcement is used, the section is over-reinforced, and concrete fails suddenly before steel yields. This case is not permitted by the ACI Code.

Two basic equations for the analysis and design of structural members are the two equations of equilibrium that are valid for any load and any section:

1. The compression force should be equal to the tension force, otherwise a section will have linear displacement plus rotation:

$$C = T \tag{3.7}$$

2. The internal nominal ultimate bending moment M_n is equal to either the compressive force C multiplied by its arm or the tension force T multiplied by the same arm:

$$M_n = C(d - z) \tag{3.8a}$$

or

$$M_n = T(d - z) \tag{3.8b}$$

($M_u = \phi M_n$ after reduction by the factor ϕ)

The use of these equations can be explained by considering the case of a rectangular section with tension reinforcement (Figure 3.7). The section may be balanced, under-reinforced, or over-reinforced depending on the percentage of steel reinforcement used.

3.9.1   The Balanced Section

Let us consider the case of a balanced section, which implies that at ultimate load the strain in concrete equals 0.003 and that of steel equals the yield stress divided by the modulus of elasticity of steel, f_y/E_s. This case is explained by the following steps:

Step 1: From the strain diagram of Figure 3.10,

$$\frac{c_b}{d - c_b} = \frac{0.003}{f_y/E_s}$$

from triangular relationships (where c_b is c for a balanced section), and by adding the numerator to the denominator,

$$\frac{c_b}{d} = \frac{0.003}{0.003 + f_y/E_s} \quad \text{or} \quad c_b = \frac{0.003}{0.003 + f_y/E_s} \times d$$

Substituting $E_s = 29 \times 10^3$ ksi,

$$c_b = \frac{87}{87 + f_y} \times d \qquad (3.9)$$

where f_y is in ksi.

Step 2: From the equilibrium equation,

$$C = T$$
$$0.85f'_c ab = A_s f_y \qquad (3.10)$$
$$a = \frac{A_s f_y}{0.85f'_c b} \qquad (3.11a)$$

where a is the depth of the compressive block, equal to $\beta_1 c$ where $\beta_1 = 0.85$ for $f'_c \le$

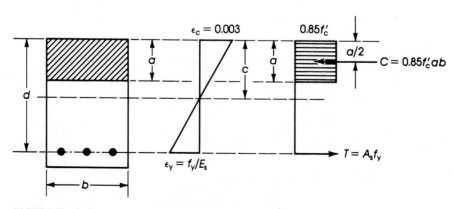

FIGURE 3.10. Rectangular balanced section.

4000 psi (27.6 MPa) and decreases linearly by 0.05 per 1000 psi (6.9 MPa) for higher concrete strengths (Figure 3.8). Since the balanced steel reinforcement ratio is used,

$$\rho_b = \frac{A_s(\text{balanced})}{bd} = \frac{A_{sb}}{bd}$$

$$A_{sb} = \rho_b bd$$

(3.12)

and substituting the value of A_{sb} in equation 3.10,

$$0.85 f'_c ab = f_y \rho_b bd$$

Therefore,

$$\rho_b = \frac{0.85 f'_c}{d f_y} a = \frac{0.85 f'_c}{d f_y} (\beta_1 c_b)$$

Substituting the value of c_b from equation 3.9, the general equation of the balanced steel ratio becomes

$$\rho_b = 0.85 \beta_1 \frac{f'_c}{f_y} \times \frac{87}{87 + f_y}$$

(3.13)

where f'_c and f_y are in ksi. The ACI Code limits the maximum percentage of reinforcement to be used to $0.75 \rho_b$; therefore,

$$\rho_{max} = 0.75 \rho_b = 0.64 \beta_1 \frac{f'_c}{f_y} \times \frac{87}{87 + f_y}$$

(3.14)

For concrete strengths of 4000 psi (28 N/mm²) or less, $\beta_1 = 0.85$, and then

$$\rho_{max} = 0.54 \frac{f'_c}{f_y} \times \frac{87}{87 + f_y}$$

(3.15)

where f'_c and f_y are in ksi.

EXAMPLE 3.1

For the section shown in Figure 3.11, calculate

    (a) the balanced steel reinforcement
    (b) the maximum reinforcement area allowed by the ACI Code
    (c) the position of the neutral axis and the depth of the equivalent compressive stress block.

Given: $f'_c = 3$ ksi, $f_y = 40$ ksi.

Solution   1. $\rho_b = 0.85 \beta_1 \dfrac{f'_c}{f_y} \times \dfrac{87}{87 + f_y}$ (3.13)

Since $f'_c < 4000$ psi, $\beta_1 = 0.85$:

$$\rho_b = (0.85)^2 \times \frac{3}{40} \times \frac{87}{87 + 40} = 0.0371$$

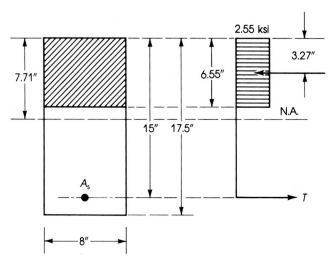

FIGURE 3.11. Example 3.1.

The area of steel reinforcement to provide a balanced condition is

$$A_{sb} = \rho_b bd = 0.0371 \times 8 \times 15 = 4.45 \text{ in.}^2$$

2. $\rho_{max} = 0.75\rho_b = 0.75 \times 0.0371 = 0.0278$

or from equation 3.15,

$$\rho_{max} = 0.54 \times \frac{f_c'}{f_y} \times \frac{87}{87 + f_y}$$

$$= 0.54 \times \frac{3}{40} \times \frac{87}{127} = 0.0278$$

$$A_{s(max)} = \rho_{max} bd$$

$$= 0.0278 \times 8 \times 15 = 3.34 \text{ in.}^2$$

or

$$A_{s(max)} = 0.75 A_{sb}$$

$$= 0.75 \times 4.45 = 3.34 \text{ in.}^2$$

3. The depth of the equivalent compressive block using $A_{s(max)}$:

$$a_{(max)} = \frac{A_{s(max)} f_y}{0.85 f_c' b} = \frac{3.34 \times 40}{0.85 \times 3 \times 8} = 6.55 \text{ in.}$$

The distance from the top fibers to the neutral axis $c = a/\beta_1$, as defined by equation 3.11. Since $f_c' < 4000$ psi, $\beta_1 = 0.85$, thus

$$c = \frac{6.55}{0.85} = 7.71 \text{ in.}$$

If the balanced steel area is used, then the position of the neutral axis will be lower than that shown and c can be calculated by equation 3.9, which is only valid for a balanced condition:

$$c_b = \frac{87}{87 + 40} \times 15 = 10.28 \text{ in.}$$ ∎

Step 3: The internal ultimate moment M_n is calculated by multiplying either C or T by the distance between them,

$$M_n = C(d - z) \tag{3.8a}$$
$$= T(d - z) \tag{3.8b}$$

For a rectangular section the distance $z = a/2$ as the line of application of the force C lies at the center of gravity of the area ab where

$$a = \frac{A_s f_y}{0.85 f_c' b}$$

Therefore,

$$M_n = C\left(d - \frac{a}{2}\right)$$

$$= T\left(d - \frac{a}{2}\right)$$

For a balanced or an under-reinforced section, $T = A_s f_y$. Then

$$M_n = A_s f_y \left(d - \frac{a}{2}\right) \tag{3.16}$$

To get the usable ultimate moment M_u, the above calculated M_n must be reduced by the capacity reduction factor ϕ. In bending, $\phi = 0.9$.

$$M_u = \phi A_s f_y \left(d - \frac{a}{2}\right) \tag{3.16a}$$

$$M_u = \phi A_s f_y \left(d - \frac{A_s f_y}{1.7 f_c' b}\right) \tag{3.16b}$$

Equation 3.16b can be written in terms of the steel percentage ρ:

$$\rho = \frac{A_s}{bd} \qquad A_s = \rho bd$$

Then,

$$M_u = \phi f_y \rho bd \left(d - \frac{\rho bd f_y}{1.7 f_c' b}\right)$$
$$= \phi \rho f_y bd^2 \left(1 - \frac{\rho f_y}{1.7 f_c'}\right) \tag{3.17}$$

Equation 3.17 can be written

$$M_u = R_u bd^2 \tag{3.18}$$

where

$$R_u = \phi \rho f_y \left(1 - \frac{\rho f_y}{1.7 f'_c}\right) \tag{3.19}$$

It is clear that R_u is a function of ρ for a given concrete strength and steel yield strength. R_u is maximum when ρ is maximum. Therefore, using ρ_{max}:

$$(R_u)_{max} = \phi \rho_{max} f_y \left(1 - \frac{\rho_{max} f_y}{1.7 f'_c}\right) \tag{3.20}$$

Similarly,

$$a_{(max)} = \frac{A_{s(max)} f_y}{0.85 f'_c b} \tag{3.11b}$$

The ratio of the equivalent compressive stress block depth a to the effective depth of the section d can be found from equation 3.10:

$$0.85 f'_c ab = A_s f_y$$

Substituting $\rho b d$ for A_s,

$$0.85 f'_c ab = \rho b d f_y$$

$$\frac{a}{d} = \frac{\rho f_y}{0.85 f'_c} \tag{3.21}$$

The maximum ratio a_{max}/d is obtained by substituting the value of ρ_{max} (equation 3.14) in equation 3.21. If

$$q = \frac{\rho f_y}{f'_c} = \frac{A_s f_y}{f'_c bd} \tag{3.22}$$

where q is a dimensionless figure, then

$$\frac{a}{d} = \frac{q}{0.85} \quad \text{or} \quad a = 1.18 qd \tag{3.23}$$

Considering equation 3.19 and substituting for q,

$$R_u = \phi \rho f_y \left(1 - \frac{\rho f_y}{1.7 f'_c}\right) \tag{3.19}$$

$$= \phi q f'_c \left(1 - \frac{q}{1.7}\right) = \phi f'_c q(1 - 0.59 q) \tag{3.24}$$

$$M_u = R_u bd^2 = \phi f'_c bd^2 q(1 - 0.59 q) \tag{3.25}$$

EXAMPLE 3.2

Determine the allowable ultimate moment and the position of the neutral axis of the rectangular section shown in Figures 3.12 and 3.13 if the reinforcement used is (a) 3 No. 6 bars, (b) 3 No. 9 bars, (c) 3 No. 10 bars. Given: $f'_c = 3$ ksi, $f_y = 40$ ksi.

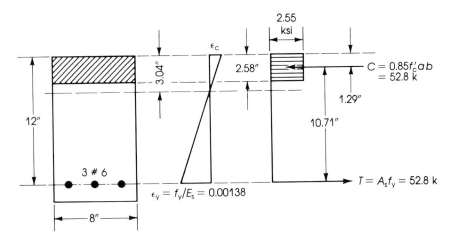

FIGURE 3.12. Example 3.2a.

Solution   1. Area of 3 No. 6 bars = 1.32 in.²

$$\rho = \frac{A_s}{bd} = \frac{1.32}{8 \times 12} = 0.0138$$

$$\rho_{max} = 0.54 \frac{f'_c}{f_y} \times \frac{87}{87 + f_y} \qquad (3.15)$$

$$= 0.54 \times \frac{3}{40} \times \frac{87}{(87 + 40)} = 0.0278$$

$$\rho < \rho_{max} < \rho_b$$

Therefore, the section is under-reinforced.

Since $\rho < \rho_{max}$, the ratio ρ should be used to calculate M_u:

$$M_u = \phi A_s f_y \left(d - \frac{a}{2} \right) \qquad (3.16a)$$

$$a = \frac{A_s f_y}{0.85 f'_c b} = \frac{1.32 \times 40}{0.85 \times 3 \times 8} = 2.58 \text{ in.}$$

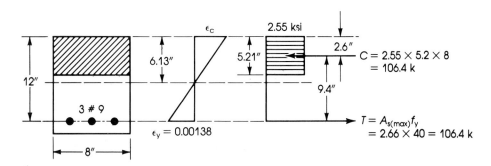

FIGURE 3.13. Example 3.2b.

$$M_u = 0.9 \times 1.32 \times 40 \left(12 - \frac{2.58}{2}\right) = 509 \text{ k·in.} = 42.4 \text{ k·ft}$$

The distance of the neutral axis from the top fibers $c = a/\beta_1$. Since $f_c' < 4000$ psi, $\beta_1 = 0.85$:

$$c = \frac{2.58}{0.85} = 3.04 \text{ in.}$$

In this example, the section is under-reinforced, which implies that the steel will yield before the concrete reaches its ultimate strength. A simple check can be made from the strain diagram (Figure 3.12). From similar triangles,

$$\frac{\varepsilon_c}{\varepsilon_y} = \frac{c}{(d - c)} \quad \text{and} \quad \varepsilon_y = \frac{f_y}{E_s} = \frac{40}{29000} = 0.00138$$

$$\varepsilon_c = \frac{3.04}{(12 - 3.04)} \times 0.00138 = 0.000468$$

which is much less than 0.003. Therefore, steel yields before concrete reaches its limiting strain of 0.003.

2. If 3 No. 9 bars are used, $A_s = 3.00$ in.^{2}

$$\rho = \frac{A_s}{bd} = \frac{3.00}{8 \times 12} = 0.0312 \qquad \rho_{max} = 0.0278 \text{ (previously calculated)}$$

Since $\rho > \rho_{max}$, then ρ_{max} should be used.

$$A_{s(max)} = \rho_{max} bd = 0.0278 \times 8 \times 12 = 2.66 \text{ in.}^2$$

$$\rho_b = \frac{\rho_{max}}{0.75} = \frac{0.0278}{0.75} = 0.0371$$

Since $\rho < \rho_b$, then the section is under-reinforced, but the maximum allowable steel area to be used as specified by the ACI Code is 2.66 in.^{2}

$$M_u = \phi A_s f_y \left(d - \frac{a}{2}\right) \tag{3.16a}$$

$$a = \frac{A_s f_y}{0.85 f_c' b} = \frac{2.66 \times 40}{0.85 \times 3 \times 8} = 5.21 \text{ in.}$$

$$M_u = 0.9 \times 2.66 \times 40 \left(12 - \frac{5.21}{2}\right) = 900 \text{ k·in.} = 75 \text{ k·ft}$$

The distance of the neutral axis from the top fibers

$$c = \frac{a}{0.85} = \frac{5.21}{0.85} = 6.13 \text{ in.}$$

Note that the utilized area of steel is 2.66 in.^{2} and not 3.00 in.^{2}.

$$\text{Utilized steel percentage} = \frac{2.66}{3.00} \times 100 = 88.7\%$$

3. If 3 No. 10 bars are used, $A_s = 3.79$ in.^{2}

$$\rho = \frac{A_s}{bd} = \frac{3.79}{8 \times 12} = 0.0395$$

$$\rho_{max} = 0.0278, \qquad \rho_b = 0.0371 \text{ (as before)}$$

Since $\rho > \rho_b$, the section is over-reinforced; $\rho > \rho_{max}$, therefore ρ_{max} should be used as specified by the ACI Code:

$$A_{s(max)} = \rho_{max} bd = 0.0278 \times 8 \times 12 = 2.66 \text{ in.}^2$$

$$M_u = 900 \text{ k·in.} = 75 \text{ k·ft (as before)}$$

$$\text{Utilized steel percentage} = \frac{2.66}{3.81} \times 100 = 70\% \qquad \blacksquare$$

From the previous example, it can be stated that a steel percentage that is less than the maximum specified by the ACI Code ($\rho_{max} = 0.75\rho_b$) will be fully utilized and consequently provides an economical solution. The section being under-reinforced, steel yields before failure of concrete, thus giving warning of impending collapse.

In example 3.2b, the section is under-reinforced, but the steel percentage used is greater than that allowed by the ACI Code, resulting in an unutilized steel percentage of 11.3 percent. Actually, the section can resist a higher ultimate moment and collapse will be due to yield of steel, but little warning is to be expected. The steel percentage in example 3.2c is high and should be avoided.

It is to be noted that the actual percentage of steel should be used and not ρ_{max} if one of the following inequalities is verified:

$$\rho < \rho_{max} \tag{3.14}$$

$$R_u < R_{u(max)} \tag{3.20}$$

$$a < a_{max}, \qquad a_{max} = \frac{A_{s(max)} f_y}{0.85 f_c' b}$$

3.9.2   Under-Reinforced Sections

A section is under-reinforced when the steel reinforcement percentage is less than that of a balanced condition ($\rho < \rho_b$). The steel reaches its yield strength f_y before the crushing of concrete. The previous equations for M_u and R_u (3.16, 3.17, 3.18, and 3.25) are still valid.

At the moment when steel starts to yield, its strain equals f_y/E_s and the strain in concrete ε_c is less than 0.003. Due to the increase of strain in steel after yielding, $\varepsilon_y > f_y/E_s$ (Figure 3.14), the deflection of the structural member will increase rapidly, causing excessive compression strain in the top fibers until concrete reaches its assumed maximum strain of 0.003. The steel bars will not fail because they undergo strain hardening; collapse will instead be the result of concrete crushing. For a given b and d, the distance of the neutral axis from the top fibers in an under-reinforced section, c_u, is less than that for the balanced case; that is, $c_u < c_b$. Consequently the depth of the equivalent stress block is less than that at the balanced condition: $a < a_b$. Then $C_u = (0.85 f_c' b)a$ is less than $C_b = (0.85 f_c' b)a_b$. If the steel percentage is increased to ρ_1 such that $\rho_1 \leq \rho_b$, then tension T will increase, which indicates that compression C has increased ($C = T$ at all times), and consequently a is increased, indicating that the neutral axis has moved downward. This means that the neutral axis in under-reinforced sections moves downward, increasing the volume of the equivalent stress block to balance the increase in the tensile force T (Figure 3.15). This is true as long as ρ is less than ρ_b.

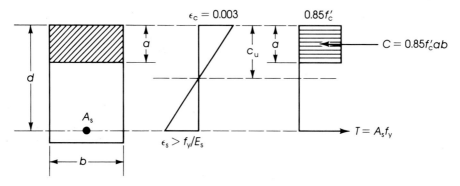

FIGURE 3.14. Under-reinforced rectangular section.

The behavior of the neutral axis and consequently the depth of the equivalent stress distribution under increase of load for a given section bd and a given steel percentage ρ is as shown in Figure 3.16. It can be seen that at small external loads, the stresses in concrete and steel are well below their ultimate strengths, and the neutral axis is quite far from the compressive fibers. With the increase of load up to the stage before yielding of steel (stage b in Figure 3.16), cracks have already lengthened, and the concrete stress distribution is not linear, but the stress in steel increases linearly as it is

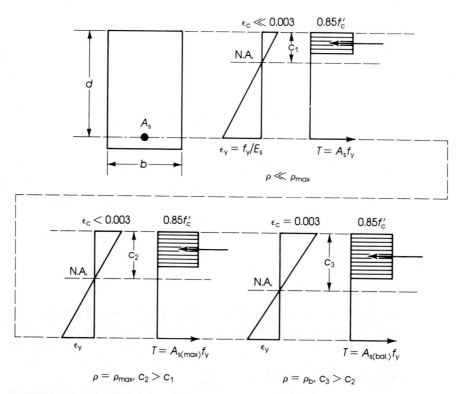

FIGURE 3.15. Movement of the neutral axis with an increase in the steel percentage ρ for under-reinforced sections.

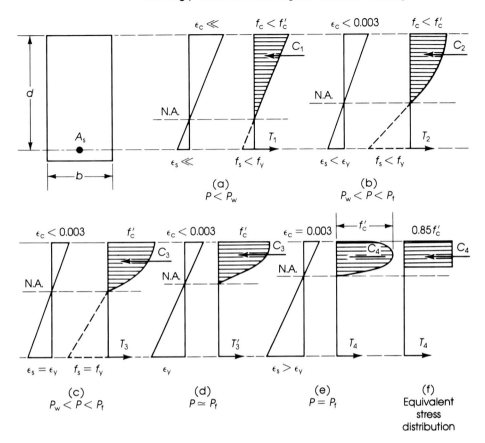

FIGURE 3.16. Movement of the neutral axis due to increasing external loading for an under-reinforced section, assuming tension failure. P is external load, P_w is working load, and P_f is failure load.

still within the elastic range. When the steel yields (stage c in Figure 3.16), the concrete does not reach its ultimate strain, and the neutral axis shifts upward. Due to this shift, the arm of the resisting moment (the distance between C and T) increases. At failure of an under-reinforced section, strain in concrete reaches its assumed maximum value of 0.003, strain in steel exceeds its yield strain ε_y, and the beam collapses due to the secondary effect of concrete crushing.

3.9.3   Over-Reinforced Sections

In over-reinforced sections (Figure 3.17), the percentage of steel used is high ($\rho > \rho_b$), and consequently the stress of steel at failure is below the yield stress ($f_s < f_y$). The steel bars behave elastically, and the stress at any load is equal to the strain multiplied by the modulus of elasticity of steel E_s. The concrete will reach its maximum compressive strength and will fail and crush suddenly.

The compressive force C equals the tensile force T at any time, that is,

$$0.85 f_c' ab = A_s f_s \qquad (3.26)$$

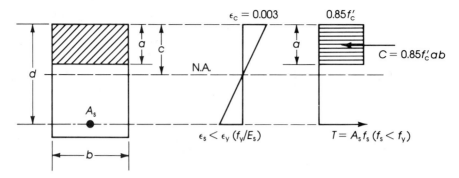

FIGURE 3.17. Over-reinforced rectangular section.

There are two unknowns, f_s and a, in the above equation.

The ultimate moment is given by

$$M_u = \phi A_s f_s \left(d - \frac{a}{2} \right)$$

or

$$M_u = \phi A_s f_s \left(d - \frac{A_s f_s}{1.7 f_c' b} \right) \qquad (3.27a)$$

The maximum steel ratio which the ACI Code allows to be used effectively is $\rho_{max} = 0.75\rho_b$; therefore equation 3.27a becomes

$$M_u = \phi A_{s(max)} f_y \left(d - \frac{A_{s(max)} f_y}{1.7 f_c' b} \right) \qquad (3.27b)$$

This was illustrated by example 3.2c.

The behavior of the neutral axis, and consequently the depth of the equivalent stress distribution under increase of load for a given section, is different from that of an under-reinforced section and is explained in Figure 3.18. The neutral axis has shifted downward due to an increase in load. The tensile force T can increase to a maximum value of $T = A_s f_y$, but concrete will not be able to withstand an equal compressive force. The compressive force will increase to its maximum value as the movement of the neutral axis downward increases the equivalent stress distribution to a maximum. Failure will be due to crushing of concrete.

In general, the equivalent stress volume ($0.85 f_c' ab$) must increase or decrease to equalize the tensile force T at all times.

3.10   Adequacy of Sections

A given section is said to be adequate if the internal resisting ultimate moment is equal to or greater than the externally applied ultimate bending moment. The section is also adequate if the steel reinforcement used is equal to or greater than that required. The following example will explain the adequacy of sections.

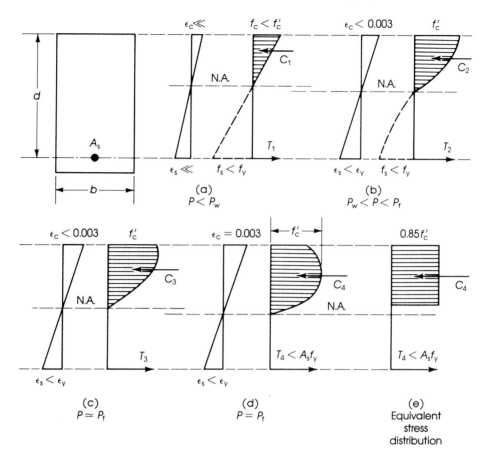

FIGURE 3.18. Movement of the neutral axis due to increasing external loading for an over-reinforced section. (Compression failure of concrete, steel does not yield.) P is external load, P_w is working load, and P_f is failure load.

EXAMPLE 3.3

An 8-ft span cantilever beam has a rectangular section and reinforcement as shown in Figure 3.19. The beam carries a dead load, including its own weight, of 0.8 k/ft and a live load of 0.7 k/ft. Using $f'_c = 4$ ksi and $f_y = 40$ ksi, check the adequacy of the section.

Solution 1. Calculate the external ultimate moment:

$$M_u = 1.4 M_D + 1.7 M_L$$

$$= 1.4 \left(\frac{W_D L^2}{2} \right) + 1.7 \left(\frac{W_L L^2}{2} \right)$$

$$= 1.4 \left(\frac{0.8}{2} \times 8^2 \right) + 1.7 \left(\frac{0.7}{2} \times 8^2 \right) = 74 \text{ k·ft} = 890 \text{ k·in.}$$

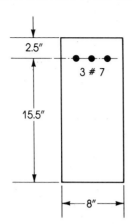
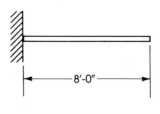

FIGURE 3.19. Example 3.3.

2. Calculate the internal ultimate moment capacity:

$$A_s = 3 \times 0.6 = 1.8 \text{ in.}^2, \qquad \rho = \frac{1.8}{8 \times 15.5} = 0.0145$$

$$\rho_{max} = 0.54 \frac{f'_c}{f_y} \times \frac{87}{87 + f_y} \qquad (3.15)$$

$$= 0.54 \times \frac{4}{40} \times \frac{87}{87 + 40} = 0.0372 > \rho$$

Therefore the section is under-reinforced; hence use $\rho = 0.0145 < \rho_{max}$.

$$M_u = \phi A_s f_y \left(d - \frac{A_s f_y}{1.7 f'_c b} \right) \qquad (3.16a)$$

$$= 0.9 \times 1.8 \times 40 \left(15.5 - \frac{1.8 \times 40}{1.7 \times 4 \times 8} \right) = 918 \text{ k·in.}$$

or calculating

$$R_u = \phi \rho f_y \left(1 - \frac{\rho f_y}{1.7 f'_c} \right) \qquad (3.17)$$

$$= 0.9 \times 0.0145 \times 40 \left(1 - \frac{0.0145 \times 40}{1.7 \times 4} \right) = 0.478 \text{ ksi}$$

$$M_u = R_u b d^2$$

$$= 0.478 \times 8 \times (15.5)^2 = 918 \text{ k·in.}$$

3. The internal ultimate moment of 918 k·in. is greater than the external applied ultimate moment of 890 k·in.; therefore the section is adequate. ∎

EXAMPLE 3.4

A simply supported beam has a span of 20 ft. If the cross-section of the beam is as shown in Figure 3.20, $f'_c = 3$ ksi, and $f_y = 40$ ksi, determine the allowable uniformly distributed service

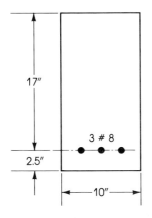
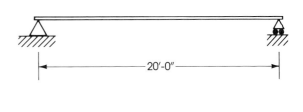

FIGURE 3.20. Example 3.4.

load on the beam. Given: $b = 10$ in., $d = 17$ in., total depth $h = 19.5$ in., and reinforced with 3 No. 8 bars ($A_s = 2.37$ in.^{2}).

1. Determine the internal ultimate moment capacity:

$$\rho = \frac{A_s}{bd} = \frac{3 \times 0.79}{10 \times 17} = 0.0139$$

$$\rho_{max} = 0.54 \frac{f'_c}{f_y} \times \frac{87}{87 + f_y} = 0.54 \left(\frac{3}{40}\right)\left(\frac{87}{87 + 40}\right) = 0.0278$$

$\rho < \rho_{max} < \rho_b$, therefore the section is under-reinforced. Use $\rho = 0.0139$, which is less than ρ_{max}.

$$M_u = \phi A_s f_y \left(d - \frac{A_s f_y}{1.7 f'_c b}\right)$$

$$= 0.9 \times 2.37 \times 40 \left(17 - \frac{2.37 \times 40}{1.7 \times 3 \times 10}\right) = 1292 \text{ k·in.} = 107.67 \text{ k·ft}$$

2. The dead load acting on the beam is self-weight:

$$w_D = \frac{10 \times 19.5}{144} \times 150 = 203 \text{ lb/ft} = 0.203 \text{ k/ft}$$

   where 150 is the weight of reinforced concrete, in pounds per cubic foot.

3. External ultimate moment:

$$M_u = 1.4 M_D + 1.7 M_L$$

$$= 1.4 \left(\frac{0.203}{8} \times 20^2\right) + 1.7 \left(\frac{W_L}{8} \times 20^2\right) = 14.21 + 85 W_L$$

   where W_L = uniform service load on the beam in k/ft.

4. Internal ultimate moment equals the external ultimate moment:

$$107.67 = 14.21 + 85 W_L \quad \text{and} \quad W_L = 1.1 \text{ k/ft}$$

   The allowable uniform service load on the beam is 1.1 k/ft. ∎

3.11   Minimum Percentage of Steel

If the external ultimate moment applied on a beam is very small, and the dimensions of the section are specified (as is sometimes required architecturally) and are larger than needed to resist the external ultimate moment, the calculation may show that no steel reinforcement is required. In this case, the maximum tensile stress due to bending moment may be equal to or less than the modulus of rupture of concrete, $f_r = 7.5\sqrt{f'_c}$. If no reinforcement is provided, sudden failure will be expected when the first crack occurs, thus giving no warning. The **ACI Code, section (10.5.1)**, specifies a minimum steel percentage equal to $200/f_y$ to be used in a reinforced concrete section. This minimum value ρ_{min} is equal to the following ratios:

For grade 40 steel ($f_y = 40$ ksi or 276 MPa), $\rho_{min} = 0.00500$
For grade 50 steel ($f_y = 50$ ksi or 345 MPa), $\rho_{min} = 0.00400$
For grade 60 steel ($f_y = 60$ ksi or 414 MPa), $\rho_{min} = 0.00333$
For grade 75 steel ($f_y = 75$ ksi or 518 MPa), $\rho_{min} = 0.00267$

The above values are not applicable to slabs of uniform thickness, which are discussed in Chapter 9.

In T-beams and joists where the web is in tension, the minimum steel ratio is computed using the width of web, according to **ACI Code, section 10.5**. Alternatively, the area of reinforcement provided at every section should be at least one-third greater than that required by analysis.

EXAMPLE 3.5

Check the design adequacy of the section shown in Figure 3.21 to resist a moment $M_u = 30$ k·ft, using $f'_c = 3$ ksi and $f_y = 40$ ksi.

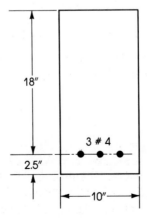

18″

3 # 4

2.5″

10″

FIGURE 3.21. Example 3.5.

Solution   1. Check ρ provided in the section:

$$\rho = \frac{A_s}{bd} = \frac{3 \times 0.2}{10 \times 18} = 0.00333$$

2. Check ρ_{min} required according to the ACI Code:

$$\rho_{min} = \frac{200}{f_y} = 0.005 > \rho = 0.00333$$

Therefore, use $\rho = \rho_{min} = 0.005 < \rho_{max} = 0.0278$:

$$A_{s(min)} = \rho_{min} bd = 0.005 \times 10 \times 18 = 0.90 \text{ in.}^2$$

Use 3 No. 5 bars ($A_s = 0.91$ in.^{2}) as 3 No. 4 bars are less than the minimum specified by the code.

3. Check ultimate moment capacity: $M_u = \phi A_s f_y \left(d - \frac{a}{2} \right)$

$$a = \frac{A_s f_y}{0.85 f_c' b} = \frac{0.91 \times 40}{0.85 \times 3 \times 10} = 1.43 \text{ in.}$$

$$M_u = 0.9 \times 0.91 \times 40 \left(18 - \frac{1.43}{2} \right) = 566 \text{ k·in.} = 47.2 \text{ k·ft}$$

This well exceeds the load of 30 k·ft.

4. Alternative solution according to the **ACI Code, section 10.5**: For 3 No. 4 bars, $A_s = 0.6$ in.^{2}

$$a = \frac{A_s f_y}{0.85 f_c' b} = \frac{0.6 \times 40}{0.85 \times 3 \times 10} = 0.94 \text{ in.}$$

$$M_u = \frac{0.9}{12} \times 0.6 \times 40 \left(18 - \frac{0.94}{2} \right) = 31.55 \text{ k·ft}$$

$$A_s \text{ required for 30 k·ft} = \frac{30}{31.55} \times 0.6 = 0.57 \text{ in.}^2$$

The minimum A_s required according to the **ACI Code, section 10.5.2**, is at least one-third greater than 0.57 in.^{2}

Minimum A_s required $= 1.33 \times 0.57 = 0.76$ in.^{2}

which exceeds the 0.6 in.^{2} provided by the No. 4 bars. Use 3 No. 5 bars, as $A_s = 0.91$ in.^{2}, greater than the 0.76 in.^{2} required. ∎

3.12   Bundled Bars

When the design of a section requires the use of a large amount of steel, for example when ρ_{max} is used, it may be difficult to fit all bars within the cross-section. The **ACI Code, section 7.6.6**, allows the use of parallel bars placed in a bundled form of two, three, or four bars as shown in Figure 3.22. Up to four bars (No. 11 or smaller) can be bundled when they are enclosed by stirrups. The same bundled bars can be used in

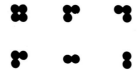

FIGURE 3.22. Bundled bar arrangement.

columns provided that they are enclosed by ties. All bundled bars may be treated as a single bar for checking the spacing and concrete cover requirements.

3.13   Rectangular Sections with Compression Reinforcement

In concrete sections proportioned to resist the bending moments resulting from external loading on a structural member, the internal moment is equal to or greater than the external moment. But a concrete section of a given width and effective depth has a maximum capacity when ρ_{max} ($= 0.75\rho_b$) is used. If external moment is greater than the internal moment capacity, more compressive and tensile reinforcement must be added.

Compression reinforcement is used when a section is limited to specific dimensions due to architectural reasons, such as a need for limited headroom in multistory buildings. Another advantage of compressive reinforcement is that long-time deflection is reduced. A third use of bars in the compressive zone is to hold stirrups, which are used to resist shear forces.

Restricted depths are frequent in ribbed slab construction, where beams are limited to the depth of the ribbed slab, resulting in what are called hidden beams (Figure 3.23).

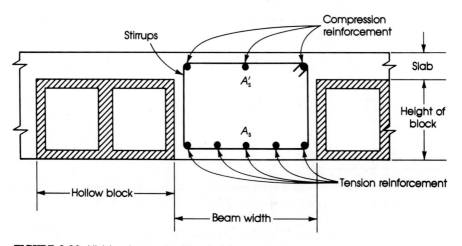

FIGURE 3.23. Hidden beam in ribbed slab construction.

Two cases of doubly reinforced concrete sections will be considered, depending upon whether compression steel yields or does not yield.

3.13.1   When Compression Steel Yields

Internal moment can simply be divided into two moments as shown in Figure 3.24: M_{u1} is the moment produced by concrete compressive force and an equivalent tension force in steel A_{s1}. M_{u2} is the additional moment produced by the compressive force in compression steel A'_s and the tension force in the additional tensile steel A_{s2}.

The moment M_{u1} is that of a singly reinforced concrete section,

$$T_1 = C_c \tag{3.28}$$

$$A_{s1}f_y = C_c = 0.85f'_c ab \tag{3.29}$$

$$a = \frac{A_{s1}f_y}{0.85f'_c b} \tag{3.30}$$

$$M_{u1} = \phi A_{s1}f_y \left(d - \frac{a}{2}\right) \tag{3.31}$$

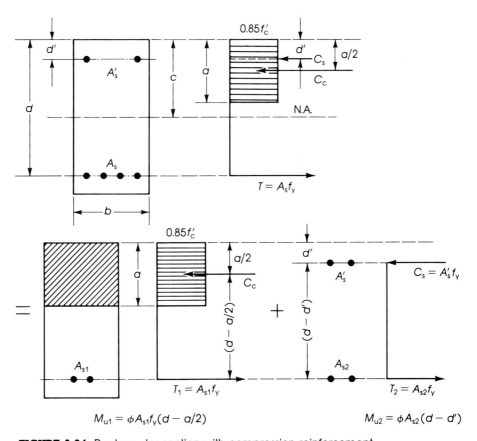

FIGURE 3.24. Rectangular section with compression reinforcement.

The restriction for M_{u1} is that $\rho_1 = A_{s1}/bd$ shall be equal to or less than ρ_{max} ($= 0.75\rho_b$) for singly reinforced concrete sections given in equation 3.14:

$$\rho_1 \le \rho_{max} \le 0.64\beta_1 \frac{f_c'}{f_y} \times \frac{87}{87 + f_y} \tag{3.32}$$

where f_y is in ksi.

Considering the moment M_{u2} and assuming that the compression steel designated as A_s' yields,

$$M_{u2} = \phi A_{s2} f_y (d - d') \tag{3.33a}$$
$$M_{u2} = \phi A_s' f_y (d - d') \tag{3.33b}$$

In this case $A_{s2} = A_s'$, producing equal and opposite forces as shown in Figure 3.24. The total resisting moment M_u is then the sum of the two moments M_{u1} and M_{u2}:

$$M_u = M_{u1} + M_{u2} = \phi[A_{s1}f_y\left(d - \frac{a}{2}\right) + A_s'f_y(d - d')] \tag{3.34}$$

The total steel reinforcement used in tension is the sum of the two steel amounts A_{s1} and A_{s2}. Therefore,

$$A_s = A_{s1} + A_{s2} \tag{3.35}$$

or

$$A_s = A_{s1} + A_s'$$

and

$$A_{s1} = A_s - A_s'$$

Then substituting $(A_s - A_s')$ for A_{s1} in equations 3.30, 3.31, and 3.34:

$$a = \frac{(A_s - A_s')f_y}{0.85f_c'b} \tag{3.36}$$

$$M_u = \phi\left[(A_s - A_s')f_y\left(d - \frac{a}{2}\right) + A_s'f_y(d - d')\right] \tag{3.37}$$

and

$$(\rho - \rho') \le \rho_{max} \le 0.64\beta_1 \times \frac{f_c'}{f_y} \times \frac{87}{87 + f_y} \tag{3.38}$$

Equation 3.38 must be fulfilled in doubly reinforced sections, which indicates that the difference between total tension steel and the compression steel should never exceed the maximum steel allowed for singly reinforced concrete sections. Failure due to the yielding of the total tensile steel will then be expected, and sudden failure of concrete is avoided.

In the above equations, it is assumed that compression steel yields. To investigate this condition, let us study the strain diagram in Figure 3.25. If compression steel yields,

$$\varepsilon_s' \ge \varepsilon_y = f_y/E_s$$

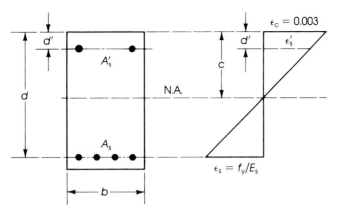

FIGURE 3.25. Strain diagram in doubly reinforced section.

Therefore, from the two triangles above the neutral axis,

$$\frac{c}{d'} = \frac{0.003}{0.003 - f_y/E_s}$$

Substitute $E_s = 29,000$ ksi and let f_y be in ksi. Then

$$\frac{c}{d'} = \frac{87}{87 - f_y}$$

and

$$c = \frac{87}{87 - f_y} \times d' \qquad (3.39)$$

From equation 3.29,

$$A_{s1}f_y = 0.85f'_cab$$

but

$$A_{s1} = A_s - A'_s$$

and

$$\rho_1 = (\rho - \rho')$$

Therefore, equation 3.29 becomes

$$(A_s - A'_s)f_y = 0.85f'_cab$$
$$(\rho - \rho')bdf_y = 0.85f'_cab$$
$$(\rho - \rho') = 0.85\frac{f'_c}{f_y} \times \frac{a}{d}$$

Also,

$$a = \beta_1 c = \beta_1 \times \frac{87}{87 - f_y} \times d'$$

as shown above. Therefore,

$$(\rho - \rho') = 0.85\beta_1 \frac{f'_c}{f_y} \times \frac{d'}{d} \times \frac{87}{87 - f_y} \qquad (3.40)$$

The quantity $(\rho - \rho')$ is the steel ratio of $(A_s - A'_s)/bd = A_{s1}/bd = \rho_1$ for any singly reinforced part.

If $(\rho - \rho')$ is greater than the value of the right-hand side in equation 3.40, then compression steel will also yield. In Figure 3.26 it can be seen that if A_{s1} is increased, T_1 and consequently C_1 will be greater and the neutral axis will shift downward, increasing the strain in the compression steel and ensuring its yield. If the tension steel used (A_{s1}) is less than the right-hand side of equation 3.40, then T_1 and C_1 consequently will be smaller, and the strain in compression steel $\varepsilon'_s < \varepsilon_y$, as the neutral axis will shift upward

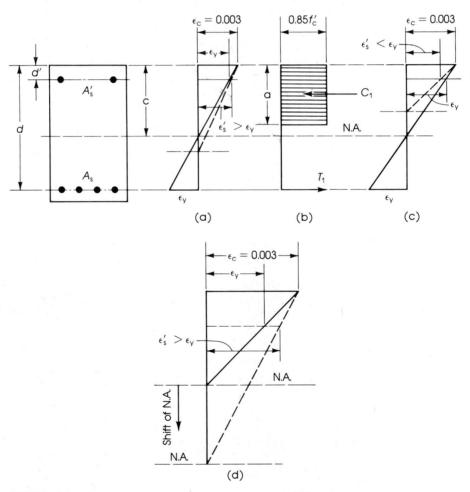

FIGURE 3.26. Yielding and non-yielding cases of compression reinforcement. Diagram (d), a close-up of (a), shows how the neutral axis responds to an increase in A_{s1}.

as shown in Figure 3.26c and compression steel will not yield. Therefore equation 3.40 can be written

$$(\rho - \rho') \geq 0.85\beta_1 \frac{f'_c}{f_y} \times \frac{d'}{d} \times \frac{87}{87 - f_y} \qquad (3.41)$$

where f_y is in ksi, and this is the condition for compression steel to yield.

EXAMPLE 3.6

A rectangular beam has a width of 10 inches and an effective depth to the centroid of tension steel bars $d = 17$ in. Tension reinforcement consists of 6 No. 8 bars in two rows; compression reinforcement consists of 2 No. 7 bars placed as shown in Figure 3.27. Calculate the ultimate moment capacity of the beam if $f'_c = 3$ ksi and $f_y = 50$ ksi.

Solution   1. Check if compression steel yields:

$$A_s = 4.71 \text{ in.}^2 \qquad \rho = A_s/bd = 4.71/(10 \times 17) = 0.0277$$
$$A'_s = 1.2 \text{ in.}^2 \qquad \rho' = A'_s/bd = 1.2/(10 \times 17) = 0.0071$$
$$A_s - A'_s = 3.51 \text{ in.}^2 \qquad (\rho - \rho') = 0.0206$$

For compression steel to yield,

$$(\rho - \rho') \geq 0.85\beta_1 \frac{f'_c}{f_y} \times \frac{d'}{d} \times \frac{87}{87 - f_y}$$

β_1 is 0.85 because $f'_c < 4000$ psi, therefore

$$\text{Right-hand side} = (0.85)^2(3/50)(2.5/17)[87/(87 - 50)] = 0.0150$$
$$(\rho - \rho') = 0.0206 > 0.0150$$

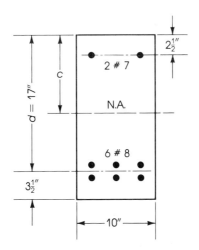 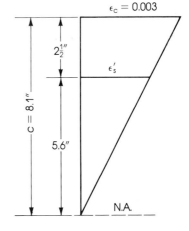

FIGURE 3.27. Example 3.6.

Therefore compression steel yields.

2. Check that $(\rho - \rho') \leq \rho_{max}$ (equation 3.15):

$$\rho_{max} = 0.54 \frac{f'_c}{f_y} \times \frac{87}{87 + f_y}$$

$$= 0.54 \times (3/50) \times 87/(87 + 50) = 0.0206$$

$$(\rho - \rho') = 0.0206 \leq \rho_{max} = 0.0206$$

3. M_u can be calculated by equation 3.37:

$$M_u = \phi \left[(A_s - A'_s)f_y \left(d - \frac{a}{2} \right) + A'_s f_y (d - d') \right]$$

where

$$a = \frac{(A_s - A'_s)f_y}{0.85 f'_c b} = \frac{3.51 \times 50}{0.85 \times 3 \times 10} = 6.88 \text{ in.}$$

$$M_u = (0.9)[3.51 \times 50(17 - 6.88/2) + 1.2 \times 50(17 - 2.5)] = 2925 \text{ k·in.}$$

$$= 243.7 \text{ k·ft}$$

4. A check to see if compression steel yields can be made now. From the strain diagram,

$$c = \frac{a}{0.85} = \frac{6.88}{0.85} = 8.09 \text{ in.}$$

$$\varepsilon'_s = \frac{5.6}{8.09} \times 0.003 = 0.00207, \qquad \varepsilon_y = f_y/E_s = 50/29000 = 0.00172$$

Because ε'_s exceeds ε_y, compression steel yields, as checked before.   ∎

3.13.2   When Compression Steel Does Not Yield

As has been explained, if

$$(\rho - \rho') < \left(0.85\beta_1 \times \frac{f'_c}{f_y} \times \frac{d'}{d} \times \frac{87}{87 - f_y} \right)$$

then the compression steel does not yield. This indicates that if a smaller amount of tension steel is used, it will yield before concrete can reach its maximum strain of 0.003, and the strain in compression steel ε'_s will not reach ε_y at failure. Yielding of compression steel will also depend on its position relative to the extreme compressive fibers d'. A higher ratio of d'/c will decrease the strain in the compressive steel ε'_s as it places compression steel A'_s nearer to the neutral axis.

If compression steel does not yield, two solutions can be tried:

1. Compression steel may be neglected completely; this approach is permitted by the ACI Code. In this case, the section is considered as singly reinforced and should be treated as before. All tension steel is considered, and ρ should not exceed ρ_{max}.
2. General analysis is made by calculating the stresses in compressive reinforcement, $f_s < f_y$, which is included in the calculation of the ultimate moment.

EXAMPLE 3.7

Determine the allowable ultimate moment of the section shown in Figure 3.28, given

$$f_c' = 2.5 \text{ ksi} \quad \text{and} \quad f_y = 50 \text{ ksi}$$

Solution

1. Calculate ρ and ρ': $A_s = 2.4$ in.^{2}, $A_s' = 1.2$ in.^{2}

$$\rho = \frac{2.4}{12 \times 16} = 0.0125, \quad \rho' = \frac{1.2}{12 \times 16} = 0.00625, \quad \rho - \rho' = 0.00625$$

2. Apply equation 3.41: $\beta_1 = 0.85$ (because $f_c' < 4000$ psi).

$$0.85\beta_1 \frac{f_c'}{f_y} \times \frac{d'}{d} \times \frac{87}{87 - f_y} = (0.85)^2 \times \frac{2.5}{50} \times \frac{3}{16} \times \frac{87}{87 - 50} = 0.0159$$

3. Compare $(\rho - \rho')$ value: $(\rho - \rho') = 0.00625 < 0.0159$. Therefore compression steel does not yield. Neglect A_s' and consider only $A_s = 2.4$ in.^{2} as singly reinforced.

4.

$$\rho_b = 0.85\,\beta_1 \frac{f_c'}{f_y} \times \frac{87}{87 + f_y} = (0.85)^2 \times \frac{2.5}{50} \times \frac{87}{137} = 0.0229$$

$$\rho_{max} = 0.75\rho_b = 0.75 \times 0.0229 = 0.0172$$

$$\rho_{(used)} = 0.0125 < \rho_{max} = 0.0172$$

The section is under-reinforced; use $\rho = 0.0125$, $A_s = 2.4$ in.^{2}

5.

$$M_u = \phi A_s f_y \left(d - \frac{a}{2}\right), \quad a = \frac{A_s f_y}{0.85 f_c' b} = \frac{2.4 \times 50}{0.85 \times 2.5 \times 12} = 4.7 \text{ in.}$$

$$M_u = 0.9 \times 2.4 \times 50(16 - 4.7/2) = 1474 \text{ k·in.} = 122.9 \text{ k·ft}$$

This is the allowable ultimate moment.

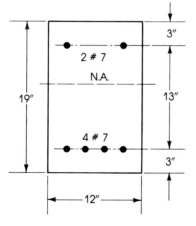
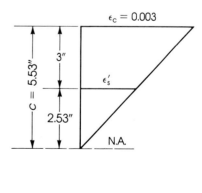

FIGURE 3.28. Example 3.7.

6. A check using the strain diagram can also prove that compression steel does not yield:

$$c = a/0.85 = 4.7/0.85 = 5.53 \text{ in.}$$

$$\varepsilon'_s = (2.53/5.53) \times 0.003 = 0.00137, \qquad \varepsilon_y = 50/29{,}000 = 0.00172$$

ε'_s is less than ε_y, so compression steel does not yield. ∎

EXAMPLE 3.8

Determine the allowable ultimate moment of the section shown in Figure 3.29, given:

$$f'_c = 5 \text{ ksi} \qquad f_y = 60 \text{ ksi}$$
$$A'_s = 1.8 \text{ in.}^2 \qquad A_s = 4.71 \text{ in.}^2$$

Solution 1. Calculate ρ and ρ':

$$\rho = \frac{A_s}{bd} = \frac{4.71}{10 \times 19} = 0.0248, \qquad \rho' = \frac{A'_s}{bd} = \frac{1.8}{10 \times 19} = 0.0095,$$
$$\rho - \rho' = 0.0152$$

2. Apply equation 3.41, assuming $\beta_1 = 0.8$ for $f'_c = 5000$ psi.

$$0.85\beta_1 \times \frac{f'_c}{f_y} \times \frac{d'}{d} \times \frac{87}{87 - f_y} = 0.85 \times 0.8 \left(\frac{5}{60}\right)\left(\frac{2}{19}\right)\left(\frac{87}{87 - 60}\right) = 0.0192$$
$$(\rho - \rho') = 0.0152 < 0.0192$$

Therefore compression steel does not yield, as $f'_s < 60$ ksi.

$$\rho_{max} = 0.75\rho_b = 0.75(0.85)(0.8)(5/60)87/(87 + 60) = 0.0252$$
$$\rho - \rho' = 0.0152 < \rho_{max}$$

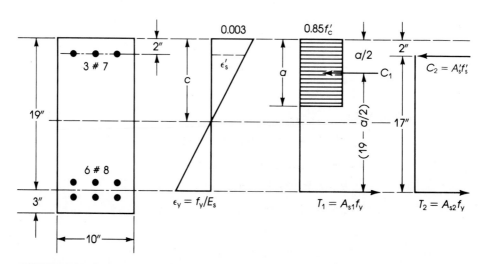

FIGURE 3.29. Example 3.8, analysis solution.

3. Calculate M_u by analysis.
   Internal forces:

$$C_1 = 0.85f'_c ab, \qquad a = \beta_1 c = 0.8c$$

$$C_1 = 0.85 \times 5(0.8c) \times 10 = 34c$$

C_2 = the force in compression steel

$$= A'_s f'_s - \text{force in displaced concrete}$$

$$= A'_s f'_s - A'_s (0.85 f'_c)$$

From strain triangles,

$$\varepsilon'_s = 0.003 \frac{(c - 2)}{c}$$

$$f'_s = E_s \varepsilon'_s \qquad \text{(since steel is in the elastic range)}$$

$$= 29,000[0.003(c - 2)/c] = 87(c - 2)/c \text{ (ksi)}$$

Therefore,

$$C_2 = 1.8[87(c - 2)/c - 0.85 \times 5] \text{ kips}$$

$$= 156.6(c - 2)/c - 7.65 \text{ kips}$$

$$T = T_1 + T_2 = (A_{s1} + A_{s2})f_y = A_s f_y$$

$$= 4.71 \times 60 = 282.6 \text{ kips}$$

4. Equating internal forces to determine the position of the neutral axis (the distance c):

$$T = C = C_1 + C_2$$

$$282.6 = 34c + 156.6(c - 2)/c - 7.65$$

$$290.25 = 34c + 156.6 - 313.2/c$$

$$c^2 - 3.93c = 9.21$$

$$(c - 1.96)^2 = 9.21 + 3.85 = 13.06$$

$$c - 1.96 = 3.61$$

$$c = 5.57 \text{ in.}$$

(The other value is negative and does not apply.)

$$a = 0.8c = 0.8 \times 5.57 = 4.46 \text{ in.}$$

5. Calculate f'_s, C_1, and C_2:

$$f'_s = 87(c - 2)/c = 87(5.57 - 2)/5.57 = 56 \text{ ksi} < 60 \text{ ksi}$$

which confirms that compression steel does not yield.

$$C_1 = 34c = 34 \times 5.57 = 189 \text{ k}$$

$$C_2 = A'_s f'_s - 7.65 = (1.8 \times 56) - 7.65 = 93 \text{ k}$$

or

$$C_2 = 156.6(c - 2)/c - 7.65$$

$$= 156.6(5.57 - 2)/5.57 - 7.65 = 93 \text{ k}$$

6. To calculate M_u, take moments about tension steel A_s:

$$M_u = \phi[C_1(d - a/2) + C_2(d - d')]$$
$$= 0.9[189(19 - 4.46/2) + 93.0(19 - 2)]$$
$$= 4275 \text{ k·in.} = 356 \text{ k·ft}$$

Another approach is to calculate the ultimate moment by neglecting the compression steel A_s' and considering the section as singly reinforced with $A_s = 4.71$ in.^{2} (6 No. 8 bars). For this case,

$$M_u = \phi A_s f_y(d - a/2)$$

using A_s only if $\rho < \rho_{max}$.

$$\rho = 0.0248 < \rho_{max} = 0.0252$$

Therefore use $A_s = 4.71$ in.^{2}

$$a = \frac{A_s f_y}{0.85 f_c' b} = \frac{4.71 \times 60}{0.85 \times 5 \times 10} = 6.64 \text{ in.}$$

$$M_u = 0.9 \times 4.71 \times 60(19 - 6.64/2) = 3988 \text{ k·in.} = 332 \text{ k·ft}$$

It can be observed that the approximation suggested by the ACI Code is acceptable in practice, although the moment determined by analysis is greater than that calculated by the approximate method. This short approach is recommended especially if the position of compression steel is well below the compression fibers, that is, if d' is relatively large, say 3 in. or more. ∎

3.14   Analysis of T- and I-Sections

3.14.1   Description

It is normal to cast concrete slabs and beams together producing a monolithic structure. Slabs have smaller thicknesses than beams. Under bending stresses, those parts of the slab on either side of the beam will be subjected to compressive stresses depending on the position of these parts relative to top fibers and relative to their distances from the beam. The part of the slab acting with the beam is called the flange, and it is indicated in Figure 3.30a by the area bt. The rest of the section confining the area $(h - t)b_w$ is called the stem or web.

In an I-section there are two flanges, a compression flange, which is actually effective, and a tension flange, which is ineffective as it lies below the neutral axis and is thus neglected completely. Therefore, the analysis and design of an I-beam is similar to that of a T-beam.

3.14.2   Effective Width

In a T-section, if the flange is very wide, the compressive stresses are at a maximum value at points adjacent to the beam and decrease approximately in a parabolic form to

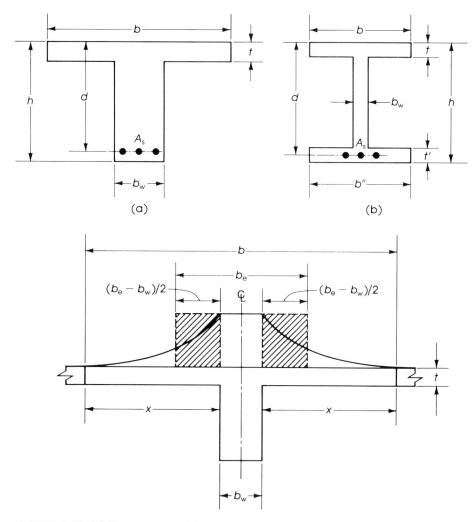

FIGURE 3.30. (a) T-section and (b) I-section, with (c) illustration of effective flange width b_e.

zero at a distance x from the face of the beam. Stresses also vary vertically from a maximum at the top fibers of the flange to a minimum at the lower fibers of the flange. This variation depends on the position of the neutral axis and the change from elastic to inelastic deformation of the flange along its vertical axis.

An equivalent stress area can be assumed to represent the stress distribution on the width b of the flange producing an equivalent flange width b_e of uniform stress (Figure 3.30(c)).

Analysis of equivalent flange widths for actual T-beams indicate that b_e is a

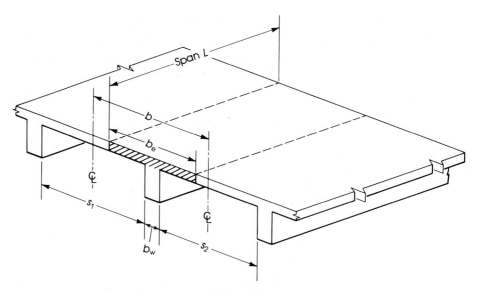

FIGURE 3.31. Effective flange width of T-beams.

function of span length of the beam.[7] Other variables that affect the effective width b_e are (Figure 3.31):

- Spacing of beams
- Width of stem (web) of beam b_w
- The relative thickness of slab with respect to the total beam depth
- End conditions of the beam (simply supported or continuous)
- The way in which the load is applied (distributed load or point load)
- The ratio of the length of the beam between points of zero moment to the width of the web and the distance between webs

The **1983 ACI Code, section 8.10.2,** prescribes the following limitations on the effective flange width b_e, considering that the span of the beam is equal to L:

1. $b_e = L/4$
2. $b_e = 16t + b_w$
3. $b_e = b$, where b is the distance between centerlines of adjacent slabs

The smallest of the above three values must be used.

These values are conservative for some cases of loading and adequate for other cases. A similar effective width of flange can be adopted for I-beam sections.

Investigations indicate that the effective compression flange increases as load is increased toward the ultimate value.[6] Under working loads, stress in the flange is within the elastic range. The restrictions on b_e mentioned above are used for both elastic and ultimate-strength methods.

A T-shaped or I-shaped section may behave as a rectangular section or a T-section. The two cases are investigated as follows.

3.14.3   T-Section Behaves as a Rectangular Section

In this case, the depth of the equivalent stress block a lies within the flange with extreme position at the level of the bottom fibers of the compression flange.

When the neutral axis lies within the flange (Figure 3.32a), the depth of the equivalent compressive distribution stress lies within the flange, producing a compressed area equal to $b_e a$. The concrete below the neutral axis is assumed ineffective and the section is considered as singly reinforced as explained earlier, with b_e replaced by b.

Therefore,

$$a = \frac{A_s f_y}{0.85 f'_c b_e}$$

and, by equation 3.16a,

$$M_u = \phi A_s f_y (d - a/2)$$

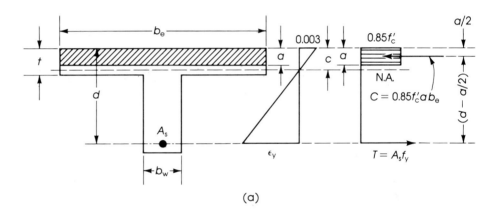

(a)

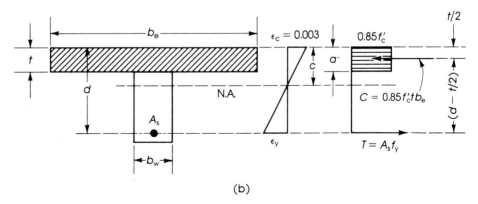

(b)

FIGURE 3.32. Rectangular section behavior (a) when the neutral axis lies within the flange and (b) when the stress distribution depth equals the slab thickness.

If the depth a is increased such that $a = t$ (this case occurs when the amount of the tension steel is increased), the neutral axis lies below the bottom fibers of the flange, resulting in a T-section. The position of the neutral axis from the top fibers, c, equals a/β_1 or t/β_1. But the ultimate moment is that of a singly reinforced concrete section:

$$M_u = \phi A_s f_y (d - t/2) \tag{3.42}$$

In this case

$$t = \frac{A_s f_y}{0.85 f'_c b_e} \quad \text{or} \quad A_s = \frac{0.85 f'_c b_e t}{f_y} \tag{3.43}$$

In this analysis, the limit of steel percentage for a singly reinforced rectangular section should apply (equation 3.14):

$$\rho = A_s / b_e d \leq \rho_{max}$$

3.14.4   Analysis of a T-Section

In this case the depth of the equivalent compressive distribution stress lies below the flange and consequently the neutral axis also lies in the web. This is due to an amount of tension steel A_s more than that calculated by equation 3.43. Part of the concrete in the web will now be effective in resisting the external moment. In Figure 3.33, the compressive force C is equal to the compression area of the flange and web multiplied by the uniform stress of $0.85 f'_c$:

$$C = 0.85 f'_c [b_e t + b_w (a - t)]$$

The position of C is at the centroid of the T-shaped compressive area, at a distance z from top fibers.

The analysis of a T-section is similar to that of a doubly reinforced concrete section considering an area of concrete $(b_e - b_w) t$ as equivalent to the compression steel area A'_s. The analysis is divided into two parts, shown in Figure 3.34:

1. A singly reinforced rectangular section $b_w d$ and steel reinforcement A_{s1}. The

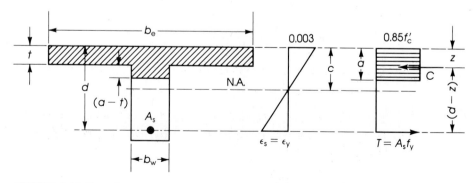

FIGURE 3.33. T-section behavior.

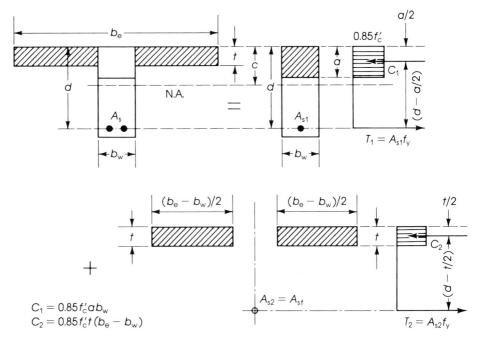

FIGURE 3.34. T-section analysis.

compressive force C_1 equals $0.85f'_c ab_w$, the tensile force T_1 equals $A_{s1}f_y$, and moment arm is equal to $(d - a/2)$.

2. A section which consists of the concrete overhanging flange sides $2 \times [(b_e - b_w)t]/2$ developing the additional compressive force (when multiplied by $0.85f'_c$) and a moment arm equal to $(d - t/2)$. A_{sf} is the area of tension steel which will develop a force equal to the compressive strength of the overhanging flanges. Therefore

$$A_{sf}f_y = 0.85f'_c(b_e - b_w)t$$

or

$$A_{sf} = \frac{0.85f'_c t(b_e - b_w)}{f_y} \tag{3.44}$$

The total steel used in the T-section A_s is $A_{s1} + A_{sf}$, or

$$A_{s1} = A_s - A_{sf} \tag{3.45}$$

The T-section is in equilibrium, $C_1 = T_1$, $C_2 = T_2$, and $C = T$; that is, $C = C_1 + C_2 = T_1 + T_2 = T$. Considering equation $C_1 = T_1$, then $A_{s1}f_y = 0.85f'_c ab_w$ or $(A_s - A_{sf})f_y = 0.85f'_c ab_w$; therefore

$$a = \frac{(A_s - A_{sf})f_y}{0.85f'_c b_w} \tag{3.46}$$

Note that b_w is used to calculate a. The ultimate moment of the section is the sum of the two moments M_{u1} and M_{u2}:

$$M_u = M_{u1} + M_{u2}$$
$$M_{u1} = \phi A_{s1} f_y (d - a/2)$$

where

$$A_{s1} = (A_s - A_{sf}) \quad \text{and} \quad a = \frac{(A_s - A_{sf})f_y}{0.85f'_c b_w}$$

$$M_{u1} = \phi(A_s - A_{sf})f_y(d - a/2)$$
$$M_{u2} = \phi A_{sf} f_y (d - t/2)$$

Therefore

$$M_u = \phi[(A_s - A_{sf})f_y(d - a/2) + A_{sf} f_y (d - t/2)] \tag{3.47}$$

The ultimate moment of a T-section or I-section can be calculated from equation 3.47.

If it is assumed that the ratio of steel in the web is ρ_w, then

$$\rho_w = A_s / b_w d$$

and that for the over-hanging flanges,

$$\rho_f = A_{sf} / b_w d$$

Then

$$A_{s1}/b_w d = \frac{A_s - A_{sf}}{b_w d} = (\rho_w - \rho_f)$$

which is the percentage of steel used in the rectangular singly reinforced stem. For a rectangular singly reinforced section, the steel percentage limit should be that specified by equation 3.14. That is, $(\rho_w - \rho_f)$ should be equal to or less than ρ_{max}:

$$(\rho_w - \rho_f) \le \rho_{max} = 0.75\rho_b \tag{3.14}$$

This condition must be met in the analysis and design of T-sections.

EXAMPLE 3.9

A series of reinforced concrete beams spaced at 7 ft, 10 in. center to center have simply supported spans of 15 ft. The beams support a reinforced concrete floor slab 4 in. thick. The dimensions and reinforcement of beams are shown in Figure 3.35; $f'_c = 3$ ksi, and $f_y = 60$ ksi. Determine the moment capacity of a typical intermediate beam.

Solution   1. Determine the effective flange width. Effective flange width is the smallest of:

$16t + b_w = (16 \times 4) + 10 = 74$ in.

Span/4 = $(15 \times 12)/4 = 45$ in.

Center to center of beams = $(7 \times 12) + 10 = 94$ in.; therefore take $b_e = 45$ in.

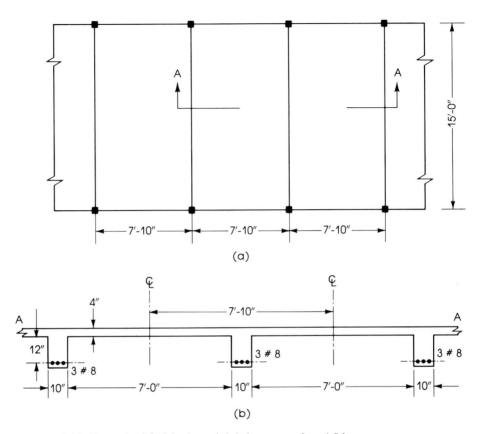

FIGURE 3.35. Example 3.9: (a) plan of slab-beam roof and (b) section A-A.

2. Check the position of neutral axis. If the section behaves as a rectangular one, then the neutral axis lies within the flange (Figure 3.36). In this case, the width of beam used is equal to 45 in.

$$a = \frac{A_s f_y}{0.85 f_c' b} \qquad A_s = 3 \times 0.79 = 2.37 \text{ in.}^2$$

$$c = \frac{A_s f_y}{0.85 f_c' b \beta_1} = \frac{2.37 \times 60}{0.85 \times 3 \times 45 \times 0.85} = 1.46 \text{ in.}$$

$$a = 0.85 c = 1.24 \text{ in.}$$

Therefore, the neutral axis lies within the flange and the section is considered rectangular.

3. Calculate the ultimate moment:

$$\rho = A_s/bd = 2.37/(45 \times 16) = 0.0033$$

$$\rho_{max} \text{ (equation 3.15)} = 0.54 \frac{f_c'}{f_y} \times \frac{87}{87 + f_y}$$

$$= 0.54 \times (3/60) \times (87/147) = 0.016$$

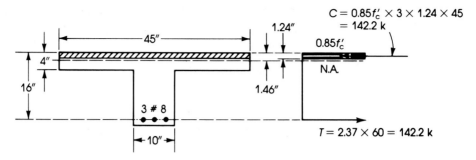

FIGURE 3.36. Example 3.9.

Therefore, use $\rho < \rho_{max}$.

$$M_u = \phi A_s f_y(d - a/2)$$
$$= 0.9 \times 2.37 \times 60(16 - 1.24/2) = 1968 \text{ k·in.} = 164 \text{ k·ft} \qquad \blacksquare$$

EXAMPLE 3.10

Calculate the ultimate moment capacity of the T-section shown in Figure 3.37; $f_c' = 3.5$ ksi, $f_y = 60$ ksi.

Solution   1. Given: $b_e = 36$ in., $b_w = 9$ in., $d = 17$ in., and $A_s = 6 \times 1 = 6$ in.^{2}; let $b = b_e$. Check the position of the neutral axis:

$$c = \frac{A_s f_y}{0.85 f_c' b \beta_1} = \frac{6 \times 60}{0.85 \times 0.85 \times 3.5 \times 36} = 3.95 \text{ in.}$$

$$a = 0.85 \times 3.95 = 3.36 \text{ in.}$$

$$c > a > 3 \text{ in.}$$

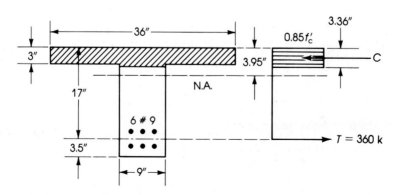

FIGURE 3.37. Example 3.10.

Therefore the neutral axis lies within the web, and a T-section analysis is necessary.

2. $A_{sf} = \dfrac{0.85f_c'(b - b_w)t}{f_y} = \dfrac{0.85 \times 3.5(36 - 9) \times 3}{60} = 4.03 \text{ in.}^2$

$A_s - A_{sf} = 6 - 4.03 = 1.97 \text{ in.}^2$

3. Check if $(\rho_w - \rho_f) \leq (\rho_{max})$:

$$\rho_w - \rho_f = \frac{A_s - A_{sf}}{b_w d} = \frac{1.97}{9 \times 17} = 0.0129$$

$$\rho_{max} = 0.54\frac{f_c'}{f_y} \times \frac{87}{87 + f_y}$$

$$= 0.54 \times (3.5/60) \times 87/(87 + 60) = 0.0186$$

$$(\rho_w - \rho_f) = 0.0129 < \rho_{max} = 0.0186$$

4. Calculate M_u using equation 3.47:

$$M_u = \phi[(A_s - A_{sf})f_y(d - a/2) + A_{sf}f_y(d - t/2)]$$

$$a = \frac{(A_s - A_{sf})f_y}{0.85f_c'b_w} = \frac{1.97 \times 60}{0.85 \times 3.5 \times 9} = 4.42 \text{ in.}$$

Therefore

$$M_u = 0.9[1.97 \times 60(17 - 4.42/2) + 4.03 \times 60(17 - 3/2)]$$

$$= 0.9[1748 + 3748] = 4946 \text{ k·in.} = 412.2 \text{ k·ft}$$

Note that a must be calculated using $(A_s - A_{sf})$ and not total steel area A_s.

Another approach to check whether the neutral axis lies within the web is to calculate the tension force $T = A_s f_y$ and compare it to the compressive force in the total flange (Figure 3.38):

$$T = A_s f_y = 6.0 \times 60 = 360 \text{ k}$$

$$C = 0.85f_c' t b_e = 0.85 \times 3.5 \times 3 \times 36 = 321.3 \text{ k}$$

Since T exceeds C, the neutral axis lies within the web, and the section behaves as a T-

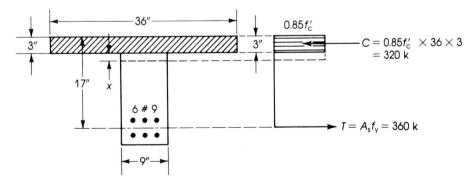

FIGURE 3.38. Example 3.10; a second approach.

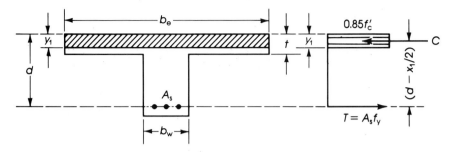

FIGURE 3.39. Analysis of the T-section when $y_1 < t$.

section. An additional area of concrete should be used to provide for the difference, that is, a compressive force of $(360 - 321.3) = 38.7$ kips. This area has a width equal to 9 in. and a depth equal to y. Therefore,

$$b_w y (0.85 f'_c) = 38.7 \text{ k}$$

$$9(y)(0.85 \times 3.5) = 38.7, \quad y = \frac{38.7}{9 \times 0.85 \times 3.5} = 1.44 \text{ in.}$$

Thus

$$a = y + t = 1.44 + 3 = 4.44 \text{ in.}$$

as calculated earlier.

Note that if C is greater than T, the concrete area provided within the compression flange will be greater than that required, and in that case the same approach may still be used. Considering $y_1 < t$,

$$(0.85 f'_c) b_e y_1 = A_s f_y$$

$$y_1 = \frac{A_s f_y}{0.85 f'_c b_e}$$

the moment arm equals $(d - y_1/2)$ and M_u equals $\phi A_s f_y (d - y_1/2)$, which is the ultimate moment of a rectangular section of a width b_e and a compressive equivalent stress block $a = y_1$ (Figure 3.39). Therefore,

$$M_u = \phi A_s f_y (d - a/2) \qquad \blacksquare$$

3.15   Dimensions of Isolated T-shaped Sections

In some cases, isolated beams with the shape of a T-section are used in which additional compression area is provided to increase the compression force capacity of sections. These sections are most commonly used as prefabricated units.

The **1983 ACI Code, section 8.10.4,** as well as the previous codes, specify the size of isolated sections of T-shape as follows:

1. Flange thickness t shall be equal to or greater than one-half of the width of the web b_w.

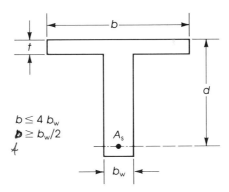

FIGURE 3.40. Isolated T-shaped sections.

2. Total flange width b shall be equal to or less than 4 times the width of the web b_w (Figure 3.40).

3.16 Inverted L-shaped Sections

In slab-beam girder floors, the end beam is called a spandrel beam. This type of floor has part of the slab on one side of the beam and is cast monolithically with the beam. The section is unsymmetrical under vertical loading (Figure 3.41). The loads on slab S_1 cause torsional moment uniformly distributed on the spandrel beam B_1. Design for torsion will be explained later. The overhanging flange width $(b - b_w)$ of a beam with the flange on one side only is limited by the **ACI Code, section 8.10.2,** to the smallest of

1. 1/12 of the span of the beam,
2. Less than or equal to 6 times the thickness of the slab,
3. Less than or equal to one-half the clear distance to the next beam.

If this is applied to the spandrel beam in Figure 3.41, then

1. $(b - 12) \leq (20 \times 12)/12 = 20$ in. (controlling case)
2. $(b - 12) \leq 6 \times 6 = 36$ in.
3. $(b - 12) \leq 3.5 \times 12 = 42$ in.

Therefore, the effective flange width $b = 20 + 12 = 32$ in., and the effective dimensions of the spandrel beam are as shown in Figure 3.41(d).

3.17 Sections of Other Shapes

Sometimes a section different from the previously defined sections is needed in special requirements of structural members. For instance, sections like those shown in Figure 3.42 may be used in the precast concrete industry. The analysis of such sections is similar to that of a rectangular section, taking into consideration the area of removed or added concrete. The best way to explain this analysis is by giving examples.

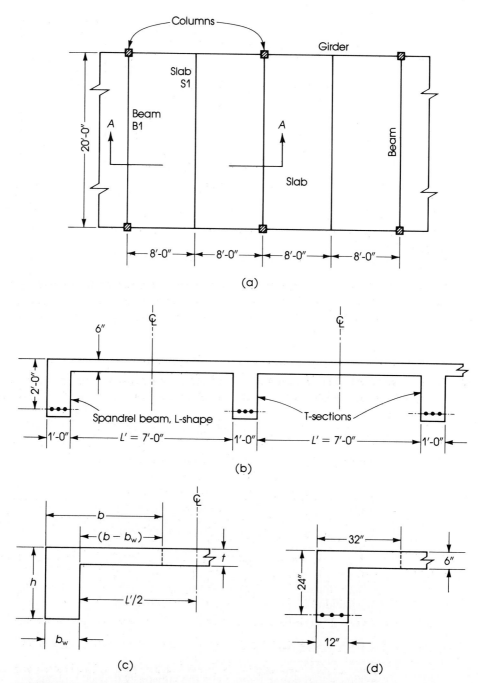

FIGURE 3.41. Slab-beam-girder floor, showing (a) plan, (b) section including spandrel beam, (c) dimensions of the spandrel beam, and (d) its effective flange width.

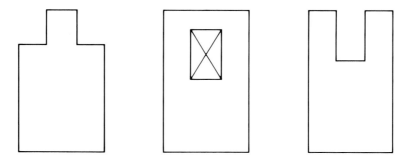

FIGURE 3.42. Sections of other shapes.

EXAMPLE **3.11**

The section shown in Figure 3.43 represents a beam in a structure containing prefabricated elements. The total width and total depth are limited to 12 in. and 18.5 in., respectively. Tension reinforcement used is 4 No. 9 bars; $f'_c = 4$ ksi and $f_y = 50$ ksi. Determine the allowable ultimate moment.

Solution

1. Check whether the section is balanced, under-reinforced, or over-reinforced. If the section is balanced, then the maximum strain of concrete is 0.003 and steel yields; that is,

$$\varepsilon_y = \frac{f_y}{E_s} = \frac{50}{29,000} = 0.001725$$

From similar triangles, find the position of the neutral axis:

$$c = \frac{0.003}{0.003 + 0.001725} \times 16 = 10.15 \text{ in.}$$

$$a = 0.85c = 0.85 \times 10.15 = 8.63 \text{ in.}$$

$C = $ total compressive force

$\quad = 0.85f'_c \times$ area of section from top fibers to a depth equal to a (8.63 in.)

$\quad = 0.85 \times 4[2(4 \times 4) + 12(8.63 - 4)] = 298 \text{ k}$

$T = A_s f_y = 4 \times 1.0 \times 50 = 200 \text{ k}$

$T < C$, thus tension steel will yield first, and the section is under-reinforced.

$$\varepsilon_s > \varepsilon_y = 0.001725, \qquad \varepsilon_c = \text{concrete maximum strain} = 0.003$$

2. Check that the steel used is equal to or less than the maximum steel specified by the ACI Code $(0.75\rho_b)$.

$$A_s \text{ (balanced)} = C_b/f_y = 298/50 = 5.96 \text{ in.}^2$$

$$A_{s(\text{max})} = 0.75A_s \text{ (balanced)} = 0.75 \times 5.96 = 4.47 \text{ in.}^2$$

$$A_s \text{ (used)} = 4 \times 1.0 = 4 \text{ in.}^2 < 4.47 \text{ in.}^2$$

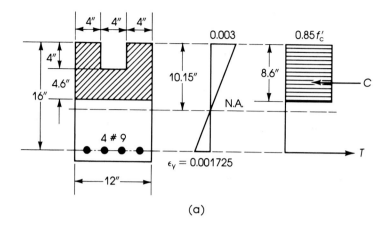

(a)

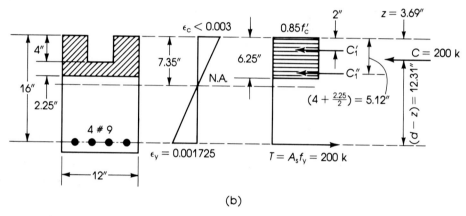

(b)

FIGURE 3.43. Example 3.11: (a) balanced and (b) under-reinforced sections.

Use the actual steel area of 4 in.^{2} and $T = 200$ k. Note that if $A_s > A_{s(max)}$, then $A_{s(max)}$ should be used and the tension force $T = A_{s(max)} f_y$.

3. Determine the position of the neutral axis based on $T = 200$ k. Since $T = C$, therefore,

$$200 = 0.85 f_c' [2(4 \times 4) + 12(a - 4)]$$

where a = depth of equivalent compressive block to produce a total compressive force of 200 k.

$$200 = 0.85 \times 4(32 + 12a - 48]$$
$$a = 6.25 \text{ in.}, \quad c = 6.25/0.85 = 7.35 \text{ in.}$$

4. Calculate M_u by taking moments of the two parts of the compressive forces (each by its arm) about tension steel.

$$C_1' = \text{compressive force on two small areas } 4 \times 4 \text{ in.}$$
$$= 0.85 \times 4[2 \times 4 \times 4] = 108.8 \text{ k}$$

$$C_1'' = \text{compressive force on area } 12 \times 2.25 \text{ in.}$$
$$= 0.85 \times 4 \times 12 \times 2.25 = 91.8 \text{ k}$$

Therefore

$$M_u = \phi[C_1'(d-2) + C_1''(d-5.12)]$$
$$= 0.9[108.8(14) + 91.8 \times 10.88] = 2270 \text{ k·in.} = 189 \text{ k·ft}$$

5. Or the resultant C_1' and C_1'' can be found:

$$C_1' + C_1'' = 108.8 + 91.8 = 200.6 \text{ k}$$

Position of resultant z from top fibers:

$$z = \frac{108.8 \times 2 + 91.8 \times 5.12}{200.6} = 3.43 \text{ in.}$$

Arm $= (d - z) = 16 - 3.43 = 12.57$ in.
$M_u = 0.9 \times 200.6 \times 12.57 = 2269$ k·in. $= 189$ k·ft ∎

EXAMPLE 3.12

Calculate the allowable ultimate moment of the section shown in Figure 3.44, $f_c' = 4$ ksi and $f_y = 60$ ksi.

Solution

1. Check if the section is balanced, under-reinforced, or over-reinforced:

For a balanced condition, compression fibers of concrete reach maximum strain of 0.003 at the same time when steel bars reach their yield stress,

$$\varepsilon_y = f_y/E_s = 60/(29 \times 10^3) = 0.00207$$

Find the position of the neutral axis:

$$c = \frac{0.003}{(0.003 + 0.00207)} \times 22 = \frac{3 \times 22}{5.07} = 13.02 \text{ in.}$$

Find stress in the compression steel:

$\varepsilon_s' = $ strain in compression steel

$$= 0.003 \times \frac{(13.02 - 3)}{13.02} = 0.00231 > \varepsilon_y = 0.00207,$$

Therefore compression steel yields if the section is balanced. Find compression and tension forces:

$$a = 0.85c = 0.85 \times 13.02 = 11.07 \text{ in.}$$
$$C_1 = \text{compression force due to concrete area to a depth } a$$
$$= 0.85f_c'[6 \times 4 + 12(11.07 - 6)] = 289 \text{ k}$$
$$C_2 = A_s'f_y = 1 \times 60 = 60 \text{ k}$$
$$\text{Total } C = C_1 + C_2 = 289 + 60 = 349 \text{ k}$$
$$T = A_s f_y = 4 \times 1 \times 60 = 240 \text{ k} < C$$

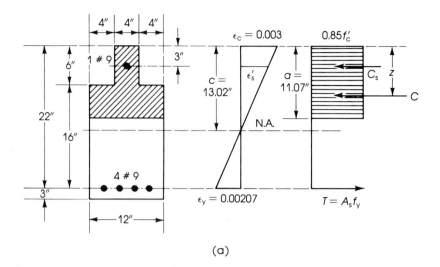

(a)

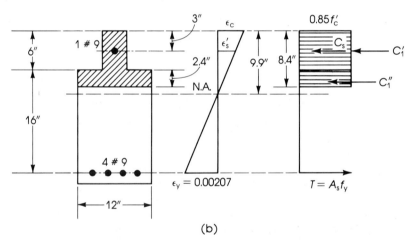

(b)

FIGURE 3.44. Example 3.12: (a) balanced condition and (b) analysis of section.

Therefore tension steel will yield long before the concrete top fibers reach maximum strain of 0.003: the section is under-reinforced.

2. Check if the tension steel used as singly reinforced is equal to or less than the maximum steel specified by the ACI Code,

$$A_{s(max)} = 0.75 A_s \text{ (balanced)}, \qquad (A_s - A'_s) = 4 - 1 = 3 \text{ in.}^2$$

A_s (balanced) as singly reinforced $= C_1/f_y = 289/60 = 4.82$ in.^{2}
$A_{s(max)} = 0.75 A_s$ (balanced) $= 0.75 \times 4.82 = 3.62$ in.^{2}
$A_{s(max)} = 3.62$ in.$^2 > (A_s - A'_s) = 3$ in.^{2}
Therefore, the section is under-reinforced using actual steel in the section.
   Note that if $(A_s - A'_s)$ is greater than $A_{s(max)}$, the latter should be used.

3. Calculate the position of neutral axis due to a maximum allowable force in tension of 240 k. Since $T = C$ at all times, and assuming that compression reinforcement yields (to be checked later):

$$T = C_1 + C_2$$
$$240 = 0.85 \times 4[6 \times 4 + 12(a - 6)] + 1 \times 60 \text{ (comp. steel)}$$

where a is the height of the concrete compressive block to produce a force of 240 k.

$$240 = 3.4[24 + 12a - 72] + 60$$
$$a = 8.41 \text{ in,} \qquad c = 8.41/0.85 = 9.9 \text{ in.}$$

Strain in compression steel

$$\varepsilon_s' = 0.003 \times \frac{(c - d')}{c} = 0.003 \times (6.9/9.9) = 0.00209 > \varepsilon_y$$

Compression steel yields.

4. Calculate M_u by taking moments of compressive forces about the tension steel:

$$C_1' = 0.85f_c' \times 6 \times 4 = 0.85 \times 4 \times 6 \times 4 = 81.6 \text{ k}$$
$$C_1'' = 0.85f_c' \times 12 \times 2.4 = 98 \text{ k}$$
$$C_s' = 1 \times (60 - 0.85 \times 4) = 56.6 \text{ k}$$
$$\begin{aligned} M_u &= \phi[C_1'(d - 6/2) + C_1''(d - 6 - 2.4/2) + C_s'(d - d')] \\ &= 0.9[81.6(22 - 3) + 98(22 - 6 - 1.2) + 56.6(22 - 3)] \\ &= 0.9[1550 + 1450 + 1075] = 3669 \text{ k·in.} = 305.7 \text{ k·ft} \end{aligned}$$ ∎

EXAMPLE 3.13

Recalculate the allowable ultimate moment in example 3.12 if the compression steel is placed $4\frac{1}{2}$ inches below the top fibers, that is, $d' = 4\frac{1}{2}$ in. (Figure 3.45).

Solution   The first two steps in the previous example for a balanced section are the same:

$$a = 11.07 \text{ in.}$$
$$C_1' = 289 \text{ k,} \qquad C_2 = 60 \text{ k,} \qquad C = 349 \text{ k}$$
$$T = 240 \text{ k}$$

1. Calculate the position of the neutral axis due to a maximum tension force of $T = 240$ k:

$$T = C_1 + C_2, \qquad a = 8.41 \text{ in.} \qquad c = 9.9 \text{ in.}$$

Check if compression steel yields:

$$\begin{aligned} \varepsilon_s' &= 0.003(c - d')/c \\ &= 0.003(9.9 - 4.5)/9.9 = 0.00164 < \varepsilon_y \end{aligned}$$

Therefore, compression steel will not yield.

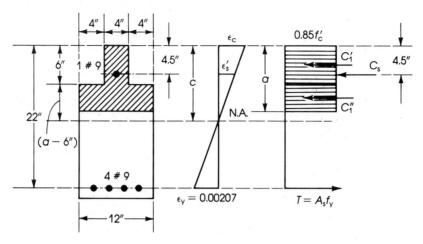

FIGURE 3.45. Example 3.13.

2. Since compression steel will not yield, it may be neglected, as explained earlier in section 3.11.

The depth of the compressive stress distribution has to be calculated:

$$T = C_1$$
$$240 = 0.85 \times 4[6 \times 4 + 12(a - 6)]$$
$$a = 9.9 \text{ in.} \qquad c = 9.9/0.85 = 11.6 \text{ in.}$$
$$C_1' = 0.85f_c' \times 6 \times 4 = 81.6 \text{ k (for the area of 6 by 4 in.)}$$
$$C_1'' = 0.85f_c' \times 12(9.9 - 6) = 159.1 \text{ k}$$

Taking moments about the tension steel:

$$M_u = 0.9[81.6(d - 6/2) + 159.1(d - 6 - 3.9/2)]$$
$$= 0.9[81.6(19) + 159.1(14.05)] = 3407 \text{ k·in.} = 284 \text{ k·ft}$$

This value is less than the ultimate moment calculated previously in example 3.12 when compression steel yields ($M_u = 305.7$ k·ft).

The ultimate moment calculated here is not the allowable ultimate moment according to the ACI Code. To calculate the allowable ultimate moment, the maximum tension steel allowed by the code is equal to $0.75\rho_b$.

$$A_s \text{ (balanced)} = 289/60 = 4.82 \text{ in.}^2$$

and

$$A_{s(max)} \text{ for the section} = 0.75 \times 4.82 = 3.62 \text{ in.}^2$$

Therefore the allowable $T = 3.62 \times 60 = 217$ k (or $T = 0.75 \times 289 = 217$ k). Then since $T = C$,

$$217 = 0.85 \times 4[6 \times 4 + 12(a - 6)], \quad a = 9.33 \text{ in.}$$

$$C_1' = 0.85 \times 4 \times 4 \times 6 = 81.6 \text{ k}$$

$$C_1'' = 0.85 \times 4 \times 12 \times 3.33 = 135.9 \text{ k}$$

$$M_u = 0.9[81.6 \times 19 + 135.9 \times 14.35] = 3151 \text{ k·in.} = 262.5 \text{ k·ft}$$

3. If compression reinforcement is to be considered in the analysis, a trial method can be used. Let $f_s' = 50$ ksi $< f_y = 60$ ksi (to be checked later); in this case, $T = 240$ k $= C_1 + C_2$

$$240 = 0.85 \times 4[6 \times 4 + 12(a - 6)] + 1 \times 50$$

$$a = 8.66 \text{ in.} \qquad c = 10.18 \text{ in.}$$

$$\varepsilon_s' = 0.003 \times \frac{(10.18 - 4.5)}{10.18} = 0.00168 < \varepsilon_y = 0.00207$$

$$f_s' = \varepsilon_s' \times E_s = 0.00168 \times 29 \times 10^3 = 48.7 \text{ ksi (close to 50 ksi)}$$

Therefore,

$$C_1' = 0.85 \times 4 \times 6 \times 4 = 81.6 \text{ k}$$

$$C_1'' = 0.85 \times 4 \times 12(8.66 - 6) = 108.5 \text{ k}$$

$$C_2 = 1 \times 50 = 50 \text{ k}$$

Taking moments about tension reinforcement:

$$M_u = 0.9[81.6(22 - 6/2) + 108.5(22 - 6 - 2.66/2) + 48.7(22 - 4.5)]$$

$$= 3595 \text{ k·in.} = 300 \text{ k·ft}$$

It can be noticed that this moment is greater than the allowable ultimate moment when compression steel was neglected (M_u was 262.5 k·ft) and less than the ultimate moment when compression steel yielded (M_u was 305.7 k·ft).

4. As an alternative to the trial procedure in step (c) above, f_s' can be calculated as a function of the distance to the neutral axis c and a quadratic equation will result:

$$\varepsilon_s' = \frac{(c - d')}{c} \times 0.003$$

$$f_s' = \varepsilon_s' E_s = 29 \times 10^3 \times 0.003 \left(\frac{c - d'}{c}\right) = 87 \left(\frac{c - 4.5}{c}\right)$$

Since $T = C$,

$$240 = 0.85 \times 4[6 \times 4 + 12(0.85c - 6)] + A_s' f_s'$$

or

$$403.2 = 34.68c + 87(c - 4.5)/c$$

Solving for c,

$$c = 10.22 \text{ in.} \qquad a = 8.69 \text{ in.}, \qquad f_s' = 48.7 \text{ ksi}, \qquad M_u = 300 \text{ k·ft}$$

The procedures explained in examples 3.11, 3.12, and 3.13 give a clear idea of how to attack the problem and the degree of work or accuracy desired. ∎

3.18 More Than One Row of Steel Bars in the Section

In the case where more than one layer of tension steel bars are presented in the section (Figure 3.46), the strain in the steel is located at the centroid of the area of steel bars. Actually, since in an under-reinforced concrete section, for instance, tension steel yields, the strain in the bottom layer will reach the yield strain ε_y while that in the top layer will not. As the load is increased, the top layer will reach its yield strain ε_y, while the strain in the bottom layer will be slightly greater than the yield strain, but the stress will still be f_y. Therefore, the above assumption that the strain in the steel is located at the centroid of the area of the steel bars is valid, and can be assumed in the analysis of reinforced concrete sections with more than one layer of steel. If in some cases the steel is placed at different depths of the section, which may not cause the yielding of the steel, the layer closest to the extreme tension fibers will be assumed to have a strain equal to $\varepsilon_y = f_y/E_s$, while the strain in the other layers can be calculated from the strain triangles of Figure 3.46.

3.19 Analysis of Sections Using Tables

Reinforced concrete sections can be analyzed and designed using the tables shown in Appendix A (for U.S. customary units) and Appendix B (for S.I. units). The tables give the values of R_u as related to the steel ratio ρ in addition to the maximum and minimum values of ρ and R_u. When the section dimensions are known, R_u is calculated, then ρ and A_s are determined from tables.

$$M_u = R_u bd^2 \quad \text{or} \quad R_u = \frac{M_u}{bd^2}$$

$$R_u = \phi\rho f_y \left(1 - \frac{\rho f_y}{1.7f_c'}\right)$$

$$A_s = \rho bd \quad \text{or} \quad \rho = \frac{A_s}{bd}$$

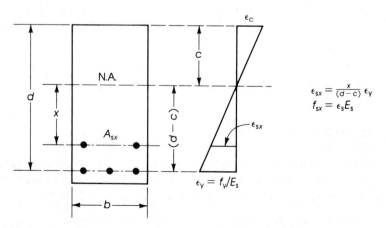

$$\varepsilon_{sx} = \frac{x}{(d-c)}\varepsilon_y$$
$$f_{sx} = \varepsilon_s E_s$$

FIGURE 3.46. Strain and stress in steel bars of different layers.

For any given value of ρ, R_u can be determined from tables, then M_u calculated. The values of ρ and R_u range between a minimum value of ρ_{min} and $R_u(min)$ to maximum values as limited by the ACI Code.

The use of the tables will reduce the calculation time appreciably. The following example explains the use of tables.

EXAMPLE 3.14

Calculate the allowable ultimate moment of the section shown in example 3.2 (Figure 3.12) using tables. Given: $b = 8$ in., $d = 12$ in., $f'_c = 3$ ksi, $f_y = 40$ ksi, and the following steel bars: (a) 3 No. 6 bars and (b) 3 No. 9 bars.

Solution

1. Using 3 No. 6 bars,

$$A_s = 1.32 \text{ in.}^2, \qquad \rho = \frac{A_s}{bd} = \frac{1.32}{8 \times 12} = 0.0138$$

From Table A.4, $\rho_{max} = 0.0278$. Since $\rho < \rho_{max}$, the section is under-reinforced. From Table A.1, for $\rho = 0.0138$, $f'_c = 3$ ksi and $f_y = 40$ ksi, get $R_u = 444$ psi (by interpolation).

$$M_u = R_u bd^2$$

The value of bd^2 can be obtained from Table A.8, or calculated. For $b = 8$ in. and $d = 12$ in., $bd^2 = 1152$ in.^{3}. Then

$$M_u = 444 \times 1152/1000 = 508 \text{ k·in.} = 42.30 \text{ k·ft}$$

Note that $R_u = 444$ psi is less than $R_u(max)$ of 783 (at the end of Table A.1), which indicates that the section is under-reinforced.

2. Using 3 No. 9 bars,

$$A_s = 3 \text{ in.}^2 \qquad \rho = \frac{3.0}{8 \times 12} = 0.0312$$

From Table A.4, $\rho_b = 0.0371$, $\rho_{max} = 0.0278$. Since $\rho < \rho_b$, the section is under-reinforced, but since $\rho > \rho_{max}$, then ρ_{max} and $R_u(max)$ must be used. From Table A.4 or Table A.1, get $R_u(max) = 783$ psi. Therefore

$$M_u = R_u(max) bd^2$$
$$= 783 \times 1152/1000 = 902 \text{ k·in.} = 75.2 \text{ k·ft}$$ ■

3.20 S.I. Examples

The following equations are some of those mentioned in this chapter, but converted to S.I. units. The other equations which are not listed here can be used for both U.S. customary and S.I. units. Note that f'_c and f_y are in MPa (N/mm^2).

$$\rho_b = 0.85\beta_1 \frac{f'_c}{f_y} \frac{600}{600 + f_y} \qquad (3.13)$$

$$\rho_{max} = 0.64\beta_1 \frac{f'_c}{f_y} \frac{600}{600 + f_y} \qquad (3.14)$$

$$\rho_{max} = 0.54 \frac{f'_c}{f_y} \frac{600}{600 + f_y} \qquad \text{when } \beta_1 = 0.85 \text{ and } f'_c \le 28 \text{ MPa}$$

$$(\rho - \rho') \ge 0.85\beta_1 \frac{f'_c}{f_y} \times \frac{d'}{d} \times \frac{600}{600 - f_y} \qquad (3.41)$$

EXAMPLE 3.15

Determine the allowable ultimate moment and the position of the neutral axis of a rectangular section which has $b = 200$ mm, $d = 300$ mm, (a) when it is reinforced with 3 bars 20 mm diameter and (b) when it is reinforced with 3 bars 28 mm diameter. Given: $f'_c = 21$ MPa and $f_y = 280$ MPa.

Solution   1. Area of 3 bars, 20 mm diameter = 942 mm².

$$\rho = \frac{A_s}{bd} = \frac{942}{200 \times 300} = 0.0157$$

$$\rho_{max} = 0.54 \frac{f'_c}{f_y} \frac{600}{600 + f_y}$$

$$= 0.54 \times \frac{21}{280} \times \frac{600}{600 + 280} = 0.0278$$

As $\rho < \rho_{max}$, the ratio ρ should be used to calculate M_u; the section is under-reinforced.

$$M_u = \phi A_s f_y \left(d - \frac{a}{2}\right)$$

$$a = \frac{A_s f_y}{0.85 f'_c b} = \frac{942 \times 280}{0.85 \times 21 \times 200} = 74 \text{ mm}$$

$$M_u = 0.9 \times 942 \times 280 \left(300 - \frac{74}{2}\right) \times 10^{-6} = 62.4 \text{ kN·m}$$

(M_u was multiplied by 10^{-6} to get the answer in kN·m.) The distance of the neutral axis from the compression fibers $c = a/\beta_1$. Since $f'_c < 28$ MPa, then $\beta_1 = 0.85$. Therefore

$$c = \frac{74}{0.85} = 87 \text{ mm}$$

2. If 3 bars 28 mm diameter are used, $A_s = 1848$ mm²

$$\rho = \frac{A_s}{bd} = \frac{1848}{200 \times 300} = 0.0308$$

$$\rho_{max} = 0.0278$$

as calculated earlier. Since $\rho > \rho_{max}$, then ρ_{max} should be used.

$$A_{s(max)} = \rho_{max}(bd) = 0.0278 \times 200 \times 300 = 1668 \text{ mm}^2$$

$$M_u = \phi A_{s(max)} f_y \left(d - \frac{a}{2}\right)$$

$$a_{(max)} = \frac{A_{s(max)} f_y}{0.85 f'_c b} = \frac{1668 \times 280}{0.85 \times 21 \times 200} = 130.8 \text{ mm}$$

$$M_u = 0.9 \times 1668 \times 280 \left(300 - \frac{130.8}{2}\right) \times 10^{-6} = 98.6 \text{ kN·m.}$$

The distance to the neutral axis

$$c(max) = \frac{a(max)}{\beta_1} = \frac{130.8}{0.85} = 154 \text{ mm}$$

$$\text{Utilized steel percentage} = \frac{1668}{1848} \times 100 = 90\%$$ ■

EXAMPLE 3.16

A 2.4-m-span cantilever beam has a rectangular section of $b = 200$ mm, $d = 390$ mm, and is reinforced with 3 bars 22 mm diameter. The beam carries a uniform dead load (including its own weight) of 12 kN/m and a uniform live load of 10.5 kN/m. Check the adequacy of the section if $f'_c = 28$ MPa and $f_y = 280$ MPa.

Solution

1. $U = 1.4D + 1.7L = 1.4 \times 12 + 1.7 \times 10.5 = 34.65 \text{ kN·m.}$

$$\text{External ultimate moment} = M_u = 1.4 M_D + 1.7 M_L$$
$$= 1.4 \left(\frac{12 \times 2.4^2}{2}\right) + 1.7 \left(\frac{10.5 \times 2.4^2}{2}\right)$$
$$= 99.8 \text{ kN·m}$$

2. Calculate the internal ultimate moment:

$$A_s = 1140 \text{ mm}^2$$

$$\rho = \frac{A_s}{bd} = \frac{1140}{200 \times 390} = 0.0146$$

$$\rho_{max} = 0.54 \frac{f'_c}{f_y} \frac{600}{600 + f_y}$$

$$= 0.54 \frac{28}{280} \times \frac{600}{600 + 280} = 0.0368$$

As $\rho < \rho_{max}$, the section is under-reinforced, and actual ρ must be used to calculate M_u.

$$M_u = \phi A_s f_y \left(d - \frac{a}{2}\right)$$

$$a = \frac{A_s f_y}{0.85 f'_c b} = \frac{1140 \times 280}{0.85 \times 28 \times 200} = 67 \text{ mm}$$

$$M_u = 0.9 \times 1140 \times 280 \left(390 - \frac{67}{2}\right) \times 10^{-6} = 102.4 \text{ kN·m}$$

3. Alternatively, calculate

$$R_u = \phi \rho f_y \left(1 - \frac{\rho f_y}{1.7 f_c'}\right)$$

$$= 0.9 \times 0.0146 \times 280 \left(1 - \frac{0.0146 \times 280}{1.7 \times 28}\right) = 3.36 \text{ MPa}$$

$$M_u = R_u b d^2 = 3.36 \times 200 \times (390)^2 \times 10^{-6} = 102.4 \text{ kN·m}$$

4. The internal ultimate moment of 102.4 kN·m is greater than the external moment of 99.8 kN·m, therefore the section is adequate. Note: R_u can be obtained from Table B.2 in Appendix B. ∎

EXAMPLE 3.17

Calculate the ultimate moment capacity of a rectangular section with the following details: $b = 250$ mm, $d = 425$ mm, $d' = 50$ mm, tension steel is 6 bars 25 mm diameter, compression steel is 2 bars 22 mm diameter, $f_c' = 21$ MPa, and $f_y = 350$ MPa.

Solution

1. Check if compression steel yields:

$$A_s = 490 \times 6 = 2940 \text{ mm}^2, \quad A_s' = 380 \times 2 = 760 \text{ mm}^2, \quad A_s - A_s' = 2180 \text{ mm}^2$$

$$\rho = \frac{A_s}{bd} = \frac{2940}{250 \times 425} = 0.0277$$

$$\rho' = \frac{A_s'}{bd} = \frac{760}{250 \times 425} = 0.0071$$

$$(\rho - \rho') = 0.0206$$

For compression steel to yield,

$$(\rho - \rho') \geq 0.85 \times 0.85 \times \frac{21}{350} \times \frac{50}{425} \times \frac{600}{600 - 350} = 0.0122$$

$$(\rho - \rho') \text{ provided} = 0.0206 > 0.0122 \text{ required}$$

Therefore, compression steel yields.

2. Check if $(\rho - \rho')$ provided $\leq \rho_{max}$ as singly reinforced:

$$\rho_{max} = 0.54 \frac{f_c'}{f_y} \frac{600}{600 + f_y} = 0.0205$$

$$(\rho - \rho') \text{ provided} < \rho_{max}$$

3. Calculate

$$M_u = \phi \left[(A_s - A_s')f_y \left(d - \frac{a}{2}\right) + A_s' f_y (d - d')\right]$$

$$a = \frac{(A_s - A_s')f_y}{0.85 f_c' b} = \frac{2180 \times 350}{0.85 \times 21 \times 250} = 171 \text{ mm}$$

$$M_u = 0.9 \left[2180 \times 350 \left(425 - \frac{171}{2} \right) + 760 \times 350(425 - 50) \right] \times 10^{-6}$$

$$= 322.9 \text{ kN·m}$$

4. Yielding of compression steel may be checked by the strain diagram, as in example 3.6. ∎

SUMMARY

SECTIONS 3.1 and 3.2

In proportioning reinforced concrete structural members, it is necessary to investigate the safety of the structure and the deflection and the width of cracks in each member under working load.

SECTION 3.3

When reinforced concrete beams are tested to failure, flexural cracks develop when the tensile stress in concrete reaches or exceeds $f_r(7.5\sqrt{f'_c})$. Shear cracks develop at a section located at about distance d from the face of the support.

SECTION 3.4

Reinforced concrete flexural members may be classified as under-reinforced, balanced, or over-reinforced, depending on the type of flexural failure. Only the first type is allowed by the ACI Code.

SECTION 3.5

Ultimate strength U shall be the critical of

$$U = 1.4D + 1.7L$$
$$U = 0.75(1.4D + 1.7L + 1.7W)$$
$$U = 0.9D + 1.3W$$

SECTION 3.6

The capacity reduction factor ϕ specified by the ACI Code is equal to

0.9 for bending or axial tension
0.85 for shear or torsion
0.75 for axial compression with or without bending (members with spirals)
0.70 for axial compression with or without bending (other than members with spirals)

SECTIONS 3.7–3.9

Analysis of sections is explained in detail.
(a) For a balanced section,

$$c_b = \left(\frac{87}{87 + f_y} d \right), \qquad a = (\beta_1 c)$$

$$\rho_b = 0.85 \beta_1 \frac{f'_c}{f_y} \times \frac{87}{87 + f_y}$$

(b) $\rho_{max} = 0.75\rho_b, \qquad A_{s(max)} = \rho_{(max)} bd$

$\beta_1 = 0.85$ for $f'_c \leq 4000$ psi and decreases linearly by 0.05 for every increase of 1000 psi in f'_c.

(c) $M_u = \phi A_s f_y \left(d - \dfrac{a}{2}\right),$ $a = \dfrac{A_s f_y}{0.85 f'_c b}$

$M_u = R_u b d^2,$ $R_u = \phi \rho f_y \left(1 - \dfrac{\rho f_y}{1.7 f'_c}\right)$

$\rho = A_s/bd$

SECTIONS 3.10–3.12

(a) A given section is said to be adequate if the internal ultimate moment equals or exceeds the externally applied ultimate moment, $\phi M_n \geq M_u$.

(b) The ACI Code specifies a minimum percentage of steel = $200/f_y$. For grades 40, 50, and 60 steels, ρ_{min} is equal to 0.005, 0.004, and 0.0033, respectively.

(c) Bundled bars are allowed by the ACI Code.

SECTION 3.13

(a) Compression steel yields when

$$(\rho - \rho') \geq 0.85\beta_1 \frac{f'_c}{f_y} \times \frac{d'}{d} \times \frac{87}{87 - f_y}   (f_y \text{ in ksi})$$

If compression steel yields, then

$$M_u = \phi \left[(A_s - A'_s)f_y \left(d - \frac{a}{2}\right) + A'_s f_y (d - d')\right]$$

$$a = \frac{(A_s - A'_s)f_y}{0.85 f'_c b}$$

and $(\rho - \rho')$ must be $\leq \rho_{max}$.

(b) When compression steel does not yield, it may be neglected in calculating M_u.

SECTION 3.14

In T- and I-sections:

(a) Effective flange width b_e is the least of: (span/4), $(16t + b_w)$, or the distance between centerlines of adjacent slabs.

(b) When the section behaves as a rectangular section ($a \leq t$), treat the section as singly reinforced.

$$M_u = \phi A_s f_y \left(d - \frac{a}{2}\right)$$

(c) When $a > t$, then

$$M_u = M_{u1} + M_{u2}$$

$$= \phi \left[(A_s - A_{sf})f_y \left(d - \frac{a}{2}\right) + A_{sf} f_y \left(d - \frac{t}{2}\right)\right]$$

$$A_{sf} = 0.85 f'_c t(b - b_w)/f_y,   a = \frac{(A_s - A_{sf})f_y}{0.85 f'_c b_w}$$

$(\rho_w - \rho_f) \leq \rho_{max}$

SECTIONS 3.15 and 3.16

Isolated T-sections and inverted L-shaped sections are treated as T-sections.

SECTION
3.17
The analysis of sections of special shape is similar to that of rectangular sections, taking into consideration the area of removed or added concrete.

SECTION
3.18
When more than one layer of tension steel bars are presented in the section, the resultant tension force is located at the centroid of the area of the bars.

SECTION
3.19
Reinforced concrete sections can be analyzed using the tables in Appendix A for U.S. customary units and Appendix B for S.I. units.

SECTION
3.20
S.I. equations and examples are presented in this section.

REFERENCES

1. E. Hognestad, N. W. Hanson, and D. McHenry: "Concrete Stress Distribution in Ultimate Strength Design," *ACI Journal,* 52, December 1955, pp. 455–79.
2. J. R. Janney, E. Hognestad, and D. McHenry: "Ultimate Flexural Strength of Prestressed and Conventionally Reinforced Concrete Beams," *ACI Journal,* 52, February 1956, pp. 601–20.
3. A. H. Mattock, L. B. Kriz, and E. Hognestad: "Rectangular Concrete Stress Distribution in Ultimate Strength Design," *ACI Journal,* 57, February 1961, pp. 875–929.
4. A. H. Mattock and L. B. Kriz: "Ultimate Strength of Nonrectangular Structural Concrete Members," *ACI Journal,* 57, January 1961, pp. 737–66.
5. *Building Code Requirements for Reinforced Concrete,* ACI 381-83, American Concrete Institute, Detroit, 1983.
6. Franco Levi: "Work of European Concrete Committee," *ACI Journal,* 57, March 1961, pp. 1049–54.
7. UNESCO, *Reinforced Concrete—An International Manual,* Butterworth, London, 1971.
8. M. N. Hassoun: "Ultimate-Load Design of Reinforced Concrete," View Point Publication, Cement and Concrete Association, London, 1981, 2nd ed.

PROBLEMS

3.1. (a) Calculate the modulus of elasticity of concrete E_c for the following types of concrete:

Density	Strength f_c'
155 pcf	4000 psi
145 pcf	3000 psi
120 pcf	2000 psi
2400 kg/m³	30 MPa
2200 kg/m³	22 MPa
2000 kg/m³	18 MPa

$$E_c = 33\,W^{1.5}\sqrt{f_c'}\ \text{(psi)}$$

$$E_c = 0.043\,W^{1.5}\sqrt{f_c'}\ \text{(S.I.)}$$

(b) Determine the modular ratio n and the modulus of rupture for each case. Tabulate your results.

$$f_r = 7.5\sqrt{f_c'} \text{ psi}, \qquad f_r = 0.62\sqrt{f_c'} \text{ MPa}$$

3.2. A standard concrete cylinder 150 by 300 mm was tested to failure and the following loads and strains were recorded:

Load kN	Strain × 10^4	Load kN	Strain × 10^4
0	0.0	318	10.0
53	1.2	371	13.6
106	2.0	424	18.0
159	3.2	477	30.0
212	5.2	424	39.0
265	7.2	371	42.0

(a) Draw the stress-strain diagram of concrete and determine the maximum stress and corresponding strain.

(b) Determine the initial modulus, secant modulus, and the modulus at a stress of 18 N/mm².

(c) Calculate the modulus of elasticity of concrete using the ACI formula for normal weight concrete and compare results.

$$E_c = 57,000\sqrt{f_c'} \text{ psi}, \qquad E_c = 4730\sqrt{f_c'} \text{ MPa}$$

3.3. Calculate ρ_b, ρ_{max}, $R_u(max)$, R_u, a/d ratio, and (a/d) max for a rectangular section that has a width $b = 12$ in. (300 mm) and an effective depth $d = 20$ in. (500 mm) for the following cases:

(a) $f_c' = 3$ ksi, $f_y = 40$ ksi, $A_s = 4$ No. 8 bars.
(b) $f_c' = 4$ ksi, $f_y = 60$ ksi, $A_s = 4$ No. 7 bars.
(c) $f_c' = 5$ ksi, $f_y = 60$ ksi, $A_s = 4$ No. 9 bars.
(d) $f_c' = 21$ MPa, $f_y = 420$ MPa, $A_s = 3 \times 28$ mm.
(e) $f_c' = 20$ MPa, $f_y = 350$ MPa, $A_s = 3 \times 29$ mm.
(f) $f_c' = 35$ MPa, $f_y = 520$ MPa, $A_s = 4 \times 25$ mm.
(g) $f_c' = 17.5$ MPa, $f_y = 280$ MPa, $A_s = 4 \times 18$ mm.

3.4. Using the ACI Code requirements, calculate the ultimate moment capacity of a rectangular section that has a width $b = 250$ mm (10 in.) and an effective depth $d = 550$ mm (22 in.) when $f_c' = 21$ MPa (3 ksi) and $f_y = 420$ MPa (60 ksi) and the steel used is

(a) 3 No. 9 bars (b) 6 No. 9 bars (c) 4 × 20 mm (d) 4 × 28 mm (e) 4 × 32 mm

3.5. A reinforced concrete simple beam has a rectangular section with a width $b = 8$ in. (200 mm) and an effective depth $d = 18$ in. (450 mm). At ultimate moment (failure) the strain in the steel was recorded and was equal to 0.0015. (The strain in concrete at failure may be assumed to be 0.003.)

(a) Check if the section is balanced, under-reinforced, or over-reinforced.

(b) Determine the steel area that will make the section balanced.

(c) Calculate the steel area provided in the section and its ultimate moment. Compare this value with the ultimate moment allowed by the ACI Code.

(d) Calculate the ultimate moment capacity of the section if the steel percentage used is $\rho = 1.4$ percent. Given: $f_c' = 3$ ksi (21 MPa) and $f_y = 50$ ksi (350 MPa)

3.6. A 10-ft (3-m) span cantilever beam has an effective cross-section (bd) of 12 in. by 24 in. (300 by 600 mm) and reinforced with 5 No. 8 (5 × 25 mm) bars. If the uniform load due to its own weight and the dead load are equal to 685 lb/ft (10 kN/m), determine the allowable uniform live load on the beam using the strength design method and ACI load factors. Given: $f'_c = 3$ ksi (21 MPa), $f_y = 60$ ksi (420 MPa)

3.7. The cross-section of a 17-ft (5-m) span simply supported beam is 10 by 28 in. (250 by 700 mm), and it is reinforced symmetrically with 8 No. 6 bars (8 × 20 mm) in two rows. Determine the allowable concentrated live load at midspan considering the total acting dead load is equal to 2.55 k/ft (37 kN/m). Given: $f'_c = 3$ ksi (21 MPa), $f_y = 40$ ksi (280 MPa)

3.8. Determine the ultimate moment capacity of the sections shown in Figure 3.47. Neglect the lack of symmetry in (b). Given: $f'_c = 4$ ksi (28 MPa), $f_y = 60$ ksi (420 MPa)

3.9. A rectangular concrete section has a width $b = 12$ in. (300 mm), an effective depth $d = 18$ in. (450 mm), and $d' = 2$ in. (50 mm). If compression steel of 2 No. 7 bars (2 × 22 mm) is used, calculate the allowable ultimate moment that can be applied on the section if the tensile steel A_s is
(a) 4 No. 7 (4 × 22 mm)   and   **(b)** 8 No. 8 (8 × 25 mm) bars. Given: $f'_c = 3$ ksi (21 MPa), $f_y = 40$ ksi (280 MPa)

3.10. A 16-ft (4.8-m) span simply supported beam has a width $b = 12$ in. (300 mm), $d = 22$ in. (550 mm), $d' = 2$ in. (50 mm), $A_s = 6$ No. 9 bars (6 × 28 mm), and $A'_s = 3$ No. 6 bars (3 × 18 mm). The beam carries a uniform dead load of 2 k/ft (30 kN/m) including its own weight. Calculate the allowable uniform live load that can be safely applied on the beam. Given: $f'_c = 3$ ksi (21 MPa), $f_y = 50$ ksi (350 MPa)

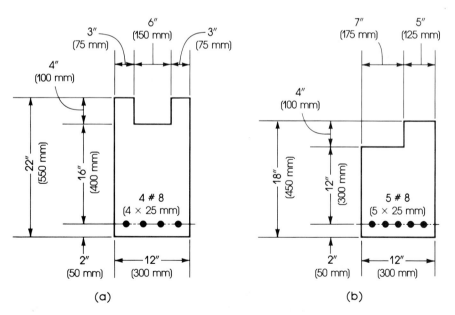

FIGURE 3.47. Problem 3.8.

3.11. Check the adequacy of a 10-ft (3-m) span cantilever beam, assuming a concrete strength $f'_c = 4$ ksi (28 MPa) and steel $f_y = 60$ ksi (420 MPa) are used. The dimensions of the beam section are $b = 10$ in. (250 mm), $d = 20$ in. (500 mm), $d' = 2$ in. (50 mm), $A_s = 6$ No. 7 (6 × 22 mm), and $A'_s = 2$ No. 5 (2 × 16 mm). The dead load on the beam, excluding its own weight, is equal to 2 k/ft (30 kN/m), and the live load equals 1.25 k/ft (20 kN/m). (Compare the internal M_u with the external ultimate moment.)

3.12. A series of reinforced concrete beams spaced at 9 ft (2.7 m) on centers are acting on a simply supported span of 18 ft (5.4 m). The beams support a reinforced concrete floor slab 4 in. (100 mm) thick. If the width of web $b_w = 10$ in. (250 mm), $d = 18$ in. (450 mm), and the beam is reinforced with 3 No. 9 bars (3 × 28 mm), determine the moment capacity of a typical interior beam. Given: $f'_c = 4$ ksi (280 MPa), $f_y = 60$ ksi (420 MPa)

3.13. Calculate the ultimate moment capacity of a T-section that has the following dimensions:

Flange width = 30 in. (750 mm)
Flange thickness = 3 in. (75 mm)
Web width = 10 in. (250 mm)
Effective depth $d = 18$ in. (450 mm)
Tension reinforcement 6 No. 8 bars (6 × 25 mm)
$f'_c = 3$ ksi (21 MPa)
$f_y = 60$ ksi (420 MPa)

3.14. Repeat problem 3.13 if $d = 24$ in. (600 mm).

3.15. Repeat problem 3.13 if the flange is an inverted L-shape with the same flange width projecting from one side only. (Neglect lack of symmetry.)

CHAPTER 4

Strength Design Method: Flexural Design of Reinforced Concrete Beams

4.1   Introduction

In the previous chapter, the analysis of different reinforced concrete sections was explained: details of the section were given, and it was required to determine the ultimate moment capacity of the section. In this chapter, the process is reversed: the external moment is given, which is equal to the internal moment, and it is required to find safe, economic, and practical dimensions of the concrete section and the area of reinforcing steel.

4.2   Rectangular Sections with Tension Reinforcement Only

From the analysis of rectangular singly reinforced sections (section 3.9), the following equations were derived, where f'_c and f_y are in ksi:

$$\rho_b = 0.85 \beta_1 \frac{f'_c}{f_y} \times \frac{87}{87 + f_y} \tag{3.13}$$

$$\rho_{max} = 0.75 \rho_b = 0.64 \beta_1 \frac{f'_c}{f_y} \times \frac{87}{87 + f_y} \tag{3.14}$$

$$(\text{for } f'_c \le 4000 \text{ psi}), \quad \rho_{max} = 0.54 \frac{f'_c}{f_y} \times \frac{87}{87 + f_y} \tag{3.15}$$

β_1 is 0.85 when $f'_c \le 4000$ psi (28 N/mm²) and decreases by 0.05 for every increase of 1000 psi (7 N/mm²) in concrete strength. The steel percentage of a balanced section ρ_b and the maximum allowable steel percentage ρ_{max} can be calculated for different values of f'_c and f_y, as shown in Table A.4 in Appendix A.

The ultimate moment equations were derived in the previous chapter in the following forms:

$$M_u = R_u b d^2 \tag{3.18}$$

where

$$R_u = \phi\rho f_y \left(1 - \frac{\rho f_y}{1.7f_c'}\right) \tag{3.19}$$

and $\phi = 0.9$, or

$$M_u = \phi b d^2 \rho f_y \left(1 - \frac{\rho f_y}{1.7f_c'}\right) \tag{3.17}$$

It can be seen that for a given ultimate moment and known f_c' and f_y, there are three unknowns in the above equations: the width b, the effective depth of the section d, and the steel ratio ρ. A unique solution is not possible unless values of two of these three unknowns are assumed. Usually ρ is assumed (using ρ_{max}, for instance), and b can also be assumed, as it may represent the width of the partition wall under that beam in some cases.

4.3   Spacing of Reinforcement and Concrete Cover

4.3.1   Specifications

Figure 4.1 shows two reinforced concrete sections. The bars are placed such that the clear spacings shall be at least equal to nominal bar diameter D but not less than 1 inch (25 mm). Vertical clear spacings between bars, in more than one layer, shall not be less than 1 inch (25 mm), according to the **ACI Code, section 7.6.** The width of the section

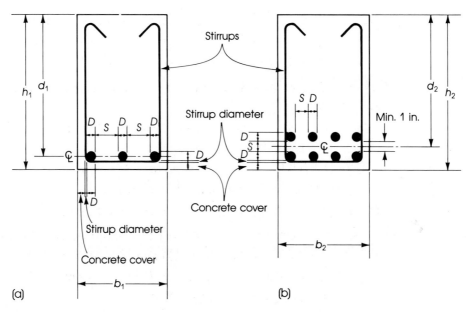

FIGURE 4.1. Spacing of steel bars (a) in one row or (b) two rows.

depends on the number n and diameter of bars used. Stirrups are placed at intervals; their diameter and spacings depend on shear requirements, to be explained later. At this stage, stirrups $\frac{3}{8}$ in. (10 mm) diameter can be assumed to calculate the width of the section. There is no need to adjust the width b if different diameters of stirrups are used. The specified minimum concrete cover for cast-in-place and precast concrete is given in the **ACI Code, section 7.7.** Concrete cover for beams and girders is equal to $1\frac{1}{2}$ in. (38 mm), and that for slabs is equal to $\frac{3}{4}$ in. (20 mm), when concrete is not exposed to weather or in contact with ground.

4.3.2   Minimum Width of Concrete Sections

The general equation for the minimum width of a concrete section can be written in the following form:

$$b_{min} = nD + (n - 1)s + 2 \times \text{stirrups diameter}$$
$$+ 2 \times \text{concrete cover} \qquad (4.1a)$$

where

n = number of bars

D = diameter of the largest bar used

s = spacing between bars (equal to D or 1 in., whichever is larger)

If the stirrups' diameter is taken equal to $\frac{3}{8}$ in. (10 mm) and concrete cover equals $1\frac{1}{2}$ in. (38 mm), then

$$b_{min} = nD + (n - 1)s + 3.75 \text{ in. (95 mm)} \qquad (4.1b)$$

This equation, if applied to the concrete sections in Figure 4.1, becomes

$$b_1 = 3D + 2S + 3.75 \text{ in. (95 mm)}$$
$$b_2 = 4D + 3S + 3.75 \text{ in. (95 mm)}$$

To clarify the use of equation 4.1, let the bars used in sections of Figure 4.1 be No. 10 (32 mm) bars. Then

$$b_1 = 5 \times 1.27 + 3.75 = 10.10 \text{ in. } (s = D), \text{ say 11 in.}$$
$$b_1 = 5 \times 32 + 95 = 255 \text{ mm, say 260 mm}$$
$$b_2 = 7 \times 1.27 + 3.75 = 12.64 \text{ in., say 13 in.}$$
$$b_2 = 7 \times 32 + 95 = 319 \text{ mm, say 320 mm}$$

If bars used are No. 6 (20 mm), the minimum widths become

$$b_1 = 3 \times 0.75 + 2 \times 1 + 3.75 = 8.0 \text{ in. } (s = 1.0 \text{ in.})$$
$$b_1 = 3 \times 20 + 2 \times 25 + 95 = 205 \text{ mm, say 210 mm}$$
$$b_2 = 4 \times 0.75 + 3 \times 1 + 3.75 = 9.75 \text{ in., say 10 in.}$$
$$b_2 = 4 \times 20 + 3 \times 25 + 95 = 250 \text{ mm}$$

The width of the concrete section shall be increased to the nearest inch.

4.3.3   Minimum Overall Depth of Concrete Sections

The effective depth d is the distance between the extreme compressive fibers of the concrete section and the centroid of the tension reinforcement. The minimum total depth is equal to d plus the distance from the centroid of the tension reinforcement to the extreme tension concrete fibers, which depends on the number of layers of the steel bars. In application to the sections shown in Figure 4.1,

$$h_1 = d_1 + D/2 + \tfrac{3}{8} \text{ in.} + \text{concrete cover} \qquad\qquad (4.2a)$$
$$= d_1 + D/2 + 1.875 \text{ in.} \quad (50 \text{ mm})$$

for one row of steel bars and

$$h_2 = d_2 + 0.5 + D + \tfrac{3}{8} \text{ in.} + \text{concrete cover} \qquad\qquad (4.2b)$$
$$= d_2 + D + 2.375 \text{ in. } (60 \text{ mm})$$

for two layers of steel bars. The overall depth shall be increased to the nearest half inch (10 mm), or better, to the nearest inch (20 mm in S.I.). For example, if $D = 1$ in. (25 mm), $d_1 = 18.9$ in. (475 mm), and $d_2 = 20.1$ in. (502 mm), then

$$\text{Minimum } h_1 = 18.9 + 0.5 + 1.875 = 21.275 \text{ in., say } 21.5 \text{ in. or } 22 \text{ in.}$$
$$(h_1 = 475 + 13 + 50) = 538 \text{ mm, say } 540 \text{ mm or } 550 \text{ mm}$$

and

$$\text{Minimum } h_2 = 20.1 + 1.0 + 2.375 = 23.475 \text{ in., say } 23.5 \text{ in. or } 24 \text{ in.}$$
$$(h_2 = 502 + 25 + 60) = 587 \text{ mm, say } 590 \text{ mm or } 600 \text{ mm}$$

If No. 9 or smaller bars are used, a practical estimate of the total depth h can be made as follows:

$$h = d + 2.5 \text{ in. } (63 \text{ mm}), \text{ for one layer of steel bars}$$
$$h = d + 3.5 \text{ in. } (88 \text{ mm}), \text{ for two layers of steel bars}$$

For more than two layers of steel bars, a similar approach may be used.

It should be mentioned that the minimum spacing between bars depends on the maximum size of coarse aggregate used in concrete. The nominal maximum size of the coarse aggregate shall not be larger than one-fifth of the narrowest dimension between sides of forms, nor one-third of the depth of slabs, nor three-fourths of the minimum clear spacing between individual reinforcing bars or bundles of bars **(ACI Code, section 3.3)**.

EXAMPLE 4.1

Design a singly reinforced rectangular section to resist an ultimate moment of 2700 k·in. using the maximum allowable steel percentage ρ_{max}. Given: $f'_c = 3$ ksi and $f_y = 40$ ksi.

Solution   For $f'_c = 3$ ksi and $f_y = 40$ ksi,

$$\rho_{max} = (0.75)(0.85)\beta_1 \frac{f'_c}{f_y} \times \frac{87}{87 + f_y}$$

and thus

$$\beta_1 = 0.85$$

$$\rho_{max} = (0.75)(0.85)^2 \left(\frac{3}{40}\right)\left(\frac{87}{87 + 40}\right) = 0.0278$$

$$(R_u)_{max} = \phi\rho_{max}f_y \left(1 - \frac{\rho_{max}f_y}{1.7f_c'}\right)$$

$$= 0.9 \times 0.0278 \times 40 \times \left(1 - \frac{0.0278 \times 40}{1.7 \times 3}\right) \text{ ksi} = 0.783 \text{ ksi}$$

Since $M_u = R_u bd^2$, therefore

$$bd^2 = M_u/R_u = 2700/0.783 = 3448 \text{ in.}^3$$

Thus for

$b = 6$ in.,	$d = 24$ in.
$b = 7$ in.,	$d = 22.2$ in.
$b = 8$ in.,	$d = 20.8$ in.
$b = 9$ in.,	$d = 19.6$ in.
$b = 10$ in.,	$d = 18.6$ in.
$b = 11$ in.,	$d = 17.7$ in.
$b = 12$ in.,	$d = 17.0$ in.

The choice of the effective depth d depends on three factors:

1. The width b required. A small width will result in a deep beam that decreases the headroom available. Furthermore, a deep narrow beam may lower the ultimate moment capacity of the structural member due to possible lateral deformation.
2. The amount of (diameter) and distribution of reinforcing steel. A narrow beam may need more than one row of steel bars, thus increasing the total depth of the section.
3. The partition or wall thickness. If cement block walls are used, usually but not necessarily, the width b is chosen to be equal to the wall thickness. Exterior walls in buildings in most cases are thicker than interior walls. The architectural plan of the structure will show the different thicknesses.

A reasonable choice of d/b ratio varies between 1.5 and 3, and the most practical values range between 2 and 2.5. It can be seen that the deeper the section, the more economical it is, as far as the quantity of concrete used. For instance, considering the unit length of a beam, the ratio of increase in volume of concrete between the last and first choices in this example is $(12 \times 17)/(6 \times 24) = 1.42$. Other ratios are shown in Table 4.1. In the same table the area of the two sides and bottom width b for a unit length of the beam is given for every choice of b and d. These values represent the approximate area of formwork.

It can be seen that deep beams require a greater area of formwork than a shallow beam. The cost of the greater concrete volume may outweigh the difference in the cost of formwork when a shallow section is chosen. However, a compromise can be found, which is, practically, that ratio d/h of 2.0 to 2.5 suggested before. A section 9 by 19.6 in. is adequate if no other restrictions are called for.

The area of the steel reinforcement A_s is equal to ρbd. The area of steel needed for the different choices of b and d are also shown in Table 4.1. Since the steel percentage required is

TABLE 4.1. Properties of sections chosen in example 4.1

b (in.)	d (in.)	Area bd (in.^{2})	Ratio $bd/6 \times 24$	Area of formwork $1 \times (b + 2d)$ (in.^{2})	Ratio $(b + 2d)/54$	A_s (in.^{2})	Ratio $A_s/4$
6	24.0	144.0	1.00	54.0	1.000	4.00	1.00
7	22.0	155.4	1.08	51.4	0.952	4.32	1.08
8	20.8	166.4	1.15	49.6	0.918	4.63	1.16
9	19.6	176.4	1.22	48.2	0.893	4.90	1.22
10	18.6	186.0	1.29	47.2	0.874	5.17	1.29
11	17.7	194.7	1.35	46.4	0.860	5.41	1.35
12	17.0	204.0	1.42	46.0	0.850	5.67	1.42

constant ($\rho_{max} = 0.0278$), then A_s is proportional to bd. The choice of a deep section will result in smaller A_s. Choosing 9 by 19.6 in. section, the area of steel A_s required is equal to 4.90 in.^{2}. From that figure, different choices of steel bars may be selected:

3 No. 11	$A_s = 4.68$ in.^{2}
4 No. 10	$A_s = 5.06$ in.^{2}
5 No. 9	$A_s = 5.00$ in.^{2}
6 No. 8	$A_s = 4.71$ in.^{2}
8 No. 7	$A_s = 4.81$ in.^{2}
11 No. 6	$A_s = 4.86$ in.^{2}
16 No. 5	$A_s = 4.91$ in.^{2}

The choice of bars will depend on two factors:

1. Adequate placement of bars in the section in one or more rows, fulfilling the restrictions of the ACI Code for minimum spacing.
2. The area of steel bars closest to the required steel area.

If 6 No. 8 bars are chosen, $A_s = 4.71$ in.^{2}, which is 0.19 in.^{2} less than the required area of 4.90 in.^{2} But since the total depth h may be increased a fraction of an inch, the actual d will be a little greater than the calculated d, consequently reducing the required A_s. Also, the 6 bars would have to be placed in two layers as 9-in. width is not sufficient. Calculating the minimum b to place 3 No. 8 bars in one layer,

$$b_{min} = 3D + 2s + 3.75 \qquad (4.1b)$$
$$= 5 + 3.75 = 8.75, \text{ say 9 in.}$$
$$h_{min} = d + D + 2.375 \qquad (4.2b)$$
$$= 19.6 + 1.0 + 2.375 = 22.975, \text{ say 23 in. (or 24 in.)}$$

Therefore, the section shown in Figure 4.2 is selected. The total depth h is chosen to be 24 in. In this case, the actual $d = 4 - 1.0 - 2.375 = 20.625$ in., which is greater than the calculated d of 19.6 in. Because of the small variation, reduction in the required steel area can be approximated by the ratio of the calculated d to the actual d. That is, A_s actually needed $= 4.90 \times 19.6/20.625 = 4.66$ in.^{2}, which is less than 4.71 in.^{2} provided. In the same way, if 22 in. total depth is chosen, then the actual d is equal to $22 - 3.375 = 18.625$ in. and the A_s actually needed $= 4.90 \times 19.6/18.625 = 5.16$ in.^{2} ∎

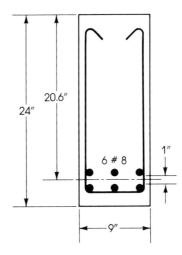

FIGURE 4.2. Example 4.1.

EXAMPLE 4.2

Solve the problem of example 4.1 using a steel percentage $\rho = \frac{1}{2}\rho_{max}$.

Solution   For $f'_c = 3$ ksi and $f_y = 40$ ksi, $\rho_{max} = 0.0278$ and $\frac{1}{2}\rho_{max} = 0.0139$.

$$R_u = \phi\rho f_y \left(1 - \frac{\rho f_y}{1.7 f'_c}\right)$$

$$= 0.9 \times 0.0139 \times 40 \left(1 - \frac{0.0139 \times 40}{1.7 \times 3}\right)$$

$$= 0.446 \text{ ksi}$$

or, from the tables in Appendix A, for $\rho = 0.0139$, and by interpolation, $R_u = 446$ psi = 0.446 ksi. Then

$$bd^2 = \frac{M_u}{R_u} = \frac{2700}{0.446} = 6054 \text{ in.}^3$$

Choosing $b = 12$ in., $d = \sqrt{504.5} = 22.5$ in.

$$A_s = \rho bd = 0.0139 \times 12 \times 22.5 = 3.75 \text{ in.}^2$$

Choose 4 No. 9 bars in one layer, $A_s = 4.00$ in.^{2}.

$$b_{min} = nD + (n-1)s + 3.75 \qquad\qquad (4.1b)$$

$$= 7 \times 1.128 + 3.75 = 11.7 \text{ in., say 12 in.}$$

$$h_{min} = d + D/2 + 1.875 \qquad\qquad (4.2a)$$

$$= 22.5 + 1.128/2 + 1.875 = 24.94 \text{ in., say 25 in.}$$

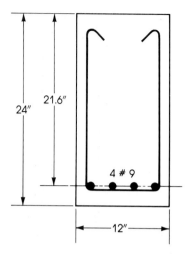

FIGURE 4.3. Example 4.2.

Since the actual A_s used is greater than the calculated A_s, a smaller depth can be adopted. Therefore take $h = 24$ in. Then

$$d = 24 - 1.128/2 - 1.875 = 21.6 \text{ in.}$$

$$A_s = 3.75 \times (22.5/21.6) = 3.91 \text{ in.}^2$$

which is less than the 4.00 in.² used (Figure 4.3). A check of the ultimate moment capacity of the section can be made, as explained earlier.

$$a = \frac{A_s f_y}{0.85 f'_c b} = \frac{4.00 \times 40}{0.85 \times 3 \times 12} = 5.23 \text{ in.}$$

$$M_u = \phi A_s f_y (d - a/2)$$
$$= 0.9 \times 4.00 \times 40(21.6 - 5.23/2) = 2734 \text{ k·in.} > 2700 \text{ k·in.} \qquad \blacksquare$$

EXAMPLE **4.3**

Find the necessary reinforcement for a given section 9 in. wide and 28 in. total depth (Figure 4.4) if it is subjected to an external ultimate moment of 2900 k·in. Given: $f'_c = 4$ ksi and $f_y = 60$ ksi.

Solution

1. Assuming one layer of No. 8 steel bars (to be checked later), then

$$d = 28 - 0.5 - 1.875 = 25.625 \text{ in. (or } d = 28 - 2.5 \text{ in.} = 25.5 \text{ in.)}$$

2. Check if the section is adequate without compression reinforcement. Compare ultimate moment capacity of the section (using ρ_{max}) with the design moment. For $f'_c = 4$ ksi and $f_y = 60$ ksi, $R_u(max) = 937$ psi $= 0.937$ ksi. The maximum $M_u = 0.937 \times 9 \times (25.625)^2 = 5538$ k·in., which, being greater than the applied moment of 2900 k·in., indicates that the section is under-reinforced.

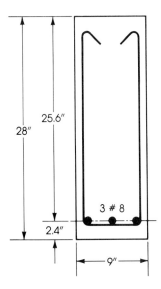

FIGURE 4.4. Example 4.3.

3. Or check that $R_u < R_u(\text{max})$:

$$R_u = \frac{M_u}{bd^2} = \frac{2900}{9 \times (25.625)^2} = 490 \text{ psi} < R_u(\text{max}) \text{ of } 937 \text{ psi}$$

Therefore, the section is under-reinforced and adequate without compression reinforcement.

4. From tables or calculations (equation 3.19), for $R_u = 490$ psi, $\rho = 0.010$.

$$A_s = \rho bd = 0.01 \times 9 \times 25.625 = 2.31 \text{ in.}^2$$

Use 3 No. 8 bars ($A_s = 2.35$ in.^{2}) in one row as assumed previously.

5. Since the yield strength $f_y > 40$ ksi, the **ACI Code, section 10.6**, specifies that it is necessary to check cracking condition, that is,

$$z = 0.6f_y\sqrt[3]{d_c A} \le 175 \text{ k/in. (interior exposure)}$$
$$\le 145 \text{ k/in. (exterior exposure)}$$

(for details, see Chapter 6, section 6.6) where

A = effective tension area of concrete surrounding and having the same centroid as the flexural tension reinforcement, divided by the number of bars, and

d_c = 2.4 in. = thickness of concrete cover measured from the extreme tension fiber to the center of the closest bar.

$$A = \frac{2 \times 2.4 \times 9}{3} = 14.4 \text{ in.}^2$$

$$z = 0.6 \times 60\sqrt[3]{2.4 \times 14.4}$$

$$= 117 \text{ k/in.} < 145 \text{ k/in.}$$

6. Another approach is to use the ultimate moment equation,

$$M_u = \phi A_s f_y(d - a/2)$$

$$= \phi A_s f_y \left(d - \frac{A_s f_y}{1.7 f'_c b} \right)$$

A second-degree equation will result, from which A_s can be computed:

$$2900 = 0.9 A_s \times 60 \left(25.625 - \frac{A_s \times 60}{1.7 \times 4 \times 9} \right)$$

$$53.7 = 25.625 A_s - 0.98 (A_s)^2$$

$$A_s = 2.34 \text{ in.}^2 \qquad \blacksquare$$

4.4   Rectangular Sections with Compression Reinforcement

A singly reinforced section has a maximum internal ultimate moment capacity when ρ_{max} of steel is used. If the applied ultimate moment is greater than the internal ultimate moment capacity, as in the case of a limited cross-section, a doubly reinforced section may be used, adding steel bars in both the compression and the tension zones. Compression steel will provide compressive force in addition to the compressive force in the concrete area.

EXAMPLE 4.4

A beam section is limited to a width $b = 10$ in. and a total depth $h = 22$ in. and has to resist an ultimate moment of 3150 k·in. Calculate the required reinforcement. Given: $f'_c = 3$ ksi and $f_y = 50$ ksi.

Solution   1. Determine the ultimate moment capacity that is allowed for the section as singly reinforced. This is done by starting with ρ_{max}. For $f'_c = 3$ ksi and $f_y = 50$ ksi,

$$\rho_{max} = 0.0206 \qquad R_u(max) = 740 \text{ psi}$$
$$M_u = R_u bd^2 \quad b = 10 \text{ in.} \quad d = 22 - 3.5 = 18.5 \text{ in.}$$

(This calculation assumes two rows of steel, to be checked later.)

$$M_u = 0.740 \times 10 \times (18.5)^2 = 2533 \text{ k·in.}$$
$$\text{Design } M_u = 3150 \text{ k·in.} > 2533 \text{ k·in.}$$

Therefore compression steel is needed to carry the difference.

2. Compute A_{s1}, M_{u1}, and M_{u2}:

$$A_{s1} = \rho_{max} bd = 0.0206 \times 10 \times 18.5 = 3.81 \text{ in.}^2$$

$$M_{u1} = 2533 \text{ k·in.}$$

$$M_{u2} = M_u - M_{u1} = 3150 - 2533 = 617 \text{ k·in.}$$

3. Calculate A_{s2} and A'_s, the additional tension and compression steel due to M_{u2}. Assume

$$d' = 2.5 \text{ in.} \qquad M_{u2} = \phi A_{s2} f_y (d - d')$$

Then,

$$A_{s2} = A'_s = \frac{M_{u2}}{\phi f_y (d - d')} = \frac{617}{0.9 \times 50(18.5 - 2.5)} = 0.86 \text{ in.}^2$$

Total tension steel is equal to A_s.

$$A_s = A_{s1} + A_{s2} = 3.81 + 0.86 = 4.67 \text{ in.}^2$$

The compression steel $A'_s = 0.86 \text{ in.}^2$

4. Check if compression steel yields:

$$\varepsilon_y = \frac{f_y}{29,000} = \frac{50}{29,000} = 0.00172$$

$$a = \frac{(A_s - A'_s)f_y}{0.85 f'_c b} = \frac{A_{s1} f_y}{0.85 f'_c b} = \frac{3.81 \times 50}{0.85 \times 3 \times 10} = 7.47 \text{ in.}$$

$$c \text{ (distance to neutral axis)} = a/\beta_1 = \frac{7.47}{0.85} = 8.80 \text{ in.}$$

ε'_s = strain in compression steel (from strain triangles)

$$= 0.003 \times \frac{(8.80 - 2.5)}{8.80} = 0.00215$$

As $\varepsilon'_s > \varepsilon_y$, compression steel yields.

5. Choose steel bars as follows:

$A_s = 4.67 \text{ in.}^2$ Choose 6 No. 8 bars ($A_s = 4.71 \text{ in.}^2$) in two rows, as assumed.

$A'_s = 0.86 \text{ in.}^2$ Choose 2 No. 6 bars ($A_s = 0.88 \text{ in.}^2$).

6. Check actual d:

$$\text{Actual } d = 22 - (1.5 + 0.375 + 1.5)$$

$$= 18.625 \text{ in.}$$

It is equal approximately to the assumed depth.

7. Since $f_y > 40$ ksi, check $z = 0.6 f_y \sqrt[3]{d_c A}$ (refer to Chapter 6, section 6.6). $d_c = 2.5$ in.,

$$A = \frac{3.375 \times 2 \times 10}{6} = 11.25 \text{ in.}^2$$

$$z = 0.6 \times 50 \sqrt[3]{2.5 \times 11.25} = 30 \times 3.04 \text{ k/in.}$$

$$= 91.2 < 145 \text{ k/in. (for an exterior member)}$$

$$< 175 \text{ k/in. (for an interior member)}$$

Therefore, the section is adequate (see Figure 4.5).

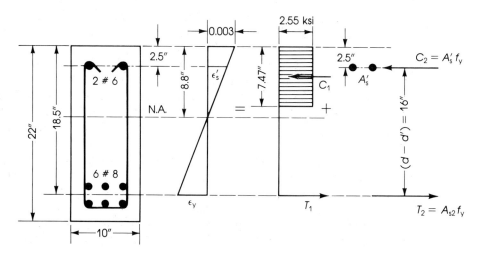

FIGURE 4.5. Example 4.4, doubly reinforced concrete section.

Notes

1. Yielding of compression steel can be checked using equation 3.41:

$$(\rho - \rho') \geq 0.85\beta_1 \times \frac{f'_c}{f_y} \times \frac{d'}{d} \times \frac{87}{87 - f_y}$$

Right-hand side $= (0.85)^2(3/50)(2.5/18.6)(87/37) = 0.0137$

Since $(\rho - \rho')$ used is equal to $\rho_{max} = 0.0206$, then $(\rho - \rho') = 0.0206 > 0.0137$. This proves that the compression steel yields.

2. In shallow beams, ε'_s may be less than ε_y, indicating that the compression steel does not yield. The stress in the compression steel is $f'_s = \varepsilon'_s E_s$. A'_s can be calculated from

$$A'_s = \frac{M_{u2}}{\phi f'_s(d - d')}$$

4.5   Design of T-Sections

In slab-beam-girder construction, the slab dimensions as well as the spacing and position of beams will be established first. The next step is to design the supporting beams, namely, the dimensions of the web and the steel reinforcement. Referring to the analysis of a T-section in the previous chapter, it can be seen that a large area of the compression flange, forming a part of the slab, is effective in resisting a great part or all of the compressive force due to bending. If the section is designed on this basis, the depth of the web will be small; consequently the moment arm is small, resulting in a large amount of tension steel, which is not favorable. Shear requirements should be met, and this usually requires quite a deep section.

In many cases, web dimensions can be known based on the flexural design of the section at the support in a continuous beam. The section at the support is subjected to a negative moment, the slab being under tension and considered not effective, and the beam width is that of the web.

If the dimensions of the web have not been established, approximate values can be obtained by considering a rectangular section with a reduced width b_r. The reduced width b_r is greater than the width of the web b_w and less than the effective flange width b. A reasonable and practical ratio of b_r/b ranges from $\frac{1}{3}$ to $\frac{2}{3}$, depending on the applied moment as well as shear requirements. If shear is high or a small amount of steel reinforcement is desired, a greater depth is needed. For shallow sections, a higher ratio is used. The next step is to estimate the effective depth using the equation $M_u = R_u b_r d^2$.

Once the dimensions of the web are chosen, the reinforcing steel area can be calculated from the equations in section 3.13.

The design of T-sections in most practical cases does not need to consider a doubly reinforced concrete section. When the width of the flange is small (as in the case of precast elements) and the effective depth d is limited, then a doubly reinforced concrete T-section may be needed.

When the neutral axis lies within the flange, the section behaves as a rectangular section with the beam width equal to the flange width.

EXAMPLE 4.5

The T-beam section shown in Figure 4.6 has a web width b_w of 10 in., a flange width b of 40 in., a flange thickness of 4 in., and an effective depth d of 13.5 in. Determine the necessary reinforcement if the applied ultimate moment is 3800 k·in. Given: $f'_c = 3$ ksi and $f_y = 60$ ksi.

Solution   1. Check the position of the neutral axis; the section may be rectangular. Assume the depth of compression block a is 4 in.; that is, $a = t = 4$ in. Then,

$$\phi M_n = \phi \times 0.85 f'_c bt(d - t/2) = 4223 \text{ k·in.}$$

The ultimate moment which the concrete flange can resist is greater than the ultimate mo-

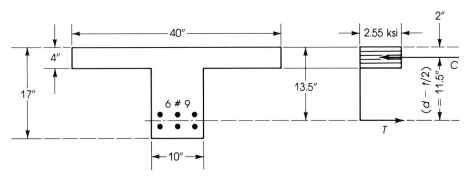

FIGURE 4.6. Example 4.5, T-section.

ment applied. Therefore the section behaves as a rectangular section and a shallower depth of compressive block is needed.

2. Determine the area of tension steel considering a rectangular section, $b = 40$ in.

$$R_u = \frac{\phi M_n}{bd^2} = \frac{3,800,000}{40 \times (13.5)^2} = 521 \text{ psi}$$

From tables in Appendix A, for $R_u = 521$ psi, $\rho = 0.0111$.

$$A_s = \rho bd = 0.0111 \times 40 \times 13.5 = 6.00 \text{ in.}^2$$

Use 6 No. 9 bars, $A_s = 6.00$ in.^{2} (in two rows).

3. Another approach can be used to determine A_s: Assume $a = t = 4$ in.

$$\text{Moment arm} = (d - a/2) = 13.5 - 2 = 11.5 \text{ in.}$$

$$A_s = \frac{M_u}{\phi f_y (\text{arm})} = \frac{3800}{0.9 \times 60 \times 11.5} = 6.12 \text{ in.}^2$$

Check

$$a = \frac{A_s f_y}{0.85 f'_c b} = \frac{6.12 \times 60}{0.85 \times 3 \times 40} = 3.6 \text{ in.}$$

$$\text{Moment arm} = 13.5 - \frac{3.6}{2} = 11.7 \text{ in.}$$

$$A_s = \frac{3800}{0.9 \times 60 \times 11.7} = 6.02 \text{ in.}^2$$

Note that one or two trials are sufficient to get the required A_s.   ∎

EXAMPLE 4.6

The floor system shown in Figure 4.7 consists of 3-in. slabs supported by 18-ft-span beams spaced at 10 ft on centers. The beams have a web width b_w of 12 in. and an effective depth d of 16.5 in. Calculate the necessary reinforcement for a typical interior beam if the ultimate applied moment is 5900 k·in., f'_c is 3 ksi, and f_y is 50 ksi.

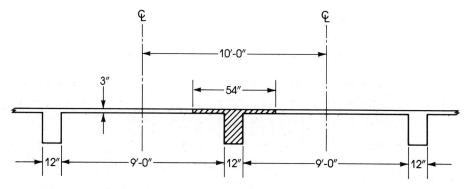

FIGURE 4.7. Example 4.6, effective flange width.

Solution   1. Find the beam flange width: Flange width is the smallest of

$$b = 16t + b_w = 3 \times 16 + 12 = 60 \text{ in.}$$

$$b = \text{span}/4 = \frac{18 \times 12}{4} = 54 \text{ in.}$$

Center to center of adjacent slabs $= 10 \times 12 = 120$ in. Use $b = 54$ in.
2. Check position of the neutral axis, assuming $a = t$.

$$
\begin{aligned}
\phi M_n \text{ (based on flange)} &= \phi \times 0.85 f'_c bt(d - t/2) \\
&= 0.9 \times 0.85 \times 3 \times 54 \times 3(16.5 - 1.5) \\
&= 5577 \text{ k·in.}
\end{aligned}
$$

As M_u applied (5900 k·in.) > 5577 k·in., the beam acts as a T-section. The neutral axis lies below the flange.
3. Find the portion of the ultimate moment taken by the overhanging portions of the flange (Figure 4.8). Area of steel required to develop a tension force balancing the compressive force in the projecting portions of the flange,

$$A_{sf} = \frac{0.85 f'_c (b - b_w)t}{f_y} = \frac{0.85 \times 3(54 - 12)3}{50} = 6.43 \text{ in.}^2$$

$$\phi M_n = M_{u1} + M_{u2}$$

That is, the sum of the ultimate moment of the web and the ultimate moment of the flanges.

$$
\begin{aligned}
M_{u2} &= \phi A_{sf} f_y (d - t/2) \\
&= 0.9 \times 6.43 \times 50(16.5 - 3/2) = 4327 \text{ k·in.}
\end{aligned}
$$

4. Calculate the ultimate moment of the web (as singly reinforced rectangular section):

$$M_{u1} = M_u - M_{u2} = 5900 - 4327 = 1573 \text{ k·in.}$$

$$R_u = \frac{M_{u1}}{bd^2} = \frac{1,573,000}{12 \times (16.5)^2} = 482 \text{ psi}$$

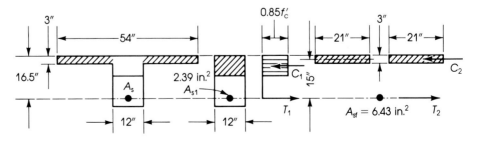

FIGURE 4.8. Analysis of example 4.6.

From tables in Appendix A: for $R_u = 482$ psi, $\rho_1 = 0.0121$:

$$A_{s1} = \rho_1 b_w d = 0.0121 \times 12 \times 16.5 = 2.39 \text{ in.}^2$$
$$= (\rho_w - \rho_f) b_w d = \rho_1 b_w d$$

Total $A_s = A_{sf} + A_{s1} = 6.43 + 2.39 = 8.82$ in.^{2}

5. Check that $(\rho_w - \rho_f) \le 0.75 \rho_b \le \rho_{max}$:

$$\rho_w - \rho_f = \rho_1 = \frac{A_{s1}}{b_w d} \qquad \rho_1 = 0.0121 \qquad \rho_{max} = 0.0206$$

Since $\rho_1 < \rho_{max}$, the above condition is satisfied; or since $R_u = 482 < R_u(max) = 740$ psi, the condition is satisfied. It can be seen that ρ_1, the steel percentage for the singly reinforced rectangular section, must always be less than or equal to ρ_{max}.

6. Another approach to the solution is as follows: Check the position of the neutral axis and obtain the ultimate moment that the whole flange can resist, M_u (flange) $= \phi M_n = 5577$ k·in., as explained above. The difference between the applied ultimate moment and that resisted by the total flange shall be balanced by the moment produced by the hatched part of the web (Figure 4.9):

$$M_u \text{ carried by web} = 5900 - 5577 = 323 \text{ k·in.}$$

Taking moments about A_s,

$$\phi b_w (a - t) \times 0.85 f_c' \left[d - \left(t + \frac{(a - t)}{2} \right) \right] = \phi M_n(\text{web})$$

$$\phi 0.85 f_c' b_w (a - t) \left[d - \left(\frac{a + t}{2} \right) \right] = M_u(\text{web})$$

$$0.9 \times 0.85 \times 3 \times 12(a - 3) \left[16.5 - \frac{a}{2} - \frac{3}{2} \right] = 323 \text{ k·in.}$$

$$(a - 3) \left(15 - \frac{a}{2} \right) = 11.73$$

$$a^2 - 33a + 114 = 0$$

$$a = 3.9 \text{ in.}$$

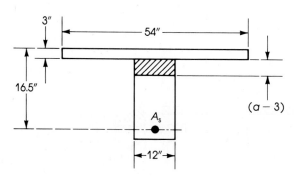

FIGURE 4.9. Alternate approach to example 4.6.

From a calculate A_{s1}:

$$a = \frac{A_{s1}f_y}{0.85f'_c b_w}$$

$$A_{s1} = 0.85 \times 3 \times 12 \times \frac{3.9}{50} = 2.39 \text{ in.}^2$$

$$A_s = A_{sf} + A_{s1} = 6.43 + 2.39 = 8.82 \text{ in.}^2$$ ■

EXAMPLE 4.7

In a slab-beam floor system, the smallest flange width was found to be 48 in., the web width adopted was 12 in., and the slab thickness was 4.0 in. (Figure 4.10). Design a T-section to resist an ultimate external moment of 867 k·ft. Given: $f'_c = 3$ ksi and $f_y = 60$ ksi.

Solution   1. Since the effective depth d is not given, a reduced flange width b_r is assumed, $b_r = 0.6b = 0.6 \times 48 = 28.8$ in. An equivalent rectangular section can be chosen with $b_r = 28.8$ in. and

$$d^2 = \frac{M_u}{R_u b_r}$$

For $f'_c = 3$ ksi, and $f_y = 60$ ksi, $R_u(\text{max}) = 702$ psi. Therefore,

$$d = \sqrt{\frac{867 \times 12}{0.702 \times 28.8}} = 22.7 \text{ in.}$$

Assume two rows of steel bars (to be checked later). Let $h = 26$ in. and $d = 26 - 3.5 = 22.5$ in.

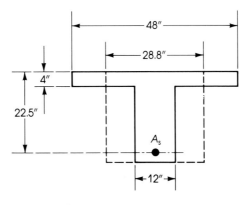

FIGURE 4.10. Example 4.7.

2. Proceed as in the previous example to calculate A_s. Check the position of the neutral axis. Let $a = t$.

$$\phi M_n \text{ (based on flange area)} = \phi 0.85 f'_c bt(d - t/2)$$
$$= 0.9 \times 0.85 \times 3.0 \times 48 \times 4(22.5 - 4/2)$$
$$= 9033 \text{ k·in.} = 753 \text{ k·ft}$$
$$M_u \text{ (applied)} = 867 \text{ k·ft} > 753 \text{ k·ft}$$

The section acts as a T-section.

3. $A_{sf} = \dfrac{0.85 f'_c (b - b_w)t}{f_y} = \dfrac{0.85 \times 3.0(48 - 12) \times 4}{60} = 6.1 \text{ in.}^2$

The ultimate moment taken by overhanging parts of the flange,

$$M_{u2} = \phi 0.85 f'_c (b - b_w)t(d - t/2)$$
$$= \phi A_{sf} f_y (d - t/2)$$
$$= 0.9 \times 6.1 \times 60 \left(22.5 - \frac{4}{2}\right) = 6752 \text{ k·in.} = 562.7 \text{ k·ft}$$

4. The ultimate moment taken by the web as a rectangular section

$$M_{u1} = M_u - M_{u2} = 867 - 562.7 = 304.3 \text{ k·ft} = 3651 \text{ k·in.}$$
$$R_u = \frac{M_{u1}}{bd^2} = \frac{3651}{12 \times (22.5)^2} = 0.601 \text{ ksi} = 601 \text{ psi}$$

From Table A.1 in Appendix A, $\rho_1 = 0.0131$.

$$A_{s1} = \rho_1 bd = 0.0131 \times 12 \times 22.5 = 3.54 \text{ in.}^2$$
$$A_s = A_{sf} + A_{s1} = 6.1 + 3.54 = 9.64 \text{ in.}^2$$

5. Check that $(\rho_w - \rho_f) = \rho_1 \leq \rho_{max}$. From tables, $\rho_{max} = 0.0161$. Because $\rho_1 = 0.0131 < \rho_{max}$, the condition is satisfied.

6. A_{s1} can be calculated by another approach, as follows:

$$M_{u1} = \phi A_{s1} f_y (d - a/2) = \phi A_{s1} f_y \left(d - \frac{A_{s1} f_y}{2 \times 0.85 f'_c b_w}\right)$$
$$3651 = 0.9 A_{s1} \times 60 \left(22.5 - \frac{A_{s1} \times 60}{2 \times 0.85 \times 3.0 \times 12}\right)$$
$$0.98 A_{s1}^2 - 22.5 A_{s1} = -(3651/540)$$
$$A_{s1}^2 - 22.96 A_{s1} = -69$$
$$(A_{s1} - 11.48)^2 = -69 + (11.48)^2 = 62.79$$
$$A_{s1} - 11.48 = \pm 7.94$$
$$A_{s1} = 3.54 \text{ in.}^2$$
$$A_s = 6.1 + 3.54 = 9.64 \text{ in.}^2$$

Use 8 No. 10 bars in two rows ($A_s = 10.12 \text{ in.}^2$).

7. If the area of steel calculated is considered high, it can be reduced by increasing the effective depth d and consequently increasing the total depth. This can be achieved in two ways:

- Choosing a smaller equivalent reduced flange width by adopting (b_r/b) less than the 0.6 assumed above.
- Find the depth necessary for the flange concrete area to take the compressive force, that is, for $a = t$. Then

$$M_u = \phi 0.85 f'_c bt(d - t/2)$$
$$867 \times 12 = 0.9 \times 0.85 \times 3.0 \times 48 \times 4(d - 2)$$

The depth d is the only unknown and equals 25.6 in.

$$R_u = \frac{M_u}{bd^2} = \frac{867 \times 12}{48 \times (25.6)^2} = 0.33 \text{ ksi} = 330 \text{ psi}$$

From tables, $\rho = 0.00665$; thus

$$A_s = 0.00665 \times 48 \times 25.6 = 8.17 \text{ in.}^2$$

This type of solution gives reasonable depth and area of steel reinforcement.

In some cases, when a large moment is applied to a T-section and the flange area is not sufficient to resist the compressive force, additional compression steel may be used in the flange, provided deflection is adequate.   ■

4.6   General Design Considerations

4.6.1   Combinations of f'_c and f_y

Previous analyses of sections were based on given values of the ultimate concrete strength f'_c and the yield strength of steel f_y.

Table 4.2 shows the variation of ρ_{max} when different combinations of f'_c and f_y are used. It is recommended here that low-strength concrete not be used with high-strength steel and that high-strength concrete not be used with low-strength steel.

If concrete with $f'_c = 6000$ psi is reinforced with steel bars of 40 ksi yield strength, the maximum allowable steel percentage ρ_{max} is equal to 4.90 percent. If this high percentage is used, the concrete section will be crowded with steel bars, causing problems in steel placement and concreting, and will increase the total cost. Also, if a small

TABLE 4.2. ρ_{max} at different f_y/f'_c ratios

		2500 psi 17.3 MPa		3000 psi 30.7 MPa		4000 psi 27.6 MPa		5000 psi 34.5 MPa		6000 psi 41.4 MPa	
f'_c											
f_y		ρ_{max} (%)	f_y/f'_c	ρ_{max} (%)	f_y/f'_c	ρ_{max} (%)	f_y/f'_c	ρ_{max} (%)	f_y/f'_c	ρ_{max} (%)	f_y/f'_c
ksi	MPa										
40	276	2.32	16	2.78	13.3	3.72	10.0	4.36	8	4.90	6.7
50	345	1.72	20	2.06	16.7	2.75	12.5	3.24	10	3.64	8.3
60	414	1.34	24	1.61	20.0	2.14	15.0	2.54	12	2.83	10.0
75	517	0.97	30	1.16	25.0	1.55	18.8	1.83	15	2.06	12.5

percentage of steel is used, the concrete will not be utilized economically. On the other hand, if concrete with $f_c' = 2500$ psi is reinforced with steel bars of 75 ksi yield strength, the maximum allowable steel percentage ρ_{max} is equal to 0.97 percent. This percentage of steel is low, thus reducing the stiffness of the member and causing excessive deflection and cracking.

It is therefore important to utilize the steel and concrete to the maximum possible degree by adopting proper combinations of concrete strength and steel yield strength. For low-strength concrete ($f_c' = 2500$ psi), a steel grade of $f_y = 40$ ksi may be used. Concrete with ultimate strength $f_c' = 3000$ psi can be used adequately with steel having $f_y = 40$ ksi or 50 ksi. Similarly, concrete with $f_c' = 4000$ psi can be adequately used with steel of $f_y = 60$ ksi. (Concrete with $f_c' = 2500$ psi should be avoided whenever possible.)

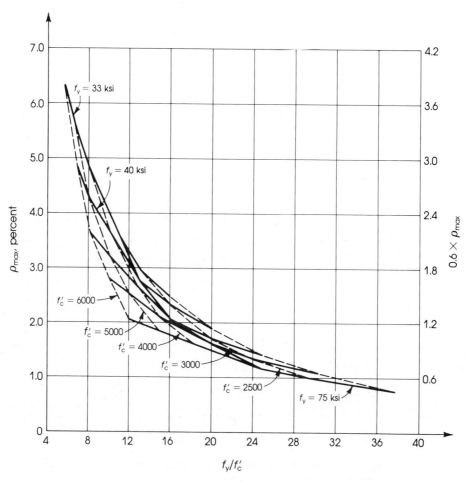

FIGURE 4.11. Maximum steel percentages ρ_{max} for various f_y/f_c' ratios.

4.6.2   Relation Between ρ_{max} and the Ratio f_y/f_c'

The variation of the maximum steel percentage ρ_{max} versus the ratio of steel yield strength to concrete ultimate strength (f_y/f_c') is plotted in Figure 4.11 for the values shown in Table 4.2. The dashed lines indicate the various strengths of concrete, while the solid lines indicate the yield strengths of steel. It can be seen that ρ_{max} increases nonlinearly with a decrease in the ratio f_y/f_c'. Extreme ends of the plotted graphs are not recommended for design purposes as they represent either high-strength steel with low-strength concrete or high-strength concrete with low-yield-strength steel. Figure 4.12 plots a smaller range of the ratio in more detail. The range of f_y/f_c' between 12 and 16 is of most practical consideration.

In the design of reinforced concrete members, three factors have to be considered besides the safety of the structural member: avoidance of sudden failure, economy, and crowding of steel.

The ACI Code permits a maximum ratio of 0.75 of the balanced steel ratio to avoid sudden failure. Using less steel than ρ_{max} is recommended in order to ensure yield of steel and the development of plastic hinges in indeterminate structures (see Chapter 16, section 16.8).

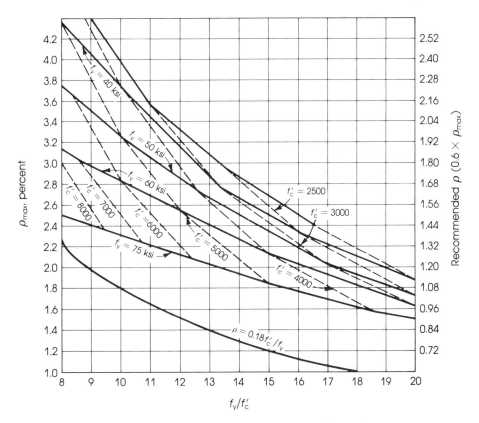

FIGURE 4.12. Maximum and recommended steel percentages for the practical range of f_y/f_c' ratios.

It was explained earlier that there can be considerable variations in the specifications and the costs of designs that meet the same ultimate moment, as a result of choosing from the full range of steel percentage specified by the ACI Code. Any design is safe as long as it complies with the code; a practical steel percentage that may be adopted whenever possible ranges between 0.5 and $0.6\rho_{max}$, as shown on the right side of Figure 4.12 for $\rho = 0.6\rho_{max}$. For example, for $f'_c = 4000$ psi and $f_y = 60$ ksi, ρ_{max} is equal to 2.14 percent, and using $0.6\rho_{max}$ gives a design percentage of 1.284 percent, or 1.85 square inches of steel per square foot of the concrete section. The weight of steel bars per unit length of the member is 6.3 pounds per cubic foot of concrete (equivalent to 100 kg/m^3 of concrete). From a practical approach, the weight of the steel used for such concrete is adequate. The ratio considered here of $\rho = 1.284$ percent (for $f'_c = 4000$ psi and $f_y = 60$ ksi) is a little higher than the equivalent steel percentage of $\rho = 0.18(f'_c/f_y) = 0.18 \times (4/60) = 1.20$ percent suggested by others. The equivalent balanced steel ratio in the working-stress design method (alternate design method) is equal to 1.4 percent.

To keep the steel bars from being crowded in the concrete section, one should adopt a steel percentage less than ρ_{max}.

Based on the above discussion, and considering the tension steel area in a singly reinforced concrete section, and to give easy, direct solutions whenever the steel percentage ρ is not known, the following suggested steel ratios ρ_s according to Table 4.3 may be used. The table lists ρ_s with the balanced steel ratio ρ_b, maximum steel ratio allowed by the ACI Code ρ_{max}, elastic steel ratio for a balanced section, and $\rho f_y/f'_c = 0.18$. It also gives the estimated weight of steel reinforcement per cubic foot of reinforced concrete.

TABLE 4.3. Comparison of the suggested steel ratio ρ_s with other steel ratios

f'_c (psi)	f_y (ksi)	ρ_b (%)	ρ_{max} (%)	ρ for $\frac{\rho f_y}{f'_c} = 0.18$ (%)	ρ for ρ_b elastic (%)	ρ_s (%)	Ratio ρ_s/ρ_{max}	Ratio ρ_s/ρ_b	Weight of ρ_s (lb/ft^3 of concrete)
2500	40	3.09	2.32	1.13	1.02	1.2	0.517	0.39	6
	50	2.29	1.72	0.90	0.90	1.2	0.700	0.52	6
3000	40	3.71	2.78	1.35	1.29	1.4	0.500	0.38	7
	50	2.75	2.06	1.08	1.11	1.2	0.583	0.44	6
	60	2.15	1.61	0.90	0.96	1.2	0.745	0.56	6
4000	40	4.96	3.72	1.80	1.88	1.6	0.430	0.32	8
	50	3.67	2.75	1.44	1.62	1.4	0.510	0.35	7
	60	2.58	2.14	1.20	1.41	1.4	0.654	0.54	7
	75	2.07	1.55	0.96	1.12	1.2	0.774	0.58	6
5000	40	5.81	4.36	2.25	2.50	1.6	0.367	0.28	8
	50	4.32	3.24	1.80	2.15	1.4	0.432	0.32	7
	60	2.54	2.54	1.50	1.87	1.4	0.551	0.55	7
	75	2.44	1.83	1.20	1.49	1.2	0.656	0.49	6

4.7   Examples Using S.I. Units

EXAMPLE 4.8

Design a singly reinforced rectangular section to resist an ultimate moment of 280 kN·m, using maximum allowable steel percentage. Given: $f_c' = 21$ N/mm², $f_y = 280$ N/mm², and $b = 220$ mm.

Solution

$$\rho_{max} = 0.64\beta_1 \frac{f_c'}{f_y}\left(\frac{600}{600 + f_y}\right)$$

$$= 0.64 \times 0.85 \times \frac{21}{280} \times \frac{600}{600 + 280} = 0.0278$$

$$R_u(max) = \phi\rho_{max}f_y\left(1 - \frac{\rho_{max}f_y}{1.7f_c'}\right)$$

$$= 0.9 \times 0.0278 \times 280\left(1 - \frac{0.0278 \times 280}{1.7 \times 21}\right) = 5.5 \text{ N/mm}^2 \text{ (MPa)}$$

$$M_u = R_u bd^2 \quad d = \sqrt{\frac{M_u}{R_u b}} = \sqrt{\frac{280 \times 10^6}{5.5 \times 220}} = 481 \text{ mm}$$

$$A_s = \rho bd = 0.0278 \times 220 \times 481 = 2940 \text{ mm}^2 = 29.4 \text{ cm}^2$$

Choose 6 bars, 25 mm diameter in two rows.

A_s provided $= 6 \times 4.9 = 29.4$ cm² $= A_s$ required

Total depth $h = d + 25$ mm $+ 60$ mm

$= 481 + 25 + 60 = 566$ mm, say 570 mm

Check minimum width:

$b_{min} = 3D + 2S + 95$ mm $= 5 \times 25 + 95 = 220$ mm   ∎

EXAMPLE 4.9

Calculate the required reinforcement for a beam that has a section of $b = 250$ mm and a total depth $h = 560$ mm to resist $M_u = 356$ kN·m. Given: $f_c' = 21$ N/mm² and $f_y = 350$ N/mm².

Solution   1. Determine the ultimate moment capacity of the section using ρ_{max}:

$$\rho_{max} = 0.64\beta_1 \frac{f_c'}{f_y} \times \frac{600}{600 + f_y} = 0.64 \times 0.85 \times \frac{21}{350} \times \frac{600}{950} = 0.0206$$

$$R_u(\text{max}) = \phi \rho_{\text{max}} f_y \left(1 - \frac{\rho_{\text{max}} f_y}{1.7 f_c'}\right)$$

$$= 0.9 \times 0.0206 \times 350 \left(1 - \frac{0.0206 \times 350}{1.7 \times 21}\right) = 5.2 \text{ N/mm}^2$$

$d = h - 85$ mm (assuming two rows of bars)

$\quad = 560 - 85 = 475$ mm

$\phi M_n = R_u bd^2 = 5.2 \times 250 \times (475)^2 \times 10^{-6} = 293.3$ kN·m

This is less than the external moment; therefore compression reinforcement is needed.

2. Calculate A_{s1}, M_{u1}, and M_{u2}:

$$A_{s1} = \rho_{\text{max}} bd = 0.0206 \times 250 \times 475 = 2446 \text{ mm}^2$$

$M_{u1} = 293.3$ kN·m (as calculated above)

$M_{u2} = M_u - M_{u1} = 356 - 293.3 = 62.7$ kN·m

3. Calculate A_{s2} and A_s' due to M_{u2}. Assume $d' = 50$ mm:

$$M_{u2} = \phi A_{s2} f_y (d - d')$$

$$62.7 \times 10^6 = 0.9 A_{s2} \times 350(465 - 50)$$

$$A_{s2} = 478 \text{ mm}^2$$

Total tension steel $= 2446 + 478 = 2924$ mm^2

The compression steel $A_s' = 478$ mm^2.

4. Compression steel yields if

$$(\rho - \rho') = \rho_1 \geq 0.85\beta_1 \times \frac{f_c'}{f_y} \times \frac{d'}{d} \times \frac{600}{600 - f_y}$$

$$\text{Right-hand side} = (0.85)^2 \times \frac{21}{350} \times \frac{50}{475} \times \frac{600}{600 - 350} = 0.0111$$

Since $(\rho - \rho') = \rho_{\text{max}} = 0.0206 > 0.0111$, therefore compression steel yields.

5. Choose steel bars as follows: For tension, choose 6 bars, 25 mm diameter. The A_s provided (2940 mm^2) is greater than A_s required (2924 mm^2). For compression steel, choose 2 bars 18 mm diameter:

$$A_s = 508 \text{ mm}^2 > 478 \text{ mm}^2$$

6. Since $f_y > 280$ N/mm, check that

$$Z \leq 0.6 f_y \sqrt[3]{d_c A} \leq 30.6 \text{ MN/m for interior members}$$

$$\leq 25.4 \text{ MN/m for exterior members}$$

$d_c = 50$ mm $A = \dfrac{85 \times 2 \times 250}{6} = 7083$ mm^2

$Z = 0.6 \times 350 \sqrt[3]{50 \times 7083} = 14{,}858$ N/mm

$\quad = 14.86$ MN/m < 25.4 MN/m

The beam can be used as an exterior as well as an interior member.   ∎

SUMMARY

Chapter 4 presents another way of using what Chapter 3 covered; thus the summary of Chapter 3 can be applied in this chapter.

Total depth $h = d + 2.5$ in. for one row of steel bars and $d + 3.5$ in. for two rows. Minimum width of concrete sections is $b_{min} = nD + (n - 1)s + 3.75$ in.

REFERENCES

1. E. Hognestad, N. W. Hanson and D. McHenry: "Concrete Stress Distribution in Ultimate Strength Design," *ACI Journal,* 52, Dec. 1955.
2. J. R. Janney, E. Hognestad and D. McHenry: "Ultimate Flexural Strength of Prestressed and Conventionally Reinforced Concrete Beams," *ACI Journal,* 52, Feb. 1956.
3. A. H. Mattock, L. B. Kriz and E. Hognestad: "Rectangular Concrete Stress Distribution in Ultimate Strength Design," *ACI Journal,* 57, Feb. 1961.
4. A. H. Mattock and L. B. Kriz: "Ultimate Strength of Nonrectangular Structural Concrete Members," *ACI Journal,* 57, Jan. 1961.
5. American Concrete Institute: "Building Code Requirements for Reinforced Concrete" (ACI 381–83), Detroit, 1983.
6. Franco Levi: "Work of European Concrete Committee," *ACI Journal,* 57, March 1961.
7. UNESCO: "Reinforced Concrete—An International Manual," Butterworth, London, 1971.
8. M. N. Hassoun: "Ultimate-Load Design of Reinforced Concrete," View Point Publication, Cement and Concrete Association, London, 1981.

PROBLEMS

4.1. Design a singly reinforced rectangular section to resist an ultimate moment of 232 k·ft (320 kN·m) if $f'_c = 3$ ksi (21 MPa), $f_y = 50$ ksi (350 MPa), and $b = 10$ in. (250 mm), using (a) ρ_{max}, (b) $\rho = 0.016$, and (c) $\rho = 0.012$.

4.2. Design a singly reinforced section to resist an ultimate moment of 186 k·ft (252 kN·m) if $b = 11$ in. (275 mm), $d = 20$ in. (500 mm), $f'_c = 3$ ksi (21 MPa), and $f_y = 40$ ksi (280 MPa).

4.3. Determine the reinforcement required for the section given in problem 4.2 when $f'_c = 4$ ksi (28 MPa) and $f_y = 60$ ksi (420 MPa).

4.4. A simply supported beam has a 20-ft span (6 m) and carries a uniform dead load of 600 lb/ft (9 kN/m) and a concentrated live load at midspan of 9 kips (40 kN) (Figure 4.13). Design the beam if $b = 12$ in. (300 mm), $f'_c = 4$ ksi (28 MPa), and $f_y = 60$ ksi (420 MPa). (Beam self-weight is not included in the dead load.)

4.5. A beam with a span of 24 ft (7.2 m) between supports has an overhanging extended part of 8 ft (2.4 m) on one side only. The beam carries a uniform dead load of 2 k/ft (30 kN/m) (including its own weight) and a uniform live load of 1.2 k/ft (18 kN/m) (Figure 4.14). Design the smallest singly reinforced rectangular section to be used for the entire beam. Select steel for positive and negative moments. Use $f'_c = 3$ ksi (21 MPa), $f_y = 60$ ksi (420 MPa), and $b = 12$ in. (300 mm). (Determine the maximum positive and the maximum

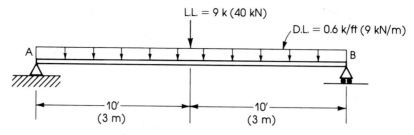

FIGURE 4.13. Problem 4.4.

negative moments by placing the live load once on the span and once on the overhanging part.)

4.6. Design a 15-ft (4.5-m) cantilever beam of uniform depth to carry a uniform dead load of 0.8 k/ft (12 kN/m) and a live load of 1 k/ft (15 kN m). Assume a beam width b = 14 in. (350 mm), f'_c = 4 ksi (28 MPa), and f_y = 60 ksi (420 MPa).

4.7. A 10-ft (3-m) cantilever beam carries a uniform dead load of 1.35 k/ft (20 kN/m) (including its own weight) and a live load of 0.67 k/ft (10 kN/m) (Figure 4.15). Design the beam using a variable depth. Draw all details of the beam and reinforcement. Given: f'_c = 3 ksi (21 MPa), f_y = 40 ksi (280 MPa), and b = 12 in. (300 mm). Assume d at the free end = 8 in. (200 mm).

4.8. Determine the necessary reinforcement for a concrete beam to resist an external ultimate moment of 290 k·ft (400 kN·m) if the width b = 12 in. (300 mm), d = 19 in. (475 mm), d' = 2.5 in. (65 mm), f'_c = 3 ksi (21 MPa), and f_y = 60 ksi (420 MPa).

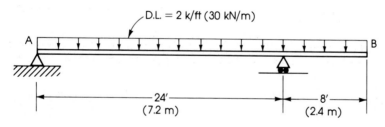

FIGURE 4.14. Problem 4.5.

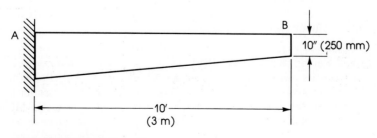

FIGURE 4.15. Problem 4.7.

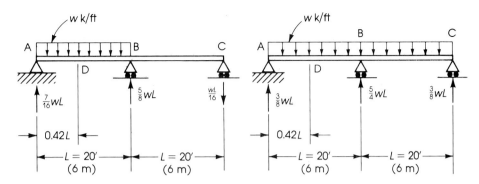

FIGURE 4.16. Problem 4.11.

4.9. Design a reinforced concrete section that can carry an ultimate moment of 260 k·ft (360 kN·m) as:
(a) Singly reinforced, $b = 10$ in. (250 mm).
(b) Doubly reinforced, 25 percent of the moment to be resisted by compression steel, $b = 10$ in. (250 mm).
(c) T-section which has a flange thickness of 3 in. (75 mm), flange width of 20 in. (500 mm), and a web width of 10 in. (250 mm), $f_c' = 3$ ksi (21 MPa), $f_y = 60$ ksi (420 MPa).
Determine the quantities of concrete and steel designed per foot length (meter length) of beams and then determine the cost of each design if the price of concrete equals \$50/cubic yard (\$67/m³) and the steel \$0.3/lb (\$0.66/kg). Tabulate and compare results.

4.10. Determine the necessary reinforcement for a T-section that has a flange width $b = 40$ in. (1000 mm), flange thickness $t = 4$ in. (100 mm), and web width $b_w = 10$ in. (250 mm) to carry an ultimate moment of 545 k·ft (750 kN·m). Given: $f_c' = 3$ ksi (21 MPa) and $f_y = 60$ ksi (420 MPa).

4.11. The two-span continuous beam shown in Figure 4.16 is subjected to a uniform dead load = 2 k/ft (including its own weight) and a uniform live load = 3 k/ft. The reactions due to two

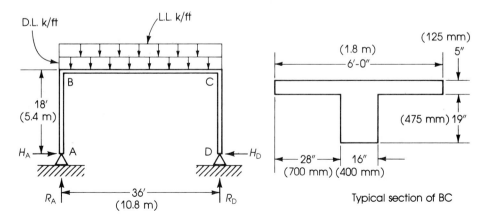

FIGURE 4.17. Problem 4.12.

different loadings are also shown. It is required to

(a) calculate the maximum negative ultimate moment at the intermediate support B, and the maximum positive ultimate moment within the span AB (at $0.42\,L$ from support A),

(b) design the critical sections at B and D, and

(c) draw the reinforcement details for the entire beam ABC. Given: $L = 20$ ft, $b = 12$ in., $h = 20.5$ in., $f_c' = 4$ ksi, and $f_y = 60$ ksi.

4.12. The two-hinged frame shown in Figure 4.17 carries a uniform dead load = 3.2 k/ft and a uniform live load = 2.4 k/ft on BC. The reactions at A and D can be evaluated as follows: $H_A = H_D = w\,L/9$, $R_A = R_D = w\,L/2$, where w = uniform load on BC. A typical cross-section of the beam BC is also shown. Using the strength design method,

(a) draw the ultimate bending moment diagram for the frame ABCD,

(b) design the beam BC for the applied ultimate moments (positive and negative), and

(c) draw the reinforcement details of BC. Given: $f_c' = 4$ ksi and $f_y = 60$ ksi.

CHAPTER

5

Working Stress Design Method: Flexural Analysis of Reinforced Concrete Beams

5.1   Assumptions

Reinforced concrete members can be analyzed and designed by a method that is based on elastic theory and the application of Hooke's Law. This method is called the working stress, straight-line, or alternate design method. The 1983 ACI Code puts little emphasis on this design concept, but it is still in use. The ACI Code of 1963 contained the following assumptions for the design of reinforced concrete structures by the working stress design method:

1. A cross-section that is plane before loading remains plane after loading. This means that strains above and below the neutral axis are proportional to the distance from the neutral axis.
2. The stress-strain relation for concrete is a straight line under service loads within the allowable working stress limits. Stresses vary as the distance from the neutral axis (except for deep beams).
3. The concrete strength in tension is neglected in a cracked section, and the steel resists all tension due to flexure. Before cracking the entire concrete area is effective.
4. The tension reinforcement is replaced in design computations with an equivalent concrete tension area equal to n times that of the reinforcing steel, where n, the modular ratio, is equal to the ratio of the moduli of elasticity of steel and concrete ($n = E_s/E_c$).
5. The modulus of elasticity of steel (E_s) is assumed to be equal to 29×10^6 psi (200,000 MPa) or 2,100,000 kgf/cm^2.

The modulus of elasticity of concrete (E_c) can be assumed equal to $33W^{1.5}\sqrt{f'_c}$ psi ($E_c = 0.1368W^{1.5}\sqrt{f'_c}$ kgf/cm^2 or $0.043W^{1.5}\sqrt{f'_c}$ MPa), where W = unit weight of concrete between 90 and 155 lb per cubic foot (1440 and 2420 kg/m^3).

5.2   Transformed Area Concept

For short-term live loads, Hooke's Law holds for both concrete and steel when stresses are below the allowable working stresses. In this case both materials are assumed to behave in an elastic manner, with unit stresses proportional to unit strains; that is, the stress is equal to the strain times the modulus of elasticity: stress $= \varepsilon E$.

The modular ratio $n = E_s/E_c$, which is used in the transformed area concept, may be used as the nearest whole number, but not less than 6. For example, when

$f'_c = 2500$ psi or 17.5 N/mm^2, $n = 10$
$f'_c = 3000$ psi or 21 N/mm^2, $n = 9$
$f'_c = 4000$ psi or 28 N/mm^2, $n = 8$
$f'_c = 5000$ psi or 35 N/mm^2, $n = 7$

The basic concept of a transformed area is that the section of steel and concrete is transformed into a homogeneous section of concrete by replacing the actual steel area with an equivalent area of concrete. Two conditions must be satisfied for the transformation process:

1. Equilibrium: the force in the steel must be equal to the force in the transformed concrete, or

$$A_s \times f_s = A_{tc} \times f_{tc} \qquad (5.1)$$

where

A_s = area of the steel in the section
f_s = the stress in the steel
A_{tc} = the transformed concrete area
f_{tc} = the stress in the transformed concrete area

2. Strain compatibility: the unit strains must be the same, or

$$f_s/E_s = f_{tc}/E_c \qquad (5.2)$$

But since $n = E_s/E_c$, then

$$f_{tc} = f_s \frac{E_c}{E_s} = f_s/n$$

and

$$f_s = n f_{tc}$$

For the transformed area of concrete,

$$f_s = n f_c \qquad (5.3)$$

and from equation 5.1:

$$A_{tc} = \frac{A_s f_s}{f_{tc}} = \frac{A_s n f_{tc}}{f_{tc}} = n A_s$$

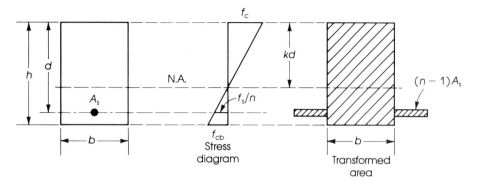

FIGURE 5.1. Transformed area for uncracked singly reinforced section.

that is, the transformed area

$$A_{\text{tc}} = nA_s \qquad (5.4)$$

For dead loads or long-term loads, the elastic stresses discussed above are complicated by the presence of shrinkage stresses and creep in the concrete producing inelastic effects, mainly in the compression steel. The actual stress in the compression steel is much greater than nf_s', where f_s' is the calculated stress in the compression steel. The ACI Code requires the use of $2n$ with compression steel.

It is to be noted that as long as the maximum tensile stress in the concrete is smaller than the modulus of rupture ($7.5\sqrt{f_c'}$), so that no tension cracks develop, the entire gross concrete area is used in computations. But when the maximum tensile stress exceeds the modulus of rupture, the cracked transformed area is used in computations. Figures 5.1, 5.2, and 5.3 and example 5.1 explain the transformed area procedure.

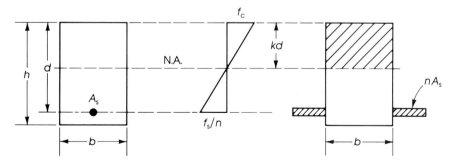

FIGURE 5.2. Transformed area for a cracked singly reinforced section.

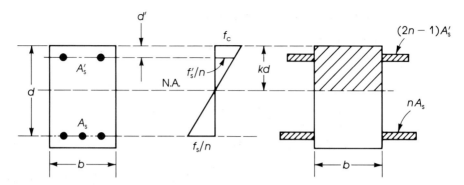

FIGURE 5.3. Transformed area for a cracked section with compression steel.

EXAMPLE 5.1

Calculate the transformed areas of the sections shown in Figures 5.1–5.3, with the following details. For all three sections, $b = 10$ in, $d = 17.5$ in., total depth $= 20$ in., tension steel is 3 No. 7 bars ($A_s = 1.8$ in.^{2}), $f'_c = 3$ ksi, $n = 9$, and the distance from the extreme compression fibers to the neutral axis, $Kd = 5$ in. For the section in Figure 5.3, $d' = 2.5$ in. and compression steel is 2 No. 6 bars, $A'_s = 0.88$ in.^{2}.

Solution

1. The section of Figure 5.1 is an uncracked section. The maximum tensile stress in concrete is less than its modulus of rupture $(7.5\sqrt{f'_c})$; thus the entire concrete section is effective. $n = E_s/E_c$; for $f'_c = 3000$ psi, n is 9.

   Net area of concrete $= bh - A_s$

   Transformed area of steel $= nA_s$

   Total transformed area $= (bh - A_s) + nA_s$
   $$= bh + (n - 1)A_s$$
   $$= 10 \times 20 + (9 - 1) \times 1.8 = 214.4 \text{ in.}^2$$

2. Figure 5.2 shows the case of a cracked section. The tensile stress in the concrete is greater than the modulus of rupture (in this example $f_r = 7.5\sqrt{3000} = 411$ psi).

   Transformed area $= b(Kd) + nA_s$
   $$= 10 \times 5 + 9 \times 1.8 = 66.2 \text{ in.}^2$$

3. Figure 5.3 shows the case of a cracked section with compression reinforcement. Use $2n$ for the compression steel only.

   Net transformed area $= [b(Kd) - A'_s] + nA_s + 2nA'_s$
   $$= b(Kd) + nA_s + (2n - 1)A'_s$$

taking into consideration the concrete displaced by the steel bars. Therefore,

$$\text{Net transformed area} = 10 \times 5 + 9 \times 1.8 + (2 \times 9 - 1) \times 0.88$$
$$= 81.16 \text{ in.}^2 \qquad \blacksquare$$

5.3   Allowable Working Stresses in Flexure

The **1983 ACI Code, Appendix B,** specifies the following allowable stresses for concrete and steel in bending.

For concrete, the allowable compressive stress $f_{ca} = 0.45f_c'$, where f_c' is the minimum specified compressive strength of a standard cylinder of concrete. Some common values are listed below.

f_c'	psi	2500	3000	3500	4000	4500	5000	
	MPa	17.3	20.7	24.1	27.6	31.0	34.5	
$f_{ca} = 0.45 f_c'$	psi	1125	1350	1575	1800	2025	2250	
	MPa	7.8	9.3	10.9	12.4	14.0	15.50	

| f_c' | (MPa) | 17.5 | 20.0 | 21.0 | 25.0 | 28.0 | 30.0 | 35.0 | 40.0 |
| $f_{ca} = 0.45 f_c'$ | (MPa) | 8.0 | 9.0 | 9.5 | 11.5 | 12.6 | 13.5 | 15.8 | 18.0 |

For steel, the allowable f_{sa} is 20 ksi (138 MPa) for grade 40 ($f_y = 40$ ksi $= 276$ MPa) and grade 50 ($f_y = 50$ ksi $= 345$ MPa) steels. For grade 60 ($f_y = 60$ ksi $= 414$ MPa) and stronger steels, allowable $f_{sa} = 24$ ksi (166 MPa). For main reinforcement $\frac{3}{8}$ in. (10 mm) or less in diameter in one-way slabs of not more than 12 ft (3.6 m) span, the allowable stress $f_{sa} = 0.5f_y \leq 30$ ksi (207 MPa). In S.I. units, allowable stresses of 140 MPa and 170 MPa are used for steel bars with yield strengths of 280 MPa and 420 MPa, respectively.

5.4   Analysis of Rectangular Sections in Bending with Tension Reinforcement

In problems of analysis, the properties of the section and the allowable working stresses are given. It is usually required either to determine the stresses applied to the section and compare them with the allowable stresses, or to determine the allowable bending moment that the section may carry. Two approaches may be used for analysis:

1. The internal couple method, in which the external bending moment is equated to the internal resisting couple. The internal couple consists of a compressive force C on one side of the neutral axis and an equal tensile force T on the other side. The arm is equal to jd, where j is less than one.

2. The flexural formula approach, in which the stress at any point is calculated by the conventional flexural formula

$$f = \frac{Mc}{I}$$

where

f = bending stress

c = distance from the considered point to the neutral axis

M = the bending moment

I = the moment of inertia of the transformed area

Figure 5.4 shows a rectangular section and the resisting internal forces C and T and the moment arm jd. The force C is equal to $\frac{1}{2}bf_cKd$, which is the volume of half a prism. The distance Kd indicates the position of the neutral axis with respect to the extreme compressive fibers.

Two main equilibrium equations must be satisfied: first, the compressive and tensile forces are equal,

$$C = T \tag{5.5}$$

and the internal moment is equal to either the compressive force or the tensile force multiplied by its arm,

$$M = C(jd) = T(jd) \tag{5.6}$$

From the above two equations, the following results can be deduced:

1. Applying the first equilibrium condition to Figure 5.4,

$$C = T \quad \text{or} \quad \tfrac{1}{2}f_cbKd = A_sf_s$$

If the percentage of steel in the section is ρ, then $\rho = A_s/bd$ or $A_s = \rho bd$.

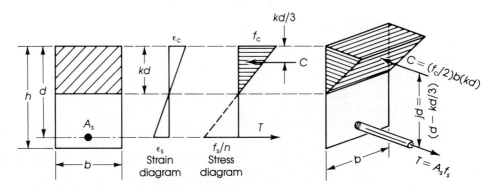

FIGURE 5.4. Internal forces in a singly reinforced rectangular section.

Therefore,

$$\tfrac{1}{2}f_cbKd = (\rho bd)f_s$$

The percentage of steel reinforcement in the section is

$$\rho = \frac{K}{2} \times \frac{f_c}{f_s} \tag{5.7}$$

2. The distance of the neutral axis from the top compressive fibers (Kd) can be established for a rectangular section by setting the moments of the transformed area about the neutral axis equal to zero:

$$\frac{b(Kd)^2}{2} - nA_s(d - Kd) = 0 \tag{5.8}$$

$$\frac{b(Kd)^2}{2} - n(\rho bd)(d - Kd) = 0$$

$$K^2 + 2n\rho K - 2n\rho = 0$$

$$K = \sqrt{n^2\rho^2 + 2n\rho} - n\rho \tag{5.9}$$

3. The value of K can be obtained from the strain diagram. From the two triangles in Figure 5.4,

$$\frac{\varepsilon_c}{\varepsilon_s} = \frac{Kd}{(d - Kd)} = \frac{K}{(1 - K)}$$

Since the strain equals stress divided by modulus of elasticity, then

$$\varepsilon_c = f_c/E_c \quad \text{and} \quad \varepsilon_s = f_s/E_s$$

therefore,

$$\frac{\varepsilon_c}{\varepsilon_s} = \frac{f_c}{f_s} \times \frac{E_s}{E_c}$$

If the ratio f_s/f_c is denoted by r and E_s/E_c by n, then

$$\frac{\varepsilon_c}{\varepsilon_s} = \frac{n}{r} = \frac{K}{(1 - K)}$$

Therefore,

$$K = \frac{n}{n + r} \tag{5.10}$$

4. The moment arm jd is equal to $d - Kd/3$, or

$$j = 1 - \frac{K}{3} \tag{5.11}$$

5. The internal moment M can be calculated using either the concrete compressive force C or the tension force T:

$$M_C = \tfrac{1}{2}f_cbKd(jd) = (\tfrac{1}{2}f_cKj)bd^2 = Rbd^2 \tag{5.12}$$

where

$$R = \tfrac{1}{2}f_c Kj$$
$$M_T = A_s f_s (jd) \qquad\qquad (5.13)$$

In the analysis procedure, it is usually easier to calculate Kd directly from equation 5.8.

EXAMPLE 5.2

Calculate the stresses in the extreme concrete compression fibers f_c and in the steel reinforcement f_s for the rectangular beam section shown in Figure 5.5 when the external moment $M = 60$ k·ft, using (1) the internal couple method and (2) the flexural formula $f = Mc/I$. Determine also the allowable moment capacity of the section. Given: $f'_c = 3$ ksi and $f_y = 40$ ksi ($f_{ca} = 1350$ psi, $f_{sa} = 20$ ksi).

Solution   1. Internal couple method: $n = 9$, $A_s = 2.37$ in.² Determine the distance of the neutral axis (N.A.) from the top compressive fibers. The sum of the moments of the effective transformed areas about the neutral axis is equal to zero. Therefore

$$b\frac{(Kd)^2}{2} - nA_s(d - Kd) = 0 \qquad\qquad (5.8)$$

$$\frac{10}{2}(Kd)^2 - 9 \times 2.37(19 - Kd) = 0$$

$$(Kd)^2 + 4.26(Kd) = 81$$

$$Kd = 7.12 \text{ in.}$$

2. The internal resisting moment $M = C(jd) = T(jd)$. But $C = \tfrac{1}{2}f_c b(Kd)$, and its distance from the top fibers is equal to $Kd/3$. Therefore

$$M = Cjd = \tfrac{1}{2}f_c b(Kd)\left(d - \frac{Kd}{3}\right)$$

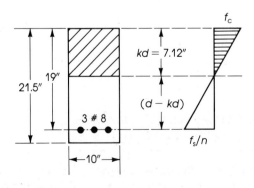

FIGURE 5.5. Example 5.2.

$$60 \times 12{,}000 = \tfrac{1}{2} \times f_c \times 10 \times 7.12\left(19 - \frac{7.12}{3}\right)$$

$$f_c = 1215 \text{ psi} < 1350 \text{ psi (the allowable concrete stress)}$$

3. $M = T(jd) = A_s f_s (jd) = A_s f_s \left(d - \dfrac{Kd}{3}\right)$

$$60 \times 12{,}000 = 2.37 \times f_s(19 - 7.12/3)$$

$$f_s = 18.25 \text{ ksi} < 20 \text{ ksi (the allowable steel stress)}$$

4. Solution using the flexural formula: The moment of inertia of the cracked transformed section,

$$I = b\frac{(Kd)^3}{3} + nA_s(d - Kd)^2$$

$$= 10 \times \frac{(7.12)^3}{3} + 9 \times 2.37(19 - 7.12)^2 = 4214 \text{ in.}^4$$

$$f_c = \frac{Mc}{I} = \frac{M(Kd)}{I} = \frac{60 \times 12{,}000 \times 7.12}{4214} = 1217 \text{ psi}$$

$$\frac{f_s}{n} = \frac{M(d - Kd)}{I} = \frac{60 \times 12{,}000(19 - 7.12)}{4214} = 2030 \text{ psi}$$

$$f_s = 2.03 \times 9 = 18.27 \text{ ksi}$$

Both f_c and f_s are less than the allowable stresses of $f_{ca} = 1350$ psi and $f_{sa} = 20$ ksi.

5. Calculate the allowable moment capacity of the section:

$$\text{Allowable } f_{ca} = 1350 \text{ psi}, \qquad \text{Allowable } f_{sa} = 20 \text{ ksi}$$

$$C = \tfrac{1}{2} f_{ca} b (Kd) = \frac{1350}{2} \times \frac{10 \times 7.12}{1000} = 48.1 \text{ k}$$

$$T = A_s f_{sa} = 2.37 \times 20 = 47.4 \text{ k}$$

Since T is less than C, the allowable stress in the steel will be reached before the concrete fibers reach their allowable stress; therefore T controls.

$$\text{Allowable } M = T(jd) = T\left(d - \frac{Kd}{3}\right)$$

$$= 47.4\left(19 - \frac{7.12}{3}\right)\Big/ 12 = 65.7 \text{ k·ft}$$

which is greater than the applied moment of 60 k·ft.   ∎

5.5   Analysis of Rectangular Sections with Compression Reinforcement

The elastic analysis of beams with compression reinforcement indicates that the compression steel has to be replaced by $(n - 1)A_s'$ units of transformed area of concrete. Shrinkage stresses and creep increase the compressive stresses, and the use of the above value does not give reasonable results. For such beams, an ultimate-strength analysis is more desirable.

The ACI Code recommends the use of $(2n - 1)A'_s$ to replace the above smaller value of $(n - 1)A'_s$. This approach was designated by P. M. Ferguson[3] as the semielastic analysis of sections with compression reinforcement. The method is fairly satisfactory for the calculation of the allowable moments and stresses.

EXAMPLE 5.3

Determine the allowable bending moment which the doubly reinforced concrete section shown in Figure 5.6 can carry. Given: $f'_c = 4$ ksi, $f_{sa} = 20$ ksi, $A_s = 7.10$ in.^{2}, $A'_s = 2.20$ in.^{2}. $f_{ca} = 1800$ psi, $n = 8$, $d = 22$ in., and $b = 13$ in.

Solution

1. Determine the position of the neutral axis:

$$nA_s = 8 \times 7.1 = 56.8 \text{ in.}^2 \qquad (2n - 1)A'_s = (16 - 1)2.20 = 33 \text{ in.}^2$$

Take moments of the transformed areas about the neutral axis and equate the sum to zero:

$$\frac{13}{2}(Kd)^2 + 33(Kd - 2.5) - 56.8(22 - Kd) = 0$$

Solve for Kd to get $Kd = 9$ in.

2. Determine if steel or concrete controls. From the stress diagram, if the steel stress is 20 ksi, then $f_{sa}/n = 20,000/8 = 2500$ psi. From triangles:

$$f_c = \frac{f_s}{n} \times \frac{(Kd)}{(d - Kd)} = 2500 \times \frac{9}{13} = 1730 \text{ psi}$$

Since $f_c = 1730$ psi is less than the allowable concrete stress of 1800 psi, steel controls as it

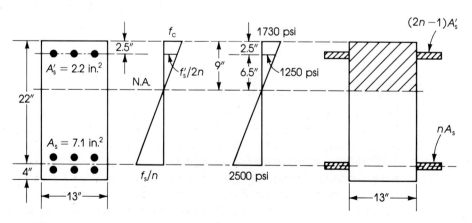

FIGURE 5.6. Example 5.3.

will reach its allowable stress of 20 ksi first. From the two triangles above the neutral axis:

$$\frac{f_s'/2n}{f_c} = \frac{6.5}{9}$$

$$\frac{f_s'}{2n} = \frac{6.5}{9} \times 1730 = 1250 \text{ psi}$$

and

$$f_s' = \frac{6.5}{9} \times f_c(2n) = \frac{6.5}{9} \times 1.730 \times 16 = 19.9 \text{ ksi}$$

which is less than the allowable stress in the steel of 20 ksi. If f_s' is found to be greater than the allowable stress, then this last value (allowable stress) must be used. Note that $2n$ is used for the calculation of the compression steel stress only.

3. Since steel controls,

$$T = A_s f_{sa} = 20 \times 7.1 = 142 \text{ k}$$

This can also be checked as follows:

$$C_1 \text{ (concrete)} = \tfrac{1}{2} \times 1.800 \times 13 \times 9 = 105.0 \text{ k}$$
$$C_2 \text{ (compression steel)} = 2.2 \times 19.9 = 43.8 \text{ k}$$
$$C = C_1 + C_2 = 148.8 \text{ k} > 142 \text{ k}$$

Therefore steel controls.

4. Calculate the allowable moment. Since $f_{sa} = 20$ ksi, $f_c = 1730$ psi, and $f_s' = 19.9$ ksi, then

$$C_1 = \tfrac{1}{2}f_c bKd = \tfrac{1}{2} \times 1.730 \times 13 \times 9 = 101.2 \text{ k}$$
$$C_2 = (2n - 1)A_s'(f_s'/2n) = 33 \times 1.250 = 41.25 \text{ k}$$

Taking the moment about the tension steel:

$$\text{Allowable bending moment} = C_1\left(d - \frac{Kd}{3}\right) + C_2(d - d')$$
$$= 101.2(22 - 9/3) + 41.25(22 - 2.5)$$
$$= 2727 \text{ k·in.} = 227.25 \text{ k·ft} \qquad \blacksquare$$

5.6   Analysis of T-Sections

Concrete slabs and beams are usually cast together forming T-shaped sections. The limitation of the effective flange width was discussed in Chapter 3, section 3.13. The effective flange width is the smallest of (1) one-fourth of the span, (2) $16t + b_w$, or (3) the distance between centerlines of adjacent slabs. The case when the neutral axis lies within the flange producing a rectangular section and the case when the neutral axis lies below the flange producing T-sections were discussed in section 3.13.

The following example will explain the analysis of T-sections by the working stress design method.

EXAMPLE 5.4

Calculate f_c and f_s of the T-beam shown in Figure 5.7 for a moment of 100 k·ft using the internal couple method. Given: $f'_c = 4$ ksi, $f_{sa} = 20$ ksi, $n = 8$, and $A_s = 3.8$ in.^{2}.

Solution

1. Determine whether the neutral axis lies within or below the flange by taking moments of areas about the bottom of the flange. For the concrete flange:

$$\text{Area} = 50 \times 4 = 200 \text{ in.}^2$$

$$\text{Moment about bottom of flange} = 200 \times 2 = 400 \text{ in.}^3$$

For the steel bars

$$nA_s = (8 \times 3.8) = 30.4 \text{ in.}^2$$

$$\text{Moment about bottom of flange} = (8 \times 3.8)18.5 = 562 \text{ in.}^3$$

This indicates that the section behaves as a T-beam because the neutral axis must lie below the flange so that the above two moments can be equal.

2. Determine the position of the neutral axis. The sum of the moments of the transformed areas about the neutral axis is equal to zero.

$$bt\left(Kd - \frac{t}{2}\right) + b_w \frac{(Kd - t)^2}{2} - nA_s(d - Kd) = 0$$

$$50 \times 4(Kd - 2) + 13 \frac{(Kd - 4)^2}{2} - 30.4(22.5 - Kd) = 0$$

$$(Kd)^2 + 27.5(Kd) - 151 = 0$$

$$Kd = 4.68 \text{ in.}$$

3. Determine the position of the resultant compressive force C. The stress in the concrete at the level of the bottom of the flange

$$f_{c1} = \frac{0.68}{4.68} f_c = 0.146 f_c$$

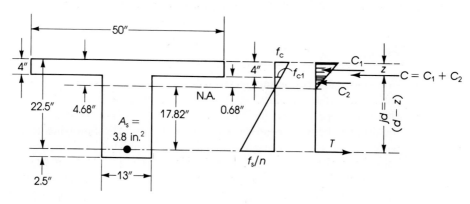

FIGURE 5.7. Example 5.4.

Find C_1, C_2, their respective moment arms, and their moments about the top fibers by subdividing the areas or stresses into parts that can be calculated easily. In this case, C_1 is taken on an area 50 in. wide and extending down to the neutral axis with a depth of 4.68 in. The force C_2 is a negative value, which is the force on the deducted area below the flange (Figure 5.8).

$$C_1 = (50 \times 4.68)\frac{f_c}{2} = 117 f_c \qquad \text{Its arm} = \frac{4.68}{3} = 1.56 \text{ in.}$$

Moment about top fibers $= 117 f_c \times 1.56 = 182.5 f_c$

$$C_2 = -(37 \times 0.68) \times \frac{0.146 f_c}{2} = -1.84 f_c \qquad \text{Its arm} = 4 + \frac{0.68}{3} = 4.23 \text{ in.}$$

Moment about top fibers $= -1.84 f_c \times 4.23 = -7.8 f_c$

Net force $C = 117 f_c - 1.84 f_c = 115.16 f_c$

Net moment about the top fibers $= 182.5 f_c - 7.8 f_c = 174.8 f_c$

If z is the distance of the resultant compressive force from the top fibers, then

$$z = \frac{174.7 f_c}{115.16 f_c} = 1.52 \text{ in.}$$

The arm $jd = d - z = 22.5 - 1.52 = 20.98$ in.
4. Determine the stresses in the concrete and steel due to the 100 k·ft external moment.

$$C = T = \frac{M}{jd} = \frac{100 \times 12}{20.98} = 57.1 \text{ k}$$

From the above calculation, $C = 115.16 f_c$. Therefore,

$$f_c = \frac{57,100}{115.16} = 496 \text{ psi} < 1800 \text{ psi}$$

$$f_s = \frac{T}{A_s} = \frac{57,100}{3.8} = 15,026 \text{ psi} < 20,000 \text{ psi}$$

Note: Sometimes when the concrete area above the neutral axis in the web is small, it may be neglected in calculating the neutral axis distance Kd, in this example the area $13(Kd - 4)$ in.^{2}. This procedure is called the approximate method. ∎

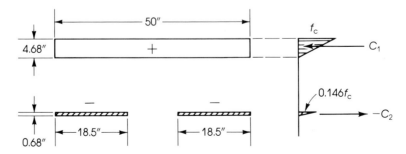

FIGURE 5.8. Transformed areas and compressive forces, example 5.4.

EXAMPLE 5.5

Determine f_c and f_s of the problem in the previous example using the flexural formula $f = Mc/I$. See Figure 5.9.

Solution   The moment of inertia of the cracked transformed section

$$I = 50 \times \frac{(4)^3}{12} + 50 \times 4 \times (2.68)^2 + \frac{13 \times (0.68)^3}{3} + 30.4 \times (17.82)^2$$

$$= 11,360 \text{ in.}^4$$

$$f_c = \frac{M(Kd)}{I} = \frac{100 \times 12,000 \times 4.68}{11,360} = 494 \text{ psi}$$

$$\frac{f_s}{n} = \frac{M(d - Kd)}{I} = \frac{100 \times 12,000 \times 17.82}{11,360} = 1882 \text{ psi}$$

$$f_s = 8 \times 1882 = 15,060 \text{ psi}$$

These results are about the same as those previously obtained values.   ■

EXAMPLE 5.6

Determine the allowable moment that can be applied on the T-section of the previous example (Figure 5.9).

Solution   Investigate first whether the concrete or steel stress controls. If the steel reaches its allowable stress of 20 ksi first, then the stress in the concrete at the top fibers is equal to

$$\frac{f_s}{n} = \frac{20,000}{8} = 2500 \text{ psi}$$

$$f_c = \frac{f_s}{n} \times \frac{Kd}{(d - Kd)} = 2500 \times \frac{4.68}{17.82} = 657 \text{ psi}$$

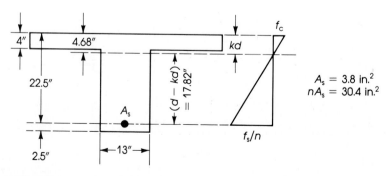

FIGURE 5.9. Examples 5.5 and 5.6.

which is lower than the allowable stress of concrete of 1800 psi. Therefore, tension controls.

$$M = A_s f_s jd$$

From the previous analysis in example 5.4, it was found that jd = 20.98 in. Therefore,

$$M = 3.8 \times 20 \times 20.98 = 1595 \text{ k·in.} = 133 \text{ k·ft}$$

Note that if only the allowable moment is required, the value of the moment arm jd must be calculated. From example 5.4, it was seen that a long procedure was done to find jd. For the calculation of the allowable moment, an approximate jd can be taken as the larger value of either $(d - t/2)$ or $0.9d$, as suggested by Ferguson.[3] In example 5.5, if jd is not known to be 20.98 in., then

$$\left(d - \frac{t}{2}\right) = \left(22.5 - \frac{4}{2}\right) = 20.5 \text{ in.}$$

or

$$0.9d = 0.9 \times 22.5 = 20.25 \text{ in.}$$

Therefore,

$$\text{Approximate allowable moment} = A_s f_s jd$$
$$= 3.8 \times 20 \times 20.5 = 1558 \text{ k·in.} = 130 \text{ k·ft} \qquad \blacksquare$$

5.7   Nonrectangular Sections in Bending

A procedure similar to that explained above can be used for the analysis of nonrectangular beams. The location of the resultant compressive force C is the main part of analysis, as the tension portion of the concrete is always neglected. For nonsymmetrical sections loaded along their principal axis, lateral deflection is expected. But if these beams are cast monolithically with the slabs, such lack of symmetry may be ignored and the bending axis will be essentially horizontal (Figure 5.10).

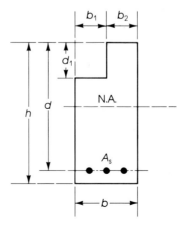

FIGURE 5.10. Nonrectangular beam section.

EXAMPLE 5.7

The section shown in Figure 5.11 is that of a 22-ft simply supported beam whose floor system is composed of precast slabs. The dead load including its own weight is equal to 900 lb/ft. It is required (a) to calculate the allowable live load on the beam, based on moment only, by the flexural equation procedure, and (b) to check results by using the internal couple method. Given: $f'_c = 5$ ksi, $f_{sa} = 20$ ksi, $n = 7$, and $A_s = 6$ in.^{2}.

Solution

1. To use the flexural formula, $f = Mc/I$, first determine the position of the neutral axis:

$$nA_s = 7 \times 6.0 = 42.0 \text{ in.}^2$$

Subdividing the compression area into three rectangles, then

$$8 \times \frac{(Kd)^2}{2} + 2 \times 3 \times \frac{(Kd - 4)^2}{2} = 42.0(22 - Kd)$$

$$7(Kd)^2 + 18(Kd) = 876$$

$$Kd = 9.97 \text{ in.}$$

2. Calculate the moment of inertia (for the three rectangles plus A_s):

$$I = 8 \times \frac{(9.97)^3}{3} + 2 \times 3 \times \frac{(5.97)^3}{3} + 42(22 - 9.97)^2 = 9177 \text{ in.}^4$$

Allowable $f_c = 0.45f'_c = 0.45 \times 5000 = 2250$ psi

3. Calculate the allowable moment:

$$M = \frac{f_c I}{c} \quad \text{or} \quad M = \frac{f_s}{n} \times \frac{I}{(d - Kd)}$$

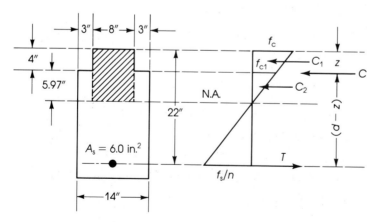

FIGURE 5.11. Example 5.7.

Internal moment based on concrete stress f_c:

$$M_c = \frac{2250 \times 9177}{9.97} = 2.07 \times 10^6 \text{ lb·in.} = 2071 \text{ k·in.}$$

Internal moment based on steel stress f_{sa} (20 ksi):

$$M_s = \frac{20}{7} \times \frac{9177}{(22 - 9.97)} = 2180 \text{ k·in.}$$

Since M_c is smaller than M_s, the internal moment based on concrete stress of 2250 psi controls. This means that the concrete will reach its allowable stress first and that the section is over-reinforced.

4. Determine the allowable live load on the beam. The external moment M is equal to the internal moment.

$$M = \frac{WL^2}{8} \times 12 = 2071 \text{ k·in.}$$

$$W = \frac{2071 \times 8}{12 \times 22 \times 22} = 2.850 \text{ k/ft} = 2850 \text{ lb/ft}$$

Therefore, the allowable live load = $2850 - 900 = 1950$ lb/ft.

5. Solution by the internal couple method. The concrete stress f_{c1} at 4 in. below the top fibers is equal to

$$f_{c1} = f_c \frac{(Kd - 4)}{Kd}$$

$$= f_c \times \frac{(9.97 - 4)}{9.97} = 0.6f_c = 0.6 \times 2250 = 1350 \text{ psi}$$

6. To calculate C and its position, the compressive concrete area is divided into a rectangle whose area is $8Kd$ and two small rectangles with areas equal to $2 \times 3(Kd - 4)$.

$$C_1 = 8 \times 9.97 \times \frac{f_c}{2} = 39.88f_c$$

Its arm from top fibers = $9.97/3 = 3.32$ in.

Moment of C_1 about top fibers = $39.88f_c \times 3.32 = 132.5f_c$

$$C_2 = 2 \times 3 \times 5.97 \times \frac{0.6f_c}{2} = 10.75f_c$$

Its arm from top fibers = $4 + 5.97/3 = 5.99$ in.

Moment of C_2 about top fibers = $10.75f_c \times 5.99 = 64.4f_c$

Total $C = C_1 + C_2 = 39.88f_c + 10.75f_c = 50.63f_c$

Total moment about top fibers = $132.5f_c + 64.4f_c = 196.9f_c$

$$z = \frac{196.9}{50.63} = 3.89 \text{ in.}$$

$$jd = (d - z) = 22 - 3.89 = 18.11 \text{ in.}$$

7. $C = 50.63 f_c = 50.63 \times 2.250 = 113.9$ k

$T = A_s f_s = 6.0 \times 20 = 120.0$ k

Since C is smaller than T, then C controls.

8. $M_c = Cjd = 113.9 \times 18.11 = 2063$ k·in.

$M_c = \dfrac{WL^2}{8} \times 12 = 2063$ k·in.

Therefore

$W = 2842$ lb/ft

and

$W_L \text{ (live load)} = 2842 - 900 = 1942$ lb/ft

which is very close to W_L calculated earlier.   ■

5.8   S.I. Examples

EXAMPLE 5.8

Calculate the stresses f_c and f_s for a rectangular section, $b = 250$ mm, $d = 475$ mm and reinforced with 3 bars 25 mm in diameter due to an external moment of 81.36 kN·m. Use (1) the internal couple method and (2) the flexural formula, then (3) determine the allowable moment. Given: $f_c' = 21$ MPa, $f_y = 280$ MPa, $f_{ca} = 9.45$ MPa, and $f_{sa} = 140$ MPa.

Solution   1. Solution using the internal couple method: given that $n = 9$, $A_s = 3 \times 490.9 = 1472.7$ mm², and $f_{ca} = 9.45$ MPa:

$$b\frac{(Kd)^2}{2} = nA_s(d - Kd)$$

$$250\frac{(Kd)^2}{2} = 9 \times 1472.7(475 - Kd)$$

$$Kd = 178 \text{ mm}$$

$$M = Cjd = \tfrac{1}{2}f_c bKd\left(d - \frac{Kd}{3}\right)$$

$$81.36 \times 10^6 \text{ (N·mm)} = \tfrac{1}{2} \times f_c \times 250(178)\left(475 - \frac{178}{3}\right)$$

Therefore

$$f_c = 8.8 \text{ MPa} < 9.45 \text{ MPa}$$

$$M = A_s f_s(jd) = A_s f_s\left(d - \frac{Kd}{3}\right)$$

$$81.36 \times 10^6 = 1472.7 f_s \left(475 - \frac{178}{3} \right)$$

$$f_s = 133 \text{ MPa} < 140 \text{ MPa}$$

2. Solution using the flexural formula:

$$I = b \frac{(Kd)^3}{3} + nA_s(d - Kd)^2$$

$$= \frac{250}{3} (178)^3 + 9 \times 1472.7(475 - 178)^2 = 1639.15 \times 10^6 \text{ mm}^4$$

$$f_c = \frac{Mc}{I} = \frac{(81.36 \times 10^6) \times (178)}{1639.15 \times 10^6} = 8.8 \text{ MPa} < 9.45 \text{ MPa}$$

$$f_s = \frac{nM(d - Kd)}{I} = \frac{9 \times 81.36 \times 10^6}{1639.15 \times 10^6} (475 - 178) = 133 \text{ MPa} < 140 \text{ MPa}$$

3. Allowable moments: given $f_{ca} = 8.45$ MPa and $f_{sa} = 140$ MPa, therefore, the compression force

$$C = \tfrac{1}{2} f_{ca} bkd$$

$$= \frac{9.45}{2} \times 250 \times 178 \times 10^{-3} = 210.26 \text{ kN}$$

$$T = A_s f_{sa} = 1472.7 \times 140 \times 10^{-3} = 206.2 \text{ kN}$$

T is less than C; thus the allowable stress in the steel will be reached before the concrete fibers reach their allowable stress: T controls.

$$\text{Allowable } M = T(jd) = T \left(d - \frac{Kd}{3} \right)$$

$$= 210.26 \left(475 - \frac{178}{3} \right) \times 10^{-3} = 87.4 \text{ kN} \cdot \text{m}$$

which is greater than the applied moment of 81.36 kN·m.   ∎

SUMMARY

SECTIONS 5.1 and 5.2   The modular ratio $n = E_s/E_c$ used in the transformed area concept is equal to 9, 8, and 7 for $f'_c = 3000$, 4000, and 5000 psi, respectively.

The transformed area $A_{tc} = nA_s$.

SECTION 5.3   The allowable stress for concrete $f_{ca} = 0.45 f'_c$. The allowable stress in steel $f_{sa} = 20$ ksi for $f_y = 40$ or 50 ksi and 24 ksi for $f_y = 60$ ksi.

SECTIONS 5.4 and 5.5   The analysis of a rectangular section may be performed either by the internal couple method or by the flexural formula $f = Mc/I$.

$$M_C = \tfrac{1}{2}f_c kj(bd^2)$$

$$M_T = A_s f_s(jd)$$

$$K = \frac{n}{n + r} \qquad j = 1 - \frac{K}{3}$$

For rectangular sections with compression reinforcement, use the transformed area $(2n - 1)A'_s$ to replace A'_s.

SECTIONS 5.6 and 5.7   The analysis of T-sections and nonrectangular sections is explained by examples.

REFERENCES

1. American Concrete Institute: *Reinforced Concrete Design Handbook (Working stress design)*, Publication SP-3, Detroit, 1965.
2. American Concrete Institute: *Building Code Requirements for Reinforced Concrete*, ACI 318-63, Detroit, 1963.
3. P. M. Ferguson: *Reinforced Concrete Fundamentals*, Wiley, New York, 1979.

PROBLEMS

5.1. Calculate the values of ρ, K, R, and j for a rectangular section when f_c and f_s are equal to the allowable stresses:
   (a) $f'_c = 2.5$ ksi, $f_y = 40$ ksi.
   (b) $f'_c = 3$ ksi, $f_y = 40$ ksi.
   (c) $f'_c = 4$ ksi, $f_y = 60$ ksi.
   (d) $f'_c = 17.5$ MPa, $f_s = 140$ MPa.
   (e) $f'_c = 21$ MPa, $f_s = 140$ MPa.

5.2. Calculate the maximum stresses in the concrete and steel for a rectangular section subjected to an applied moment of 72 k·ft (96 kN·m). The section has a width $b = 11$ in. (275 mm), an effective depth $d = 16.5$ in. (410 mm), total depth $h = 19$ in. (475 mm), and is reinforced with 4 No. 8 bars (4×25 mm). Given: $f'_c = 4$ ksi (280 MPa), $f_{sa} = 24$ ksi (170 MPa), and $n = 8$.

5.3. A rectangular section has a width $b = 8$ in. (200 mm), an effective depth $d = 20$ in. (500 mm), and is reinforced with 6 No. 6 bars (6×20 mm). Calculate the maximum stresses in the steel and concrete if a moment $M = 53$ k·ft (72 kN·m) is applied to the section using (a) the internal couple method and (b) the flexural formula. Determine the allowable moment on the section. Given: $f'_c = 3$ ksi (21 MPa), $f_{sa} = 20$ ksi (140 MPa), and $n = 9$.

5.4. A 10-ft (3-m) span cantilever beam carries a uniform load of 4 k/ft (60 kN/m) and has the section shown in Figure 5.12. Calculate the bending stresses in the concrete and steel if $f'_c = 4$ ksi (28 MPa), $f_{sa} = 20$ ksi (140 MPa), and $n = 8$.

5.5. A 15-ft (4.5-m) span simply supported beam carries a uniform load of 2 k/ft (30 kN/m) and a concentrated load of 6 k (27 kN) at midspan. The section of the beam is shown in Figure 5.13. Calculate the bending stresses in the concrete and steel using the internal couple method, and compare with the allowable stresses. Given: $f'_c = 4$ ksi (280 MPa), $f_{sa} = 22$ ksi (155 MPa), and $n = 8$.

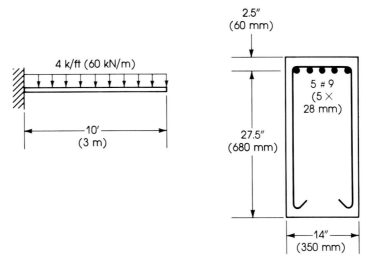

FIGURE 5.12. Problem 5.4.

5.6. Determine the allowable moment on the section of problem 5.5 and calculate the additional uniform load that can be applied on the beam, if the concentrated load P is removed.

5.7. If an 8-ft (2.4-m) span cantilever beam has the section given earlier in problem 5.2, determine the allowable uniform load that can be applied on the beam.

5.8. Determine the allowable bending moment on a rectangular section if b = 12 in. (300 mm), d = 22 in. (550 mm), d' = 2.5 in. (65 mm), A_s = 8 No. 7 bars (8 × 22 mm), A'_s = 3 No. 8 bars (3 × 25 mm), f'_c = 3 ksi (21 MPa), and f_{sa} = 20 ksi (140 MPa).

5.9. A T-section has a flange width b = 40 in. (1000 mm), a flange thickness t = 4 in. (100 mm), a web width b_w = 10 in. (250 mm), d = 19.2 in. (480 mm), and is reinforced with 6 No. 8 bars (6 × 25 mm). For M = 72.5 k·ft (100 kN·m), calculate the maximum stresses in concrete and steel by (a) the internal couple method and (b) the flexural formula. Determine also the allowable moment on the section. Given: f'_c = 3 ksi (21 MPa), f_{sa} = 22 ksi (155 MPa), and n = 9.

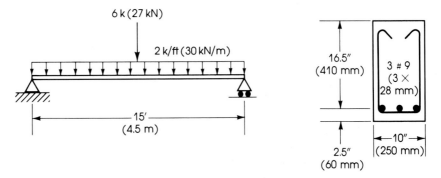

FIGURE 5.13. Problem 5.5.

5.10 and 5.11. Calculate the resisting moment of the sections shown in Figures 5.14 and 5.15. Use $f'_c = 4$ ksi (280 MPa) and $f_{sa} = 24$ ksi (170 MPa).

5.12 and 5.13. Determine the resisting moment of the rectangular sections with compression reinforcement shown in Figures 5.16 and 5.17. Use $f'_c = 4$ ksi (280 MPa) and $f_{sa} = 24$ ksi (170 MPa).

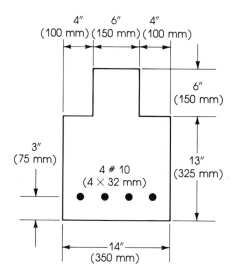

FIGURE 5.14. Problem 5.10.

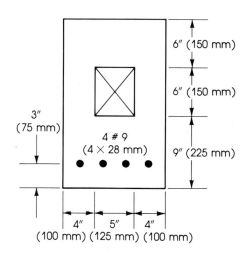

FIGURE 5.15. Problem 5.11.

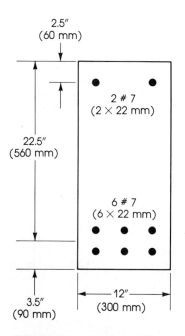

FIGURE 5.16. Problem 5.12.

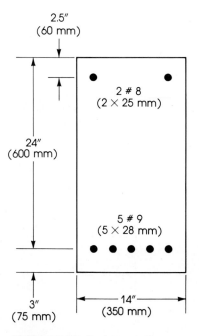

FIGURE 5.17. Problem 5.13.

5.14. Compute the resisting moment of the irregular section with compression reinforcement shown in Figure 5.18. Use $f'_c = 3$ ksi (210 MPa) and $f_{sa} = 20$ ksi (140 MPa).

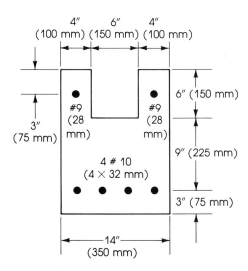

FIGURE 5.18. Problem 5.14.

Serviceability of Reinforced Concrete Beams: Deflection and Control of Cracking

CHAPTER 6

6.1   Deflection of Reinforced Concrete Members

Two important factors affect the serviceability of reinforced concrete flexural members: cracks and deflection. Adequate stiffness of members is necessary to prevent excessive cracks and deflection.

The widespread use of the strength design method, taking into consideration the nonlinear relationship between stress and strain in concrete, has resulted in smaller sections than those designed by the elastic theory. The ACI Code recognizes the use of steel up to a yield strength of 80 ksi (560 MPa) and the use of high-strength concrete. The use of high-strength steel and concrete results in smaller sections and a reduction in the stiffness of the flexural member and consequently increases its deflection.

The permissible deflection is governed by many factors, such as the type of the building, the appearance of the structure, the presence of plastered ceilings, partitions, the damage expected due to excessive deflection, and type and magnitude of live load.

The **ACI Code, section 9.5,** specifies minimum thickness for one-way flexural members and one-way slabs as shown in Table A.6 in Appendix A. The values are for members not supporting or attached to partitions or other construction likely to be damaged by large deflections. The minimum thicknesses indicated in Table A.6 are used for members made of normal weight concrete, $W = 145$ pcf (2320 kg/m³), and for steel reinforcement with yield strengths as mentioned in the tables. The values are modified for cases of lightweight concrete or a steel yield strength different from 60 ksi as follows:

- For lightweight concrete having unit weights in the range of 90–120 pcf (1440–1920 kg/m³), the values in the tables for $f_y = 60$ ksi (420 MPa) shall be multiplied by the greater of $(1.65 - 0.005W_c)$ or 1.09, where W_c is the unit weight of concrete in pounds per cubic foot.
- For yield strength of steel different from 60 ksi (420 MPa), the values in the tables for 60 ksi shall be multiplied by $(0.4 + f_y/100)$ where f_y is in ksi.

6.1.1   Instantaneous Deflection

The deflection of structural members is mainly due to the dead load plus a fraction or all of the live load. The deflection that occurs immediately upon the application of the load is called the immediate or instantaneous deflection. Under sustained loads, the deflection increases appreciably with time. Various methods are available for computing deflections in statically determinate and indeterminate structures. The instantaneous deflection calculations are based on the elastic behavior of the flexural members. The elastic deflection Δ is a function of the load W, span L, moment of inertia I, and the modulus of elasticity of the material E:

$$\Delta = f\left(\frac{WL}{EI}\right) \tag{6.1}$$

In general, the deflection can be expressed as follows:

$$\Delta = \alpha\left(\frac{WL^3}{EI}\right) \tag{6.2}$$

where

$\qquad W$ = total load on the span

$\qquad \alpha$ = a coefficient that depends on the degree of fixity at the supports, the variation of moment of inertia along the span, and the distribution of load

For example, the maximum deflection of a uniformly loaded simply supported beam is

$$\Delta = \frac{5WL^3}{384EI}$$

where

$\qquad W$ = is the total load on the span

$\qquad$ = wL (uniform load per unit length $\times$ span).

Deflections of beams with different loadings and different end conditions as a function of the load, span, and EI are given in Appendix C.

Since W and L are known, the problem then is to calculate the modulus of elasticity E and the moment of inertia I of the reinforced concrete member, or the flexural stiffness of the member EI.

6.1.2   Modulus of Elasticity

The **ACI Code, section 8.5**, specifies that the modulus of elasticity of concrete E_c may be taken as:

$$E_c = 33W_c^{1.5}\sqrt{f_c'} \text{ psi} \tag{6.3}$$

for values of W_c between 90 and 155 pounds per cubic foot. For normal weight concrete ($W_c = 145$ pcf),

$$E_c = 57{,}400\sqrt{f_c'} \text{ psi} \tag{6.4}$$

The secant modulus of elasticity of concrete was discussed in detail in Chapter 2. The stress-strain curve of concrete is nonlinear, and the secant modulus is an approximate value for the modulus of elasticity of concrete as adopted by the ACI Code.

The modulus of elasticity is usually determined by the short-term loading of a concrete cylinder. In actual members, creep due to sustained loading, at least for the dead load, affects the modulus on the compression side of the member. For the tension side, the modulus in tension is assumed to be the same as in compression when the stress magnitude is low. At high stresses the modulus decreases appreciably. Furthermore, the modulus varies along the span due to the variation of moments and shear forces.

6.1.3   Moment of Inertia

The moment of inertia, in addition to the modulus of elasticity, determines the stiffness of the flexural member. Under small loads, the produced maximum moment will be small, and the tension stresses at the extreme tension fibers will be less than the modulus of rupture of concrete; in this case the gross transformed cracked section will be effective in providing the rigidity. At working loads or higher, flexural tension cracks are formed. At the cracked section, the position of the neutral axis is high, while at sections midway between cracks along the beam, the position of the neutral axis is lower (nearer to the tension steel). In both locations only the transformed cracked sections are effective in determining the stiffness of the member; therefore, the effective moment of inertia varies considerably along the span. At maximum bending moment, the concrete is cracked, and its portion in the tension zone is neglected in the calculations of moment of inertia. Near the points of inflection the stresses are low, and the entire section may be uncracked. For this situation and in the case of beams with variable depth, exact solutions are complicated.

Figure 6.1(a) shows the load-deflection curve of a reinforced concrete beam tested to failure. The beam is a simply supported 17-ft span and loaded by two concentrated loads 5 ft apart, symmetrical about the centerline. The beam cross-section is 8 by 12 in. and reinforced as shown in the figure. The beam was subjected to two cycles of loading; in the first (curve cy 1), the load-deflection curve was a straight line up to a load $P = 1.7$ k when cracks started to occur in the beam. The line (a) represents the load-deflection relationship using a moment of inertia for the uncracked transformed section. It can be seen that the actual deflection of the beam under loads less than the cracking load, based on a homogeneous uncracked section, is very close to the calculated deflection (line a). Curve cy 1 represents the actual deflection curve when the load is increased to about half the ultimate load. The slope of the curve, at any level of load, is less than the slope of line (a) because cracks developed, and the cracked part of the concrete section reduces the stiffness of the beam. The load was then released, and a residual deflection was observed at midspan. Once cracks have developed, the assumption of uncracked section behavior under small loads does not hold.

In the second cycle of loading, the deflection (curve c) increased at a rate greater than that of line (a) because the resistance of the concrete tension fibers is lost. When the load is increased, the load-deflection relationship is represented by curve cy 2. If the load in the first cycle is increased up to the ultimate load, curve cy 1 will take the path

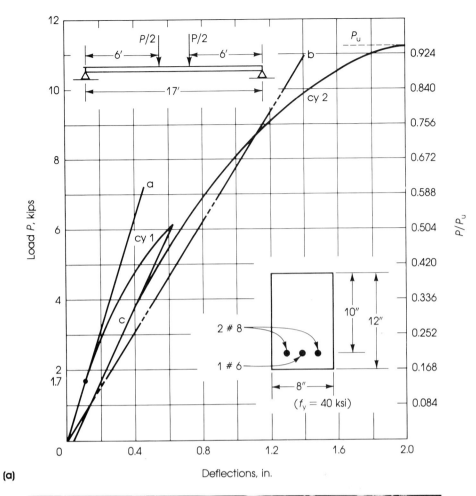

(a)

(b)

FIGURE 6.1. (a) Experimental and theoretical load-deflection curves for a beam of the section and load illustrated, and (b) deflection of a reinforced concrete beam.

cy 2 at about 0.6 of the ultimate load. Curve (c) represents the actual behavior of the beam for any additional loading or unloading cycles.

Line (b) represents the load-deflection relationship based on a cracked transformed section; it can be seen that the deflection calculated on that basis differs from the actual deflection. The **1983 ACI Code, section 9.5,** presents an equation to determine the effective moment of inertia used in calculating deflection in flexural members. The effective moment of inertia given by the ACI Code (formula 9.7) is based on the expression proposed by Branson[3] and calculated as follows:

$$I_e = \left(\frac{M_{cr}}{M_a}\right)^3 I_g + \left[1 - \left(\frac{M_{cr}}{M_a}\right)^3\right] I_{cr} \leq I_g \tag{6.5}$$

where

I_e = effective moment of inertia

M_{cr} = cracking moment $\left(\dfrac{f_r I_g}{y_t}\right)$ (6.6)

    f_r = modulus of rupture of concrete ($7.5\sqrt{f_c'}$ psi, $0.623\sqrt{f_c'}$ MPa)   (6.7)

    M_a = maximum moment in member at stage for which deflection is being computed

    I_g = moment of inertia of gross concrete section about the centroidal axis, neglecting the reinforcement

    I_{cr} = moment of inertia of cracked transformed section

    y_t = distance from centroidal axis of cross-section, neglecting steel, to extreme fiber tension

The following limitations are specified by the code:

- For continuous spans, the effective moment of inertia may be taken as the average of the moment of inertia of the critical positive and negative moment sections.
- For lightweight concrete, the modulus of rupture f_r to be used in equation 6.6 is equal to

$$f_r = 7.5\left(\frac{f_{ct}}{6.7}\right) \quad \text{where} \quad \frac{f_{ct}}{6.7} \leq \sqrt{f_c'} \tag{6.8a}$$

where f_{ct} is the average splitting tensile strength, in psi. When f_{ct} is not known, f_r may be taken as follows: For all-lightweight concrete,

$$f_r = 5.6\sqrt{f_c'} \tag{6.8b}$$

For sand-lightweight concrete

$$f_r = 6.4\sqrt{f_c'} \tag{6.8c}$$

- For prismatic members, I_e may be taken as the value obtained from equation 6.5 at midspan for simple and continuous spans and at the support section for cantilevers **(ACI Code, section 9.5.2.4).**

Note that I_e computed by equation 6.5 provides a transition between the upper and lower bounds of the gross moment of inertia I_g and the cracked moment of inertia I_{cr} as a function of the level of M_{cr}/M_a. Heavily reinforced concrete members may have an effective moment of inertia I_e very close to that of a cracked section I_{cr}, while flanged members may have an effective moment of inertia close to the gross moment of inertia I_g.

For continuous beams, the **ACI Code Commentary, section 9.5,** presents an approximate value of the average I_e for prismatic or nonprismatic members for somewhat improved results as follows: For beams with both ends continuous,

$$\text{Average } I_e = 0.70I_m + 0.15(I_{e1} + I_{e2}) \tag{6.9}$$

For beams with one end continuous,

$$\text{Average } I_e = 0.85I_m + 0.15(I_{con.}) \tag{6.10}$$

where

$$I_m = \text{midspan } I_e$$
$$I_{e1}, I_{e2} = I_e \text{ at beam ends}$$
$$I_{con.} = I_e \text{ at continuous end}$$

Moment envelopes should be used in computing both positive and negative values of I_e. In the case of a beam subjected to a single heavy concentrated load, only the midspan I_e should be used.

6.2   Long-Time Deflection

Deflection of reinforced concrete members continues to increase under sustained load, though more slowly with time. Shrinkage and creep are the cause of this additional deflection, which is called long-time deflection.[1] It is influenced mainly by temperature, humidity, age at time of loading, curing, quantity of compression reinforcement, and magnitude of the sustained load. (Shrinkage and creep were discussed in Chapter 2.) The **ACI Code, section 9.5,** suggests that unless values are obtained by a more comprehensive analysis, the additional long-term deflection for both normal and lightweight concrete flexural members shall be obtained by multiplying the immediate deflection by the factor

$$\lambda = \frac{T}{1 + 50\rho'} \tag{6.11}$$

where

$\rho' = A_s'/bd$ for the section at midspan of a simply supported or continuous beams or at the support of a cantilever beam

T = time-dependent factor for sustained loads that may be taken equal to 1.0, 1.2, 1.4, and 2.0 for periods of 3, 6, 12, and 60 months, respectively

TABLE 6.1. Multipliers for long-time deflections

Period, months	1	3	6	12	18	24	36	48	60
T	0.5	1.0	1.2	1.4	1.6	1.7	1.8	1.9	2.0

The factor λ is used to compute deflection caused by the dead load and the portion of the live load that will be sustained for a sufficient period to cause significant time-dependent deflections. The factor λ is a function of the material property, represented by T, and a section property, represented by $(1 + 50\rho')$. In equation 6.11, the effect of compression reinforcement is related to the area of concrete rather than the ratio of compression to tension steel, as was used in the 1977 ACI Code.

The **ACI Code Commentary, section 9.5,** presents a curve to estimate T for periods less than 60 months. These values are estimated as shown in Table 6.1.

The total deflection is equal to the immediate deflection plus the additional long-time deflection. For instance, the total additional long-time deflection of a flexural beam with $\rho' = 0.01$ at a 5-year period, according to the ACI Code, is equal to λ times the immediate deflection, where $\lambda = 2/(1 + 50 \times 0.01) = 1.33$.

6.3   Allowable Deflection

Deflection shall not exceed the following values according to the **ACI Code, section 9.5:**

- $L/180$ for immediate deflection due to live load for flat roofs not supporting elements that are likely to be damaged,
- $L/360$ for immediate deflection due to live load for floors not supporting elements likely to be damaged,
- $L/480$ for the part of the total deflection which occurs after attachment of elements, that is, the sum of the long-time deflection due to all sustained loads and the immediate deflection due to any additional live load, for floors or roofs supporting elements likely to be damaged, and
- $L/240$ for the part of the total deflection occurring after elements are attached, for floors or roofs not supporting elements likely to be damaged.

6.4   Deflection Due to Combinations of Loads

If a beam is subjected to different types of loads (uniform, nonuniform, or concentrated loads) or subjected to end moments, the deflection may be calculated for each type of loading or force applied on the beam separately and the total deflection calculated by superposition. This means that all separate deflections are added up algebraically to get the total deflection. The deflections of beams under individual loads are shown in the tables of Appendix C. The following examples illustrate the technique.

EXAMPLE 6.1

Calculate the instantaneous midspan deflection for the simply supported beam shown in Figure 6.2, which carries a uniform dead load of 0.4 k/ft and a live load of 0.6 k/ft, in addition to a concentrated dead load of 5 kips at midspan. Given: $f'_c = 4$ ksi, $f_y = 60$ ksi, $b = 13$ in., $d = 21$ in., and total depth = 24 in.

Solution

1. Check minimum depth according to the ACI Code, Table A.6.

$$\text{Minimum total depth} = L/16 = \frac{40 \times 12}{16} = 30 \text{ in.}$$

Total thickness used = 24 in. < 30 in.; therefore deflection must be checked.

2. The deflection at midspan due to a distributed load, from tables in Appendix C, is

$$\Delta_1 = \frac{5wL^4}{384E_cI_e}$$

The deflection at midspan due to a concentrated load

$$\Delta_2 = \frac{PL^3}{48E_cI_e}$$

Since w, P, and L are known, it is required to determine the modulus of elasticity E_c and the effective moment of inertia I_e.

3. The modulus of elasticity of concrete

$$E_c = 57,400 \sqrt{f'_c}$$
$$= 57,400 \sqrt{4000} = 3.63 \times 10^6 \text{ psi}$$

4. The effective moment of inertia is equal to

$$I_e = \left(\frac{M_{cr}}{M_a}\right)^3 I_g + \left[1 - \left(\frac{M_{cr}}{M_a}\right)^3\right] I_{cr} \le I_g$$

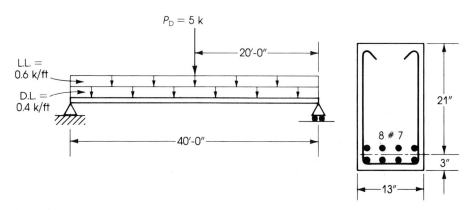

FIGURE 6.2. Example 6.1.

Determine values of all terms on the right-hand side:

$$M_a = \frac{wL^2}{8} + \frac{PL}{4} = \frac{(0.6 + 0.4)}{8}(40)^2 \times 12 + \frac{5 \times 40}{4} \times 12 = 3000 \text{ k·in.}$$

$$I_g = \frac{bh^3}{12} \text{ (for concrete section only)} = 14{,}976 \text{ in.}^4$$

$$M_{cr} = \frac{f_r I_g}{y_t}$$

$$y_t = h/2 = 12 \text{ in.}$$

$$f_r = 7.5\sqrt{f_c'} = 7.5\sqrt{4000} = 474 \text{ psi}$$

$$M_{cr} = \frac{0.474 \times 14{,}976}{12} = 592 \text{ k·in.}$$

The moment of inertia of the cracked transformed area I_{cr} is calculated as follows: Determine the position of the neutral axis for a cracked section by equating the moments of the transformed area about the neutral axis to zero, letting $x = Kd$ = distance to the neutral axis:

$$\frac{bx^2}{2} - nA_s(d - x) = 0$$

$$n = \frac{E_s}{E_c} = 8.0 \qquad A_s = 4.8 \text{ in.}^2$$

$$\frac{13}{2}x^2 - (8)(4.8)(21 - x) = 0$$

$$x^2 + 5.9x - 124 = 0$$

$$x = 8.6 \text{ in.}$$

$$I_{cr} = \frac{bx^3}{3} + nA_s(d - x)^2 = \frac{13(8.6)^3}{3} + 38.4(21 - 8.6)^2 = 8660 \text{ in.}^4$$

With all terms calculated,

$$I_e = \left(\frac{592}{3000}\right)^3 \times 14{,}976 + \left[1 - \left(\frac{592}{3000}\right)^3\right] \times 8660 = 8708 \text{ in.}^4$$

5. Calculate the deflections from the different loads:

$$\Delta_1 \text{ (due to distributed load)} = \frac{5wL^4}{384E_cI_e}$$

$$= \left(\frac{5}{384}\right) \times \left(\frac{1000}{12}\right) \times \frac{(40 \times 12)^4}{3.63 \times 10^6 \times 8708} = 1.82 \text{ in.}$$

$$\Delta_2 \text{ (due to concentrated load)} = \frac{PL^3}{48E_cI_e}$$

$$= \frac{5000 \times (40 \times 12)^3}{48 \times 3.63 \times 10^6 \times 8708} = 0.36 \text{ in.}$$

Total immediate deflection $= \Delta_1 + \Delta_2 = 1.82 + 0.36 = 2.18$ in.

6. Compare them with the allowable deflection: The immediate deflection due to a uniform

live load of 0.6 k/ft is equal to 0.6(1.82) = 1.09 in. If the member is part of a floor construction not supporting or attached to partitions or other elements likely to be damaged by large deflection, the allowable immediate deflection due to live load is equal to

$$L/360 = (40 \times 12)/360 = 1.33 \text{ in.} > 1.09 \text{ in.}$$

If the member is part of a flat roof and similar to the above, the allowable immediate deflection due to live load is $L/180 = 2.67$ in. Both allowable values are greater than the actual deflection of 1.09 in. due to the uniform applied live load. ∎

EXAMPLE 6.2

Determine the long-time deflection of the beam in the previous example if the time-dependent factor equals 2.0.

Solution

1. The sustained load causing long-time deflection is that due to dead load, consisting of a distributed uniform load of 0.4 k/ft and a concentrated load of 5 k at midspan.

   Deflection due to uniform load = 0.4 × 1.82 = 0.728 in.

   Deflection is a linear function of load w, all other values (L, E_c, I_e) being the same.

   Deflection due to concentrated load = 0.36 in.

   Total immediate deflection due to sustained loads = 0.728 + 0.36 = 1.088 in.

2. For additional long-time deflection, the immediate deflection is multiplied by the factor λ according to the ACI Code:

   $$\lambda = \frac{T}{1 + 50\rho'} = \frac{2}{1 + 0}$$

   In this problem $A_s' = 0$; therefore λ = 2.0.

   Additional log-time deflection = 2 × 1.088 = 2.176 in.

3. Total long-time deflection is the immediate deflection plus additional long-time deflection: 2.18 + 2.176 = 4.356 in.
4. Deflection due to live load plus additional long-time deflection due to shrinkage and creep is 1.088 + 2.176 = 3.264 in. ∎

EXAMPLE 6.3

Calculate the instantaneous and 1-year long-time deflection at the free end of the cantilever beam shown in Figure 6.3. The beam has a 20-ft span and carries a uniform dead load of 0.4 k/ft, a uniform live load of 0.4 k/ft, a concentrated dead load P_D of 3 k at the free end, and a concentrated live load P_L of 4 k placed at 10 ft from the fixed end. Given: $f_c' = 4$ ksi, $f_y = 60$ ksi, $b = 12$ in., $d = 21.5$ in., and total depth of section = 25 in. (Tension steel is 6 No. 8 bars and compression steel is 2 No. 8 bars.)

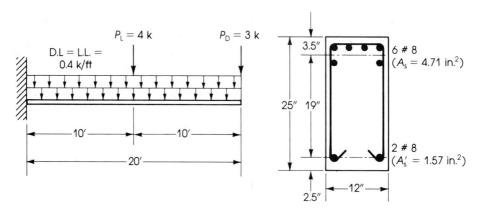

FIGURE 6.3. Example 6.3.

Solution   1. Minimum depth $= L/8 = 20/8 = 2.5$ ft $= 30$ in., which is greater than 25 in. used. There-fore, deflection must be checked. The maximum deflection of a cantilever beam is at the free end. From tables in Appendix C, the deflection at the free end is as follows:
Deflection due to distributed load

$$\Delta_1 = \frac{wL^4}{8EI}$$

Deflection due to a concentrated dead load at the free end

$$\Delta_2 = \frac{P_DL^3}{3EI}$$

Deflection due to a concentrated live load at $l = 10$ ft from the fixed end is maximum at the free end:

$$\Delta_3 = \frac{P_L(l)^2}{6EI}(3L - l)$$

2. The modulus of elasticity of normal weight concrete

$$E_c = 57,400\sqrt{f'_c} = 57,400\sqrt{4000} = 3.63 \times 10^6 \text{ psi}$$

3. Maximum moment at the fixed end

$$M_a = \frac{wL^2}{2} + P_D \times 20 + P_L \times 10$$

$$= \frac{(0.4 + 0.4)(400)}{2} + 3 \times 20 + 4 \times 10 = 260 \text{ k·ft}$$

4. $I_g =$ gross moment of inertia (concrete only)

$$= \frac{bh^3}{12} = \frac{12 \times (25)^3}{12} = 15,625 \text{ in.}^4$$

5. $M_{cr} = \dfrac{f_r I_g}{y_t} = \dfrac{(7.5\sqrt{4000}) \times 15,625}{25/2} = 593 \text{ k·in.} = 49.41 \text{ k·ft}$

6. Determine the position of the neutral axis, then determine the moment of inertia of the cracked transformed section. Take moments of areas about the neutral axis and equate to zero. Use $n = 8$ to calculate the transformed area of A_s and use $(2n - 1) = 15$ to calculate the transformed area of A_s' (see Chapter 5, section 5.5).

$$b\frac{(Kd)^2}{2} + (2n - 1)A_s'(Kd - 2.5) - nA_s(d - Kd) = 0$$

For this section, $Kd = 7.97$ in.

$$I_{cr} = \frac{b}{3}(Kd)^3 + (2n - 1)A_s'(Kd - d')^2 + nA_s(d - Kd)^2 = 9627.4 \text{ in.}^4$$

7. Effective moment of inertia

$$I_e = \left(\frac{M_{cr}}{M_a}\right)^3 I_g + \left[1 - \left(\frac{M_{cr}}{M_a}\right)^3\right] I_{cr} \leq I_g$$

$$= \left(\frac{49.41}{260}\right)^3 \times 15{,}625 + \left[1 - \left(\frac{49.41}{260}\right)^3\right] \times 9627.4 = 9668 \text{ in.}^4$$

8. Determine the components of the deflection:

$$\Delta_1 \text{ (due to uniform load of 0.8 k/ft)} = \frac{800}{12} \times \frac{(20 \times 12)^4}{8 \times 3.63 \times 10^6 \times 9668} = 0.786 \text{ in.}$$

$$\Delta_1 \text{ (due to dead load)} = 0.786 \times \frac{0.4}{0.8} = 0.393 \text{ in.}$$

$$\Delta_2 \text{ (due to concentrated dead load) at free end} = \frac{3000(20 \times 12)^3}{3 \times 3.63 \times 10^6 \times 9668}$$

$$= 0.392 \text{ in.}$$

Δ_3 (due to concentrated live load at 10 ft from fixed end) =

$$\frac{4000(10 \times 12)^2 \times (3 \times 20 \times 12 - 10 \times 12)}{6 \times 3.63 \times 10^6 \times 9668} = 0.164 \text{ in.}$$

The total immediate deflection

$$\Delta = \Delta_1 + \Delta_2 + \Delta_3$$
$$= 0.786 + 0.392 + 0.164 = 1.342 \text{ in.}$$

9. For additional long-time deflection the immediate deflection is multiplied by the factor λ. For a 1-year period, $T = 1.4$.

$$\rho' = A_s'/bd = 1.57/(12 \times 21.5) = 0.0061$$

$$\lambda = \frac{1.4}{1 + 50 \times 0.0061} = 1.073$$

Total immediate deflection Δ_s due to sustained load (here only the dead load of 0.4 k/ft and $P_D = 3$ k at free end):

$$\Delta_s = 0.393 + 0.392 = 0.785 \text{ in.}$$

Additional long-time deflection = $1.073 \times 0.785 = 0.842$ in.

10. Total long-time deflection is the immediate deflection plus long-time deflection due to shrinkage and creep.

$$\text{Total } \Delta = 1.342 + 0.842 = 2.184 \text{ in.} \qquad \blacksquare$$

EXAMPLE 6.4

Calculate the instantaneous midspan deflection of beam AB in Figure 6.4, which has a span of 32 ft. The beam is continuous over several supports of different span lengths. The absolute bending moment diagram and cross-sections of the beam at midspan and supports are also shown. The beam carries a uniform dead load of 4.2 k/ft and a live load of 3.6 k/ft. Given: $f_c' = 3$ ksi, $f_y = 60$ ksi, and $n = 9.2$.

$$\text{Moment at midspan, } M_D = 192 \text{ k·ft} \qquad M_{(D+L)} = 480 \text{ k·ft}$$
$$\text{Moment at left support A, } M_D = 179 \text{ k·ft} \qquad M_{(D+L)} = 420 \text{ k·ft}$$
$$\text{Moment at right support B, } M_D = 216 \text{ k·ft} \qquad M_{(D+L)} = 542 \text{ k·ft}$$

Solution

1. The beam AB is subjected to a positive moment that causes a deflection downward at midspan, and negative moments at the two ends causing a deflection upward at midspan. As was explained earlier, the deflection is a function of the effective moment of inertia I_e. In a continuous beam, I_e to be used is the average value for the positive and negative moment regions. Therefore, three sections will be considered, the section at midspan and the sections at the two supports.

2. Calculate I_e:
   For the gross area of all sections, $Kd = 13.5$ in., and $I_g = 114,300$ in.[4]. Also, $f_r = 7.5\sqrt{f_c'} = 410$ psi and $E_c = 57,400\sqrt{f_c'} = 3.15 \times 10^6$ for all sections. The values of Kd, I_{cr}, and M_{cr} for each cracked section, I_e for dead load only (using M_a of dead load) and I_e for dead and live loads (using M_a for dead and live loads) are calculated and tabulated below:

Section	Kd (in.)	I_{cr} (in.[4])	M_{cr} (k·ft)	I_e (in.[4]) (Dead load)	I_e (in.[4]) (D + L)
Midspan	6.67	48,550	159.4	86,160	50,960
Support A	10.9	34,930	289.3	114,300	60,880
Support B	12.6	44,860	289.3	114,300	55,415

Note that when the beam is subjected to dead load only and the ratio M_{cr}/M_a is greater than 1.0, I_e is equal to I_g.

3. Calculate average I_e from equation 6.9:

$$I_{e1} \text{ (average)} = 0.7(50,960) + 0.15(60,880 + 55,415) = 53,116 \text{ in.}^4$$

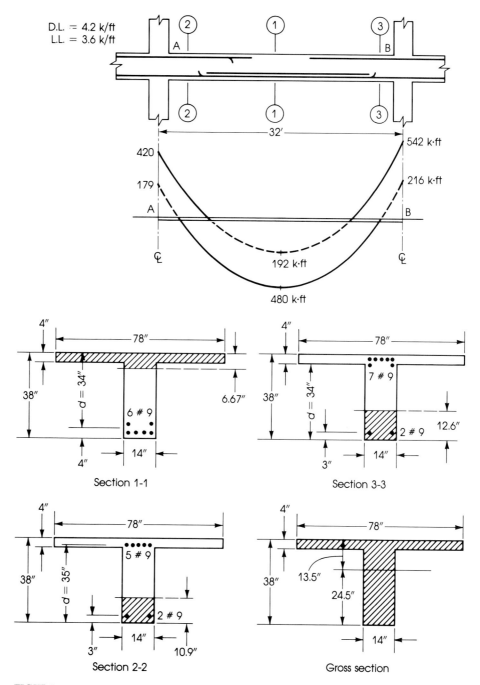

FIGURE 6.4. Example 6.4, deflection of a continuous beam.

For dead and live loads,

$$\text{Average } I_e \text{ for end sections} = \tfrac{1}{2}(60{,}880 + 55{,}415) = 58{,}150 \text{ in.}^4$$

$$I_{e2} \text{ (average)} = \tfrac{1}{2}(50{,}960 + 58{,}150) = 54{,}550 \text{ in.}^4$$

For dead load only,

$$\text{Average } I_e \text{ for end sections} = 114{,}300 \text{ in.}^4$$

$$I_{e3} \text{ (average)} = \tfrac{1}{2}(86{,}160 + 114{,}300) = 100{,}230 \text{ in.}^4$$

4. Calculate immediate deflection at midspan:

$$\Delta_1 \text{ (due to uniform load)} = (5wL^4)/(384EI_e) \text{ (downward)}$$
$$\Delta_2 \text{ (due to a moment at A, } M_A) = -(M_A L^2)/(16EI_e) \text{ (upward)}$$
$$\Delta_3 \text{ (due to a moment at B, } M_B) = -(M_B L^2)/(16EI_e) \text{ (upward)}$$
$$\text{Total deflection } \Delta = \Delta_1 - \Delta_2 - \Delta_3$$

The dead load deflection for a uniform dead load of 4.2 k/ft and taking $M_A(\text{D.L.}) = 179$ k·ft, $M_B(\text{D.L.}) = 216$ k·ft, and $I_{e3} = 100{,}230$ in.^{4}, then substituting in the above equations,

$$\Delta = 0.314 - 0.063 - 0.075 = 0.176 \text{ in. (downward)}$$

The deflection due to combined dead and live loads is found by taking (D + L) load = 7.8 k/ft, $M_A = 420$ k·ft, $M_B = 542$ k·ft, and $I_{e2} = 54{,}550$ in.^{4}:

$$\Delta = 1.071 - 0.270 - 0.349 = 0.452 \text{ in. (downward)}$$

The immediate deflection due to live load only is $0.542 - 0.176 = 0.276$ in. (downward). If the limiting permissible deflection $= L/480 = (32 \times 12)/480 = 0.8$ in., then the section is adequate.

There are a few points to mention about the results.

- If I_e of the midspan section only is used ($I_e = 50{,}960$ in.^{4}), then the deflection due to dead plus live loads is calculated by multiplying the obtained value in step 4 above by the ratio of the two I_e:

$$\Delta(\text{D} + \text{L}) = 0.452 \times \left(\frac{54{,}550}{50{,}960}\right) = 0.484 \text{ in.}$$

The difference is small, about 7 percent on the conservative side.
- If I_{e1} (average) is used ($I_{e1} = 53{,}116$ in.^{4}), then $\Delta(\text{D} + \text{L}) = 0.471$ in. The difference is small, about 4 percent on the conservative side.
- It is believed that it is more convenient to use I_e at midspan section unless a more rigorous solution is required. ∎

EXAMPLE 6.5 (S.I. UNITS)

Calculate the instantaneous and 5-year long-time deflection for a 12-m span simply supported beam which carries a uniform dead load of 15 kN/m and a live load of 10 kN/m. Given: $f'_c = 28$ MPa, $f_y = 420$ MPa, $n = 8$, $b = 300$ mm, $h = 700$ mm, $d = 620$ mm. Reinforcement is 6 bars × 25 mm in two rows ($A_s = 2940$ mm^2) in the tension zone.

Solution   1. Check minimum thickness according to the ACI Code:

$$\frac{L}{16} = \frac{12 \times 1000}{16} = 750 \text{ mm} > 700 \text{ mm used}$$

Therefore check deflection.

2. Maximum deflection occurs at midspan:

$$\Delta = \frac{5wL^4}{384EI}$$

3. $E_c = 4730\sqrt{f'_c} = 4730\sqrt{28} = 25{,}030 \text{ N/mm}^2$
4. The effective moment of inertia I_e is equal to

$$I_e = \left(\frac{M_{cr}}{M_a}\right)^3 I_g + \left[1 - \left(\frac{M_{cr}}{M_a}\right)^3\right] I_{cr} \le I_g$$

Determine values for the terms on the right-hand side:

$$M_a = \frac{wL^2}{8} = \frac{(15 + 10)}{8} \times (12)^2 = 450 \text{ kN·m}$$

$$I_g = \frac{bh^3}{12} = \frac{300}{12} \times (700)^3 = 85.75 \times 10^8 \text{ mm}^4 \text{ (concrete only)}$$

$$M_{cr} = \frac{f_r I_g}{y_t}$$

$$f_r = 0.623\sqrt{28} = 3.31 \text{ N/mm}^2 \qquad y_t = \frac{700}{2} = 350 \text{ mm}$$

$$M_{cr} = \frac{3.31 \times 85.75 \times 10^8}{350} \times 10^{-6} = 81 \text{ kN·m}$$

Calculate the moment of inertia of the cracked transformed section I_{cr}. Determine the location of the neutral axis first, then set moments of the transformed area about the neutral axis equal to zero.

$$\frac{b}{2}(Kd)^2 - nA_s(d - Kd) = 0$$

$$\frac{300}{2}(Kd)^2 - 8 \times 2940(620 - Kd) = 0 \qquad Kd = 243 \text{ mm}$$

$$I_{cr} = \frac{b}{3}(Kd)^3 + nA_s(d - Kd)^2$$

$$= \frac{300}{3}(243)^3 + 8 \times 2940(620 - 243)^2 = 47.78 \times 10^8 \text{ mm}^4$$

Having determined all values, calculate I_e:

$$I_e = \left(\frac{81}{450}\right)^3 \times 85.75 \times 10^8 + \left[1 - \left(\frac{81}{450}\right)^3\right] \times 47.78 \times 10^8 = 48.0 \times 10^8 \text{ mm}^4$$

5. Calculate deflection:

$$\Delta = \frac{5}{384} \times \left(\frac{25 \times 1000}{1000}\right) \times \frac{(12 \times 1000)^4}{25{,}030 \times 48.0 \times 10^8} = 56 \text{ mm}$$

6. Compare with allowable deflection:

The immediate deflection due to live load

$$\Delta_{L.L.} = \frac{10}{25} \times 56 = 22.4 \text{ mm}$$

For a floor construction attached to elements likely to be damaged by large deflection, the allowable deflection

$$\Delta = \frac{L}{360} = \frac{12 \times 1000}{360} = 33.3 \text{ mm}$$

This is greater than 22.4 mm, the deflection due to live load; thus the beam is adequate.

7. For long-time deflection:

$$\Delta_{D.L.} \text{ (sustained load)} = 56 - 22.4 = 33.6 \text{ mm}$$

Calculate the factor λ for additional long-time deflection. For a 5-year period, $T = 2.0$.

$$\lambda = \frac{T}{1 + 50\rho'} \qquad \rho' = \frac{A_s'}{bd}$$

Since $A_s' = 0$, then $\lambda = 2.0$. Therefore, additional long-time deflection is $2 \times 33.6 = 67.2$ mm, and the total deflection is $56 + 67.2 = 123.2$ mm.   ■

6.5   Cracks in Flexural Members

The study of crack formation, behavior of cracks under increasing load, and control of cracking is necessary for proper design of reinforced concrete structures. In flexural members, cracks develop under working loads, and since concrete is weak in tension, reinforcement is placed in the cracked tension zone to resist the tension force produced by the external loads.

Flexural cracks develop when the stress at the extreme tension fibers exceeds the modulus of rupture of concrete. With the use of high-strength reinforcing bars, excessive cracking may develop in reinforced concrete members. The use of high-tensile steel has many advantages, yet the development of undesirable cracks seems to be inevitable. Wide cracks may allow corrosion of the reinforcement or leakage of water structures and may spoil the appearance of the structure.

A crack is formed in concrete when a narrow opening of indefinite dimension has developed in the concrete beam as the result of internal tensile stresses. These internal stresses may be due to one or more of the following:

- external forces such as direct axial tension, shear, flexure, or torsion,
- shrinkage,
- creep, or
- internal expansion resulting from a change of properties of the concrete constituents.

In general, cracks may be divided into two main types, secondary and main cracks.

6.5.1   Secondary Cracks

Secondary cracks, very small cracks that develop in the first stage of cracking, are produced by the internal expansion and contraction of the concrete constituents and by low flexural tension stresses due to the self-weight of the member and any other dead loads. There are three types of secondary cracks:

Shrinkage Cracks.   These are important cracks because they affect the pattern of cracking that is produced by loads in flexural members. When they develop, they form a weak path in the concrete. When load is applied, cracks start to appear at the weakest sections, such as along the reinforcing bars. The number of cracks formed is limited by the amount of shrinkage in concrete and the presence of restraints. Shrinkage cracks are difficult to control.

Secondary Flexural Cracks.   Usually these are widely spaced, and each crack does not influence the formation of others.[8] They are expected to occur under low loads such as dead loads. When a load is applied gradually on a simple beam, tensile stress develops at the bottom fibers, and when it exceeds the flexural tensile stress of concrete, cracks start to develop. They widen gradually and extend toward the neutral axis. It is difficult to predict the sections at which secondary cracks start because concrete is not a homogeneous, isotropic material.

Saliger[9] and Billig[10] estimated the steel stress just before cracking to be from about 6000 to 7000 psi (42 to 49 MPa). An initial crack width of the order of 0.001 in. (0.025 mm) is expected at the extreme concrete tensile fibers. Once cracks are formed, the tensile stress of concrete at the cracked section decreases to zero, and the steel bars take all the tensile force. At this moment, some slip occurs between the steel bars and the concrete due to the differential elongation of concrete and steel and extends to a section where the concrete and steel strains are equal. Figure 6.5 shows the typical stress distribution between cracks in a member under axial tension.

Corrosion Secondary Cracks.   These cracks form when moisture containing deleterious agents such as sodium chloride, carbon dioxide, and dissolved oxygen penetrates the concrete surface, corroding the steel reinforcement.[11] The oxide compounds formed by deterioration of steel bars occupy a larger volume than the steel and exert mechanical pressure that perpetuates extensive cracking.[12,13] This type of cracking may be severe enough to result in eventual failure of the structure. The failure of a roof in Muskegan, Michigan, in 1955 due to the corrosion of steel bars was reported by Shermer.[13] The extensive cracking and spalling of concrete in the San Mateo-Hayward Bridge in California within seven years was reported by Stratfull.[12] Corrosion cracking may be forestalled by using proper construction methods and high-quality concrete. More details are discussed by Evans[14] and Mozer et al.[15]

6.5.2   Main Cracks

Main cracks develop at a later stage than secondary cracks. They are caused by the difference in strains in steel and concrete at the section considered. The behavior of main cracks changes at two different stages. At low tensile stresses in steel bars, the

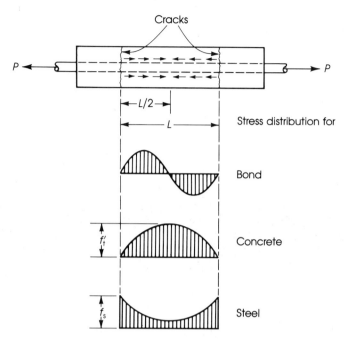

FIGURE 6.5. Typical stress distribution between cracks.

number of cracks increases while the widths of cracks remain small; as tensile stresses are increased, an equilibrium stage is reached. When stresses are further increased, the second stage of cracking develops, and crack widths increase without any significant increase in the number of cracks. Usually one or two cracks start to widen more than the others, forming critical cracks (Figures 6.6 and 6.7).[2,16]

Main cracks in beams and axially tensioned members have been studied by many investigators; prediction of the width of cracks and crack control were among the problems studied. These will be discussed here as well as the requirements of the ACI Code.

Crack Width.   Crack width and crack spacing, according to existing crack theories, depend on many factors, which include steel percentage, its distribution in the concrete section, steel flexural stress at service load, concrete cover, and properties of the concrete constituents. Different equations for predicting the width and spacing of cracks in reinforced concrete members were presented at the Symposium on Bond and Crack Formation in Reinforced Concrete in Stockholm, Sweden, in 1957. Chi and Kirstein[16] in 1958 presented equations for the crack width and spacing as a function of an effective area of concrete around the steel bar: a concrete circular area of diameter equal to four times the diameter of the bar was used to calculate crack width. Other equations were presented over the next decade.[17-23]

FIGURE 6.6. Two main cracks, about midspan, in a reinforced concrete beam tested to failure.

FIGURE 6.7. Spacing of cracks in a reinforced concrete beam.

An empirical formula presented by Kaar and Mattock[20] as a result of extensive research to estimate the maximum crack width W at the steel level is

$$W(\text{max}) = 0.115 f_s \sqrt[4]{A} \times 10^{-6} \qquad (6.12)$$

where

A = area of concrete surrounding each bar, equal to the effective tension area of concrete having the same centroid as that of the reinforcing bars divided by the number of bars, in square inches

f_s = the stress in the steel bars

At the tension concrete face, Kaar and Mattock suggested the following formula:

$$W(\text{max}) = 0.115 R f_s \sqrt[4]{A} \times 10^{-6} \qquad (6.13)$$

where R is the ratio of distances from the neutral axis to the tension face and to the steel centroid, and A and f_s are as defined above.

Gergely and Lutz[23] presented the following formula for the limiting crack width, which is adopted by the ACI Code:

$$W = 0.076 R f_s \sqrt[3]{A d_c} \times 10^{-6} \qquad (6.14)$$

where

R, A, and f_s are as defined above

d_c = thickness of concrete cover measured from the extreme tension fiber to the center of the closest bar

The value of R can be taken approximately equal to 1.2; therefore equation 6.14 becomes

$$W = 0.0912 f_s \sqrt[3]{A d_c} \times 10^{-6} \text{ in} \qquad (6.15)$$

Note that f_s is in psi and W in inches.

The mean ratio of maximum crack width to average crack width was found to vary between 1.5 and 2.0 as reported by many investigators. An average value of 1.75 may be used.

In S.I. units (mm and MPa), equation 6.14 is

$$W = 11.0 R f_s \sqrt[3]{A d_c} \times 10^{-6} \qquad (6.16)$$

and in metric units (cm and kgf/cm²),

$$W = 10.8 R f_s \sqrt[3]{A d_c} \times 10^{-6} \qquad (6.17)$$

Permissible Crack Width. This is a necessary standard, for the formation of cracks in reinforced concrete members is unavoidable. Hairline cracks occur even in carefully designed and constructed structures. Cracks are usually measured at the face of the concrete, but actually they are related to crack width at the steel level, where corrosion is expected. The permissible crack width is also influenced by aesthetic and appearance requirements. The naked eye can detect a crack about 0.006 in. (0.15 mm) wide, depending on the surface texture of concrete. Different values for permissible crack width at the steel level have been suggested by many investigators, ranging from 0.010 to 0.016 in. (0.25–0.40 mm) for interior structures and from 0.006 to 0.010 in. (0.15–0.25 mm) for exterior exposed structures. The ACI Code of 1963, section 1508-b, allowed an average crack width at the concrete surface of 0.015 in. for interior members and 0.010 in. for exterior members, with maximum values of about 0.026 and 0.017 in., respectively. These values were considered high, and the ACI Code of 1983 set a limiting crack width of 0.016 in. (0.40 mm) for interior members and 0.013 in. (0.32 mm) for exterior members.

EXAMPLE 6.6

Determine the limiting crack width expected in a simply supported beam if the steel stress at service load is 30 ksi and if the beam cross-section is as shown in Figure 6.8.

Solution   The limiting crack width in beams, according to equation 6.15, is

$$W = 0.0912f_s \sqrt[3]{Ad_c} \times 10^{-6} \text{ in.}$$

1. For the section shown in Figure 6.8(a), $d_c = 2.5$ in. and the effective tension area of concrete can be determined by computing the centroid of the steel reinforcement. In this beam section, the centroid of 3 No. 9 bars is 2.5 in. from the bottom fibers. If a concrete depth of 5 in. is used, then the centroid of the concrete area of 10 by 5 in. is the same as the

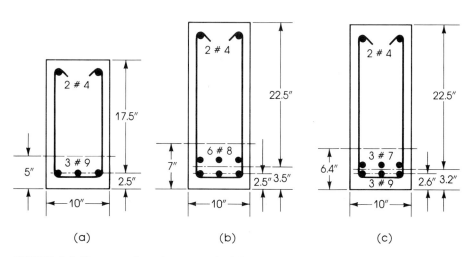

(a)   (b)   (c)

FIGURE 6.8. Three sections for example 6.6.

centroid of the steel reinforcement. Therefore the effective tension area of concrete for one bar,

$$A = \frac{10 \times 5}{3} = 16.67 \text{ in.}^2$$

$$W = 0.0912 \times 30,000\sqrt[3]{16.67 \times 2.5} \times 10^{-6} = 0.0095 \text{ in.}$$

This is less than the allowable limiting values of 0.016 in. and 0.013 in. for interior and exterior members, respectively.

2. For the section shown in Figure 6.8(b), $d_c = 2.5$ in. and the steel bars are placed in two layers. If the minimum clear distance between the two layers of 1 in. is used, then the centroid of the steel bars is 3.5 in. from the bottom fibers and the height of the effective tension concrete area is equal to $2 \times 3.5 = 7$ in. Therefore, for one bar,

$$A = \frac{10 \times 7}{6} = 11.67 \text{ in.}^2$$

$$W = 0.0912 \times 30,000\sqrt[3]{11.67 \times 2.5} \times 10^{-6} = 0.0085 \text{ in.}$$

This value of crack width is adequate (less than 0.013 in.).

3. For the section shown in Figure 6.8(c), $d_c = 2.6$ in. and the centroid of the steel bars is 3.2 in. from the bottom fibers. The effective tension concrete area is equal to 64 in.^{2} for all bars. Since there are two different bar diameters, the code specifies that the number of bars can be computed as the total steel area divided by the area of the largest bar used. The area of a No. 9 bar is 1.0 in.^{2}, and of a No. 7 bar is 0.6 in.^{2} Therefore the number of No. 9 bars to be used is 4.8. For one bar,

$$A = \frac{10 \times 6.4}{4.8} = 13.3 \text{ in.}^2$$

$$W = 0.0912 \times 30,000\sqrt[3]{13.3 \times 2.6} \times 10^{-6} = 0.009 \text{ in.}$$ ∎

Crack Control. Control grows in importance with the use of high-strength steel in reinforced concrete members, as larger cracks develop under working loads because of the high allowable stresses. Control of cracking depends on the permissible crack width: it is always preferable to have a large number of fine cracks rather than a small number of large cracks. Secondary cracks are minimized by controlling the total amount of cement paste, water-cement ratio, permeability of aggregate and concrete, rate of curing, shrinkage, and end restraint conditions.

The factors involved in controlling main cracks are the reinforcement stress, bond characteristics of reinforcement, the distribution of reinforcement, the diameter of the steel bars used, steel percentage, concrete cover, and properties of concrete constituents. Any improvement in the above factors will help in reducing the width of cracks.

6.6   ACI Code Requirements

To control cracks in reinforced concrete members, the **1983 ACI Code, Chapter 10,** specifies the following:

1. Only deformed bars are permitted as main reinforcement.
2. Tension reinforcement should be well distributed in the zones of maximum tension (**section 10.6.3**).
3. When the flange of the section is under tension, part of the main reinforcement should be distributed over the effective flange width or one-tenth of the span, whichever is smaller. Some longitudinal reinforcement has to be provided in the outer portion of the flange (**section 10.6.6**).
4. The design yield strength of reinforcement should not exceed 80 ksi (560 MPa) (**section 9.4**). In Europe, satisfactory structures have been built with design yield strength greater than 80 ksi.
5. Limiting crack width is equal to 0.016 in. (0.4 mm) for interior members and 0.013 in. (0.32 mm) for exterior members.
6. When the design yield strength of steel f_y is greater than 40 ksi (280 MPa), the cross-section of maximum positive and maximum negative moments should be proportioned such that

$$Z = f_s\sqrt[3]{Ad_c} \le 175 \text{ kip/in. (31 kN/mm) for interior members} \quad (6.18)$$
$$\le 145 \text{ kip/in. (26 kN/mm) for exterior members} \quad (6.19)$$

(**section 10.6.4**) where f_s = steel flexural stress at service load (ksi), and may be taken as $0.6f_y$. A and d_c are the effective tension area of concrete and thickness of concrete cover, respectively. The above equations are based on the expression suggested by Gergely and Lutz[20] mentioned earlier:

$$W = 0.076Rf_s\sqrt[3]{Ad_c} \times 10^{-6} \text{ in. } (f_s \text{ in psi}) \quad (6.14)$$

Here R is the ratio of distances to the neutral axis from the extreme concrete tension fibers and from the centroid of the main reinforcement. Considering limiting crack widths of 0.016 and 0.013 in., and taking f_s in ksi,

$$Z = f_s\sqrt[3]{Ad_c} = \frac{W \times 1000}{0.0912}$$

$$Z_i = \frac{0.016 \times 1000}{0.0912} = 175 \text{ k/in. for interior members}$$

$$Z_e = \frac{0.013 \times 1000}{0.0912} = 145 \text{ k/in. for exterior members}$$

When f_s is not known, it can be assumed equal to $0.6f_y$. Therefore

$$Z_i = 0.6f_y\sqrt[3]{Ad_c} \le 175 \text{ k/in. (31 kN/mm)} \quad (6.20)$$
$$Z_e = 0.6f_y\sqrt[3]{Ad_c} \le 145 \text{ k/in. (26 kN/mm)} \quad (6.21)$$

7. For slabs, R can be considered equal to 1.35, and substituting the value of R in the above expressions,

$$Z_i \le 155 \text{ k/in. (27 kN/mm) for interior members} \quad (6.22)$$
$$Z_e \le 129 \text{ k/in. (23 kN/mm) for exterior members} \quad (6.23)$$

8. For relatively deep girders, with depth equal to or greater than 3 ft (900 mm), light reinforcement should be added near the vertical faces in the tension zone

to control cracking in the web. About 10 percent of the main steel may be used, and bars should be spaced not farther than 12 in. (300 mm) or web width, whichever is smaller. If this additional steel is not provided, only a few wide cracks may extend into the web **(ACI Code, section 10.6.7)**. The zone of maximum tension will contain numerous fine and well-distributed cracks.[24]

EXAMPLE **6.7**

Design a simply supported beam of 24 ft span to carry an ultimate uniform load of 4 k/ft. Given: $f'_c = 4$ ksi, $f_y = 75$ ksi, and $b = 10$ in.

Solution

1. Choosing a steel percentage $\rho = 0.8$ percent, then $R_u = 490$ psi (from tables in Appendix A). External ultimate moment,

$$M_u = \frac{WL^2}{8} = \frac{4 \times 24 \times 24}{8} \times 12 = 3456 \ \text{k·in.}$$

$$d^2 = \frac{M_u}{R_u b} = \frac{3456}{0.490 \times 10} = 705.3 \ \text{in.}^2$$

$$d = 26.5 \ \text{in.}$$

$$A_s = 0.008 \times 10 \times 26.5 = 2.12 \ \text{in.}^2$$

Choose 3 No. 8 bars ($A_s = 2.35$ in.²)

Total depth $h = 26.5 + 2.5 = 29$ in.   (Figure 6.9)

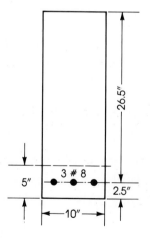

FIGURE 6.9. Example 6.7.

2. Since the yield strength of steel is greater than 40 ksi, crack width limitation must be checked by equation 6.20 or 6.21, depending on whether the beam is an interior or exterior member.

$$Z = 0.6 f_y \sqrt[3]{A d_c} \leq 175 \text{ k/in.}$$

$$\leq 145 \text{ k/in.}$$

$$A = \frac{10 \times 5}{3} = 16.67 \text{ in.}^2 \quad \text{and} \quad d_c = 2.5 \text{ in.}$$

$$Z = 0.6 \times 75 \sqrt[3]{16.67 \times 2.5} = 155.7 \text{ k/in.}$$

This value of Z is less than Z_i of 175 k/in., which indicates that the beam can be used as an interior member. The value 155.7 is greater than Z_e of 145 k/in., which indicates that the beam cannot be used as an exterior member. In other words, the expected limiting crack width is greater than 0.013 in. but less than 0.016 in. The crack width can be checked as follows:

$$W = 0.0912 \times 0.6 f_y \sqrt[3]{A d_c} \times 10^{-6} \text{ in.}$$

$$= 0.0912 \times 0.6 \times 75000 \sqrt[3]{16.67 \times 2.5} \times 10^{-6}$$

$$= 0.0142 \text{ in.}$$

3. If the beam is to be used as an exterior member, then either a better steel distribution or a different section with greater beam width may be used. If 4 No. 7 bars are used ($A_s = 2.41 \text{ in.}^2$), then

$$A = \frac{10 \times 5}{4} = 12.5. \text{ in.}^2 \quad \text{and} \quad d_c = 2.5 \text{ in.}$$

$$Z = 0.6 \times 75 \sqrt[3]{12.5 \times 2.5} = 141.75 \text{ k/in.}$$

which is less than the 145 k/in. required, and the beam can be used externally. ∎

EXAMPLE 6.8 (S.I. UNITS)

Design a simply supported beam of 7.2 m span to carry an ultimate uniform load of 60 kN/m. Given: $f_c' = 28$ MPa, $f_y = 520$ MPa, and $b = 250$ mm.

Solution

1. Choose a steel percentage $\rho = 0.8$ percent, which is less than ρ_{max} of 1.55 percent, $R_u = 3.45$ MPa ($R_u(\text{max}) = 6.1$ MPa) from Table B.2 in Appendix B.

$$\text{External ultimate moment} = \frac{UL^2}{8} = \frac{60 \times (7.2)^2}{8} = 388.8 \text{ kN·m}$$

$$d^2 = \frac{M_u}{R_u b} = \frac{388.8 \times 10^6}{3.45 \times 250}$$

$$d = 671 \text{ mm}$$

$$A_s = \rho b d = 0.008 \times 250 \times 671 = 1343 \text{ mm}^2$$

Choose 3 bars × 25 mm (A_s = 1370 mm²); the total depth is 671 + 65 = 736 mm, say 740 mm.

2. Since the yield strength of steel is greater than 280 MPa, then check crack limitations:

$$Z = 0.6f_y \sqrt[3]{Ad_c} \leq 31 \text{ kN/mm for interior members}$$

$$\leq 26 \text{ kN/mm for exterior members}$$

$$A = \frac{250 \times 65 \times 2}{3} = 10{,}834 \text{ mm}^2 \qquad d_c = 65 \text{ mm}$$

$$Z = 0.6 \times 520 \sqrt[3]{10{,}834 \times 65} = 27{,}758 \text{ N/mm} = 27.76 \text{ kN/mm}$$

which indicates that the beam can be used as an interior member only. The value 27.76 kN/mm is greater than Z_e of 26 kN/mm, which indicates that the beam cannot be used as an exterior member; in other words, the expected crack width W is greater than 0.32 mm but less than 0.4 mm.

$$W = 11.0Rf_s \sqrt[3]{Ad_c} \times 10^{-6} \text{ mm}$$

$$= 11 \times 1.2 \times (0.6 \times 520) \sqrt[3]{10{,}834 \times 65} \times 10^{-6} = 0.36 \text{ mm}$$

3. If the beam is to be used as an exterior member, and 4 bars of 22 mm are used, then

$$A = \frac{250 \times 65 \times 2}{4} = 8125 \text{ mm}^2 \qquad d_c = 65 \text{ mm}$$

$$Z = 0.6 \times 520 \times \sqrt[3]{8125 \times 65} = 25{,}218 \text{ N/mm} = 25.22 \text{ kN/mm}$$

which is less than the 26 kN/mm required. The beam can be used now as an exterior as well as an interior member. ∎

SUMMARY

SECTION 6.1

(a) Deflection $\Delta = \alpha \left(\dfrac{WL^3}{EI} \right)$ $\qquad\qquad$ (6.2)

$$\Delta = \frac{5WL^3}{384EI}$$

for a simply supported beam subjected to a uniform total load = $W = wL$.

$$E_c = 33w^{1.5} \sqrt{f_c'} = 57{,}400 \sqrt{f_c'}$$

for normal weight concrete.

(b) Effective moment of inertia

$$I_e = \left(\frac{M_{cr}}{M_a} \right)^3 I_g + \left[1 - \left(\frac{M_{cr}}{M_a} \right)^3 \right] I_{cr} \leq I_g \qquad\qquad (6.5)$$

$$M_{cr} = f_r \times I_g/y_t \quad \text{and} \quad f_r = 7.5\sqrt{f_c'}$$

SECTION 6.2

The deflection of reinforced concrete members continues to increase under sustained load.

Additional long-time deflection $= \lambda \times$ Instantaneous deflection

$$\lambda = \frac{T}{1 + 50\rho'}$$

where

$\rho' = A_s'/bd$

$T = 1.0, 1.2, 1.4,$ and 2.0 for periods of 3, 6, 12, and 60 months, respectively

SECTIONS 6.3 and 6.4

(a) The allowable deflection varies between $L/180$ and $L/480$.

(b) Deflections for different types of loads may be calculated for each type of loading separately and then added algebraically to obtain the total deflection.

SECTION 6.5

(a) Cracks are classified as secondary cracks (shrinkage, corrosion, or secondary flexural cracks) and main cracks.

(b) Maximum crack width,

$$W(\max) = 0.076Rf_s \sqrt[3]{Ad_c} \times 10^{-6} \tag{6.14}$$

Approximate values for R, f_s, and d_c are

$R = 1.2$ for beams and 1.35 slabs

$d_c = 2.5$ in.

$f_s = 0.6 f_y$

(c) The limiting crack width is 0.016 in. for interior members and 0.013 in. for exterior members.

SECTION 6.6

When the design yield strength of steel f_y is greater than 40 ksi, the crack width must be checked as follows:

$$Z = f_s \sqrt[3]{Ad_c} \leq 175 \text{ k/in. for interior members} \tag{6.18}$$

$$\leq 145 \text{ k/in. for exterior members} \tag{6.19}$$

REFERENCES

1. Wei-Wen Yu and G. Winter: "Instantaneous and Longtime Deflections of Reinforced Concrete Beams under Working Loads," *ACI Journal*, 57, July 1960.
2. M. N. Hassoun: "Evaluation of Flexural Cracks in Reinforced Concrete Beams," *Journal of Engineering Sciences*, 1, No. 1, January 1975, College of Engineering, Riyadh, Saudi Arabia.
3. D. W. Branson: "Instantaneous and Time-Dependent Deflections of Simple and Continuous Reinforced Concrete Beams, Part 1," Alabama Highway Research Report No. 7, August 1963.
4. ACI Committee 435: "Allowable Deflections," *ACI Journal*, 65, June 1968.

5. ACI Committee 435: "Deflections of Continuous Beams," *ACI Journal,* 70, December 1973.

6. ACI Committee 435: "Variability of Deflections of Simply Supported Reinforced Concrete Beams," *ACI Journal,* 69, January 1972.

7. Dan E. Branson: *Deformation of Concrete Structures,* McGraw-Hill, New York, 1977.

8. American Concrete Institute: *Causes, Mechanism and Control of Cracking in Concrete,* ACI Publication SP-20, Detroit, 1968.

9. R. Saliger: "High Grade Steel in Reinforced Concrete," *Proceedings 2nd Congress of International Association for Bridge and Structural Engineering,* Berlin-Munich, 1936.

10. K. Billig: *Structural Concrete,* St. Martins Press, New York, 1960.

11. P. E. Halstead: "The Chemical and Physical Effects of Aggressive Substances on Concrete," *The Structural Engineer,* 40, 1961.

12. R. E. Stratfull: "The Corrosion of Steel in a Reinforced Concrete Bridge," *Corrosion,* 13, No. 3, March 1957.

13. C. L. Shermer: "Corroded Reinforcement Destroys Concrete Beams," *Civil Engineering,* 26, December 1956.

14. V. R. Evans: *An Introduction to Metallic Corrosion,* Edward Arnold Publishers, London, 1948.

15. J. D. Mozer, A. Bianchini and C. Kesler: "Corrosion of Steel Reinforcement in Concrete," University of Illinois Dept. of Theoretical and Applied Mechanics Report No. 259, April 1964.

16. M. Chi and A. F. Kirstein: "Flexural Cracks in Reinforced Concrete Beams," *ACI Journal,* 54, April 1958.

17. R. C. Mathy and D. Watstein: "Effect of Tensile Properties of Reinforcement on the Flexural Characteristics of Beams," *ACI Journal,* 56, June 1960.

18. F. Levi: "Work of European Concrete Committee," *ACI Journal,* 57, March 1961.

19. E. Hognestad: "High Strength Bars as Concrete Reinforcement," *Journal of the Portland Cement Association, Development Bulletin,* 3, No. 3, September 1961.

20. P. H. Karr and A. H. Mattock: "High Strength Bars as Concrete Reinforcement (Control of Cracking), Part 4," *PCA Journal,* 5, January 1963.

21. B. B. Broms: "Crack Width and Crack Spacing in Reinforced Concrete Members," *ACI Journal,* 62, October 1965.

22. G. D. Base, J. B. Reed and H. P. Taylor: "Discussion on 'Crack and Crack Spacing in Reinforced Concrete Members,' " *ACI Journal,* 63, June 1966.

23. P. Gergely and L. A. Lutz: "Maximum Crack Width in Reinforced Concrete Flexural Members," in *Causes, Mechanism and Control of Cracking in Concrete,* ACI Publication SP-20, 1968.

24. ACI Committee 224: "Control of Cracking in Concrete Structures," *ACI Journal,* 69, December 1972.

PROBLEMS

6.1 to 6.4. For the beams shown in Figures 6.10 to 6.13, design the critical sections using the maximum steel ratio allowed by the ACI Code (ρ_{max}), and calculate the instantaneous and long-term deflections. Use $f'_c = 4$ ksi (28 MPa), $f_y = 60$ ksi (420 MPa), and $b = 12$ in. (300 mm). Assume that 20 percent of the live loads are sustained, dead loads include estimated self-weight of beams, and time-dependent factor $T = 2.0$.

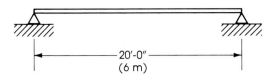

FIGURE 6.10. Problem 6.1. D.L. = 1.5 k/ft (22.5 kN/m) and L.L. = 2.2 k/ft (33 kN/m).

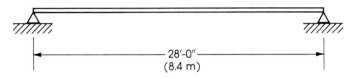

FIGURE 6.11. Problem 6.2. D.L. = 2 k/ft (30 kN/m) and L.L. = 1.4 k/ft (21 kN/m).

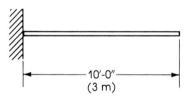

FIGURE 6.12. Problem 6.3. D.L. = 3 k/ft (45 kN/m) and L.L. = 2 k/ft (30 kN/m).

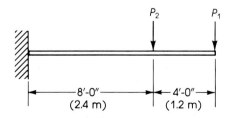

FIGURE 6.13. Problem 6.4. D.L. = 2 k/ft (30 kN/m) and concentrated live loads are P_1 = 10 k (45 kN) and P_2 = 16 k (72 kN).

6.5. Design the critical section of the beam shown in Figure 6.11 as doubly reinforced, considering that compression steel resists 20 percent of the maximum bending moment. Then calculate the maximum instantaneous and long-term deflections.

6.6. Design a 28-ft (8.4-m) simply supported beam to carry an ultimate load (including self-weight) of 5 k/ft (75 kN/m). Calculate the maximum crack width and check the cracking limits set by the ACI Code. Use $f'_c = 4$ ksi (28 MPa), $f_y = 75$ ksi (520 MPa), $b = 12$ in. (300 mm), and a steel ratio $\rho = \frac{1}{2}\rho_b$.

6.7 to 6.10. Compute the anticipated maximum crack width of the beams in problems 6.1 to 6.4. Check the cracking limits set by the ACI Code.

Bond Stress and Development Length

7.1 Introduction

The joint behavior of steel and concrete in a reinforced concrete member is based on the fact that a bond is maintained between the two materials after the concrete hardens. If a straight bar of round section is embedded in concrete, a considerable force is required to pull the bar out of the concrete. If the embedded length of the bar is long enough, the steel bar may yield, leaving some length of the bar in the concrete. The bonding force is friction between steel and concrete. The bond is caused by adhesion and by the grip of the concrete on the steel resulting from the shrinkage of the concrete in setting.[1]

Bonding is influenced mainly by the roughness of the steel surface area, the concrete mix, shrinkage, and the cover of concrete. Deformed bars give a better bond than plain bars. Rich mixes have greater adhesion than weak mixes. An increase in the concrete cover will improve the ultimate bond stress of a steel bar.[2]

7.2 Flexural Bond

Consider a length dx of a beam subjected to uniform loading. Let the moment produced on one side be M_1 and on the other side M_2, with M_1 being greater than M_2. The moments will produce internal compression and tension forces as shown in Figure 7.1. Since M_1 is greater than M_2, T_1 is greater than T_2, and consequently C_1 is greater than C_2.

At any section, $T = M/jd$, where jd is the moment arm:

$$T_1 - T_2 = dT = dM/jd$$

But

$$T_1 = T_2 + u\Sigma O dx$$

197

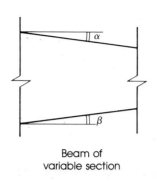
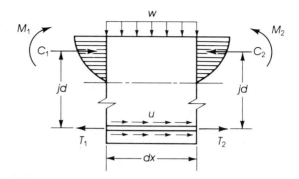

Beam of
variable section

FIGURE 7.1. Flexural bond.

where u is the average bond stress and ΣO is the sum of perimeters of bars in the section at the tension side. Therefore,

$$T_1 - T_2 = u\Sigma Odx = dM/jd$$

or

$$u = \frac{dM}{dx} \times \frac{1}{jd\Sigma O}$$

The rate of change of the moment with respect to x is the shear, or

$$dM/dx = V$$

Therefore,

$$u = \frac{V}{jd\Sigma O} \qquad (7.1)$$

The value u is the average bond stress; and for practical calculations, j can be taken approximately equal to 0.87:

$$u = \frac{V}{0.87d\Sigma O}$$

In the strength-design method, the nominal bond strength is reduced by the capacity reduction factor $\phi = 0.85$. Thus

$$U_u = \frac{V_u}{\phi(0.87)d\Sigma O} \qquad (7.2)$$

If the beam is of variable depth, then the effective depth d is not constant and it is a function of x.

$$T = \frac{M}{jd}$$

and

$$dT = \frac{jd(dM) - Md(jd)}{(jd)^2} = u\Sigma Odx$$

or

$$u = \frac{dM}{dx}\left(\frac{1}{jd\Sigma O}\right) - \frac{M}{(jd)^2\Sigma O}\left[\frac{d(jd)}{dx}\right]$$

$\frac{d(jd)}{dx}$ is the slope and $\frac{dM}{dx} = V$

Therefore,

$$u = \frac{V}{jd\Sigma O} \pm \frac{M}{(jd)^2\Sigma O}(\tan\alpha + \tan\beta)$$

where α and β are the angles that the top and bottom surfaces of the beam make with the horizontal. If j is taken equal to 0.87,

$$u = \frac{V}{0.87d\Sigma O} \pm \frac{M}{(0.87d)^2\Sigma O}(\tan\alpha + \tan\beta) \qquad (7.3)$$

The minus sign is used when the depth d *increases* as the moment *decreases,* and the plus sign is used when the depth *decreases* as the moment *increases.*

In the strength-design method, and if the top surface is horizontal ($\alpha = 0$), then

$$U_u = \frac{V_u}{\phi(0.87)d\Sigma O} \pm \frac{M_u}{\phi(0.87d)^2\Sigma O}(\tan\beta) \qquad (7.4)$$

7.3 Development Length

If a steel bar is embedded in concrete as shown in Figure 7.2, and is subjected to a tension force T, then this force will be resisted by the bond stress between the steel bar and the concrete. The maximum tension force is equal to $A_s f_y$, where A_s is the area of the steel bar. This force is resisted by another internal force of magnitude $U_u O l_d$, where U_u is the ultimate average bond stress, l_d is the embedded length of the bar, and O is the perimeter of the bar (πD). The two forces must be equal for equilibrium:

$$A_s f_y = U_u O l_d$$

or

$$l_d = \frac{A_s f_y}{U_u O}$$

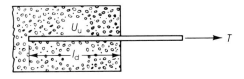

FIGURE 7.2. Anchorage or development length of a bar.

For a combination of bars,

$$l_d = \frac{A_s f_y}{U_u \Sigma O}$$

(7.5)

The length l_d is the minimum permissible anchorage length and is called the development length. For the working-stress design method,

$$l_d = \frac{A_s f_s}{u \Sigma O}$$

(7.6)

where f_s = allowable stress in the steel and u = allowable bond stress. For one bar of diameter D,

$$l_d = \frac{\pi D^2 f_s}{4u(\pi D)} = \frac{D f_s}{4u}$$

(7.7)

For ultimate-strength design, equation 7.7 becomes

$$l_d = \frac{D f_y}{4 U_u}$$

(7.8)

The ACI Code of 1983 adopted the development length concept to replace the flexural bond system. The flexural bond concept emphasized the computation of the nominal maximum bond stresses, whereas the development length concept considers an average bond stress along the bar. The development length l_d specified in the 1983 Code is based upon the 1963 Code's permissible stresses, with the length increased about 20 percent. Therefore, from equation 7.8:

$$l_d = \frac{1.2 D f_y}{4 U_u}$$

(7.9)

where U_u is the permissible ultimate bond stress.

It can be mentioned here that compression steel bars sustain higher bond stresses than bars in tension. Also, bond stresses in top bars are about 70 percent of the bond stresses in the bottom bars. Thus the development length of top bars should be 40 percent more than that of bottom bars. This is because of the tendency of the wet concrete to settle away from the top bars by gravity before hardening takes place.

The reduction factor ϕ is not used in the specified development length as that includes an allowance for reduction of strength.

7.4   Critical Sections for Bond in Flexural Members

The critical sections for development of reinforcement in flexural members are

- at points of maximum stress
- at points where tension bars within the span are terminated or bent
- at the face of the support
- at points of inflection at which moment changes signs

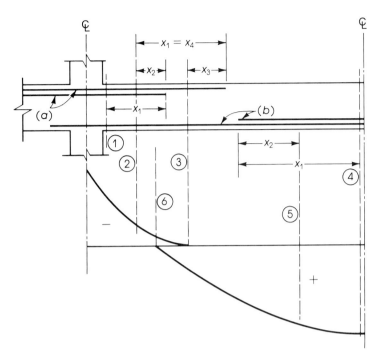

FIGURE 7.3. Critical sections (circled numbers) and development lengths (x_1–x_4).

The critical sections for a typical uniformly loaded continuous beam are shown in Figure 7.3. The sections and the relative development lengths are explained as follows:

1. Three sections are critical for the negative moment reinforcement:

    Section 1 is at the face of the support, where the negative moment as well as stress are at maximum. Two development lengths, x_1 and x_2, must be checked.

    Section 2 is the section where part of the negative reinforcement bars can be terminated. To develop full tensile force, the bars should extend a distance x_2 before they can be terminated. Once part of the bars are terminated, the remaining bars develop maximum stress.

    Section 3 is at the point of inflection. The bars shall extend a distance x_3 beyond section 3; x_3 must be equal to or greater than the effective depth d, 12 bar diameters, or $\frac{1}{16}$ clear span, whichever is greater. At least one-third of the total reinforcement provided for negative moment at the support shall be extended a distance x_3 beyond the point of inflection, according to the **ACI Code, section 12.12.3.**

2. Three sections are critical for positive moment reinforcement:

    Section 4 is that of maximum positive moment and maximum stresses. Two development lengths, x_1 and x_2, have to be checked. The length x_1 is the development length l_d specified by the **ACI Code, section 12.11,** as will be mentioned later. The length x_2 is equal to or greater than d or 12 bar diameters.

Section 5 is where part of the positive reinforcement bars may be terminated. To develop full tensile force, the bars should extend a distance x_2. The remaining bars will have a maximum stress due to the termination of part of the bars. At the face of support, section 1, at least one-fourth of the positive moment reinforcement in continuous members shall extend along the same face of the member into the support, according to the **ACI Code, section 12.11.1.** For simple members, at least one-third of the reinforcement shall extend into the support.

At points of inflection, section 6, limits are according to **section 12.11.3 of the ACI Code.**

7.5   Development Length of Deformed Bars

7.5.1   Development Length in Tension (ACI Code, Section 12.2)

For No. 11 or smaller bars,

$$l_d = 0.04 A_b f_y / \sqrt{f_c'} \geq 0.0004 d_b f_y \qquad (7.10)$$

where

A_b = area of an individual bar, in.^{2}

d_b = nominal diameter of bar or wire, in.

f_c' = the ultimate compressive strength of concrete, psi

f_y = yield strength of reinforcement, psi

For deformed wire,

$$l_d = 0.03 d_b f_y / \sqrt{f_c'} \qquad (7.11)$$

For No. 14 bars,

$$l_d = 0.085 f_y / \sqrt{f_c'}$$

and for No. 18 bars,

$$l_d = 0.11 f_y / \sqrt{f_c'}$$

The above values shall be multiplied by:

- 1.4 for top bars
- $\left(2 - \dfrac{60,000}{f_y}\right)$ for bars with $f_y > 60,000$ psi   (7.12)
- 0.8 for reinforcement bars spaced laterally at least 6 in. on centers, and at least 3 in. from the edge of bar to the face of the member
- A_s(required)/A_s(provided) ≤ 1.0, if reinforcement provided in a structural member is in excess of that required
- 0.75 for bars enclosed within a spiral not less than $\frac{1}{4}$ in. in diameter and not more than 4 in. in pitch

In no case shall the development length l_d be less than 12 in. (300 mm). In S.I. units, equation 7.10 becomes

$$l_d = 0.019 A_b f_y / \sqrt{f'_c} \geq 0.057 d_b f_y \qquad (7.13)$$

Equation 7.11 becomes

$$l_d = 0.36 d_b f_y / \sqrt{f'_c} \qquad (7.14)$$

The factor of equation 7.12 becomes

$$\left(2 - \frac{420}{f_y}\right) \qquad (7.15)$$

Units are in mm and MPa.

7.5.2 Development Length in Compression (ACI Code, Section 12.3)

For all bars,

$$l_d = 0.02 d_b f_y / \sqrt{f'_c} \geq 0.0003 d_b f_y \qquad (7.16)$$

but not less than 8 in. (200 mm). l_d may be reduced by the ratio of the required area to the area provided of steel reinforcement. In S.I. units, equation 7.16 becomes

$$l_d = 0.24 d_b f_y / \sqrt{f'_c} \geq 0.0427 d_b f_y \qquad (7.17)$$

7.5.3 Derivations

The ACI Code's basic expressions for the development length l_d are derived by using equation 7.9 and a set of values of ultimate bond stress capacity U_u for different bar sizes. For example, No. 11 or smaller bars have

$$U_u = \frac{9.5 \sqrt{f'_c}}{d_b} \leq 800 \text{ psi}$$

Substituting this value in equation 7.9,

$$l_d = \frac{1.2 f_y d_b}{4 U_u} = \frac{1.2 f_y d_b}{4 \left(\dfrac{9.5 \sqrt{f'_c}}{d_b}\right)} = 0.04 A_b f_y / \sqrt{f'_c} \qquad (7.10)$$

For No. 14 bars, $d_b = 1.69$ in., $U_u = 6\sqrt{f'_c}$, and

$$l_d = \frac{1.2 f_y (1.69)}{4(6\sqrt{f'_c})} = 0.085 f_y / \sqrt{f'_c}$$

For No. 18 bars, $d_b = 2.26$ in., $U_u = 6\sqrt{f'_c}$, and

$$l_d = \frac{1.2 f_y (2.26)}{4(6\sqrt{f'_c})} = 0.11 f_y / \sqrt{f'_c}$$

For wires, $U_u = 10\sqrt{f'_c}$ and

$$l_d = 0.03 f_y d_b / \sqrt{f'_c} \qquad\qquad (7.11)$$

For all compression bars, $U_u = 13\sqrt{f'_c} \leq 800$ psi.

Tables A.10 and B.9 in the appendixes give the development lengths for different bar diameters and concrete and steel strengths based on the previous equations.

7.5.4   Standard Hooks (ACI Code, Sections 12.5 and 7.1)

The term *standard hook* is given to any bent shape shown in Figure 7.4. A hook is used at the end of a bar when its straight embedment length is less than the necessary development length l_d. Thus the full capacity of the bar can be maintained in the shortest distance of embedment. The minimum diameter of bend, measured on the inside of the main bar of a standard hook of stirrups and ties of No. 3 to No. 5 bars, is 4 bar diameters. For all other main bars and steel grades, the minimum inside diameter of bend D_b is as follows:

• For No. 3 to No. 8 (10–25 mm), $D_b = 6d_b$
• For No. 9 to No. 11 (28, 32, and 36 mm), $D_b = 8d_b$
• For No. 14 and No. 18 (43 and 58 mm), $D_b = 10d_b$

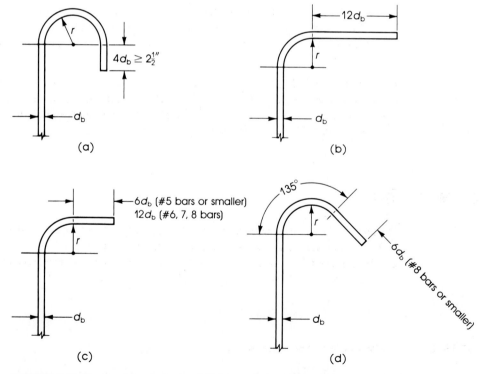

FIGURE 7.4. Standard hooks (a and b) for main bars and (c and d) stirrups and ties.

The inside diameters of bends for stirrups and ties are as follows:

- For No. 3 bars (10 mm), D_b = 1.5 in. (38 mm)
- For No. 4 bars (12 mm), D_b = 2 in. (50 mm)
- For No. 5 bars (16 mm), D_b = 2.5 in. (63 mm)

All other bars are treated similar to the main bars.

The **1983 ACI Code, section 12.5.2**, specifies a basic development length l_{hb} for grade 60 hooked bar (f_y = 60 ksi) as follows:

$$l_{hb} = 1200 d_b / \sqrt{f'_c} \tag{7.18}$$
$$l_{hb} = 100 d_b / \sqrt{f'_c} \text{ (S.I. units)}$$

Based on different conditions, the basic development length l_{hb} must be multiplied by one of the following factors:

1. When bars with yield strength other than 60 ksi are used, the basic development length l_{hb} is multiplied by $f_y/60$, where f_y is in ksi.
2. When No. 11 or smaller bars are used, and the hook is enclosed vertically or horizontally within stirrups or ties spaced not less than 3 times the diameter of the hooked bar, the basic development length is multiplied by 0.8.
3. When No. 11 or smaller bars are used and the side concrete cover, normal to the plane of the hook, is not less than 2.5 in., the basic development length is multiplied by 0.7. The same factor applies for a 90° hook when the concrete cover on bar extension beyond the hook is not less than 2 in.
4. When a bar anchorage is not required, the basic development length for the reinforcement in excess of that required is multiplied by the ratio A_s(required)/A_s(provided).

The development length l_{dh} of a standard hook for deformed bars in tension is equal to the basic development length l_{hb} multiplied by the applicable modified factor but must not be less than $8d_b$ or 6 in., whichever is greater. It is recommended that the development length should be doubled when plain bars are used.

EXAMPLE 7.1

The section of a simply supported beam, 2 ft from the face of the support, has an effective depth d = 18 in. and a width b = 12 in. and is reinforced with 4 No. 7 bars. The ultimate shearing force at this section is 59 k. Check the flexural bond at the given section. Given: f'_c = 3 ksi and f_y = 40 ksi.

Solution   The allowable ultimate bond stress

$$U_u = \frac{9.5\sqrt{f'_c}}{D} \leq 800 \text{ psi}$$

For No. 7 bars, the diameter is 0.875 in. and the perimeter is 2.75 in. Therefore,

$$U_u = \frac{9.5\sqrt{3000}}{0.875} = 600 \text{ psi}$$

The ultimate flexural bond stress

$$U_u = \frac{V_u}{\phi j d \Sigma O}$$

$$= \frac{59 \times 1000}{0.85 \times 0.87 \times 18 \times (4 \times 2.75)} = 404 \text{ psi} < 600 \text{ psi} \qquad \blacksquare$$

EXAMPLE 7.2

A continuous beam has the bar details shown in Figure 7.5. The bending moments for maximum positive and negative moments are also shown. It is necessary to check the development lengths at all critical sections. Given: $f'_c = 4$ ksi, $f_y = 40$ ksi, $b = 11$ in., $d = 18$ in., and span $L = 24$ ft.

Solution   The critical sections are (a) at the face of the support for tension and compression reinforcement (section 1), (b) at points where tension bars are terminated within the span (sections 2 and 5), (c) at points of inflection (sections 3 and 6), and (d) at midspan (section 4).

1. Development lengths for negative moment reinforcement, from Figure 7.5, are as follows: 3 No. 9 bars are terminated at a distance $x_1 = 3$ ft from the face of the support, while the other three bars extend to a distance of 4'-9" (57 in.) from the face of the support.

    The development length for x_1 (tension bars)

$$l_d = 0.04 A_b f_y / \sqrt{f'_c} = 0.04 \times 1 \times 40,000 / \sqrt{4000}$$

$$= 1600/63 = 25.5 \text{ in.}$$

Being top bars, required $l_d = 1.4 \times 25.5 = 35.7$ in.

$$\text{Minimum } l_d = 0.0004 d_b f_y = 0.0004 \times 1.128 \times 40,000 = 18 \text{ in.}$$

but not less than 12 in. (For No. 9 bar, area $A_b = 1$ in.^{2}, diameter $d_b = 1.128$ in.) Therefore, $x_1 = 35.7$ in. controls.

$$x_1 \text{ provided } = 3'-0'' = 36 \text{ in.}$$

The development length x_2 shall extend beyond the point where 3 No. 9 bars are not needed, either d or $12 d_b$ (18 in. or 14 in.). Therefore, $x_2 = 18$ in. as provided.

The distance x_4 provided $= 57 - 18 = 39$ in. $\geq x_1$. The required development length $x_4 = 35.7$ in., similar to x_1, which is less than the length provided. Total length $y \geq x_1 + (x_1 - x_2) \geq (2x_1 - x_2)$.

Beyond the point of inflection (section 3), 3 No. 9 bars extend a length x_3. Three out of 6 No. 9 bars in the section are more than what is required by the ACI Code, which requires that one-third of the bars should extend beyond the point of inflection. The development length x_3 is the greatest value of

$$d = 18 \text{ in.}$$

$$12 d_b = 12 \times 1.128 = 13.5 \text{ in.}$$

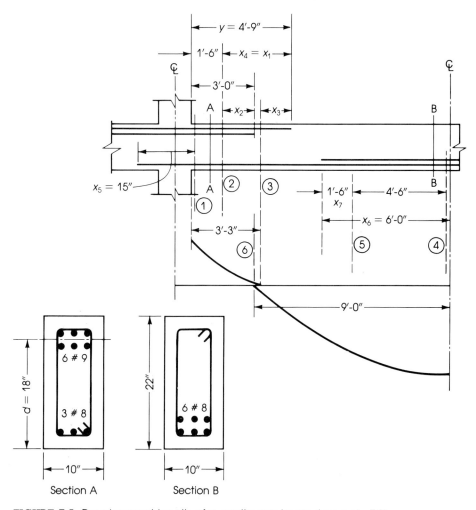

FIGURE 7.5. Development length of a continuous beam (example 7.2).

or

$$\frac{L}{16} = \frac{24 \times 12}{16} = 18 \text{ in.}$$

x_3 provided $= 18$ in., which is adequate.

2. Compressive reinforcement at the face of the support (section 1):
   The development length x_5 is equal to

$$l_d = 0.02 d_b f_y / \sqrt{f_c'} = 0.02 \times 1 \times 40{,}000 / \sqrt{4000} = 12.7 \text{ in.}$$
$$\text{Minimum } l_d = 0.0003 d_b f_y = 0.0003 \times 1 \times 40{,}000 = 12 \text{ in.}$$

but not less than 8 in. The length 12.7 in. controls. (For No. 8 bar, $d_b = 1$ in.) l_d provided $= 15$ in., which is greater than the 12.7 in. required.

3. Development length for positive moment reinforcement:

Three No. 8 bars extend 6 ft beyond the centerline, while the other bars extend to the support. The development length x_6 from the centerline is the greatest of

$$l_d = 0.04 A_b f_y / \sqrt{f_c'} = 0.04 \times 0.8 \times 40{,}000 / \sqrt{4000} = 20.5 \text{ in.}$$

$$\text{Minimum } l_d = 0.0004 d_b f_y = 0.0004 \times 1 \times 40{,}000 = 16 \text{ in.}$$

but not less than 12 in. x_6 provided = 6 ft = 72 in. > 20.5 in.

The length x_7 is equal to d or $12 d_b$, that is, 18 in. or $12 \times 1 = 12$ in. The x_7 provided = 18 in., which is adequate.

The actual position of the termination of bars within the span can be determined by the moment resistance diagram, as will be explained later. ∎

7.6   Splices

Steel bars that are used as reinforcement in structural members are fabricated in lengths of 20, 40, and 60 ft (6, 12, and 18 m), depending on the bar diameter, transportation facilities, and other reasons. Bars are usually tailored according to the reinforcement details of the structural members. When some bars are short, it is necessary to splice them by lapping the bars a sufficient distance to transfer stress through the bond from one bar to the other.

Splices may be made by lapping or welding or with mechanical devices that provide positive connection between bars. Lap splices should not be used for bars larger than No. 11 (36 mm). For noncontact lap splices in flexural members, bars should not be spaced transversely farther apart than one-fifth the required length or 6 in. (150 mm). An approved welded splice is one in which the bars are butted and welded to develop in tension at least 125 percent of the specified yield strength of the bar. The **ACI Code, section 12.14,** also specifies that full positive connections must develop in tension or compression at least 125 percent of the specified yield strength of the bar.

The minimum splice lengths in tension and compression according to the ACI Code are given as a function of the development length l_d explained in Section 7.5, equations 7.10 to 7.17. Splices in tension and compression as specified by the **ACI Code, sections 12.15 and 12.16,** are as follows:

For splices in tension:

- Minimum splice length is 12 in. (300 mm).
- Splices in the region of maximum moment should be avoided where possible, otherwise the splice is anchored for the full yield strength of the bar. If more than half of the bars are spliced, the lap length must not be less than $1.7 l_d$, but if half of the bars or less are spliced, the lap length must not be less than $1.3 l_d$.
- In regions of low calculated stresses, f_s is less than $0.5 f_y$ or the ratio $[A_s(\text{provided})/A_s(\text{required})] \geq 2.0$, and if not more than $\frac{3}{4}$ of the bars are spliced, then the lap length must not be less than $1.0 l_d$. If more than $\frac{3}{4}$ of the bars are spliced, the lap length must not be less than $1.3 l_d$.
- In tension tie members, the splices must be staggered and have a lap splice of twice the development length, $2 l_d$.

- The above values shall be increased by 20 percent for a 3-bar bundle and 33 percent for a 4-bar bundle.

For splices in compression:

- The minimum length of a lap splice when $f'_c \geq 3000$ psi (21 MPa) is

$$\text{Lap} = 0.02 d_b f_y / \sqrt{f'_c}$$

but

$$\text{Lap} \geq 0.0005 d_b f_y \text{ when } f_y \leq 60 \text{ ksi,}$$
$$\geq (0.0009 f_y - 24) d_b \text{ when } f_y > 60 \text{ ksi, or}$$
$$\geq 12 \text{ in. (300 mm)}$$

- When f'_c is less than 3000 psi (21 MPa), the lap length in compression must be increased by one-third.
- When bars of different size are lap-spliced, splice length must be the larger of l_d of the larger bar or the splice length of the smaller bar **(ACI Code, section 12.16.2)**.
- In tied compression members, where area of ties is greater than $0.0015hs$, where h is the overall thickness of the member and s is the tie spacing, a lap length of 0.83 of that specified in the first category above can be used with a minimum lap length of 12 in. (300 mm) **(ACI Code, section 12.16.3)**.
- When longitudinal bars in columns are offset, the slope of the inclined portion of the bar with the axis of the column should not exceed 1 to 6. When column faces are offset 3 in. or more, splices of vertical bars should be made by separate dowels adjacent to the offset face.

7.7　Moment Resistance Diagram (Bar Cutoff Points)

The moment capacity of a beam is a function of its effective depth d, width b, and the steel area, for given strengths of concrete and steel. For a given beam, with constant width and depth, the amount of reinforcement can be varied according to the variation of the bending moment along the span. It is a common practice to cut off the steel bars where they are no longer needed to resist the flexural stresses. In some other cases, as in continuous beams, positive moment steel bars may be bent up, usually at 45°, to provide tensile reinforcement for the negative moments over the supports.

It was mentioned earlier that the ultimate moment capacity of an under-reinforced concrete beam at any section is

$$M_u = \phi A_s f_y (d - a/2) \tag{7.19}$$

and for the working-stress design method,

$$M = A_s f_s (jd) \tag{7.20}$$

The lever arm $(d - a/2)$ or jd varies for sections along the span as the amount of reinforcement varies; however, the variation in the lever arm along the beam length is small and is never less than the value obtained at the section of maximum bending moment. Thus, it may be assumed that the moment capacity of any section is proportional to the tensile force or the area of the steel reinforcement, assuming proper anchorage lengths are provided.

To determine the position of the cut-off or bent points, the moment diagram due to external loading is drawn first. A moment resistance diagram is also drawn on the same graph, indicating points where some of the steel bars are no longer required. The ultimate moment resistance of one bar M_{ub} is

$$M_{ub} = \phi A_{sb} f_y (d - a/2) \tag{7.21}$$

where

$$a = \frac{A_s f_y}{0.85 f_c' b} \quad \text{and} \quad A_{sb} = \text{area of one bar}$$

The intersection of the moment resistance lines with the external bending moment diagram indicates the theoretical points where each bar can be terminated. To illustrate the above discussion, Figure 7.6 shows a uniformly loaded simple beam, its cross section, and the bending moment diagram. The bending moment curve is a parabola with a maximum moment at midspan of 2400 k·in. Since the beam is reinforced with 4 No. 8 bars, the ultimate moment resistance of one bar is

$$M_{ub} = \phi A_{sb} f_y (d - a/2)$$
$$a = \frac{A_s f_y}{0.85 f_c' b} = \frac{4 \times 0.79 \times 50}{0.85 \times 3 \times 12} = 5.2 \text{ in.}$$
$$M_{ub} = 0.9 \times 0.79 \times 50(20 - 5.2/2) = 620 \text{ k·in.}$$

The ultimate moment resistance of 4 bars is thus 2480 k·in., which is greater than the external moment of 2400 k·in. If the moment diagram is drawn to scale on the base line A–A, it can be seen that one bar can be terminated at point a, a second bar at point b, the third bar at point c, and the fourth bar at the support end A. These points are the theoretical positions for the termination of the bars. However, it is necessary to develop part of the strength of the bar by bond, as explained earlier. The ACI Code specifies that every bar should be continued at least a distance equal to the effective depth d of the beam or 12 bar diameters, whichever is greater, beyond the theoretical points a, b, and c. The code also specifies (section 12.11.1) that at least one-third of the positive moment reinforcement must be continued to the support for simple beams. Therefore, for the example discussed here, two bars must extend into the support, and the moment resistance diagram M_{uR} is shown by the dark line, which must enclose the external bending moment diagram at all points.

For continuous beams, the bars are bent at the required points and used to resist the negative moments at the supports. At least one-third of the total reinforcement provided for the negative moment at the support must be extended beyond the inflection points a distance not less than the effective depth, 12 bar diameters, or $\frac{1}{16}$ the clear span, whichever is greatest (ACI Code, section 12.12.3).

Bent bars are also used to resist part of the shear stresses in beams, as will be explained in Chapter 8. The moment resistance diagram for a continuous beam is shown in Figure 7.7.

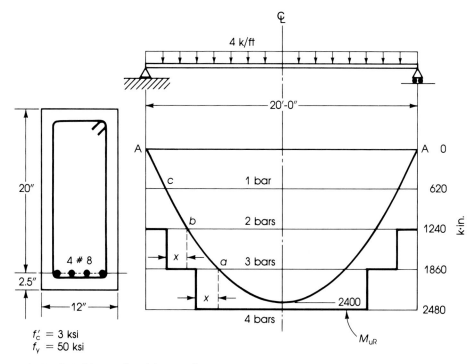

FIGURE 7.6. Moment resistance diagram.

EXAMPLE 7.3

For the simply supported beam shown in Figure 7.8, (a) design the beam for the given ultimate loads and draw the moment resistance diagram, and (b) show the points where the reinforcing bars can be terminated. Given: beam width $b = 10$ in., steel percentage $\rho = 0.013$,
$f'_c = 2.5$ ksi, and $f_y = 50$ ksi.

Solution   For $\rho = 1.3$ percent, which is less than ρ_{max} of 1.72 percent, $R_u = 496$ psi. Maximum bending
moment at midspan,

$$M_u(\text{max}) = 132.5 \text{ k·ft}$$
$$M_u = R_u bd^2$$

Therefore

$$d^2 = \frac{132.5 \times 12{,}000}{496 \times 10} = 320.56$$
$$d = 17.9 \text{ in.}$$
$$A_s = 0.013 \times 10 \times 17.9 = 2.33 \text{ in.}^2$$

Choose 4 No. 7 bars ($A_s = 2.4$ in.²). Total depth $h = 17.9 + 0.5 + 1.875 = 20.3$ in. Let $h = 21$ in.; then actual $d = 21 - 1.875 - 0.875 = 18.25$ in.

$$M_{uR} = \phi A_s f_y (d - a/2)$$

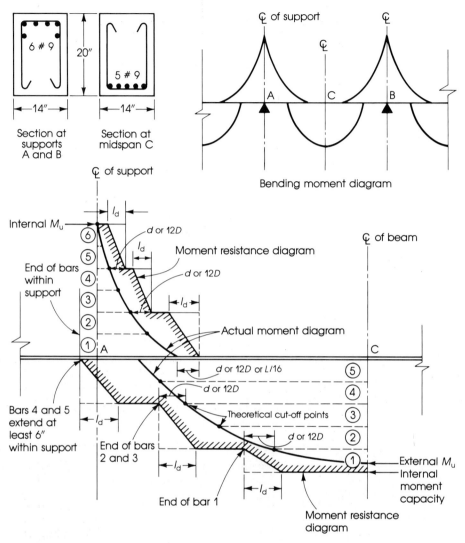

FIGURE 7.7. Sections and bending moment diagram (top) and moment resistance diagram (bottom) of a continuous beam. Bar diameter is signified by D.

$$\text{where } a = \frac{\text{actual } A_s f_y}{0.85 f'_c b} = \frac{2.4 \times 50}{0.85 \times 2.5 \times 10} = 5.65 \text{ in.}$$

$$M_{uR} \text{ (for one bar)} = 0.9 \times 0.6 \times 50(18.25 - 5.65/2)$$

$$= 416.6 \text{ k·in.} = 34.7 \text{ k·ft}$$

$$M_{uR} \text{ (for four bars)} = 4 \times 34.7 = 138.8 \text{ k·ft}$$

Details of the bending moment resistance diagram and the position of bent bars are shown in Figure 7.9. Note that bars are bent at a distance $d = 18$ in. beyond the points where, theoretically, the bars are not needed. ∎

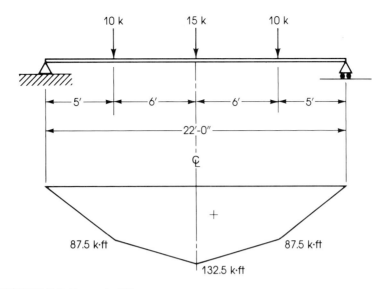

FIGURE 7.8. Example 7.3.

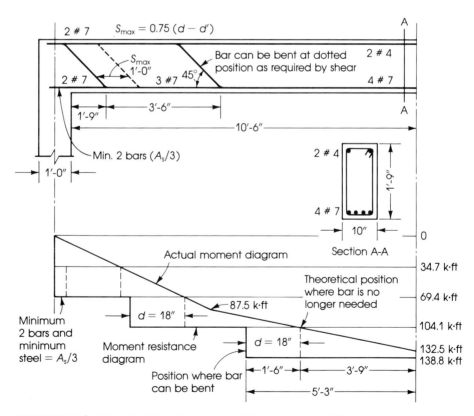

FIGURE 7.9. Details of reinforcing bars and the moment resistance diagram, example 7.3.

SUMMARY

SECTIONS
7.1 to 7.3   Bonding is influenced mainly by the roughness of the steel surface area, the concrete mix, shrinkage, and the cover of concrete. In general,

$$l_d = \frac{A_s f_y}{U_u \Sigma O} \tag{7.5}$$

For one bar of diameter D,

$$l_d = \frac{1.2 D f_y}{4 U_u} \tag{7.9}$$

SECTION
7.4   The critical sections for the development of reinforcement in flexural members are:

- at points of maximum stress
- at points where tension bars are terminated within the span
- at the face of the support
- at points of inflection

SECTION
7.5   (a) Development length in tension for No. 11 or smaller bars is

$$l_d = 0.04 A_b f_y / \sqrt{f'_c} \geq 0.0004 d_b f_y \geq 12 \text{ in.} \tag{7.10}$$

   l_d shall be multiplied by 1.4 for top bars.

(b) Development length in compression for all bars

$$l_d = 0.02 d_b f_y / \sqrt{f'_c} \geq 0.0003 d_b f_y \geq 8 \text{ in.} \tag{7.16}$$

(c) For all main bars, the minimum diameter of bends in hooks are

- For No. 3 to No. 8, $6 d_b$
- For No. 9 to No. 11, $8 d_b$

The basic development length l_{hb} of a standard hook is

$$l_{hb} = 1200 d_b / \sqrt{f'_c} \tag{7.18}$$

SECTION
7.6   (a) For splices in tension, the minimum splice length is 12 in. If more than half the bars are spliced, the lap length must be $\geq 1.7 l_d$. If less than half the bars are spliced, the lap length must be $\geq 1.3 l_d$.

(b) For splices in compression, the minimum lap length when $f'_c \geq 3$ ksi is

$$0.02 d_b f_y / \sqrt{f'_c} \geq 0.0005 d_b f_y \geq 12 \text{ in.}$$

SECTION
7.7   The ultimate moment resistance of one bar in a flexural member is

$$M_{ub} = \phi A_{sb} f_y \left(d - \frac{a}{2} \right)$$

REFERENCES

1. L. A. Lutz and P. Gergely: "Mechanics of Bond and Slip of Deformed Bars in Concrete," *ACI Journal,* 68, April 1967.
2. ACI Committee 408: "Bond Stress—The State of the Art," *ACI Journal,* 63, November 1966.
3. ACI Committee 408: "Opportunities in Bond Research," *ACI Journal,* 67, November 1970.
4. Y. Goto: "Cracks Formed in Concrete around Deformed Tensioned Bars," *ACI Journal,* 68, April 1971.
5. T. D. Mylrea: "Bond and Anchorage," *ACI Journal,* 44, March 1948.
6. E. S. Perry and J. N. Thompson: "Bond Stress Distribution on Reinforcing Steel in Beams and Pullout Specimens," *ACI Journal,* 63, August 1966.
7. C. O. Orangum, J. O. Jirsa and J. E. Breen: "A Reevaluation of Test Data on Development Length and Splices," *ACI Journal,* 74, March 1977.
8. J. Minor and J. O. Jirsa: "Behavior of Bent Bar Anchorage," *ACI Journal,* 72, April 1975.

PROBLEMS

7.1. A continuous beam has the typical steel reinforcement details shown in Figure 7.10; the sections at midspan and at the face of the support of a typical interior span are also shown.

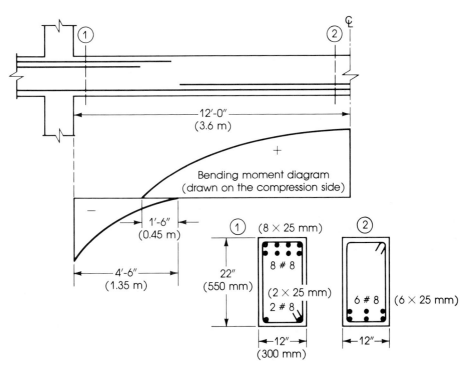

FIGURE 7.10. Problem 7.1.

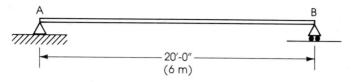

FIGURE 7.11. Problem 7.2. D.L. = 1.5 k/ft (22.5 kN/m), L.L. = 2.2 k/ft (33 kN/m).

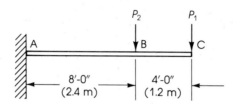

FIGURE 7.12. Problem 7.3. D.L. = 2 k/ft (30 kN/m); L.L. (concentrated loads only) is P_1 = 10 k (45 kN), P_2 = 16 k (72 kN).

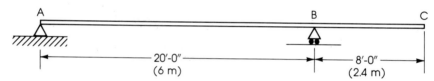

FIGURE 7.13. Problem 7.4 D.L. = 6 k/ft (90 kN/m), L.L. = 4 k/ft (60 kN/m).

It is required to check the development lengths of the reinforcing bars at all critical sections. Use f'_c = 4 ksi (28 MPa) and f_y = 60 ksi (420 MPa).

7.2. Design the beam shown in Figure 7.11 using ρ_{max}. Draw the moment resistance diagram and indicate where the reinforcing bars can be terminated. The beam carries a uniform dead load of 1.5 k/ft (22.5 kN/m) and a live load of 2.2 k/ft (33 kN/m). Use f'_c = 4 ksi (28 MPa), f_y = 60 ksi (420 MPa), and b = 12 in. (300 mm).

7.3. Design the beam shown in Figure 7.12 using a steel ratio ρ = $1/2\rho_b$. Draw the moment resistance diagram and indicate the cutoff points. Use f'_c = 3 ksi (21 MPa), f_y = 60 ksi (420 MPa), and b = 12 in. (300 mm).

7.4. Design the section at support B of the beam shown in Figure 7.13 using ρ_{max}. Adopting the same dimensions of the section at B for the entire beam ABC, determine the reinforcement required for part AB and draw the moment resistance diagram for the beam ABC. Use f'_c = 4 ksi (28 MPa), f_y = 60 ksi (420 MPa), and b = 12 in. (300 mm).

Shear and Diagonal Tension

8.1　Introduction

The general formula for the shear stress in a homogeneous beam is

$$v = \frac{VQ}{Ib} \qquad (8.1)$$

where

V = total shear at the section considered

Q = statical moment about neutral axis of that portion of cross-section lying between a line through the point in question parallel to the neutral axis and nearest face, upper or lower, of the beam

I = moment of inertia of cross-section about the neutral axis

b = width of beam at the given point

Consider a simply supported beam of rectangular section loaded with a uniform load (Figure 8.1). Take any portion of the beam dx and determine the stresses f_{c1} and f_{c2} on each face (Figure 8.2).

$$f_{c1} = \frac{M_1 K d}{I} \qquad f_1 \text{ at any point} = \frac{M_1 y}{I}$$

$$f_{c2} = \frac{M_2 K d}{I} \qquad f_2 \text{ at any point} = \frac{M_2 y}{I}$$

$$C_1 = \int_A f_1 \, dA \qquad C_2 = \int_A f_2 \, dA$$

Since M_2 is greater than M_1, as shown in Figure 8.1, then $C_2 > C_1$ and $f_{c2} > f_{c1}$. The portion of the beam dx is in equilibrium and, therefore, a shear stress develops along the face a–a₁ (Figure 8.2). If v is the shear stress per unit area, then the shearing force

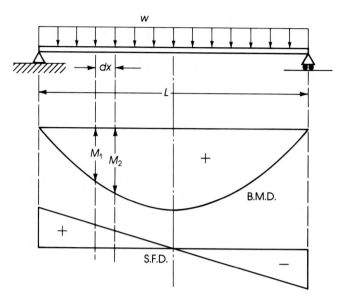

FIGURE 8.1. Bending moment and shearing force diagrams for a simple beam.

developed equals $vb\ dx$. Since the sum of horizontal forces equals 0, then

$$C_2 - C_1 = vb\ dx$$

$$\int_A f_2\ dA - \int_A f_1\ dA = vb\ dx$$

or

$$\int_A \frac{M_2}{I} y\ dA - \int_A \frac{M_1}{I} y\ dA = vb\ dx$$

$$\left(\frac{M_2}{I} - \frac{M_1}{I}\right) \int_A y\ dA = vb\ dx$$

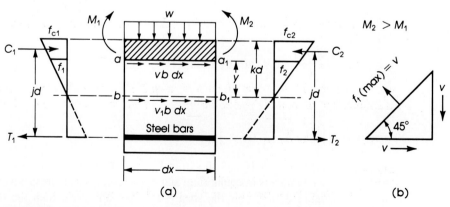

FIGURE 8.2. (a) Analysis of forces and (b) case of pure shear.

The value $\int_{Ay} y\, dA$ is the first area moment Q of the part above the point considered about the neutral axis, and $M_2 - M_1 = dM$. Therefore,

$$Q \frac{dM}{I} = vb\, dx \quad \text{or} \quad v = \frac{Q}{Ib} \times \frac{dM}{dx}$$

But the rate of change of moment with respect to x is equal to the shearing force V. Then

$$v = \frac{VQ}{Ib} \tag{8.1}$$

From Figure 8.2, if the total compressive force on one face is C_1 and the total compressive force (down to the neutral axis) on the other face is C_2, and the average shearing force is equal to $v_a b\, dx$, then

$$C_2 - C_1 = v_a b\, dx = dC$$

The compressive force C, at any section, is equal to M/jd. Therefore

$$dC = dM/jd \quad \text{or} \quad dM/jd = v_a b\, dx$$

where v_a is the average shear stress:

$$v_a = \frac{dM}{dx} \times \frac{1}{bjd}$$

But since $dM/dx = V$, therefore

$$v_a = \frac{V}{bjd} \tag{8.2}$$

It can be assumed that where normal stress is low, a case of pure shear may be considered. In this case the maximum tensile stress acting at 45° equals the shear stress (Figure 8.2 (b)). Such an analysis considers the shear stress as representing the diagonal tension. But since the diagonal tension effects are complex and the combined normal and shear stresses must be taken into consideration, the j value in equation 8.2 is eliminated, and the average shear stress is then

$$v_a = \frac{V}{bd} \tag{8.3}$$

If the ultimate shearing force is considered, then the ultimate shear stress v_u is

$$v_u = \frac{V_u}{bd} \tag{8.4}$$

For the combined action of shear and normal stresses at any point in a beam, the maximum and minimum diagonal tension (principal stresses) are given by the equation

$$f_p = \frac{f}{2} \pm \sqrt{\left(\frac{f}{2}\right)^2 + v^2} \tag{8.5}$$

where

f = intensity of normal stress due to bending

v = shear stress

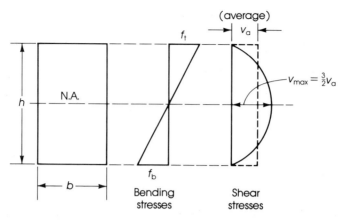

FIGURE 8.3. Bending and shear stresses in a homogeneous beam, according to elastic theory.

The direction of the maximum diagonal tension (principal stress) is given by the relation

$$\tan 2\alpha = 2v/f$$

where α = the angle between the diagonal tension and the horizontal direction.

Shear failure is most likely to occur where the shear forces are maximum, generally near the supports of the member. The first evidence of impending failure is the formation of diagonal cracks.

The distribution of bending and shear stresses according to elastic theory for a homogeneous rectangular beam is as shown in Figure 8.3. The bending stresses are calculated from the flexural formula $f = Mc/I$, while the shear stress at any point is calculated by the shear formula of equation 8.1. The maximum shear stress is at the neutral axis and is equal to 1.5 v_a (average shear), where $v_a = V/bh$. The shear stress curve is parabolic. From the shear formula $v = VQ/Ib$,

$$Q = b\left(\frac{h}{2}\right)\left(\frac{h}{4}\right) = \frac{bh^2}{8}$$

h = total depth of the section, and

$$I = \frac{bh^3}{12}$$

Therefore,

$$v = \frac{V}{b} \times \frac{bh^2}{8} \times \frac{12}{bh^3} = 1.5\frac{V}{bh} = 1.5v_a$$

8.2   Shear Stresses in Reinforced Concrete Beams

For a singly reinforced concrete beam, the distribution of shear stress above the neutral axis is a parabolic curve. Below the neutral axis, the maximum shear stress is main-

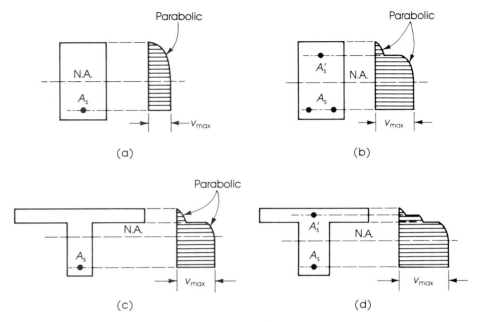

FIGURE 8.4. Distribution of shear stresses in reinforced concrete beams: (a) singly reinforced, (b) doubly reinforced, (c) T-section, (d) T-section with compression steel.

tained down to the level of the tension steel, because there is no change in the tensile force down to this point and the concrete in tension is neglected. The shear stress below the tension steel is zero (Figure 8.4). For doubly reinforced and T-sections, the distribution of shear stresses is as shown in Figure 8.4.

EXAMPLE 8.1

(a) Calculate the maximum shear stress in a rectangular beam reinforced with 3 No. 8 bars, for a shearing force of 10 kips.
(b) Determine the average shear stress. Given: $b = 10$ in., $d = 19$ in., $f'_c = 3$ ksi, $f_y = 40$ ksi, and $n = 9$.

Solution 1. Determine the position of the neutral axis:

$$b \frac{(Kd)^2}{2} = nA_s(d - Kd)$$

$$10 \frac{(Kd)^2}{2} = 9 \times 2.37 \, (19 - Kd)$$

$$(Kd)^2 + 4.26 \, (Kd) = 81$$

$$(Kd) = 7.12 \text{ in.} \text{ (similar to example 5.1)}$$

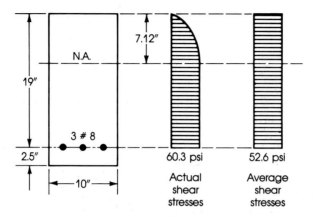

FIGURE 8.5. Shear stress distribution (example 8.1).

2. $v = VQ/Ib$

$$Q = b \frac{(Kd)^2}{2} = \frac{10}{2} (7.12)^2 = 254 \text{ in.}^3$$

$$I = b \frac{(Kd)^3}{3} + nA_s(d - Kd)^2$$

$$= \frac{10}{3} (7.12)^3 + 9 \times 2.37(19 - 7.12)^2 = 4214 \text{ in.}^4$$

3. v_{max} is at the level of the neutral axis.

$$v_{max} = \frac{VQ}{Ib} = \frac{10,000 \times 254}{4214 \times 10} = 60.3 \text{ psi}$$

4. The average shear stress = V/bd.

$$v_a = \frac{10 \times 1000}{10 \times 19} = 52.6 \text{ psi} \qquad \text{(Figure 8.5)}$$ ■

EXAMPLE **8.2**

For the section shown in Figure 8.6, determine the shear stresses at different levels and draw the shear stress distribution for a shearing force of 100 k. Given: $f'_c = 4$ ksi, $f_s = 20$ ksi, and $n = 8$.

Solution   1. Determine transformed areas:

For compression steel, $(2n - 1)A'_s = (16 - 1) \times 0.88 = 13.2 \text{ in.}^2$

For 2 No. 9 bars tension steel, $nA_s = 8 \times 2 = 16 \text{ in.}^2$

For 5 No. 9 bars tension steel, $nA_s = 8 \times 5 = 40 \text{ in.}^2$

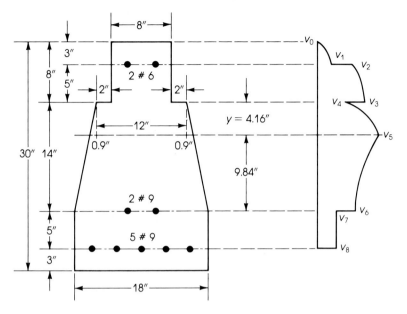

FIGURE 8.6. Example 8.2.

2. Take moments about the neutral axis (Figure 8.6):

$$40(19 - y) + 16(14 - y) = 13.2(y + 5) + 8 \times 8(y + 4)$$

$$+ 12 \times y \times \frac{y}{2} + 2\left[\frac{y}{2}\left(\frac{3}{14}\,y\right)\left(\frac{y}{3}\right)\right]$$

$$y^3 + 84y^2 + 1865y - 9268 = 0$$
$$y = 4.16 \text{ in.}$$

3. Calculate the moment of inertia of the section:

$$I = 40(14.84)^2 + 16(9.84)^2 + 13.2(9.16)^2 + 64(12.16 - 4.0)^2$$

$$+ \frac{(8)^4}{12} + \frac{12(4.16)^3}{3} + 2\left[\frac{0.9}{12}\,(4.16)^3\right]$$

$$= 8809 + 1549 + 1107 + 4261 + 341 + 288 + 10.8 = 16{,}366 \text{ in.}^4$$

4. Determine $v = \dfrac{VQ}{Ib}$ at any level:

$$v = \frac{100{,}000}{16{,}366}\left(\frac{Q}{b}\right)$$

The various values can be arranged as in Table 8.1.   ■

TABLE 8.1. Shear and web reinforcement formulas

Distance from top (in.)	Shear stress designation	b (in.)	Q (in.³)	Shear stress $v = \left(\dfrac{100,000}{16,366}\right)\dfrac{Q}{b}$ (psi)
0	v_0	8	0	0
3	v_1	8	8(3)(10.66) = 256	196
3	v_2	8	256 + 13.2(9.16) = 377	288
8	v_3	8	13.2(9.16) + 64(8.16) = 644	492
8	v_4	12	644	328
12.16 (at N.A.)	v_5	13.8	$644 + 12(4.16)(2.08)$ $+ 2\left(\dfrac{0.9}{2}\right)(4.16)\left(\dfrac{4.16}{3}\right) = 753$	333
22	v_6	18	753	256
22	v_7	18	40(14.84) = 594	202
27	v_8	18	594	202

8.3   Behavior of Beams without Shear Reinforcement

Concrete is weak in tension and the beam will collapse if proper reinforcement is not provided. The tensile stresses develop in beams due to axial tension, bending, shear, torsion, or a combination of these forces. The location of cracks in the concrete beam depends on the direction of principal stresses. For the combined action of normal stresses and shear stresses, maximum diagonal tension may occur at about a distance d from the face of the support.

The behavior of reinforced concrete beams with and without shear reinforcement tested under increasing load was discussed in Chapter 3, section 3.3. In the tested beams, vertical flexural cracks developed at the section of maximum bending moment when the tensile stresses in concrete exceeded the modulus of rupture of concrete, $f_r = 7.5 \sqrt{f'_c}$. Inclined cracks in the web developed at a later stage, at a location very close to the support.

An inclined crack occurring in a beam that was previously uncracked is generally referred to as a web-shear crack. If the inclined crack starts at the top of an existing flexural crack and propagates into the beam, the crack is referred to as flexural-shear crack (Figure 8.7). Web-shear cracks occur in beams with thin webs in regions with high shear and low moment. They are relatively uncommon cracks and may occur near the inflection points of continuous beams or adjacent to the supports of simple beams.

Flexural-shear cracks are the most common type found in reinforced concrete beams. A flexural crack extends vertically into the beam, then the inclined crack forms starting from the top of it when shear stresses develop in that region. In regions of high shear stresses, beams must be reinforced by stirrups or by bent bars to produce ductile beams that do not rupture at a failure. To avoid a shear failure before a bending failure, a greater factor of safety must be provided against a shear failure. The ACI Code specifies a capacity reduction factor ϕ of 0.85 for shear and 0.9 for flexure.

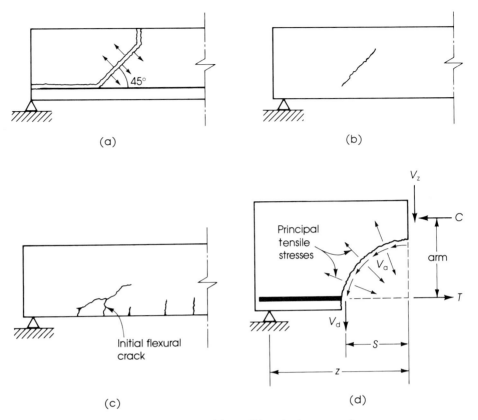

FIGURE 8.7. Shear failure: (a) general form, (b) web-shear crack, (c) flexural-shear crack, (d) analysis of forces involved in shear. V_a is interface shear, V_z is shear resistance, and V_d is dowel force.

Shear resistance in reinforced concrete members is developed by a combination of the following mechanisms[2] (Figure 8.7):

- shear resistance of the uncracked concrete V_z[3]
- interface shear transfer V_a due to aggregate interlock tangentially along the rough surfaces of the crack[3]
- arch action[4]
- dowel action V_d due to the resistance of the longitudinal bars to the transverse shearing force[5]

In addition to the above forces, shear reinforcement increases the shear resistance by V_s, which depends on the diameter and spacings of stirrups used in the concrete member. If shear reinforcement is not provided in a rectangular beam, the proportions of the shear resisted by the various mechanisms are 20–40 percent by V_z, 35–50 percent by V_a, and 15–25 percent by V_d.[6]

8.4   Moment Effect on Shear Strength

In simply supported beams under uniformly distributed load, the midspan section is subjected to a large bending moment and small shear, whereas sections near the ends are subjected to large shear and small bending moments. The shear and moment values are both high near the intermediate supports of a continuous beam. At a location of large shear force and small bending moment, there will be little flexural cracking, and an average shear stress v is equal to V/bd. The diagonal tensile stresses are inclined at about $45°$ and are numerically equal to v. Diagonal cracks can be expected when the diagonal tensile stress in the vicinity of the neutral axis reaches or exceeds the tensile strength of concrete.

The shear strength varies between $3.5\sqrt{f_c'}$ and $5\sqrt{f_c'}$. After completing a large number of beam tests on shear and diagonal tension,[1] it was found that in regions with large shear and small moment, diagonal tension cracks were formed at an average shear stress of

$$v_{cr} = \frac{V_{cr}}{bd} = 3.5\sqrt{f_c'} \tag{8.6}$$

In locations where shear force and bending moments are high, flexural cracks are formed first. At a later stage, some cracks bend in a diagonal direction when the diagonal tension stress at the upper end of such cracks exceeds the tensile strength of concrete. Given the presence of large moments on a beam, for which adequate reinforcement is provided, the nominal shear stress at which diagonal tension cracks develop is given by

$$v_{cr} = \frac{V_{cr}}{bd} = 1.9\sqrt{f_c'} \tag{8.7}$$

This value is a little more than half of the value in equation 8.6 when bending moment is very small. This means that large bending moments reduce shear stresses. The following equation has been suggested to predict the nominal shear stress at which a diagonal crack is expected[1]:

$$v_{cr} = \frac{V}{bd} = 1.9\sqrt{f_c'} + 2500\rho\frac{Vd}{M} \le 3.5\sqrt{f_c'} \tag{8.8}$$

The **1983 ACI Code, section 11.3.2,** adopted this equation for the nominal ultimate shear stress to be resisted by concrete:

$$v_c = 1.9\sqrt{f_c'} + 2500\rho_\omega\frac{V_u d}{M_u} \le 3.5\sqrt{f_c'}$$

$$v_c = 0.16\sqrt{f_c'} + 17.2\rho_\omega\frac{V_u d}{M_u} \le 0.29\sqrt{f_c'} \quad \text{(S.I.)} \tag{8.9}$$

and the nominal shear strength

$$V_c = v_c b_w d$$

where

$$\rho_\omega = \frac{A_s}{b_w d}$$

b_w being the web width in T-sections or the width of a rectangular section

V_u and M_u = ultimate shearing force and bending moment occurring simultaneously on the considered section

The value of $V_u d/M_u$ must not exceed 1.0 in equation 8.9. The shearing force equals the shear stress times the effective concrete area, or $V_c = v_c b_w d$.

If M_u is large in equation 8.9, the second term becomes small and v_c approaches $1.9\sqrt{f'_c}$. If M_u is small, the second term becomes large and the upper limit of $3.5\sqrt{f'_c}$ controls. As an alternate to equation 8.9, the **ACI Code, section 11.3.1**, permits evaluating the shear strength of concrete as follows:

$$v_c = 2\sqrt{f'_c} \quad \text{and} \quad V_c = (2\sqrt{f'_c})b_w d$$
$$v_c = 0.17\sqrt{f'_c} \quad \text{and} \quad V_c = (0.17\sqrt{f'_c})b_w d \quad \text{(S.I.)} \tag{8.10}$$

For members subjected simultaneously to axial force and shear, the magnitude of the diagonal tension is modified. Axial compression will decrease the tensile principal stresses caused by shear, thus increasing the concrete shear capacity. Axial tension, on the other hand, will increase the principal stresses caused by shear, thus reducing the net concrete shear capacity. To compute the nominal shear strength of concrete V_c for a cross-section subjected to axial force N_u and shear V_u simultaneously, the **ACI Code, section 11.3.2**, presents the following equations:

1. In the case of an axial compression force N_u:

$$V_c = \left(1.9\sqrt{f'_c} + 2500\rho_\omega \frac{V_u d}{M_m}\right)b_w d \tag{8.11}$$

where

$$\rho_\omega = \frac{A_s}{b_w d}$$

$$M_m = M_u - N_u\left(\frac{4h - d}{8}\right)$$

h = overall depth

$V_u d/M_u$ may be greater than 1.0, but V_c must not exceed

$$V_c = (3.5\sqrt{f'_c})\sqrt{1 + \frac{N_u}{500A_g}}\,b_w d \tag{8.12}$$

where A_g = the gross area in in.^{2}.

Alternatively, V_c may be computed by

$$V_c = \left(2 + 0.001\frac{N_u}{A_g}\right)\sqrt{f'_c}\,b_w d \tag{8.13}$$

2. In the case of an axial tensile force N_u,

$$V_c = \left(2 + 0.004\frac{N_u}{A_g}\right)\sqrt{f'_c}\,b_w d \tag{8.14}$$

where N_u is to be taken negative.

It is to be emphasized that torsional stresses must be taken into consideration when they exist on the beam cross-section, in addition to the stresses due to shear forces (see Chapter 15).

8.5   Beams with Web Reinforcement

The common types of web reinforcement are

- stirrups, which can be placed either perpendicular to the longitudinal reinforcement or inclined to the longitudinal reinforcement, usually making a 45° angle and welded to the main reinforcement
- bent bars, which are part of the longitudinal reinforcement, bent up (where they are no longer needed) at an angle of 30° to 60°, usually at 45°
- combinations of stirrups and bent bars

The shear strength of a reinforced concrete beam is increased by the use of web reinforcement. Prior to the formation of diagonal tension cracks, web reinforcement contributes very little to the shear resistance. After diagonal cracks have developed, web reinforcement augments the shear resistance of a beam, and a redistribution of internal forces occurs at the cracked section. When the amount of web reinforcement provided is small, failure due to yielding of web steel may be expected, but if the amount of web reinforcement is too high, a shear-compression failure may be expected, which should be avoided.

Concrete, stirrups, and bent bars act together to resist transverse shear. The concrete, by virtue of its high compressive strength, acts as the diagonal compression member of a lattice girder system where the stirrups act as vertical tension members. The diagonal compression force is such that its vertical component is equal to the tension force in the stirrup. Bent-up reinforcement acts also as tension members in a truss, as shown in Figure 8.8.

In general, the contribution of shear reinforcement to the shear strength of a reinforced concrete beam can be described as follows:[2]

- It resists part of the shear V_s.
- It increases the magnitude of the interface shear V_a (Figure 8.7) by resisting the growth of the inclined crack.
- It increases the dowel force V_d (Figure 8.7) in the longitudinal bars.
- The confining action of the stirrups on the compression concrete may increase its strength.
- The confining action of stirrups on the concrete increases the rotation capacity of plastic hinges that develop in indeterminate structures at ultimate load, and increases the length over which yielding takes place.[7]

The total nominal shear strength of beams with web reinforcement V_n is due partly to the shear strength attributed to the concrete V_c and partly to the shear strength contributed by the shear reinforcement V_s:

$$V_n = V_c + V_s \qquad\qquad (8.15)$$

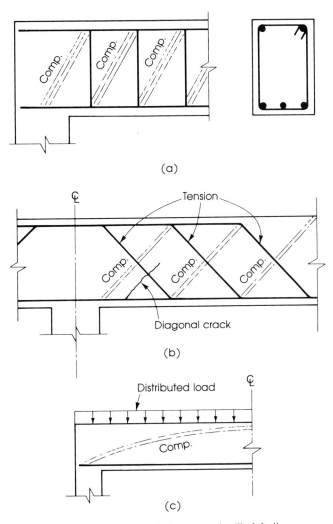

(a)

(b)

(c)

FIGURE 8.8 Truss action of web reinforcement with (a) stirrups,
(b) bent bars, and (c) tension steel.

The shear force V_u produced by factored loads must be less than or equal to the total nominal shear strength V_n, or

$$V_u \leq \phi V_n \leq \phi(V_c + V_s) \tag{8.16}$$

where $V_u = 1.4V_D + 1.7V_L$ and $\phi = 0.85$.

An expression for V_s may be developed from the truss analogy (Figure 8.9). For a 45° crack and a series of inclined stirrups or bent bars, the vertical shear force V_s resisted by shear reinforcement is equal to the sum of the vertical components of the tensile forces developed in the inclined bars. Therefore

$$V_s = nA_v f_y \sin \alpha \tag{8.17}$$

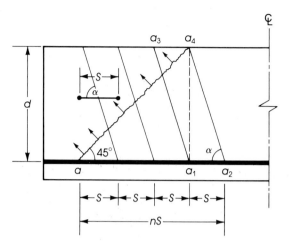

FIGURE 8.9. Factors in inclined shear reinforcement.

where A_v = area of shear reinforcement with a spacing s and f_y = yield strength of shear reinforcement. But ns is the distance a–a_1–a_2:

$$d = a_1\text{–}a_4 = a\text{–}a_1 \tan 45° \text{ (from triangle a–}a_1\text{–}a_4\text{)}$$

and

$$d = a_1\text{–}a_4 = a_1\text{–}a_2 \tan \alpha \text{ (from triangle }a_1\text{–}a_2\text{–}a_4\text{)}$$
$$ns = a\text{–}a_1 + a_1\text{–}a_2$$
$$= d(\cot 45° + \cot \alpha) = d(1 + \cot \alpha)$$

or

$$n = \frac{d}{s}(1 + \cot \alpha)$$

Substituting this value in equation 8.17,

$$V_s = \frac{A_v f_y d}{s} \sin \alpha(1 + \cot \alpha)$$

$$V_s = \frac{A_v f_y d}{s}(\sin \alpha + \cos \alpha) \tag{8.18}$$

But $V_s = v_s bd$. Therefore, in terms of shear stress,

$$v_s = \frac{A_v f_y}{bs}(\sin \alpha + \cos \alpha) \tag{8.19}$$

For the case of vertical stirrups, $\alpha = 90°$ and

$$V_s = \frac{A_v f_y d}{s} \quad \text{or} \quad s = \frac{A_v f_y d}{V_s} \tag{8.20}$$

In the case of T-sections, b is replaced by the width of web b_w in all shear equations. In terms of shear stress,

$$v_s = \frac{V_s}{bd} = \frac{A_v f_y}{b_w s} \qquad (8.21)$$

When $\alpha = 45°$, equation 8.18 becomes

$$V_s = 1.414 \frac{A_v f_y d}{s} \quad \text{or} \quad s = \frac{1.414 A_v f_y d}{V_s} \qquad (8.22)$$

For a single bent bar or group of parallel bars in one position, the shearing force resisted by steel is

$$V_s = A_v f_y \sin \alpha \quad \text{or} \quad A_v = \frac{V_s}{f_y \sin \alpha} \qquad (8.23)$$

For $\alpha = 45°$,

$$A_v = 1.414 \frac{V_s}{f_y} \qquad (8.24)$$

8.6   ACI Code Design Requirements

8.6.1   Critical Section for Nominal Shear Strength Calculation

The **ACI Code, section 11.1.2,** permits taking the critical section for nominal shear strength calculation at a distance d from the face of the support. This recommendation is based on the fact that the first inclined crack is likely to form within the shear span of the beam and some distance d away from the support. The distance d is also based on experimental work and appeared in the testing of the beams discussed in Chapter 3.

The code also specifies that shear reinforcement must be provided between the face of the support and the distance d, using the same reinforcement adopted for the critical section.

8.6.2   Minimum Area of Shear Reinforcement

To ensure that the stirrups will have sufficient strength to resist the diagonal tension in concrete, the **ACI Code, section 11.5.5,** requires all stirrups to have a minimum shear reinforcement area A_v equal to

$$A_v = \frac{50 b_w s}{f_y} \qquad (8.25)$$

where b_w = width of web and s = spacing of stirrups. The minimum amount of shear reinforcement is required whenever V_u exceeds $\phi V_c/2$, except in

- slabs and footings
- concrete floor joist construction

- beams where the total depth does not exceed 10 in., 2.5 times the flange thickness for T-shaped flanged sections, or one-half the web width, whichever is greatest

It is a common practice to increase the depth of a slab, footing, or shallow beam to increase its shear capacity. Stirrups may not be effective in shallow members because their compression zones have relatively small depths and may not satisfy the anchorage requirements of stirrups. For beams that are not shallow, reinforcement is not required when V_u is less than $\phi V_c/2$.

8.6.3   Maximum Shear Carried by Web Reinforcement V_s

To prevent a shear-compression failure where the concrete may crush due to high shear and compressive stresses in the critical region on top of a diagonal crack, the **ACI Code, section 11.5.6.8**, requires that V_s not exceed $(8\sqrt{f_c'})b_w d$. This is equivalent to a maximum stress $v_s \leq 8\sqrt{f_c'}$.

8.6.4   Maximum Spacing of Stirrups

To ensure that a diagonal crack will always be intersected by at least one stirrup, the **ACI Code, section 11.5.4.2**, requires that the spacings between stirrups not exceed $d/2$, provided that $V_s \leq (4\sqrt{f_c'})b_w d$. This is based on the assumption that a diagonal crack develops at 45° and extends a horizontal distance of about d. In regions of high shear where V_s exceeds $(4\sqrt{f_c'})b_w d$, the maximum spacing between stirrups must not exceed $d/4$. This limitation is necessary to ensure that the diagonal crack will be intersected by at least three stirrups. When V_s exceeds the maximum value of $8\sqrt{f_c'}$, the above limitation of maximum stirrup spacing does not apply, and the dimensions of the concrete cross-section should be increased.

Another limitation for the maximum spacing of stirrups may also be obtained from the condition of minimum area of shear reinforcement. A minimum A_v is obtained when the spacing s is maximum, or maximum $s = A_v f_y/50b_w$. The least value of all maximum spacings must be adopted. The ACI Code maximum spacing requirement ensures closely spaced stirrups that hold the longitudinal tension steel in place within the beam, thereby increasing their dowel capacity V_d (Figure 8.7).

8.6.5   Yield Strength of Shear Reinforcement

The **ACI Code, section 11.5.2**, requires that the yield strength of shear reinforcement not exceed 60 ksi (420 MPa). The reason behind this limitation is to limit the crack width caused by the diagonal tension and to ensure that the sides of the crack remain in close contact to improve the interface shear transfer V_a (Figure 8.7).

8.6.6   Anchorage of Stirrups

The **ACI Code, section 12.13**, requires that web reinforcement must be carried as high as possible into the compression zone. Ends of single legs, simple U-stirrups, or multiple

U-stirrups must be anchored by adding standard hooks or embedment $d/2$ above or below mid-depth on the compression side of the member for a development length $l_d \geq 24d_b \geq 12$ in., or by bending stirrups around the longitudinal steel bars to reduce the high local bearing stress under the hook. The ACI Code also requires that each bend in the continuous portion of a simple U-stirrup enclose a longitudinal bar.

8.6.7   Stirrups Adjacent to the Support

The **ACI Code, section 11.1.2,** specifies that shear reinforcement provided between the face of the support and the critical section at a distance d from it may be designed for the same shear V_u at the critical section. It is common practice to place the first stirrup at a distance $s/2$ from the face of the support, where s is the spacing calculated by equation 8.20 for V_u at the critical section.

8.6.8   Effective Length of Bent Bars

Only the center three-fourths of the inclined portion of any longitudinal bar shall be considered effective for shear reinforcement. This means that the maximum spacing of bent bars is $0.75(d - d')$. From Figure 8.10,

$$\text{Effective length of bent bar} = \frac{0.75(d - d')}{\sin 45°} = 0.75(1.414)(d - d')$$

The maximum spacing s, the horizontal projection of the effective length,

$$s_{max} = 0.707[0.75(1.414)(d - d')] = 0.75(d - d')$$

The ACI Code requirements for the design of shear reinforcement are summarized in Table 8.2.

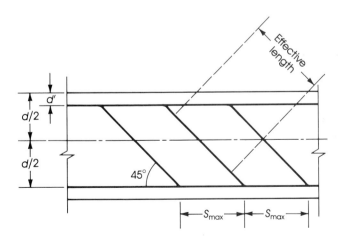

FIGURE 8.10. Effective length and spacing of bent bars.

TABLE 8.2. Shear and web reinforcement formulas

U.S. customary units	*S.I. units*
$V_u = v_u b_w d$	$V_u = v_u b_w d$

(Maximum design V_u is at a distance d from the face of the support)

$$v_c = 2.0\sqrt{f_c'} \quad \text{or}$$

$$v_c = 1.9\sqrt{f_c'} + \left(2500\rho_w \frac{V_u d}{M_u}\right)$$

$$v_c = 0.17\sqrt{f_c'} \quad \text{or}$$

$$v_c = 0.16\sqrt{f_c'} + \left(17.2\rho_w \frac{V_u d}{M_u}\right)$$

where

$$\rho_w = A_s/b_w d$$
$$V_u d/M_u \le 1.0$$
$$v_c \le 3.5\sqrt{f_c'}$$
$$V_c = v_c b_w d$$

where

$$\rho_w = A_s/b_w d$$
$$V_u d/M_u \le 1.0$$
$$v_c \le 0.29\sqrt{f_c'}$$
$$V_c = v_c b_w d$$

Vertical stirrups

$$A_v = \frac{v_s b_w s}{f_y} = \frac{V_s s}{f_y d}$$

$$s = \frac{A_v f_y}{v_s b_w} = \frac{A_v f_y d}{V_s}$$

$$\phi V_s = V_u - \phi V_c$$

$$V_s = v_s b_w d$$

$$A_v = \frac{v_s b_w s}{f_y} = \frac{V_s s}{f_y d}$$

$$s = \frac{A_v f_y}{v_s b_w} = \frac{A_v f_y d}{V_s}$$

$$\phi V_s = V_u - \phi V_c$$

$$V_s = v_s b_w d$$

Series of bent bars or inclined stirrups

$$A_v = \frac{V_s s}{f_y d(\sin \alpha + \cos \alpha)}$$

$$A_v = \frac{V_s s}{f_y d(\sin \alpha + \cos \alpha)}$$

For $\alpha = 45°$,

$$A_v = \frac{0.7 V_s s}{f_y d}$$

$$s = \frac{1.414 A_v f_y d}{V_s}$$

For $\alpha = 45°$,

$$A_v = \frac{0.7 V_s s}{f_y d}$$

$$s = \frac{1.414 A_v f_y d}{V_s}$$

For a single bent bar or group of bars, parallel and bent in one position

$$A_v = \frac{V_s}{f_y \sin \alpha}$$

$$A_v = \frac{V_s}{f_y \sin \alpha}$$

TABLE 8.2. (continued)

U.S. customary units	S.I. units
For $\alpha = 45°$,	For $\alpha = 45°$,

$$A_v = \frac{1.414 V_s}{f_y} \qquad\qquad A_v = \frac{1.414 V_s}{f_y}$$

$$V_s \leq (3.0\sqrt{f_c'})b_w d \qquad\qquad V_s \leq (0.25\sqrt{f_c'})b_w d$$

$$\phi V_s = V_u - \phi V_c \qquad\qquad \phi V_s = V_u - \phi V_c$$

$$\text{Minimum } A_v = \frac{50 b_w s}{f_y} \qquad\qquad \text{Minimum } A_v = \frac{0.35 b_w s}{f_y}$$

$$\text{Maximum } s = \frac{A_v f_y}{50 b_w} \qquad\qquad \text{Maximum } s = \frac{A_v f_y}{0.35 b_w}$$

For vertical web reinforcement

Maximum $s = d/2 \leq 24$ in.	Maximum $s = d/2 \leq 600$ mm
if $V_s \leq 4.0\sqrt{f_c'}\,(b_w d)$	if $V_s \leq 0.33\sqrt{f_c'}\,(b_w d)$
Maximum $s = d/4$	Maximum $s = d/4$
if $V_s > 4.0\sqrt{f_c'}\,(b_w d)$	if $V_s > 0.33\sqrt{f_c'}\,(b_w d)$
$V_s \leq 8\sqrt{f_c'}\,(b_w d)$	$V_s \leq 0.67\sqrt{f_c'}\,(b_w d)$

Otherwise increase the dimensions of the section

EXAMPLE 8.3

A 20-ft span simply supported beam carries uniformly distributed dead and live loads of 3 k/ft and 1.3 k/ft, respectively. The dimensions of the beam section and the steel reinforcement are as shown in Figure 8.11. Check the section for shear and design the necessary web reinforcement. Given: $f_c' = 3$ ksi and $f_y = 40$ ksi.

Solution

1. Ultimate load $= 1.4D + 1.7L$
   $$= 1.4 \times 3 + 1.7 \times 1.3 = 6.41 \text{ k/ft}$$

2. Ultimate shearing force at support $= 6.41 \times 10 = 64.1$ k
3. Maximum design shear V_u is at a distance d from the support:

$$V_{ud} \text{ (at distance } d) = 64.1 - \frac{22}{12} \times 6.41 = 52.3 \text{ k}$$

$$V_u = \phi(V_c + V_s) \qquad \text{where } \phi = 0.85$$

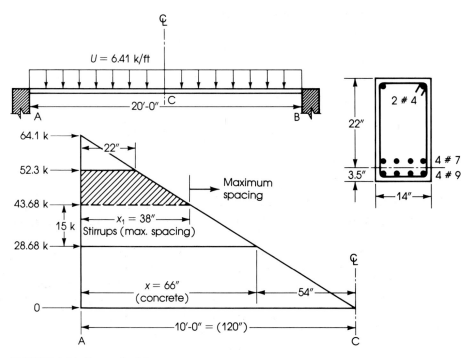

FIGURE 8.11. Example 8.3.

4. The nominal shear strength provided by concrete

$$V_c = (2\sqrt{f_c'})bd = 2\sqrt{3000}\,(14 \times 22) = 33.74 \text{ k}$$
$$\phi V_c = 0.85 \times 33.74 = 28.68 \text{ k}$$

5. Distance from the face of the support at which shear strength ϕV_c is 28.68 k is equal to

$$x = \frac{(64.1 - 28.68)}{64.1}(10) = 5.5 \text{ ft} = 66 \text{ in.}$$

   from the triangles of the shear diagram.

6. Nominal shear strength to be provided by web reinforcement:

$$\phi V_s = V_u - \phi V_c = 52.3 - 28.68 = 23.62 \text{ k}$$
$$V_s = \frac{23.62}{0.85} = 27.78 \text{ k}$$

7. Design of stirrups:
   Choose No. 3 stirrups, 2 branches (legs),

$$A_v = 2 \times 0.11 = 0.22 \text{ in.}^2$$

   The two vertical legs of the stirrups will resist shear.

$$\text{Spacing } s = \frac{A_v f_y d}{V_s} = \frac{0.22 \times 40 \times 22}{27.78} = 6.96 \text{ in.}$$

Check maximum spacing of stirrups:

Maximum $s = d/2 = 22/2 = 11$ in.

or

$$s_{max} = \frac{A_v f_y}{50 b_w} = \frac{0.22 \times 40,000}{50 \times 14} = 12.6 \text{ in.}$$

Check for maximum spacing $(d/4)$ $V_s \leq (4\sqrt{f_c'})b_w d$:

$$(4\sqrt{f_c'})b_w d = (4\sqrt{3000}) \times 14 \times 22 = 67.5 \text{ k}$$
$$V_s = 27.78 \text{ k} < 67.5 \text{ k}$$

Therefore, use No. 3 stirrups spaced at 6.5 in. Note that 6.5 in. is the minimum value of all spacings. A spacing of 7 in. may be used, which is very close to the calculated spacing of 6.96 in.

8. The web reinforcement, No. 3 stirrups at 6.5 in., will only be needed for a distance of 22 in. from the face of the support. After that, the shear force V_s decreases to zero at a distance $x = 66$ in. (see Figure 8.11). Therefore, the spacing of the stirrups may be increased as required by the shear force diagram. It is not practical to provide stirrups at different spacing, due to extra workmanship and cost. One simplification is to determine the distance from the face of the support where maximum spacing can be used, and then only two different spacings may be adopted.

Maximum spacing $= d/2 = 11$ in.

$$V_s \text{ (for } s_{max} = 11 \text{ in.)} = \frac{A_v f_y d}{V_s} = \frac{0.22 \times 40 \times 22}{11} = 17.6 \text{ k}$$

$$\phi V_s = 0.85 \times 17.6 = 15 \text{ k}$$

Distance from the face of the support where s_{max} (11 in.) can be used (from the triangles of shear forces):

$$x_1 = [64.1 - (28.68 + 15.0)]\left(\frac{10}{64.1}\right) = 3.18 \text{ ft} = 38 \text{ in.}$$

Then, for 38 in. from the face of each support, use No. 3 stirrups at 6.5 in. and for the rest of the beam, minimum web reinforcement (or stirrups with maximum spacings) can be used. In this way two different spacings are used giving a reasonable as well as practical solution. The first stirrup is usually placed at a distance $s/2$ from the face of the support.

9. Distribution of stirrups: place the first stirrup at $s/2 = 3.25$ in., say 3 in. Place 6 stirrups at 6.5 in. This adds up to 42 in., more than the necessary 38 in. Place 7 stirrups at 11 in. for another 77 in. The total distance is 119 in. Place the last stirrup at midspan; that is, put the seventh of the 11-in.-spaced stirrups at 12 in., giving a total distance of 120 in. up to the midspan of the beam. A similar arrangement is used on the other side of the beam. It can be seen from Figure 8.11 that for a distance of 54 in. from the centerline of the beam no web reinforcement is required, but the ACI Code specifies a minimum web reinforcement in reinforced concrete beams as calculated in this example. Figure 8.12 shows the distribution of stirrups. The actual area of web reinforcement needed (excluding minimum web rein-

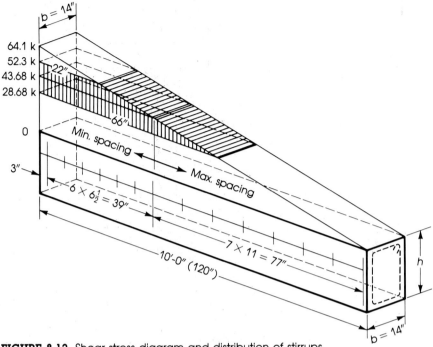

FIGURE 8.12. Shear stress diagram and distribution of stirrups (example 8.3).

forcement) is equal to the volume of the hatched parts divided by f_y:

$$A_v = \frac{V_s s}{f_y d}$$

$$\text{Total } A_v = \frac{V_s}{f_y d} \sum s = \frac{\text{Area of hatched part } (\Sigma V_s s)}{f_y d}$$

$$\text{Total } A_v \text{ required} = \frac{22 \times 27.78 + 1/2 \times 44 \times 27.78}{40 \times 22} = 1.4 \text{ in.}^2$$

If No. 3 stirrups are used each of which has $A_v = 0.22$ in.², then the number of stirrups actually required is 1.4/0.22 = 6.36, say 7 stirrups, compared to the total number used of 13. The additional stirrups cover the minimum web reinforcement specified by the ACI Code, and they provide additional shear resistance (see Figure 8.13). ■

8.7   Shear Resistance Diagram

A shear resistance diagram may be drawn showing the required V_s and the shear resisted by the provided stirrups. In Figure 8.13, the excess shear which must be resisted by web reinforcement is plotted versus the distance from the face of the support. The results of example 8.3 are shown in Figure 8.13 to illustrate the shear resistance and spacing diagrams.

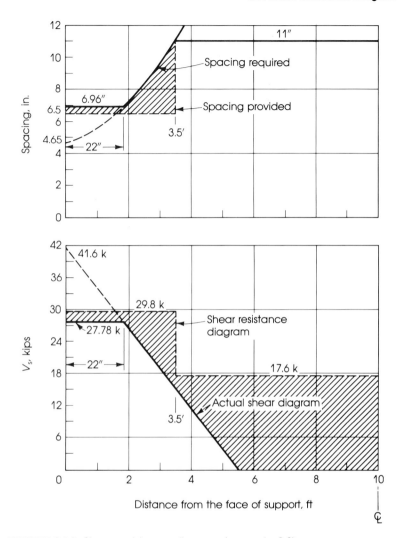

FIGURE 8.13. Shear resistance diagram (example 8.3).

In the shear resistance diagram, the maximum shear V_s is equal to 27.78 k at a distance d from the face of the support. Then the shear V_s decreases to zero at a distance 5.5 ft from the face of the support (see Figure 8.11). Two spacings of the stirrups were adopted, one at 6.5 in. for a distance of 42 in. (3.5 ft), and the second at 11 in. for the other part of half the beam up to midspan. The shear resistance values of the stirrups can be calculated from equation 8.20:

$$V_s = \frac{A_v f_y d}{s} = \frac{0.22 \times 40 \times 22}{s} = \frac{193.6}{s}$$

$$\phi V_s = \frac{164.6}{s}$$

For $s = 6.5$ in., $V_s = 29.8$ k, $\phi V_s = 25.33$ k ($\phi = 0.85$), and for $s = 11$ in., $\phi V_s = 15$ k and $V_s = 17.6$ k, which is plotted in Figure 8.13. It can be seen that the additional shear resistance provided is presented by the hatched area, between the actual shear stress required and the shear resistance provided. From the diagram, it is obvious that other spacing arrangements may be used as long as the actual shear lies within the shear resistance diagram. The top part of Figure 8.13 shows the curve relating the spacings of the stirrup to the distance from the face of the support. Once the required spacing is calculated at some section, at the support for instance, then the required spacings elsewhere can be determined easily.

At the face of the support, the spacing is

$$s_0 = \frac{164.6}{\phi V_s} = \frac{164.6}{35.42} = 4.65 \text{ in.}$$

For a uniform load, ϕV_s is a linear function of the distance from the point of zero (ϕV_s), 5.5 ft from the face of the support. Therefore, the spacings at 0.5, 1, 1.5, 2, 3, 4, and 5 ft from the face of the support are calculated as follows:

$$s_{0.5} = s_0 \times \frac{5.5}{5.0} = 4.65 \times \frac{5.5}{5} = 5.1 \text{ in.}$$

$$s_1 = 4.65 \times \frac{5.5}{4.5} = 5.7 \text{ in.} \qquad s_{1.5} = 4.65 \times \frac{5.5}{4} = 6.4 \text{ in.}$$

$$s_2 = 4.65 \times \frac{5.5}{3.5} = 7.3 \text{ in.} \qquad s_3 = 4.65 \times \frac{5.5}{2.5} = 10.3 \text{ in.}$$

$$s_4 = 4.65 \times \frac{5.5}{1.5} = 17.0 \text{ in.} \qquad s_5 = 4.65 \times \frac{5.5}{0.5} = 51 \text{ in.}$$

These values of the spacing are plotted in the upper part of Figure 8.13. Minimum spacing required is that for $\phi V_s = 23.62$ k, which is 6.96 in., and maximum spacing is 11 in. ($d/2$), as explained in example 8.3. The provided spacings are 6.5 in. and 11 in., respectively, which are below or the same as the required spacings. Any variation in spacing is acceptable as long as the value s is below the "spacing required" curve, and as long as the shear resistance (in the lower part of Figure 8.13) is greater than the actual shear force.

In most cases, it is unnecessary to plot Figure 8.13 unless shear is an important factor in the design and different spacings are adopted.

EXAMPLE 8.4

A rectangular beam is to be designed to carry an ultimate shear force of 25 k. Determine the minimum beam cross-section if controlled by shear, and neglect the shearing force resisted by the minimum web reinforcement specified by the ACI Code. Given: $f'_c = 4$ ksi, $f_y = 60$ ksi, and $b = 11$ in.

Solution

Since the shearing force controls the design of the beam section, then the concrete will resist the total shear. Ultimate shear stress provided by concrete

$$v_c = 2\sqrt{f'_c} = 126 \text{ psi}$$

The ultimate shear stress on the section

$$v_u = V_u/\phi bd$$

or

$$bd = V_u/\phi v_u = V_u/\phi v_{uc} = \frac{25,000}{0.85 \times 126} = 235 \text{ in.}^2$$

$$d = 235/11 = 21.3 \text{ in.}$$

Therefore, a section 11 by 24 in. will be adequate. Use minimum stirrups according to ACI Code requirements, as explained earlier. ∎

EXAMPLE 8.5

Design the necessary web reinforcement for a 14-ft simply supported beam which carries a uniform ultimate load of 8 k/ft and an ultimate concentrated load at midspan of 20 k. The beam section is limited to $b = 12$ in. and $d = 20$ in. Given: $f'_c = 3$ ksi and $f_y = 40$ ksi.

Solution

1. Shearing force at the support is $8 \times 7 + 10 = 66$ k.
2. Maximum design shear V_u is at a distance d from the support,

$$V_u = 66 - 8 \left(\frac{20}{12}\right) = 52.67 \text{ k}$$

3. Allowable shear stress

$$v_c = 2\sqrt{f'_c} = 2\sqrt{3000} = 109 \text{ psi}$$

Shear strength provided by concrete

$$\phi V_c = \phi v_c bd = 0.85 \times 0.109 \times 12 \times 20 = 22.24 \text{ k}$$

4. Shear force at midspan is 10 k.
5. Distance from the face of the support at which shear force is 22.24 k:

$$x = \frac{(66 - 22.24)}{(66 - 10)} (84) = 65.4 \text{ in. (Figure 8.14)}$$

6. Maximum shear force to be resisted by web reinforcement

$$\phi V_s = 52.67 - 22.24 = 30.43 \text{ k}$$

$$V_u = \phi(V_c + V_s) \qquad V_s = \frac{30.43}{0.85} = 35.76 \text{ k}$$

7. Design of stirrups:
   Choose No. 3 stirrups, two branches, $A_v = 2 \times 0.11 = 0.22$ in.^{2}

$$\text{Spacing } s = \frac{A_v f_y d}{V_s} = \frac{0.22 \times 40 \times 20}{35.76} = 4.92 \text{ in.}$$

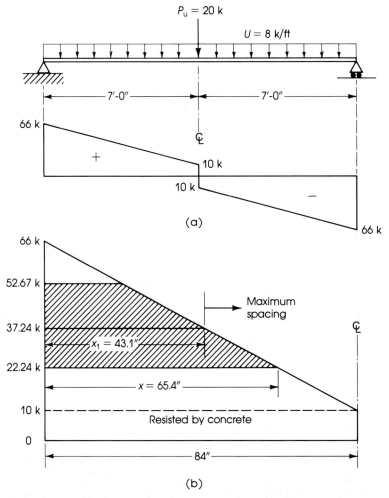

FIGURE 8.14. (a) Shear force diagram and (b) calculation of reinforcement spacing (example 8.5).

Check maximum spacing of stirrups:

Maximum $s = d/2 = 20/2 = 10$ in.

or

$$s = \frac{A_v f_y}{50 b_w} = \frac{0.22 \times 40,000}{50 \times 12} = 14.6 \text{ in.}$$

Check for maximum spacing of $d/4$:

$$V_s = 35.76 \text{ k} < 4\sqrt{f_c'}\, b_w d < 4\sqrt{3000} \times 12 \times 20 < 52.6 \text{ k}$$

Therefore, the use of No. 3 stirrups spaced at 4.5 in. is adequate.

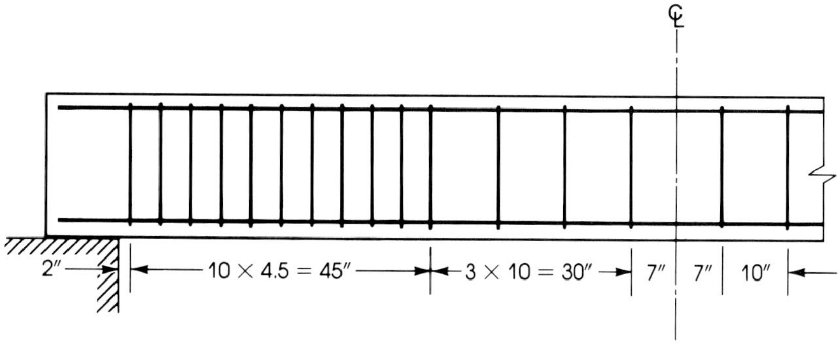

FIGURE 8.15. Distribution of stirrups (example 8.5).

8. Determine the distance from the face of the support at which maximum spacing of 10 in. can be used:

$$V_s \text{ (for maximum } s) = \frac{A_v f_y d}{s} = \frac{0.22 \times 40 \times 20}{10} = 17.6 \text{ k}$$

Distance from the face of the support where $s = 10$ in. can be used ($\phi V_s = 15$ k):

$$X_1 = \frac{(66 - 37.24)}{(66 - 10)} \times 84 = 43.1 \text{ in.}$$

Thus, for 43.1 in. from the face of each support, No. 3 stirrups at 4.5-in. spacings will be used and for the rest of the beam, No. 3 stirrups at 10 in. will be adequate.
9. Distribution of stirrups (Figure 8.15):
Place the first stirrup at a distance $s/2$ from the face of the support, at 2.0 in. Place 10 stirrups at 4.5 in. The total distance is 47.0 in., greater than the 43.1 in. required. Place 3 stirrups at 10 in., for a total distance of 77.0 in. The distance left to midspan is 7.0 in.; therefore add another stirrup at midspan. Always adjust the number of stirrups according to the spacings as the last stirrup may be placed at midspan of the beam. ∎

EXAMPLE 8.6

Redesign the beam of example 8.3 using a combination of stirrups and bent bars. Given: $f'_c = 3$ ksi, $f_y = 40$ ksi, $b = 14$ in., $d = 22$ in., and A_s is 4 No. 9 and 4 No. 7 bars.

Solution   The calculations for the shear forces will be the same as in example 8.3.
1. Uniform ultimate load = 6.41 k/ft

$$V_u \text{ (at support)} = 64.1 \text{ k}$$
$$V_u \text{ (at a distance } d \text{ from the face of the support)} = 52.3 \text{ k}$$

ϕV_c (shear resisted by concrete) = 28.68 k

V_s (shear to be resisted by web reinforcement) = 27.78 k

Other calculated values are shown in Figure 8.16.

2. Design of stirrups: In the design for shear, it is desirable to have stirrups spaced uniformly all over the beam for practical reasons. The additional shear may be covered by bent bars, which are chosen from the main reinforcement of the beam and at points where they are no longer needed for moment resistance.

   Choose No. 3 stirrups (10 mm) spaced at 11 in., with the first stirrup placed at 3 in. from the face of the support.

$$A_v = 2 \times 0.11 = 0.22 \text{ in.}^2$$

Shear strength of web reinforcement:

$$V_s \text{ (stirrups)} = \frac{A_v f_y d}{s} = \frac{0.22 \times 40 \times 22}{11} = 17.6 \text{ k}$$

The shear force which must be resisted by bent bars is $27.78 - 17.6 = 10.18$ k. Maximum spacing of stirrups:

$$s_{max} = d/2 = 22/2 = 11 \text{ in.}$$

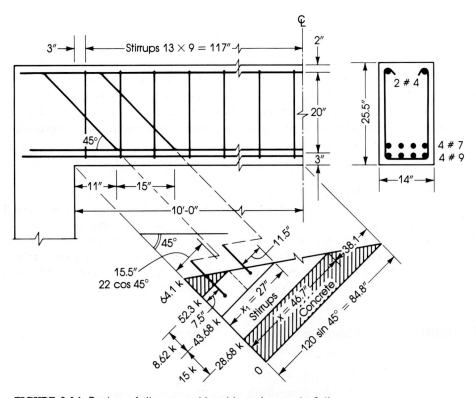

FIGURE 8.16. Design of stirrups and bent bars (example 8.6).

or

$$s_{max} = \frac{A_v f_y}{50 b_w} = \frac{0.22 \times 40{,}000}{50 \times 14} = 12.6 \text{ in.}$$

Spacing of 11 in. is adequate.

$$\text{Number of stirrups} = 2 \left(1 + \frac{117}{11} \right) = 24 \text{ stirrups}$$

3. Bent bars:

$$x_1 = \frac{(64.1 - 43.68)}{64.1} (84.8) = 27 \text{ in. (Figure 8.16)}$$

For $\alpha = 45°$, $\sin 45° = 0.707$.

$$A_v = \frac{V_s s}{f_y d (\sin \alpha + \cos \alpha)} = \frac{V_s}{f_y d} (0.707 s)$$

The shear diagram can be drawn on a line inclined 45° from the horizontal, and A_v is equal to the volume of the shear strength diagram ($V_s \times 0.707 s$) divided by $f_y d$ (Figure 8.16). The volume can be divided into two parts, both having the same maximum shear. The first part is a rectangle 8.62 by 15.5 and the second part is a triangle $(27 - 15.5)$ by 8.62 by $\frac{1}{2}$. The width of the beam is 14 in. Therefore

$$A_v = \frac{(8.62 \times 15.5) + (8.62 \times 11.5) \times 1/2}{40 \times 22} = 0.21 \text{ in.}^2$$

From the main reinforcement of the beam, 2 No. 7 bars can be chosen ($A_s = 1.2 \text{ in.}^2$) and can be bent at the center of gravity of the two divisions as shown in Figure 8.16. The first bar is bent at a distance 7.5 in. from the face of the support on the inclined line (11 in. on the horizontal projection). The bars resist a diagonal tension equal to a width of 15 in. on the inclined force diagram and on a beam width of 14 in.; that is, diagonal tension resisted by the bars

$$A_s f_y = 1.2 \times 40 = 48 \text{ k}$$

The total diagonal tension force to be resisted by bent bars is 10.18 k (see the above calculations). Since the shear resistance provided by the bent bars is greater than that required, one bent bar or smaller bars (if available in the section) may be used.

It is recommended to draw the shear diagram to scale and measure all the dimensions required. A moment resistance diagram can also be drawn to determine the points where bent bars are no longer needed for bending. The ACI Code gives the following equations to calculate the spacing s or A_v for bent bars placed in series:

$$s = \frac{A_v f_y d}{V_s} (\sin \alpha + \cos \alpha)$$

For $\alpha = 45°$,

$$s = \frac{1.41 A_v f_y d}{V_s} \quad \text{and} \quad A_v = \frac{0.7 V_s s}{f_y d}$$

In this example, let the spacing of the inclined bars be 15 in.; then

$$A_v = \frac{0.7 \times 10.18 \times 15}{40 \times 22} = 0.12 \text{ in.}^2$$

If No. 7 bars, bent at 15 in. spacing, are chosen for illustration, then A_v provided is 1.2 in.^{2} and is greater than A_v required.

One should be careful in placing the first bar to avoid bending it into the supporting columns. In this example, it will be recommended to start the first bent bar at a distance $0.75d = 0.75 \times 22 = 16.5$ in. from the face of the support and bent at 45°. ∎

8.8   Shear Stresses in Members of Variable Depth

The shear stress v is a function of the effective depth d; therefore, shear stresses vary along a reinforced concrete beam with variable depth.[10] In such a beam (Figure 8.17), consider a small element dx. The compression force C at any section is equal to the moment divided by its arm,

$$C = \frac{M}{y}$$

The first derivative of C is

$$dC = \frac{y\,dM - M\,dy}{y^2}$$

If C_1 is greater than C_2,

$$C_1 - C_2 = dC = vb\,dx$$

$$vb\,dx = \frac{y\,dM - M\,dy}{y^2}$$

$$= \frac{dM}{y} - \frac{M}{y^2}\,dy$$

$$v = \frac{1}{yb}\left(\frac{dM}{dx}\right) - \frac{M}{by^2}\left(\frac{dy}{dx}\right)$$

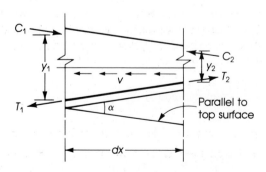

FIGURE 8.17. Shear stress in a beam with variable depth.

Since $y = jd$, dM/dx is equal to the shearing force V and $d(jd)/dx$ is the slope,

$$v = \frac{V}{bjd} - \frac{M}{b(jd)^2}\left[\frac{d}{dx}(jd)\right]$$

and

$$v = \frac{V}{bjd} \pm \frac{M}{b(jd)^2}(\tan \alpha) \tag{8.26}$$

where V and M are the external shear and moment, respectively, and α is the slope angle of one face of the beam relative to the other face. The plus sign is used when the beam depth decreases as the moment increases, while the minus sign is used when the depth increases as the moment increases or decreases. The above formula is used for small slopes such that the angle α is less than or equal to 30°.

A simpler form of equation 8.26 can be used by eliminating the j value,

$$v = \frac{V}{bd} \pm \frac{M}{bd^2}(\tan \alpha) \tag{8.27}$$

For the strength design method, the following equation may be used:

$$v_u = \frac{V_u}{\phi bd} \pm \frac{M_u}{\phi bd^2}(\tan \alpha) \tag{8.28}$$

or the shearing force

$$\phi V_n = V_u \pm \frac{M_u}{d}(\tan \alpha) \tag{8.29}$$

Figure 8.18 shows a cantilever beam with a concentrated load P at the free end. The moment and the depth d increase toward the support. In this case a negative sign is used in equations 8.27 and 8.28. Similarly, a negative sign is used for section t in the simply supported beam shown, and a positive sign is used for section Z, where moment increases as the depth decreases.

In many cases, the variation in the depth of beams occurs on parts of the beams near their supports (Figure 8.18). Tests[11] on beams with variable depth indicate that beams with greater depth at the support fail mainly by shear compression. Beams with smaller depth at the support fail generally by instability type of failure, caused by the propagation of the major crack in the beam upward and then horizontally to the beam's top section. Tests[11] also indicate that for beams with variable depth (Figure 8.18) with an inclination α of about 10°, and subjected to shear and flexure, the concrete shear strength V_{cv} may be computed by

$$V_{cv} = V_c(1 + \tan \alpha) \tag{8.30}$$

where

V_{cv} = shear strength of beam with variable depth,

V_c = **ACI Code equation 11.6** (equation 8.9 in this book)

$$= \left(1.9\sqrt{f_c'} + 2500\rho_\omega \frac{V_u d_s}{M_u}\right) b_w d_s \leq 3.5\sqrt{f_c'}\, b_w d_s$$

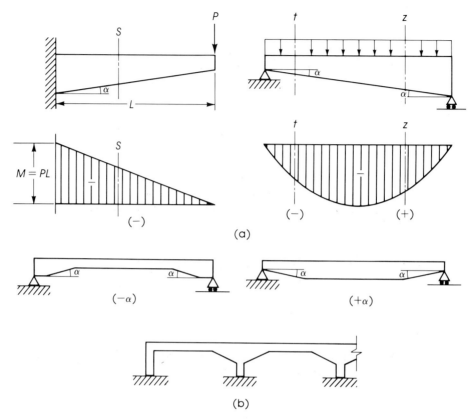

FIGURE 8.18. Beams with variable depth: (a) moment diagrams and (b) typical forms.

α = inclination of beam at the support, considered positive for beams of small depth at the support and negative for beams with greater depth at the support (Figure 8.18)

d_s = effective depth of the beam at the support

The simplified **ACI Code equation 11.3** can also be used to compute V_c:

$$V_c = (2\sqrt{f_c'})\, b_w d_s \tag{8.31}$$

Tests also indicate that the variation in the depth of beams has no influence on the shear capacity of vertical stirrups.

EXAMPLE 8.7

Design the cantilever beam shown in Figure 8.19 under the ultimate loads applied if the total depth at the free end is 12 in. and it increases toward the support. Use a steel percentage $\rho = 1.5$ percent, $f_c' = 3$ ksi, $f_y = 50$ ksi, and $b = 10$ in.

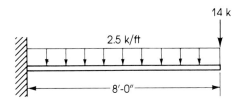

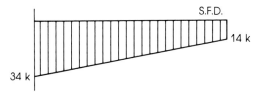

FIGURE 8.19. Example 8.7, with bending moment diagram (middle) and shear force diagram (bottom).

Solution

1. M_u (at the support) $= \dfrac{2.5}{2} \times (8)^2 \times 12 + 14 \times 8 \times 12 = 2304$ k·in.

2. For $\rho = 1.5$ percent, $R_u = 576$ psi (from tables).

$$d = \sqrt{\frac{M}{R_u b}} = \sqrt{\frac{2304}{0.576 \times 10}} = 20 \text{ in.}$$

$$A_s = 0.015 \times 10 \times 20 = 3 \text{ in.}^2 \quad \text{(use 4 No. 8 or 3 No. 9 bars)}$$

3. Design for shear:

Maximum shear at the support is $14 + 20 = 34$ k. Since the beam section is variable, moment effect shall be considered; and since the beam depth increases as the moment increases, a minus sign will be used in equation 8.28.

$$v_u = \frac{V_u}{\phi b d} - \frac{M_u}{\phi b d^2} (\tan \alpha)$$

To find $\tan \alpha$, let d at the free end $= 10$ in., and d at the support $= 20$ in.:

$$\tan \alpha = \frac{(20 - 10)}{8 \times 12} = 0.104$$

$$v_u \text{ (at the support)} = \frac{34,000}{(0.85 \times 10 \times 20)} - \frac{2304 \times 1000 \times 0.104}{[0.85 \times 10 \times (20)^2]} = 130 \text{ psi}$$

4. Shear stress at the free end $= V_u/\phi bd$ $(M_u = 0)$

$$= \frac{14{,}000}{0.85 \times 10 \times 10} = 165 \text{ psi}$$

5. Since at a distance (d) from the face of the support, the effective depth is less than 20 in., try $d = 18$ in.

$$d \text{ (at 18 in. from support)} = 10 + \frac{78}{96} \times 10 = 18.1 \text{ in.}$$

$$V_u = 34 - 2.5 \times \frac{18}{12} = 30 \text{ k}$$

$$M_u \text{ (at 18 in. from support)} = 14 \times 78 + \frac{2.5}{12} \times \frac{(78)^2}{2} = 1726 \text{ k·in.}$$

$$v_u = \frac{30{,}000}{0.85 \times 10 \times 18.1} - \frac{1726 \times 1000 \times 0.104}{0.85 \times 10 \times (18.1)^2} = 130 \text{ psi}$$

6. At midspan (48 in. from the support):

$$d = 15 \text{ in.}$$

$$V_u = 14 + 10 = 24 \text{ k}$$

$$M_u = 14 \times 48 + \frac{2.5}{12} \times \frac{(48)^2}{2} = 912 \text{ k·in.}$$

$$v_u = \frac{24{,}000}{0.85 \times 10 \times 15} - \frac{912 \times 1000 \times 0.104}{0.85 \times 10 \times (15)^2} = 139 \text{ psi}$$

Similarly, at 6 ft from the support (2 ft from the free end),

$$d = 12.5 \text{ in.} \qquad V_u = 19 \text{ k} \qquad M_u = 396 \text{ k·in.} \qquad v_u = 149 \text{ psi}$$

At 1 ft from the free end,

$$d = 11.25 \text{ in.} \qquad V_u = 16.5 \text{ k} \qquad M_u = 183 \text{ k·in.} \qquad v_u = 155 \text{ psi}$$

These values are shown in Figure 8.20.

7. Shear stress resisted by concrete $= 2\sqrt{f_c'} = 2\sqrt{3000} = 109$ psi
   Minimum shear stress to be resisted by web reinforcement,

$$v_{us} = 165 - 109 = 56 \text{ psi}$$

(v_u and consequently v_{us} have already been increased by the ratio $1/\phi$ in equation 8.28)

8. Choose No. 3 stirrups, 2 branches.

$$A_v = 2 \times 0.11 = 0.22 \text{ in.}^2$$

$$s(\text{required}) = \frac{A_v f_y}{v_s b_w} = \frac{0.22 \times 50{,}000}{56 \times 10} = 19.6 \text{ in.}$$

$$s_{max}(\text{for } d/2) = 10 \text{ in. to 5 in. at the free end}$$

$$s_{max}(\text{for minimum } A_v) = \frac{A_v f_y}{50 b_w} = \frac{0.22 \times 50{,}000}{50 \times 10} = 22 \text{ in.}$$

9. Check for maximum spacing $(d/4)$:

$$v_{us} \leq 4\sqrt{f_c'}$$

$$4\sqrt{f_c'} = 4\sqrt{3000} = 218 > 56 \text{ psi}$$

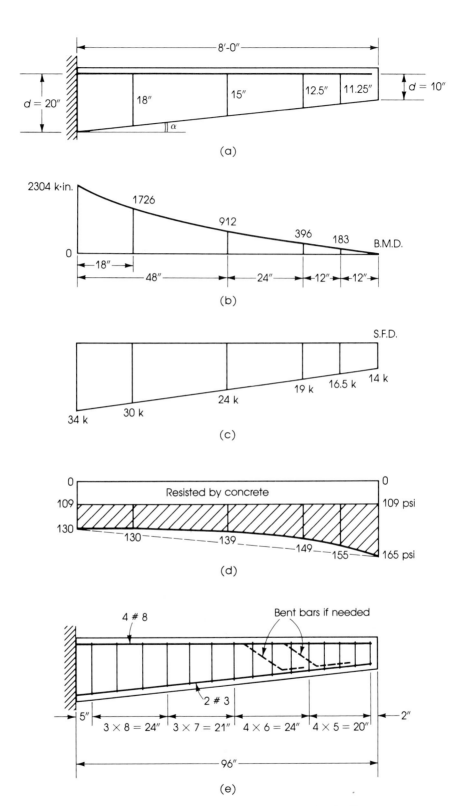

FIGURE 8.20. Web reinforcement for a beam of variable depth (example 8.7).

10. Distribution of stirrups (distances from the free end):

$$
\begin{aligned}
\text{1st stirrup at 2 in.} &= \ \ 2 \text{ in.}\\
\text{4 stirrups at 5 in.} &= 20 \text{ in.}\\
\text{4 stirrups at 6 in.} &= 24 \text{ in.}\\
\text{3 stirrups at 7 in.} &= 21 \text{ in.}\\
\text{3 stirrups at 8 in.} &= \underline{24 \text{ in.}}\\
&\ \ \ 91 \text{ in.}
\end{aligned}
$$

There is 5 in. left to the face of the support.

Bent bars may be used to take the excess shear near the free end and they may be placed as shown dotted in Figure 8.20(e).

The same design may be performed using forces instead of the stress approach used in this example. ∎

8.9   S.I. Examples

The general design requirements for web reinforcement according to the ACI Code are summarized in Table 8.2, which gives the necessary design equations in both U.S. customary and S.I. units. The following example will show the design of shear reinforcement using S.I. units.

EXAMPLE 8.8

A 6-m clear span simply supported beam carries a uniform dead load of 45 kN/m and a live load of 20 kN/m (Figure 8.21). The dimensions of the beam section are: $b = 350$ mm, $d = 550$ mm. The beam is reinforced with 4 bars 25 mm diameter in one row. It is required to design the necessary web reinforcement. Given: $f'_c = 21$ MPa and $f_y = 280$ MPa.

Solution

1. Ultimate load $= 1.4D + 1.7L$

$$= 1.4 \times 45 + 1.7 \times 20 = 97 \text{ kN/m}$$

2. Ultimate shear force at the face of the support

$$V_u = 97 \times \frac{6}{2} = 291 \text{ kN}$$

3. Maximum design shear at a distance d from the face of the support:

$$V_u(\text{at } d \text{ distance}) = 291 - 0.55 \times 97 = 237.65 \text{ kN}$$

4. The nominal shear strength provided by concrete

$$V_c = (0.17 \sqrt{f'_c})\, bd = (0.17 \sqrt{21}) \times 350 \times 550 = 150 \text{ kN}$$

$$V_u = \phi V_c + \phi V_s \qquad \phi V_c = 0.85 \times 150 = 127.5 \text{ kN}$$

$$\phi V_s = 237.65 - 127.5 = 110.15 \text{ kN}$$

$$V_s = \frac{110.15}{0.85} = 129.6 \text{ kN}$$

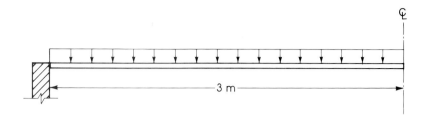

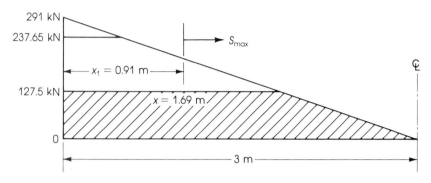

FIGURE 8.21. Example 8.8.

5. Distance from the face of the support at which $\phi V_c = 127.5$ kN

$$x = \frac{(291 - 127.5)}{291}(3) = 1.69 \text{ m (from triangles)}$$

6. Design of stirrups:

   Choose stirrups 10 mm diameter, 2 branches.

   $$A_v = 2 \times 78.5 = 157 \text{ mm}^2$$

   $$\text{Spacing } s = \frac{A_v f_y d}{V_s} = \frac{157 \times 280 \times 550}{129.6 \times 10^3} = 186.6 \text{ mm}$$

   say, 180 mm. Check maximum spacing of stirrups:

   $$\text{Maximum } s = \frac{d}{2} = \frac{550}{2} = 275 \text{ mm}$$

   or

   $$s_{max} = \frac{A_v f_y}{0.35b} = \frac{157 \times 280}{0.35 \times 350} = 359 \text{ mm}$$

   Check for maximum spacing of ($d/4$):

   $$\text{If } V_s \leq (0.33\sqrt{f_c'})bd, \quad s_{max} = \frac{d}{2}$$

   $$\text{If } V_s > (0.33\sqrt{f_c'})bd, \quad s_{max} = \frac{d}{4}$$

$$bd(0.33\sqrt{f'_c}) = 0.33\sqrt{21} \times 350 \times 550 = 291.1 \text{ kN}$$

Actual $V_s = 129.6 \text{ kN} < 291.1 \text{ kN}$

Therefore, s_{max} is limited to $d/2 = 275$ mm.

7. The web reinforcement, stirrups 10 mm diameter, spaced at 180 mm, will only be needed for a distance $d = 0.55$ m from the face of the support. Beyond that, the shear stress V_s decreases to zero at a distance $x = 1.69$ m when $\phi V_c = 127.5$ kN. It is not practical to provide stirrups at many different spacings. One simplification is to find out the distance from the face of support where maximum spacing can be used, and then only two different spacings may be adopted.

$$\text{Maximum spacing} = \frac{d}{2} = 275 \text{ mm}$$

$$V_s \text{ (for } s_{max} = 275 \text{ mm)} = \frac{A_v f_y d}{s} = \frac{157 \times 0.280 \times 550}{275} = 87.9 \text{ kN}$$

$$\phi V_s = 87.9 \times 0.85 = 74.7 \text{ kN}$$

Distance from the face of support where $s_{max} = 275$ mm can be used (from the triangles):

$$x_1 = \frac{291 - (127.5 + 74.7)}{291} (3) = 0.91 \text{ m}$$

Then, for 0.91 m from the face of support, use stirrups 10 mm diameter at 180 mm, and for the rest of the beam, minimum stirrups (with maximum spacings) can be used.

8. Distribution of stirrups:

Place first stirrup at $\frac{s}{2} = \frac{180}{2} = $ 90 mm

5 stirrups at 180 mm = 900 mm

Total = 990 mm = 0.99 m > 0.91 m

7 stirrups at 270 mm = 1890 mm

Total = 2880 mm = 2.88 m < 3 m

The last stirrup is $(3 - 2.88) = 0.12$ m = 120 mm from the centerline of the beam, which is adequate. A similar stirrup distribution applies to the other half of the beam, giving a total number of stirrups of 26.

The other examples in this chapter can be worked out in a similar way using the equations given in Table 8.2.   ∎

SUMMARY

SECTIONS 8.1 and 8.2

The shear stress in a homogeneous beam is $v = VQ/Ib$. The distribution of the shear stress above the neutral axis in a singly reinforced concrete beam is parabolic. Below the neutral axis the maximum shear stress is maintained down to the level of the steel bars.

SECTION 8.3

The development of shear resistance in reinforced concrete members occurs by

• shear resistance of the uncracked concrete
• interface shear transfer

- arch action
- dowel action

SECTION 8.4

The shear stress at which a diagonal crack is expected is

$$v_c = \frac{V}{bd} = \left(1.9\sqrt{f_c'} + 2500\rho_\omega \frac{V_u d}{M_u}\right) \le 3.5\sqrt{f_c'}$$

The nominal shear strength

$$V_c = v_c b_w d = 2\sqrt{f_c'}\, b_w d$$

SECTIONS 8.5 and 8.6

(a) The common types of web reinforcement are stirrups (perpendicular or inclined to the main bars), bent bars, or combinations of stirrups and bent bars.

$$V_u = \phi V_n = \phi V_c + \phi V_s \quad \text{and} \quad V_s = \frac{1}{\phi}(V_u - \phi V_c)$$

(b) The ACI Code design requirements are summarized in Table 8.2.

SECTIONS 8.7 and 8.8

(a) A shear resistance diagram may be drawn as shown in Figure 8.13, when shear is a major design force acting on the structure.
(b) For members with variable depth,

$$v_u = \frac{V_u}{\phi bd} \pm \frac{M_u}{\phi bd^2}(\tan\alpha) \tag{8.28}$$

REFERENCES

1. "Report of the ACI-ASCE Committee 326," *ACI Journal*, 59, 1962.
2. ACI-ASCE Committee 426: "The Shear Strength of Reinforced Concrete Members," *ASCE Journal, Structural Division*, June 1973.
3. R. C. Fenwick and T. Paulay: "Mechanisms of Shear Resistance of Concrete Beams," *ASCE Journal, Structural Division*, October 1968.
4. G. N. Kani: "Basic Facts Concerning Shear Failure," *ACI Journal*, 63, June 1966.
5. D. W. Johnson and P. Zia: "Analysis of Dowel Action," *ASCE Journal, Structural Division*, 97, May 1971.
6. H. P. Taylor: "The Fundamental Behavior of Reinforced Concrete Beams in Bending and Shear," in *Shear in Reinforced Concrete*, vol. 1, American Concrete Institute Special Publication 42, Detroit, 1974.
7. A. L. L. Baker: *Limit-State Design of Reinforced Concrete*, Cement and Concrete Association, London, 1970.
8. P. E. Regan and M. H. Khan: "Bent-up Bars as Shear Reinforcement," in *Shear in Reinforced Concrete*, vol. 1, American Concrete Institute Special Publication 42, Detroit, 1974.
9. J. G. MacGregor: "The Design of Reinforced Concrete Beams for Shear," in *Shear in Reinforced Concrete*, vol. 2, American Concrete Institute Special Publication 42, Detroit, 1974.
10. German Code of Practice, DIN 1045.
11. S. Y. Debaiky and E. I. Elniema: "Behavior and Strength of Reinforced Concrete Beams in Shear," *ACI Journal*, 79, May–June 1982.

PROBLEMS

8.1. Calculate the maximum shear stress in a rectangular concrete section that is reinforced with 4 No. 7 bars (4 × 22 mm) and subjected to a shearing force of 15 kips (67 kN). Determine also the average shear stress. Given: $b = 11$ in. (275 mm), $d = 20$ in. (500 mm), $f'_c = 3$ ksi (21 MPa), $f_y = 40$ ksi (280 MPa), and $n = 9.0$.

8.2. An 18-ft (5.4-m) span simply supported beam carries a uniform dead load of 4 k/ft (60 kN/m) and a live load of 1.5 k/ft (22 kN/m). The beam has a width $b = 12$ in. (300 mm), a depth $d = 24$ in. (600 mm), and is reinforced with 3 No. 9 plus 3 No. 8 bars (3 × 28 mm + 3 × 25 mm). Check the section for shear and design the necessary web reinforcement. Given: $f'_c = 3$ ksi (21 MPa) and $f_y = 40$ ksi (280 MPa).

8.3. For the beam given in problem 8.2, draw the shear resistance diagram based on the web reinforcement chosen.

8.4. A rectangular beam is to be designed to carry an ultimate shearing force of 75 kips (335 kN). Determine the minimum beam section if controlled by shear, using the minimum web reinforcement as specified by the ACI Code. (Refer to Table 8.2 for minimum A_v.) Given: $f'_c = 4$ ksi (28 MPa), $f_y = 40$ ksi (280 MPa), and $b = 16$ in. (400 mm).

8.5. Redesign problem 8.2, using a combination of stirrups and bent bars. Use No. 3 stirrups (10 mm diameter) placed at maximum spacings allowed by the ACI Code.

8.6. Design the critical sections of an 11-ft (3.3-m) span simply supported beam for bending moment and shearing forces using ρ_{max} and the strength design method. Given: $f'_c = 3$ ksi (21 MPa), $f_y = 60$ ksi (420 MPa), and $b = 10$ in. (250 mm). Dead load is 2.75 k/ft (40 kN/m), live load is 1.375 k/ft (20 kN/m).

8.7. A cantilever beam of 7.4-ft (2.20-m) span carries a uniform dead load of 685 lb/ft (10 kN/m) and a concentrated live load of 18 k (80 kN) at a distance of 3 ft (0.9 m) from the face of the support. Design the beam for moment and shear using the strength design method. Given: $f'_c = 3$ ksi (21 MPa), $f_y = 60$ ksi (420 MPa), $b = 8$ in. (200 mm), and use $\rho = \frac{3}{4}\rho_{max}$.

8.8. Design the necessary web reinforcement for a 14-ft (4.2-m) simply supported beam that carries an ultimate uniform load of 5 k/ft (90 kN/m) and an ultimate concentrated load at midspan $P_u = 24$ k (108 kN). The beam has a width $b = 14$ in. (350 mm), a depth $d = 16$ in. (400 mm), and is reinforced with 4 No. 8 bars (4 × 25 mm). Given $f'_c = 4$ ksi (28 MPa), $f_y = 60$ ksi (420 MPa).

8.9. Redesign the web reinforcement of the beam in problem 8.8 if the uniform ultimate load of 6 k/ft (90 kN/m) is due to dead load and the ultimate concentrated load $P_u = 24$ k (108 kN) is due to a moving live load. Change the position of the live load to cause maximum shear at the support and at midspan.

8.10. Design a cantilever beam that has a span of 9 ft (2.7 m) to carry an ultimate triangular load which varies from zero load at the free end to maximum load of 8 k/ft (120 kN/m) at the face of the support. The beam shall have a variable depth, with minimum depth at the free end of 10 in. (250 mm) increasing linearly toward the support. Use maximum steel percentage ρ_{max} for flexural design. Given: $f'_c = 3$ ksi (21 MPa), $f_y = 60$ ksi (420 MPa) for flexural reinforcement, $f_y = 40$ ksi (280 MPa) for stirrups, and $b = 11$ in. (275 mm).

CHAPTER

9

Design of One-Way Slabs

9.1 Introduction

Reinforced concrete slabs are constructed to provide flat surfaces, usually horizontal, in building floors and roofs, bridges, and other types of structures. The slab may be supported by masonry or reinforced concrete walls, by columns, by reinforced concrete beams usually poured monolithically with the slab, by structural steel beams, or by the ground. The depth of the slab is usually very small compared to its span.

9.2 Types of Slabs

Slabs can be classified structurally as follows:

1. **One-way slabs.** If a slab is supported on two opposite sides only, it will bend or deflect in a direction perpendicular to the supported edges. The structural action is one-way, and the loads are carried by the slab in the deflected short direction. This type of slab is called a one-way slab (Figure 9.1(a)). If the slab is supported on four sides, and the ratio of the long side to the short side is equal to or greater than 2, most of the load (about 95 percent or more) is carried in the short direction, and one-way action is considered for all practical purposes (Figure 9.1(b)). If the slab is made of reinforced concrete with no voids, then it is called a one-way solid slab.

2. **One-way joist floor system.** This type of slab is also called a ribbed slab. It consists of a floor slab, usually 2–4 in. (50–100 mm) thick, supported by reinforced concrete ribs (or joists). The ribs are usually tapered and uniformly spaced at distances that do not exceed 30 in. (750 mm). The ribs are supported on girders that rest on columns. The spaces between the ribs may be formed using removable steel or fiberglass form fillers (pans), which may be used many times (Figure 9.2). In some ribbed slabs, the space between ribs may be filled with permanent fillers to provide a horizontal slab

257

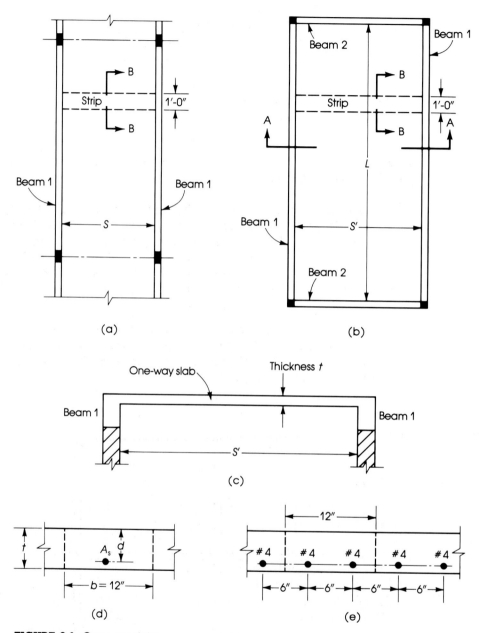

FIGURE 9.1. One-way slabs.

soffit. Different materials are used as fillers such as hollow lightweight or normal weight concrete blocks, hollow clay tile blocks, or any lightweight material.

3. **Two-way slabs.** When the slab is supported on four sides and the ratio of the long side to the short side is less than 2, the slab will deflect in double curvature in both

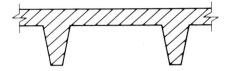

FIGURE 9.2. One-way ribbed slab.

directions. The floor load is carried in two directions to the four beams and the slab is called a two-way slab (Figure 9.3).

4. Two-way ribbed slabs and the waffle slab system. This type of slab consists of a floor slab with a length-to-width ratio less than 2. The thickness of the slab is usually 2 to 4 in. (50–100 mm) and supported by ribs (or joists) in two directions. The ribs are arranged in each direction at spacings of about 20–30 in. (500–750 mm), producing square or rectangular shapes (Figure 9.4). The ribs can also be arranged at 45° or 60° from the centerline of slabs, producing architectural shapes at the soffit of the slab. In two-way ribbed slabs, different systems can be adopted:

- A two-way rib system with voids between the ribs is obtained by using special removable and usable forms (pans) normally square in shape. The ribs are supported on four sides by girders that rest on columns. This type is called a two-way ribbed (joist) slab system.

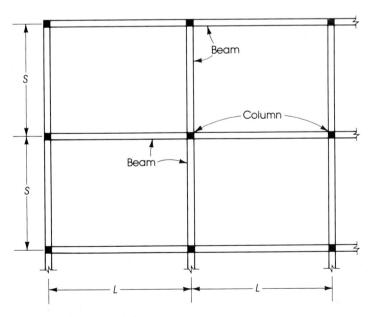

FIGURE 9.3. Two-way slab.

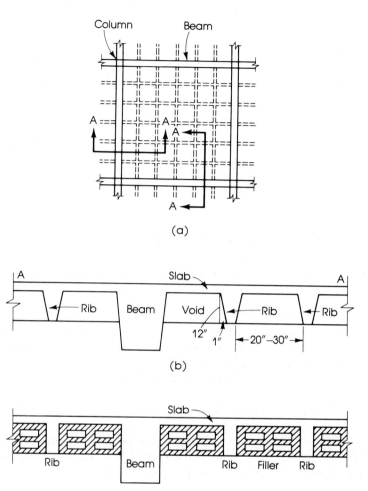

FIGURE 9.4. (a) Two-way ribbed slab, (b) cross-section without fillers, and (c) cross-section with fillers.

- A two-way rib system with permanent fillers between ribs that produce horizontal slab soffits. The fillers may be of hollow lightweight or normal weight concrete or any other lightweight material. The ribs are supported by girders on four sides in turn supported by columns. This type is also called a two-way ribbed (joist) slab system or a hollow-block two-way ribbed system.
- A two-way rib system with voids between the ribs has the ribs continuing in both directions without supporting beams and resting directly on columns through solid panels above the columns. This type is called a waffle slab system.

5. **Flat slabs.** A flat slab is a two-way slab reinforced in two directions that usually does not have beams or girders, and the loads are transferred directly to the

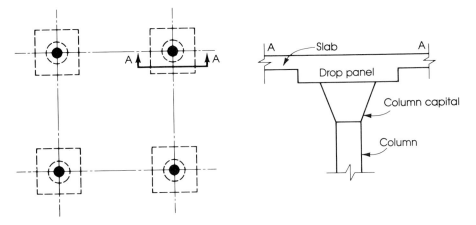

FIGURE 9.5. Flat slab floor with drop panel.

supporting columns. The column tends to punch through the slab, which can be treated by three methods:

- Using a drop panel and a column capital (Figure 9.5).
- Using a drop panel without a column capital. The concrete panel around the column capital should be thick enough to withstand the diagonal tensile stresses arising from the punching shear.
- Using a column capital without drop panel (Figure 9.6).

    6. Flat plate floors.   A flat plate floor is a two-way slab system consisting of a uniform slab that rests directly on columns and does not have beams or column capitals (Figure 9.7). In this case the column tends to punch through the slab, producing diagonal tensile stresses. Therefore, a general increase in the slab thickness is required or special reinforcement (as shearhead reinforcement) is used.

    7. Slabs resting directly on ground.   These are commonly used in basements or ground floors, sidewalks, and mostly in roads, highways, and airport runways. These slabs may be designed by empirical methods for simple cases, like basement floors, or can be analyzed by methods developed for beams on elastic foundations.

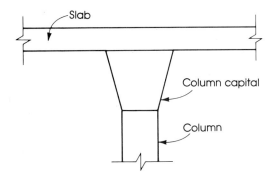

FIGURE 9.6. Flat slab floor section without drop panel.

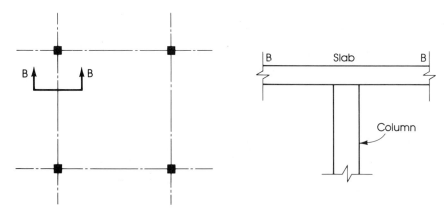

FIGURE 9.7. Flat plate floor.

8. Lift slabs.   These are specially constructed flat-plate slabs. The columns are fixed in place before casting begins. Sleeves or collars that fit loosely around the column are embedded in the concrete. The bottom level is poured first and serves as the pouring bed for the following levels until all slabs are poured. When cured, the slabs are lifted by hydraulic jacks to the desired level, where the collars are fixed to the columns.

9.3   Design of One-Way Solid Slabs

As mentioned earlier, in a one-way slab the ratio of the length of the slab to its width is greater than 2. Nearly all the loading is transferred in the short direction, and the slab may be treated as a beam. A unit strip of slab, usually 1 ft (or 1 m) at right angles to the supporting girders, is considered as a rectangular beam. The beam has a unit width with a depth equal to the thickness of the slab and a span length equal to the distance between the supports. A one-way slab thus consists of a series of rectangular beams placed side by side (Figure 9.1).

If the slab is one span only and rests freely on its supports, the maximum positive moment M for a uniformly distributed load of w psf is $M = (wL^2)/8$, where L is the span length between the supports. If the same slab is built monolithically with the supporting beams or is continuous over several supports, the positive and negative moments are calculated either by elastic analysis or by moment coefficients as for continuous beams. The **ACI Code, section 8.3,** permits the use of moment and shear coefficients in the case of two or more approximately equal spans (Figure 9.8). This condition is met when the larger of two adjacent spans does not exceed the shorter span by more than 20 percent. For uniformly distributed loads, the unit live load shall not exceed three times the unit dead load. When these conditions are not satisfied, elastic analysis is required. In elastic analysis, the negative bending moments at the centers of the supports are calculated. The value that should be considered in the design is the negative moment at the face of the support. To obtain this value, reduce from the maximum moment value at the center

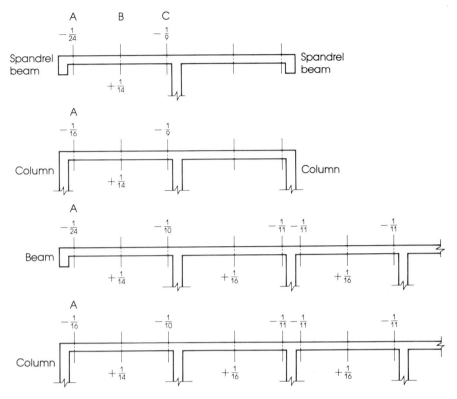

FIGURE 9.8. Moment coefficients for continuous beams and slabs (ACI Code, section 8.3).

Notes: 1. Spans are approximately equal: Longer span ≤ 1.2 shorter span.
2. Loads are uniformly distributed.
3. L.L./D.L. ≤ 3.
4. For slabs with spans ≤ 10 ft, negative bending moment at face of all supports $= (\frac{1}{12}) w_u l_n^2$.
5. For unrestrained discontinuous end at A, coefficient = 0 at A and $= +\frac{1}{11}$ at B.
6. Shearing force at C $= 1.15 \frac{w_u l_n}{2}$ and at face of all other support $= \frac{1}{2} w_u l_n$.
7. $M_u =$ (coefficient)$(w_u l_n^2)$ and $l_n =$ clear span.

of the support a quantity equal to $Vb/3$, where V is the shearing force calculated from the analysis and b is the width of the support:

$$M_f \text{ (at face of the support)} = M_c \text{ (at centerline of support)} - \frac{Vb}{3}$$

In addition to moment, diagonal tension and development length of bars should also be checked for proper design. The ACI coefficients to compute the maximum and minimum moments in continuous slabs are shown in Figure 9.8.

9.4   Design Limitations of the ACI Code

1. A typical imaginary strip 1 ft (or 1 m) wide is assumed.
2. The minimum thickness of one-way slabs using grade 60 steel according to the **ACI Code, Table 9.5(a)**, is as follows:

   $L/20$ for simply supported slabs

   $L/24$ for one end continuous slabs

   $L/28$ for both ends continuous slabs

   $L/10$ for a cantilever slab

3. Strength design as well as working stress design methods are permitted.
4. Deflection is to be checked when the slab supports are attached to construction likely to be damaged by large deflections. Deflection limits are set by the **ACI Code, section 9.5(b)**.
5. It is preferable to choose slab depth to the nearest $\frac{1}{2}$ in. (or 10 mm).
6. Shear should be checked, although it does not usually control.
7. Concrete cover in slabs shall be not less than $\frac{3}{4}$ in. (20 mm) at surfaces not exposed to weather or ground.
8. In structural slabs of uniform thickness, the minimum amount of reinforcement in the direction of the span shall not be less than that required for shrinkage and temperature reinforcement **(ACI Code, section 7.12)**.
9. The principal reinforcement shall be spaced not farther apart than three times the slab thickness, nor more than 18 in. **(ACI Code, section 7.6.5)**.
10. Straight bar systems may be used in both tops and bottoms of continuous slabs. An alternative bar system of straight and bent (trussed) bars placed alternately may also be used.
11. In addition to main reinforcement, steel bars at right angles to the main reinforcement must be provided. This additional steel is called secondary, distribution, shrinkage, or temperature reinforcement.

9.5   Temperature and Shrinkage Reinforcement

Concrete shrinks as the cement paste hardens, and a certain amount of shrinkage is usually anticipated. If a slab is left to move freely on its supports, it can contract to accommodate the shrinkage. However, slabs and other members are joined rigidly to other parts of the structure, causing a certain degree of restraint at the ends. This results in tension stresses known as shrinkage stresses. A decrease in temperature may have an effect similar to shrinkage. Since concrete is weak in tension, the temperature and shrinkage stresses are likely to cause hairline cracks. Reinforcement is placed in the slab to counteract contraction and distribute the cracks uniformly. As the concrete shrinks, the steel bars are subjected to compression.

Reinforcement for shrinkage and temperature stresses normal to the principal reinforcement should be provided in a structural slab in which the principal reinforcement extends in one direction only. The **ACI Code, section 7.12.2**, specifies the following minimum steel ratios: for slabs in which grade 40 or 50 deformed bars are used, $\rho = 0.2$ percent, and for slabs in which grade 60 deformed bars or welded wire fabric are used, $\rho = 0.18$ percent. In no case shall such reinforcement be placed farther apart than five times the slab thickness, nor more than 18 in.

9.6   Reinforcement Details

In continuous one-way slabs, the steel area of the main reinforcement is calculated for all critical sections, at midspans, and at supports. The choice of bar diameter and detailing depends mainly on the steel areas, spacing requirements, and development length. Two bar systems may be adopted.

In the straight bar system (Figure 9.9), straight bars are used for top and bottom reinforcement in all spans. This is mainly recommended in slabs with small thickness, about 6 in. or less. The time to produce straight bars is less than that required to produce bent bars.

In the bent bar or trussed system, straight and bent bars are placed alternately in the floor slab. The location of bent points should be checked for flexural, shear, and development length requirements. For normal loading in buildings, the bar details at the end and interior spans of one-way solid slabs may be adopted as shown in Figure 9.9.

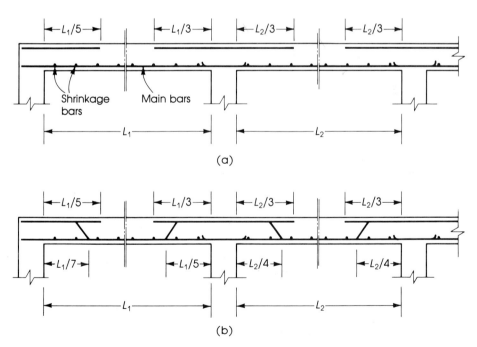

FIGURE 9.9. Reinforcement details in continuous one-way slabs: (a) straight bars and (b) bent bars.

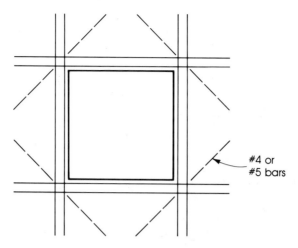

FIGURE 9.10. Reinforcement around openings.

9.7   Openings in Slabs

Openings in floor slabs are usually provided for service ducts and sometimes for lift shafts. The openings are trimmed by special beams or reinforcement so that the design strength of the slab is not affected. Diagonal cracks may develop at the corners of the opening, and special reinforcement should be considered.

When the openings are small (as for ducts) and no beams are provided, additional reinforcement must be provided. The normal reinforcement in both directions of the slab is cut off at the opening and additional reinforcement of cross-sectional area at least equal to the cut bars that would have crossed the opening is placed along each side of the opening. Diagonal reinforcement across the four corners of the opening is usually provided using No. 4 or No. 5 (12 or 16 mm) bars but not less than the diameter of the main bars (Figure 9.10).

9.8   Distribution of Loads from One-Way Slabs to Supporting Beams

In one-way floor slab systems, the loads from slabs are transferred to the supporting beams along the long ends of the slabs. The beams transfer their loads in turn to the supporting columns.

From Figure 9.11, it can be seen that beam B_2 carries loads from two adjacent slabs. Considering a 1-ft length of beam, the load transferred to the beam is equal to the area of a strip 1 ft wide and S ft in length multiplied by the intensity of load on the slab. This load produces a uniformly distributed load on the beam,

$$U_B = U_S \times S$$

The uniform load on the end beam B_1 is half the load on B_2, as it supports a slab from one side only.

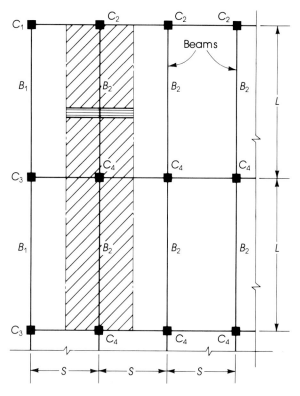

FIGURE 9.11. Distribution of loads on beams.

The load on column C_4 is equal to the reaction of two beams B_2 from both ends:

Load on column $C_4 = U_B L = U_S L S$

The load on column C_3 is one-half the load on column C_4, as it supports loads from slabs on one side only. Similarly, the load on columns C_2 and C_1 are:

$$\text{Load on } C_2 = U_S \frac{L}{2} S = \text{Load on } C_3$$

$$\text{Load on } C_1 = U_S \frac{L}{2} \frac{S}{2}$$

From the above analysis, it can be seen that each column carries loads from slabs surrounding the column and up to the centerline of adjacent slabs: up to $L/2$ in the long direction and $S/2$ in the short direction.

EXAMPLE 9.1

Calculate the ultimate moment capacity of a one-way slab that has an effective depth $d = 6$ in. and is reinforced with No. 4 bars spaced at 6 in. Given: $f'_c = 3$ ksi and $f_y = 40$ ksi.

Solution

1. The slab is reinforced with No. 4 bars spaced at 6 in. For a 1-ft strip of slab, the width $b = 12$ in. and the effective depth $d = 6$ in. The steel area provided in 12-in. width is equal to the area of 2 No. 4 bars, $A_s = 0.39$ in.^{2}

2. The percentage of steel used in a 1-ft width.

$$\rho = \frac{A_s}{bd} = \frac{0.39}{12 \times 6} = 0.0054$$

3. Compare ρ used with ρ_{max} allowed by the ACI Code.

$$\rho_{max} = 0.75\rho_b = 0.75\left(0.85\beta_1 \frac{f'_c}{f_y} \frac{87}{87 + f_y}\right)$$

Since $f'_c < 4000$ psi, then $\beta_i = 0.85$.

$$\rho_{max} = 0.75 \times 0.85 \times 0.85 \times \frac{3}{40} \times \frac{87}{87 + 40} = 0.0278$$

ρ (used) $= 0.0054 < \rho_{max}$, so use ρ.

4. Compare with ρ_{min}:

$$\rho_{min} = \frac{200}{f_y} = \frac{200}{40,000} = 0.005 \text{ (flexure)}$$

ρ_{min} for shrinkage reinforcement $= 0.002$

ρ (used) $= 0.0054 > \rho_{min}$, so use ρ.

5. The section is treated as a singly reinforced rectangular section of width $b = 12$ in. Therefore,

$$M_u = \phi A_s f_y\left(d - \frac{a}{2}\right)$$

$$a = \frac{A_s f_y}{0.85 f'_c b} = \frac{0.39 \times 40}{0.85 \times 3 \times 12} = 0.51 \text{ in.}$$

$$M_u = 0.9 \times 0.39 \times 40\left(6 - \frac{0.51}{2}\right) = 80.7 \text{ k·in.} = 6.72 \text{ k·ft} \qquad \blacksquare$$

EXAMPLE 9.2

Repeat the previous problem if the slab is reinforced with No. 6 bars spaced at 7 in.

Solution

1. The slab is reinforced with No. 6 bars spaced at 7 in. The area of one No. 6 bar is 0.44 in.^{2} To calculate the steel area in a 1-ft width, the area of one No. 6 bar is divided by the spacing of 7 in. to get the average steel area per inch width of slab.

$$A_s(\text{in 12-in. width}) = \frac{0.44}{7} \times 12 = 0.754 \text{ in.}^2/\text{ft}$$

2. Compare the steel percentage provided with ρ_{max} and ρ_{min} already calculated in the previous example:

$$\rho \text{ (used)} = \frac{0.754}{12 \times 6} = 0.0105$$

ρ (used) is less than $\rho_{max} = 0.0278$ and greater than ρ_{min}; therefore, use ρ provided in the slab to calculate its ultimate moment.

3. $$M_u = \phi A_s f_y \left(d - \frac{a}{2} \right)$$

$$a = \frac{A_s f_y}{0.85 f'_c b} = \frac{0.754 \times 40}{0.85 \times 3 \times 12} = 0.99 \text{ in.}$$

$$M_u = 0.9 \times 0.754 \times 40 \left(6 - \frac{0.99}{2} \right) = 149.3 \text{ k·in.} = 12.44 \text{ k·ft} \qquad \blacksquare$$

EXAMPLE 9.3

Determine the allowable uniform live load that can be applied on the slab of the previous example if the slab spans 10 ft between simple supports and carries a uniform dead load of 200 lb/sq. ft.

Solution

1. The ultimate moment capacity of the slab per foot width was calculated in the previous example, $M_u = 12.44$ k·ft.
2. The ultimate bending moment of a simply supported one-way slab,

$$M_u = \frac{UL^2}{8}$$

Therefore,

$$12.44 = U \frac{(10)^2}{8}$$

$$U = 0.995 \text{ k/ft}^2 = 995 \text{ psf (lb/ft}^2)$$

Note that the unit of U is k/ft², which indicates a unit length of the slab along the span and a unit width ($b = 1$ ft).

3. The ultimate load $U = 1.4 \, D + 1.7 \, L$ or

$$995 = 1.4 \times 200 + 1.7 \, L$$

$$L = \frac{1}{1.7} (995 - 1.4 \times 200) = 420.6 \text{ psf} \qquad \blacksquare$$

EXAMPLE 9.4

Design a 12-ft simply supported solid one-way slab to carry a uniform ultimate load of 840 psf including its own weight. Given: $f'_c = 3$ ksi, $f_y = 40$ ksi, and ACI specifications.

Solution   1. Minimum thickness of one-way simply supported slab is $L/25$ when $f_y = 40$ ksi (Table A.6 in Appendix A).

$$\text{Minimum depth} = \frac{L}{25} = \frac{12 \times 12}{25} = 5.76 \text{ in.}$$

If a 6-in. slab is adopted, deflection need not be checked, But if a 5-in. slab is assumed (which is less than 5.75 in.), deflection must be checked, as explained in Chapter 6.

2. $M_u = \dfrac{UL^2}{8} = \dfrac{0.84(12)^2}{8} \times 12 = 181.6$ k·in.

(per foot width of slab)

3. Choose a reinforcement percentage of 1.4 percent.

$$R_u = \phi \rho f_y \left(1 - \frac{\rho f_y}{1.7 f_c'}\right) = 0.9 \times 0.014 \times 40,000 \left(1 - \frac{0.014 \times 40}{1.7 \times 3}\right)$$

$$= 450 \text{ psi} = 0.45 \text{ ksi (or from Table A.1 in Appendix A)}$$

$$M_u = R_u b d^2$$

$$181.6 = 0.45(12)\, d^2$$

Therefore,

$$d = 5.8 \text{ in.}$$

$$A_s = \rho b d = 0.014 \times 12 \times 5.8 = 0.975 \text{ in.}^2$$

Use No. 6 bars spaced at 5 in. (actual A_s used = 1.06 in.²).

4. Total depth = $d + \frac{1}{2}$ bar diameter + cover

$$t = 5.8 + \tfrac{3}{8} + \tfrac{3}{4} = 6.925 \text{ in.}$$

say, 7 in. This total depth is greater than the 5.75 in. required, and therefore deflection need not be checked.

$$\text{Actual } d = 7 - \tfrac{3}{8} - \tfrac{3}{4} = 5.875 \text{ in.}$$

5. Check shear requirement: determine the shear at a distance d from the face of the support. Since the width of support is not given, calculate V_u from one end of the span:

$$V_u = 0.84 \times \left(\frac{12}{2} - \frac{5.875}{12}\right) = 4.63 \text{ kips}$$

Allowable shear stress $v_c = 2\sqrt{f_c'} = 2\sqrt{3000} = 109$ psi

Allowable ultimate shear force = $\phi v_c b d = 0.85 \times 109 \times 12 \times 5.875$

$$= 6530 \text{ lb} = 6.53 \text{ k}$$

Therefore the section is adequate.

6. Determine secondary (shrinkage) reinforcement: the percentage of reinforcement required in the transverse direction (when $f_y = 40$ ksi) is 0.2 percent.

$$A_s = 0.002bt = 0.002 \times 12 \times 7 = 0.168 \text{ in.}^2$$

Use No. 3 bars spaced at 8 in. (A_s used = 0.17 in.²). ∎

EXAMPLE 9.5

The cross-section of a continuous one-way solid slab in a building is shown in Figure 9.12. The slabs are supported by beams that span 24 ft between simple supports. The dead load on the slabs is that due to self-weight plus 60 psf; the live load is 120 psf. Design the continuous slab and draw a detailed section. Given: $f'_c = 3$ ksi and $f_y = 40$ ksi.

Solution
1. The minimum thickness of the first slab is $L/30$, as one end is continuous and the second end is discontinuous (Table A.6 in Appendix A). The distance between centers of beams may be considered the span L, here equal to 12 ft.

$$\text{Minimum total depth} = \frac{L}{30} = \frac{12 \times 12}{30} = 4.8 \text{ in.}$$

$$\text{Minimum total depth for interior span} = \frac{L}{35} = 4.1 \text{ in.}$$

Assume a uniform thickness of 5 in., which is greater than 4.8 in.; and therefore it is not necessary to check deflection.
2. Calculate loads and moments in a unit strip:

$$\text{Dead load} = \text{weight of slab} + 60 \text{ psf}$$

$$= \left(\frac{5}{12} \times 150\right) + 60 = 122.5 \text{ psf}$$

(Note that reinforced concrete weighs 150 pcf.)

$$\text{Ultimate load } U = 1.4D + 1.7L = 1.4 \times 122.5 + 1.7 \times 120 = 375.5 \text{ psf}$$

The clear span is 11.0 ft. The ultimate moment in the first span is over the support and equals $UL^2/10$.

$$M_u = \frac{U \times (11)^2}{10} = 0.3755 \times \frac{121}{10} = 4.54 \text{ k·ft} = 54.5 \text{ k·in.}$$

3. Assume $\rho = 1.4$ percent and $R_u = 450$ psi $= 0.45$ ksi, as calculated in example 9.4. This is less than ρ_{max} of 0.0278 and greater than ρ_{min} of 0.005.

$$d = \sqrt{\frac{M_u}{R_u b}} = \sqrt{\frac{54.5}{0.45 \times 12}} = 3.18 \text{ in.}$$

$$A_s = \rho bd = 0.014 \times 12 \times 3.18 = 0.53 \text{ in.}^2$$

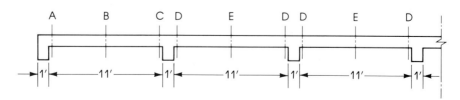

FIGURE 9.12. Example 9.5.

Choose No. 5 bars, which can be checked later.

Total depth $= d + \frac{1}{2}$ bar diameter $+$ cover $= 3.18 + \frac{5}{16} + \frac{3}{4} = 4.25$ in.

Use slab thickness of 5 in., as assumed earlier.

Actual d used $= 5 - \frac{3}{4} - \frac{5}{16} = 3.9$ in.

4. Moments and steel reinforcement required at other sections using $d = 3.9$ in. are as follows:

Location	Moment coefficient	M_u (k·in.)	$R_u = M_u/bd^2$ (psi)	ρ (%)	A_s (in.²)
A	$-\frac{1}{24}$	22.7	Small	0.50	0.23
B	$+\frac{1}{14}$	38.9	213	0.65	0.30
C	$-\frac{1}{10}$	54.5	300	0.90	0.44
D	$-\frac{1}{11}$	49.6	271	0.80	0.38
E	$+\frac{1}{16}$	34.1	187	0.55	0.26

5. To choose the bar diameter and spacing, it is recommended to investigate first the sections at C and B.

For the section at C, $A_s = 0.44$ in.². Choose No. 4 bars spaced at 5 in. (A_s actually used $= 0.45$ in.²). All bars are bent from adjacent slabs.

For the section at B, $A_s = 0.3$ in.²; choose No. 3 straight bars spaced at 10 in. and bent No. 4 bars spaced at 10 in. This means that there is a bar every 5 in., one No. 3 straight and one No. 4 bent bar placed alternately. The bent bars contribute to half the steel reinforcement required in the negative zone at C. The area of steel is 0.13 in.² (from No. 3 bars at 10 in.) plus 0.24 in.² (from No. 4 bars at 10 in.) totaling 0.37 in.², which is greater than that required at section B.

At section A, only No. 4 bent bars are provided at the top; A_s provided $= 0.24$ in.², which is greater than the 0.23 in.² required.

The above arrangement can be used only for the first two spans to provide the required A_s of 0.44 in.² at section C. For the rest of the spans, the spacings can be increased to 6 in. and 12 in. (instead of 5 in. and 10 in. spacings, respectively). These spacings are adequate as they provide $A_s = 0.39$ in.² at section D (No. 4 bars spaced at 6 in.), which is greater than the 0.38 in.² required. At section E, A_s provided will be 0.31 in.². It is obvious that this arrangement can be done because the moments at interior spans are less than those of the first span.

The steel bars should be checked for proper development length and location of bent-up bars, as explained in Chapter 7. In continuous one-way slabs, the arrangement as shown in Figure 9.13 may be used. It is also practical to use the straight bar system in continuous slabs to reduce labor cost.

Note that actual d provided (when No. 4 bars are used) has increased to 4 in. Revision of calculations is not necessary.

6. Maximum shear occurs at the exterior face of the second support (at section C).

$$V_u \text{ (at C)} = 1.15 \frac{UL_n}{2}$$

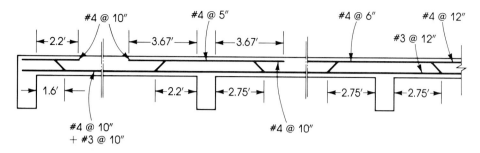

FIGURE 9.13. Reinforcement details of example 9.5.

where U = ultimate load and

L_n = clear span

$$V_u = 1.15 \times \frac{0.3755}{2} \times 11 = 2.375 \text{ kips/ft of width}$$

Allowable ultimate shear stress $v_c = 2\sqrt{f_c'} = 2\sqrt{3000} = 109$ psi

Allowable ultimate shear force $= \phi v_c b d = 0.85 \times 109 \times 12 \times 3.9$

$$= 4366 \text{ lb} = 4.3 \text{ k/ft}$$

which is greater than 2.375 kip/ft. Therefore the section is adequate.   ■

EXAMPLE 9.6

Determine the uniform ultimate load on an intermediate beam supporting the slabs of example 9.5. Calculate also the axial load on an interior column.

Solution
1. The uniform ultimate load per foot length on an intermediate beam is equal to the ultimate uniform load on slab multiplied by S, the short dimension of the slab. Therefore,

$$U \text{ (beam)} = U \text{ (slab)} \times S = 0.3755 \times 12 = 4.5 \text{ k/ft length of beam}$$

The weight of the web of the beam shall be added to the above value. Span of beam = 24 ft.

$$\text{Estimated total depth} = \frac{L}{20} \times 0.8 = \left(\frac{24}{20} \times 0.8\right) \times 12 = 11.5 \text{ in., say, 12 in.}$$

Slab thickness is 5 in. and height of web is $12 - 5 = 7$ in.

$$\text{Ultimate weight of beam web} = \left(\frac{7}{12} \times 150\right) \times 1.4 = 122.5 \text{ lb/ft}$$

$$\text{Total uniform load on beam} = 4.5 + 0.1225 = 4.62 \text{ k/ft}$$

2. Axial load on an interior column:

$$P_u = 4.62 \times 24 \text{ ft} = 110.9 \text{ k}$$ ■

9.9   One-Way Joist Floor System

The one-way joist floor system consists of hollow slabs with a total depth greater than that of solid slabs. The system is most economical for buildings where superimposed loads are small and spans are relatively large, such as schools, hospitals, and hotels. The concrete in the tension zone is ineffective; therefore this area is left open between ribs or filled with lightweight material to reduce the self-weight of the slab.

The design procedure and requirements of ribbed slabs follow the same steps as that of rectangular and T-sections explained in Chapter 3. The following points may be noticed in the design of one-way ribbed slabs:

1. Ribs are usually tapered and uniformly spaced at about 16 to 30 in. (400 to 750 mm). They are usually formed by using pans (molds) 20 in. (500 mm) wide and 6 to 20 in. (150 to 500 mm) deep, depending on the design requirement. The standard increment in depth is 2 in. (50 mm).
2. The ribs shall not be less than 4 in. (100 mm) wide and must have a depth of not more than 3.5 times the width.
3. Shear strength V_c provided by concrete for the ribs may be taken 10 percent greater than that for beams. This is mainly due to the interaction between the slab and the closely spaced ribs **(ACI Code, section 8.11.8.).**
4. The thickness of the slab on top of the ribs is usually 2 to 4 in. (50 to 100 mm) and contains minimum reinforcement (shrinkage reinforcement). This thickness shall not be less than $\frac{1}{12}$ of the clear span between ribs or 2 in. (50 mm).
5. The ACI coefficients to calculate moments in continuous slabs can be used for continuous ribbed slab design.
6. There are additional practical limitations, which can be summarized as follows:
   - The minimum width of the rib is $\frac{1}{3}$ of the total depth or 4 in. (100 mm), whichever is greater.
   - Minimum slab thickness is 2 in. or $\frac{1}{10}$ of spacing between ribs, whichever is greater.
   - Secondary reinforcement in the slab in the transverse directions of ribs should not be less than the shrinkage reinforcement or $\frac{1}{5}$ of the area of the main reinforcement in the ribs.
   - Secondary reinforcement parallel to the ribs shall be placed in the slab and spaced at distances not more than half of the spacings between ribs.
   - If the live load on the ribbed slab is less than 3 kN/m² (60 psf) and the span of ribs exceeds 5 m (17 ft), a secondary transverse rib should be provided at midspan (its direction is perpendicular to the direction of main ribs) and reinforced with the same amount of steel as the main ribs. Its top reinforcement shall not be less than half of the main reinforcement in the tension zone.
   - If the live load exceeds 3 kN/m² (60 psf) and the span of ribs varies between 4 and 7 m (13 and 23 ft), one transverse rib must be provided as indicated above. If the span exceeds 7 m (23 ft), at least two transverse ribs at one-third span must be provided with reinforcement, as explained above.

EXAMPLE 9.7

Design an interior rib of a concrete joist floor system with the following description: Span of rib = 20 ft (simply supported), Dead load (excluding own weight) = 10 psf, Live load = 70 psf, f'_c = 3 ksi, and f_y = 40 ksi.

Solution

1. Design of the slab: Assume a top slab thickness of 2 in. and fixed to ribs that have a clear spacing of 20 in. No fillers are used. Self-weight of slab = $\frac{2}{12}$ × 150 = 25 psf.

$$\text{Total D.L.} = 25 + 10 = 35 \text{ psf}$$

$$U = 1.4D + 1.7L = 1.4 \times 35 + 1.7 \times 70 = 168 \text{ psf}$$

$$M_u = \frac{UL^2}{12} \text{ (slab is assumed fixed to ribs)}$$

$$= \frac{0.168}{12}\left(\frac{20}{12}\right)^2 = 0.0389 \text{ k·ft} = 0.467 \text{ k·in.}$$

Considering that the moment in slab will be carried by plain concrete only, the 1977 ACI Code allows flexural tensile strength of $f_t = 5\sqrt{f'_c}$, with a capacity reduction factor $\phi = 0.65$. $f_t = 5\sqrt{3000} = 274$ psi.

$$\text{Flexural tensile strength} = \frac{Mc}{I} = \phi f_t$$

$$I = \frac{bh^3}{12} = \frac{12(2)^3}{12} = 8 \text{ in.}^4 \qquad c = \frac{h}{2} = \frac{2}{2} = 1 \text{ in.}$$

Therefore,

$$M = \phi f_t \frac{I}{c} = 0.65 \times 0.274 \times \frac{8}{1} = 1.42 \text{ k·in.}$$

This value is greater than M_u = 0.467 k·in., and the slab is adequate.

For shrinkage reinforcement,

$$A_s = \frac{0.2}{100} \times 12 \times 2 = 0.048 \text{ in.}^2$$

Use No. 3 bars spaced at 12 in. laid transverse to the direction of the ribs. Welded wire fabric may be economically used for this low amount of steel reinforcement. Use similar shrinkage reinforcement No. 3 bars spaced at 12 in. laid parallel to the direction of ribs, one bar on top of each rib and one bar in the slab between ribs.

2. Calculate moment in an interior rib:

$$\text{Minimum depth} = \frac{L}{25} = \frac{20 \times 12}{25} = 9.6 \text{ in.}$$

say, 10 in. The total depth of rib and slab is 10 + 2 = 12 in. Assume a rib width of 4 in. at the lower end that tapers to 6 in. at the level of the slab (Figure 9.14). The average width is 5 in. Note that the increase in the rib width using removable forms has a ratio of about 1 horizontal to 12 vertical.

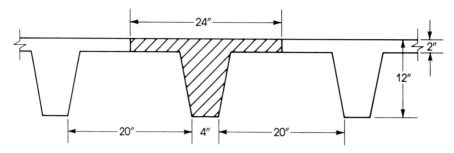

FIGURE 9.14. Example 9.7.

$$\text{Weight of rib} = \frac{5}{12} \times \frac{10}{12} \times 150 = 52 \text{ lb/ft length of rib}$$

The rib carries a load from $(20 + 4)$ in. width slab plus its own weight:

$$U = \frac{24}{12} \times 168 + (1.4 \times 52) = 409 \text{ lb/ft}$$

$$M_u = \frac{UL^2}{8} = \frac{0.409}{8} (20)^2 \times 12 = 245 \text{ k·in.}$$

3. Design the rib:

The total depth is 12 in. Assuming No. 5 bars and concrete cover of $\frac{3}{4}$ in., the effective depth d is $12 - \frac{3}{4} - \frac{5}{16} = 10.94$ in.

Check the moment capacity of the flange:

$$M_u \text{ (flange)} = \phi C \left(d - \frac{t}{2} \right)$$

where

$$C = 0.85 f_c' bt$$

$$M_u = 0.9(0.85 \times 3 \times 24 \times 2) \left(10.94 - \frac{2}{2} \right) = 1095 \text{ k·in.}$$

The moment capacity of the flange is greater than the applied moment; thus the rib acts as a rectangular section with $b = 24$ in., and the depth of the equivalent compressive block a is less than 2 in.

$$M_u = \phi A_s f_y \left(d - \frac{a}{2} \right) = \phi A_s f_y \left(d - \frac{A_s f_y}{1.7 f_c' b} \right)$$

$$245 = 0.9 A_s \times 40 \left(10.94 - \frac{A_s \times 40}{1.7 \times 3 \times 24} \right)$$

$$A_s = 0.62 \text{ in.}^2$$

$$a = \frac{A_s f_y}{0.85 \times f_c' b} = 0.4 \text{ in.} < 2 \text{ in.}$$

Use 2 No. 5 bars, $A_s = 0.62$ in.² Under normal loads, ribbed slabs act as rectangular sections with $a \leq t/2$. Approximate A_s can be calculated using $a = t/3 = \frac{2}{3}$ in.

$$M_u = \phi A_s f_y \left(d - \frac{a}{2} \right)$$

$$245 = 0.9 A_s \times 40 \left(10.94 - \frac{2}{6} \right)$$

$$A_s = 0.64 \text{ in.}^2$$

which is very close to the above calculated value. The value of a can be revised to get more accurate A_s.

In continuous beams, one bar may be bent over the supports, and its area shall be considered in the total A_s required in the negative moment zone. Straight bar system is mostly preferred.

4. Calculate shear in rib:

The allowable shear strength of the web of the rib

$$\phi V_c = \phi(1.1) \times 2\sqrt{f'_c} b_w d$$
$$= 0.85 \times 1.1 \times 2\sqrt{3000} \times 5 \times 10.94 = 5603 \text{ lb}$$

The ultimate shear at a distance d from the support

$$V_u = 409 \left(10 - \frac{10.94}{12} \right) = 3717 \text{ lb}$$

This is less than the shear capacity of the rib. Minimum stirrups may be used, and in this case an additional No. 4 bar will be placed within the slab above the rib to hold the stirrups in place.

It is advisable to add one transverse rib at midspan perpendicular to the direction of the ribs having the same reinforcement as that of the main ribs.   ■

SUMMARY

SECTIONS 9.1 and 9.2 Slabs are of different types, one-way (solid or joist floor systems) and two-way (solid, ribbed, waffle, flat slabs, and flat plates).

SECTIONS 9.3 and 9.4
(a) The ACI Code moment and shear coefficients for continuous one-way slabs are given in Figure 9.8.
(b) The minimum thickness of one-way slabs using grade 60 steel is $L/20$, $L/24$, $L/28$, and $L/10$ for simply supported, one end continuous, both ends continuous, and cantilever slabs, respectively.

SECTION 9.5 The minimum steel ratios ρ_{min} in slabs are 0.002 in. for slabs in which grade 40 or grade 50 bars are used, and 0.0018 in. for slabs in which deformed bars of grade 60 are used. Maximum spacings between bars ≤ 5 times slab thickness ≤ 18 in.

(a) Reinforcement details are shown in Figure 9.9.

(b) When openings are provided in slabs, additional reinforcement around the openings must be provided.

(c) Distribution of loads from one-way slabs to the supporting beams is shown in Figure 9.11.

The design procedure of ribbed slabs is similar to that of rectangular and T-sections. The width of ribs must be ≥4 in., while the depth must be ≤3.5 times the width. The minimum thickness of the top slab is 2 in. or ≥1/12 of the clear span between ribs.

REFERENCES

1. Concrete Reinforcing Steel Institute: *CRSI Handbook,* Chicago, 1984.
2. Portland Cement Association: *Continuity in Concrete Building Frames,* Chicago, 1959.

PROBLEMS

9.1. Calculate the ultimate moment capacity of a 5-in. (125-mm) thick one-way solid slab reinforced with No. 5 (16-mm) bars spaced at 4 in. (100 mm). Given: $f'_c = 4$ ksi (28 MPa) and $f_y = 60$ ksi (420 MPa).

9.2. A 16-ft (4.8-m) span simply supported slab carries a uniform dead load of 200 psf (10 kN/m^2) (excluding its own weight). The slab has a uniform thickness of 7 in. (175 mm) and is reinforced with No. 6 (20-mm) bars spaced at 5 in. (125 mm). Determine the allowable uniformly distributed load that can be applied on the slab if $f'_c = 3$ ksi (21 MPa) and $f_y = 60$ ksi (420 MPa).

9.3. Design a 10-ft (3-m) cantilever slab to carry a uniform total dead load of 120 psf (5.8 kN/m^2) and a concentrated live load at the free end of 2 k/ft (30 kN/m), when $f'_c = 4$ ksi (28 MPa) and $f_y = 60$ ksi (420 MPa).

9.4. A 6-in. (150-mm) solid one-way slab carries a uniform dead load of 160 psf (7.8 kN/m^2) (including its own weight) and a live load of 80 psf (3.9 kN/m^2). The slab spans 12 ft (3.6 m) between 10-in. (250-mm) wide simple supports. Determine the necessary slab reinforcement using $f'_c = 4$ ksi (28 MPa) and $f_y = 50$ ksi (350 MPa).

9.5. Repeat problem 9.3 using a variable section with a minimum total depth at the free end of 4 in. (100 mm).

9.6. Design a continuous one-way solid slab supported on beams spaced at 14 ft (4.2 m) on centers. The width of beams is 12 in. (300 mm), leaving clear slab spans of 13 ft (3.9 m). The slab carries a uniform dead load of 100 psf (4.8 kN/m^2) (including self-weight of slab) and a live load of 120 psf (5.8 kN/m^2). Use $f'_c = 3$ ksi (21 MPa), $f_y = 40$ ksi (280 MPa), and the ACI coefficients. Show two arrangements for the bars: (a) using straight bars for all top and bottom reinforcement, and (b) using the trussed (bent) bar system.

9.7. Repeat problem 9.6 using equal clear spans of 10 ft (3 m), $f'_c = 3$ ksi (21 MPa), and $f_y = 60$ ksi (420 MPa).

9.8. Repeat problem 9.6 using $f'_c = 4$ ksi (28 MPa) and $f_y = 60$ ksi (420 MPa).

9.9. Design an interior rib of a concrete joist floor system with the following description: Span of ribbed slab is 18 ft (5.4 m) between simple supports; uniform dead load (excluding self-weight) is 15 psf (720 N/m²); live load is 100 psf (4.8 kN/m²); support width is 14 in. (350 mm); f'_c = 3 ksi (21 MPa) and f_y = 60 ksi (420 MPa). Use 30-in. (750-mm) wide removable pans.

9.10. Repeat problem 9.9 using 20-in. (500-mm) wide removable pans.

9.11. Use the information given in problem 9.9 to design a continuous ribbed slab with three equal spans of 18 ft (5.4 m) each.

10 Axially Loaded Columns

10.1 Axial Compression

A column is defined by the **1983 ACI Code, section 2.1,** as a member used primarily to support axial compressive loads having a ratio of height to the least lateral dimension of 3 or greater. A column which is subjected to pure axial load does not exist. Reinforced concrete beams, floors, and columns are cast monolithically and this situation produces some moment on the columns due to end restraint. Perfect alignment of columns is not possible; thus there will be eccentric loadings on columns. Therefore, it can be said that axially loaded columns are those with very small eccentricities.

Concrete is an inexpensive construction material and has a high compressive strength, which makes it economical to use for compression members. However, steel reinforcement must be provided in columns to resist any bending moment that may exist and to reduce the concrete section, as steel carries a much greater force in compression than concrete.

10.2 Types of Columns

The main types of reinforced concrete columns are as follows:

1. *Tied columns* have longitudinal bars braced by closed ties at intervals.
2. *Spiral columns* have longitudinal bars braced with a spaced helix or spiral.
3. *Composite columns* contain structural steel or cast iron members encased with concrete and may also contain longitudinal bars and spirals.
4. *Combination columns* contain structural steel members wrapped with wire and encased in at least 2.5 in. (60 mm) of concrete.
5. *Steel pipe columns* are filled with concrete.

The different types of columns are shown in Figure 10.1. Column sections may be square, rectangular, circular, L-shaped, octagonal, or any desired shape with an adequate side width.

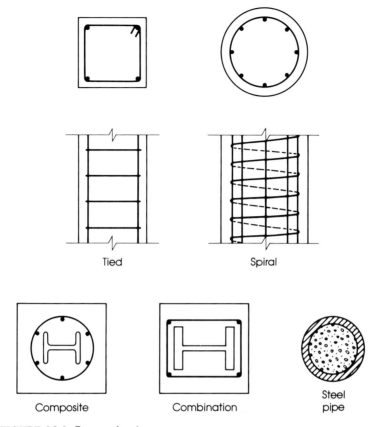

FIGURE 10.1. Types of columns.

Ties and spirals are provided in columns to serve two main functions: to hold the longitudinal bars in position in the forms during concreting, and to prevent the slender, highly stressed longitudinal bars from buckling and bursting the concrete cover.

10.3   Behavior of Axially Loaded Columns

Tests on reinforced concrete columns under axial loads were made as early as 1900. In 1911, Whitney at the University of Wisconsin realized that the unit stress in the concrete and in the steel could not be calculated under sustained loading; creep was not known at that time. Since 1930, experimental investigations at Lehigh University and at the University of Illinois have been performed on columns. The results indicate that there is no definite relationship between stresses based on elastic theory assumptions and the actual stresses in an axially loaded column. The true ratio of steel stress to concrete stress depends on the amount of shrinkage, creep, and the loading condition of the member. Design procedures based on strength design were suggested to replace the elastic design of columns.

When an axial load is applied to a reinforced concrete column, the concrete can be considered to behave elastically up to a low stress of about $\frac{1}{3}f'_c$. If the load on the column is increased to reach its ultimate strength, the concrete will reach the maximum strength and the steel will reach its yield strength f_y. The maximum strength of concrete under ultimate load is assumed to be $0.85f'_c$, as was explained in Chapter 3. The ultimate nominal load capacity of the column can be written as follows:

$$P_o = 0.85f'_c A_n + A_{st}f_y \qquad (10.1)$$

where

$$A_n \text{ and } A_{st} = \text{the net concrete and total steel compressive areas,}$$
$$\text{respectively}$$
$$A_n = A_g - A_{st}$$

and

$$A_g = \text{gross concrete area}$$

Two different types of failure occur in columns, depending on whether ties or spirals are used. For a tied column, the concrete fails by crushing and shearing outward, the longitudinal steel bars fail by buckling outward between ties, and the column failure occurs suddenly, much like the failure of a concrete cylinder.

A spiral column undergoes a marked yielding, followed by considerable deformation before complete failure. The concrete in the outer shell fails and spalls off. The concrete inside the spiral is confined and provides little strength before the initiation of column failure. A hoop tension develops in the spiral, and for a closely spaced spiral the steel may yield. A sudden failure is not expected. Figure 10.2 shows typical load deformation curves for tied and spiral columns. Up to point a, both columns behave similarly. At point a, the longitudinal steel bars of the column yield and the spiral column shell spalls off. After the ultimate load is reached, a tied column fails suddenly (curve b), while a spiral column deforms appreciably before failure (curve c).

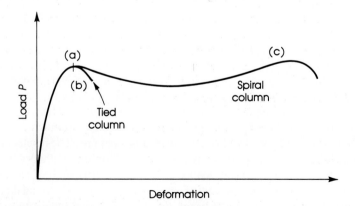

FIGURE 10.2. Behavior of tied and spiral columns.

10.4   Safety Provisions

The load factors used for columns are the same as described in Chapter 3 for the strength design method. The ultimate strength of a column is also reduced by the reduction factor ϕ. This factor has different values, depending on the designed structural member, and covers the variation in material strength, workmanship, dimensions, control, and degree of supervision.

For axially as well as eccentrically loaded columns, the 1983 ACI Code sets the reduction factors at $\phi = 0.70$ for tied columns and $\phi = 0.75$ for spirally reinforced columns. The difference of 0.05 between the two values shows the additional ductility of spirally reinforced columns.

The reduction factor for columns is much lower than those for flexure ($\phi = 0.9$) and shear ($\phi = 0.85$). This is because in axially loaded columns, the strength depends mainly on the concrete compression strength, while the strength of members in bending is less affected by the variation of concrete strength, especially in the case of an under-reinforced section. Furthermore, the concrete in columns is subjected to more segregation than in the case of beams. Columns are cast vertically in long, narrow forms, while the concrete in beams are cast in shallow, horizontal forms. Also, the failure of a column in a structure is more critical than that of a floor beam.

10.5   Limitations

The ACI Code presents the following limitations for reinforcement in compression members:

1. The minimum longitudinal steel percentage is 1 percent and the maximum percentage is 8 percent of the gross area of the section **(ACI Code, section 10.9.1)**. Minimum reinforcement is necessary to provide resistance to bending, which may exist, and to reduce the effects of creep and shrinkage of the concrete under sustained compressive stresses. Practically, it is very difficult to fit more than 8 percent of steel reinforcement into a column and maintain sufficient space for concrete to flow between bars.
2. A minimum of four bars is required for tied circular and rectangular members and six bars for circular members enclosed by spirals **(ACI Code, section 10.9.2)**. For other shapes, one bar should be provided at each corner and proper lateral reinforcement must be provided. For tied triangular columns, at least three bars are required. Bars shall not be located at a distance greater than 6 in. clear on either side from a laterally supported bar.
3. The minimum ratio of spiral reinforcement ρ_s is limited to

$$\rho_s = 0.45 \left(\frac{A_g}{A_c} - 1\right)\frac{f'_c}{f_y} \quad \textbf{(ACI Code equation 10.5)} \qquad (10.2)$$

where

A_g = gross area of section

A_c = area of core of spirally reinforced column measured to the outside diameter of spiral

f_y = yield strength of spiral reinforcement $\leq$ 60 ksi **(ACI Code, section 10.9.3)**

4. Minimum diameter of spirals is $\frac{3}{8}$ in., and their clear spacing should not be more than 3 in. nor less than 1 in., according to the **ACI Code, section 7.10.4**. Splices may be provided by welding or lapping the spiral bars by 48 diameters or a minimum of 12 in.

5. Ties for columns must have a minimum diameter of $\frac{3}{8}$ in. to enclose longitudinal bars of No. 10 size or smaller, and a minimum diameter of $\frac{1}{2}$ in. for larger bar diameters.

6. Spacing of ties shall not exceed the smallest of 48 times the tie diameter, 16 times the longitudinal bar diameter, or the least dimension of the column.

The code does not give restrictions on the size of columns to allow wider utilization of reinforced concrete columns in smaller sizes.

10.6   Spiral Reinforcement

Spiral reinforcement in compression members prevents a sudden crushing of concrete and buckling of longitudinal steel bars. It has the advantage of producing a tough column that undergoes gradual and ductile failure. The minimum spiral ratio required by the ACI Code is meant to provide an additional compressive capacity to compensate for the spalling of the column shell. The strength contribution of the shell is

$$P_u(\text{shell}) = 0.85 f'_c (A_g - A_c) \tag{10.3}$$

where A_g is the gross concrete area and A_c is the core area (Figure 10.3).

From tests on spirally reinforced columns, it is reported that spiral steel is at least twice as effective as longitudinal bars; therefore the strength contribution of spiral equals $2\rho_s A_c f_y$, where ρ_s is the ratio of volume of spiral reinforcement to total volume of core.

If the strength of the column shell is equated to the spiral strength contribution, then

$$0.85 f'_c (A_g - A_c) = 2\rho_s A_c f_y$$

or

$$\rho_s = 0.425 \left(\frac{A_g}{A_c} - 1 \right) \frac{f'_c}{f_y} \tag{10.4}$$

The ACI Code adopted a minimum ratio of ρ_s according to the following equation:

$$\rho_s = 0.45 \left(\frac{A_g}{A_c} - 1 \right) \frac{f'_c}{f_y} \tag{10.2}$$

The design relationship of spirals may be obtained as follows (Figure 10.3):

$$
\begin{aligned}
\rho_s &= \frac{\text{Volume of spiral in one loop}}{\text{Volume of core for a spacing } S} \\
&= \frac{a_s \pi (D_c - d_s)}{\left(\dfrac{\pi D_c^2}{4} \right) S} = \frac{4 a_s (D_c - d_s)}{D_c^2 S}
\end{aligned}
\tag{10.5}
$$

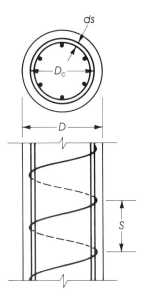

FIGURE 10.3. Dimensions of a column spiral.

where

    a_s = area of spiral reinforcement

    D_c = diameter of the core measured to the outside diameter of spiral

    D = diameter of the column

    d_s = diameter of the spiral

    S = spacing of the spiral

10.7   Design Equations

The ultimate nominal load capacity of a member under axial compression was given in equation 10.1. Since a perfect axially loaded column does not exist, some eccentricity occurs on the column section. The ACI Code does not specify a minimum eccentricity, but it specifies that the maximum nominal load P_o should be multiplied by 0.8 for tied columns and 0.85 for spirally reinforced columns. Introducing the capacity reduction factor ϕ, the allowable ultimate loads on columns according to the **ACI Code, section 10.3.5,** are as follows:

$$P_u = \phi P_n = \phi(0.80)[0.85 f_c'(A_g - A_{st}) + A_{st}f_y] \tag{10.6}$$

for tied columns, and

$$P_u = \phi P_n = \phi(0.85)[0.85 f_c'(A_g - A_{st}) + A_{st}f_y] \tag{10.7}$$

for spiral columns, where

A_g = gross concrete area

A_{st} = total steel compressive area

ϕ = 0.7 for tied columns and 0.75 for spirally reinforced columns

Equations 10.6 and 10.7 may be written as a function of the gross steel ratio:

$$\rho_g = \frac{A_{st}}{A_g}$$

Therefore, equations 10.6 and 10.7 become

$$P_u = (0.8)\phi A_g[0.85f'_c + \rho_g(f_y - 0.85f'_c)] \qquad (10.8)$$

for tied columns, and

$$P_u = (0.85)\phi A_g[0.85f'_c + \rho_g(f_y - 0.85f'_c)] \qquad (10.9)$$

for spirally reinforced columns. These last two equations are used in design problems when A_g is required and a steel ratio ρ_g is assumed.

Earlier ACI Codes specified minimum eccentricities for the design of axially loaded columns of 0.10h for tied columns and 0.05h for spirally reinforced columns, but not less than 1 in. The minimum eccentricity was measured with respect to either axis, and the term h represents the total depth of a rectangular column or the outside diameter of a round column.

10.8   Axial Tension

Concrete will not crack as long as stresses are below its tensile strength; in this case both concrete and steel resist the tensile stresses. But when the tension force exceeds the tensile strength of concrete (about one-tenth of the compressive strength), cracks develop across the section, and the entire tension force is resisted by steel. The ultimate load that the member can carry is that due to tension steel only:

$$T_n = A_{st}f_y \qquad (10.10)$$

and

$$T_u = \phi A_{st}f_y \qquad (10.11)$$

where ϕ = 0.9 for axial tension.

Tie rods in arches and similar structures are subjected to axial tension. Under working loads, the concrete cracks and the steel bars carry the whole tension force. The concrete acts as a fire and corrosion protector. Special provisions must be taken for water structures, as in the case of water tanks. In such designs, the concrete is not allowed to crack under the tension caused by the fluid pressure.

10.9   Long Columns

The equations developed in this chapter for the strength of axially loaded members are for short columns. In the case of long columns, the load capacity of the column is reduced by a reduction factor.

A long column is one with a high slenderness ratio h'/r, where h' is the effective height of the column and r is the radius of gyration. The design of long columns is explained in detail in Chapter 12.

10.10   Arrangement of Vertical Bars and Ties in Columns

Figure 10.4 shows the arrangement of longitudinal bars in tied columns and the distribution of ties. Ties shown in dotted lines are required when the clear distance on either side from laterally supported bars exceeds 6 in. The minimum concrete cover in columns is 1.5 in.

EXAMPLE 10.1

Determine the allowable ultimate axial load on a 12-in.-square short tied column reinforced with 4 No. 8 bars. Ties are No. 3 bars spaced at 12 in. Given: $f'_c = 4$ ksi and $f_y = 60$ ksi.

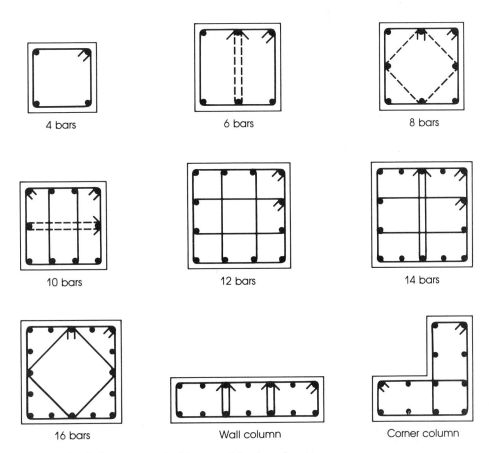

FIGURE 10.4. Arrangement of bars and ties in columns.

Solution   1. Since the column is short, there is no reduction in the capacity of the column due to slenderness or buckling effect.

2. Using equation 10.6,

$$P_u = \phi 0.8[0.85 f_c'(A_g - A_{st}) + A_{st}f_y]$$

For the tied column, $\phi = 0.7$, $A_g = 12 \times 12 = 144$ in.^{2}, and $A_{st} = 4 \times 0.79 = 3.16$ in.^{2}. Therefore,

$$P_u = 0.7 \times 0.8[(0.85 \times 4) \times (144 - 3.16) + 3.16 \times 60] = 374.3 \text{ k}$$

3. Steel percentage

$$\rho = \frac{4 \times 0.79}{12 \times 12} = 0.022$$

which exceeds the 0.01 minimum steel ratio and is less than ρ_{max} of 0.08.

4. Check tie spacing:
The minimum tie diameter is $\frac{3}{8}$ in. Spacing is the smallest of 48 tie diameters, 16 bar diameters, or the least column side:

$$S_1 = 48 \times \tfrac{3}{8} = 18 \text{ in.} \quad S_2 = 16 \times 1 = 16 \text{ in.} \quad S_3 = 12 \text{ in.}$$

Spacing used is 12 in., which is adequate.   ∎

EXAMPLE 10.2

Determine the allowable ultimate load on the short tied rectangular column section shown in Figure 10.5. Given: $f_c' = 3$ ksi and $f_y = 40$ ksi.

Solution   1. For a short tied column,

$$P_u = \phi \times 0.8[0.85 f_c'(A_g - A_{st}) + A_{st}f_y]$$

Here $\phi = 0.7$, $A_{st} = 6 \times 1 = 6$ in.^{2}, and $A_g = 16 \times 12 = 192$ in.^{2}.

$$P_u = 0.7 \times 0.8[0.85 \times 3(192 - 6) + 6 \times 40] = 400 \text{ k}$$

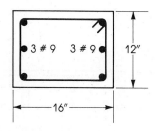

3 # 9   3 # 9   12"

16"

FIGURE 10.5. Example 10.2

2. Steel ratio = $(6 \times 1)/(12 \times 16) = 0.031$, which is greater than the 0.01 minimum reinforcement and less than $\rho_{max} = 0.08$.
3. Check spacing of No. 3 ties, as in example 10.1.   ∎

EXAMPLE 10.3

Design a square tied short column for an ultimate axial load of 436 kips. Use $f'_c = 3$ ksi, $f_y = 40$ ksi, and $\rho = 0.02$.

Solution
1. Using equation 10.8,

$$436 = P_u = 0.8 \times 0.7 A_g[0.85 \times 3 + 0.02(40 - 2.55)]$$

$$A_g = \frac{436}{0.56(2.55 + 0.75)} = 236 \text{ in.}^2$$

Column side $= \sqrt{236} = 15.36$ in.

Choose a square column 16 by 16 in. ($A_g = 256$ in.^{2}):

$$A_{st} = 0.02 \times 236 = 4.72 \text{ in.}^2$$

2. Since a greater area of the concrete section is adopted, the steel percentage may be reduced by substituting again in equation 10.6 using $P_u = 436$ and $A_g = 256$ in.^{2}

$$P_u = \phi \times 0.8[0.85 f'_c (A_g - A_{st}) + A_{st} f_y]$$
$$436 = 0.7 \times 0.8[0.85 \times 3(256 - A_{st}) + A_{st} \times 40]$$
$$A_{st} = 3.36 \text{ in.}^2$$

Use 6 No. 7 bars ($A_s = 3.61$ in.^{2}).
3. Design of ties:
Choosing No. 3 ties, the spacing is the least of (1) $16 \times \frac{7}{8} = 14$ in., (2) $48 \times \frac{3}{8} = 18$ in., (3) 16 in. Use No. 3 ties spaced at 14 in. Distances between bars are 5.5 in. < 6 in. Therefore no additional ties are required. The section is shown in Figure 10.6.   ∎

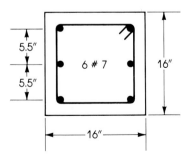

FIGURE 10.6. Example 10.3.

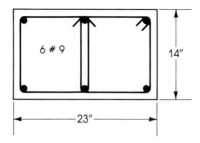

FIGURE 10.7. Example 10.4.

EXAMPLE 10.4

A rectangular tied column carries an axial dead load of 200 k and an axial live load of 240 k and has an effective height of 22 ft. Design the column using $f'_c = 4$ ksi, $f_y = 50$ ksi, the least side of the column $b = 14$ in., and assuming a reduction factor of 0.9 due to column height.

Solution

1. $P_u = 1.4D + 1.7L = 1.4 \times 200 + 1.7 \times 240 = 688$ k
2. Adjusting for the column height reduction factor, the load for an equivalent short column is

$$P_u(\text{short}) = \frac{688}{0.90} = 764.4 \text{ k}$$

3. Assuming a steel percentage of 2 percent and using equation 10.8:

$$P_u = (0.8)\phi A_g[0.85f'_c + \rho(f_y - 0.85f'_c)]$$
$$764.4 = 0.7 \times 0.8A_g[0.85 \times 4 + 0.02(50 - 3.4)]$$
$$A_g = \frac{764.4}{0.56(3.4 + 0.932)} = 315.1 \text{ in.}^2$$

For $b = 14$ in., column side $= 315.1/14 = 22.5$ in. Choose a column section 23 by 14 in. ($A_g = 322$ in.^{2})

$$A_s = 0.02 \times 315.1 = 6.3 \text{ in.}^2$$

4. Since A_g has increased, reduce A_{st} by using equation 10.6:

$$764.4 = 0.7 \times 0.8[0.85 \times 4(322 - A_{st}) + A_{st} \times 50]$$
$$A_{st} = 5.8 \text{ in.}^2$$

Choose 6 No. 9 bars ($A_s = 6$ in.^{2}).

5. Choose No. 3 ties spaced at 14 in., as shown in Figure 10.7. ∎

EXAMPLE 10.5

A circular short column has a diameter of 16 in. and is reinforced with 6 No. 8 bars enclosed by a No. 3 spiral spaced at 1.5 in. If $f'_c = 5$ ksi and $f_y = 60$ ksi, determine the allowable ultimate axial load on the column according to the ACI Code requirements (Figure 10.8).

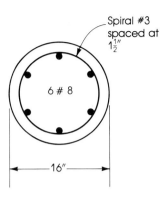

Spiral #3
spaced at
$1\frac{1}{2}''$

6 # 8

—16"—

FIGURE 10.8. Example 10.5.

Solution

1. For a short column,

$$P_u = (0.85)\phi[0.85f'_c(A_g - A_{st}) + A_{st}f_y]$$

where $\phi = 0.75$ for spiral columns

$$A_{st} = 6 \times 0.79 = 4.74 \text{ in.}^2$$

$$A_g = \frac{\pi}{4} \times (16)^2 = 201 \text{ in.}^2$$

$$P_u = 0.75 \times 0.85[0.85 \times 5(201 - 4.74) + 4.74 \times 60] = 713 \text{ k}$$

2. The steel percentage

$$\rho = \frac{4.74}{201}(100) = 2.36 \text{ percent}$$

which is more than the 1 percent minimum reinforcement and less than ρ_{max} of 8 percent.

3. Check adequacy of spirals:
   Minimum spiral ratio

$$\rho_s = 0.45\left(\frac{A_g}{A_c} - 1\right)\frac{f'_c}{f_y} \tag{10.2}$$

$$A_g = \frac{\pi}{4} \times (16)^2$$

$$A_c = \frac{\pi}{4}(13)^2$$

where A_c = the area of the core, with 1.5 in. concrete cover. Therefore,

$$\rho_s(\text{min}) = 0.45\left[\frac{(16)^2}{(13)^2} - 1\right]\left(\frac{5}{60}\right) = 0.0193$$

Spiral ratio provided

$$\rho_s = \frac{4a_s(D_c - d_s)}{(D_c)^2 S} \tag{10.5}$$

$a_s = 0.11$ (for No. 3 bar), $D_c = 13$ in., $d_s = 0.375$ in., and $S = 1.5$ in. Therefore,

$$\rho_s = \frac{4 \times 0.11(13 - 0.375)}{1.50 \times (13)^2} = 0.0219$$

which is greater than the minimum ratio of 0.0193. Spacing of spirals used is 1.50 in., which is between the minimum spacing of 1 in. and the maximum of 3 in.

The allowable ultimate axial load on the column, then, is 713 k. ∎

EXAMPLE 10.6

Design a rectangular tied short column to carry an ultimate axial load of 1900 kN. Use $f'_c = 28$ MPa, $f_y = 420$ MPa, and column width $b = 300$ mm.

Solution   1. Using equation 10.8,

$$P_u = 0.8\phi A_g[0.85f'_c + \rho_g(f_y - 0.85f'_c)]$$

Assuming a steel percentage of 2 percent,

$$1900 \times 10^3 = 0.8 \times 0.7A_g[0.85 \times 28 + 0.02(420 - 0.85 \times 28)]$$
$$A_g = 106{,}950 \text{ mm}^2$$

For $b = 300$ mm, the other side of the rectangular column is $106{,}950/300 = 356.5$ mm. Therefore, use a section of 300 by 350 mm ($A_g = 105{,}000$ mm^2).

2. $A_s = 0.02 \times 106{,}950 = 2139$ mm^2

Choose 6 bars, 22 mm in diameter ($A_s = 2280$ mm^2).

3. Check the actual axial load capacity of the section using equation 10.6:

$$P_u = 0.8\phi[0.85f'_c(A_g - A_{st}) + A_{st}f_y]$$
$$= 0.8 \times 0.7[0.85 \times 28(105{,}000 - 2280) + 2280 \times 420] \times 10^{-3}$$
$$= 1905.3 \text{ kN}$$

This meets the required P_u of 1900 kN.

4. Choose ties 10 mm in diameter. Spacing is the least of (1) $16 \times 22 = 352$ mm, (2) $48 \times 10 = 480$ mm, or (3) 300 mm. Choose 10-mm ties spaced at 300 mm. ∎

SUMMARY

SECTIONS 10.1 to 10.4	Columns may be tied, spirally reinforced, or composite.

$\phi = 0.7$ for tied columns

$\phi = 0.75$ for spirally reinforced columns

SECTIONS 10.5 and 10.6   Minimum ratio of spirals

$$\rho_s = 0.45 \left(\frac{A_g}{A_c} - 1\right)\frac{f'_c}{f_y} \tag{10.2}$$

$$\rho_s = \frac{4a_s(D_c - d_s)}{D_c^2 S} \qquad (10.5)$$

The minimum diameter of spirals is $\frac{3}{8}$ in., and their clear spacings should be not more than 3 in. nor less than 1 in.

SECTION
10.7

For tied columns:

$$P_u = \phi P_n = 0.8\phi[0.85f_c'(A_g - A_{st}) + A_{st}f_y] \qquad (10.6)$$

or

$$P_u = \phi P_n = 0.8\phi A_g[0.85f_c' + \rho_g(f_y - 0.85f_c')] \qquad (10.8)$$

For spiral columns:

$$P_u = \phi P_n = 0.85\phi[0.85f_c'(A_g - A_{st}) + A_{st}f_y] \qquad (10.7)$$

or

$$P_u = \phi P_n = 0.85\phi A_g[0.85f_c' + \rho_g(f_y - 0.85f_c')] \qquad (10.9)$$

where $\rho_g = A_{st}/A_g$.

SECTIONS
10.8 to 10.10

(a) For axial tension,

$$T_u = \phi A_{st}f_y \qquad (\phi = 0.9) \qquad (10.11)$$

(b) Arrangements of vertical bars and ties in columns are shown in Figure 10.4.

REFERENCES

1. ACI Committee 315: *Manual of Standard Practice for Detailing Reinforced Concrete Structures*, 6th ed., American Concrete Institute, Detroit, 1974.
2. Concrete Reinforcing Steel Institute: *CRSI Handbook,* 2nd ed., Chicago, 1984.
3. B. Bresler and P. H. Gilbert: "Tie Requirements for Reinforced Concrete Columns," *ACI Journal,* 58, November 1961.
4. J. F. Pfister: "Influence of Ties on the Behavior of Reinforced Concrete Columns," *ACI Journal,* 61, May 1964.
5. N. G. Bunni: "Rectangular Ties in Reinforced Concrete Columns," *Publication No. SP-50,* American Concrete Institute, Detroit, 1975.
6. Ti-Huang: "On the Formula for Spiral Reinforcement." *ACI Journal,* 61, March 1964.
7. American Concrete Institute: *Design Handbook, Vol. 2, Columns,* ACI Publication SP-17a, Detroit, 1978.
8. M. N. Hassoun: *Design Tables of Reinforced Concrete Members,* Cement and Concrete Association, London, 1978.

PROBLEMS

10.1. Determine the allowable ultimate axial load on a rectangular short tied column 12 by 18 in. (300 by 450 mm) reinforced with 6 No. 9 (6 × 28 mm) bars. Ties are No. 3 (10 mm diameter) and are spaced at 12 in. (300 mm). Given: $f_c' = 4$ ksi (28 MPa), $f_y = 60$ ksi (420 MPa).

10.2. Repeat problem 10.1 for a circular short tied column, 15 in. (375 mm) diameter.

10.3. Repeat problem 10.2 if the round column has adequate spirals instead of ties.

10.4. Design a square tied short column to carry an axial dead load of 180 k (800 kN) and an axial live load of 160 k (712 kN). Determine the required ties and draw the cross-section showing the bar arrangement. Use $f'_c = 5$ ksi (35 MPa), $f_y = 60$ ksi (420 MPa), and a steel percentage of 3 percent.

10.5. Repeat problem 10.4 if the section is rectangular with $b = 10$ in. (250 mm) and the steel percentage is 2 percent.

10.6. Repeat problem 10.4 by choosing the smallest possible square tied column. (Increase the steel percentage to 8 percent).

10.7. Design a spirally reinforced circular short column to carry an ultimate axial load of 620 k (2760 kN). Design the necessary spirals and draw a detailed section. Use $f'_c = 4$ ksi (28 MPa), $f_y = 60$ ksi (420 MPa), and steel percentage of 2.5 percent.

10.8. Repeat problem 10.7 using a steel percentage of 4 percent.

10.9. Repeat problem 10.4 if the column is long and causes a reduction in column capacity of 15 percent.

10.10. Repeat problem 10.7 if the column is long and causes a reduction in column capacity of 10 percent.

11

Members in Compression and Bending

11.1 Introduction

A structural member subjected to a direct force will be subjected to a bending moment if the direct force acts eccentrically on the section of the structural member. The force may be a compression or a tension force. In practice, members that are part of a building frame or arches are subjected to combined axial loads and bending moments. For example, Figure 11.1 shows a two-hinged portal frame that carries a uniform ultimate load on BC. The bending moment is drawn on the tension side of the frame. The columns AB and CD are subjected to an axial compressive force and a bending moment. The ratio of the moment to the axial force is usually defined as the eccentricity e, where $e = M_n/P_n$. Therefore, e represents the distance from the plastic centroid of the section to the point of application of the load. The plastic centroid is obtained by determining the location of the resultant force produced by the steel and the concrete, assuming that both are stressed in compression to f_y and $0.85f'_c$, respectively. For symmetrical sections, the plastic centroid coincides with the centroid of the section. For nonsymmetrical sections, the plastic centroid is determined by taking moments about an arbitrary axis, as explained below.

EXAMPLE 11.1

Determine the plastic centroid of the section shown in Figure 11.2. Given: $f'_c = 4$ ksi and $f_y = 60$ ksi.

Solution
1. It is assumed that the concrete is stressed in compression to $0.85f'_c$:

$$F_c = \text{force in concrete} = (0.85f'_c)A_g$$
$$= (0.85 \times 4) \times 14 \times 20 = 952 \text{ k}$$

F_c is located at the centroid of the concrete section (at 10 in. from axis A–A).

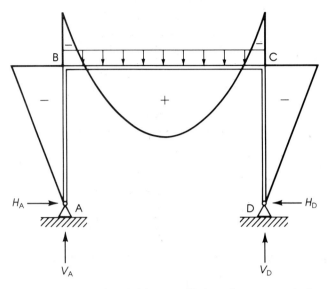

FIGURE 11.1. Two-hinged portal frame with bending moment diagram drawn on the tension side.

2. Forces in steel bars:

$$F_{s1} = A_{s1}f_y = 4 \times 60 = 240 \text{ k}$$
$$F_{s2} = A_{s2}f_y = 3 \times 60 = 180 \text{ k}$$

3. Take moments about A–A:

$$x = \frac{(952 \times 10) + (240 \times 2.5) + (180 \times 17.5)}{952 + 240 + 180} = 9.67 \text{ in.}$$

Therefore, the plastic centroid lies at 9.67 in. from axis A–A.

4. If $A_{s1} = A_{s2}$ (symmetrical section), then $x = 10$ in. from axis A–A. ∎

11.2   Load-Moment Interaction Diagram

When a normal force is applied on a short reinforced concrete column, the following cases may arise, according to the location of the normal force with respect to the plastic centroid.

1. *Axial Compression* (P_o): This is a theoretical case assuming that a large axial load is acting at the plastic centroid: $e = 0$ and $M_n = 0$. Failure of the column occurs by crushing of the concrete and yielding of steel bars. This is represented by P_o on the curve of Figure 11.3.
2. *Maximum Allowable Axial Load P_n [P_n(max)]*: This is the case of a normal force acting on the section with minimum eccentricity. According to the ACI Code, P_n(max) = $0.80P_o$ for tied columns and $0.85P_o$ for spirally reinforced

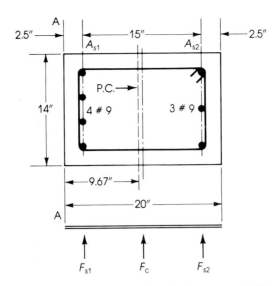

FIGURE 11.2. Plastic centroid (P.C.) of section in example 11.1.

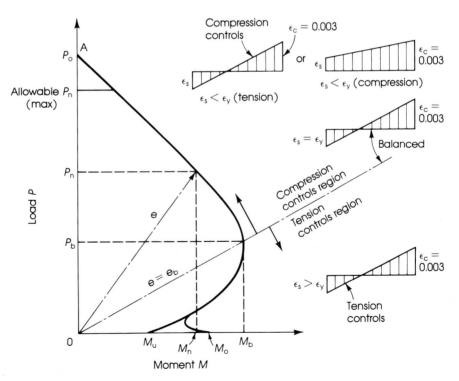

FIGURE 11.3. Load-moment strength interaction diagram showing ranges of cases discussed in text.

columns, as explained in Chapter 10. In this case, failure occurs by crushing of the concrete ($\varepsilon_c = 0.003$) and the yielding of steel bars.

3. *Compression Controls:* This is the case of a large axial load acting at a small eccentricity. The range of this case varies between a maximum value of $P_n = P_n(\text{max})$ to a minimum value of $P_n = P_b$ (balanced load). Failure occurs by crushing of the concrete on the compression side with a strain of 0.003, while the stress in the steel bars (on the tension side) is less than the yield strength $f_y(\varepsilon_s < \varepsilon_y)$. In this case $P_n > P_b$ and $e < e_b$.

4. *Balanced Condition* (P_b): A balanced condition is reached when the compression strain in the concrete reaches 0.003 and the strain in the tensile reinforcement reaches $\varepsilon_y = f_y/E_s$ simultaneously, and failure of concrete occurs at the same time as the steel yields. The moment that accompanies this load is called the balanced moment M_b, and the relevant balanced eccentricity is $e_b = M_b/P_b$.

5. *Tension Controls:* This is the case of a small axial load with large eccentricity, that is, a large moment. Before failure, tension occurs in a large portion of the section, causing the tension steel bars to yield before actual crushing of the concrete. At failure, the strain in the tension steel is greater than the yield strain ε_y, while the strain in the concrete reaches 0.003. The range of this case extends from the balanced to the case of pure flexure (Figure 11.3). When tension controls, $P_n < P_b$ and $e > e_b$.

6. *Pure Flexure:* The section in this case is subjected to a bending moment M_o, while the axial load $P_n = 0$. Failure occurs as in a beam subjected to bending moment only. The eccentricity is assumed to be at infinity. Note that radial lines from the origin represent constant ratios of $M_n/P_n = e =$ eccentricity of the load P_n from the plastic centroid.

Cases 1 and 2 were discussed in Chapter 10, and case 6 was discussed in detail in Chapter 3. The other cases will be discussed in this chapter.

11.3   Design Assumptions for Columns

The **ACI Code, section 10.2,** adopted the following design assumptions:

1. Strains in concrete and steel are proportional to the distance from the neutral axis.
2. Equilibrium of forces and strain compatibility must be satisfied.
3. The maximum usable compressive strain in concrete = 0.003.
4. Strength of concrete in tension can be neglected.
5. The stress in the steel $f_s = \varepsilon E_s \le f_y$.
6. The concrete stress block may be taken as a rectangular shape with concrete stress = $0.85f'_c$ and extending from the extreme compressive fibers a distance $a = \beta_1 c$, where c is the distance to the neutral axis and β_1 is 0.85 when $f'_c \le$ 4000 psi (28 MPa) and decreases by 0.05 for each 1000 psi (7 MPa) above 4000 psi (28 MPa), but not less than 0.65.

11.4   Safety Provisions

The safety provisions for load factors in the design of columns are as given in Chapter 3. For gravity and wind loads:

$$U = 1.4D + 1.7L$$
$$U = 0.75(1.4D + 1.7L + 1.7W)$$

or

$$U = 0.9D + 1.3W$$

The most critical ultimate load should be used.

The strength reduction factor ϕ to be used for columns may vary according to the following cases:

1. When $P_u = \phi P_n \geq 0.1 f'_c A_g$, then

$$\phi = 0.70 \text{ for tied columns} \tag{11.1}$$
$$\phi = 0.75 \text{ for spirally reinforced columns} \tag{11.2}$$

This case occurs generally when compression controls. A_g is the gross area of the concrete section.
2. Between values of $0.1 f'_c A_g$ and zero, P_u lies in the tension control zone and ϕ is larger than when compression controls. The **ACI Code, section 9.3.2,** specifies that for members in which f_y does not exceed 60 ksi, with symmetrical reinforcement and with the distance between compression and tension steel $(h - d' - d_s)$ or $(d - d')$ not less than $0.7h$ (h = total depth of section), the value of ϕ is

$$\phi = 0.9 - \frac{2.0P_u}{f'_c A_g} \geq 0.70 \text{ for tied columns} \tag{11.3}$$

$$\phi = 0.9 - \frac{1.5P_u}{f'_c A_g} \geq 0.75 \text{ for spirally reinforced columns} \tag{11.4}$$

(see Figure 11.4). For other reinforced members, ϕ may be increased linearly to 0.9 as ϕP_n decreases from $0.1 f'_c A_g$ or ϕP_b (whichever is smaller) to zero.
3. When $P_u = 0$, then $\phi = 0.9$ because it is a case of pure flexure.

11.5   Balanced Condition—Rectangular Sections

A balanced condition occurs in a column section when a load is applied on the section and produces, at ultimate strength, a strain of 0.003 in the compressive fibers of concrete and a strain $\varepsilon_y = f_y/E_s$ in the tension steel bars simultaneously. This is a special case where the neutral axis can be determined from the strain diagram with known extreme values. When the ultimate applied eccentric load is greater than P_b, compression controls; if it is smaller than P_b, tension controls in the section.

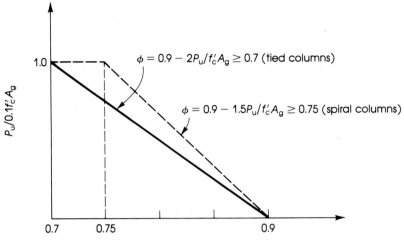

FIGURE 11.4. Variation in ϕ when tension controls.

The analysis of a balanced column section can be explained by referring to Figure 11.5:

1. Let c equal the distance from the extreme compressive fibers to the neutral axis. From the strain diagram

$$\frac{c_b(\text{balanced})}{d} = \frac{0.003}{0.003 + \dfrac{f_y}{E_s}} \quad (\text{where } E_s = 29{,}000 \text{ ksi})$$

Therefore,

$$c_b = \frac{87d}{87 + f_y} \quad (\text{where } f_y \text{ is in ksi}) \tag{11.5}$$

The depth of the equivalent compressive block

$$a_b = \beta_1 c_b = \left(\frac{87}{87 + f_y}\right)\beta_1 d \tag{11.6}$$

where $\beta_1 = 0.85$ for $f'_c \le 4000$ psi (28 MPa) and decreases by 0.05 for each 1000-psi (7-MPa) increase in f'_c.

2. From equilibrium, the sum of the horizontal forces equals zero:

$$P_b - C_c - C_s + T = 0 \tag{11.7}$$

where

$$C_c = 0.85f'_c ab \quad \text{and} \quad T = A_s f_y$$

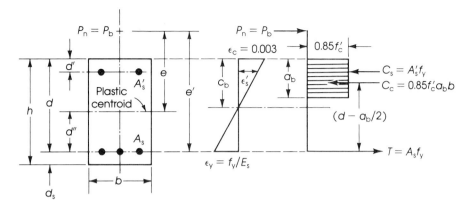

FIGURE 11.5. Balanced condition (rectangular section).

If compression steel yields, then

$$C_s = A'_s(f_y - 0.85f'_c)$$

taking the displaced concrete into account. Therefore equation 11.7 becomes

$$P_b = 0.85f'_c ab + A'_s(f_y - 0.85f'_c) - A_s f_y \qquad (11.8)$$

3. The eccentricity e_b is measured from the plastic centroid and e' is measured from the centroid of the tension steel, $e' = e + d''$ (in this case $e = e_b + d''$), where d'' is the distance from the plastic centroid to the centroid of the tension steel. The value of e_b can be determined by taking moments about the plastic centroid.

$$P_b e_b = C_c \left(d - \frac{a}{2} - d''\right) + C_s(d - d' - d'') + Td'' \qquad (11.9)$$

or

$$P_b e_b = M_b = 0.85f'_c ab \left(d - \frac{a}{2} - d''\right)$$
$$+ A'_s(f_y - 0.85f'_c)(d - d' - d'') + A_s f_y d'' \quad (11.10)$$

The balanced eccentricity e_b can then be calculated as follows:

$$e_b = \frac{M_b}{P_b}$$

EXAMPLE **11.2**

Determine the balanced compressive forces P_b, e_b, and M_b for the section shown in Figure 11.6. Given: $f'_c = 4$ ksi and $f_y = 60$ ksi.

Solution
1. For a balanced condition, the strain in the concrete = 0.003 and the strain in the tension steel is

$$\varepsilon_y = \frac{f_y}{E_s} = \frac{60}{29,000} = 0.00207$$

2. Locate the neutral axis:

$$c_b = \frac{87}{87 + f_y} d = \frac{87}{87 + 60} (19.5) = 11.54 \text{ in.}$$

$$a_b = 0.85 c_b = 0.85 \times 11.54 = 9.81 \text{ in.}$$

3. Check if compression steel yields. From the strain diagram

$$\frac{\varepsilon_s'}{0.003} = \frac{c - d'}{c} = \frac{11.54 - 2.5}{11.54}$$

$$\varepsilon_s' = 0.00235$$

which exceeds ε_y of 0.00207; thus compression steel yields.

4. Calculate the forces acting on the section:

$$C_c = 0.85 f_c' ab = 0.85 \times 4 \times 9.81 \times 14 = 467 \text{ k}$$

$$T = A_s f_y = 4 \times 60 = 240 \text{ k}$$

$$C_s = A_s'(f_y - 0.85 f_c') = 4(60 - 3.4) = 226.4 \text{ k}$$

5. Calculate P_b and e_b:

$$P_b = C_c + C_s - T = 467 + 226.4 - 240 = 453.4 \text{ k}$$

From equation 11.9,

$$M_b = P_b e_b = C_c \left(d - \frac{a}{2} - d''\right) + C_s(d - d' - d'') + Td''$$

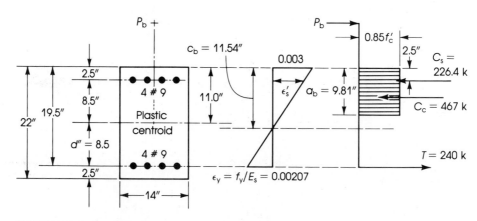

FIGURE 11.6. Example 11.2, balanced condition.

The plastic centroid is at the centroid of the section and $d'' = 8.5$ in.

$$M_b = 453.4 e_b = 467 \left(19.5 - 8.5 - \frac{9.81}{2} \right)$$

$$+ \ 226.4(19.5 - 8.5 - 2.5) + 240 \times 8.5$$

$$= 6810.8 \ \text{k·in.} = 567.6 \ \text{k·ft}$$

$$e_b = \frac{M_b}{P_b} = \frac{6810.8}{453.4} = 15.0 \ \text{in.} \qquad \blacksquare$$

11.6 Rectangular Sections Under Eccentric Force, General Case

There are two basic equations of equilibrium which can be used in the analysis of columns under eccentric loadings: (1) The sum of the horizontal or vertical forces = 0, and (2) the sum of moments about any axis = 0. Referring to Figure 11.7, the following equations may be established.

1. $\qquad P_n - C_c - C_s + T = 0$ $\qquad\qquad\qquad\qquad$ (11.11)

   where

   $C_c = 0.85 f'_c ab$
   $C_s = A'_s(f'_s - 0.85 f'_c)$ (if compression steel yields, then $f'_s = f_y$)
   $T = A_s f_s$ (if tension steel yields, then $f_s = f_y$)

2. Taking moments about A_s:

$$P_n e' - C_c \left(d - \frac{a}{2} \right) - C_s(d - d') = 0 \qquad\qquad (11.12)$$

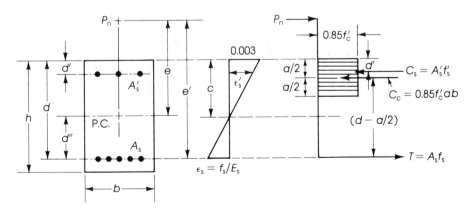

FIGURE 11.7. General case, rectangular section.

or

$$P_ne' - C_c\left(d - \frac{a}{2}\right) - A_s'f_s'(d - d') = 0$$

The quantity $e' = d + d''$, and $e' = \left(e + d - \frac{h}{2}\right)$ for symmetrical reinforcement.

$$A_s' = \frac{P_ne' - C_c\left(d - \frac{a}{2}\right)}{f_s'(d - d')} \qquad (11.13)$$

(Use $f_s' = f_y$ if compression steel yields.)

$$P_n = \frac{1}{e'}\left[C_c\left(d - \frac{a}{2}\right) + C_s(d - d')\right] \qquad (11.14)$$

If $A_s' = 0$, then

$$P_ne' - C_c\left(d - \frac{a}{2}\right) = 0 \qquad (11.15)$$

3. Taking moments about A_s':

$$P_n[e' - (d - d')] + C_c\left(\frac{a}{2} - d'\right) - T(d - d') = 0$$

$$P_n(e' + d' - d) + C_c\left(\frac{a}{2} - d'\right) - A_sf_s(d - d') = 0 \qquad (11.16)$$

(Use $f_s = f_y$ if tension steel yields.)

$$A_s = \frac{1}{f_s(d - d')}\left[P_n(e + d' - d) + C_c\left(\frac{a}{2} - d'\right)\right] \qquad (11.17)$$

$$P_n = \frac{-C_c\left(\frac{a}{2} - d'\right) + T(d - d')}{(e' + d' - d)} \qquad (11.18)$$

4. Taking moments about C_c:

$$P_n\left[e' - \left(d - \frac{a}{2}\right)\right] - T\left(d - \frac{a}{2}\right) - C_s\left(\frac{a}{2} - d'\right) = 0 \qquad (11.19)$$

$$P_n = \frac{T\left(d - \frac{a}{2}\right) + C_s\left(\frac{a}{2} - d'\right)}{(e' + a/2 - d)} \qquad (11.20)$$

If $A_s = A_s'$ and $f_s = f_s' = f_y$, then

$$P_n = \frac{A_sf_y(d - d')}{(e' + a/2 - d)} \qquad (11.21)$$

$$A_s = A_s' = \frac{P_n\left(e' + \dfrac{a}{2} - d\right)}{f_y(d - d')} \tag{11.22}$$

11.7   Strength of Columns When Compression Controls

If the compressive applied force P_n exceeds the balanced force P_b, or the eccentricity $e = M_n/P_n$ is less than e_b, compression failure is expected. In this case compression controls, and the strain in the concrete will reach 0.003 while the strain in the steel is less than ε_y (Figure 11.8). A large part of the column will be in compression. The neutral axis moves toward the tension steel, increasing the compression area, and therefore the distance to the neutral axis c is greater than the balanced c_b. It can be noticed from Figure 11.8 that the strain in the compression steel will exceed that strain ε_y for a balanced section, which indicates that the compression steel has yielded.

Because it is difficult to predict if compression or tension controls whenever a section is given, it can be assumed that compression controls when $e' < d$ or $e < \frac{2}{3}d$, which should be checked later. The ultimate eccentric load P_n can be calculated using the principles of statics. The analysis of column sections when compression controls can be performed by calculating the distance to the neutral axis for a balanced section c_b and then choosing a value of c a little greater than c_b. Calculate a, C_c, and P_n; then check the assumed c or a. This is a trial method, and the results will converge to the required value of P_n.

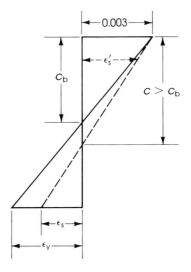

FIGURE 11.8. Strain diagram when compression controls. When $\varepsilon_s < \varepsilon_y$, then $c > c_b$ and $\varepsilon_s' > \varepsilon_y$.

EXAMPLE 11.3

Determine the nominal compressive strength P_n for the section given in example 11.2 if $e = 10$ in. (see Figure 11.9).

Solution

1. Since $e = 10$ in. $< \frac{2}{3}d = 13$ in. (or $e' = 18.5$ in. $< d$), assume compression controls. This assumption will be checked later.

2. Calculate the distance to the neutral axis for a balanced section c_b:

$$c_b = \frac{87}{87 + f_y} d = \frac{87}{87 + 60} (19.5) = 11.54 \text{ in.}$$

3. From the equations of equilibrium,

$$P_n = C_c + C_s - T \tag{11.11}$$

where

$$C_c = 0.85 f'_c ab = 0.85 \times 4 \times 14a = 47.6a$$
$$C_s = A'_s(f_y - 0.85f'_c) = 4(60 - 0.85 \times 4) = 226.4 \text{ k.}$$

Assume compression steel yields (this assumption will be checked later).

$$T = A_s f_s = 4f_s \qquad (f_s < f_y)$$
$$P_n = 47.6a + 226.4 - 4f_s \tag{A}$$

4. Taking moments about A_s:

$$P_n = \frac{1}{e'} \left[C_c \left(d - \frac{a}{2} \right) + C_s(d - d') \right] \tag{11.14}$$

$$e' = e + d - \frac{h}{2} \text{ (for symmetrical reinforcement)}$$

$$= 10 + 19.5 - \frac{22}{2} = 18.5 \text{ in.}$$

$$P_n = \frac{1}{18.5} \left[47.6a \left(19.5 - \frac{a}{2} \right) + 226.4(19.5 - 2.5) \right]$$

$$P_n = 50.17a - 1.29a^2 + 208 \tag{B}$$
$$47.6a = P_n - 226.4 + 4f_s \tag{A}$$

5. Assume $c = 13.4$ in., which exceeds c_b (11.54 in.).

$$a = 0.85 \times 13.4 = 11.39 \text{ in.}$$

Substitute $a = 11.39$ in equation B:

$$P_n = 50.17 \times 11.39 - 1.29(11.39)^2 + 208 = 612.1 \text{ k}$$

6. Calculate ε_s from the strain diagram when $c = 13.4$ in.:

$$\varepsilon_s = \left(\frac{19.5 - 13.4}{13.4} \right) (0.003) = 0.001366 < \varepsilon_y$$

$$f_s = \varepsilon_s E_s = 0.001366 \times 29{,}000 = 39.6 \text{ ksi} < f_y$$

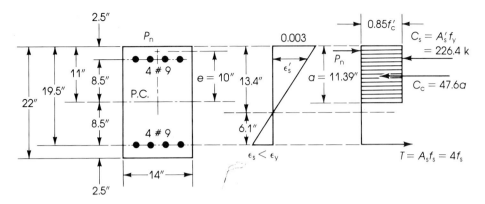

FIGURE 11.9. Example 11.3. Compression controls.

7. Substitute $P_n = 612.1$ k and $f_s = 39.6$ ksi in equation A above to calculate a (or substitute a and f_s to check P_n):

$$47.6a = 612.1 - 226.4 + 4 \times 39.6$$

$$a = 11.4 \text{ in.}$$

which is very close to the assumed $a = 11.39$ in. Then $P_n = 612.1$ k.

8. Check if compression steel yields: From the strain diagram,

$$\varepsilon_s' = \frac{13.4 - 2.5}{13.4}(0.003) = 0.00244 > \varepsilon_y = 0.00207$$

Compression steel yields, as assumed.

9. $P_n = 612.1$ k is greater than $P_b = 453.4$ k, and $e = 10$ in. $< e_b = 15$ in., both calculated in the previous example, indicating that compression controls, as assumed. ∎

11.7.1   Numerical Analysis Solution

The analysis of columns when compression controls can also be performed by reducing the calculations into one cubic equation in the form

$$A\alpha^3 + B\alpha^2 + C\alpha + D = 0$$

and then solving for α by a numerical method. The solution can be obtained rapidly by using a programmable calculator. From the equations of equilibrium

$$P_n = C_c + C_s - T \tag{11.11}$$

$$P_n = (0.85f_c'ab) + A_s'(f_y - 0.85f_c') - A_sf_s \tag{11.11a}$$

Taking moments about the tension steel A_s:

$$P_n = \frac{1}{e'}\left[C_c\left(d - \frac{a}{2}\right) + C_s(d - d')\right] \tag{11.14}$$

$$= \frac{1}{e'}\left[0.85f_c'ab\left(d - \frac{a}{2}\right) + A_s'(f_y - 0.85f_c')(d - d')\right] \tag{11.14a}$$

From the strain diagram,

$$\varepsilon_s = \left(\frac{d - c}{c}\right)(0.003) = \frac{(d - a/\beta_1)}{a/\beta_1}(0.003)$$

The stress in the tension steel is

$$f_s = \varepsilon_s E_s = 29,000\varepsilon_s = \frac{87}{a}(\beta_1 d - a)$$

Substituting this value of f_s in equation 11.11a and equating equations 11.11a and 11.14a and simplifying:

$$(0.85f_c'b/2)a^3 + [0.85f_c'b(e' - d)]a^2$$
$$+ [A_s'(f_y - 0.85f_c')(e' - d + d') + 87A_se']a - 87A_se'\beta_1 d = 0$$

This is a cubic equation in terms of a:

$$Aa^3 + Ba^2 + Ca + D = 0$$

where

$$A = 0.85f_c'\frac{b}{2}$$

$$B = 0.85f_c'b(e' - d)$$
$$C = A_s'(f_y - 0.85f_c')(e' - d + d') + 87A_se'$$
$$D = -87A_se'\beta_1 d$$

The solution of the cubic equation can be obtained by using the well-known Newton-Raphson method. This method is very powerful for finding a root of $f(x) = 0$. It uses a simple technique and the solution converges rapidly.
    Let

$$f(a) = Aa^3 + Ba^2 + Ca + D$$

The first derivative

$$f'(a) = 3Aa^2 + 2Ba + C$$

Assume any initial value of a, say a_0, and compute the next value

$$a_1 = a_0 - \frac{f(a_0)}{f'(a_0)}$$

Then use the obtained value a_1 in the same way to get

$$a_2 = a_1 - \frac{f(a_1)}{f'(a_1)}$$

Repeat the same steps to get the answer up to the desired accuracy. In the case of the analysis of columns when compression controls, the value a is greater than the balanced a (a_b). Therefore start with $a_0 = a_b$ and repeat twice to get reasonable results.

Applying this method to the previous example:

$$A = 0.85 \times 4 \times \frac{14}{2} = 23.8$$

$$B = 0.85 \times 4 \times 14(18.5 - 19.5) = -47.6$$
$$C = 4(60 - 0.85 \times 4)(18.5 - 19.5 + 2.5) + 87 \times 4 \times 18.5 = 6777.6$$
$$D = -87 \times 4 \times 18.5 \times (0.85 \times 19.5) = -106{,}710$$

For a balanced section, $c_b = 11.54$ in. and $a_b = 0.85 \times 11.54 = 9.81$ in.

$$f(a) = 23.8a^3 - 47.6a^2 + 6777.6a - 106{,}710$$
$$f'(a) = 71.4a^2 - 95.2a + 6777.6$$

Let $a_o = a_b = 9.81$ in.

$$a_1 = 9.81 - \frac{f(9.81)}{f'(9.81)} = 9.81 - \frac{-22{,}334}{12{,}715} = 11.566 \text{ in.}$$

Let $a_1 = 11.566$; then

$$a_2 = 11.566 - \frac{f(11.566)}{f'(11.566)} = 11.566 - \frac{2136}{15{,}228} = 11.43 \text{ in.}$$

This value of a is similar to that obtained earlier in example 11.3. Substitute the value of a in equation 11.11a or 11.14a to get $P_n = 612$ k.

11.8   Whitney Equation (Compression Controls)

An approximate equation was suggested by Whitney to estimate the nominal compressive strength of short columns when compression controls, as follows:[15]

$$P_n = \frac{bh f_c'}{\left(\dfrac{3he}{d^2}\right) + 1.18} + \frac{A_s' f_y}{\left(\dfrac{e}{d - d'}\right) + 0.5} \qquad (11.23)$$

This equation can only be used when the reinforcement is symmetrically placed in single layers parallel to the axis of bending.

EXAMPLE 11.4

Determine the nominal compressive strength P_n for the section given in example 11.3 using the same eccentricity $e = 10$ in. and compare results.

Solution   1. Properties of the section shown in Figure 11.9: $b = 14$ in., $h = 22$ in., $d = 19.5$ in., $d' = 2.5$ in., $A_s' = 4.0$ in.^{2}, and $(d - d') = 17$ in.

2. Apply the Whitney equation:

$$P_n = \frac{14 \times 22 \times 4}{\dfrac{3 \times 22 \times 10}{(19.5)^2} + 1.18} + \frac{4 \times 60}{\left(\dfrac{10}{17}\right) + 0.5} = 643 \text{ k}$$

3. P_n calculated by the Whitney equation is not a conservative value in this example, and the value of $P_n = 643$ k is greater than the more exact value of 612.1 k calculated by statics in example 11.3. ■

EXAMPLE 11.5

Calculate P_n for the section given in example 11.3 assuming a straight-line relationship between P_o and P_b in the interaction load-moment diagram of Figure 11.3 (see Figure 11.10).

Solution

1. The balanced force P_b and eccentricity e_b of this section were calculated earlier:

$$P_b = 453.4 \text{ k} \quad e_b = 15 \text{ in.} \quad M_b = 567.6 \text{ k·ft}$$

2. Calculate P_o (theoretical axial load):

$$P_o = 0.85 f'_c A_g + A_{st}(f_y - 0.85 f'_c)$$
$$= (0.85 \times 4) \times (14 \times 22) + (8 \times 1)(60 - 0.85 \times 4) = 1500 \text{ k}$$

3. From triangular relations (Figure 11.10):

$$\frac{1500 - P_n}{1500 - 453.4} = \frac{P_n e}{M_b} = \frac{10 P_n}{567.6 \times 12}$$
$$P_n = 591.35 \text{ k}$$

which is a conservative value of P_n. It is always less than P_n calculated by statics ($P_n = 612.1$ k). ■

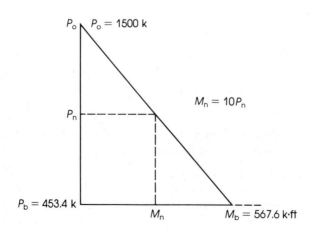

FIGURE 11.10. Example 11.5.

11.9   Strength of Columns When Tension Controls

When a column is subjected to an eccentric force with large eccentricity e, tension controls. The column section fails due to the yielding of steel and crushing of concrete when the strain in the steel exceeds ε_y ($\varepsilon_y = f_y/E_s$). In this case the ultimate capacity P_n will be less than P_b, or the eccentricity $e = M_n/P_n$ is greater than the balanced eccentricity e_b. Since it is difficult in some cases to predict if tension or compression controls, it can be assumed (as a guide) that tension controls when the eccentricity $e > d$ or $e' > 1.5d$. This assumption should be checked later.

The general equations of equilibrium developed in the general case earlier will be used to calculate the ultimate nominal strength of the column. This is illustrated in the following example.

EXAMPLE   11.6

Determine the nominal compressive strength P_n for the section given in example 11.2 if $e = 20$ in. (see Figure 11.11).

Solution
1. Since $e = 20$ in. is greater than $d = 19.5$ in., then assume that tension controls (to be checked later).
2. The strain in the tension steel ε_s will be greater than ε_y and its stress is f_y. Assume that compression steel yields ($f'_s = f_y$), which should be checked later.
3. From the equations of equilibrium:

$$P_n = C_c + C_s - T \tag{11.11}$$

where

$$C_c = 0.85 f'_c ab = 0.85 \times 4 \times 14a = 47.6a$$
$$C_s = A'_s(f_y - 0.85f'_c) = 4(60 - 0.85 \times 4) = 226.4 \text{ k}$$
$$T = A_s f_y = 4 \times 60 = 240 \text{ k}$$

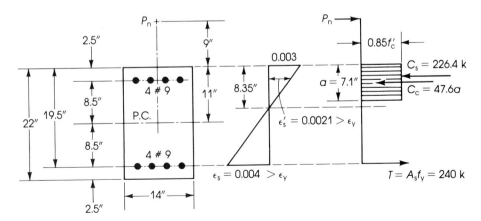

FIGURE 11.11. Example 11.6. Tension controls.

Therefore,

$$P_n = 47.6a + 226.4 - 240 = (47.6a - 13.6) \qquad (A)$$

4. Taking moments about A_s:

$$P_n = \frac{1}{e'}\left[C_c\left(d - \frac{a}{2}\right) + C_s(d - d')\right] \qquad (11.14)$$

$$e' = e + d - \frac{h}{2} \text{ (for symmetrical reinforcement)}$$

$$= 20 + 19.5 - \frac{22}{2} = 28.5 \text{ in.}$$

$$P_n = \frac{1}{28.5}\left[47.6a\left(19.5 - \frac{a}{2}\right) + 226.4 \times 17\right]$$

or

$$P_n = 32.56a - 0.835a^2 + 135.0 \qquad (B)$$

5. Substituting the value of P_n from equation A into equation B:

$$P_n = (47.6a - 13.6) = 32.56a - 0.835a^2 + 135.0$$

or

$$a^2 + 18a = 178.0$$
$$(a + 9)^2 = 178 + 81 = 259$$
$$a + 9 = 16.1$$
$$a = 7.1 \text{ in.}$$

6. From equation A,

$$P_n = 47.6 \times 7.1 - 13.6 = 324.4 \text{ k}$$

$$M_n = P_n e = 324.4 \times \frac{20}{12} = 540.67 \text{ k·ft}$$

7. Check if compression steel has yielded:

$$c = \frac{a}{0.85} = \frac{7.1}{0.85} = 8.35 \text{ in.}$$

$$\varepsilon_y = \frac{60}{29,000} = 0.00207$$

$$\varepsilon_s' = \frac{(8.35 - 2.5)}{8.35} \times 0.003 = 0.0021 > \varepsilon_y$$

Compression steel yields. Check strain in tension steel:

$$\varepsilon_s = \left(\frac{19.5 - 8.35}{8.35}\right) \times 0.003 = 0.004 > \varepsilon_y$$

If compression steel does not yield, use f_s' as calculated from $f_s' = \varepsilon_s' E_s$ and revise the calculations.

8. From example 11.2, it was calculated that $P_b = 453.4$ k and $e_b = 15$ in. In this example $P_n < P_b$ and $e > e_b$; therefore, tension controls. ∎

11.10 Approximate Equation for Rectangular Sections When Tension Controls

The ultimate strength capacity P_n of a rectangular short column subjected to an eccentric force with large eccentricity (tension controls) can be estimated using the approximate following equation:[15]

$$P_n = 0.85 f_c' bd \left[\rho'm' - \rho m + \left(1 - \frac{e'}{d}\right) \right.$$
$$\left. + \sqrt{\left(1 - \frac{e'}{d}\right)^2 + 2\left(\frac{e'}{d}\right)(\rho m - \rho'm') + 2\rho'm'\left(1 - \frac{d'}{d}\right)} \right] \quad (11.24)$$

where

$$m = \frac{f_y}{0.85 f_c'}$$

$$m' = (m - 1)$$

$$\rho = \frac{A_s}{bd}$$

$$\rho' = \frac{A_s'}{bd}$$

$$e' = e + d''$$

When $A_s = A_s'$ or $\rho = \rho'$, then

$$P_n = 0.85 f_c' bd \left[\left(-\rho + 1 - \frac{e'}{d}\right) \right.$$
$$\left. + \sqrt{\left(1 - \frac{e'}{d}\right)^2 + 2\rho m' \left(1 - \frac{d'}{d}\right) + 2\rho \frac{e'}{d}} \right] \quad (11.25)$$

When no compression reinforcement is used ($A_s' = 0$ or $\rho' = 0$):

$$P_n = 0.85 f_c' bd \left[\left(-\rho m + 1 - \frac{e'}{d}\right) + \sqrt{\left(1 - \frac{e'}{d}\right)^2 + \frac{2e'\rho m}{d}} \right] \quad (11.26)$$

EXAMPLE 11.7

Calculate P_n for the section given in example 11.6, using the approximate equation.

Solution   1. Since $A_s = A_s' = 4$ in.^{2} in the given section, then use equation 11.25:

2. $m = \dfrac{f_y}{0.85 f'_c} = \dfrac{60}{0.85 \times 4} = 17.65$

$m' = (m - 1) = 16.65$

$\rho = \dfrac{A_s}{bd} = \dfrac{4}{14 \times 19.5} = 0.01465$

$e' = e + d - \dfrac{h}{2} = 20 + 19.5 - \dfrac{22}{2} = 28.5$ in.

$d' = 2.5$ in.

$$P_n = 0.85 \times 4 \times 14 \times 19.5 \left[\left(-0.01465 + 1 - \dfrac{28.5}{19.5} \right) \right.$$
$$\left. + \sqrt{ \left(1 - \dfrac{28.5}{19.5} \right)^2 + 2 \times 0.01465 \times 16.65 \left(1 - \dfrac{2.5}{19.5} \right) + 2 \times 0.01465 \times \dfrac{28.5}{19.5} } \right]$$
$$= 928.2[-0.4762 + \sqrt{0.8253}] = 928.2 \times 0.3491 = 324 \text{ k}$$

This value of P_n is similar to that obtained by statics in the previous example. ∎

11.11   Strength of Columns with Circular Sections, Balanced Condition

The values of the balanced force P_b and the balanced moment M_b for circular sections can be determined using the equations of equilibrium, as was done in the case of rectangular sections. The bars in a circular section are arranged in such a way that their distance from the axis of plastic centroid varies depending on the number of bars in the section. The main problem is to find the depth of the compressive block a and the stresses in the reinforcing bars. The following example explains the analysis of circular sections under balanced conditions. A similar procedure can be adopted to analyze sections when tension or compression controls.

EXAMPLE 11.8

Determine the balanced load P_b and the balanced moment M_b for the 16-in.-diameter circular section in Figure 11.12. Given: $f'_c = 4$ ksi and $f_y = 60$ ksi.

Solution   1. Since the reinforcement bars are symmetrical about the axis A–A passing through the center of the circle, then the plastic centroid lies on that axis.
2. Determine the location of the neutral axis:

$$d = 13.1 \text{ in.} \quad \varepsilon_y = \dfrac{f_y}{E_s} \quad (E_s = 29{,}000 \text{ ksi})$$

$$\dfrac{c_b}{d} = \dfrac{0.003}{0.003 + \varepsilon_y} = \dfrac{0.003}{0.003 + \dfrac{f_y}{E_s}} = \dfrac{87}{87 + f_y}$$

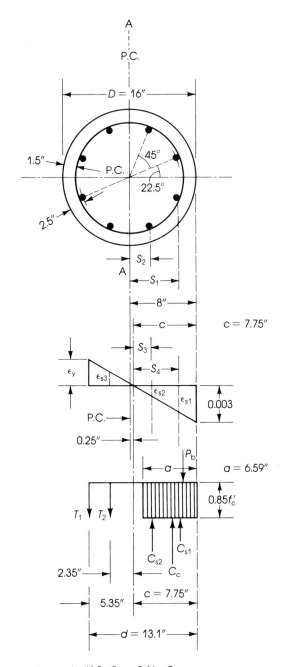

FIGURE 11.12. Example 11.8. Bars: 8 No. 9

$$S = 8 - 2.5 = 5.5 \text{ in.}$$
$$S_1 = S \cos 22.5° = 5.1 \text{ in.}$$
$$S_2 = S \cos 67.5° = 2.1 \text{ in.}$$
$$d = 8 + 5.1 = 13.1 \text{ in.}$$
$$S_3 = 1.85 \text{ in.}$$
$$S_4 = 4.85 \text{ in.}$$

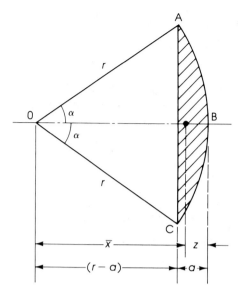

FIGURE 11.13. Properties of circular segments.

$$c_b = \frac{87}{87 + 60} (13.1) = 7.75 \text{ in.}$$

$$a_b = 0.85 \times 7.75 = 6.59 \text{ in.}$$

3. Calculate the properties of a circular segment (Figure 11.13):

$$\text{Area of segment} = r^2(\alpha - \sin \alpha \cos \alpha) = \frac{r^2}{2} (2\alpha - \sin 2\alpha) \qquad (11.27)$$

Location of centroid $\bar{x}$ (from the circle center O):

$$\bar{x} = \frac{2}{3} \frac{(r \sin^3 \alpha)}{(\alpha - \sin \alpha \cos \alpha)} = \frac{2}{3} \frac{(r \sin^3 \alpha)}{(\alpha - 1/2 \sin^2 2\alpha)} \qquad (11.28)$$

$$Z = r - \bar{x} \qquad (11.29)$$

$$r \cos \alpha = r - a \quad \text{or} \quad \cos \alpha = \left(1 - \frac{a}{r}\right) \qquad (11.30)$$

$$\cos \alpha = \left(1 - \frac{6.59}{8}\right) = 0.17625$$

and $\alpha = 79.85°$, $\sin \alpha = 0.984$, and $\alpha = 1.394$ radians.

$$\text{Area of segment} = (8)^2(1.394 - 0.984 \times 0.17625) = 78.12 \text{ in.}^2$$

$$\bar{x} = \left(\frac{2}{3}\right) \frac{8(0.984)^3}{(1.394 - 0.984 \times 0.17625)} = 4.16 \text{ in.}$$

$$Z = r - \bar{x} = 8 - 4.16 = 3.84 \text{ in.}$$

4. Calculate the compressive force C_c:

$$C_c = 0.85 f_c' \times \text{area of segment}$$
$$= 0.85 \times 4 \times 78.12 = 265.6 \text{ k}$$

It acts at 2.78 in. from the center of the column.

5. Calculate the strains, stresses, and forces in the tension and the compression steel. Determine the strains from the strain diagram. For T_1

$$\varepsilon = \varepsilon_y = 0.00207 \quad f_s = f_y = 60 \text{ ksi}$$
$$T_1 = 2 \times 60 = 120 \text{ k}$$

For T_2

$$\varepsilon_{s3} = \frac{2.35}{5.35} \varepsilon_y = \frac{2.35}{5.35} \times 0.00207 = 0.00091$$
$$f_{s3} = 0.00091 \times 29{,}000 = 26.4 \text{ ksi}$$
$$T_2 = 26.4 \times 2 = 52.8 \text{ k}$$

For C_{s1}

$$\varepsilon_{s1} = \frac{4.85}{7.75} \times 0.003 = 0.00188$$
$$f_{s1} = 0.00188 \times 29{,}000 = 54.5 \text{ ksi} < 60 \text{ ksi}$$
$$C_{s1} = 2(54.5 - 3.4) = 102.2 \text{ k}$$

For C_{s2}

$$\varepsilon_{s2} = \frac{1.85}{7.75} \times 0.003 = 0.000716$$
$$f_{s2} = 0.000716 \times 29{,}000 = 20.8 \text{ ksi}$$
$$C_{s2} = 2(20.8 - 3.4) = 34.8 \text{ k}$$

The stresses in the compression steel have been reduced to take into account the concrete displaced by the steel bars.

6. The balanced force $P_b = C_c + C_s - T$:

$$P_b = 265.6 + (102.2 + 34.8) - (120 + 52.8) = 230 \text{ k}$$

7. Take moments about the plastic centroid (axis A–A through the center of the section) for all forces:

$$M_b = P_b e_b = [C_c \times 4.16 + C_{s1} \times 5.1 + C_{s2} \times 2.1 + T_1 \times 5.1 + T_2 \times 2.1]$$
$$= 2422.1 \text{ k·in.} = 201.9 \text{ k·ft}$$

$$e_b = 10.5 \text{ in.} \qquad\qquad \blacksquare$$

11.12 Strength of Circular Columns When Compression Controls

The circular section under eccentric load can be analyzed in a similar way as was done for rectangular sections. The procedure requires the calculation of the strains and forces, and their locations on the section, and compression controls when P_n exceeds P_b or the eccentricity e is less than e_b.

An approximate equation to estimate P_n in a circular section when compression controls was suggested by Whitney:[15]

$$P_n = \frac{A_g f'_c}{\left[\frac{9.6De}{(0.8D + 0.67D_s)^2} + 1.18\right]} + \frac{A_{st} f_y}{\left(\frac{3e}{D_s} + 1\right)} \qquad (11.31)$$

where

A_g = gross area of the section

D = external diameter

D_s = diameter measured through the centroid of the bar arrangement

A_{st} = total vertical steel area

e = eccentricity measured from the plastic centroid

EXAMPLE 11.9

Calculate the nominal compressive strength P_n for the section of the previous example using the Whitney equation if the eccentricity $e = 6$ in.

Solution   1. $e = 6$ in. is less than $e_b = 8.82$ in. calculated earlier; thus compression controls.
2. Using the Whitney equation:

$$A_g = \frac{\pi}{4}D^2 = \frac{\pi}{4}(16)^2 = 201.1 \text{ in.}^2$$

$$D = 16 \text{ in.} \quad D_s = 16 - 5 = 11.0 \text{ in.} \quad A_{st} = 8 \times 1 = 8 \text{ in.}^2$$

$$P_n = \frac{(201.1 \times 4)}{\left[\frac{(9.6 \times 16 \times 6)}{(0.8 \times 16 + 0.67 \times 11)^2} + 1.18\right]} + \frac{8 \times 60}{\left(\frac{3 \times 6}{11}\right) + 1}$$

$$= 415.5 \text{ k}$$

$$M_n = P_n e = 415.5 \times \frac{6}{12} = 207.8 \text{ k·ft} \qquad \blacksquare$$

11.13   Approximate Equation for Circular Sections When Tension Controls

The procedure to analyze circular sections using the principles of statics is a lengthy one because of the arrangement and geometry of the section. An approximate equation to estimate P_n for circular sections when tension controls is suggested by Whitney-Hognestad as follows:[15]

$$P_n = 0.85f'_cD^2 \left[\sqrt{\left(\frac{0.85e}{D} - 0.38\right)^2 + \frac{\rho_g mD_s}{2.5D}} - \left(\frac{0.85e}{D} - 0.38\right) \right] \quad (11.32)$$

where

D = diameter of the concrete section

$m = \dfrac{f_y}{0.85f'_c}$

$\rho_g = \dfrac{A_{st}}{A_g}$ = ratio of total steel

D_s = diameter measured through centroid of bars

EXAMPLE 11.10

Calculate P_n for the section of example 11.8 using the Whitney-Hognestad equation if the eccentricity $e = 12$ in.

Solution   1. $e = 12$ in. is greater than $e_b = 8.82$ in., which indicates that tension controls.
2. Using the Whitney-Hognestad equation 11.32:

$D = 16$ in. $D_s = 11.0$ in.

$m = \dfrac{60}{0.85 \times 4} = 17.65$ $\rho_g = \dfrac{A_{st}}{A_g} = \dfrac{8}{201.1} = 0.0398$

$$P_n = 0.85 \times 4(16)^2 \left[\sqrt{\left(\frac{0.85 \times 12}{16} - 0.38\right)^2 + \left(\frac{0.0398 \times 17.65 \times 11}{2.5 \times 16}\right)} \right.$$
$$\left. - \left(\frac{0.85 \times 12}{16} - 0.38\right) \right] = 219.25 \text{ k}$$

$$M_n = P_n e = 219.25 \times \frac{12}{12} = 219.25 \text{ k·ft}$$ ∎

11.14   Design of Columns Under Eccentric Loading

In the previous sections, analysis of columns under eccentric loading was discussed where the column section was given and the nominal compressive strength was required. In the following sections, design of columns will be explained for cases when compression controls and when tension controls.

11.14.1   When Compression Controls

This is the case when P_n is greater than P_b or e is less than e_b. The following example will explain the design procedure.

EXAMPLE 11.11

Design a rectangular short tied column for an axial force of 180 k and a bending moment of 55 k·ft due to dead load; and for an axial load of 120 k and a bending moment of 30 k·ft due to live load. Given: $f'_c = 3$ ksi, $f_y = 40$ ksi, and $b = 14$ in.

Solution

1. Ultimate axial load $= P_u = 1.4P_D + 1.7P_L$

$$P_u = 1.4 \times 180 + 1.7 \times 120 = 456 \text{ k}$$

$$M_u = 1.4M_D + 1.7M_L = 1.4 \times 55 + 1.7 \times 30 = 128 \text{ k·ft}$$

2. Nominal strength:

$$P_n = \frac{P_u}{\phi} = \frac{456}{0.7} = 651.4 \text{ k}$$

$$M_n = \frac{M_u}{\phi} = \frac{128}{0.7} = 182.86 \text{ k·ft}$$

$$e = \frac{M_n}{P_n} = \frac{182.86 \times 12}{651.4} = 3.37 \text{ in.}$$

3. Assume a steel percentage of 3 percent (minimum 1 percent) and $A_s = A'_s$; that is,

$$\rho = \rho' = 1.5 \text{ percent}$$

4. Since $e = 3.37$ in. is small, assume compression controls.
5. Determine the balanced condition requirements: For a balanced section,

$$c = \frac{87}{87 + f_y} d$$

and

$$a_b = \frac{87}{87 + f_y} (0.85d)$$

$$a_b = \frac{87}{87 + 40} (0.85d) = 0.5823d$$

$$P_b = C_c + C_s - T = 0.85f'_c \, a_b b + A'_s f_y - A_s f_y$$

neglecting the concrete displaced by the compression steel. Since $A_s = A'_s$, then

$$P_b = 0.85 \times 3 \times 0.5823d \times 14 = 20.79d$$

If $P_b = P_n = 651.4$ k, then

$$d = \frac{651.4}{20.79} = 31.3 \text{ in.}$$

This means that if the effective depth d used is less than 31.4 in., compression may control; and if d is greater than 31.3 in., tension may control.

6. An approximate estimate of the total cross-sectional area when compression controls can be obtained as follows:

$$A_g = \frac{P_n}{0.85 f'_c} = \frac{651.4}{0.85 \times 3} = 256 \text{ in.}^2$$

Assume a section of 14×20 in. ($A_g = 280$ in.^{2}).

$$A_s = A'_s = 0.015 bh = 0.015 \times 14 \times 20 = 4.2 \text{ in.}^2$$

Choose 4 No. 9 bars on the compression side and the same bars on the tension side ($A_s = A'_s = 4.0$ in.^{2}). Let $h = 20$ in., $d = 17.5$ in., and $d' = 2.5$ in. This is an assumed section which will be checked now.

7. Since $d = 17.5$ in. is less than $\frac{2}{3} d_b = 20.87$ in., and the eccentricity $e = 3.37$ in. is greater than $e_{\min} = 0.1h = 0.1 \times 20 = 2.0$ in., compression controls.

8. Check P_n using the Whitney equation (symmetrical reinforcement is used in single layer), equation 11.23:

$$P_n = \frac{14 \times 20 \times 3}{\left(\dfrac{3 \times 20 \times 3.37}{(17.5)^2}\right) + 1.18} + \frac{4 \times 40}{\left(\dfrac{3.37}{15} + 0.5\right)} = 677.3 \text{ k}$$

This is greater than the required P_n of 651.4 k.

$$M_n = P_n e = 677.3 \times \frac{3.37}{12} = 190.2 \text{ k·ft}$$

which is greater than the 182.86 k·ft required.

9. Check balanced condition:

$$a_b = 0.5823 d = 0.5823 \times 17.5 = 10.2 \text{ in.}$$
$$C_c = 0.85 f'_c a_b d = 0.85 \times 3 \times 10.2 \times 14 = 364.1 \text{ k}$$
$$C_s = A'_s (f_y - 0.85 f'_c) = 4(40 - 2.55) = 149.8 \text{ k}$$
$$T = A_s f_y = 4 \times 40 = 160 \text{ k}$$
$$P_b = C_c + C_s - T = 364.1 + 149.8 - 160 = 353.9 \text{ k}$$

$P_n > P_b$; thus compression controls.

10. Section may be checked by statics, as explained in example 11.3.

11. Design of ties: Choose No. 3 ties spaced at the least of $b = 14$ in. (least column side), 16 bar diameter $= 16 \times \frac{9}{8} = 18$ in., or 48 tie diameter $= 48 \times \frac{3}{8} = 18$ in. Add 2 No. 5 bars at mid-depth of the section, so that reinforcement bars will not be spaced more than 12 in. apart (Figure 11.14). ∎

11.14.2 When Tension Controls

Tension controls when the nominal column strength P_n is less than P_b, as explained in section 11.9. It can be assumed, as a guide, that tension controls when the eccentricity $e > d$ or $e' > 1.5d$. Two cases will be considered: the case when no compression reinforcement is used ($A'_s = 0$) and the case when tension and compression reinforcement are provided in the section.

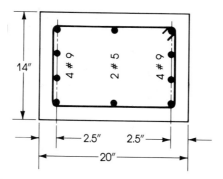

FIGURE 11.14. Example 11.11.

For the case when $A_s' = 0$, proceed as follows: From Figure 11.15(a), $e' = e + d''$. Taking moments about the compressive force C_c:

$$P_n \left[e' - \left(d - \frac{a}{2} \right) \right] = T \left(d - \frac{a}{2} \right) = A_s f_y \left(d - \frac{a}{2} \right)$$

$$P_n e' - P_n \left(d - \frac{a}{2} \right) = A_s f_y \left(d - \frac{a}{2} \right)$$

$$A_s = \frac{P_n e'}{f_y \left(d - \frac{a}{2} \right)} - \frac{P_n}{f_y}$$

$$A_s = \frac{M_n'}{f_y \left(d - \frac{a}{2} \right)} - \frac{P_n}{f_y} \qquad (11.33)$$

where $M_n' = P_n e'$ = moment of the force P_n about the tension steel A_s. The first term represents the area of steel A_s'' for a section subjected to simple bending only with

$$M_n' = A_s'' f_y \left(d - \frac{a}{2} \right)$$

or

$$A_s = A_s'' - \frac{P_n}{f_y} = (\rho'' bd) - \frac{P_n}{f_y} \qquad (11.34)$$

where ρ'' is the steel percentage for $M_n' = P_n e'$. The capacity reduction ϕ varies between 0.7 and 0.9 for pure bending. The tables in Appendix A or B can be used to calculate A_s''. Note that $e' > e$ and $M_n' > M_n$, consequently a larger section is expected when M_n' is used.

$$P_n = \frac{P_u}{\phi} \quad \text{and} \quad M_n = \frac{M_u}{\phi}$$

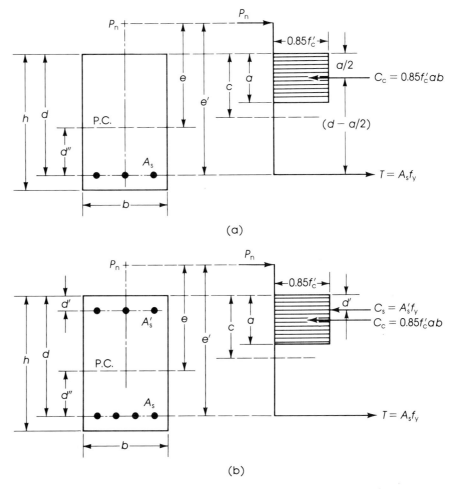

FIGURE 11.15. Sections of columns where tension controls: (a) $A_s' = 0$, and (b) with compression and tension reinforcement.

For the case when compression and tension steel are used in the section (Figure 11.15(b)) and taking moments about the tension steel A_s:

$$P_n e' - C_c \left(d - \frac{a}{2} \right) - C_s(d - d') = 0$$

where

$$C_c = 0.85 f_c' ab \text{ and } C_s = A_s' f_y \text{ (if compression steel yields)}$$

$$A_s' = \frac{P_n e' - C_c \left(d - \frac{a}{2} \right)}{f_y(d - d')}$$

or

$$A_s' = \frac{M_n' - M_{nl}}{f_y(d - d')} \tag{11.35}$$

where

$$M'_n = P_n e' \text{ and}$$

$$M_{nl} = C_c \left(d - \frac{a}{2}\right) = A_{sl} f_y \left(d - \frac{a}{2}\right) \tag{11.36}$$

which is the moment capacity of the section as singly reinforced due to simply bending only. Since tension steel yields, then using f_y in the above equation is adequate.

Taking moments about the compression steel A'_s:

$$P_n[e' - (d - d')] + C_c \left(\frac{a}{2} - d'\right) - T(d - d') = 0$$

$$P_n e' - [P_n(d - d')] + C_c \left(\frac{a}{2} - d'\right) = A_s f_y(d - d')$$

Writing $(a/2 - d')$ in the form $(d - d') - (d - a/2)$, then

$$A_s = \left[\frac{P_n e'}{f_y(d - d')} - \frac{P_n}{f_y}\right] + \frac{C_c}{f_y(d - d')} \left[(d - d') - \left(d - \frac{a}{2}\right)\right]$$

$$A_s = \left[\frac{M'_n}{f_y(d - d')} - \frac{P_n}{f_y}\right] + \frac{C_c}{f_y} - \frac{C_c \left(d - \frac{a}{2}\right)}{f_y(d - d')}$$

But

$$M_{nl} = C_c \left(d - \frac{a}{2}\right) = A_{sl} f_y \left(d - \frac{a}{2}\right) \tag{11.36}$$

Therefore,

$$A_s = \left(\frac{C_c}{f_y} - \frac{P_n}{f_y}\right) + \frac{M'_n - M_{nl}}{f_y(d - d')} \tag{11.37}$$

$$A_s = \left(A_{sl} - \frac{P_n}{f_y}\right) + \frac{M'_n - M_{nl}}{f_y(d - d')} \tag{11.40}$$

or

$$A_s = \left(\rho_l bd - \frac{P_n}{f_y}\right) + \frac{M'_n - M_{nl}}{f_y(d - d')} \tag{11.41}$$

where $M_{nl} = C_c(d - a/2) = A_{sl} f_y(d - a/2)$ is the equation of a reinforced section subjected to simple bending.

$$\frac{C_c}{f_y} = \frac{T_l}{f_y} = A_{sl} = \rho_l bd$$

which is the steel ratio for M_{nl} ($\rho_l \leq \rho_{max}$).

Notes

1. The above equations assume that compression steel yields. If it does not yield, the actual stress in the compression steel must be used in equation 11.35:

$$A'_s = \frac{M'_n - M_{nl}}{f'_s(d - d')}$$

2. Yielding of compression steel can be checked using the tables in Appendix A for U.S. customary units and Appendix B for S.I. units.
3. Check if $\rho_l \leq \rho_{max}$.

EXAMPLE 11.12

Design a rectangular section for a short tied column to carry an axial ultimate load $P_u = 30$ k and an ultimate moment $M_u = 180$ k·ft. Given: $f'_c = 3$ ksi, $f_y = 40$ ksi, and $b = 10$ in. Do not use compression steel in the section ($A'_s = 0$).

Solution

1. $$e = \frac{M_u}{P_u} = \frac{180 \times 12}{30} = 72 \text{ in.}$$

   The eccentricity e is quite large, and since the expected d is less than 72 in., tension may be assumed to control and must be checked later.

2. Since d is not known, determine an approximate d from the given moment, considering an assumed reduction factor $\phi = 0.7$:

$$M_n = \frac{M_u}{\phi} = \frac{180}{0.7} = 257.1 \text{ k·ft}$$

$$R = \rho f_y \left(1 - \frac{\rho f_y}{1.7 f'_c}\right) \qquad (3.19)$$

For $\rho = 0.014$, $R = 500$ psi $= 0.5$ ksi (or use tables in Appendix A).

$$d = \sqrt{\frac{M_n}{Rb}} = \sqrt{\frac{257.1 \times 12}{0.5 \times 10}} = 24.8 \text{ in.}$$

Letting $h = $ total depth $= 25$ in., then $d = 25 - 2.5 = 22.5$ in.

3. Check the assumed value of ϕ for a tied column.

$$\phi = 0.9 - \frac{2P_u}{f'_c A_g} \geq 0.7 \qquad (11.3)$$

$$= 0.9 - \frac{2 \times 30}{3 \times (10 \times 25)} = 0.82$$

$$\frac{d - d'}{h} = \frac{20}{25} = 0.8 > 0.7$$

$$M_n = \frac{180}{\phi} = \frac{180}{0.82} = 219.5 \text{ k·ft}$$

$$P_n = \frac{30}{\phi} = \frac{30}{0.82} = 36.6 \text{ k}$$

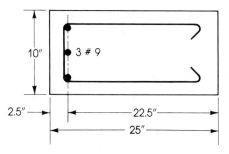

FIGURE 11.16. Example 11.12.

4. Calculate A_s: Let

$$e' = e + d'' = e + d - \frac{h}{2} \text{ (approximately)}$$

$$= 72 + 22.5 - \frac{25}{2} = 82 \text{ in.}$$

$$M'_n = P_n e' = 36.6 \times \frac{82.0}{12} = 250.1 \text{ k·ft}$$

$$R = \frac{M'_n}{bd^2} = \frac{250.1 \times 12,000}{10 \times (22.5)^2} = 593 \text{ psi}$$

The steel ratio ρ can be calculated from

$$R = \rho f_y \left(1 - \frac{\rho f_y}{1.7 f'_c} \right)$$

or from tables in Appendix A. The values of R in Tables A.1 to A.3 are multiplied by 0.9. Therefore, to use these tables, multiply the calculated $R = 593$ psi by 0.9; then $R_u = 0.9R = 534$ psi. From Table A.1, for $R_u = 534$ psi, $\rho'' = 0.017$.

$$A_s = \rho' bd - \frac{P_n}{f_y}$$

$$= 0.017 \times 10 \times 22.5 - \frac{36.6}{40} = 2.91 \text{ in.}^2$$

Choose 3 No. 9 bars ($A_s = 3.00$ in.^{2}) (Figure 11.16).

$$\rho \text{ used} = \frac{3.0}{10 \times 22.5} = 0.0133 < \rho_{max} = 0.0278$$

Minimum $\rho = 0.01$.

5. Calculate the balanced force P_b:

$$\phi P_b = \phi [0.85 f'_c ab + A'_s (f_y - 0.85 f'_c) - A_s f_y]$$

$$a_b = \frac{87}{87 + 40} \times 0.85 \times 22.5 = 13.1 \text{ in.} \qquad A'_s = 0$$

$$\phi P_b = 0.7[0.85 \times 3 \times 13.1 \times 10 - 3.0 \times 40] = 150 \text{ k}$$

$$0.1 f'_c A_g = 0.1 \times 3 \times 10 \times 25 = 75 \text{ k} < P_b \qquad (\phi = 0.82)$$

$$P_u = 30 \text{ k} < 75 \text{ k}$$

Therefore, tension controls.

6. The section can be checked for P_n using equation 11.26. ∎

EXAMPLE 11.13

The section of a short tied beam-column is limited to a width $b = 11$ in. and a total depth $h = 23$ in. The column is subjected to an ultimate moment $M_u = 300$ k·ft and an ultimate axial load $P_u = 30$ k. Determine the necessary reinforcement if $f'_c = 3$ ksi and $f_y = 50$ ksi.

Solution

1. The eccentricity

$$e = \frac{M_u}{P_u} = \frac{300 \times 12}{30} = 120 \text{ in.}$$

This is a large eccentricity when compared with a section of $h = 23$ in. It will be assumed that tension controls.

2. Determine the capacity reduction factor ϕ:

$$\phi = 0.9 - \frac{2 P_u}{f'_c A_g} \geq 0.7 \tag{11.3}$$

$$= 0.9 - \frac{2 \times 30}{3 \times (11 \times 23)} = 0.82$$

$$\frac{d - d'}{h} = \frac{18}{23} = 0.78 > 0.7$$

$$M_n = \frac{300}{0.82} = 365.9 \text{ k·ft}$$

$$P_n = \frac{30}{0.82} = 36.5 \text{ k}$$

3. Assume two rows of steel bars and assume $d = 23 - 3.5 = 19.5$ in. Let e' equal the distance from P_n to the centroid of tension steel:

$$e' = e + d - \frac{h}{2} = 120 + 19.5 - 11.5 = 128.0 \text{ in.}$$

$$M'_n = P_n e' = 36.5 \times 128.0 = 4672 \text{ k·in.}$$

$$R = \frac{M'_n}{bd^2} = \frac{4672 \times 1000}{11 \times (19.5)^2} = 1117 \text{ psi}$$

$$R_{max} \text{ for singly reinforced beam section} = \rho_{max} f_y \left(1 - \frac{\rho_{max} f_y}{1.7 f'_c} \right) \tag{3.19}$$

ρ_{max} is 0.0206 for $f'_c = 3$ ksi and $f_y = 50$ ksi and thus $R_{max} = 822$ psi. These values can be obtained from Table A.1 in Appendix A. Note that R_u values in the tables are already multiplied by $\phi = 0.9$ for flexural design, or $R_u = 0.9R$.

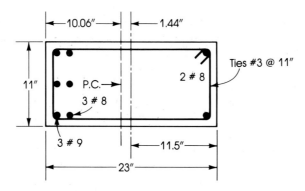

FIGURE 11.17. Example 11.13.

$$R_{max} = \frac{R_u(max)}{0.9} = \frac{740}{0.9} = 822 \text{ psi}$$

The value of R_{max} is less than $R = 1117$ psi of the section. This indicates that reinforcement on the compression side of the section must be provided.

4. Let $d' = 2.5$ in.

$$M_{nl} = R_{max} bd^2 = 0.822 \times 11 \times (19.5)^2 = 3438 \text{ k·in.} = 286.5 \text{ k·ft}$$

$$A_s' = \frac{M_n' - M_{nl}}{f_y(d - d')} = \frac{4672 - 3438}{50(17)} = 1.45 \text{ in.}^2 \qquad (11.35)$$

Use 2 No. 8 bars ($A_s = 1.58$ in.^{2}).

$$A_s = \left(\rho_{max}bd - \frac{P_n}{f_y}\right) + \frac{M_n' - M_{nl}}{f_y(d - d')} \qquad (11.39)$$

$$A_s = \left(0.0206 \times 11 \times 19.5 - \frac{36.5}{50}\right) + 1.45 = 5.14 \text{ in.}^2$$

Use 3 No. 9 plus 3 No. 8 bars in two rows ($A_s = 5.35$ in.^{2}), as in Figure 11.17.

5. Check if compression steel yields:

$$\rho - \rho' = \rho_{max} = 0.0206$$

$$\frac{d'}{d} = \frac{2.5}{19.5} = 0.128$$

The yielding of compression steel can be checked from a strain diagram along the depth of the section, or from Table A.9 in Appendix A: $(\rho - \rho')_{min}$ for compression steel to yield is 0.014 (by interpolation); $(\rho - \rho')$ used $= \rho_{max}$, which is much greater than 0.014. Therefore compression steel yields.

6. Design column ties: Use No. 3 ties spaced at the least of width of column = 11.0 in., 16 bar diameters = $16 \times \frac{8}{8}$ = 16.0 in., or 48 tie diameters = $48 \times \frac{3}{8}$ = 18.0 in. Use No. 3 ties spaced at 11.0 in.

Notes

1. Since compression steel differs from the tension steel, $e' = e + d''$, which is slightly different from $e' = e + d - h/2$ assumed but will not change the steel requirement. The plastic

centroid lies at 1.44 in. from the centroid of the concrete section, as shown in Figure 11.17. Actual e' is $120 + 6.56 = 126.56$ in. compared to $e' = 128$ in. used. M_n' calculated using $e' = 128$ in. will thus be on the conservative side.

2. This method of design allows for choosing between variable amounts of steel reinforcement in both tension and compression sides.
3. Flexibility of design can be provided, for example, by adopting a smaller steel ratio for the singly reinforced part $(\rho - \rho')$, that is, assuming $(\rho - \rho') < \rho_{max}$, and consequently increasing the compression reinforcement.
4. When tension controls, the ratio of the compression to the tension reinforcement A_s'/A_s may vary from 1.0 to a minimum of zero. The choice of the ratio depends on the dimensions of the sections, the location of $e = M_n/P_n$ on the interaction diagram, and on any other restrictions. It is practical to provide reinforcement on the compression side of the section. This can be achieved by using $(\rho - \rho')$ less than ρ_{max} assumed in the above example, or using $A_s' = 0.5A_s$.
5. The section can be checked for P_n using equation 11.26.
6. If the depth of the section is not known, calculate $M_n = M_u/0.7$ and determine an approximate depth from the flexural design using M_n. ∎

EXAMPLE 11.14

Assuming that $\rho_1 = \rho - \rho' = 0.75\rho_{max}$ used in the previous example, calculate the tension and compression reinforcement required to carry the given loads.

Solution

1. Using $\rho_1 = \rho - \rho' = 0.75\rho_{max} = 0.75 \times 0.0206 = 0.01545$,

$$\text{Design } R = 0.75R_{max} = 0.75 \times 822 = 616.5 \text{ psi}$$

$$M_{nl} = Rbd^2 = 0.6165 \times 11 \times (19.5)^2 = 2579 \text{ k·in.}$$

2.

$$A_s' = \frac{M_n' - M_{nl}}{f_y(d - d')} = \frac{4672 - 2579}{50 \times 17} = 2.46 \text{ in.}^2$$

$$A_s = \left(\rho_1 bd - \frac{P_n}{f_y}\right) + \frac{M_n' - M_{nl}}{f_y(d - d')}$$

$$= \left(0.01545 \times 11 \times 19.5 - \frac{36.5}{50}\right) + 2.46 = 5.04 \text{ in.}^2$$

$$\text{Ratio } \frac{A_s'}{A_s} = \frac{2.46}{5.04} = 0.49$$

Use 5 No. 9 bars ($A_s = 5.0$ in.²) in two rows in the tension side and 2 No. 10 bars ($A_s = 2.53$ in.²) in the compression side.

3. Since ρ_1 used here $= 0.01545$ is greater than 0.014 (see the previous example), then compression steel yields. Otherwise calculate ε_s' from a strain diagram and if it is less than ε_y, calculate $f_s' = \varepsilon_s' E_s$ and use it in the above equation. ∎

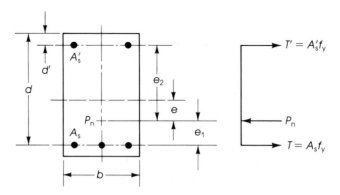

FIGURE 11.18. Section subjected to tension force with small eccentricity.

11.15   Design of Sections Subjected to Tension Force with Small Eccentricity

This case occurs when the tension force P_n lies within the distance between the steel reinforcement. The eccentricity e is less than $d/2$. Since the concrete tensile strength is neglected, then the total force is resisted by T and T' (Figure 11.18). The reinforcements A_s and A'_s are inversely proportional to their distance from the point of application of the normal force.

$$P_n = T + T' \tag{11.40}$$

Taking moments about A'_s,

$$P_n e_2 = A_s f_y (d - d')$$

or

$$A_s = \frac{P_n e_2}{f_y(d - d')} \tag{11.41}$$

Taking moments about A_s,

$$P_n e_1 = A'_s f_y (d - d')$$

$$A'_s = \frac{P_n e_1}{f_y(d - d')} \tag{11.42}$$

The capacity reduction factor ϕ in flexure and axial tension is 0.9:

$$P_u = \phi P_n = 0.9 P_n$$

EXAMPLE   **11.15**

Determine the necessary reinforcement for the section shown in Figure 11.19 to resist an ultimate tensile force $P_u = 207$ k acting at a distance of 3 in. from the centroid of the concrete section. Given: $f'_c = 3$ ksi and $f_y = 60$ ksi.

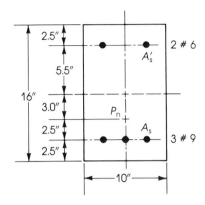

FIGURE 11.19. Example 11.15.

Solution

$$P_n = \frac{P_u}{\phi} = \frac{207}{0.9} = 230 \text{ k}$$

$$e_1 = 2.5 \text{ in.} \quad e_2 = 8.5 \text{ in.} \quad d - d' = 11.0 \text{ in.}$$

$$A_s = \frac{P_n e_2}{f_y(d - d')} = \frac{230 \times 8.5}{60 \times 11} = 2.96 \text{ in.}^2$$

Use 3 No. 9 bars ($A_s = 3.00 \text{ in.}^2$).

$$A_s' = \frac{P_n e_1}{f_y(d - d')} = \frac{230 \times 2.5}{60 \times 11} = 0.87 \text{ in.}^2$$

Use 2 No. 6 bars ($A_s = 0.88 \text{ in.}^2$). ∎

11.16 Design of Columns Using Charts and Tables

The design procedures explained earlier are based on the principles of statics. In practical design, the structural engineer uses charts, tables, and computer programs to design all types of columns with different arrangements of steel reinforcement. These charts and tables are published by the American Concrete Institute (ACI), the Concrete Reinforcing Steel Institute (CRSI), and the Portland Cement Association (PCA). Programs for the design of columns can be prepared by the student or the engineer using programmable hand calculators or computers. The main purpose is to provide easy, accurate, and fast design of structural members. The use of charts can be illustrated in the following examples, using the sample charts shown in Figures 11.20 and 11.21 and the sample column tables of Appendix A prepared by the author.

EXAMPLE 11.16

Determine the ultimate strength P_u of the short tied column shown in Figure 11.22 (p. 339) acting at an eccentricity $e = 12$ in., using charts. Given: $f_c' = 5$ ksi and $f_y = 60$ ksi.

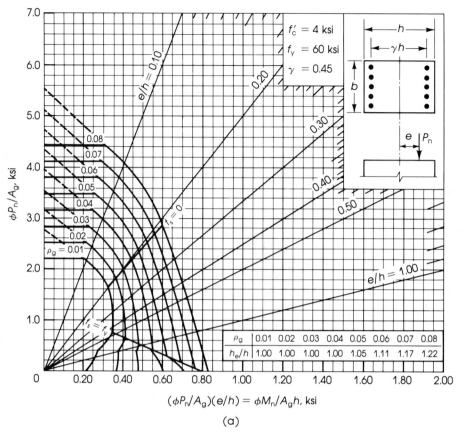

$$(\phi P_n/A_g)(e/h) = \phi M_n/A_g h, \text{ ksi}$$

(a)

FIGURE 11.20. Load-moment strength interaction diagram for rectangular columns where $f'_c = 4$ ksi, $f_y = 60$ ksi, and (a) $\gamma = 0.45$, (b) $\gamma = 0.60$, (c) $\gamma = 0.75$, and (d) $\gamma = 0.90$. Courtesy of American Concrete Institute.[7]

Solution   1. Properties of the section:

$$h = 24 \text{ in.}, \quad \gamma h = 24 - 5 = 19 \text{ in.} = \text{ distance between } A_s \text{ and } A'_s$$

$$\gamma = \frac{19}{24} = 0.79 \qquad \rho = \frac{A_{st}}{bh} = \frac{8 \times 1.27}{14 \times 24} = 0.03$$

$$\frac{e}{h} = \frac{12}{24} = 0.5 \qquad A_g = bh$$

2. From charts (Figure 11.21),

$$\frac{P_u}{bh} = 1.5 \quad \text{for} \quad \gamma = 0.75$$

$$= 1.65 \quad \text{for} \quad \gamma = 0.9$$

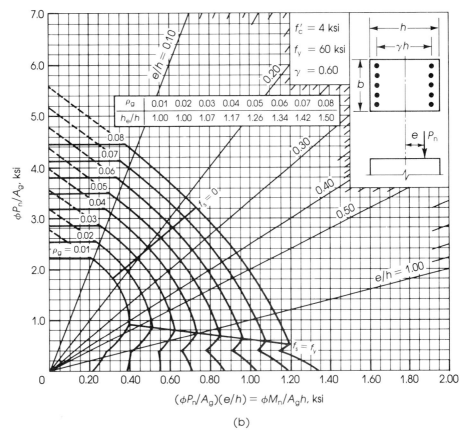

$$f_c' = 4 \text{ ksi}$$
$$f_y = 60 \text{ ksi}$$
$$\gamma = 0.60$$

ρ_g	0.01	0.02	0.03	0.04	0.05	0.06	0.07	0.08
h_e/h	1.00	1.00	1.07	1.17	1.26	1.34	1.42	1.50

$$(\phi P_n/A_g)(e/h) = \phi M_n/A_g h, \text{ ksi}$$

(b)

FIGURE 11.20 (continued).

Therefore, by interpolation for $\gamma = 0.79$:

$$\frac{P_u}{bh} = 1.56 \quad \text{and} \quad P_u = 1.56 \times 14 \times 24 = 524 \text{ k}$$

EXAMPLE 11.17

Determine the necessary reinforcement for a short tied column to carry an ultimate load of 525 k and an ultimate moment of 350 k·ft. The column section has a width of $b = 14$ in. and a total depth $h = 20$ in. Given: $f_c' = 4$ ksi, $f_y = 60$ ksi, and the charts of Figure 11.20.

Solution   1. The eccentricity

$$e = \frac{M_u}{P_u} = \frac{350 \times 12}{525} = 8 \text{ in.}$$

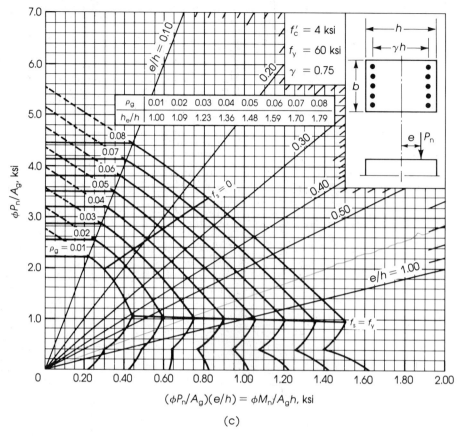

FIGURE 11.20 (continued).

Let $d = 20 - 2.5 = 17.5$ in. and the distance between tension and compression steel

$$\gamma h = 20 - (2.5 + 2.5) = 15.0 \text{ in.}$$

$$\gamma = \frac{15}{20} = 0.75 \qquad \frac{e}{h} = \frac{8}{20} = 0.40$$

$$\frac{P_u}{bh} = \frac{525}{14 \times 20} = 1.88$$

2. From graphs, for $\frac{P_u}{bh} = 1.88$, $\frac{e}{h} = 0.40$, and $\gamma = 0.75$,

$$\rho = 0.044$$

$$A_s = 0.044 \times 14 \times 20 = 12.32 \text{ in.}^2$$

Choose 8 No. 11 bars ($A_s = 12.48$ in.^{2}), 4 bars on each short side, and use No. 4 ties spaced at 14 in.

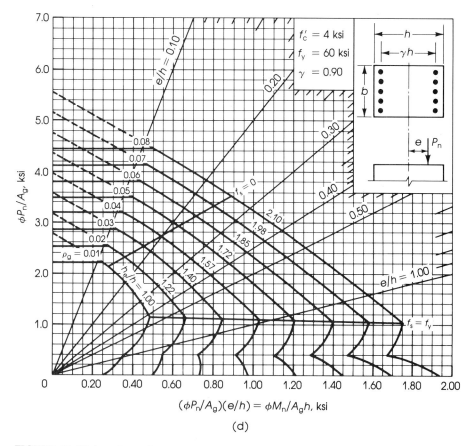

FIGURE 11.20 (continued).

EXAMPLE 11.18

Repeat the same problem of example 11.17 using the column tables of Appendix A.

Solution   Given: $P_u = 525$ k, $M_u = 350$ k·ft, $b = 14$ in., $h = 20$ in., $f'_c = 4$ ksi, and $f_y = 60$ ksi. Required: A_s and A'_s.

1. Determine the eccentricity

$$e = \frac{M_u}{P_u} = \frac{350 \times 12}{525} = 8 \text{ in.}$$

From the column tables of $f'_c = 4$ ksi, and for the section 14 by 20 in., choose the reinforcement A_s and A'_s (on the short sides of the column) corresponding to a load $P_u \geq$ 525 k under an eccentricity $e = 8$ in.

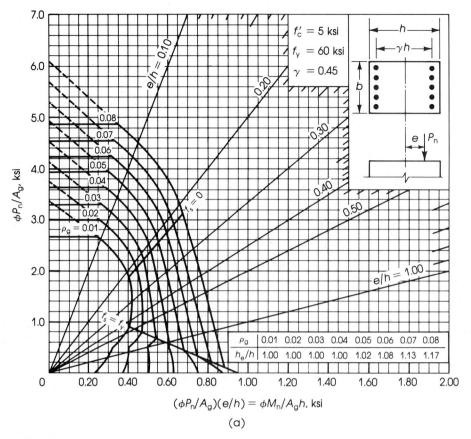

FIGURE 11.21. Load-moment strength interaction diagram for rectangular columns where $f'_c = 5$ ksi, $f_y = 60$ ksi, and (a) $\gamma = 0.45$, (b) $\gamma = 0.60$, (c) $\gamma = 0.75$, and (d) $\gamma = 0.90$. Courtesy of American Concrete Institute.[7]

2. For vertical bars distributed along the short side, if 3 No. 14 bars are selected, $P_u = 545$ k. If 4 No. 11 bars are selected, $P_u = 529$ k. Therefore, choose 4 No. 11 bars (in one row) on each short side. Total steel area used (8 No. 11) = 12.32 in.^{2} and the steel percentage $\rho = 12.32/(14 \times 20) = 0.044$ or 4.4 percent. ∎

EXAMPLE 11.19

Design a circular column section to resist the ultimate load and moment given in the previous example. Use $f'_c = 4$ ksi, $f_y = 60$ ksi, and the tables of Appendix A.

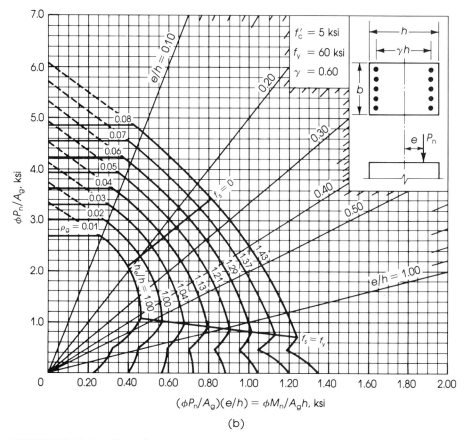

$f'_c = 5$ ksi
$f_y = 60$ ksi
$\gamma = 0.60$

$e/h = 0.10$
0.20
0.30
0.40
0.50

$f_s = 0$

$e/h = 1.00$

$f_s = f_y$

$\rho_g = 0.01$
0.02
0.03
0.04
0.05
0.06
0.07
0.08

$h_e/h = 1.00$
1.00
1.04
1.13
1.21
1.29
1.37
1.43

$\phi P_n/A_g$, ksi

$(\phi P_n/A_g)(e/h) = \phi M_n/A_g h$, ksi

(b)

FIGURE 11.21 (continued).

Solution   Given: $P_u = 525$ k and $M_u = 350$ k·ft. Eccentricity $e = M_u/P_u = (350 \times 12)/525 = 8$ in.
From the column tables of circular sections, the following may be selected:

Diameter (in.)	Steel bars	P_u (kips)	ρ (%)
24	6 No. 10	528	1.68
24	8 No. 9	534	1.77
24	10 No. 8	533	1.74
22	6 No. 14	525	3.55
22	10 No. 11	557	4.10
20	6 No. 18	553	7.64
20	10 No. 14	546	7.16

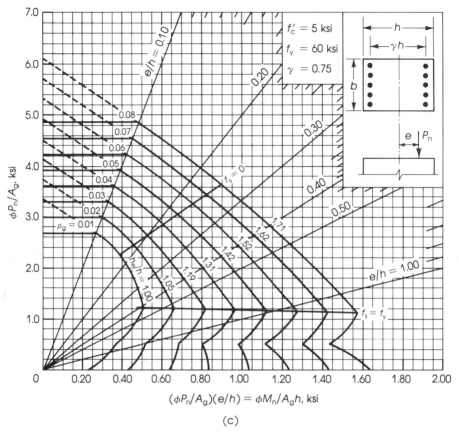

FIGURE 11.21 (continued).

A smaller column section needs a higher steel ratio. If no limitations exist, choose a 22-in. column reinforced with 6 No. 14 bars, or 24-in. column reinforced with 6 No. 10 bars. ■

11.17   Biaxial Bending

The analysis and design of columns under eccentric loading was discussed earlier in this chapter considering a uniaxial case. This means that the load P_n was acting along the y-axis (Figure 11.23) causing a combination of axial load P_n and a moment about the x-axis $= M_{nx} = P_n e_y$; or acting along the x-axis (Figure 11.24) with an eccentricity e_x, causing a combination of an axial load P_n and a moment $M_{ny} = P_n e_x$.

If the load P_n is acting anywhere such that its distance from the x-axis is e_y and its distance from the y-axis is e_x, then the column section will be subjected to a combination of forces: an axial load P_n, a moment about the x-axis $= M_{nx} = P_n e_y$, and a moment about the y-axis $= M_{ny} = P_n e_x$ (Figure 11.25). The column section in this case is said to be subjected to biaxial bending. The analysis and design of columns under this combina-

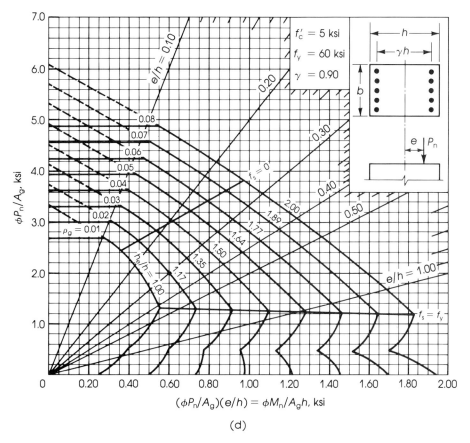

FIGURE 11.21 (continued).

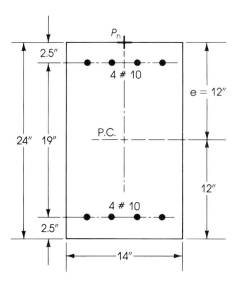

FIGURE 11.22. Example 11.16.

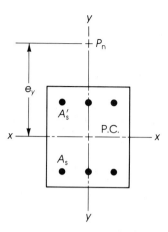

FIGURE 11.23. Uniaxial bending with load P_n along the y-axis with eccentricity e_y.

tion of forces is not simple when the principles of statics are used. The neutral axis is at an angle with respect to both axes, and lengthy calculations are needed to determine the location of the neutral axis, strains, concrete compression area, and internal forces and their point of application. Therefore it was necessary to develop practical solutions to estimate the strength of columns under axial load and biaxial bending. The formulas developed relate the response of the column in biaxial bending to its uniaxial strength about each major axis.

The biaxial bending strength of an axially loaded column can be represented by a three-dimensional interaction curve, as shown in Figure 11.26. The surface is formed by a series of uniaxial interaction curves drawn radially from the P_n axis. The curve M_{ox}

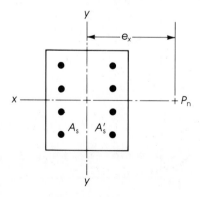

FIGURE 11.24. Uniaxial bending with load P_n along the x-axis with eccentricity e_x.

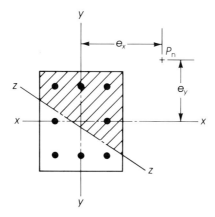

FIGURE 11.25. Biaxial bending.

represents the interaction curve in uniaxial bending about the x-axis, and the curve M_{oy} represents the curve in uniaxial bending about the y-axis. The plane at constant axial load P_n shown in Figure 11.26 represents the contour of the bending moment M_n about any axis.

Different shapes of columns may be used to resist axial loads and biaxial bending. Circular, square, or rectangular column cross-sections may be used with equal or unequal bending capacities in the x and y directions.

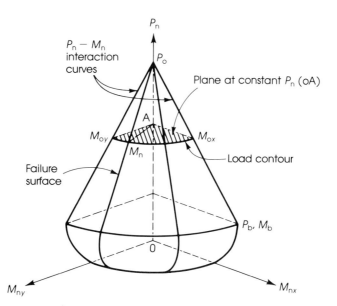

FIGURE 11.26. Biaxial interaction surface.

11.18   Circular Columns with Uniform Reinforcement

Circular columns with reinforcement distributed uniformly about the perimeter of the section have almost the same moment capacity in all directions. If a circular column is subjected to biaxial bending about the x- and y-axes, the biaxial ultimate moment M_u can be calculated using either of the following equations:

$$M_u = \sqrt{(M_{ux})^2 + (M_{uy})^2} \tag{11.43}$$

and

$$e = \sqrt{(e_x)^2 + (e_y)^2} \tag{11.44}$$

where

$M_{ux} = P_u e_y$ = uniaxial moment about the x-axis

$M_{uy} = P_u e_x$ = uniaxial moment about the y-axis

$M_u = P_u e$ = uniaxial moment capacity of the section about any axis

In circular columns, a minimum of six bars should be used, and these should be uniformly distributed in the section.

11.19   Square and Rectangular Columns Under Biaxial Bending

Square or rectangular columns with unequal bending moments about their major axes will require a different amount of reinforcement in each direction. An approximate method of analysis of such sections was developed by Boris Bresler and is called the Bresler reciprocal method.[9,12] According to this method, the load capacity of the column under biaxial bending can be determined by using the following expression:

$$\frac{1}{P_u} = \frac{1}{P_{ux}} + \frac{1}{P_{uy}} - \frac{1}{P_{uo}} \tag{11.45}$$

or

$$\frac{1}{P_n} = \frac{1}{P_{nx}} + \frac{1}{P_{ny}} - \frac{1}{P_{no}} \tag{11.46}$$

where

P_u = load capacity under biaxial bending

P_{ux} = uniaxial load capacity when the load acts at an eccentricity e_y and $e_x = 0$

P_{uy} = uniaxial load capacity when the load acts at an eccentricity e_x and $e_y = 0$

P_{uo} = pure axial load when $e_x = e_y = 0$

$$P_n = \frac{P_u}{\phi} \quad P_{nx} = \frac{P_{ux}}{\phi} \quad P_{ny} = \frac{P_{uy}}{\phi} \quad P_{no} = \frac{P_{uo}}{\phi}$$

The uniaxial load strengths P_{nx}, P_{ny}, and P_{no} can be calculated according to the equations and methods given earlier in this chapter or by using the appropriate charts. After that, they are substituted in equation 11.46 to calculate P_n.

The Bresler equation is valid for all cases when P_n is equal to or greater than $0.10P_{no}$. When P_n is less than $0.10P_{no}$, the axial force may be neglected and the section can be designed as a member subjected to pure biaxial bending according to the following equation:

$$\frac{M_{ux}}{M_x} + \frac{M_{uy}}{M_y} \leq 1.0 \qquad (11.47)$$

or

$$\frac{M_{nx}}{M_{ox}} + \frac{M_{ny}}{M_{oy}} \leq 1.0 \qquad (11.48)$$

where

$M_{ux} = P_u e_y =$ design moment about the x-axis

$M_{uy} = P_u e_x =$ design moment about the y-axis

M_x and $M_y =$ uniaxial moment capacities about the x- and y-axes

$$M_{nx} = \frac{M_{ux}}{\phi} \quad M_{ny} = \frac{M_{uy}}{\phi} \quad M_{ox} = \frac{M_x}{\phi} \quad M_{oy} = \frac{M_y}{\phi}$$

The Bresler equation is not recommended when the section is subjected to axial tension loads.

11.20 Bresler Load Contour Method

In this method, the failure surface shown in Figure 11.26 is cut at a constant value of P_n, giving the related values of M_{nx} and M_{ny}. The general nondimensional expression for the load contour method is

$$\left(\frac{M_{nx}}{M_{ox}}\right)^{\alpha_1} + \left(\frac{M_{ny}}{M_{oy}}\right)^{\alpha_2} = 1.0 \qquad (11.49)$$

Bresler indicated that the exponent α can have the same value in both terms of the above expression ($\alpha_1 = \alpha_2$). Furthermore, he indicated that the value of α varies between 1.15 and 1.55 and can be assumed to be 1.5 for rectangular sections. For square sections, α varies between 1.5 and 2.0, and an average value of $\alpha = 1.75$ may be used for practical designs. When the reinforcement is uniformly distributed around the four faces in square columns, α may be assumed to be 1.5.

$$\left(\frac{M_{nx}}{M_{ox}}\right)^{1.5} + \left(\frac{M_{ny}}{M_{oy}}\right)^{1.5} = 1.0 \qquad (11.50)$$

The 1972 British Code CP110 assumes $\alpha = 1.0, 1.33, 1.67,$ and 2.0 when the ratio $P_u/1.1P_{uo}$ is equal to 0.2, 0.4, 0.6, and ≥ 0.8, respectively.

11.21   PCA Load Contour Method

The load contour approach, proposed by the Portland Cement Association (PCA), is an extension of the method developed by Bresler. In this approach, which is also called Parme method,[11] a point B on the load contour (of a horizontal plane at a constant P_n shown in Figure 11.26) is defined such that the biaxial moment capacities M_{nx} and M_{ny} are in the same ratio as the uniaxial moment capacities M_{ox} and M_{oy}, that is

$$\frac{M_{nx}}{M_{ny}} = \frac{M_{ox}}{M_{oy}}$$

or

$$\frac{M_{nx}}{M_{ox}} = \frac{M_{ny}}{M_{oy}} = \beta$$

The ratio β is shown in Figure 11.27 and represents that constant portion of the uniaxial moment capacities which may be permitted to act simultaneously on the column section.

For practical design, the load contour shown in Figure 11.27 may be approximated by two straight lines AB and BC. The slope of line AB $= (1 - \beta)/\beta$ and the slope of line BC $= \beta/(1 - \beta)$. Therefore when

$$\frac{M_{ny}}{M_{oy}} > \frac{M_{nx}}{M_{ox}}$$

then

$$\frac{M_{ny}}{M_{oy}} + \frac{M_{nx}}{M_{ox}}\left(\frac{1-\beta}{\beta}\right) = 1 \tag{11.51}$$

and when

$$\frac{M_{ny}}{M_{oy}} < \frac{M_{nx}}{M_{ox}}$$

then

$$\frac{M_{nx}}{M_{ox}} + \frac{M_{ny}}{M_{oy}}\left(\frac{1-\beta}{\beta}\right) = 1 \tag{11.52}$$

The actual value of β depends on the ratio P_n/P_o as well as the material and properties of the cross-section. For lightly loaded columns, β will vary from 0.55 to 0.7. An average value of $\beta = 0.65$ can be used for design purposes.

When uniformly distributed reinforcement is adopted along all faces of rectangular columns, the ratio M_{oy}/M_{ox} is approximately b/h, where b and h are the width and total depth of the rectangular section, respectively. Substituting this ratio in equations 11.51 and 11.52,

$$M_{ny} + M_{nx}\left(\frac{b}{h}\right)\left(\frac{1-\beta}{\beta}\right) \simeq M_{oy} \tag{11.53}$$

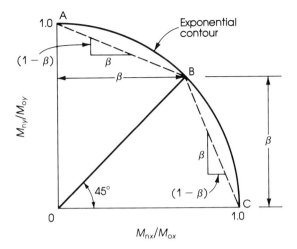

FIGURE 11.27. Nondimensional load contour at constant P_n (straight-line approximation).

and

$$M_{nx} + M_{ny} \left(\frac{h}{b}\right)\left(\frac{1 - \beta}{\beta}\right) \simeq M_{ox} \tag{11.54}$$

For $\beta = 0.65$ and $h/b = 1.5$, then

$$M_{oy} \simeq M_{ny} + 0.36 M_{nx} \tag{11.55}$$

and

$$M_{ox} \simeq M_{nx} + 0.80 M_{ny} \tag{11.56}$$

From the above presentation, it can be seen that direct explicit equations for the design of columns under axial load and biaxial bending are not available. Therefore, the designer should have enough experience to make an initial estimate of the section using the values of P_n, M_{nx}, and M_{ny} and the uniaxial equations, then check the adequacy of the column section using the equations for biaxial bending.

EXAMPLE 11.20

The section of a short tied column is reinforced with 8 No. 10 bars distributed as shown in Figure 11.28. Determine the allowable ultimate load on the section P_u if it acts at $e_x = 8$ in. and $e_y = 12$ in. Use $f'_c = 5$ ksi and $f_y = 60$ ksi.

Solution   1. Determine the uniaxial load capacity P_{ux} about the x-axis:

$$\gamma h = 24 - 5 = 19 \text{ in.} \quad \gamma = \frac{19}{24} = 0.792$$

$$\frac{e_y}{h} = \frac{12}{24} = 0.5$$

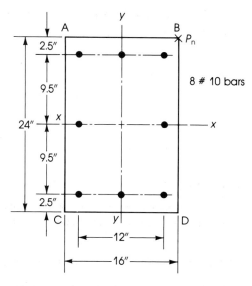

FIGURE 11.28. Example 11.20.

Considering 3 No. 10 bars at AB and 3 No. 10 bars at CD to calculate P_{ux}, then the steel ratio

$$\rho = \frac{6 \times 1.27}{16 \times 24} = 0.02$$

From graphs of Figure 11.21:

$$\text{For } \gamma = 0.75, \frac{P_u}{bh} = 1.3$$

$$\text{For } \gamma = 0.9, \frac{P_u}{bh} = 1.4$$

Therefore by interpolation and for $\gamma = 0.792$,

$$\frac{P_u}{bh} = 1.328$$

$$P_{ux} = 1.328 \times 16 \times 24 = 510 \text{ k}$$

2. Determine the uniaxial load capacity P_{uy} about the y-axis: let

$$\gamma h = 16 - 4 = 12 \text{ in.} \quad \gamma = \frac{12}{16} = 0.75$$

(normally $\gamma h = 11$ in.)

$$\frac{e_x}{b} = \frac{8}{16} = 0.5$$

Considering 3 No. 10 bars at BD and 3 No. 10 bars at AC for bending about the y-axis to calculate P_{uy}, the steel ratio

$$\rho = \frac{6 \times 1.27}{16 \times 24} = 0.02$$

From charts: For $\gamma = 0.75$,

$$\frac{P_u}{bh} = 1.25$$

Therefore

$$P_{uy} = 1.25 \times 16 \times 24 = 480 \text{ k}$$

3. Determine $P_{uo} = \phi P_{uo} = \phi(0.85 f'_c A_g + A_{st} f_y)$. From the chart, it is clear that P_{ux} and P_{uy} are greater than the balanced loads. Therefore use $\phi = 0.7$ when compression controls $(P_u > 0.1 f'_c A_g)$.

$$P_{uo} = 0.7(0.85 \times 5 \times 16 \times 24 + 8 \times 1.27 \times 60) = 1569 \text{ k}$$

4. Use the Bresler equation and multiply by 100 to simplify calculations:

$$\frac{1}{P_u} = \frac{1}{P_{ux}} + \frac{1}{P_{uy}} - \frac{1}{P_{uo}}$$

$$\frac{100}{P_u} = \frac{100}{510} + \frac{100}{480} - \frac{100}{569} = 0.3407$$

Therefore $P_u = 293.5$ k and $P_n = \dfrac{293.5}{0.7} = 419.3$ k.   ▪

EXAMPLE 11.21

Determine the nominal ultimate load P_n for the column section of the previous example using the PCA (or Parme) method.

Solution

1. Assume $\beta = 0.65$. From the previous example, the uniaxial load capacities in the direction of x- and y-axes were calculated:

$$P_{ux} = 510 \text{ k} \qquad P_{uy} = 499 \text{ k}$$

Since

$$\frac{P_u}{f'_c A_g} > 0.1 \quad \text{or} \quad P_u > 0.1 f'_c A_g$$

the capacity reduction factor $\phi = 0.7$.

$$P_{nx} = \frac{510}{0.7} = 728.6 \text{ k} \qquad P_{ny} = \frac{499}{0.7} = 712.9 \text{ k}$$

2. The moment capacity of the section about the x-axis

$$M_{nx} = P_n e_y = P_n \times 12 \text{ k·in.}$$

The moment capacity of the section about the y-axis

$$M_{oy} = P_{ny} e_x = 712.9 \times 8 \text{ k·in.}$$

3. Let the nominal load capacity = P_n. The nominal design moment on the section about the x-axis

$$M_{nx} = P_n e_y = P_n \times 12 \text{ k·in.}$$

and that about the y-axis

$$M_{ny} = P_n e_x = 8 P_n$$

4. Check if $M_{ny}/M_{oy} > M_{nx}/M_{ox}$:

$$\frac{8 P_n}{712.9 \times 8} > \frac{12 P_n}{728.6 \times 12} \quad \text{or} \quad \frac{P_n}{712.9} > \frac{P_n}{728.6}$$

Since $712.9 < 728.6$, then $M_{ny}/M_{oy} > M_{nx}/M_{ox}$. Therefore use equation 11.51.

5. $$\frac{8 P_n}{712.9 \times 8} + \frac{12 P_n}{728.6 \times 12}\left(\frac{1 - 0.65}{0.65}\right) = 1$$

(multiply by 1000 to simplify calculations)

$$1.4 P_n + 0.739 P_n = 1000$$

$$P_n = 467 \text{ k}$$

$$P_u = \phi P_n = 326.9 \text{ k} \quad (\phi = 0.7)$$

Note that P_u is greater than that value of 300.5 k obtained by the Bresler reciprocal method (equation 11.45) in the previous example.

6. If $\beta = 0.6$ is assumed, then

$$\frac{P_n}{712.9} + \frac{P_n}{728.6}\left(\frac{1 - 0.6)}{0.6}\right) = 1$$

$P_n = 431$ k and $P_u = 0.7 \times 431 = 301.7$ k, which is approximately equal to that of the previous example. ∎

EXAMPLE 11.22

Design a rectangular section for a short tied column to carry an ultimate axial load $P_u = 350$ k and a biaxial bending moment $M_{ux} = 210$ k·ft and $M_{uy} = 140$ k·ft. Use $f'_c = 5$ ksi, $f_y = 60$ ksi, and $b = 15$ in.

Solution   1. Assume $\phi = 0.7$ and calculate the nominal design forces.

$$P_n = \frac{P_u}{\phi} = \frac{350}{0.7} = 500 \text{ k}$$

$$M_{nx} = \frac{210}{0.7} = 300 \text{ k·ft}$$

$$M_{ny} = \frac{140}{0.7} = 200 \text{ k·ft}$$

$$e_x = \frac{M_{ny}}{P_n} = \frac{200 \times 12}{500} = 4.8 \text{ in.}$$

$$e_y = \frac{M_{nx}}{P_n} = \frac{300 \times 12}{500} = 7.2 \text{ in.}$$

2. Assume $\beta = 0.65$ and estimate the equivalent uniaxial bending moment using equations 11.53 and 11.54:

$$M_{ox} = M_{nx} + M_{ny} \left(\frac{h}{b}\right)\left(\frac{1-\beta}{\beta}\right) \tag{11.54}$$

Let the ratio $h/b = 1.5$. Therefore

$$M_{ox} = 300 + 200 \times 1.5 \times \left(\frac{1-0.65}{0.65}\right) = 461 \text{ k·ft}$$

The equivalent uniaxial eccentricity

$$e_y = \frac{M_{ox}}{P_n} = \frac{461 \times 12}{500} = 11.0 \text{ in.}$$

$$M_{oy} = M_{ny} + M_{nx} \left(\frac{b}{h}\right)\left(\frac{1-\beta}{\beta}\right) \tag{11.53}$$

$$= 200 + 300 \left(\frac{1}{1.5}\right)\left(\frac{1-0.65}{0.65}\right) = 308 \text{ k·ft}$$

The equivalent uniaxial eccentricity

$$e_x = \frac{M_{oy}}{P_n} = \frac{308 \times 12}{500} = 7.4 \text{ in.}$$

3. Estimate the dimension of the column section from M_{ox} about the x-axis = 461 k·ft: For $f'_c = 5$ ksi, $f_y = 60$ ksi, and from Table A.3 in Appendix A, $\rho_{max} = 2.52$ percent. Choose $\rho = 2$ percent $< \rho_{max}$ and $R_u = 927$ psi.

$$d = \sqrt{\frac{M_u}{R_u b}} = \sqrt{\frac{461 \times 12}{0.927 \times 15}} = 20 \text{ in. (in the long direction)}$$

Check d_s, the effective depth in the short direction, using M_{oy}:

$$d_s = \sqrt{\frac{308 \times 12}{0.927 \times 22}} = 13.46 \text{ in.}$$

Then use a concrete section 15 by 22 in.

4. The steel reinforcement can be determined using the charts of Figures 11.20 and 11.21:

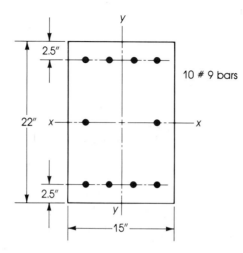

FIGURE 11.29. Example 11.22.

For $P_n = 500$ k, $M_{ox} = 461$ k·ft, $e_y = 11$ in., $d - d' = 17$ in., and

$$\frac{e}{h} = \frac{11}{22} = 0.5 \qquad \gamma = \frac{17}{22} = 0.77$$

$$\frac{P_u}{bh} = \frac{500}{15 \times 22} = 1.5$$

From charts, the steel ratio $\rho = 0.03$.

$$A_{st} = 0.03 \times 22 \times 15 = 9.9 \text{ in.}^2$$

Use 10 No. 9 bars distributed as shown in Figure 11.29.

5. Check the adequacy of the section using the Bresler reciprocal equation (11.45):

$$M_{nx} = 300 \text{ k·ft} \qquad e_y = 7.2 \text{ in.}$$
$$M_{ny} = 200 \text{ k·ft} \qquad e_x = 4.8 \text{ in.}$$

Determine the uniaxial load capacity P_{ux} about the x-axis:

$$d - d' = 17 \text{ in.} \qquad \gamma = \frac{17}{22} = 0.77$$

$$\frac{e}{h} = \frac{7.2}{22} = 0.33 \qquad \rho = \frac{8 \times 1}{15 \times 22} = 0.0242$$

From graphs (Figures 11.20 and 11.21),

$$\frac{P_{ux}}{bh} = 1.8$$

$$P_{ux} = 1.8 \times 15 \times 22 = 594 \text{ k}$$

Determine uniaxial load capacity P_{uy} about the y-axis:

$$\gamma = \frac{10}{15} = 0.67 \qquad \frac{e}{h} = \frac{4.8}{15} = 0.32$$

The steel ratio

$$\rho = \frac{6 \times 1}{15 \times 22} = 0.0182$$

$$\frac{P_{uy}}{bh} = 1.65$$

$$P_{uy} = 1.65 \times 15 \times 22 = 544 \text{ k}$$

Determine $P_{uo} = \phi P_{no} = \phi(0.85 f'_c A_g + A_s f_y)$

$$P_{uo} = 0.7(0.85 \times 5 \times 15 \times 22 + 10 \times 60) = 1402 \text{ k}$$

$$\frac{1}{P_u} = \frac{1}{P_{ux}} + \frac{1}{P_{uy}} - \frac{1}{P_{uo}}$$

$$\frac{100}{P_u} = \frac{100}{594} + \frac{100}{544} - \frac{100}{1402} = 0.2808$$

$$P_u = 356 \text{ k}$$

which is greater than the applied ultimate load of 350 k. Therefore, the section is adequate.   ∎

11.22 S.I. Examples

EXAMPLE 11.23

Determine the balanced compressive forces P_b, e_b, and M_b for the section shown in Figure 11.30. Use $f'_c = 28$ MPa and $f_y = 420$ MPa ($b = 350$ mm, $d = 490$ mm).

Solution   1. For a balanced condition, the strain in the concrete = 0.003 and the strain in the tension steel = $\varepsilon_y = f_y/E_s = 420/200,000 = 0.0021$ where $E_s = 200,000$ MPa.

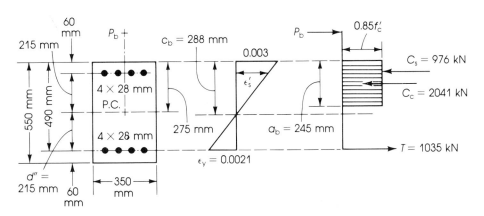

FIGURE 11.30. Example 11.23.

2. Locate the neutral axis depth c_b:

$$c_b = \left(\frac{600}{600 + f_y}\right) d \text{ (where } f_y \text{ in MPa)}$$

$$= \left(\frac{600}{600 + 420}\right)(490) = 288 \text{ mm}$$

$$a_b = 0.85c_b = 0.85 \times 288 = 245 \text{ mm}$$

3. Check if compression steel yields. From the strain diagram:

$$\frac{\varepsilon_s'}{0.003} = \frac{c - d'}{c} = \frac{288 - 60}{288}$$

$$\varepsilon_s' = 0.00238 > \varepsilon_y$$

Therefore, compression steel yields.

4. Calculate the forces acting on the section:

$$C_c = 0.85f_c'ab = \frac{0.85}{1000} \times 28 \times 245 \times 350 = 2041 \text{ kN}$$

$$T = A_s f_y = 2464 \text{ mm}^2 \times 0.420 = 1035 \text{ kN}$$

$$C_s = A_s'(f_y - 0.85f_c') = \frac{2464 \text{ mm}^2}{1000}(420 - 0.85 \times 28) = 976 \text{ kN}$$

5. Calculate P_b and M_b:

$$P_b = C_c + C_s - T = 2041 + 976 - 1035 = 1982 \text{ kN}$$

From equation 11.9:

$$M_b = P_b e_b = C_c\left(d - \frac{a}{2} - d''\right) + C_s(d - d' - d'') + Td''$$

The plastic centroid is at the centroid of the section and $d'' = 215$ mm.

$$M_b = 2041\left(490 - \frac{245}{2} - 215\right) + 976(490 - 60 - 215) + 1035 \times 215$$

$$= 743.6 \text{ kN·m}$$

$$e_b = \frac{M_b}{P_b} = \frac{743.6}{1982} = 0.375 \text{ m} = 375 \text{ mm} \qquad \blacksquare$$

SUMMARY

(a) The plastic centroid can be obtained by determining the location of the resultant force produced by the steel and the concrete, assuming both are stressed in compression to f_y and $0.85f_c'$, respectively.

(b) On a load-moment interaction diagram the following cases of analysis are developed:

1. Axial compression P_o.
2. Maximum allowable axial load $P_n(\text{max}) = 0.8P_o$ (for tied columns) and $P_n(\text{max}) = 0.85P_o$ (for spiral columns).

3. Compression controls when $P_n > P_b$ or $e < e_b$.
4. Balanced condition, P_b and M_b.
5. Tension controls when $P_n < P_b$ or $e > e_b$.
6. Pure flexure.

SECTION 11.4

(a) When $P_u = \phi P_n \geq 0.1 f'_c A_g$:

$$\phi = 0.7 \text{ for tied columns and } 0.75 \text{ for spiral columns}$$

(b) When $0.1 f'_c A_g > P_u > 0$:

$$\phi = 0.9 - \frac{2P_u}{f'_c A_g} \geq 0.7 \quad \text{(for tied columns)}$$

$$\phi = 0.9 - \frac{1.5P_u}{f'_c A_g} \geq 0.75 \quad \text{(for spiral columns)}$$

SECTION 11.5

For a balanced section,

$$c_b = \frac{87d}{87 + f_y} \quad \text{and} \quad a_b = \beta_1 c_b$$

$$\beta_1 = 0.85 \text{ for } f'_c \leq 4 \text{ ksi}$$

$$P_b = C_c + C_s - T$$

$$= 0.85 f'_c ab + A'_s(f_y - 0.85 f'_c) - A_s f_y$$

$$M_b = P_b e_b = C_c \left(d - \frac{a}{2} - d'' \right) + C_s(d - d' - d'') + Td''$$

$$e_b = M_b / P_b$$

SECTION 11.6

The equations for the general analysis of rectangular sections under eccentric forces are summarized.

SECTIONS 11.7 and 11.8

Examples for the case when compression controls are given.

SECTIONS 11.9 and 11.10

Examples for the case when tension controls are given.

SECTIONS 11.11 to 11.13

Equations to compute the strength of columns with circular sections are given. The cases of a balanced section when compression controls and when tension controls are explained.

SECTION 11.14

This section gives examples of the design of columns when compression controls and when tension controls.

SECTION 11.15

This section gives an example of the design of sections subjected to a tension force.

SECTION   This section gives examples of the design of columns using charts and tables.
11.16

SECTIONS   Biaxial bending:
11.17 to 11.21 (a) For circular columns with uniform reinforcement,

$$M_u = \sqrt{(M_{ux})^2 + (M_{uy})^2}$$
$$e = \sqrt{(e_x)^2 + (e_y)^2}$$

(b) For square and rectangular sections,

$$\frac{1}{P_n} = \frac{1}{P_{nx}} + \frac{1}{P_{ny}} - \frac{1}{P_{no}}$$

$$\frac{M_{nx}}{M_{ox}} + \frac{M_{ny}}{M_{oy}} \leq 1.0$$

(c) In the Bresler load contour method,

$$\left(\frac{M_{nx}}{M_{ox}}\right)^{1.5} + \left(\frac{M_{ny}}{M_{oy}}\right)^{1.5} = 1.0$$

(d) In the PCA load contour method,

$$M_{ny} + M_{nx}\left(\frac{b}{h}\right)\left(\frac{1-\beta}{\beta}\right) = M_{oy}$$

$$M_{nx} + M_{ny}\left(\frac{h}{b}\right)\left(\frac{1-\beta}{\beta}\right) = M_{ox}$$

REFERENCES

1. B. Brester: *Reinforced Concrete Engineering,* Vol. 1, Wiley, New York, 1974.
2. E. O. Pfrang, C. P. Siess and M. A. Sozen: "Load-Moment-Curvature Characteristics of Reinforced Concrete Cross-Sections," *ACI Journal,* 61, July 1964.
3. F. E. Richart, J. O. Draffin, T. A. Olson and R. H. Heitman: "The Effect of Eccentric Loading, Protective Shells, Slenderness Ratio, and Other Variables in Reinforced Concrete Columns," *Bulletin No. 368,* Engineering Experiment Station, University of Illinois, Urbana, 1947.
4. N. G. Bunni: "Rectangular Ties in Reinforced Concrete Columns," in *Reinforced Concrete Columns,* Publication No. SP-50, American Concrete Institute, 1975.
5. C. S. Whitney: "Plastic Theory of Reinforced Concrete," *Transactions ASCE,* 107, 1942.
6. Concrete Reinforcing Steel Institute: *CRSI Handbook,* Chicago, 1984.
7. American Concrete Institute: *Design Handbook, Vol. 2, Columns,* Publication SP-17a, Detroit, 1978.
8. Portland Cement Association: "Ultimate Load Tables for Circular Columns," Chicago, 1960.
9. B. Bresler: "Design Criteria for Reinforced Concrete Columns," *ACI Journal,* 57, November 1960.

10. R. Furlong: "Ultimate Strength of Square Columns under Biaxially Eccentric Loads," *ACI Journal,* 57, March 1961.
11. A. L. Parme, J. M. Nieves and A. Gouwens: "Capacity of Reinforced Rectangular Columns Subjected to Biaxial Bending," *ACI Journal,* 63, September 1966.
12. J. F. Fleming and S. D. Werner: "Design of Columns Subjected to Biaxial Bending," *ACI Journal,* 62, March 1965.
13. M. N. Hassoun: "Ultimate-Load Design of Reinforced Concrete," *View Point Publication,* Cement and Concrete Association, London, 1981.
14. M. N. Hassoun: *Design Tables of Reinforced Concrete Members,* Cement and Concrete Association, London, 1978.
15. American Concrete Institute: *Building Code Requirements for Reinforced Concrete (ACI 318-63),* Detroit, 1963.

PROBLEMS

11.1. Determine the balanced compressive force P_b, the moment M_b, and the balanced eccentricity for a rectangular section, $b = 12$ in. (300 mm), total depth $h = 20$ in. (500 mm), and reinforced with 3 No. 10 (3 × 32 mm) bars on the two opposite short sides. Given: $f'_c = 4$ ksi (28 MPa) and $f_y = 60$ ksi (420 MPa).

11.2. Determine the nominal compressive strength P_n for the section of problem 11.1 if the eccentricity $e = 18$ in. (450 mm). Calculate P_n using the approximate formula and compare results.

11.3. Repeat problem 11.2 if $e = 5$ in. (12.5 mm). Calculate P_n using the Whitney equation and compare results.

11.4. Calculate P_n for a circular section with diameter = 20 in. (500 mm) and reinforced with 8 No. 9 (8 × 28 mm) bars when $e = 6$ in. (150 mm), $f'_c = 4$ ksi (28 MPa), and $f_y = 60$ ksi (420 MPa).

11.5. Repeat problem 11.4 when $e = 16$ in. (400 mm).

11.6. Design a rectangular short tied column for an axial load = 120 k (535 kN) and a bending moment $M = 30$ k·ft (40 kN·m) due to D.L., and an axial load of 140 k (625 kN) and a moment $M = 20$ k·ft (27 kN·m) due to L.L. Use $f'_c = 4$ ksi (28 MPa), $f_y = 60$ ksi (420 MPa), and $b = 12$ in. (300 mm).

11.7. Design a rectangular short tied column for an axial ultimate load $P_u = 60$ k (267 kN) and an ultimate moment $M_u = 200$ k·ft (270 kN·m). Use $f'_c = 5$ ksi (35 MPa), $f_y = 60$ ksi (420 MPa), $b = 12$ in. (300 mm), and A'_s about $\frac{1}{2}A_s$.

11.8. Design a rectangular tied short column for $P_u = 80$ k (356 kN) and $M_u = 280$ k·ft (380 kN·m). Use $f'_c = 4$ ksi (28 MPa), $f_y = 60$ ksi (420 MPa), $b = 14$ in. (350 mm), and $A'_s = 0$.

11.9. Repeat problem 11.6 using $f'_c = 5$ ksi (35 MPa) and $f_y = 60$ ksi (420 MPa).

11.10. Repeat problem 11.7 using $f'_c = 4$ ksi (28 MPa), $f_y = 60$ ksi (420 MPa), and $b = 14$ in. (350 mm).

11.11. Repeat problem 11.8 using $f'_c = 5$ ksi (35 MPa) and $f_y = 60$ ksi (420 MPa).

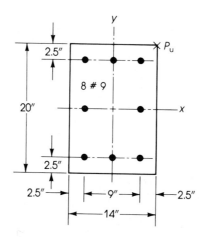

FIGURE 11.31. Problems 11.13 and 11.14.

11.12. Design a rectangular tied short column using the tables in Appendix A for (a) $P_u = 721$ k (500 kN), $M_u = 215$ k·ft (20 kN·m), and $b = 14$ in. (350 mm), using $f'_c = 4$ ksi (28 MPa) and $f_y = 60$ ksi (420 MPa), and for (b) $P_u = 25$ k (110 kN), $M_u = 75$ k·ft (102 kN·m), and $b = 14$ in. (350 mm).

11.13. The section of a short tied column is shown in Figure 11.31. Determine the allowable ultimate load on the section P_u if it acts at $e_x = 7$ in. and $e_y = 10$ in. Use $f'_c = 4$ ksi, $f_y = 60$ ksi, and the Bresler equation.

11.14. Repeat problem 11.13 using the PCA (Parme) equation.

11.15. Design a rectangular section for a short tied column to carry an ultimate axial load $P_u = 295$ k and a biaxial bending moment $M_{ux} = 180$ k·ft and $M_{uy} = 120$ k·ft. Use $f'_c = 4$ ksi, $f_y = 60$ ksi, and $b = 14$ in.

12

Long Columns

12.1   Introduction

In the analysis and design of short columns discussed in the previous two chapters, it was assumed that buckling, elastic shortening, and secondary moment due to lateral deflection had minimal effect on the ultimate strength of the column, thus these were not included in the design procedure. However, when the column is long, these factors must be considered: the extra length will cause a reduction in the column strength that varies with the column effective height, width of the section, the slenderness ratio, and the column end conditions.

A column with a high slenderness ratio will have a considerable reduction in strength, while a low slenderness ratio means that the column is relatively short and the reduction in strength may not be significant. The slenderness ratio is the ratio of the column height l to the radius of gyration r, where $r = \sqrt{I/A}$, I being the moment of inertia of the section and A the sectional area.

For a rectangular section of width b and a depth h (Figure 12.1), $I_x = bh^3/12$ and $A = bh$. Therefore, $r_x = \sqrt{I/A} = 0.288h$ (or approximately, $r_x = 0.3h$). Similarly $I_y = hb^3/12$ and $r_y = 0.288b$ (or approximately $0.3b$). For a circular column with diameter D, $I_x = I_y = \pi D^4/64$ and $A = \pi D^2/4$ and, therefore, $r_x = r_y = 0.25D$.

In general, columns may be considered:

- Long with a relatively high slenderness ratio, where lateral bracing or shear walls are required.
- Long with a relatively medium slenderness ratio that causes a reduction in the column strength. Lateral bracing may not be required.
- Short where the slenderness ratio is relatively small, causing a slight reduction in strength, as discussed in previous chapters.

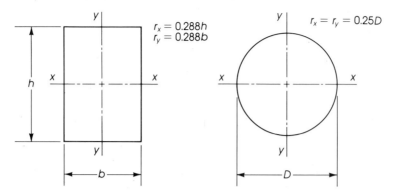

FIGURE 12.1. Rectangular and circular sections of columns, with radii of gyration r.

12.2   Effective Column Length Kl_u

The slenderness ratio l/r can be calculated accurately when the effective length of the column (Kl_u) is used. This effective length is a function of two main factors.

1. The unsupported length l_u represents the unsupported height of the column between two floors. It is measured as the clear distance between slabs, beams, or any structural member providing lateral support to the column. In a flat slab system with column capitals, the unsupported height of the column is measured from the top of the lower floor slab to the bottom of the column capital. If the column is supported with a deeper beam in one direction than in the other direction, l_u should be calculated in both directions (about the x- and the y-axes) of the column section. The critical (greater) value must be considered in the design.
2. The effective length factor K represents the ratio of the distance between points of zero moment in the column and the unsupported height of the column in one direction. For example, if the unsupported length of a column hinged at both ends on which sidesway is prevented is l_u, the points of zero moment will be at the top and bottom of the column, that is, at the two hinged ends. Therefore the factor $K = l_u/l_u = 1.0$. If a column is fixed at both ends and sidesway is prevented, the points of inflection (points of zero moment) are at $l_u/4$ from each end. Therefore, $K = 0.5l_u/l_u = 0.5$ (Figure 12.2). To evaluate the proper value of K, two main cases will be considered.

   When structural frames are braced, the frame, which consists of beams and columns, is braced against sidesway by shear walls, rigid bracing, or lateral support from an adjoining structure. The ends of the columns will stay in position, and lateral translation of joints is prevented. The range of K in braced frames is always equal to or less than 1.0. The **ACI Code, section 10.11.2**, recommends the use of $K = 1.0$ for braced frames.

   When the structural frames are unbraced, the frame is not supported against sidesway, and it depends on the stiffness of the beams and columns to

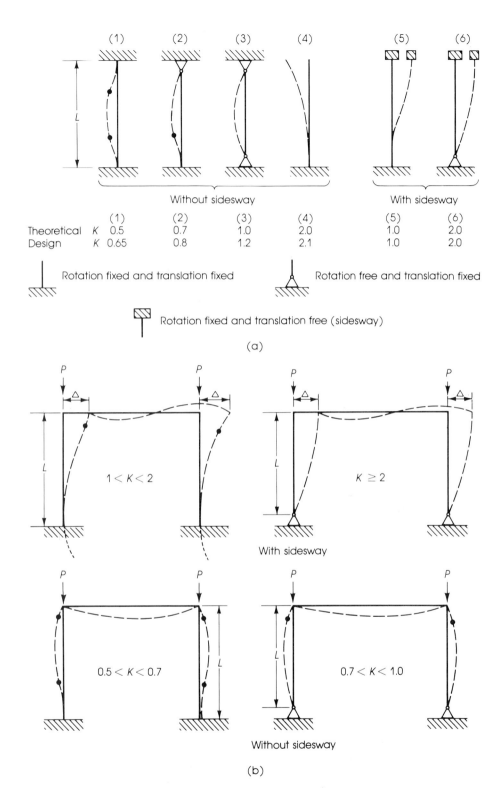

FIGURE 12.2. (a) Effective lengths of columns and length factor K and (b) effective lengths and K for portal frame columns.

prevent lateral deflection. Joint translations are not prevented, and the frame sways in the direction of lateral loads. The range of K for different columns and frames is given in Figure 12.2, considering the two cases when sidesway is prevented or not prevented.

12.3   Alignment Charts

The effective length of columns can be estimated by using the alignment chart shown in Figure 12.3.[10] To find the effective length factor K, it is necessary first to calculate the end restraint factors ψ_A and ψ_B at the top and bottom of the column, respectively, where

$$\psi = \frac{\Sigma EI/l \text{ of columns}}{\Sigma EI/l \text{ of beams}} \quad \text{(both in the plane of bending)} \qquad (12.1)$$

The ψ factor at one end shall include all columns and beams meeting at the joint. For a hinged end, ψ is infinite and may assumed to be 10.0. For a fixed end, ψ is zero and may assumed to be 1.0. Those assumed values may be used because a perfect frictionless hinge, as well as perfectly fixed ends, cannot exist in reinforced concrete frames.

The procedure to estimate K is to calculate ψ_A for the top end of the column and ψ_B for the bottom end of the column. Plot ψ_A and ψ_B on the alignment chart of Figure 12.3 and connect the two points to intersect the middle line, which indicates the K value. Two nomograms are shown, one for braced frames where sidesway is prevented, and the second for unbraced frames where sidesway is not prevented. The development of the charts is based on the assumptions that (1) the structure consists of symmetrical rectangular frames, (2) the girder moment at a joint is distributed to columns according to their relative stiffnesses, and (3) all columns reach their critical loads at the same time.

12.4   Member Stiffness *EI*

The stiffness of a structural member is equal to the modulus of elasticity E times the moment of inertia I of the section. The values of E and I for reinforced concrete members can be estimated as follows.

1. The modulus of elasticity of concrete was discussed in Chapter 2; the ACI Code gives the following expression:

$$\begin{aligned} E_c &= 33w^{1.5}\sqrt{f'_c} \\ &= 57,400\sqrt{f'_c} \end{aligned}$$

for normal weight concrete (or $57,000\sqrt{f'_c}$ as suggested by the code). The modulus of elasticity of steel $E_s = 29 \times 10^6$ psi.

2. For reinforced concrete members, the moment of inertia I varies along the member depending on the degree of cracking and the percentage of reinforcement in the section considered.

To evaluate the factor ψ, EI must be calculated for beams and columns. For this purpose, EI can be estimated as follows: For flexural members (beams), the cracked

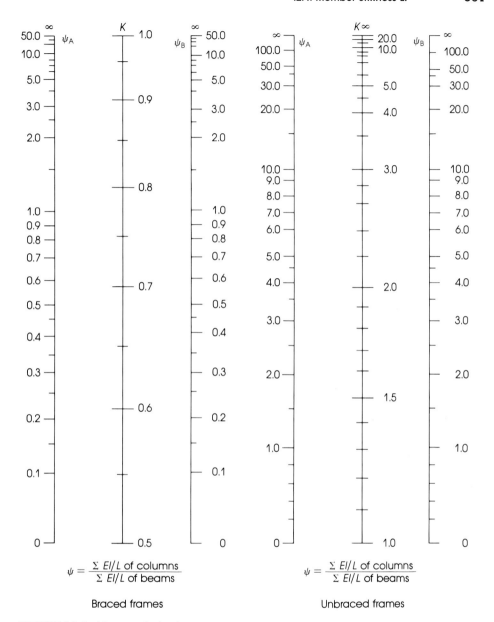

$$\psi = \frac{\Sigma \ EI/L \text{ of columns}}{\Sigma \ EI/L \text{ of beams}}$$

Braced frames

$$\psi = \frac{\Sigma \ EI/L \text{ of columns}}{\Sigma \ EI/L \text{ of beams}}$$

Unbraced frames

FIGURE 12.3. Alignment chart.

transformed area can be used. For columns, the gross moment of inertia may be used or the following approximate equation:

$$EI = 0.2E_cI_g + E_sI_s \qquad (12.2)$$

where I_g = the gross moment of inertia, neglecting A_s, and I_s = moment of inertia of the steel reinforcement. An approximate method to evaluate I is to consider the gross moments of inertia of the column section and 0.5 times the gross moment of inertia of the flexural member.[10]

12.5   Limitation of the Slenderness Ratio

The **ACI Code, section 10.11**, recommends the following limitations between short and long columns in braced frames:

1. The effect of slenderness may be neglected and the column can be designed as a short column when

$$\frac{Kl_u}{r} < 34 - \frac{12M_{1b}}{M_{2b}} \qquad\qquad (12.3)$$

   where M_{1b} and M_{2b} are the factored end moments of the column and M_{2b} is greater than M_{1b} **(ACI Code, section 10.11.4)**.
2. The ratio M_{1b}/M_{2b} is considered positive if the member is bent in single curvature and negative for double curvature (Figure 12.4).
3. If the member is subjected to transverse loads causing the moment within the column length to be greater than the end moments, then the ratio M_{1b}/M_{2b} can be considered equal to 1. Consequently, the moment magnifier coefficient $C_m = 1.0$ (to be defined later).
4. If there are no calculated moments at the ends of the columns, the ratio $M_{1b}/M_{2b} = 1.0$ and the moment magnifier coefficient $C_m = 1.0$.
5. If the factored column moments are zero or $e = M_u/P_u < e_{min}$, the value of M_{2b} should be calculated using the minimum eccentricity according to the **ACI Code (section 10.11.5.4)** as follows:

$$e_{min} = (0.6 + 0.03h)$$

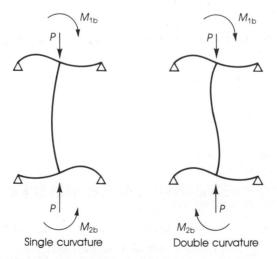

Single curvature   Double curvature

FIGURE 12.4. Single and double curvatures.

Therefore,

$$M_u = M_{2b} = P_u(0.6 + 0.03h)$$

In columns in unbraced frames, the effect of the slenderness ratio may be neglected when

$$\frac{Kl_u}{r} \le 22 \tag{12.4}$$

(ACI Code, section 10.11.4.2)

12.6   The Moment Magnifier Method (ACI Code Method)

In general, compression members may be subjected to lateral deflections that cause secondary moments. If the secondary moment M' is added to the applied moment on the column M_a, the final moment is $M = M_a + M'$. An approximate method to estimate the final moment M is to multiply the applied moment M_a by a factor called the magnifying moment factor δ, which must be equal to or greater than 1.0, or $M_{max} = \delta M_a$ and $\delta \ge 1.0$. The moment M_a is obtained from the elastic structural analysis using factored loads, and it is the maximum moment that acts on the column at either end or within the column if transverse loadings are present. To calculate the moment magnifier factor δ,

1. Determine if the frame is braced against sidesway and find the unsupported length l_u and the effective length factor K.
2. Calculate the member stiffness EI using the larger value of the two following equations **(ACI Code, section 10.11.5)**:

$$EI = \frac{0.2E_cI_g + E_sI_s}{1 + \beta_d} \tag{12.5}$$

(Note that in equation 12.2, β_d is assumed $= 0$ in the calculation of the factor ψ) or

$$EI = \frac{0.4E_cI_g}{1 + \beta_d} \tag{12.6}$$

where

$E_c = 57{,}000\sqrt{f'_c}$

$E_s = 29 \times 10^6$ psi

I_g = gross moment of inertia of the section about the axis considered, neglecting A_s

I_s = moment of inertia of the reinforcing steel

$\beta_d = \dfrac{\text{maximum factored dead load moment}}{\text{maximum factored total load moment}} = \dfrac{1.4D}{1.4D + 1.7L}$

3. Determine the Euler buckling load P_c:

$$P_c = \frac{\pi^2 EI}{(Kl_u)^2} \quad \textbf{(ACI Code equation 10.9)} \qquad (12.7)$$

Use the values of EI, K, and l_u as calculated from steps 1 and 2.

4. Calculate the value of the factor C_m to be used in the equation of the moment magnifier factor. For braced members without transverse loads,

$$C_m = 0.6 + 0.4 \frac{M_{1b}}{M_{2b}} \geq 0.4 \quad \text{and} \quad M_{1b} < M_{2b} \qquad (12.8)$$

where M_{1b} and M_{2b} are the end moments. Note that $C_m = 1.0$ for all unbraced frames, and for braced frames with transverse loading, as explained above.

5. Calculate the moment magnifier factor δ_b

$$\delta_b = \frac{C_m}{1 - \left(\dfrac{P_u}{\phi P_c}\right)} \geq 1.0 \quad \textbf{(ACI Code equation 10.7)} \qquad (12.9)$$

and

$$\delta_s = \frac{1.0}{1 - \left(\dfrac{\Sigma P_u}{\phi \Sigma P_c}\right)} \geq 1.0 \quad \textbf{(ACI Code equation 10.8)} \qquad (12.10)$$

where

δ_b = moment magnification factor for frames braced against sidesway

δ_s = moment magnification factor for frames not braced against sidesway

ΣP_u and ΣP_c = summations for all axial and critical loads of columns in one story

The factor δ_b reflects the effects of the curvature between the ends of the compression member, while the factor δ_s reflects the effect of the lateral drift resulting from lateral and gravity loads.

6. Design the compression member using the axial factored load P_u from a conventional frame analysis and a magnified factored moment M_c computed as follows:

$$M_c = \delta_b M_{2b} + \delta_s M_{2s} \quad \textbf{(ACI Code equation 10.6)} \qquad (12.11)$$

where M_{2b} and M_{2s} are the values of the larger factored end moment due to loads that result in no sidesway and sidesway, respectively. For frames braced against sidesway, $\delta_s = 0$ and $M_c = \delta_b M_{2b}$. For frames not braced against sidesway, both δ_b and δ_s must be computed. Therefore, $P_n = P_u/\phi$ and $M_n = M_c/\phi$.

Note that in equation 12.11, the ACI Code expresses the column secondary moments of the unbraced frames as the sum of the magnification due to (1) nonsway moments due to gravity loads, represented by δ_b, and (2) sway moments due to wind and other lateral loads, represented by δ_s. When the structural frame is not braced by shear walls or bracing elements, δ_b and δ_s must be evaluated and used to calculate M_c. In this case, where the structure is subjected to gravity and lateral loads, reduced load factors are used, as specified by the **ACI Code, section 9.2,** and given in Chapter 3. The columns of the structural frame must also be checked for gravity loads effect using $(1.4D + 1.7L)$ load factors. Both cases are explained in example 12.4.

7. If Kl_u/r is greater than 100, **ACI Code, section 10.11.4,** requires an exact analysis to be made based on forces and moments determined from the analysis of the structure.

EXAMPLE 12.1

The column section shown in Figure 12.5 carries an axial load $P_D = 125$ k and a moment $M_D = 105$ k·ft due to dead load and an axial load $P_L = 112$ k and a moment $M_L = 96$ k·ft due to live load. The column is part of a frame which is braced against sidesway and bent in single curvature about its major axis. The unsupported length of the column $l_u = 19$ ft and the moments at both ends of the column are equal. Check the adequacy of the column using $f'_c = 4$ ksi and $f_y = 60$ ksi.

Solution

1. Calculate ultimate loads:

$$P_u = 1.4P_D + 1.7P_L = 1.4 \times 125 + 1.7 \times 112 = 365.4 \text{ k}$$

$$M_u = 1.4M_D + 1.7M_L = 1.4 \times 105 + 1.7 \times 96 = 310.2 \text{ k·ft}$$

$$e = \frac{M_u}{P_u} = \frac{310.2 \times 12}{365.4} = 10.2 \text{ in.}$$

2. Check if the column is long: Since the frame is braced against sidesway, assume $K = 1.0$, and $r = 0.3h = 0.3 \times 22 = 6.6$ in., $l_u = 19$ ft.

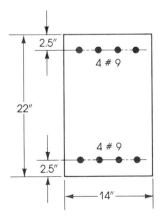

FIGURE 12.5. Example 12.1.

$$\frac{Kl_u}{r} = \frac{1 \times 19 \times 12}{6.6} = 34.5$$

For braced columns, if $Kl_u/r \le 34 - 12\, M_{1b}/M_{2b}$, slenderness effect may be neglected. Given end moments $M_{1b} = M_{2b}$ and M_{1b}/M_{2b} positive for single curvature,

$$\text{Right-hand side} = 34 - 12\, \frac{M_{1b}}{M_{2b}} = 34 - 12 \times 1 = 22$$

Since $Kl_u/r = 34.5 > 22$, slenderness effect must be considered.

3. Calculate the moment magnifier δ_b as follows ($\delta_s = 0$). Calculate C_m:

$$C_m = 0.6 + 0.4\, \frac{M_{1b}}{M_{2b}} \ge 0.4$$

$$= 0.6 + 0.4 \times 1 = 1.0$$

Calculate E_c:

$$E_c = 57{,}000 \sqrt{f_c'} = 57{,}000\sqrt{4000} = 3605 \text{ ksi}$$

$$E_s = 29{,}000 \text{ ksi}$$

The moment of inertia

$$I_g = \frac{14(22)^3}{12} = 12{,}422 \text{ in.}^4$$

$$A_s = A_s' = 4.0 \text{ in.}^2$$

$$I_{se} = 2 \times 4.0 \left(\frac{22-5}{2}\right)^2 = 578 \text{ in.}^4$$

The dead-load moment ratio

$$\beta_d = \frac{\text{design dead load moment}}{\text{total load moment}} = \frac{1.4 \times 105}{310.2} = 0.474$$

The stiffness

$$EI = \frac{0.2 E_c I_g + E_s I_{se}}{1 + \beta_d}$$

$$= \frac{(0.2 \times 3605 \times 12{,}422) + (29{,}000 \times 578)}{1 + 0.474} = 17.45 \times 10^6 \text{ k·in.}^2$$

Calculate P_c:

$$P_c = \frac{\pi^2 EI}{(Kl_u)^2} = \frac{\pi^2 \times 17.45 \times 10^6}{(12 \times 19)^2} = 3313 \text{ k}$$

The moment magnifier factor

$$\delta_b = \frac{C_m}{1 - \left(\dfrac{P_u}{\phi P_c}\right)}$$

$$= \frac{1.0}{1 - \left(\dfrac{365.4}{0.7 \times 3313}\right)} = 1.19$$

4. Calculate the design moment $\delta_b M_n$:

$$\text{Required } P_n = \frac{P_u}{\phi} = \frac{365.4}{0.7} = 522 \text{ k}$$

$$\text{Required } M_n = \frac{M_u}{\phi} = \frac{310.2}{0.7} = 443.1 \text{ k·ft}$$

$$\text{Design moment } \delta_b M_n = 1.19 \times 443.1 = 527.3 \text{ k·ft}$$

$$\text{Design eccentricity } e = \frac{527.3 \times 12}{522} = 12.12 \text{ in.}$$

5. Determine the nominal load capacity of the section using $e = 12.12$ in. according to the procedure explained in example 11.3:

$$P_n = 47.6a + 226.4 - 4f_s \qquad (A)$$

$$e' = e + d - \frac{h}{2} = 12.12 + 19.5 - \frac{22}{2} = 20.62 \text{ in.}$$

$$P_n = \frac{1}{20.62}\left[47.6a\left(19.5 - \frac{a}{2}\right) + 226.4(19.5 - 2.5)\right]$$

$$= 45a - 1.15a^2 + 186.6 \qquad (B)$$

Solving for a from equations A and B,

$$a = 10.6 \text{ in.} \quad \text{and} \quad P_n = 535 \text{ k}$$

The load capacity P_n is greater than the required load of 522 k; therefore, the section is adequate.

6. The nominal load capacity P_n can also be calculated from graphs, as explained in Chapter 11.
7. If the section is not adequate, increase the steel reinforcement in the section.   ∎

EXAMPLE 12.2

Check the adequacy of the column in the previous example if the unsupported length $l_u = 10$ ft. Determine the maximum allowable nominal load on the column.

Solution

1. Applied loads are $P_n = 522$ k and $M_n = 443.1$ k.
2. Check if the column is long: $l_u = 10$ ft, $r = 0.3h = 0.3 \times 22 = 6.6$ in., and $K = 1.0$ (frame is braced against sidesway).

$$\frac{Kl_u}{r} = \frac{1 \times (10 \times 12)}{6.6} = 18.2$$

Check if $Kl_u/r \leq 34 - 12M_{1b}/M_{2b}$

$$\text{Right-hand side} = 34 - 12 \times 1 = 22$$

$$\frac{Kl_u}{r} = 18.2 < 22$$

Therefore slenderness effect can be neglected.

3. Determine the nominal load capacity of the short column as explained in example 11.3. From example 11.3, the nominal compressive strength $P_n = 612.1$ k, which is greater than the required load of 522 k, because the column is short.
4. P_n can also be calculated using graphs. ∎

EXAMPLE 12.3

Check the adequacy of the column in example 12.1 if the frame is unbraced against sidesway, the end restraint factors ψ_A (top) = 0.8 and ψ_B (bottom) = 2.0, and the unsupported length $l_u = 16$ ft.

Solution

1. Determine the value of K from the alignment chart (Figure 12.3) for unbraced frames. Connect the values of $\psi_A = 0.8$ and $\psi_B = 2.0$ to intersect the K line at $K = 1.4$.

$$\frac{Kl_u}{r} = \frac{1.4 \times (16 \times 12)}{6.6} = 40.7$$

2. For unbraced frames, if $Kl_u/r \leq 22$, the column can be designed as a short column. Since actual $Kl_u/r = 40.7 > 22$, the slenderness effect must be considered.
3. Calculate the moment magnifier δ, given $C_m = 1.0$, $K = 1.4$, $EI = 17.45 \times 10^6$ k·in.² (from example 12.1), and

$$P_c = \frac{\pi^2 EI}{(Kl_u)^2} = \frac{\pi \times 17.45 \times 10^6}{(1.4 \times 16 \times 12)^2} = 2384 \text{ k}$$

$$\delta_b = \frac{C_m}{1 - \left(\dfrac{P_u}{\phi P_c}\right)} = \frac{1.0}{1 - \left(\dfrac{365.4}{0.7 \times 2384}\right)} = 1.28$$

4. From example 12.1, the applied loads are $P_u = 365.4$ k and $M_u = 310.2$ k·ft or

$$P_n = \frac{P_u}{\phi} = \frac{365.4}{0.7} = 522 \text{ k}$$

and

$$M_n = \frac{M_u}{\phi} = \frac{310.2}{0.7} = 443.1 \text{ k·ft}$$

The design moment = $\delta_b M_n = 1.28 \times 443.1 = 567.2$ k·ft. The design eccentricity

$$e = \frac{\delta_b M_n}{P_n} = \frac{567.2 \times 12}{522} = 13.0 \text{ in.}$$

5. The requirement now is to check the adequacy of a short column for $P_n = 522$ k, $\delta_b M_n = 567.2$ k·ft, and $e = 13.0$ in. The procedure is explained in examples 11.3 and 12.1.
6. From example 11.3:

$$P_n = 47.6a + 226.4 - 4f_s \tag{A}$$

$$e' = e + d - \frac{h}{2} = 13 + 19.5 - \frac{22}{2} = 21.5 \text{ in.}$$

$$P_n = \frac{1}{21.5} \left[47.6a \left(19.5 - \frac{a}{2} \right) + 226.4(19.5 - 2.5) \right]$$

$$= 43.16a - 1.1a^2 + 179 \qquad\qquad (B)$$

$$a = 10.4 \text{ in.}$$

Thus $c = 12.24$ in. and $P_n = 508$ k. This load capacity of the column is less than the required P_n of 522 k. Therefore, the section is not adequate.

7. Increase steel reinforcement to 4 No. 10 bars on each side and repeat calculations. $P_n = 568$ k.
8. The steel reinforcement required for $P_n = 522$ k and $\delta_b M_n = 567.2$ k·ft can be determined using design graphs as explained in section 11.19 of Chapter 11.
9. P_n can be calculated using the Whitney equation. ∎

EXAMPLE 12.4

Design an interior square column of the first story of an 8-story office building. Clear height of the first floor is 16 ft, while the height of all other floors is 11 ft. The building layout is in 24 bays (Figure 12.6) and the columns are not braced against sidesway. The loads acting on a first floor interior column due to gravity and wind are as follows:

Axial dead load = 380 k

Axial live load = 140 k

Axial wind load = 0 k

Dead load moments are 32 k·ft (top) and 54 k·ft (bottom)

Live load moments are 20 k·ft (top) and 36 k·ft (bottom)

Wind load moments are 60 k·ft (top) and 60 k·ft (bottom)

EI/l for beams = 360×10^3 k·in.

Use $f'_c = 5$ ksi, $f_y = 60$ ksi, and the ACI Code requirements. Assume exterior column load = $\frac{2}{3}$ interior column load and corner column load = $\frac{1}{3}$ interior column load.

Solution   1. Calculate ultimate forces using load combinations. For gravity loads:

$$P_u = 1.4D + 1.7L = 1.4(380) + 1.7(140) = 770 \text{ k}$$

$$M_u = M_{2b} = 1.4M_D + 1.7M_L = 1.4(54) + 1.7(36) = 136.8 \text{ k·ft}$$

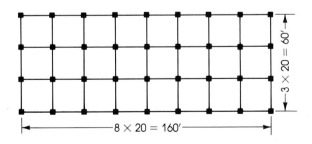

FIGURE 12.6. Example 12.4.

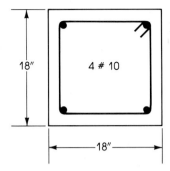

FIGURE 12.7. Column cross-section, example 12.4.

For gravity plus wind load:

$$P_u = 0.75(1.4D + 1.7L + 1.7W)$$
$$= 0.75[1.4(380) + 1.7(140 + 0)] = 577.5 \text{ k}$$
$$M_{ub} = M_{2b} = 0.75(1.4M_D + 1.7M_L) = 0.75[1.4(54) + 1.7(36)] = 102.6 \text{ k·ft}$$
$$M_{us} = M_{2s} = 0.75(1.7M_W) = 0.75(1.7 \times 60) = 76.5 \text{ k·ft}$$

Other combinations are not critical:

$$P_u = 0.9D + 1.3W = 0.9(380) + 1.3(0) = 342 \text{ k}$$
$$M_{2b} = 0.9M_D = 0.9(54) = 48.6 \text{ k·ft}$$
$$M_{2s} = 1.3M_W = 1.3(60) = 78 \text{ k·ft}$$

2. Select a preliminary section of column based on gravity load combination using tables or charts. Select a section 18 by 18 in. reinforced by 4 No. 10 bars (Figure 12.7). From charts of Figure 11.21 and for $P_u = 770$ k, the allowable ultimate moment $M_u = 206$ k·ft, which is greater than 136.8 k·ft.
3. Check Kl_u/r:

$$I_g = \frac{(18)^4}{12} = 8748 \text{ in.}^4 \quad E_c = 4.03 \times 10^6 \text{ psi}$$

For 16-ft column,

$$\frac{EI_g}{l_c} = \frac{8748(4.03 \times 10^6)}{16 \times 12} = 183.6 \times 10^6$$

For 11-ft column,

$$\frac{EI_g}{l_c} = \frac{8748(4.03 \times 10^6)}{11 \times 12} = 267.1 \times 10^6$$

$$\psi \text{ (top)} = \psi \text{ (bottom)} = \frac{\Sigma(EI/l_c)}{\Sigma(EI/l_b)} = \frac{183.6 + 267.1}{2(0.5 \times 360)} = 1.25$$

From the chart (Figure 12.3), $K = 1.37$ for unbraced frame and 0.8 for braced frame.

$$\frac{Kl_u}{r} = \frac{1.37(16 \times 12)}{0.3 \times 18} = 48.7$$

which is more than 22 and less than 100. Therefore, slenderness ratio must be considered.

4. Compute P_c:

$$E_c = 4.03 \times 10^3 \text{ ksi} \qquad E_s = 29 \times 10^3 \text{ ksi}$$

$$I_g = 8748 \text{ in.}^4 \qquad I_{se} = 5.06 \left(\frac{13}{2}\right)^2 = 214 \text{ in.}^4$$

$$\beta_d = \frac{1.4 M_D}{M_u} = \frac{1.4(54)}{136.8} = 0.55$$

$$EI = (0.2 E_c I_g + E_s I_{se})/(1 + \beta_d)$$

$$= \frac{0.2(4.03 \times 10^3 \times 8748) + 29 \times 10^3(214)}{1 + 0.55} = 8.55 \times 10^6 \text{ k·in.}^2$$

For calculation of δ_s, $\beta_d = 0$ and $EI = 8.55 \times 10^6(1.55) = 13.25 \times 10^6 \text{ k·in.}^2$

$$P_c = \frac{\pi^2 EI}{(Kl_u)^2} = \frac{\pi^2(8.55 \times 10^6)}{(0.8 \times 16 \times 12)^2} = 3577 \text{ k} \quad \text{(braced)}$$

$$= \frac{\pi^2(13.25 \times 10^6)}{(1.37 \times 16 \times 12)^2} = 1890 \text{ k} \quad \text{(unbraced)}$$

5. Compute δ_b for gravity load combination:

$$P_u = 770 \text{ k}$$

$$M_{2b} = 1.4(54) + 1.7(36) = 136.8 \text{ k·ft} \quad \text{(bottom)}$$

$$M_{1b} = 1.4(32) + 1.7(20) = 78.8 \text{ k·ft} \quad \text{(top)}$$

$$C_m = 0.6 + 0.4 \frac{M_{1b}}{M_{2b}} = 0.6 + 0.4 \left(\frac{78.8}{136.8}\right) = 0.83$$

$$\delta_b = \frac{C_m}{1 - \left(\dfrac{P_u}{\phi P_c}\right)} = \frac{0.83}{1 - \left(\dfrac{770}{0.7 \times 3577}\right)} = 1.20$$

Check minimum $M_{2b} = P_u(0.6 + 0.03h)$:

$$M_{2b} = 770(0.6 + 0.03 \times 18)/12 = 73 \text{ k·ft} < 136.8 \text{ k·ft}$$

$$M_c = \delta_b M_{2b} = 1.2(136.8) = 164.2 \text{ k·ft}$$

For $P_u = 770$ k, the allowable M_u on the section $= 206$ k·ft > 164.2 k·ft; therefore the section is adequate.

6. Compute δ_b and δ_s considering wind loading. Loads are $P_u = 577.5$ k, $M_{2b} = 102.6$ k·ft, and $M_{2s} = 76.5$ k·ft.

$$\delta_b = \frac{C_m}{1 - \left(\dfrac{P_u}{\phi P_c}\right)} = \frac{0.83}{1 - \left(\dfrac{577.5}{0.7 \times 3577}\right)} = 1.08 > 1.0$$

$$\delta_s = \frac{1.0}{1 - \left(\dfrac{\Sigma P_u}{\phi \Sigma P_c}\right)}$$

For one floor in the building, the number of interior columns = 14, exterior columns = 18, corner columns = 4.

$$\Sigma P_u = 14(557.5) + 18\left(\frac{2}{3} \times 577.5\right) + 4\left(\frac{1}{3} \times 577.5\right) = 15{,}785 \text{ k}$$

$$\Sigma P_c = 14(1890) + 22\left(\frac{2}{3} \times 1890\right) = 54{,}180 \text{ k}$$

$$\delta_s = \frac{1.0}{1 - \left(\dfrac{15{,}785}{0.7 \times 54{,}180}\right)} = 1.71$$

$$M_c = \delta_b M_{2b} + \delta_s M_{2s} = 1.08(102.6) + 1.71(76.5) = 241.6 \text{ k·ft}$$

From the chart (Figure 11.21), and for $P_u = 577.5$ k, $\rho = 0.015$, the allowable $M_u = 260$ k·ft > 241.6 k·ft. Therefore, the column section is adequate.   ■

SUMMARY

SECTIONS 12.1 to 12.3

(a) Radius of gyration $r = \sqrt{I/A}$

$r = 0.3h$ for rectangular sections and $0.25D$ for circular sections.

(b) The effective column length $= Kl_u$. For braced frames, $K = 1.0$; for unbraced frames, K varies as shown in Figure 12.2.

(c) K can be determined from the alignment chart (Figure 12.3).

SECTION 12.4

Member stiffness is EI: ($E_c = 33w^{1.5}\sqrt{f'_c}$)

Approximate $I = I_g$ (for columns), and $I = 0.5I_g$ for flexural members (beams); $I_g = bh^3/12$.

SECTION 12.5

The effect of slenderness may be neglected when

(a) $\quad \dfrac{Kl_u}{r} \leq 22$ for unbraced frames $\qquad\qquad\qquad (12.4)$

(b) $\quad \dfrac{Kl_u}{r} \leq \left(34 - \dfrac{12M_{1b}}{M_{2b}}\right)$ for braced columns $\qquad (12.3)$

where M_{2b} and M_{1b} are the end moments and $M_{2b} > M_{1b}$.

SECTION 12.6

(a) The moment magnifier factors δ_b and δ_s are

$$\delta_b = \frac{C_m}{1 - \left(\dfrac{P_u}{\phi P_c}\right)} \geq 1.0 \qquad\qquad\qquad (12.9)$$

$$\delta_s = \frac{1.0}{1 - \dfrac{\Sigma P_u}{\phi \Sigma P_c}} \geq 1.0 \qquad\qquad\qquad (12.10)$$

(b) $\qquad M_c = \delta_b M_{2b} + \delta_s M_{2s}$ $\qquad\qquad\qquad$ (*12.11*)

$$P_c = \frac{\pi^2 EI}{(Kl_u)^2}$$ $\qquad\qquad\qquad$ (*12.7*)

(c) $\qquad C_m = \left(0.6 + 0.4\,\frac{M_{1b}}{M_{2b}}\right) \geq 0.4$ $\qquad\qquad\qquad$ (*12.8*)

(d) *EI* is the larger of:

$$EI = \frac{0.4 E_c I_g}{1 + \beta_d}$$ $\qquad\qquad\qquad$ (*12.6*)

or

$$EI = \frac{0.2 E_c I_g + E_s I_s}{1 + \beta_d}$$ $\qquad\qquad\qquad$ (*12.5*)

$$\beta_d = \frac{1.4D}{(1.4D + 1.7L)}$$

REFERENCES

1. B. B. Broms: "Design of Long Reinforced Concrete Columns," *Journal of Structural Division ASCE,* 84, July 1958.
2. J. G. MacGregor, J. Breen and E. O. Pfrang: "Design of Slender Concrete Columns," *ACI Journal,* 67, January 1970.
3. American Concrete Institute: *Design Handbook, Vol. 2, Columns,* ACI Publication SP-17a, 1978, Detroit.
4. W. F. Chang and P. M. Ferguson: "Long Hinged Reinforced Concrete Columns," *ACI Journal,* 60, January 1963.
5. R. W. Furlong and P. M. Ferguson: "Tests of Frames with Columns in Single Curvature," *Symposium on Reinforced Columns,* American Concrete Institute SP-13, 1966.
6. J. G. MacGregor and S. L. Barter: "Long Eccentrically Loaded Concrete Columns Bent in Double Curvature," *Symposium on Reinforced Concrete Columns,* American Concrete Institute SP-13, 1966.
7. R. Green and J. E. Breen: "Eccentrically Loaded Concrete Columns Under Sustained Load," *ACI Journal,* 66, November 1969.
8. B. G. Johnston: *Guide to Stability Design for Metal Structures,* 3rd ed., Wiley, New York, 1976.
9. T. C. Kavanagh: "Effective Length of Framed Columns," *Transactions ASCE,* 127, Part II, 1962.
10. American Concrete Institute: "Commentary on Building Code Requirements for Reinforced Concrete," ACI 318-83, 1983.
11. R. G. Drysdale and M. W. Huggins: "Sustained Biaxial Load on Slender Concrete Columns," *Journal of Structural Division ASCE,* 97, May 1971.
12. M. N. Hassoun: *Ultimate Load Design of Reinforced Concrete,* Cement and Concrete Association, London, 1981.

PROBLEMS

12.1. The column section in Figure 12.8 carries an axial load P_D = 120 k and a moment M_D = 110 k·ft due to dead load and an axial load P_L = 95 k and a moment M_L = 100 k·ft due to live load. The column is part of a frame, braced against sidesway, and bent in single curvature about its major axis. The unsupported length of the column l_u = 18 ft and the moments at both ends are equal. Check the adequacy of the section using f'_c = 4 ksi and f_y = 60 ksi.

12.2. Repeat problem 12.1 if l_u = 12 ft.

12.3. Repeat problem 12.1 if the frame is unbraced against sidesway and the end restraint factors ψ (top) = 0.7 and ψ (bottom) = 1.8 and the unsupported height l_u = 14 ft.

12.4. Design a 20-ft-long rectangular tied column for an axial load P_D = 200 k and a moment M_D = 60 k·ft due to dead load and an axial load P_L = 120 k and a moment M_L = 40 k·ft due to live load. The column is bent in single curvature about its major axis, braced against sidesway, and the end moments are equal. The end restraint factors are ψ (top) = 2.5 and ψ (bottom) = 1.4. Use f'_c = 5 ksi, f_y = 60 ksi, and b = 15 in.

12.5. Design the column in problem 12.4 if the column length = 10 ft.

12.6. Repeat problem 12.4 if the column is unbraced against sidesway.

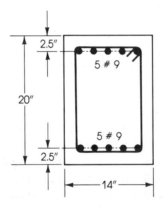

FIGURE 12.8. Problem 12.1.

13

Footings

13.1   Introduction

Footings are structural members used to support columns and walls and to transmit and distribute their loads to the soil in such a way that

- the load bearing capacity of the soil is not exceeded,
- excessive settlement, differential settlement, or rotations are prevented, and
- adequate safety against overturning or sliding is maintained.

When a column load is transmitted to the soil by the footing, the soil becomes compressed. The amount of settlement depends on many factors, such as the type of soil, the load intensity, the depth below ground level, and the type of footing. If different footings of the same structure have different settlements, new stresses develop in the structure. Excessive differential settlement may lead to the damage of nonstructural members in the buildings, even failure of the affected parts.

Vertical loads are usually applied at the centroid of the footing. If the resultant of the applied loads does not coincide with the centroid of the bearing area, a bending moment develops. In this case, the pressure on one side of the footing will be greater than the pressure on the other side, causing higher settlement on one side and a possible rotation of the footing.

If the bearing soil capacity is different under different footings, for example if the footings of a building are partly on soil and partly on rock, a differential settlement will occur. It is usual in such cases to provide a joint between the two parts to separate them, allowing for independent settlement.

The depth of the footing below the ground level is an important factor in the design of footings. This depth should be determined from soil tests, which should provide reliable information on safe bearing capacity at different layers below ground level. Soil test reports specify the allowable bearing capacity to be used in the design. The depth of footing below ground level may be calculated by Rankine's formula:

$$h = \frac{P}{W}\left(\frac{1 - \sin\phi}{1 + \sin\phi}\right)^2 \qquad\qquad (13.1)$$

where

h = depth below ground level

P = soil pressure (psf)

W = unit weight of soil

ϕ = angle of repose

In cold areas where freezing occurs, frost action may cause heaving or subsidence. It is necessary to place footings below freezing depth to avoid movements.

13.2   Types of Footings

Different types of footings may be used to support building columns or walls. The most common types are as follows:

1. *Wall footings* are used to support structural walls that carry loads from other floors, or to support nonstructural walls. They have a limited width and a continuous length under the wall (Figure 13.1). Wall footings may have one thickness, be stepped, or have a sloped top.
2. *Isolated or single footings* are used to support single columns (Figure 13.2). They may be square, rectangular, or circular. Again, the footing may be of uniform thickness, stepped, or have a sloped top. This is one of the most economical types of footings and it is used when columns are spaced at relatively long distances.
3. *Combined footings* (Figure 13.3) usually support two columns, or three columns not in a row. The shape of the footing in plan may be rectangular or trapezoidal, depending on column loads. Combined footings are used when two columns are so close that single footings cannot be used, or when one column is located at or near a property line.
4. *Cantilever or strap footings* (Figure 13.4) consist of two single footings connected with a beam or a strap and support two single columns. They are used when one footing supports an eccentric column and the nearest adjacent footing lies at quite a distance from it. This type replaces a combined footing and is more economical.
5. *Continuous footings* (Figure 13.5) support a row of three or more columns. They have limited width and continue under all columns.
6. *Raft or mat foundations* (Figure 13.6) consist of one footing, usually placed under the entire building area, and support the columns of the building. They are used when
   - the soil bearing capacity is low
   - column loads are heavy
   - single footings cannot be used
   - piles are not used
   - differential settlement must be reduced through the entire footing system

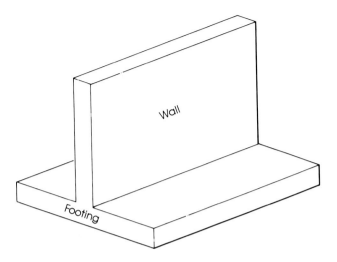

FIGURE 13.1. Wall footing.

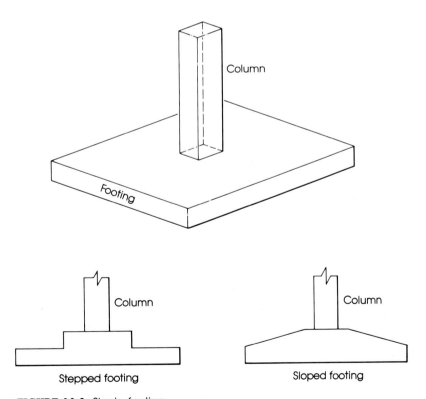

Stepped footing

Sloped footing

FIGURE 13.2. Single footing.

378

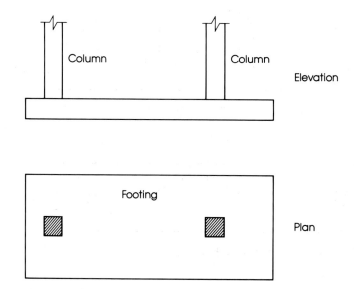

FIGURE 13.3. Combined footing.

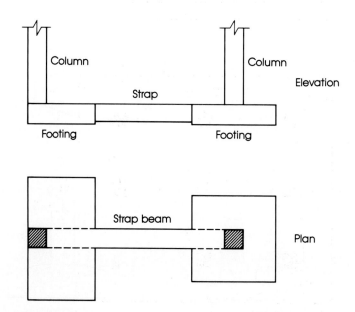

FIGURE 13.4. Strap footing.

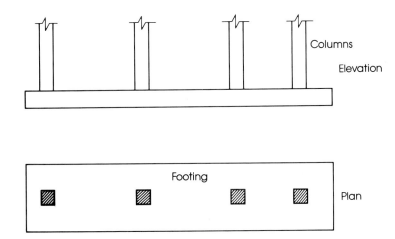

FIGURE 13.5. Continuous footing.

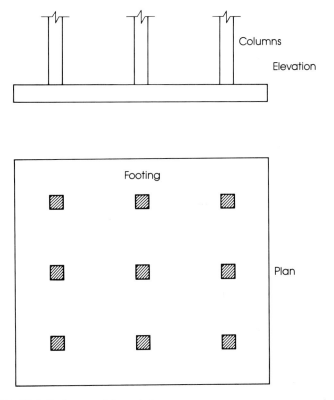

FIGURE 13.6. Raft or mat foundation.

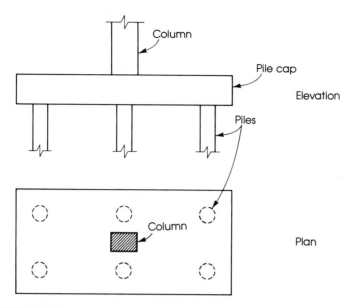

FIGURE 13.7. Pile cap footing.

7. *Pile caps* (Figure 13.7) are thick slabs used to tie a group of piles together, and to support and transmit column loads to the piles.

13.3   Distribution of Soil Pressure

Figure 13.8 shows a footing supporting a single column. When the column load P is applied on the centroid of the footing, a uniform pressure is assumed to develop on the soil surface below the footing area. However, the actual distribution of soil pressure is not uniform but depends on many factors, especially the composition of the soil and the degree of flexibility of the footing.

For example, the distribution of pressure on cohesionless soil (sand) under a rigid footing is shown in Figure 13.9. The pressure is maximum under the center of the footing and decreases toward the ends of the footing. The cohesionless soil tends to move from

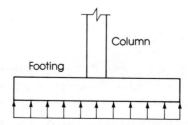

FIGURE 13.8. Distribution of soil pressure assuming uniform pressure.

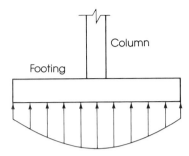

FIGURE 13.9. Soil pressure distribution in cohesionless soil (sand).

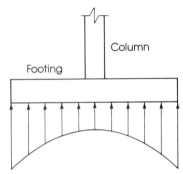

FIGURE 13.10. Soil pressure distribution in cohesive soil (clay).

the edges of the footing causing a reduction in pressure, while the pressure increases around the center to satisfy equilibrium conditions. If the footing is resting on a cohesive soil like clay, the pressure under the edges is greater than at the center of the footing (Figure 13.10). The clay near the edges has a strong cohesion with the adjacent clay surrounding the footing, causing the nonuniform pressure distribution.

Referring to Figure 13.8, when the load P is applied, the part of the footing below the column tends to settle downward. The footing will tend to take a uniform curved shape, causing an upward pressure on the projected parts of the footing. Each part acts as a cantilever and must be designed for both bending moments and shearing forces. The design of footings will be explained in detail later.

13.4   Design Considerations

Footings must be designed to carry the column loads and transmit them to the soil safely while satisfying code limitations. The design procedure must take the following strength requirements into consideration:

- the area of the footing based on the allowable bearing soil capacity
- two-way shear or punching shear

- one-way shear
- bending moment and steel reinforcement required
- bearing capacity of columns at their base
- dowel requirements
- development length of bars
- differential settlement

These strength requirements will be explained in the following sections.

13.4.1   Size of Footings

The area of the footing can be determined from the actual external loads (unfactored forces and moments) such that the allowable soil pressure is not exceeded. In general, for vertical loads:

$$\text{Area of footing} = \frac{\text{total load (including self-weight)}}{\text{allowable soil pressure}}$$

where total load is the working (unfactored) design load. Once the area is determined, an ultimate soil pressure is obtained by dividing the factored load, $P_u = 1.4D + 1.7L$, by the area of the footing. This is required to design the footing by the strength design method.

$$q_u = \frac{P_u}{\text{area of footing}} \tag{13.2}$$

13.4.2   Two-Way Shear (Punching Shear)

The **ACI Code, section 11.11.2,** allows a shear strength V_c of footings without shear reinforcement for two-way shear action as follows:

$$V_c = \left(2 + \frac{4}{\beta_c}\right)\sqrt{f'_c}\,b_o d \leq 4\sqrt{f'_c}\,b_o d \tag{13.3}$$

where

β_c = ratio of long side to short side of rectangular area

b_o = perimeter of the critical section taken at $d/2$ from the loaded area (column section)

d = effective depth of footing

This shear is a measure of the diagonal tension caused by the effect of the column load on the footing. Inclined cracks may occur in the footing at a distance $d/2$ from the face of the column on all sides. The footing will fail as the column tries to punch out part of the footing, as shown in Figure 13.11. It can be noticed from equation 13.3 that $V_c = 4\sqrt{f'_c}b_o d$ whenever

$$\beta_c = \frac{\text{footing long side}}{\text{footing short side}} \leq 2$$

When $\beta_c > 2$, the allowable V_c is reduced.

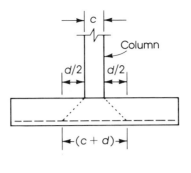

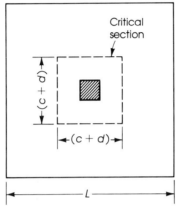

FIGURE 13.11. Punching shear (two-way).

The two-way shearing force V_u and the effective depth d required (if shear reinforcement is not provided) can be calculated as follows (refer to Figure 13.11):

1. Assume d.
2. Determine b_o: $b_o = 4(c + d)$ for square columns, where one side $= c$; $b_o = 2(c_1 + d) + 2(c_2 + d)$ for rectangular columns of sides c_1 and c_2.
3. The shearing force V_u acts at a section that has a length $b_o = 4(c + d)$ or $[2(c_1 + d) + 2(c_2 + d)]$ and a depth d; the section is subjected to a vertical downward load P_u and a vertical upward pressure q_u (equation 13.2). Therefore,

$$V_u = P_u - q_u(c + d)^2 \text{ for square columns} \tag{13.4a}$$

$$= P_u - q_u(c_1 + d)(c_2 + d) \text{ for rectangular columns} \tag{13.4b}$$

4. Allowable $\phi V_c = 4\phi \sqrt{f'_c}\, b_o d$ (introducing $\phi = 0.85$). Let $V_u = \phi V_c$. Then

$$d = \frac{V_u}{4\phi \sqrt{f'_c}\, b_o} \tag{13.5}$$

If d is not close to the assumed d, revise your assumption.

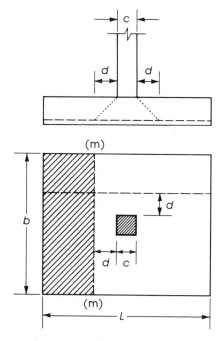

FIGURE 13.12. One-way shear.

13.4.3   One-Way Shear

For footings with bending action in one direction, the critical section is located at a distance d from the face of the column. The diagonal tension at section m–m in Figure 13.12 can be checked as was done before in beams. The allowable shear in this case is equal to

$$\phi V_c = 2\phi \sqrt{f_c'} bd \qquad (13.6)$$

where b = width of section m–m. The ultimate shearing force at section m–m can be calculated as follows:

$$V_u = q_u b \left(\frac{L}{2} - \frac{c}{2} - d \right) \qquad (13.7)$$

where b = side of footing parallel to section m–m.

If no shear reinforcement is to be used, then d can be checked, assuming $V_u = \phi V_c$:

$$d = \frac{V_u}{2\phi \sqrt{f_c'} b} \qquad (13.8)$$

13.4.4   Flexural Strength and Footing Reinforcement

The critical sections for moment occur at the face of the column (section n–n, Figure 13.13). The bending moment in each direction of the footing must be checked and the

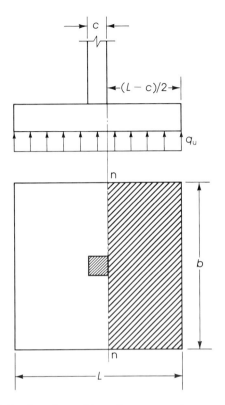

FIGURE 13.13. Critical section of bending moment.

appropriate reinforcement must be provided. In square footings, the bending moments in both directions are equal. To determine the reinforcement required, the depth of the footing in each direction may be used. As the bars in one direction rest on top of the bars in the other direction, the effective depth d varies with the diameter of the bars used. The value of d_{min} may be adopted.

The depth of the footing is often controlled by the shear, which requires a depth greater than that required by the bending moment. The steel reinforcement in each direction can be calculated in the case of flexural members as follows:

$$A_s = \frac{M_u}{\phi f_y(d - a/2)} \tag{13.9}$$

Another approach is to calculate $R_u = M_u/bd^2$ and determine the steel percentage required ρ from tables in Appendix A. In footings, the value of a in equation 13.9 is small, and A_s can be calculated by assuming that a is between 1 and 2 in. Determine A_s, then check if assumed a is close to the calculated a ($a = A_s f_y/0.85 f'_c b$). If it is not close, use the calculated a again in the equation to get the revised A_s. An approximate value for $(d - a/2)$ may be assumed equal to $0.9d$.

The minimum steel percentage requirement in flexural members is equal to $200/f_y$. However, the **ACI Code, section 10.5.3,** indicates that for structural slabs of uniform thickness, the minimum area and maximum spacing of steel bars in the direction of

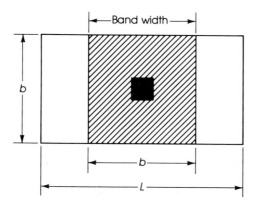

FIGURE 13.14. Band width for reinforcement distribution.

bending shall be as required for shrinkage and temperature reinforcement. This last minimum steel requirement is very small and a higher minimum reinforcement ratio is recommended, but not greater than $200/f_y$.

The reinforcement in one-way footings and two-way footings must be distributed across the entire width of the footing. In the case of two-way rectangular footings, the **ACI Code, section 15.4.4,** specifies that in the long direction reinforcement must be distributed uniformly along the width of the footing. In the short direction a certain ratio of the total reinforcement in this direction must be placed uniformly within a band width equal to the length of the short side of the footing according to

$$\frac{\text{Reinforcement in band width}}{\text{Total reinforcement in the short direction}} = \frac{2}{\beta + 1} \qquad (13.10)$$

where

$$\beta = \frac{\text{long side of footing}}{\text{short side of footing}}$$

The band width must be centered on the centerline of the column (Figure 13.14). The remaining reinforcement in the short direction must be uniformly distributed outside the band width. This remaining reinforcement percentage shall not be less than that required for shrinkage and temperature.

When structural steel columns or masonry walls are used, then the critical sections for moments in footings are taken at halfway between the middle and the edge of masonry walls, and halfway between the face of the column and the edge of the steel base place **(ACI Code, section 15.4.2).**

13.4.5   Bearing Capacity of Column at Base

The loads from the column act on the footing at the base of the column, on an area equal to the area of the column cross-section. Compressive forces are transferred to the footing directly by bearing on the concrete. Tensile forces must be resisted by reinforcement, neglecting any contribution by concrete.

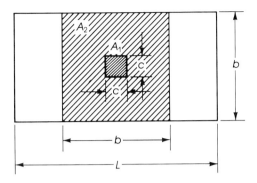

FIGURE 13.15. Bearing areas on footings. $A_1 = c^2$, $A_2 = b^2$.

Forces acting on the concrete at the base of the column must not exceed the bearing strength of concrete as specified by the **ACI Code, section 10.15:**

$$\text{Bearing strength } N_1 = \phi(0.85 f'_c A_1) \qquad (13.11)$$

where $\phi = 0.7$ and $A_1 =$ the bearing area of the column. The value of the bearing strength given in equation 13.11 may be multiplied by a factor $\sqrt{A_2/A_1} \leq 2.0$ for bearing on footings when the supporting surface is wider on all sides than the loaded area. Here A_2 is the area of the part of the supporting footing that is geometrically similar to and concentric with the loaded area (Figure 13.15). Since $A_2 > A_1$, the factor $\sqrt{A_2/A_1}$ is greater than unity, indicating that the allowable bearing strength is increased because of the lateral support from the footing area surrounding the column base. The modified bearing strength

$$N_2 \leq \phi(0.85 f'_c A_1)\sqrt{A_2/A_1} \qquad (13.12)$$
$$N_2 \leq 2\phi(0.85 f'_c A_1) \qquad (13.13)$$

If the calculated bearing force is greater than either N_1 or N_2, reinforcement must be provided to transfer the excess force. This is achieved by providing dowels or extending the column bars into the footing. If the calculated bearing force is less than either N_1 or N_2, then minimum reinforcement must be provided. The **ACI Code, section 15.8.2,** indicates that the minimum area of the dowel reinforcement is at least $0.005 A_g$, but not less than 4 bars, where A_g is the gross area of the column section of the supported member. The minimum reinforcement requirements apply also to the case when the calculated bearing forces are greater than N_1 and N_2.

13.4.6   Dowels in Footings

It was explained in the previous section that dowels are required in any case, even if the bearing strength is adequate. The ACI Code specifies a minimum steel ratio $\rho = 0.005$ of the column section as compared to $\rho = 0.01$ as minimum reinforcement for the column itself. The minimum number of dowel bars needed is four; these may be placed at the four corners of the column. The dowel bars are usually extended into the footing, bent at their ends, and tied to the main footing reinforcement.

The dowel diameter shall not exceed the diameter of the longitudinal bars in the columns by more than 0.15 in. This requirement is necessary to insure proper action of and tying between the column and footing.

The development length of the dowels must be checked to determine proper transfer of the compression force into the footing.

13.4.7   Development Length of the Reinforcing Bars

The critical sections for checking the development length of the reinforcing bars are the same as those for bending moments. The development length for compression bars was given in Chapter 7:

$$l_d = 0.02 f_y d_b / \sqrt{f'_c} \qquad (7.16)$$

but not less than $0.0003 f_y d_b \geq 8$ in. For other values refer to Chapter 7. Dowel bars must also be checked for proper development length.

13.4.8   Differential Settlement (Balanced Footing Design)

Footings usually support the following loads:

- dead loads from the substructure and superstructure
- live load resulting from materials or occupancy
- weight of materials used in backfilling
- wind loads

Each footing in a building is designed to support the maximum load that may occur on any column due to the critical combination of loadings, using the allowable soil pressure.

The dead load, and maybe a small portion of the live load (called the "usual" live load), may act continuously on the structure. The rest of the live load may occur at intervals and on some parts of the structure only, causing different loadings on columns. Consequently, the pressure on the soil under different footings will vary according to the loads on the different columns, and differential settlement will occur under the various footings of one structure. Since partial settlement is inevitable, the problem turns out to be the amount of differential settlement that the structure can tolerate. The amount of differential settlement depends on the variation in the compressibility of the soils, the thickness of the compressible material below foundation level, and the stiffness of the combined footing and superstructure. Excessive differential settlement results in cracking of concrete and damage to claddings, partitions, ceilings, and finishes.

Differential settlement may be expressed in terms of angular distortion of the structure. Bjerrum[5] indicated the danger limits of distortion for some conditions as in Table 13.1.

For practical purposes it can be assumed that the soil pressure under the effect of sustained loadings is the same for all footings, thus causing equal settlements. The sustained load (or the usual load) can be assumed equal to the dead load plus a percentage of the live load, which occurs very frequently on the structure. Footings then are proportioned for these sustained loads to produce the same soil pressure under all footings. In no case is the allowable soil bearing capacity to be exceeded under the dead load plus the maximum live load for each footing. Example 13.5 explains the procedure

TABLE 13.1. Danger limits of distortion

Angular distortion	Behavior of the structure
1/600	Limit of danger for frames with diagonals.
1/500	Safe limit for buildings where cracking is not permissible.
1/300	Limit where first cracking in panel walls is to be expected.
1/250	Limit where tilting of high rigid buildings becomes visible.
1/150	Limit where structural damage of buildings is to be feared.

for calculating the areas of footings, taking into consideration the effect of differential settlement.

13.5   Plain Concrete Footings

Plain concrete footings may be used to support masonry walls or other light loads and transfer them to the supporting soil. The 1977 **ACI Code, section 15.11,** allows the use of plain concrete pedestals and footings on soil, provided that the design stresses shall not exceed the following:

1. Maximum flexural stress in tension is $\leq 5\phi \sqrt{f_c'}$ (where $\phi = 0.65$).
2. Maximum stress in one-way shear (beam action) $\leq 2\phi \sqrt{f_c'}$ (where $\phi = 0.85$).
3. Maximum shear stress in two-way action $\leq 4\phi \sqrt{f_c'}$ (where $\phi = 0.85$).
4. Maximum compressive strength shall not exceed the concrete bearing strength. The minimum thickness of plain concrete footings shall not be less than 8 in. The critical sections for bending moments are at the face of the column or wall. The critical sections for one-way shear and two-way shear action are at distances d and $d/2$ from the face of the column or wall, respectively. Although plain concrete footings do not require steel reinforcement, it will be advantageous to provide shrinkage reinforcement in the two directions of the footing.
5. Stresses due to factored loads are computed assuming a linear distribution in concrete.
6. The effective depth d must be taken equal to the overall thickness minus 3 in. Whenever possible, it is advisable to use reinforced concrete footings and to avoid the use of plain concrete footings.

13.6   General Requirements for Footings Design

The following points illustrate the general requirements for the design of footings.

1. A site investigation is required to determine the chemical and physical properties of the soil.

2. Determine the magnitude and distribution of loads from the superstructure.
3. Establish the criteria and the tolerance for the total and differential settlements of the structure.
4. Determine the most suitable and economic type of foundation.
5. Determine the depth of footings below the ground level and the method of excavation.
6. Establish the allowable bearing pressure to be used in design.
7. Determine the pressure distribution beneath the footing based on its width.
8. Perform a settlement analysis.

EXAMPLE 13.1

Design a reinforced concrete footing to support a 20-in.-wide concrete wall carrying a dead load = 26 k/ft, including the weight of the wall, and a live load = 20 k/ft. The bottom of the footing is 6 ft below final grade. Use f'_c = 4 ksi, f_y = 40 ksi, and an allowable soil pressure = 5 ksf.

Solution

1. Calculate the effective soil pressure. Assume a total depth of footing = 20 in. Weight of footing = 20/12 × 150 = 250 psf. Weight of the soil fill on top of the footing, assuming that soil weighs 100 lb/ft³ = (6 − 20/12) × 100 = 433 psf. Effective soil pressure at the bottom of the footing = 5000 − 250 − 433 = 4317 psf = 4.32 ksf.

2. Calculate width of the footing for a 1-ft length of the wall:

$$\text{Width of footing} = \frac{\text{total load}}{\text{effective soil pressure}} = \frac{26 + 20}{4.32} = 10.7 \text{ ft, use 11 ft}$$

3. Net upward pressure = $\frac{\text{ultimate load}}{\text{footing width}}$ (for 1-ft length).

Ultimate load = $1.4D + 1.7L = 1.4 \times 26 + 1.7 \times 20 = 70.4$ k

Net upward pressure $q_u = \frac{70.4}{11.0} = 6.4$ ksf

4. Check the assumed depth for shear requirements. The concrete cover in footings = 3 in., and assume No. 8 bars, then $d = 20 − 3.5 = 16.5$ in. The critical section for one-way shear is at a distance d from the face of the wall:

$$V_u = q_u \left(\frac{L}{2} - d - \frac{c}{2}\right) = 6.4 \left(\frac{11}{2} - \frac{16.5}{12} - \frac{20}{2 \times 12}\right) = 21.06 \text{ k}$$

Allowable one-way shear = $2\sqrt{f'_c} = 2\sqrt{4000} = 126.5$ psi

Required $d = \frac{V_u}{\phi(2\sqrt{f'_c})(b)} = \frac{21.06 \times 1000}{0.85(126.5)(12)} = 16.32$ in.

b = 1-ft length of footing = 12 in.

Total depth = 16.32 + 3.5 = 19.82 in., say 20 in.

Actual d = 20 − 3.5 = 16.5 in.   (as assumed)

5. Calculate the bending moment and steel reinforcement. The critical section is at the face of the wall:

$$M_u = \frac{1}{2} q_u \left(\frac{L}{2} - \frac{c}{2} \right)^2 = \frac{6.4}{2} \left(\frac{11}{2} - \frac{20}{24} \right)^2 = 70 \text{ k·ft}$$

$$R_u = \frac{M_u}{bd^2} = \frac{70 \times 12,000}{12(16.5)^2} = 257 \text{ psi}$$

From Table A.1 in Appendix A, for $R_u = 257$ psi, $f'_c = 4$ ksi, and $f_y = 40$ ksi, the steel percentage $\rho = 0.0075$. Minimum steel percentage for flexural members

$$\rho_{min} = \frac{200}{f_y} = \frac{200}{40,000} = 0.5 \text{ percent}$$

Percentage of shrinkage reinforcement = 0.2 percent (for $f_y = 40$ ksi). Therefore, use $\rho = 0.0075$ as calculated, which is greater than the minimum $\rho = 0.002$ required according to the **ACI Code, section 10.5.3.**

$$A_s = 0.0075 \times 12 \times 16.5 = 1.48 \text{ in.}^2$$

Use No. 8 bars spaced at 6 in. ($A_s = 1.57$ in.^{2}).

6. Check development length. Required $l_d = 20$ in. for bottom bars (from tables in Appendix A). l_d provided is the distance from the section of maximum moment at the face of the wall to the edge of the footing minus 3 in. (concrete cover).

$$l_d \text{ (provided)} = \frac{L}{2} - \frac{c}{2} - 3 = \frac{11 \times 12}{2} - \frac{20}{2} - 3 = 53 \text{ in.}$$

This is greater than the 20 in. required.

7. Shrinkage reinforcement must be provided in the longitudinal direction (minimum $\rho = 0.002$).

$$A_s = 0.002 \times 12 \times 20 = 0.48 \text{ in.}^2$$

Use No. 5 bars spaced at 7 in. ($A_s = 0.53$ in.^{2}). Details of the section are shown in Figure 13.16. ■

EXAMPLE 13.2

Design a plain concrete footing to support a 16-in.-thick concrete wall. The loads on the wall consist of 16 k/ft dead load (including the self-weight of wall) and a 10 k/ft live load. The base of the footing is 4 ft below final grade. Use $f'_c = 3$ ksi and an allowable soil pressure = 5 ksf.

Solution

1. Calculate the effective soil pressure. Assume a total depth of footing = 28 in.

$$\text{Weight of footing} = \frac{28}{12} \times 145 = 338 \text{ psf}$$

Weight of soil on top of the footing, assuming that soil weighs 100 pcf, is $(4 - 2.33) \times 100 = 167$ psf.

$$\text{Effective soil pressure} = 5000 - 338 - 167 = 4495 \text{ psf}$$

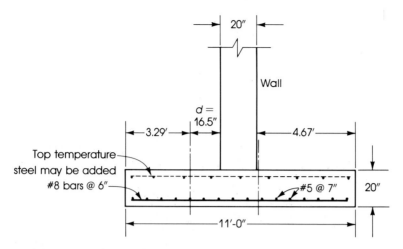

FIGURE 13.16. Wall footing, example 13.1.

2. Calculate width of footing for 1-ft length of the wall:

$$\text{Width of footing} = \frac{\text{total load}}{\text{effective soil pressure}}$$

$$= \frac{16 + 10}{4.495} = 5.79 \text{ ft, say, 6.0 ft (Figure 13.17)}$$

$b = 1$ ft

3. $U = 1.4D + 1.7L = 1.4 \times 16 + 1.7 \times 10 = 39.4$ k/ft

Net upward pressure $q_u = 39.4/6 = 6.57$ ksf

4. Check bending stresses. The critical section is at the face of the wall. For 1-ft length of wall and footing,

$$M_u = \frac{1}{2} q_u \left(\frac{L}{2} - \frac{c}{2}\right)^2 = \frac{1}{2} \times 6.57 \left(\frac{6}{2} - \frac{16}{2 \times 12}\right)^2 = 17.88 \text{ k·ft}$$

Let the effective depth $= 28 - 3 = 25$ in., assuming that the bottom 3 in. is not effective.

$$I_g = \frac{bd^3}{12} = \frac{12}{12}(25)^3 = 15{,}625 \text{ in.}^4$$

$$\text{Flexural tensile stress} = f_t = \frac{M_u c}{I} = \frac{(17.88 \times 12{,}000)}{15{,}625}\left(\frac{25}{2}\right) = 172 \text{ psi}$$

5. Allowable flexural tensile stress $= 5\phi\sqrt{f_c'} = 5 \times 0.65\sqrt{3000} = 178$ psi, which is greater than the flexural tensile stress f_t of 172 psi.

6. Check shear stress. Critical section is at a distance $d = 25$ in. from the face of the wall.

$$V_u = q_u \left(\frac{L}{2} - \frac{c}{2} - d\right) = 6.57 \left(\frac{6}{2} - \frac{16}{2 \times 12} - \frac{25}{12}\right) = 1.65 \text{ k}$$

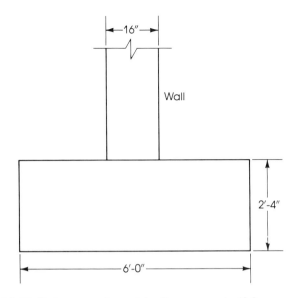

FIGURE 13.17. Plain concrete wall footing, example 13.2.

$$\text{Shear stress } v_u = \frac{V_u}{\phi bd} = \frac{1.65 \times 1000}{0.85 \times 12 \times 25} = 6.5 \text{ psi}$$

$$\text{Allowable shear } v_u = 2\sqrt{f_c'} = 2\sqrt{3000} = 110 \text{ psi}$$

Therefore, the section is adequate. It is advisable to use minimum reinforcement in this type of footing or to use a reinforced concrete footing. ∎

EXAMPLE 13.3

Design a square single footing to support an 18-in.-square tied column reinforced with 8 No. 9 bars. The column carries an unfactored axial dead load of 245 k and an axial live load of 200 k. The base of the footing is 4 ft below final grade and the allowable soil pressure = 5 ksf. Use f_c' = 4 ksi and f_y = 40 ksi.

Solution

1. Calculate the effective soil pressure. Assume a total depth of footing = 2 ft. Weight of footing = 2 × 150 = 300 psf. Weight of soil on top of the footing (assuming the weight of soil = 100 pcf) = (4 − 2) × 100 = 200 psf.

    Effective soil pressure = 5000 − 300 − 200 = 4500 psf

2. Calculate area of footing:

    Actual loads = $D + L$ = 245 + 200 = 445 k

    $$\text{Area of footing} = \frac{445}{4.5} = 98.9 \text{ ft}^2$$

    Side of footing = 9.94 ft, say, 10 ft (Figure 13.18)

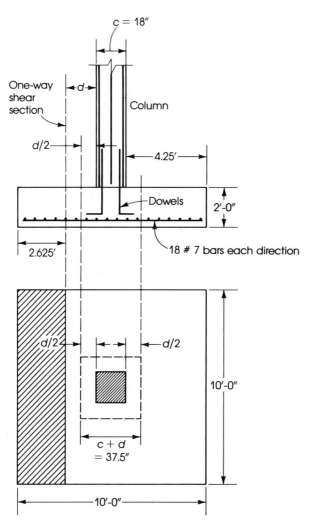

FIGURE 13.18. Square footing, example 13.3.

3. Net upward pressure = ultimate load/area of footing.

$$P_u = 1.4D + 1.7L = 1.4 \times 245 + 1.7 \times 200 = 683 \text{ k}$$

$$\text{Net upward pressure } q_u = \frac{683}{10 \times 10} = 6.83 \text{ ksf}$$

4. Check depth due to two-way shear. If no shear reinforcement is used, two-way shear determines the critical footing depth required. For an assumed total depth = 24 in., calculate d to the centroid of the top layer of the steel bars to be placed in the two directions within the footing. Let bars to be used be No. 8 bars for d calculation.

$$d = 24 - 3 \text{ (cover)} - 1.5 \text{ (bar diameters)} = 19.5 \text{ in.}$$
$$b_o = 4(c + d) = 4(18 + 19.5) = 150 \text{ in.}$$

$$c + d = 18 + 19.5 = 37.5 \text{ in.} = 3.125 \text{ ft}$$

$$V_u = P_u - q_u(c + d)^2 = 683 - 6.83(3.125)^2 = 616.3 \text{ k}$$

$$\text{Required } d = \frac{V_u}{4\phi \sqrt{f_c'} b_o} = \frac{616.3 \times 1000}{4 \times 0.85 \sqrt{4000} \times 150} = 19.1 \text{ in.}$$

This is less than 19.5 in.; thus the assumed depth is adequate. Two or more trials may be needed to reach an acceptable d very close to the assumed one.

5. Check depth due to one-way shear action: The critical section is at a distance d from the face of the column.

$$\text{Distance from edge of footing} = \left(\frac{L}{2} - \frac{c}{2} - d\right) = 2.625 \text{ ft}$$

$$V_u = 6.83 \times (2.625)(10) = 179.3 \text{ k}$$

$$\text{Shear capacity of the section} = 2\phi \sqrt{f_c'} bd$$

$$= 2 \times 0.85 \sqrt{4000} \times (10 \times 12) \times 19.5 \times 10^{-3}$$

$$= 251.6 \text{ k}$$

This is greater than the applied shear of 179.3 k; thus the depth is adequate. The depth may be calculated and compared with the assumed depth of 19.5 in. as an alternative approach.

6. Calculate the bending moment and steel reinforcement. The critical section is at the face of the column.

$$\text{Distance from edge of footing} = \left(\frac{L}{2} - \frac{c}{2}\right) = 5 - \frac{1.5}{2} = 4.25 \text{ ft}$$

$$M_u = \frac{1}{2} q_u \left(\frac{L}{2} - \frac{c}{2}\right)^2 b = \frac{1}{2} \times 6.83(4.25)^2 \times 10 = 617 \text{ k·ft}$$

$$M_u = \phi A_s f_y \left(d - \frac{a}{2}\right)$$

and

$$A_s = \frac{M_u}{\phi f_y(d - a/2)}$$

To calculate A_s, assume $a = 1.0$ in., calculate A_s, then check the assumed $a = A_s f_y / 0.85 f_c' b$, or calculate $R_u = M_u/bd^2$ and determine the steel percentage ρ from the tables in Appendix A.

$$A_s = \frac{617 \times 12}{0.9 \times 40 \left(19.5 - \frac{1.0}{2}\right)} = 10.82 \text{ in.}^2$$

Check

$$a = \frac{10.82 \times 40}{0.85 \times 4 \times (10 \times 12)} = 1.06 \text{ in.}$$

which is very close to the assumed a; therefore use $A_s = 10.82$ in.^{2}. Minimum steel reinforcement required is $0.002 \times 120 \times 24 = 5.76$ in.^{2}, which is less than A_s calculated.

Choose 18 No. 7 bars (A_s = 10.82 in.^{2}) each direction of the footing, one layer spread and tied on top of the other.

$$\text{Spacing of bars} = \frac{120 - 6}{17} = 6.7 \text{ in.}$$

(leaving 3 in. cover from each edge of the footing).

7. Check bearing stress at the base of the column.

$$\text{Bearing strength } N_1 = \phi(0.85 f_c' A_1) \tag{13.11}$$

$$N_1 = 0.7(0.85 \times 4 \times 18 \times 18) = 771 \text{ k}$$

This is greater than P_u = 683 k applied on the column; therefore, bearing stress is adequate. (N_2 is not critical.) Minimum area of dowels required, according to the ACI Code, is $0.005 A_1 = 0.005(18)^2 = 1.62$ in.^{2}. Minimum number of bars is 4. Use 4 No. 9 bars at the corners of the column to extend to the footing a development length

$$l_d \text{ in compression} = 0.02 f_y d_b / \sqrt{f_c'}$$

$$= \frac{0.02 \times 40{,}000 \times 1.128}{\sqrt{4000}} = 14.3 \text{ in.}$$

but not less than $l_d = 0.0003 f_y d_b = 13.5$ in. ≥ 8 in. Therefore, extend the bars a distance 19 in. and bend the bars to be laid and tied properly on top of the main bars.

Note that if l_d required is greater than the depth of the footing, then either smaller-sized dowels or a pedestal may be used. The same l_d is needed within the column if dowels are to extend into the footing and into the column.

8. The development length required for the No. 7 bars of the main reinforcement is l_d = 15 in. The development length provided equals the distance from the face of the column to 3 in. before the edge of the footing = $(L/2 - c/2) - 3$ in. = 48 in. This is greater than the 15 in. required.

Details of the footing are shown in Figure 13.18.   ■

EXAMPLE 13.4

Design a rectangular footing for the column of the previous example if one side of the footing is limited to 8.5 ft.

Solution

1. The design procedure for rectangular footings is similar to that of square footings, taking into consideration the forces acting on the footing in each direction separately.

2. From the previous example, the area of the footing required is 98.9 ft.^{2}

$$\text{Length of footing} = \frac{98.9}{8.5} = 11.63 \text{ ft, say, } 11.75 \text{ ft (Figure 13.19)}$$

$$\text{Footing dimensions} = 8.5 \times 11.75 \text{ ft}$$

3. P_u = 683 k. Thus

$$\text{Net upward pressure} = \frac{683}{8.5 \times 11.75} = 6.84 \text{ ksf}$$

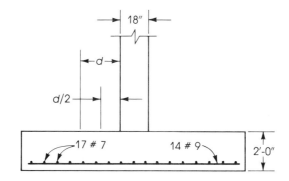

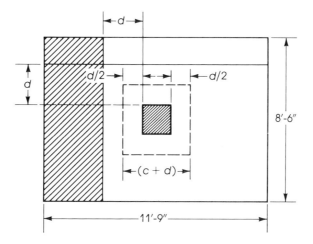

FIGURE 13.19. Rectangular footing, example 13.4.

4. Check the depth due to one-way shear. The critical section is at a distance d from the face of the column. In the longitudinal direction,

$$V_u = \left(\frac{L}{2} - \frac{c}{2} - d\right) \times q_u b$$

$$= \left(\frac{11.75}{2} - \frac{1.5}{2} - \frac{19.5}{12}\right) \times 6.84 \times 8.5 = 203.5 \text{ k}$$

This shear controls. In the short direction, $V_u = 150.7$ k (not critical).

$$\text{Required } d = \frac{V_u}{2\phi \sqrt{f_c'} b} = \frac{203.5 \times 1000}{2 \times 0.85 \sqrt{4000} \times (8.5 \times 12)} = 18.6 \text{ in.}$$

d provided $= 19.5$ in. > 18.6 in.

5. Check the depth for two-way shear action (punching shear). The critical section is at a distance $d/2$ from the face of the column on four sides.

$$b_o = 4(18 + 19.5) = 150 \text{ in.}$$

$$(c + d) = 18 + 19.5 = 37.5 \text{ in.} = 3.125 \text{ ft}$$

$$\beta_c = \frac{11.75}{8.5} = 1.38 < 2 \quad \text{use } V_c = 4\phi\sqrt{f_c'}b_o d$$

$$V_u = P_u - q_u(c + d)^2 = 683 - 6.84(3.125)^2 = 616.2 \text{ k}$$

$$\text{Required } d = \frac{V_u}{4\phi\sqrt{f_c'}b_o} = \frac{616.2 \times 1000}{4 \times 0.85\sqrt{4000} \times 150} = 19.1 \text{ in.} < 19.5 \text{ in.}$$

6. Design steel reinforcement in the longitudinal direction. The critical section is at the face of the support. Distance from the edge of the footing

$$\frac{L}{2} - \frac{c}{2} = \frac{11.75}{2} - \frac{1.5}{2} = 5.125 \text{ ft}$$

$$M_u = \frac{1}{2} \times 6.84(5.125)^2 \times 8.5 = 763.5 \text{ k·ft}$$

Let $a = 1.5$ in.

$$A_s = \frac{M_u}{\phi f_y(d - a/2)} = \frac{763.5 \times 12}{0.9 \times 40\left(19.5 - \dfrac{1.5}{2}\right)} = 13.5 \text{ in.}^2$$

Check a:

$$a = \frac{A_s f_y}{0.85 f_c' b} = \frac{13.5 \times 40}{0.85 \times 4 \times (8.5 \times 12)} = 1.56 \text{ in.}$$

which is very close to the assumed a.

$$\text{Minimum } A_s = 0.002 \times 24 \times 8.5 \times 12 = 4.9 \text{ in.}^2 < 13.5 \text{ in.}^2$$

Use 14 No. 9 bars ($A_s = 14.0$ in.^{2}).

7. Design steel reinforcement in the short direction.

$$\text{Distance from face of column to edge of footing} = \frac{8.5}{2} - \frac{1.5}{2} = 3.5 \text{ ft}$$

$$M_u = \frac{1}{2} \times 6.84\,(3.5)^2 \times 11.75 = 492.3 \text{ k·ft}$$

Let $a = 1.0$ in.

$$A_s = \frac{492.3 \times 12}{0.9 \times 40\left(19.5 - \dfrac{1.0}{2}\right)} = 8.64 \text{ in.}^2 > A_s \text{ (min)}$$

Check a:

$$a = \frac{8.64 \times 40}{0.85 \times 4 \times (11.75 \times 12)} = 0.7 \text{ in.}$$

Check again; for $a = 0.7$ in., $A_s = 8.6$ in.^{2} (small variation). Use 15 No. 7 bars ($A_s = 9.0$ in.^{2}).

$$\frac{\text{Reinforcement in band width}}{\text{Total reinforcement}} = \frac{2}{\beta + 1} = \frac{2}{\left(\dfrac{11.75}{8.5}\right) + 1} = 0.84$$

Number of bars in 8.5-ft band $= 15 \times 0.84 = 12.6$ bars, say 13 bars. Number of bars left on each side $= (15 - 13)/2 = 1$ bar. Therefore, place 13 No. 7 bars within the 8.5-ft central part, then 2 No. 7 bars ($A_s = 1.2$ in.^{2}) within $(11.75 - 8.5)/2 = 1.625$ ft on each side of the band.

$$\text{Minimum } A_s \text{ required} = 0.002 \times 24 \times (1.625 \times 12) = 0.94 \text{ in.}^2 < 1.2 \text{ in.}^2$$

Details of reinforcement are shown in Figure 13.19.

8. Check the bearing stress at the base of the column, as explained in the previous example. Use 4 No. 9 dowel bars.
9. Development length of the main reinforcement $l_d = 17.5$ in. for No. 7 bars and 25 in. for No. 9 bars.

$$l_d \text{ (long direction)} = \left(\frac{l}{2} - \frac{c}{2} - 3 \text{ in.}\right) = 58.5 \text{ in.} > 25 \text{ in.}$$

$$l_d \text{ (short direction)} = 39 \text{ in.} > 17.5 \text{ in.} \qquad \blacksquare$$

EXAMPLE 13.5

Determine the footing areas required for equal settlement (balanced footing design) if the usual live load is 20 percent for all footings. The footings are subjected to dead loads and live loads as indicated below. The allowable net soil pressure is 6 ksf.

	Footing no.				
	1	2	3	4	5
Dead load	120 k	180 k	140 k	190 k	210 k
Live load	150 k	220 k	200 k	170 k	240 k

Solution

1. Determine the footing which has the largest ratio of live load to dead load. In this example, footing 3's ratio of 1.43 is higher than the other ratios.
2. Calculate the usual load for all footings. The usual load is the dead load plus the portion of live load that most commonly occurs on the structure. In this example,

$$\text{Usual load} = \text{D.L.} + 0.2 \text{ L.L.}$$

The values of the usual loads are shown in the table below.

3. Determine the area of the footing which has the highest ratio of L.L./D.L.:

$$\text{Area of footing 3} = \frac{\text{D.L.} + \text{L.L.}}{\text{allowable soil pressure}} = \frac{140 + 200}{6} = 56.7 \text{ ft}^2$$

The usual soil pressure under footing 3 is

$$\frac{\text{Usual load}}{\text{Area of footing}} = \frac{180}{56.7} = 3.18 \text{ ksf}$$

4. Calculate the area required for each footing by dividing its usual load by the usual soil pressure of footing 3. The areas are tabulated in the table below. For footing 1, for example, the required area = 150/3.18 = 47.2 ft².

5. Calculate the maximum soil pressure under each footing:

$$q_{max} = \frac{D + L}{area} \leq 6 \text{ ksf (allowable soil pressure)}$$

	Footing no.				
	1	2	3	4	5
$\dfrac{\text{Live load}}{\text{Dead load}}$	1.25	1.22	1.43	0.90	1.14
Usual load = D.L. + 0.2 L.L	150 k	224 k	180 k	224 k	258 k
Area required = $\dfrac{\text{usual load}}{3.18 \text{ ksf}}$	47.2 ft²	70.4 ft²	56.7 ft²	70.4 ft²	81.1 ft²
Max. soil pressure $= \dfrac{D + L}{area}$	5.72 ksf	5.68 ksf	6.00 ksf	5.11 ksf	5.55 ksf

13.7   Combined Footings

When a column is located near a property line, part of the single footing might extend into the neighboring property. To avoid this situation, the column may be placed on one side or edge of the footing, causing eccentric loading. This may not be possible under certain conditions, and sometimes it is not an economical solution. A better design can be achieved by combining the footing with the nearest internal column footing, forming a combined footing. The center of gravity of the combined footing coincides with the resultant of the loads on the two columns.

Another case where combined footings become necessary is when the soil is poor and the footing of one column overlaps the adjacent footing. The shape of the combined footing may be rectangular or trapezoidal (Figure 13.20). When the load of the external column near the property line is greater than the load of the interior column, a trapezoidal footing is necessary to keep the centroid of footing in line with the resultant of the two column loads. In most other cases a rectangular footing may be advantageous.

The length and width of the combined footing are chosen to the nearest 3 in., which may cause a small variation in the uniform pressure under the footing, but it can be tolerated. For a uniform upward pressure, the footing will deflect as shown in Figure 13.21. The **ACI Code, section 15.10,** does not provide a detailed approach for the design of combined footings. The design, in general, is based on an empirical approach.

A simple method of analysis is to treat the footing as a beam in the longitudinal direction, loaded with uniform upward pressure q_u. For the transverse direction, it is assumed that the column load is spread over a width under the column equal to the column width plus d on each side, whenever that is available. In other words, the column load acts on a beam under the column within the footing which has a maximum

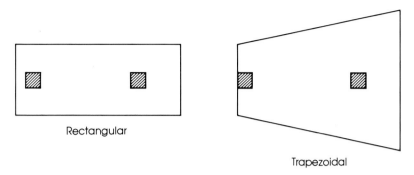

Rectangular

Trapezoidal

FIGURE 13.20. Combined footings.

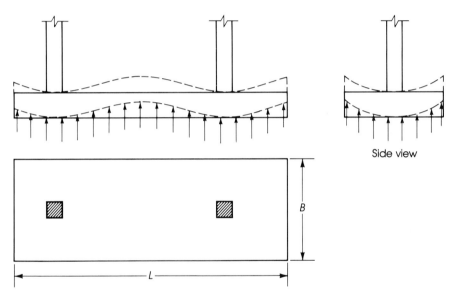

Side view

B

L

FIGURE 13.21. Upward deflection of a combined footing in two directions.

width of $(c + 2d)$ and a length equal to the short side of the footing (Figure 13.22). A smaller width, down to $(c + d)$, may be used. The next example will explain the design method in detail.

EXAMPLE 13.6

Design a rectangular combined footing to support two columns as shown in Figure 13.23. The edge column I has a section 16 by 16 in. and carries a D.L. = 180 k and a L.L. = 120 k. The interior column II has a section 20 by 20 in. and carries a D.L. = 250 k and a L.L. = 140 k.

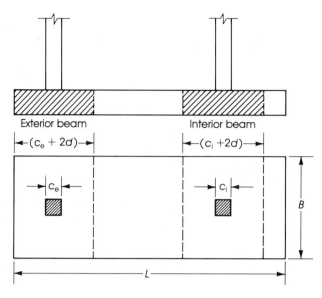

FIGURE 13.22. Analysis of combined footing in the transverse direction.

The net allowable soil pressure is 5 ksf. Design the footing using f'_c = 4 ksi, f_y = 40 ksi, and the ACI strength design method.

Solution

1. Determine the location of the resultant of the column loads. Take moments about the center of the exterior column I:

$$\bar{x} = \frac{(250 + 140) \times 16}{(250 + 140) + (180 + 120)} = 9 \text{ ft from column 1}$$

Distance of the resultant from the property line is

9 + 2 = 11.0 ft

Length of footing = 2 × 11 = 22.0 ft

In this case the resultant of column loads will coincide with the resultant of the upward pressure on the footing.

2. Determine the area of the footing. Assume the footing total depth is 40 in. (d = 40 − 4.5 = 35.5 in.).

Total actual (working) loads = 300 + 390 = 690 k

$$\text{Net upward pressure} = 5000 - \left(\frac{40}{12} \times 150\right) = 4500 \text{ psf}$$

(Soil fill is small in this example and is neglected.)

$$\text{Required area} = \frac{690}{4.5} = 153.3 \text{ ft}^2$$

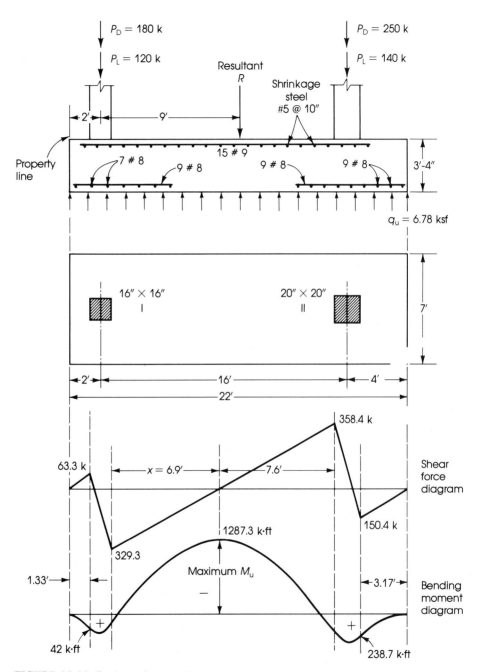

FIGURE 13.23. Design of a combined footing, example 13.6.

$$\text{Width of footing} = \frac{153.3}{22} = 6.97 \text{ ft, say, 7 ft}$$

Choose a footing 22 by 7 ft (area = 154 ft^2).

3. Determine the ultimate upward pressure using ultimate loads:

$$P_{u1} \text{ (column I)} = 1.4 \times 180 + 1.7 \times 120 = 456 \text{ k}$$

$$P_{u2} \text{ (column II)} = 1.4 \times 250 + 1.7 \times 140 = 588 \text{ k}$$

The net ultimate soil pressure

$$q_u = \frac{456 + 588}{154} = 6.78 \text{ ksf}$$

4. Draw the factored shearing force diagram as for a beam of $L = 22$ ft supported on two columns and subjected to an upward pressure of 6.78 ksf $\times$ 7 (width of footing) = 47.5 k/ft (per foot length of footing).

$$V_u \text{ (at outer face column I)} = 47.5 \left(2 - \frac{8}{12}\right) = 63.3 \text{ k}$$

$$V_u \text{ (at interior face column I)} = 456 - 47.5 \left(2 + \frac{8}{12}\right) = 329.3 \text{ k}$$

$$V_u \text{ (at outer face column II)} = 47.5 \left(4 - \frac{10}{12}\right) = 150.4 \text{ k}$$

$$V_u \text{ (at interior face column II)} = 588 - \left(4 - \frac{10}{12}\right) \times 47.5 = 358.4 \text{ k}$$

Find the point of zero shear x: Distance between interior faces of columns I and II

$$= 16 - \frac{8}{12} - \frac{10}{12} = 14.5 \text{ ft}$$

$$x = \frac{329.3}{(329.3 + 358.4)} (14.5) = 6.9 \text{ ft}$$

5. Draw the factored moment diagram considering the footing as a beam of $L = 22$ ft supported by the two columns. The uniform upward pressure = 47.5 k/ft.

$$M_{u1} \text{ (at outer face column I)} = 47.5 \frac{(1.33)^2}{2} = 42 \text{ k·ft}$$

$$M_{u2} \text{ (at outer face column II)} = 47.5 \frac{(3.17)^2}{2} = 238.7 \text{ k·ft}$$

The maximum moment occurs at zero shear:

Maximum M_u (calculated from column I side)

$$= 456 \left(6.9 + \frac{8}{12}\right) - \frac{47.5}{2} \left(6.9 + \frac{8}{12} + 2\right)^2 = 1276.8 \text{ k·ft}$$

Maximum M_u (from column II side)

$$= 588 \left(7.6 + \frac{10}{12}\right) - \frac{47.5}{2} \left(7.6 + \frac{10}{12} + 4\right)^2 = 1287.3 \text{ k·ft}$$

The moments calculated from both sides of the footings are close enough, and $M_u(\text{max}) = 1287.3$ k·ft may be adopted. This variation occured mainly because of the adjustment of the length and width of the footing.

6. Check the depth for one-way shear. Maximum shear occurs at a distance $d = 35.5$ in. from the interior face of column II (Figure 13.23).

$$V_u = 358.4 - \frac{35.5}{12} \times 47.5 = 217.9 \text{ k}$$

$$d = \frac{V_u}{\phi(2\sqrt{f_c'})b} = \frac{217.9 \times 1000}{0.85 \times 2\sqrt{4000} \times (7 \times 12)} = 24.1 \text{ in.}$$

The effective depth provided is 35.5 in. > 24.1 in.; thus the footing is adequate and shear reinforcement is not required.

7. Check depth for two-way shear (punching shear). For the interior column:

$$b_o = 4(c + d) = \frac{4}{12}(20 + 35.5) = 18.5 \text{ ft}$$

$$c + d = \frac{20 + 35.5}{12} = 4.625 \text{ ft}$$

Shear at a section $d/2$ from all sides of the column,

$$V_u = P_{u2} - q_u(c + d)^2 = 588 - 6.78(4.625)^2 = 433 \text{ k}$$

$$d = \frac{V_u}{\phi(4\sqrt{f_c'})b_o} = \frac{443,000}{0.85 \times 4\sqrt{4000} \times (18.5 \times 12)} = 9.3 \text{ in.}$$

which is not critical. Exterior column is also not critical.

8. Check the depth for moment and determine the required reinforcement in the long direction.

Maximum bending moment $= 1287.3$ k·ft

$$R_u = \frac{M_u}{bd^2} = \frac{1287.3 \times 12,000}{(7 \times 12)(35.5)^2} = 146 \text{ psi}$$

From tables in Appendix A, the steel percentage $\rho = 0.41$ percent. Use a minimum steel ratio

$$\rho = \frac{200}{f_y} = \frac{200}{40,000} = 0.005$$

$$A_s = 0.005 \times 84 \times 35.5 = 14.91 \text{ in.}^2$$

Use 15 No. 9 bars ($A_s = 15$ in.^{2}).

$$\text{Spacing of bars} = \frac{84 - 6 \text{ (concrete cover)}}{14 \text{ (no. of spacings)}} = 5.6 \text{ in.}$$

The total depth of 40 in. may be reduced using a higher steel ratio and using shear reinforcement in the shape of multiple loop stirrups within the footing. The main reinforcement bars are extended between the columns at the top of the footing, leaving a concrete cover of 3 in.

Place minimum reinforcement at the bottom of the projecting ends of the footing beyond the columns to take care of the positive moments. Extend the bars a development length $l_d = 25$ in. beyond the side of the column.

$$\text{Minimum shrinkage reinforcement} = 0.002 \times 84 \times 40.0 = 6.72 \text{ in.}^2$$

Use 9 No. 8 bars ($A_s = 7.11$ in.^{2}).

The development length required for the main top bars is $1.4l_d = 1.4 \times 25 = 35$ in. beyond the point of maximum moment. Development lengths provided to both columns are greater than 35 in.

9. For reinforcement in the short direction, calculate the bending moment in the short (transverse) direction, as in the case of single footings. The reinforcement under each column is to be placed within a maximum band width equal to the column width plus twice the effective depth d of the footing (Figure 13.24).

Reinforcement under the exterior column I:

Band width = 16 in. (column width)

+ 16 in. (on exterior side of column) + 35.5 in. (d)

= 67.5 in. = 5.625 ft

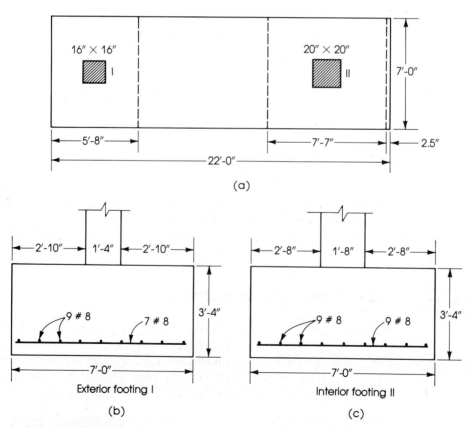

FIGURE 13.24. Design of combined footing, transverse direction: (a) plan, (b) exterior footing, and (c) interior footing.

Net upward pressure in the short direction under column I is

$$\frac{P_{u1}}{\text{width of footing}} = \frac{456}{7} = 65.1 \text{ k/ft}$$

Distance from the free end to the face of the column is

$$\frac{7}{2} - \frac{8}{12} = 2.83 \text{ ft}$$

$$M_u \text{ (at face of column I)} = \frac{65.1}{2} (2.83)^2 = 260.7 \text{ k·ft}$$

$$R_u = \frac{M_u}{bd^2} = \frac{260.7 \times 12{,}000}{(5.625 \times 12)(35.5)^2} = 36.8 \text{ psi}$$

The steel percentage $\rho <$ minimum ρ for shrinkage reinforcement ratio of 0.002.

$$A_s \text{ (min)} = 0.002 \times (5.625 \times 12)(40) = 5.4 \text{ in.}^2$$

Use 7 No. 8 bars ($A_s = 5.5$ in.^{2}) placed within the band width of 67.5 in. (say, 68 in.).
   Reinforcement under the interior column II:

Band width $= 20 + 35.5 + 35.5 = 91$ in. $= 7.6$ ft

$$\text{Net upward pressure} = \frac{P_{u2}}{\text{width of footing}} = \frac{588}{7} = 84 \text{ k/ft}$$

$$\text{Distance to face of column} = \frac{7}{2} - \frac{10}{12} = 2.67 \text{ ft}$$

$$M_u \text{ (at face of column II in short direction)} = \frac{1}{2} \times 84(2.67)^2 = 299.4 \text{ k·ft}$$

$$R_u = \frac{299.4 \times 12{,}000}{91 \times (35.5)^2} = 31.3 \text{ psi}$$

which is very small. Use minimum shrinkage reinforcement ratio of 0.002.

$$A_s = 0.002 \times 91 \times 40 = 7.28 \text{ in.}^2$$

Use 9 No. 8 bars ($A_s = 7.11$ in.^{2}) placed within the band width of 91 in. under column II, as shown in Figure 13.23, or 10 No. 8 bars.
   Development length l_d of No. 8 bars in the short direction is 20 in. The development lengths provided of 2.85 ft and 2.67 ft are greater than 20 in. (1.7 ft). ∎

13.8   Eccentrically Loaded Footings

When a column transmits axial loads only, the footing can be designed such that the load acts at the centroid of the footing, producing uniform pressure under the footing. However, in some cases, the column transmits an axial load and a bending moment, as in the case of the footings of fixed end frames. The pressure q that develops on the soil will not be uniform and can be evaluated from the following equation:

$$q = +\frac{P}{A} \pm \frac{Mc}{I} \geq 0 \tag{13.14}$$

where A and $I =$ area and moment of inertia of the footing, respectively.

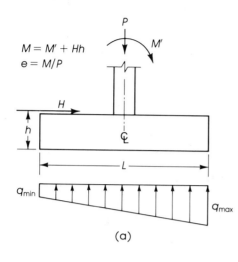

(a) Case 1: $e < L/6$, $q_{max \atop min} =$

$$\frac{P}{A} \pm \frac{M_c}{I} = \frac{P}{LB} \pm \frac{6M}{BL^2}$$

(b) Triangular soil pressure ($e = L/6$)

$$q_{min} = 0 = \frac{P}{LB} - \frac{6M}{BL^2}$$

$$q_{max} = \frac{P}{LB} + \frac{6M}{BL^2} = \frac{2P}{LB}$$

(c) Triangular soil pressure ($e > L/6$)

$$x = (L - y)/3 = L/2 - e$$
$$P = q_{max}(3x/2)\,B$$
$$q_{max} = 4P/3B(L - 2e)$$

(d) Footing moved a distance e from the axis of the column. Maximum moment occurs at section n–n. This is a case of uniform soil pressure.

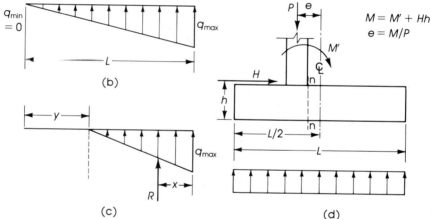

FIGURE 13.25. Single footing subjected to eccentric loading. L = length of footing, B = width, and h = height.

Different soil conditions exist depending on the magnitudes of P, M, and allowable soil pressure. The different design conditions are shown in Figure 13.25.

13.9   Footings Under Biaxial Moment

In some cases, a footing may be subjected to an axial force and biaxial moments about its x- and y-axes; such a footing may be needed for a factory crane that rotates 360°. The footing then must be designed for the critical loading.

Referring to Figure 13.26, if the axial load P acts at a distance e_x from the y-axis and e_y from the x-axis, then

$$M_x = Pe_y \quad \text{and} \quad M_y = Pe_x$$

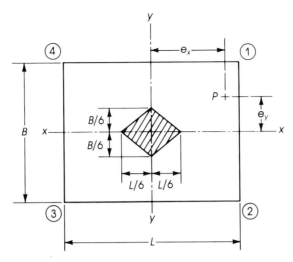

FIGURE 13.26. Footing subjected to P and biaxial moment. If $e_x <$ $L/6$ and $e_y < B/6$, footing will be subjected to upward soil pressure on all bottom surface (nonuniform pressure).

The soil pressure at corner 1

$$q_{max} = \frac{P}{A} + \frac{M_x c_y}{I_x} + \frac{M_y c_x}{I_y} \qquad (13.15)$$

At corner 2,

$$q_2 = \frac{P}{A} - \frac{M_x c_y}{I_x} + \frac{M_y c_x}{I_y} \qquad (13.16)$$

At corner 3,

$$q_3 = \frac{P}{A} - \frac{M_x c_y}{I_x} - \frac{M_y c_x}{I_y} \qquad (13.17)$$

At corner 4,

$$q_4 = \frac{P}{A} + \frac{M_x c_y}{I_x} - \frac{M_y c_x}{I_y} \qquad (13.18)$$

Again, note that the allowable soil pressure must not be exceeded and the soil cannot take any tension; that is, $q_i \geq 0$.

EXAMPLE   13.7

A 12 by 24 in. column of an unsymmetrical shed shown in Figure 13.27(a) is subjected to an axial load $P_D = 220$ k and a moment $M_D = 180$ k·ft due to dead load, and an axial load $P_L = 165$ k and a moment $M_L = 140$ k·ft due to live load. The base of the footing is 5 ft below final

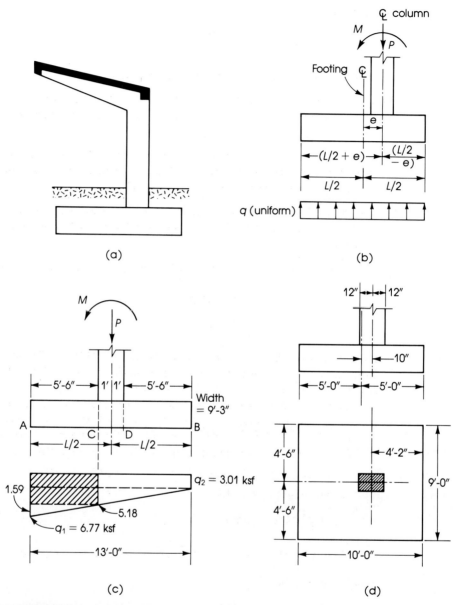

FIGURE 13.27. Example 13.7.

grade, and the allowable soil bearing pressure is 5 ksf. Design the footing using $f'_c = 4$ ksi and $f_y = 40$ ksi.

Solution   The footing is subjected to an axial load and a moment

$$P = 220 + 165 = 385 \text{ k}$$
$$M = 180 + 140 = 320 \text{ k·ft}$$

The eccentricity

$$e = \frac{M}{P} = \frac{320 \times 12}{385} = 9.97 \text{ in.}$$

say, 10 in. The footing may be designed by two methods:

Method 1: Move the center of the footing a distance $e = 10$ in. from the center of the column. In this case the soil pressure will be considered uniformly distributed under the footing (Figure 13.27(b)).

Method 2: The footing is placed concentric with the center of the column. In this case, the soil pressure will be trapezoidal or triangular (Figure 13.27(c)), and the maximum and minimum values can be calculated as shown in Figure 13.25.

The application of the two methods to example 13.7 can be explained briefly as follows:
1. For the first method, assume 20 in. footing ($d = 16.5$ in.), and assume the weight of soil is 100 pcf.

$$\text{Net upward pressure} = 5000 - \frac{20}{12} \times 150 \text{ (footing)} - \left(5 - \frac{20}{12}\right) \times 100$$

$$= 4417 \text{ psf} = 4.42 \text{ ksf}$$

$$\text{Area of footing} = \frac{385}{4.42} = 87.1 \text{ ft}^2$$

Assume a footing width of 9 ft; then the footing length = 87.1/9 = 9.7 ft, say, 10 ft. Choose a footing 9 by 10 ft and place the column eccentrically, as shown in Figure 13.27(d). The center of the footing is 10 in. away from the center of the column.
2. The design procedure now is similar to that for a single footing. Check the depth for two-way and one-way shear action. Determine the bending moment at the face of the column for the longitudinal and transverse directions. Due to the eccentricity of the footing, the critical section will be on the left face of the column in Figure 13.27(d).

$$\text{Distance to end of footing} = (5 \times 12) - 2 = 58 \text{ in.} = 4.833 \text{ ft}$$

$$P_u = 1.4D + 1.7L = 1.4 \times 220 + 1.7 \times 165 = 588.5 \text{ k}$$

$$q_u = \frac{588.5}{9 \times 10} = 6.54 \text{ ksf}$$

$$\text{Maximum } M_u = (6.54 \times 9) \times \frac{(4.833)^2}{2} = 687.4 \text{ k·ft (in 9-ft width)}$$

In the transverse direction,

$$M_u = (6.54 \times 10) \times \frac{(4)^2}{2} = 523.2 \text{ k·ft}$$

Revise the assumed depth if needed and choose the required reinforcement in both directions of the footing, as was explained in the single footing example.
3. For the second method, calculate the area of the footing in the same way as explained in the first method, then calculate the maximum soil pressure and compare it with that allowable using actual loads.

$$\text{Total load } P = 385 \text{ k}$$

$$\text{Size of footing} = 10 \times 9 \text{ ft}$$

Since the eccentricity

$$e = 10 \text{ in.} < \frac{L}{6} = \frac{10}{6} \times 12 = 20 \text{ in.}$$

then the shape of the upward soil pressure is trapezoidal. Calculate the maximum and minimum soil pressure:

$$q_{max} = \frac{P}{LB} + \frac{6M}{BL^2} = \frac{385}{10 \times 9} + \frac{6 \times 320}{9(10)^2}$$

$$= 4.28 + 2.13 = 6.41 \text{ ksf} > 4.42 \text{ ksi}$$

The footing is not safe. Try a footing 9.25×13 ft (Area $= 120.25$ ft^2).

$$q_{max} = \frac{385}{120.25} + \frac{6 \times 320}{9.25(13)^2} = 3.20 + 1.22 = 4.22 \text{ ksf} < 4.42 \text{ ksi}$$

$$q_{min} = 3.2 - 1.22 = 1.98 \text{ ksf}$$

4. Calculate the ultimate upward pressure using factored loads:

$$P_u = 1.4 \times 220 + 1.7 \times 165 = 588.5 \text{ k}$$

$$M_u = 1.4 \times 180 + 1.7 \times 140 = 490 \text{ k·ft}$$

Area of footing $= 9.25 \times 13 = 120.25$ ft^2.

$$q_u(max) = \frac{P_u}{LB} + \frac{6M_u}{BL^2} = \frac{588.5}{120.25} + \frac{6 \times 490}{(9.25)(13)^2}$$

$$= 4.89 + 1.88 = 6.77 \text{ ksf}$$

$$q_u(min) = 4.89 - 1.88 = 3.01 \text{ ksf} \qquad \text{(Figure 13.27(c))}$$

5. The maximum bending moment in the footing occurs at the left face of the column (section c).

$$q_u \text{ (section c)} = 3.01 + \frac{7.5}{13}(6.77 - 3.01) = 5.18 \text{ ksf}$$

To take moments about the face of the column, divide the upward pressure acting on AC into a rectangle of a pressure $= 5.18$ ksf and a triangle of pressure $= 6.77 - 5.18 = 1.59$ ksf.

$$M_u \text{ (at face of column)} = 5.18 \frac{(5.5)^2}{2} \times 9.25 + \frac{1.59 \times 5.5}{2}\left(\frac{2}{3} \times 5.5\right) \times 9.25$$

$$= 873 \text{ k·ft}$$

For a depth $d = 16.5$ in.,

$$R_u = \frac{M_u}{bd^2} = \frac{873 \times 12,000}{(9.25 \times 12)(16.5)^2} = 347 \text{ psi}$$

The steel percentage $\rho = 0.0107$:

$$A_s = 0.0107 \times (9.25 \times 12) \times 16.5 = 19.6 \text{ in.}^2$$

Choose 20 No. 9 bars. Similarly, calculate the bending moment at the other face of the column (section D), and the number of bars can be reduced when extended to the other end of the footing.

6. Check one-way and two-way shear and other requirements, as explained earlier.   ■
7. The second method of design is not economical.

13.10   Slabs on Ground

A concrete slab laid directly on ground may be subjected to

- uniform load over its surface, producing small internal forces, or
- nonuniform or concentrated loads, producing some moments and shearing forces. Tensile stresses develop, and cracks will occur in some parts of the slab.

Tensile stresses are generally induced by a combination of

- contraction due to temperature and shrinkage, restricted by the friction between the slab and the subgrade, causing tensile stresses
- warping of the slab
- loading conditions
- settlement

Contraction joints may be formed to reduce the tensile stresses in the slab. Expansion joints may be provided in thin slabs up to a thickness of 10 in.

Basement floors in residential structures may be made of 4 to 6 in. concrete slabs reinforced in both directions with a wire fabric reinforcement. In warehouses, slabs may be 6 to 12 in. thick, depending on the loading on the slab. Reinforcement in both directions must be provided, usually in the form of wire fabric reinforcement. Basement floors are designed to resist upward earth pressure and any water pressure. If the slab rests on very stable or incompressible soils, then differential settlement is negligible. In this case the slab thickness will be a minimum if no water table exists. Columns in the basement will have independent footings. If there is any appreciable differential settlement, the floor slab must be designed as a stiff raft foundation.

13.11   Footings on Piles

When the ground consists of soft material for a great depth, and its bearing capacity is very low, it is not advisable to rest the footings directly on the soil. It may be better to transmit the loads through piles to a deep stratum that is strong enough to bear the loads or to develop sufficient friction around the surface of the piles.

There is a large variety of piles used for foundations. The choice depends on ground conditions, presence of ground water, function of the pile, and cost. Piles may be made of concrete, steel, or timber.

In general, a pile cap (or footing) is necessary to distribute the load from a column to the heads of a number of piles. The cap should be of sufficient size to accommodate deviation in the position of the pile heads. The caps are designed as beams spanning between the pile heads and carrying concentrated loads from columns. When the column is supported by two piles, the cap may be designed as a reinforced concrete truss of a triangular shape.

The **ACI Code, section 15.2,** indicates that computations for moments and shears for footings on piles may be based on the assumption that the reaction from any pile is concentrated at the pile center. The base area of the footing or number of piles shall be determined from the unfactored forces and moments.

The minimum concrete thickness above the reinforcement in a pile footing is limited to 12 in. **(ACI Code, section 15.7).** For more design details of piles and pile footings, refer to books on foundation engineering or to the Concrete Reinforcing Steel Institute Handbook.

13.12   S.I. Equations

The following equations are the S.I. equivalents to the equations given in this chapter. The equations that do not appear here may be used for both U.S. customary and S.I. units. Note that $\sqrt{f_c'}$ (psi) is equivalent to $0.083\sqrt{f_c'}$ (MPa).

$$V_c = 0.083\left(2 + \frac{4}{\beta_c}\right)\sqrt{f_c'}b_o d \le 0.33\sqrt{f_c'}b_o d \tag{13.3}$$

$$V_c = 0.33\sqrt{f_c'}b_o d \quad \text{when} \quad \beta_c \le 2$$

$$d = \frac{V_u}{0.33\phi\sqrt{f_c'}b_o} \tag{13.5}$$

$$\phi V_c = 0.17\phi\sqrt{f_c'}bd \tag{13.6}$$

$$d = \frac{V_u}{0.17\phi\sqrt{f_c'}b} \tag{13.8}$$

$$\text{Minimum flexural steel percentage} = \frac{1.38}{f_y}$$

SUMMARY

SECTIONS 13.1 and 13.2   Footings are structural members used mainly to support columns and walls. The most common types of footings are wall, single, combined, strap, and continuous footings. Pile caps and raft foundations are also used to support columns.

SECTION 13.3   The actual distribution of soil pressure under footings is not uniform, but in practical design problems it is assumed to be uniform.

SECTION 13.4

(a) Area of footing = total actual load divided by the allowable soil pressure. Also,

$$q_u = P_u/(\text{area of footing}) \tag{13.2}$$

(b) For two-way shear (punching shear):

$$V_c = \left(2 + \frac{4}{\beta_c}\right)\sqrt{f'_c}b_o d \leq 4\sqrt{f'_c}b_o d \tag{13.3}$$

$$V_u = P_u - q_u(c + d)^2 \tag{13.4}$$

$$d = V_u/(4\phi\sqrt{f'_c})b_o \tag{13.5}$$

(c) For one-way shear:

$$\phi V_c = 2\phi\sqrt{f'_c}bd \tag{13.6}$$

$$d = V_u/(2\phi\sqrt{f'_c})b \tag{13.8}$$

(d) Flexural strength can be determined as follows:

$$M_u = \phi A_s f_y \left(d - \frac{a}{2}\right)$$

The value of a varies between 1 and 2 in., and thus A_s can be determined assuming a trial $a = 1$ in. or $d - (a/2) = 0.9d$. In rectangular footings, a certain ratio of reinforcement in the short direction should concentrate within a band equal to the short side. This ratio is equal to $2/(\beta + 1)$, where $\beta =$ long side of footing/short side.

(e) Bearing strength $N_1 = \phi(0.85f'_c A_1)$ (13.11)

$$\phi = 0.70 \quad \text{and} \quad A_1 = \text{bearing area of column}$$

$$N_2 = N_1\sqrt{\frac{A_2}{A_1}} \leq 2N_1$$

(f) The ACI Code specifies a minimum steel area for column dowels of $0.005A_g$. The development length for compression bars is $l_d = 0.02f_y d_b/\sqrt{f'_c} \geq 0.0003f_y d_b \geq 8$ in.

(g) Footings can be proportioned for balanced design under sustained loads, assuming the same soil pressure under all footings. Example 13.5 explains this method.

SECTIONS 13.5 and 13.6

(a) Plain concrete footings may be used to support compression members.

(b) A site investigation is required before starting the design of footings.

SECTION 13.7

An example of the design of a combined footing is given in this section.

SECTIONS 13.8 and 13.9

(a) When a column transmits an axial load and a bending moment, then

$$q = \frac{P}{A} \pm \frac{Mc}{I} \geq 0$$

(b) For footings under biaxial moment:

$$q(\max) = \frac{P}{A} + \frac{M_x c_y}{I_x} + \frac{M_y c_x}{I_y}$$

SECTION 13.10   A concrete slab laid directly on ground may be subjected to uniform loads, concentrated loads, temperature effect, warping, and settlement.

SECTION 13.11   A pile cap is necessary to distribute the load from the column to the heads of a number of piles. The cap is designed as a beam spanning the pile heads.

REFERENCES

1. R. W. Furlong: "Design Aids for Square Footings," *ACI Journal,* 62, March 1965.
2. R. Peck, W. Hanson and T. Thornburn: *Foundation Engineering,* 2nd ed., Wiley, New York, 1974.
3. ACI Committee 436: "Suggested Design Procedures for Combined Footings and Mats," *ACI Journal,* 63, October 1966.
4. F. Kramrisch and P. Rogers: "Simplified Design of Combined Footings," *Journal of Soil Mechanics and Foundation Division, ASCE,* 85, October 1961.
5. L. S. Blake (ed.): *Civil Engineer's Reference Book,* 3rd ed., Butterworths, London, 1975.

PROBLEMS

13.1. Design a reinforced concrete footing to support a 14-in. (350-mm) concrete wall that carries a uniform dead load = 15 k/ft (225 kN/m) and a live load = 8 k/ft (120 kN/m). The bottom of the footing is 4 ft (102 m) from the final grade. Use f'_c = 3 ksi (21 MPa), f_y = 40 ksi (280 MPa), and an allowable soil pressure = 4 ksf (190 kN/m²). Assume soil weight = 100 pcf.

13.2. Repeat problem 13.1 if the dead load = 28 k/ft (420 kN/m) and the live load = 20 k/ft (300 kN/m), using f'_c = 4 ksi (28 MPa) and f_y = 60 ksi (420 MPa).

13.3. Design a plain concrete footing to support a 16-in. (400-mm) thick masonry wall carrying a uniform dead load = 6 k/ft (90 kN/m) and a live load = 4 k/ft (60 kN/m). Use f'_c = 3 ksi (21 MPa) and an allowable soil pressure of 3 ksf (140 kN/m²).

13.4. Design a square single footing to support a 16-in. (400-mm) square tied column reinforced with No. 9 (28 mm) bars to carry an axial dead load = 160 k (720 kN) and a live load = 140 k (630 kN). Use f'_c = 4 ksi (28 MPa), f_y = 40 ksi (280 MPa), and an allowable soil pressure = 4 ksf (190 kN/m²). The bottom of the footing is 6 ft (1.8 m) below final grade.

13.5. Repeat problem 13.4 if the dead load = 100 k (450 kN) and the live load = 80 k (360 kN), f'_c = 3 ksi (21 MPa), and f_y = 40 ksi (280 MPa).

13.6. Repeat problem 13.4 if the dead load = 220 k (980 kN) and the live load = 180 k (810 kN), and the column section is 16 by 16 in. (400 by 400 mm).

13.7. Repeat problem 13.6 if the column has a circular section of 16 in. (400 mm) diameter.

13.8. Redesign the footing in problem 13.4 as a rectangular footing with the length equal to 1.5 times the width.

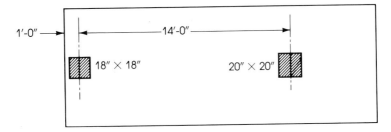

1'-0" →   ┤←————14'-0"————→│

18" × 18"   20" × 20"

FIGURE 13.28. Problem 13.10.

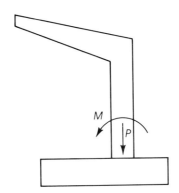

M
P

FIGURE 13.29. Problem 13.12.

13.9. Redesign the footing of problem 13.6 if the column has a section 14 by 20 in. (350 by 500 mm) and the footing is rectangular with the length equal to 1.33 times the width.

13.10. Design a rectangular combined footing to support the two columns shown in Figure 13.28. The center of the exterior column is 1 ft (0.3 m) away from the property line and 14 ft (4.2 m) from the center of the interior column. The exterior column is square with 18 in. (450 mm) side, reinforced with No. 8 (25 mm) bars, and carries an axial dead load = 160 k (720 kN) and a live load = 140 k (630 kN). The interior column is square with 20 in. (500 mm) side, reinforced with No. 9 (28 mm) bars, and carries an axial dead load = 240 k (1080 kN) and a live load = 150 k (675 kN). The bottom of the footing is 5 ft (1.5 m) below final grade. Use f'_c = 4 ksi (28 MPa), f_y = 50 ksi (350 MPa), and an allowable soil pressure = 5 ksf (240 kN/m²).

13.11. Determine the footing areas required for a balanced footing design (equal settlement approach) if the usual load = 25 percent for all footings. The allowable soil pressure = 5 ksf (240 kN/m²) and the dead and live loads are as follows:

	Footing no.					
	1	2	3	4	5	6
Dead load	130 k	220 k	150 k	180 k	200 k	240 k
Live load	160 k	220 k	210 k	180 k	220 k	200 k

13.12. The 12 by 20 in. (300 by 500 mm) column of the frame shown in Figure 13.29 is subjected to an axial load $P_D = 200$ k (900 kN) and a moment $M_D = 120$ k·ft (165 kN·m) due to dead load and an axial load $P_L = 160$ k (720 kN) and a moment $M_L = 110$ k·ft (150 kN·m) due to live load. The base of the footing is 4 ft (1.2 m) below final grade. Design the footing using $f'_c = 4$ ksi (28 MPa), $f_y = 40$ ksi (280 MPa), and an allowable soil pressure $= 4$ ksf (190 kN/m²). Use a uniform pressure and eccentric footing approach.

13.13. Repeat problem 13.12 if both the column and the footing have the same centerline (concentric case).

13.14. Determine the size of a square or round footing for the case of problem 13.13, assuming that the loads and moments on the footing are for a rotating crane fixed at its base.

14 Retaining Walls

14.1 Types of Retaining Walls

Retaining walls are structural members used to provide stability for soil or other materials and to prevent them from assuming their natural slope. In this sense, the retaining wall maintains unequal levels of earth on its two faces. The retained material on the higher level exerts a force on the retaining wall that may cause its overturning or failure. Retaining walls are used in bridges as abutments, in buildings as basement walls, and in embankments. They are also used to retain liquids, as in water tanks and sewage treatment tanks. Various types of commonly used retaining walls are shown in Figure 14.1:

1. *Gravity walls* usually consist of plain concrete or masonry and depend entirely on their own weight to provide stability against the thrust of the retained material. These walls are proportioned so that tensile stresses do not develop in the concrete or masonry due to the exerted forces on the wall. The practical height of a gravity wall does not exceed 10 ft.

2. *Semi-gravity walls* are gravity walls that have a wider base to improve the stability of the wall and to prevent the development of tensile stresses in the base. Light reinforcement is sometimes used in the base or stem to reduce the large section of the wall.

3. *The cantilever retaining wall* is a reinforced concrete wall that is generally used for heights from 10 to 25 ft. It is the most common type of retaining structure because of economy and simplicity of construction. It consists of three cantilever beams: (1) the vertical wall or the stem, (2) the toe, which is the part of the base pressed down in the soil, and (3) the heel, which is the part of the base that tends to be lifted away from the soil.

4. *Counterfort retaining walls:* A cantilever retaining wall higher than 20 ft develops a relatively large bending moment at the base of the stem, which makes the design of such walls uneconomical. One solution in this case is to introduce

419

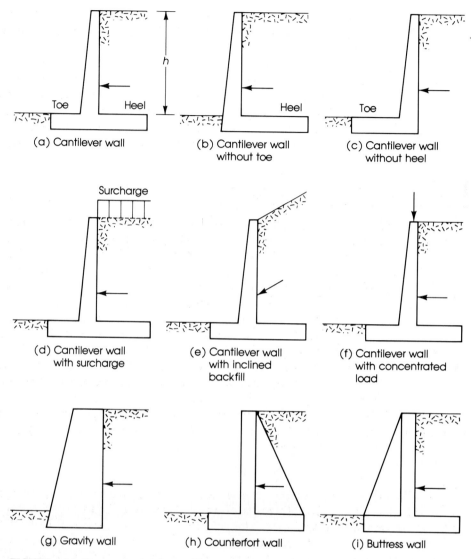

FIGURE 14.1. Types of retaining walls.

transverse walls (or counterforts) that tie the stem and the base together at intervals. The counterforts act as tension ties supporting the vertical walls. Economy is achieved because the stem is designed as a continuous slab spanning horizontally between counterforts, while the heel is designed as a slab supported on three sides.

5. *The buttressed retaining wall* is similar to the counterfort wall, but in this case the transverse walls are located on the opposite, visible side of the stem and act in compression. The design of such walls becomes economical for heights greater than 20 ft. They are not popular because of the exposed buttresses.

6. *Bridge abutments* are retaining walls that are supported at the top by the bridge deck. The wall may be assumed fixed at the base and simply supported at the top.

7. *Anchored retaining walls* are mostly used to retain earth during the excavation of building sites several floors below ground level. The wall is anchored back into a large mass of earth or concrete by rods or prestressed strands.

8. *Basement walls* resist earth pressure from one side of the wall and span vertically from the basement floor slab to the first floor slab. The wall may be assumed fixed at the base and simply supported or partially restrained at the top.

14.2   Forces on Retaining Walls

Retaining walls are generally subjected to gravity loads and to earth pressure due to the retained material on the wall. Gravity loads due to the weights of the materials are well defined and can be calculated easily and directly. The magnitude and direction of the earth pressure on a retaining wall depends on the type of soil retained and on other factors, and cannot be determined as accurately as gravity loads. Several references on soil mechanics[1,2] explain the theories and procedure to determine the soil pressure on retaining walls. The stability of retaining walls and the effect of dynamic reaction on walls are discussed in references 3 and 4. Equations developed by Rankine and Coulomb[1,6] will be used here to calculate the earth pressure on walls.

Granular materials such as sand behave differently from cohesive materials such as clay or from any combination of both types of soils. Although the pressure intensity of soil on a retaining wall is complex, it is common to assume a linear pressure distribution on the wall. The pressure intensity increases with depth linearly, and its value is a function of the height of the wall and the weight and type of soil. The pressure intensity p at a depth h below the earth's surface may be calculated as follows:

$$p = Cwh \qquad (14.1)$$

where w = unit weight of soil, and C = the coefficient that depends on the physical properties of soil. The value of the coefficient C varies from 0.3 for loose granular soil such as sand to about 1.0 for cohesive soil such as wet clay.

If the retaining wall is assumed absolutely rigid, a case of earth pressure at rest develops. Under soil pressure, the wall may deflect or move a small amount from the earth, and active soil pressure develops, as shown in Figure 14.2. If the wall moves toward the soil, a passive soil pressure develops. Both the active and passive soil pressures are assumed to vary linearly with the depth of wall (Figure 14.2). For dry, granular, noncohesive materials, the assumed linear pressure diagram is fairly satisfactory; cohesive soils or saturated sands behave in a different, nonlinear manner. Therefore, it is very common to use granular materials as backfill to provide an approximately linear pressure diagram and also to provide for the release or drainage of water from behind the wall.

For a linear pressure, the active and passive pressure intensities are determined as follows:

$$P_a = C_a wh \qquad (14.2)$$

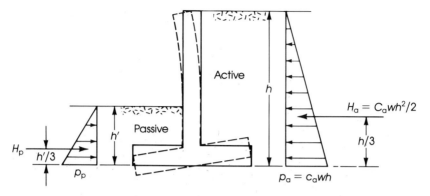

FIGURE 14.2. Active and passive earth pressure.

$$P_p = C_p wh \tag{14.3}$$

where C_a and C_p are the approximate coefficients of the active and passive pressures, respectively. The values $C_a w$ and $C_p w$ are referred to as the equivalent fluid pressures.

14.3   Theories of Earth Pressure

The two theories most commonly used in the calculation of earth pressure are those of Rankine and Coulomb.[1]

14.3.1   Rankine's Theory of Earth Pressure

In 1857, William J. M. Rankine presented a theory to calculate the earth pressure on retaining structures. He assumed that the retaining wall yields a sufficient amount to develop a state of plastic equilibrium in the soil mass at the wall surface. The rest of the soil remains in the state of elastic equilibrium. The theory applies mainly to a homogeneous, incompressible, cohesionless soil and neglects the friction between soil and wall. The active and passive soil pressures are calculated as follows:

Active soil pressure.   Active soil pressure develops when the retaining wall deflects away from the backfill under the lateral soil pressure. The active soil pressure at a depth h on a retaining wall with a horizontal backfill is determined as follows:

$$P_a = C_a wh = wh \left(\frac{1 - \sin \phi}{1 + \sin \phi} \right) \tag{14.4}$$

where

$$C_a = \frac{1 - \sin \phi}{1 + \sin \phi}$$

ϕ = angle of internal friction of the soil

Total active pressure $H_a = \dfrac{wh^2}{2} \left(\dfrac{1 - \sin \phi}{1 + \sin \phi} \right)$ (14.5)

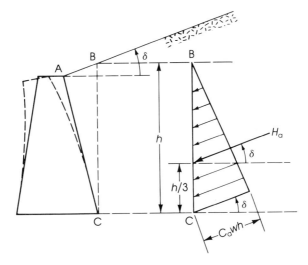

FIGURE 14.3. Active soil pressure with surcharge.

The resultant H_a acts at $h/3$ from the base (Figure 14.2). When the earth is surcharged at an angle δ to the horizontal, then

$$C_a = \cos \delta \left(\frac{\cos \delta - \sqrt{\cos^2 \delta - \cos^2 \phi}}{\cos \delta + \sqrt{\cos^2 \delta - \cos^2 \phi}} \right) \tag{14.6}$$

$$P_a = C_a wh \quad \text{and} \quad H_a = C_a \frac{wh^2}{2}$$

The resultant H_a acts at $h/3$ and is inclined at an angle δ to the horizontal (Figure 14.3). The values of C_a expressed by equation 14.6 for different values of δ and ϕ are shown in Table 14.1.

Passive soil pressure. Passive soil pressure develops when the retaining wall moves against and compresses the soil. The passive soil pressure at a depth h on a retaining wall with horizontal backfill is determined as follows:

$$P_p = C_p wh = wh \left(\frac{1 + \sin \phi}{1 - \sin \phi} \right) \tag{14.7}$$

TABLE 14.1. Values of C_a

δ	\multicolumn{7}{c}{ϕ (Angle of internal friction)}
	28°	30°	32°	34°	36°	38°	40°
0°	0.361	0.333	0.307	0.283	0.260	0.238	0.217
10°	0.380	0.350	0.321	0.294	0.270	0.246	0.225
20°	0.461	0.414	0.374	0.338	0.306	0.277	0.250
25°	0.573	0.494	0.434	0.385	0.343	0.307	0.275
30°	0	0.866	0.574	0.478	0.411	0.358	0.315

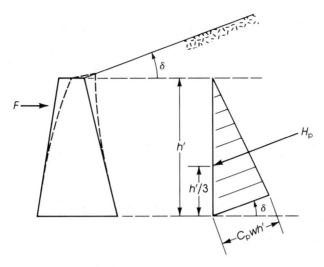

FIGURE 14.4. Passive soil pressure with surcharge.

where

$$C_p = \frac{1 + \sin \phi}{1 - \sin \phi}$$

The total passive pressure $H_p = \frac{wh^2}{2}\left(\frac{1 + \sin \phi}{1 - \sin \phi}\right)$ (14.8)

The resultant H_p acts at $h'/3$ from the base (Figure 14.2). When the earth is surcharged at an angle δ to the horizontal, then

$$C_p = \cos \delta \left(\frac{\cos \delta + \sqrt{\cos^2\delta - \cos^2\phi}}{\cos \delta - \sqrt{\cos^2\delta - \cos^2\phi}}\right)$$ (14.9)

$$P_p = C_p wh$$

and

$$H_p = C_p \frac{wh^2}{2}$$

H_p acts at $h'/3$ and is inclined at an angle δ to the horizontal (Figure 14.4). The values of C_p expressed by equation 14.9 for different values of δ and ϕ are shown in Table 14.2.

14.3.2   Coulomb's Theory of Earth Pressure

In Coulomb's theory, the active soil pressure is assumed to be the result of the tendency of a wedge of soil to slide against the surface of a retaining wall. Hence Coulomb's theory is referred to as the wedge theory. While it takes into consideration the friction of the soil on the retaining wall, it assumes that the surface of sliding is a plane, whereas in reality it is slightly curved. The error in this assumption is negligible in calculating the

TABLE 14.2. Values of C_p

	ϕ (Angle of internal friction)						
δ	28°	30°	32°	34°	36°	38°	40°
0°	2.77	3.00	3.25	3.54	3.85	4.20	4.60
10°	2.55	2.78	3.02	3.30	3.60	3.94	4.32
20°	1.92	2.13	2.36	2.61	2.89	3.19	3.53
25°	1.43	1.66	1.90	2.14	2.40	2.68	3.00
30°	0	0.87	1.31	1.57	1.83	2.10	2.38

active soil pressure. Coulomb's equations to calculate the active and passive soil pressures are as follows[6]:

The active soil pressure is

$$P_a = C_a wh$$

where

$$C_a = \frac{\cos^2(\phi - \theta)}{\cos^2\theta \cos(\theta + \beta)\left[1 + \sqrt{\dfrac{\sin(\phi + \beta)\sin(\phi - \delta)}{\cos(\theta + \beta)\cos(\theta - \delta)}}\right]^2} \qquad (14.10)$$

where

ϕ = angle of internal friction of soil

θ = angle of the soil pressure surface from the vertical

β = angle of friction along the wall surface (angle between soil and concrete)

δ = angle of surcharge to the horizontal

The total active soil pressure

$$H_a = C_a\frac{wh^2}{2} = P_a\frac{h}{2}$$

When the wall surface is vertical, $\theta = 0°$, and if $\beta = \delta$, then C_a in equation 14.10 reduces to equation 14.6 of Rankine.

Passive soil pressure is

$$P_p = C_p wh' \quad \text{and} \quad H_p = \left(\frac{wh'^2}{2}\right) C_p = P_p\frac{h'}{2}$$

where

$$C_p = \frac{\cos^2(\phi + \theta)}{\cos^2\theta \cos(\theta - \beta)\left[1 - \sqrt{\dfrac{\sin(\phi + \beta)\sin(\phi + \delta)}{\cos(\theta - \beta)\cos(\phi - \delta)}}\right]^2} \qquad (14.11)$$

The values of ϕ and w vary with the type of backfill used. As a guide, common values of ϕ and w are given in Table 14.3.

TABLE 14.3. Values of w and ϕ

Type of backfill	Unit weight w		Angle of internal friction ϕ
	pcf	kg/m³	
Soft clay	90–120	1440–1920	0°–15°
Medium clay	100–120	1600–1920	15°–30°
Dry loose silt	100–120	1600–1920	27°–30°
Dry dense silt	110–120	1760–1920	30°–35°
Loose sand and gravel	100–130	1600–2100	30°–40°
Dense sand and gravel	120–130	1920–2100	25°–35°
Dry loose sand, well graded	115–130	1840–2100	33°–35°
Dry dense sand, well graded	120–130	1920–2100	42°–46°

14.4   Earth Pressure Due to Submerged Soil

When the soil is saturated, the pores of the permeable soil are filled with water, which exerts hydrostatic pressure. In this case the buoyed unit weight of soil must be used. The buoyed unit weight (or submerged unit weight) is a reduced unit weight of soil and equals w minus the weight of water displaced by the soil. The effect of the hydrostatic water pressure must be included in the design of retaining walls subjected to a high water table and submerged soil. The value of the angle of internal friction may be used as shown in Table 14.3.

14.5   Effect of Surcharge

Different types of loads are often imposed on the surface of the backfill behind a retaining wall. If the load is uniform, an equivalent height of soil h_s may be assumed acting on the wall to account for the increased pressure. For the wall shown in Figure 14.5, the horizontal pressure due to the surcharge is constant throughout the depth of the retaining wall.

$$h_s = w_s/w \qquad (14.12)$$

where

h_s = equivalent height of soil (ft)

w_s = pressure of the surcharge (psf)

w = unit weight of soil (pcf)

The total pressure is

$$H_a = H_{a1} + H_{a2} = C_a w \left(\frac{h^2}{2} + h_s \right) \qquad (14.13)$$

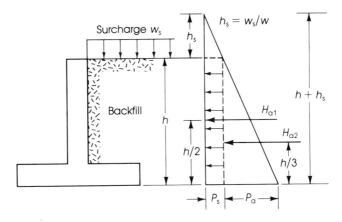

FIGURE 14.5. Surcharge effect under a uniform load.

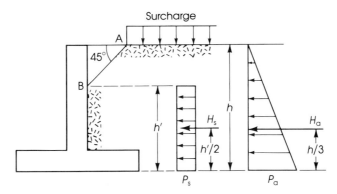

FIGURE 14.6. Surcharge effect under a partial uniform load at a distance from the wall.

In the case of a partial uniform load acting at a distance from the wall, only a portion of the total surcharge pressure affects the wall (Figure 14.6).

It is a common practice to assume that the effective height of pressure due to partial surcharge is h' measured from point B to the base of the retaining wall.[1] The line AB forms an angle of 45° with the horizontal.

In the case of a wheel load acting at a distance from the wall, the load is to be distributed over a specific area, which is usually defined by known specifications such as AASHTO and AREA[4] specifications.

14.6   Friction on the Retaining Wall Base

The horizontal component of all forces acting on a retaining wall tends to push the wall in a horizontal direction. The retaining wall base must be wide enough to resist the sliding of the wall. The coefficient of friction to be used is that of soil on concrete for coarse granular soils and the shear strength of cohesive soils.[4] The coefficients of friction μ that may be adopted for different types of soil are as follows:

- coarse-grained soils without silt $\mu = 0.55$
- coarse-grained soils with silt $\mu = 0.45$
- silt $\mu = 0.35$
- sound rock $\mu = 0.60$

The total frictional force F on the base to resist the sliding effect is

$$F = \mu R + H_p \qquad (14.14)$$

where

μ = the coefficient of friction

R = the vertical force acting on the base

H_p = passive resisting force

The factor of safety against sliding is

$$\text{Factor of safety} = \frac{F}{H_{ah}} = \frac{\mu R + H_p}{H_{ah}} \geq 1.5 \qquad (14.15)$$

where H_{ah} = the horizontal component of the active pressure H_a. The factor of safety against sliding should not be less than 1.5 if the passive resistance H_p is neglected, and not less than 2.0 if H_p is taken into consideration.

14.7   Stability Against Overturning

The horizontal component of the active pressure H_a tends to overturn the retaining wall about the point O on the toe (Figure 14.7). The overturning moment is equal to $M_o = H_a h/3$. The weight of the concrete and soil tends to develop a balancing moment or rightening moment to resist the overturning moment. The balancing moment for the case of the wall shown in Figure 14.7 is equal to

$$M_b = w_1 x_1 + w_2 x_2 + w_3 x_3 = \Sigma wx$$

The factor of safety against overturning is

$$\text{Factor of safety} = \frac{M_b}{M_o} = \frac{\Sigma wx}{Hh/3} \geq 2.0 \qquad (14.16)$$

This factor of safety should not be less than 2.0.

The resultant of all forces acting on the retaining wall R_A intersects the base at point C (Figure 14.7). In general, the point C does not coincide with the center of the base L, thus causing eccentric loading on the footing. It is desirable to keep the point C within the middle third to get the whole footing under soil pressure. (The case of a footing under eccentric load was discussed in Chapter 13, section 13.8).

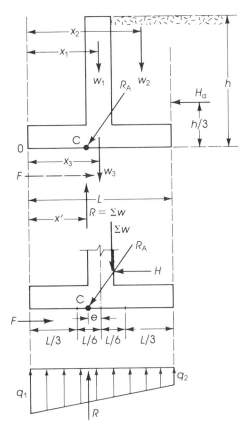

FIGURE 14.7. Overturning of a cantilever retaining wall.

14.8   Proportions of Retaining Walls

The design of a retaining wall begins with a trial section and approximate dimensions. The assumed section is then checked for stability and structural adequacy. The following rules may be used to determine the approximate sizes of the different parts of a cantilever retaining wall.

1. *Height of the wall:* The overall height of the wall is equal to the difference in elevation required plus 3 to 4 ft, which is the estimated frost penetration depth in northern states.

2. *Thickness of the stem:* The intensity of the pressure increases with the depth of the stem and reaches its maximum value at the base level. Consequently the maximum bending moment and shear in the stem occur at its base. The stem base thickness may be estimated as 1/12 to 1/10 of the height h. The thickness at the top of the stem may be assumed to be 8 to 12 in. Since retaining walls are designed for active earth pressure, causing a small deflection of the wall, it is advisable to provide the face of the wall with a batter (taper) of $\frac{1}{4}$ in. per foot

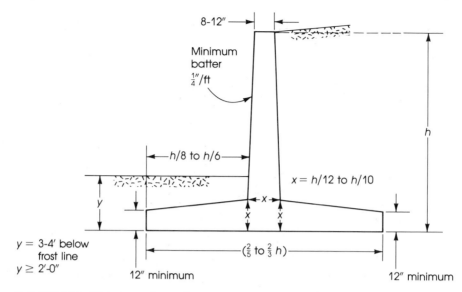

FIGURE 14.8. Trial proportions of a cantilever retaining wall.

of height h to compensate for the forward deflection. For short walls up to 10 ft high, a constant thickness may be adopted.

3. *Length of the base:* An initial estimate for the length of the base of 2/5 to 2/3 of the wall height h may be adopted.

4. *Thickness of the base:* The base thickness below the stem is estimated as the same thickness of the stem at its base, that is, 1/12 to 1/10 of the wall height. A minimum thickness of about 12 in. is recommended. The wall base may be of uniform thickness or tapered to the ends of the toe and heel, where the bending moment is zero.

The approximate initial proportions of a cantilever retaining wall are shown in Figure 14.8.

14.9   Drainage

The earth pressure discussed in the previous sections does not include any hydrostatic pressure. If water accumulates behind the retaining wall, the water pressure must be included in the design. Surface or underground water may seep into the backfill and develop the case of submerged soil. To avoid hydrostatic pressure, drainage should be provided behind the wall. If well-drained cohesionless soil is used as a backfill, the wall can be designed for earth pressure only. The drainage system may consist of one or a combination of the following:

• Weep holes in the retaining wall of 4 in. or more in diameter and spaced about 5 ft on centers horizontally and vertically (Figure 14.9(a)).

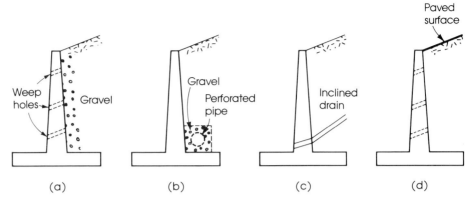

FIGURE 14.9. Drainage systems.

- Perforated pipe 8 in. in diameter laid along the base of the wall and surrounded by gravel (Figure 14.9(b)).
- Blanketing or paving the surface of the backfill with asphalt to prevent seepage of water from the surface.
- Any other method to drain surface water.

EXAMPLE **14.1**

The trial section of a semi-gravity plain concrete retaining wall is shown in Figure 14.10. It is required to check the safety of the wall against overturning, sliding, and bearing pressure under the footing. Given: weight of back fill = 120 pcf, angle of internal friction $\phi = 35°$, coefficient of friction between concrete and soil $\mu = 0.5$, allowable soil pressure = 3 ksf, and $f'_c = 3$ ksi.

Solution

1. Using the Rankine equation:

$$C_a = \frac{1 - \sin \phi}{1 + \sin \phi} = \frac{1 - 0.574}{1 + 0.574} = 0.271$$

The passive pressure on the toe is that for a height of 1 ft, which is small and can be neglected.

$$H_a = \frac{C_a w h^2}{2} = \frac{0.271}{2} (120)(11)^2 = 1968 \text{ lb}$$

H_a acts at a distance $h/3 = 11/3 = 3.67$ ft from the base.

2. Overturning moment $M_o = 1.968 \times 3.67 = 7.22$ k·ft.

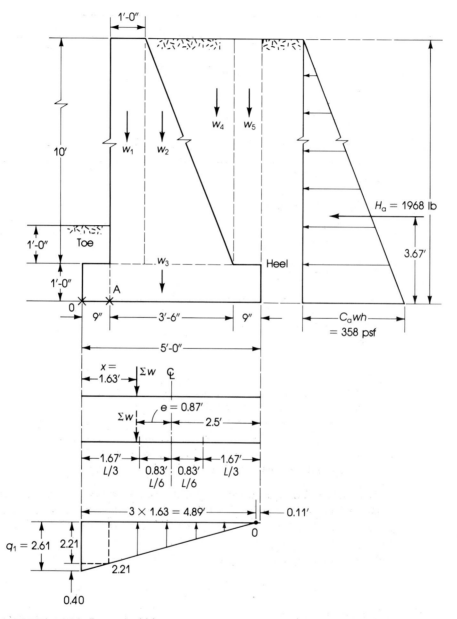

FIGURE 14.10. Example 14.1.

3. Calculate the balancing moment taken about the toe end O (Figure 14.10):

Weight (lb)	Arm (ft)	Moment (k·ft)
$w_1 = 1 \times 10 \times 145 = 1450$	1.25	1.81
$w_2 = \frac{1}{2} \times 2.5 \times 10 \times 145 = 1812$	2.60	4.71
$w_3 = 5 \times 1 \times 145 = 725$	2.50	1.81
$w_4 = \frac{1}{2} \times 2.5 \times 10 \times 120 = 1500$	3.42	5.13
$w_5 = \frac{9}{12} \times 10 \times 120 = 900$	4.625	4.16
$\Sigma w = R = 6.39$ k		$M_b = \Sigma M = 17.62$ k·ft

4. Factor of safety against overturning is

$$\frac{17.62}{7.22} = 2.44 > 2.0$$

5. Force resisting sliding, $F = \mu R$:

$$F = 0.5 \times 6.39 = 3.2 \text{ k}$$

Factor of safety against sliding

$$\frac{F}{H_a} = \frac{3.2}{1.97} = 1.62 > 1.5$$

6. Calculate soil pressure under the base.
Distance of the resultant from toe end O:

$$X = \frac{M_b - M_o}{R} = \frac{\Sigma M - H_a h/3}{R} = \frac{17.62 - 7.22}{6.39} = 1.63 \text{ ft}$$

$$e = 2.50 - 1.63 = 0.87 \text{ ft}$$

The resultant R acts just outside the middle third of the base and has an eccentricity $e = 0.87$ ft from the center of the base compared to 0.83 ft on Figure 14.10. For 1-ft lengths of the footing, the effective length of footing $= 3 \times 1.63 = 4.89$ ft.

$$I = \frac{1 \times (4.89)^3}{12} = 9.75 \text{ ft}^4$$

Area $= 1 \times 4.89 = 4.89 \text{ ft}^2$

For a triangular pressure:

$$\frac{1}{2} q_1 \times 4.89 = R = 6.39 \text{ k}$$

Therefore, $q_1 = 2.61$ ksf < 3 ksf, and $q_2 = 0$. Soil pressure is adequate.

7. Check bending stress at point A of the toe. Soil pressure at A is

$$q_A = \frac{4.14}{4.89} \times 2.61 = 2.21 \text{ ksf}$$

$$M_A = 2.21 \times \frac{(0.75)^2}{2} + \left(0.4 \times 0.75 \times \frac{1}{2}\right) \times \left(\frac{2}{3} \times 0.75\right) = 0.70 \text{ k·ft}$$

$$\text{Flexural stress in the concrete} = \frac{Mc}{I} = \frac{(0.70 \times 12{,}000) \times 6}{\frac{12}{12}(12)^3} = 29 \text{ psi}$$

This is a very low stress.

$$\text{Modulus of rupture } f_r = 7.5\sqrt{f'_c} = 7.5\sqrt{3000} = 410 \text{ psi}$$

The factor of safety against cracking = 410/29 = 14; therefore, the wall section is adequate. ■

EXAMPLE 14.2

Design a cantilever retaining wall to support a bank of earth 16.5 ft high. The top of the earth is to be level with a surcharge of 330 psf. Given: Weight of backfill = 110 pcf, angle of internal friction $\phi = 35°$, coefficient of friction between concrete and soil $\mu = 0.5$, coefficient of friction between soil layers $\mu = 0.7$, allowable soil bearing capacity = 4 ksf, $f'_c = 3$ ksi, and $f_y = 40$ ksi.

Solution

1. Using the Rankine equation:

$$C_a = \frac{1 - \sin\phi}{1 + \sin\phi} = \frac{1 - 0.574}{1 + 0.574} = 0.271$$

or from Table 14.1, for $\delta = 0$ and by interpolation for $\phi = 35°$, $C_a = 0.271$.

2. Determine the dimensions of the retaining wall using the approximate relationships shown in Figure 14.8.

- Height of wall: Allowing 3 ft for frost penetration, the height of the wall becomes $h = 16.5 + 3 = 19.5$ ft.
- Base thickness: Assume base thickness $0.08h = 0.08 \times 19.5 = 1.56$ ft, say, 1.5 ft. Height of stem = $19.5 - 1.5 = 18$ ft.
- Base length: The base length varies between $0.4h$ and $0.67h$. Assuming an average value of $0.53h$, then the base length equals $0.53 \times 19.5 = 10.3$ ft, say, 10.5 ft. The projection of the base in front of the stem varies between $0.17h$ and $0.125h$. Assume a projection of $0.17h = 0.17 \times 19.5 = 3.3$ ft, say, 3.5 ft.
- Stem thickness: The maximum stem thickness is at the bottom of the wall and varies between $0.08h$ and $0.1h$. Choose a maximum stem thickness equal to that of the base = 1.5 ft. Select a practical minimum thickness of the stem at the top of

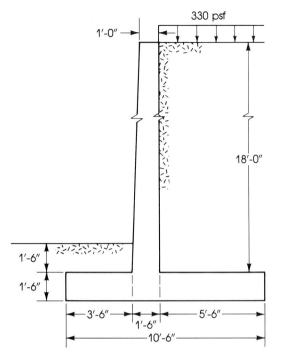

FIGURE 14.11. Example 14.2, trial configuration of retaining wall.

the wall, of 1.0 ft. The minimum batter of the face of the wall is $\frac{1}{4}$ in./ft. For an 18-ft-high wall, the minimum batter $= \frac{1}{4} \times 18 = 4.5$ in., which is less than the $1.5 - 1.0 = 0.5$ ft (6 in.) provided. The trial dimensions of the wall are shown in Figure 14.11.

3. Factor of safety against overturning:

   Calculate the actual unfactored forces acting on the retaining wall, first those acting to overturn the wall:

   $$h_s \text{ (due to surcharge)} = \frac{w_s}{w} = \frac{330}{110} = 3 \text{ ft}$$

   $$p_1 = C_a w h_s = 0.271 \times (110 \times 3) = 90 \text{ psf}$$

   $$p_2 = C_a w h = 0.271 \times (110 \times 19.5) = 581 \text{ psf}$$

   $$H_{a1} = 90 \times 19.5 = 1755 \text{ lb} \qquad \text{arm} = \frac{19.5}{2} = 9.75 \text{ ft}$$

   $$H_{a2} = \frac{1}{2} \times 581 \times 19.5 = 5665 \text{ lb} \qquad \text{arm} = \frac{19.5}{3} = 6.5 \text{ ft}$$

   Overturning moment $= 1.755 \times 9.75 + 5.665 \times 6.5 = 53.93 \text{ k·ft}$

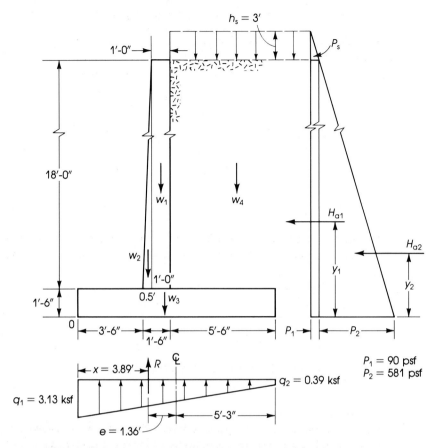

FIGURE 14.12. Example 14.2, forces acting on retaining wall.

Next, the balancing moment against overturning (see Figure 14.12):

Force (lb)	Arm (ft)	Moment (k·ft)
$w_1 = 1 \times 18 \times 150 = 2700$	4.50	12.15
$w_2 = \frac{1}{2} \times 18 \times \frac{1}{2} \times 150 = 675$	3.83	2.59
$w_3 = 10.5 \times 1.5 \times 150 = 2363$	5.25	12.41
$w_4 = 5.5 \times 21 \times 110 = 12{,}705$	7.75	98.46
$\Sigma w = R = 18.44 \text{ k}$		$\Sigma M = 125.61 \text{ k·ft}$

$$\text{Factor of safety against overturning} = \frac{125.61}{53.93} = 2.33 > 2.00$$

A stability check may be performed using the critical factored loads according to the **ACI Code, section 9.2.4:**

$$U = 0.9D + 1.7H \quad \text{or} \quad U = 1.4D + 1.7(L + H)$$

where H = horizontal earth pressure.

4. Calculate the base soil pressure. Take moments about the toe end O (Figure 14.12) to determine the location of the resultant R of the vertical forces.

$$x = \frac{\Sigma M - \Sigma Hy}{R} = \frac{\text{balancing } M - \text{overturning } M}{R}$$

$$= \frac{125.61 - 53.93}{18.44} = 3.89 \text{ ft} > \frac{10.5}{3} \text{ or } 3.5 \text{ ft}$$

The eccentricity

$$e = \frac{10.5}{2} - 3.89 = 1.36 \text{ ft}$$

The resultant R acts within the middle third of the base and has an eccentricity $e = 1.36$ ft from the center of the base. For a 1-ft length of the footing, area = $10.5 \times 1 = 10.5$ ft^2.

$$I = 1 \times \frac{(10.5)^3}{12} = 96.47 \text{ ft}^4$$

$$q_1 = \frac{R}{A} + \frac{(Re)c}{I} = \frac{18.44}{10.5} + \frac{(18.44 \times 1.36) \times 5.25}{96.47}$$

$$= 1.76 + 1.37 = 3.13 \text{ ksf} < 4 \text{ ksf}$$

$$q_2 = 1.76 - 1.37 = 0.39 \text{ ksf}$$

Soil pressure is adequate.

5. Calculate the factor of safety against sliding. In this example, the passive pressure against sliding is neglected unless a key is provided. A minimum factor of safety of 1.5 must be maintained.

$$\text{Force causing sliding} = H_{a1} + H_{a2} = 1.76 + 5.67 = 7.43 \text{ k}$$

$$\text{Resisting force} = \mu R = 0.5 \times 18.44 = 9.22 \text{ k}$$

$$\text{Factor of safety against sliding} = \frac{9.22}{7.43} = 1.24 < 1.5$$

The resistance provided does not give an adequate safety against sliding. In this case, a key should be provided to develop a passive pressure large enough to resist the excess force that causes sliding. Another function of the key is to provide sufficient development length for the dowels of the stem. The key is therefore placed such that its face is about 6 in. from the back face of the stem (Figure 14.13). In the calculation of the passive pressure, the top foot of the earth at the toe side is usually neglected, leaving a height of 2 ft in this example. Assume a key depth $t = 1.5$ ft and a width $b = 1.5$ ft,

$$C_p = \frac{1 + \sin \phi}{1 - \sin \phi} = \frac{1}{C_a} = \frac{1}{0.271} = 3.69$$

$$H_p = \frac{1}{2} C_p w(h' + t)^2 = \frac{1}{2} \times 3.69 \times 110(2 + 1.5)^2 = 2486 \text{ lb}$$

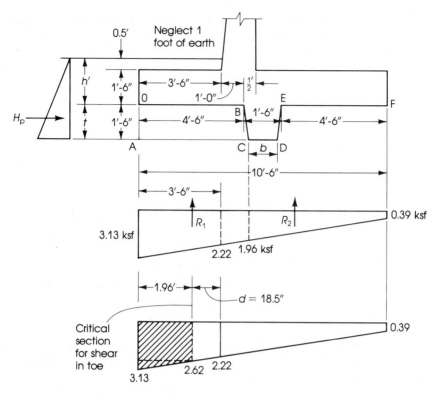

FIGURE 14.13. Footing details, example 14.2.

The sliding may occur now on the surfaces AC, CD, and EF (Figure 14.13). The sliding surface AC lies within the soil layers with a coefficient of internal friction $= \tan \phi = \tan 35° = 0.7$, while the surfaces CD and EF are those between concrete and soil with a coefficient of internal friction $= 0.5$, as given in this example. The frictional resistance $F = \mu_1 R_1 + \mu_2 R_2$.

$$R_1 = \text{reaction on AC} = \left(\frac{3.13 + 1.96}{2}\right) \times 4.5 = 11.44 \text{ k}$$

$$R_2 = R - R_1 = 18.44 - 11.44 = 7.0 \text{ k}$$

or

$$R_2 = \text{reaction on CDF} = \left(\frac{1.96 + 0.39}{2}\right) \times 6 = 7.05 \text{ k}$$

$$F = 0.7(11.44) + 0.5(7.00) = 11.50 \text{ k}$$

Total resisting force $= F + H_p = 11.50 + 2.49 = 13.99 \text{ k}$

Factor of safety against sliding $= \dfrac{13.99}{7.43} = 1.9 \quad \text{or} \quad \dfrac{11.5}{7.43} = 1.55 > 1.5$

The factor is greater than 1.5, which is recommended when passive resistance against sliding is not included.

6. Design the wall (stem). The design of the different reinforced concrete structural elements can be performed now using the ACI Code strength design method.

(a) Main reinforcement: The lateral forces applied to the wall are calculated using a load factor of 1.7. The critical section for bending moment is at the bottom of wall, height = 18 ft.

Ultimate forces:

$$P_1 = 1.7(C_a w h_s) = 1.7(0.271 \times 110 \times 3) = 152 \text{ lb}$$

$$P_2 = 1.7(C_a w h) = 1.7(0.271 \times 110 \times 18) = 912 \text{ lb}$$

$$H_{a1} = 0.152 \times 18 = 2.74 \text{ k} \qquad \text{arm} = \frac{18}{2} = 9 \text{ ft}$$

$$H_{a2} = \frac{1}{2} \times 0.912 \times 18 = 8.21 \text{ k} \qquad \text{arm} = \frac{18}{3} = 6 \text{ ft}$$

$$M_u \text{ (at bottom of wall)} = 2.74 \times 9 + 8.21 \times 6 = 73.92 \text{ k·ft}$$

Total depth used = 18 in., b = 12 in., and

$$d = 18 - 2 \text{ (concrete cover)} - 0.5 \text{ (half bar diameter)} = 15.5 \text{ in.}$$

$$R_u = \frac{M_u}{bd^2} = \frac{73.92 \times 12,000}{12(15.5)^2} = 308 \text{ psi}$$

From Table A.1 in Appendix A, the steel ratio (for R_u = 308 psi, f_c' = 3 ksi, f_y = 40 ksi) is ρ = 0.0092 < ρ_{max} = 0.0278. For economy, it is a common practice to use an approximate $\rho \leq 0.5\rho_{max}$.

$$A_s = 0.0092 \times 12 \times 15.5 = 1.71 \text{ in.}^2$$

Use No. 8 vertical bars spaced at 5.5 in. (A_s used = 1.71 in.²). Minimum vertical A_s required according to the **ACI Code, section 14.3.2**, is

$$A_s(\text{min}) = 0.0015 \times 12 \times 18 = 0.32 \text{ in.}^2 < 1.71 \text{ in.}^2$$

Since the moment decreases along the height of the wall, A_s may be reduced according to the moment requirements. A moment resistance diagram may be drawn to determine the steel bars required and their cut-off points. It is also practical to use one type of spacing for the lower half and a second spacing for the upper half of the wall. Some designers prefer three types of spacings at each third of the wall height. To calculate the moment at mid-height of the wall, 9 ft from the top:

$$P_1 = 1.7(0.271 \times 110 \times 3) = 152 \text{ lb}$$

$$P_2 = 1.7(0.271 \times 110 \times 9) = 456 \text{ lb}$$

$$H_{a1} = 0.152 \times 9 = 1.37 \text{ k} \qquad \text{arm} = 9/2 = 4.5 \text{ ft}$$

$$H_{a2} = \frac{1}{2} \times 0.456 \times 9 = 2.1 \text{ k} \qquad \text{arm} = 9/3 = 3 \text{ ft}$$

$$M_u = 1.37 \times 4.5 + 2.1 \times 3 = 12.47 \text{ k·ft}$$

$$\text{Total depth at mid-height of wall} = \frac{12 + 18}{2} = 15 \text{ in.}$$

$$d = 15 - 2 - 0.5 = 12.5 \text{ in.}$$

$$R_u = \frac{M_u}{bd^2} = \frac{12.47 \times 12,000}{12 \times (12.5)^2} = 80 \text{ psi}$$

$\rho = 0.0024$ and $A_s = 0.0024 \times 12 \times 12.5 = 0.36 \text{ in.}^2$

$A_s(\text{min}) = 0.0015 \times 12 \times 15 = 0.27 \text{ in.}^2 < 0.36 \text{ in.}^2$

Use No. 5 vertical bars spaced at 5.5 in., similar to the lower vertical steel bars in the wall.

(b) Temperature and shrinkage reinforcement: Minimum horizontal reinforcement at the base of the wall according to **ACI Code, section 14.3.3**:

$$A_s(\text{min}) = 0.0025 \times 12 \times 18 = 0.54 \text{ in.}^2 \text{ (for the bottom third)}$$

$$A_s(\text{min}) = 0.0025 \times 12 \times 15 = 0.45 \text{ in.}^2 \text{ (for the upper two-thirds)}$$

Since the front face of the wall is mostly exposed to temperature changes, use half to two-thirds of the horizontal bars at the external face of the wall and place the balance at the internal face **(ACI Code, section 14.3.4)**.

$$0.5 A_s = 0.5 \times 0.54 = 0.27 \text{ in.}^2$$

Use No. 4 horizontal bars spaced at 8 in. ($A_s = 0.29 \text{ in.}^2$) at both the internal and external surfaces of the wall. Use No. 4 vertical bars spaced at 12 in. at the front face of the wall to support the horizontal temperature and shrinkage reinforcement.

(c) Dowels for the wall vertical bars: The anchorage length of No. 8 bars into the footing must be at least 23 in. (Table A.11, Appendix A). Use an embedment length $= 2$ ft into the footing and key below the stem.

(d) Design for shear: The critical section for shear is at a distance $d = 15.5$ in. from the bottom of the stem. At this section, the distance from the top equals $18 - 15.5/12 = 16.7$ ft.

$$P_1 = 152 \text{ lb}$$

$$P_2 = 1.7(0.271 \times 110 \times 16.7) = 846 \text{ lb}$$

$$H_{a1} = 0.152 \times 16.7 = 2.54 \text{ k}$$

$$H_{a2} = \frac{1}{2} \times 0.846 \times 16.7 = 7.07 \text{ k}$$

Total $H = 2.54 + 7.07 = 9.61$ k

$$\phi V_c = \phi(2\sqrt{f_c'})bd = \frac{0.85 \times 2}{1000} \times \sqrt{3000} \times 12 \times 15.5$$

$$= 17.32 \text{ k} > 9.61 \text{ k}$$

7. Design of the heel: A load factor of 1.4 is used to calculate the factored bending moment and shearing force due to the backfill and concrete, while a load factor of 1.7 is used for the surcharge. The upward soil pressure is neglected as it will reduce the effect of the backfill and concrete on the heel. Referring to Figure 14.12, the total load on the heel:

$$V_u = 1.4[(18 \times 5.5 \times 110) + (1.5 \times 5.5 \times 150)] + 1.7(3 \times 5.5 \times 110)$$
$$= 20.0 \text{ k}$$

$$M_u \text{ (at face of wall)} = V_u \times \frac{5.5}{2}$$

$$= 20 \times 2.75 = 55 \text{ k·ft}$$

The critical section for shear is usually at a distance d from the face of the wall when the reaction introduces compression into the end region of the member (ACI Code, section 11.1.2). In this case, the critical section will be considered at the face of the wall, because tension and not compression develops in the concrete.

$$V_u = 20.0 \text{ k}$$

$$\phi V_c = \phi(2\sqrt{f_c'})bd = \frac{0.85 \times 2}{1000} \times \sqrt{3000} \times 12 \times 14.5 = 16.2 \text{ k}$$

ϕV_c is less than V_u of 20.0 k, and the section must be increased by the ratio 20.0/16.2 or shear reinforcement must be provided.

$$\text{Required } d = \frac{20.0}{16.2} \times 14.5 = 17.9 \text{ in.}$$

Total thickness required $= 17.9 + 3.5 = 21.4$ in.

Use a base thickness of 22 in., $d = 18.5$ in.

$$R_u = \frac{M_u}{bd^2} = \frac{55 \times 12,000}{12 \times (18.5)^2} = 161 \text{ psi}$$

Steel ratio $\rho = 0.0046$, which is less than

$$\rho_{min} = \frac{200}{f_y} = 0.005$$

$$A_s = 0.005 \times 12 \times 18.5 = 1.11 \text{ in.}^2$$

Use No. 8 bars spaced at 8 in. ($A_s = 1.18$ in.²). The development length for the No. 8 top bars equals $1.4l_d = 32$ in. (Table A.11). Therefore, the bars must be extended 3 ft into the toe of the base.

Temperature and shrinkage reinforcement in the longitudinal direction is not needed in the heel or toe, but it may be preferable to use minimal amounts of reinforcement in that direction, say No. 4 bars spaced at 12 in.

8. Design of the toe: The toe of the base acts as a cantilever beam subjected to upward pressures, as calculated in step 4. The factored soil pressure is obtained by multiplying the service load soil pressure by a load factor of 1.7 as it is primarily caused by the lateral forces. The critical section for the bending moment is at the front face of the stem. The critical section for shear is at a distance d from the front face of the stem, since the reaction in the direction of shear introduces compression into the toe.

Referring to Figure 14.13, the toe is subjected to an upward pressure from the soil and downward pressure due to self-weight of the toe slab.

$$V_u = 1.7 \left(\frac{3.13 + 2.62}{2}\right) \times 1.96 - 1.4 \left(\frac{22}{12} \times 0.150\right) \times 1.96 = 8.83 \text{ k}$$

This is less than ϕV_c of 16.2 k calculated for the heel in step 7.

$$M_u = 1.7 \left[\frac{2.22}{2} \times (3.5)^2 + (3.13 - 2.22) \times 3.5 \times 0.5 \left(\frac{2}{3} \times 3.5\right)\right]$$

$$- 1.4 \left[\left(\frac{22}{12} \times 0.150\right) \times \frac{(3.5)^2}{2}\right] = 27.10 \text{ k·ft}$$

$$R_u = \frac{M_u}{bd^2} = \frac{27.1 \times 12,000}{12 \times (18.5)^2} = 79 \text{ psi}$$

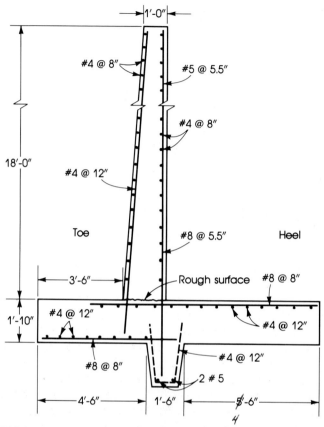

FIGURE 14.14. Reinforcement details, example 14.2.

R_u is very low, and $\rho < \rho_{\min}$.

$$A_s = \rho_{\min}bd = 0.005 \times 12 \times 18.5 = 1.11 \text{ in.}^2$$

Use No. 8 bars spaced at 8 in. ($A_s = 1.18$ in.^{2}), similar to steel of the heel. The development length required for No. 8 bars as bottom bars is 23 in. Extend the bars into the heel a distance of 2 ft measured from the face of the wall. The final reinforcement details are shown in Figure 14.14.

9. Shear keyway between wall and footing: In the construction of retaining walls, the footing is cast first and then the wall is cast on top of the footing at a later date. A construction joint is used at the base of the wall. The joint surface takes the form of a keyway, as shown in Figure 14.15, or is left in a very rough condition (Figure 14.14). The joint must be capable of transmitting the stem shear into the footing.

10. Proper drainage of the backfill is essential in this design as the earth pressure used is for drained backfill. Weep holes should be provided in the wall, of 4 in. diameter and spaced at 5 ft in the horizontal and vertical directions. ∎

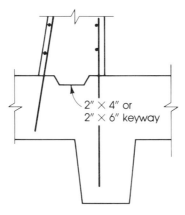

FIGURE 14.15. Keyway details.

14.10   Basement Walls

A common case of walls retaining backfill is basement walls of buildings. These walls span vertically between the basement floor slab and the first floor slab. Two possible cases of design should be investigated for a basement wall.

First, when the wall only has been built on top of the basement floor slab, the wall will be subjected to lateral earth pressure with no vertical loads except its own weight. The wall in this case acts as a cantilever, and adequate reinforcement should be provided for a cantilever wall design. This case can be avoided by installing the basement and the first floor slabs before backfilling against the wall.

Second, when the first floor and the other floor slabs have been constructed and the building is fully loaded, the wall in this case will be designed as a propped cantilever wall subjected to earth pressure and to vertical load.

For an angle of internal friction of 35°, the coefficient of active pressure $C_a = (1 - \sin 35°)/(1 + \sin 35°) = 0.271$. The horizontal earth pressure at the base $p_a = C_a wh$. For $w = 110$ pcf and an average height of a basement $h = 10$ ft, then

$$p_a = 0.271 \times 0.110 \times 10 = 0.3 \text{ ksf}$$

$$H_a = 0.271 \times 0.110 \times \frac{100}{2} = 1.49 \text{ k/ft of wall}$$

H_a acts at $h/3 = 10/3 = 3.33$ ft from the base.

An additional pressure must be added to allow for a surcharge of about 200 psf on the ground behind the wall. The equivalent height to the surcharge

$$h_s = \frac{200}{110} = 1.82 \text{ ft}$$

$$p_s = C_a wh_s = 0.271 \times 0.110 \times 1.82 = 0.054 \text{ ksf}$$

$$H_s = (C_a wh_s) \times h = 0.054 \times 10 = 0.54 \text{ k/ft of wall}$$

H_s of the surcharge acts at $h/2 = 5$ ft from base.

In the above calculations, it is assumed that the backfill is dry, but it is necessary to investigate the presence of water pressure behind the wall. The maximum water pressure occurs when the whole height of the basement wall is subjected to water pressure, and $p_w = wh = 62.5 \times 10 = 625$ psf.

$$H_w = \frac{wh^2}{2} = 0.625 \times 5 = 3.125 \text{ k/ft of wall}$$

This maximum pressure may not be present continuously behind the wall. Therefore, if the ground is intermittently wet, a percentage of the above pressure may be adopted, say 50 percent:

$$p_w/2 = \frac{0.625}{2} = 0.31 \text{ ksf}$$

$$H'_w = H_w/2 = (0.5\,wh)\,\frac{h}{2} = \frac{3.125}{2} = 1.56 \text{ k/ft of wall}$$

H'_w acts at $h/3 = 10/3 = 3.33$ ft from the base. Water may be prevented from collecting against the wall by providing drains at the lower end of the wall.

In addition to drainage, a waterproofing or dampproofing membrane must be laid or applied to the external face of the wall.

EXAMPLE 14.3

Determine the thickness and necessary reinforcement for the basement retaining wall shown in Figure 14.16. Given: weight of backfill = 110 pcf, angle of internal friction = 35°, $f'_c = 3$ ksi, and $f_y = 40$ ksi.

Solution

1. The wall spans vertically and will be considered as fixed at the bottom end and propped at the top. Consider a span $L = 9.75$ ft as shown in Figure 14.16. For the above data, the different lateral pressures on one foot length of the wall are:

   Due to active soil pressure, $p_a = 0.3$ ksf and $H_a = 1.49$ k
   Due to water pressure, $p_w = 0.31$ ksf and $H_w = 1.56$ k
   Due to surcharge, $p_s = 0.054$ ksf and $H_s = 0.54$ k

   H_a and H_w are due to triangular loadings, while H_s is due to uniform loading. Referring to Figure 14.16 and using moment coefficients of a propped beam subjected to triangular and uniform loads,

   $$M_u = 1.7(H_a + H_w)\frac{L}{7.5} + 1.7H_s\frac{L}{8}$$

   $$= 1.7\left(\frac{3.05}{7.5} \times 9.75 + 0.54 \times \frac{9.75}{8}\right) = 7.87 \text{ k·ft}$$

   $$R_B = 1.7\left(\frac{3.05}{3} + \frac{0.54}{2}\right) - \frac{7.87}{9.75} = 1.38 \text{ k}$$

   $$R_A = 4.73 \text{ k}$$

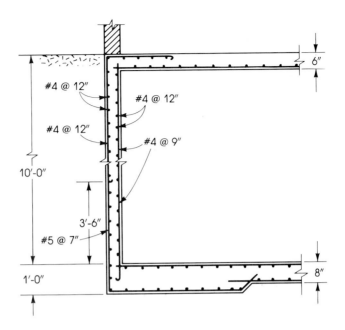

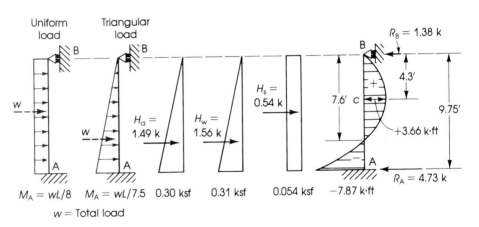

FIGURE 14.16. Basement wall, example 14.3.

Maximum positive bending moment within the span occurs at the section of zero shear.

$$V_u = 1.38 - 1.7(0.054x) - 1.7\left(0.063\frac{x^2}{2}\right) = 0$$

$$x = 4.3 \text{ ft}$$

$$M_c = 1.38 \times 4.3 - 1.7\left[\frac{0.054}{2}(4.3)^2 + \frac{0.27}{2}\frac{(4.3)^2}{3}\right] = +3.66 \text{ k·ft}$$

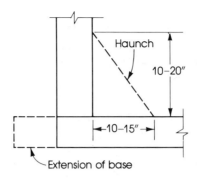

FIGURE 14.17. Adjustment of wall base, example 14.3.

2. Assuming 0.01 steel ratio, $R_u = 332$ psi:

$$d = \sqrt{\frac{M_u}{R_u b}} = \sqrt{\frac{7.87 \times 12}{0.332 \times 12}} = 4.87 \text{ in.}$$

Total depth $= 4.87 + 1.5$ (concrete cover) $+ 0.25 = 6.62$ in.

Use 7-in. slab, $d = 5.25$ in.

$$R_u = \frac{M_u}{bd^2} = \frac{7.87 \times 12,000}{12 \times (5.25)^2} = 286 \text{ psi}$$

Steel ratio $\rho = 0.0085$ and $A_s = 0.0085 \times 12 \times 5.25 = 0.53$ in.^{2}. Use No. 5 bars spaced at 7 in. ($A_s = 0.53$ in.^{2}).

3. For the positive moment, $M_c = 3.66$ k·ft:

$$R_u = \frac{3.66 \times 12,000}{12 \times (5.25)^2} = 133 \text{ psi}$$

$$\rho = 0.004 > 0.002 \ (\rho_{min})$$

$$A_s = 0.004 \times 12 \times 5.25 = 0.25 \text{ in.}^2$$

Use No. 4 bars spaced at 9 in. ($A_s = 0.26$ in.^{2}).

4. Zero moment occurs at a distance of 7.6 ft from the top and 2.15 ft from the base. The development length of No. 5 bars is 12 in. Therefore, extend the main No. 5 bars to a distance $= 2.15 + 1 = 3.15$ ft, say, 3.5 ft above the base, then use No. 4 bars spaced at 12 in. at the exterior face. For the interior face use No. 4 bars spaced at 9 in. throughout.

5. Longitudinal reinforcement: Use minimum steel ratio of 0.002 **(ACI Code, section 14.3.3)**. $A_s = 0.0025 \times 7 \times 12 = 0.21$ in.^{2}, so use No. 4 bars spaced at 12 in. on each side of the wall.

6. If the bending moment at the base of the wall is quite high, it may require a thick wall slab, say, 12 in. or more. In this case a haunch may be adopted as shown in Figure 14.17. This solution will reduce the thickness of the wall, as it will be designed for the moment at the section exactly above the haunch.

7. The basement slab may have a thickness greater than the wall thickness and may be extended outside the wall by about 10 in. or more as required.   ▪

SUMMARY

SECTIONS 14.1 to 14.3

(a) A retaining wall maintains unequal levels of earth on its two faces. The most common types of retaining walls are: gravity, semi-gravity, cantilever, counterfort, buttressed, and basement walls.

(b) For a linear pressure, the active and passive pressure intensities are:

$$P_a = C_a wh \quad \text{and} \quad P_p = C_p wh$$

According to Rankine's theory:

$$C_a = \left(\frac{1 - \sin \phi}{1 + \sin \phi}\right) \quad \text{and} \quad C_p = \left(\frac{1 + \sin \phi}{1 - \sin \phi}\right)$$

Values of C_a and C_p for different values of ϕ and δ are given in Tables 14.1 and 14.2.

SECTIONS 14.4 and 14.5

(a) When soil is saturated, the submerged unit weight must be used to calculate earth pressure. The hydrostatic water pressure must also be considered.

(b) A uniform surcharge on a retaining wall causes an additional pressure height, $h_s = w_s/w$.

SECTIONS 14.6 to 14.8

(a) The total frictional force F to resist sliding effect is:

$$F = \mu R + H_p \tag{14.14}$$

$$\text{Factor of safety against sliding} = \frac{F}{H_{ah}} \geq 1.5 \tag{14.15}$$

(b)

$$\text{Factor of safety against overturning} = \frac{M_b}{M_o} = \frac{\Sigma wx}{H_h/3} \geq 2.0 \tag{14.16}$$

(c) Approximate dimensions of a cantilever retaining wall are shown in Figure 14.8.

SECTIONS 14.9 and 14.10

(a) To avoid hydrostatic pressure on a retaining wall, a drainage system should be used that consists of weep holes, perforated pipe, or any other adequate device.

(b) Basement walls in buildings may be designed as propped cantilever walls subjected to earth pressure and vertical loads. This case occurs only if the first floor slab has been constructed. A surcharge of 200 psf may be adopted.

REFERENCES

1. K. Terzaghi and R. B. Peck: *Soil Mechanics in Engineering Practice,* Wiley, New York, 1968.
2. W. C. Huntington: *Earth Pressures and Retaining Walls,* Wiley, New York, 1957.
3. G. P. Fisher and R. M. Mains: "Sliding Stability of Retaining Walls," *Civil Engineering,* July 1952.
4. "Retaining Walls and Abutments," *AREA Manual,* vol. 1, American Railway and Engineering Association, Chicago, 1958.
5. M. S. Ketchum: *The Design of Walls, Bins and Grain Elevators,* 3rd ed., McGraw-Hill, New York, 1949.

6. W. C. Teng: *Foundation Design,* Prentice-Hall, New York, 1962.

7. J. E. Bowles: *Foundation Analysis and Design,* McGraw-Hill, New York, 1980.

8. American Concrete Institute: *ACI Reinforced Concrete Detailing Manual,* ACI 315-74, Detroit, 1974.

PROBLEMS

14.1 and 14.2. Check the adequacy of the retaining walls shown in Figures 14.18 and 14.19 with regard to overturning, sliding, and the allowable soil pressure. Given: weight of backfill = 110 pcf (1760 kg/m^3), angle of internal friction ϕ = 30°, coefficient of friction between concrete and soil μ = 0.5, allowable soil pressure = 3.5 ksf (170 kN/m^2), f'_c = 3 ksi (21 MPa).

14.3. Design the cantilever retaining wall shown in Figure 14.20. The top of the backfill is level without any surcharge. Given: weight of backfill = 120 pcf (1920 kg/m^3), angle of internal friction ϕ = 35°, coefficient of friction between concrete and soil μ = 0.52, coefficient of friction between soil layers = 0.70, allowable soil pressure = 4 ksf (200 kN/m^2), f'_c = 3 ksi (21 MPa), and f_y = 40 ksi.

14.4. Repeat problem 14.3 assuming a surcharge of 300 psf.

14.5. Repeat problem 14.3 assuming that the backfill slopes at 30° to the horizontal.

14.6. Determine the approximate dimensions of a cantilever retaining wall to support a bank of earth 16 ft (4.8 m) high. Assume that frost penetration depth is 4 ft (1.2 m). Check the safety of the retaining wall against overturning and sliding only. Given: weight of backfill = 120 pcf (1920 kg/m^3), angle of internal friction = 33°, coefficient of friction between concrete and soil = 0.45, coefficient of friction between soil layers = 0.65, and allowable soil pressure = 4 ksf (200 kN/m^2).

14.7. Design a cantilever retaining wall to support a bank of earth 12 ft high, as shown in Figure 14.21. The top of the backfill is to be level without surcharge. Assume a frost penetration

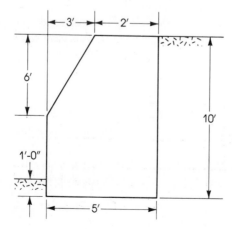

FIGURE 14.18. Problem 14.1, gravity wall.

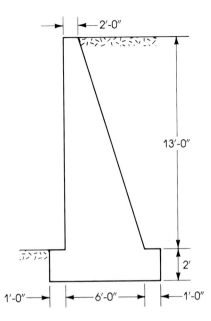

FIGURE 14.19. Problem 14.2, semi-gravity wall.

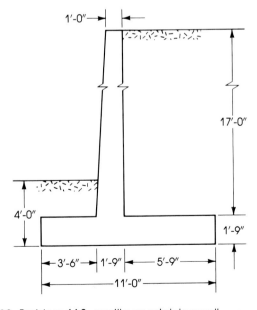

FIGURE 14.20. Problem 14.3, cantilever retaining wall.

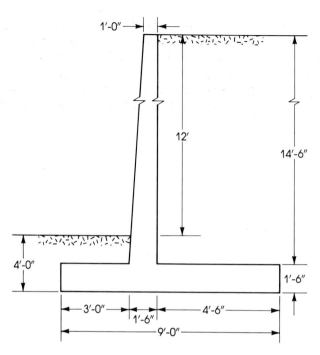

FIGURE 14.21. Problem 14.7, cantilever retaining wall.

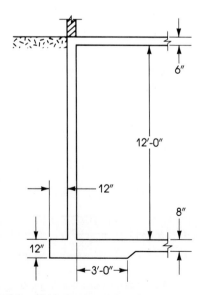

FIGURE 14.22. Problem 14.8, basement wall.

depth of 4 ft (1.2 m). Given: weight of backfill = 110 pcf (1760 kg/m³), angle of internal friction = 35°, coefficient of friction between concrete and soil = 0.55, and allowable soil pressure = 4 ksf (200 kN/m²). Use f'_c = 3 ksi (21 MPa) and f_y = 40 ksi (280 MPa).

14.8. Determine the thickness and necessary reinforcement for the basement wall shown in Figure 14.22. The weight of backfill = 120 pcf (1920 kg/m³) and the angle of internal friction ϕ = 30°. Assume a surcharge of 400 psf (20 kN/m²) and use f'_c = 3 ksi (21 MPa) and f_y = 60 ksi (420 MPa).

15

Design for Torsion

15.1   Introduction

Torsional stresses develop in a beam section when a moment acts on that section parallel to its surface. Such moments, called torsional moments, cause a rotation in the structural member and cracking on its surface, usually in the shape of a spiral. To illustrate torsional stresses, let a torque T be applied on a circular cantilever beam made of elastic homogeneous material, as shown in Figure 15.1. The torque will cause a rotation of the beam. Point B moves to point B' at one end of the beam while the other end is fixed. The angle θ is called the angle of twist. The plane AO'OB will be distorted to the shape AO'OB'. Assuming that all longitudinal elements have the same length, the shear strain

$$\gamma = \frac{(\mathrm{BB'})}{L} = \frac{r\theta}{L}$$

where L = the length of the beam and r = the radius of the circular section.

In reinforced concrete structures, members may be subjected to torsional moments when they are curved in plan, support cantilever slabs, act as spandrel beams (end beams), or are part of a spiral stairway.

Structural members may be subjected to pure torsion only or, as in most cases, subjected simultaneously to shearing forces and bending moments.

Example 15.1 illustrates the different forces that may act at different sections of a cantilever beam.

EXAMPLE 15.1

Calculate the forces acting at sections 1, 2, and 3 of the cantilever beam shown in Figure 15.2. The beam is subjected to a vertical force $P_1 = 15$ k and a horizontal force $P_2 = 12$ k acting at

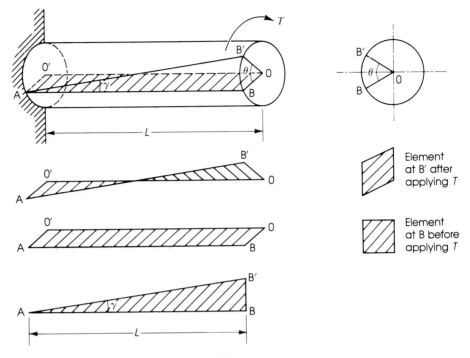

FIGURE 15.1. Torque applied to a cantilever beam.

C, and a horizontal force $P_3 = 20$ k acting at B and perpendicular to the direction of the force P_2.

Solution Let N = normal force, V = shearing force, M = bending moment, and T = torsional moment. The forces are as listed below.

Section	N (k)	M_x (k·ft)	M_y (k·ft)	V_x (k)	V_y (k)	T (k·ft)
1	0	−135 (15 × 9)	+108 (12 × 9)	+12	+15	0
2	−12 compression	0	+108	+20	+15	135 (15 × 9)
3	−12 compression	−180	+348	+20	+15	135 (15 × 9)

If P_1, P_2, and P_3 are ultimate loads ($P_u = 1.4P_D + 1.7P_L$), then the values in the table will be the ultimate design forces. ∎

454   Chapter 15: Design for Torsion

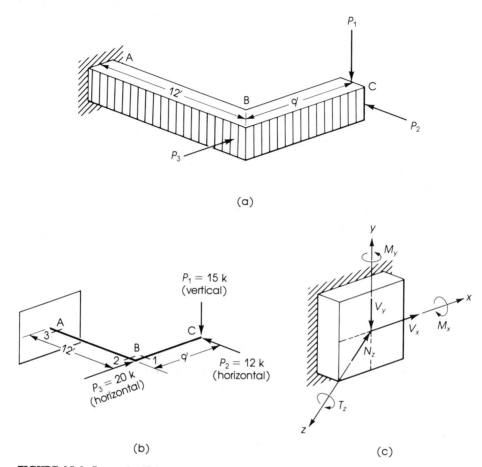

FIGURE 15.2. Example 15.1.

15.2   Torsional Moments in Circular Beams

It was shown in example 15.1 that forces can act on building frames, causing torsional moments. If a concentrated load P is acting at point C in the frame ABC shown in Figure 15.3(a), it develops a torsional moment in beam AB = $T = PZ$ acting at D. When D is at midspan of AB, then the torsional design moment in AD equals that in DB = $\frac{1}{2}T$. If a cantilever slab is supported by the beam AB in Figure 15.3(b), the slab causes a uniform torsional moment m_t along AB. This uniform torsional moment is due to the load on a unit width strip of the slab. If S = the width of the cantilever slab and w = load on the slab (psf), then $m_t = wS^2/2$ k·ft/ft of beam AB. The maximum torsional design moment in beam AB = $T = \frac{1}{2}m_t L$ acting at A and B. Other cases of loading are explained in Table 15.1. In general, the distribution of torsional moments in beams has the same shape and numerically has the same values as the shear diagrams for beams subjected to a load m_t or T.

TABLE 15.1. Torsion diagrams

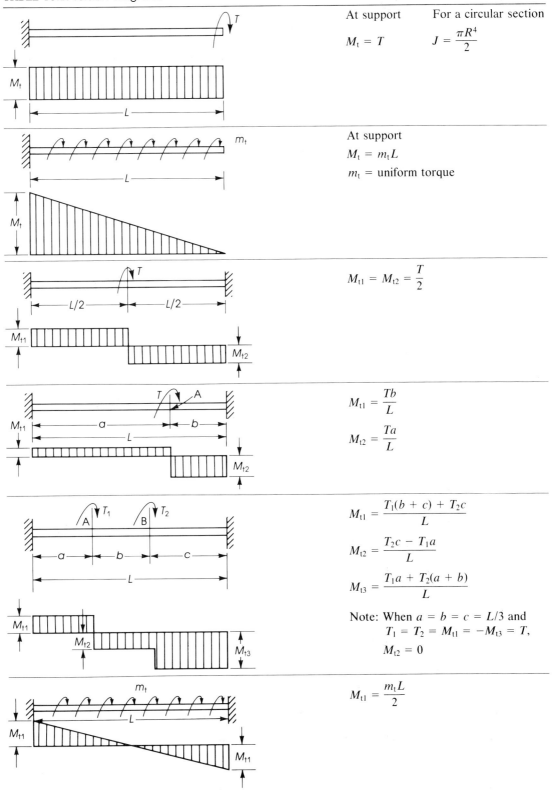

At support   For a circular section

$$M_t = T \qquad J = \frac{\pi R^4}{2}$$

At support

$$M_t = m_t L$$

m_t = uniform torque

$$M_{t1} = M_{t2} = \frac{T}{2}$$

$$M_{t1} = \frac{Tb}{L}$$

$$M_{t2} = \frac{Ta}{L}$$

$$M_{t1} = \frac{T_1(b + c) + T_2 c}{L}$$

$$M_{t2} = \frac{T_2 c - T_1 a}{L}$$

$$M_{t3} = \frac{T_1 a + T_2(a + b)}{L}$$

Note: When $a = b = c = L/3$ and
$T_1 = T_2 = M_{t1} = -M_{t3} = T$,
$M_{t2} = 0$

$$M_{t1} = \frac{m_t L}{2}$$

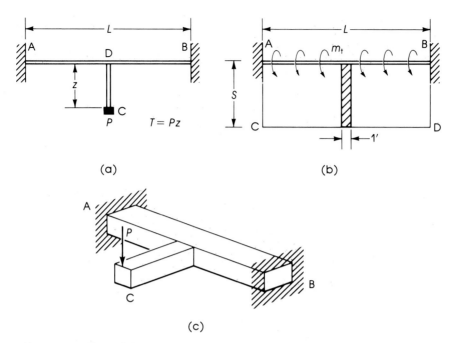

FIGURE 15.3. Torsional moments on AB.

15.3   Elastic Torsional Stresses

Considering the cantilever beam with circular section of Figure 15.1, the torsional moment T will cause a shearing force dV perpendicular to the radius of the section. From the conditions of equilibrium, the external torsional moment is resisted by an internal torque equal to and opposite to T. If dV is the shearing force acting on the area dA (Figure 15.4), then the magnitude of the torque $T = \int r\,dV$. Let the shearing stress = v, then

$$dV = v\,dA \quad \text{and} \quad T = \int rv\,dA$$

The maximum elastic shear occurs at the external surface of the circular section at radius R. The shear stress at any radius $r < R$ is equal to $v = v_{max}\,r/R$. Considering a ring at radius r with thickness dr, then the torque T can be evaluated by taking moments about the center O for the ring area:

$$dT = (2\pi r\,dr)vr$$

where $(2\pi r\,dr)$ is the area of the ring and v is the shear stress in the ring. Thus

$$T = \int_{r=0}^{r=R} (2\pi r\,dr)vr = \int_{r=0}^{r=R} 2\pi r^2 v\,dr \qquad (15.1)$$

For a hollow section with internal radius R_1,

$$T = \int_{r=R_1}^{r=R} 2\pi r^2 v\,dr \qquad (15.2)$$

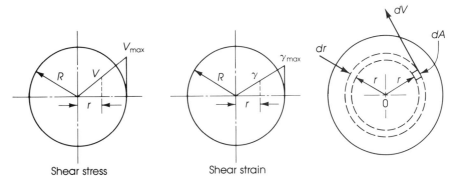

FIGURE 15.4. Torque in circular sections.

For a solid section, using equation 15.2 and using $v = v_{max}\, r/R$:

$$T = \int_{r=0}^{R} 2\pi r^2 \left(v_{max}\, \frac{r}{R} \right) dr = \frac{2\pi}{R}\, v_{max} \int_{r=0}^{R} r^3\, dr$$

$$= \frac{2\pi}{R}\, v_{max} \times \frac{R^4}{4} = \frac{\pi}{2}\, v_{max} R^3$$

or

$$v_{max} = \frac{2T}{\pi R^3} \tag{15.3}$$

The polar moment of inertia of a circular section $J = \pi R^4/2$. Therefore, the shear stress can be written as a function of the polar moment of inertia J as follows:

$$v_{max} = \frac{TR}{J} \tag{15.4}$$

15.4   Torsional Moment in Rectangular Sections

The determination of the stresses in noncircular members subjected to torsional loading is not as simple as that for circular sections. However, results obtained from the theory of elasticity[1] indicate that the maximum shearing stress for rectangular sections can be calculated as follows:

$$v_{max} = \frac{T}{\alpha x^2 y} \tag{15.5}$$

where

T = the applied torque

x = the short side of the rectangular section

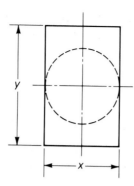

FIGURE 15.5. Rectangular section.

y = the long side of the rectangular section

α = coefficient that depends on the ratio of y/x; its value is given below

$\dfrac{y}{x}$	*1.0*	*1.2*	*1.5*	*2.0*	*3*	*4*	*5*	*10*	∞
α	0.208	0.219	0.231	0.246	0.267	0.282	0.291	0.312	0.333

The maximum shearing stress occurs along the centerline of the longer side y (Figure 15.5).

For members composed of rectangles, such as T-, L-, or I-sections, the value of α can be assumed equal to $\frac{1}{3}$, and the section may be divided into several rectangular components having a long side y_i and a short side x_i. The maximum shearing stress can be calculated from

$$v_{max} = \frac{3T}{\Sigma x_i^2 y_i} \qquad (15.6)$$

where $\Sigma x_i^2 y_i$ is the value obtained from the rectangular components of the section. When $y/x \leq 10$, a more exact expression may be used:

$$v_{max} = \frac{3T}{\Sigma x^2 y \left(1 - 0.63\dfrac{x}{y}\right)} \qquad (15.7)$$

15.5   ACI Code Analysis

In the analysis of torsion on concrete beams at working load levels, the concrete will be assumed to be cracked.

The 1983 ACI Code, which adopted the strength design method of analysis, uses an approximate expression to calculate the nominal torsional moment:

$$T_n = \frac{1}{3} v_t \Sigma x^2 y \tag{15.8}$$

and

$$T_u \leq \phi T_n \quad \textbf{(ACI Code, section 11.6.5)} \tag{15.9}$$

which is similar to equation 15.6 given above. Introducing the reduction factor ϕ, the ultimate torsional stress can be calculated as follows:

$$v_{tu} = \frac{3T_u}{\phi \Sigma x^2 y} \tag{15.10}$$

where $\phi = 0.85$ and x and y are the short and long sides of each of the rectangular components of the section.

Following are several cases for deriving $\Sigma x^2 y$.

1. For rectangular sections, $x^2 y$ can be calculated directly using the short dimension x and the long dimension y (Figure 15.5).
2. For T-, L-, or I-sections, the compound section is divided into rectangles. The **ACI Code, section 11.61,** limits the effective overhang width of the flange thickness or $y_2 \leq 3x_2$ in the T-section shown in Figure 15.6. The rectangle $(x_1 y_1)$ is the largest rectangle and extends from the top of the flange to the bottom of the web. If the flange is thicker than the web, then the main rectangle extends along the flange. The same principle applies to other solid sections of rectangular components.
3. Hollow sections are permitted by the **ACI Code, section 11.61,** to resist torsional moments based on the following three limitations (Figure 15.7):

    • When the wall thickness $h \geq x/4$:
      In this case, the same procedure as for solid sections may be used for the rectangular components of the hollow sections, or $v_{tu} = 3T_u/\phi \Sigma x^2 y$ (equation 15.10).

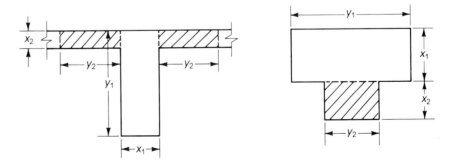

FIGURE 15.6. T-sections divided into rectangular components.

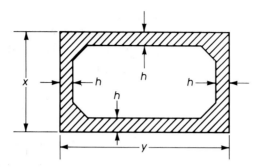

FIGURE 15.7. Hollow section.

- When the wall thickness $x/10 < h < x/4$:
  In this case, the section may be treated as solid rectangles except that $\Sigma x^2 y$ must be multiplied by $4h/x$, or

$$v_{tu} = \frac{3T_u}{\phi \Sigma x^2 y}\left(\frac{x}{4h}\right) \tag{15.11}$$

- When the wall thickness $h < x/10$:
  In this case, the wall stiffness must be considered because of the flexibility of the walls and the possibility of local buckling. Such sections should be avoided. When hollow box sections are used, a fillet is required at the corners of the box, as shown in Figure 15.7.

EXAMPLE 15.2

Calculate the ultimate torsional stress v_{tu} for the four sections shown in Figure 15.8. The applied ultimate torque $T_u = 36$ k·ft.

Solution   1. For the rectangular section (Figure 15.8(a)),

$$v_{tu} = \frac{3T_u}{\phi x^2 y}$$

$$\phi = 0.85 \quad\quad x = 12 \text{ in. and } y = 20 \text{ in.}$$

$$v_{tu} = \frac{3(36 \times 12,000)}{0.85(12)^2(20)} = 529 \text{ psi}$$

2. For the T-section (Figure 15.8(b)), the section can be divided into a main rectangle $x_1 y_1 = 10 \times 25$ in. and two small rectangles on each side of the flange $(x_y y_2)$. The effective width of the flange rectangle $= 3h = 3 \times 5 = 15$ in. or $(50 - 10)/2 = 20$ in. Use $y_2 = 15$ in. < 20 in. and $x_2 = 5$ in.

$$\Sigma x^2 y = (10)^2 \times 25 + 2(5)^2 \times 15 = 3250 \text{ in.}^3$$

$$v_{tu} = \frac{3(36 \times 12,000)}{0.85 \times 3250} = 469 \text{ psi}$$

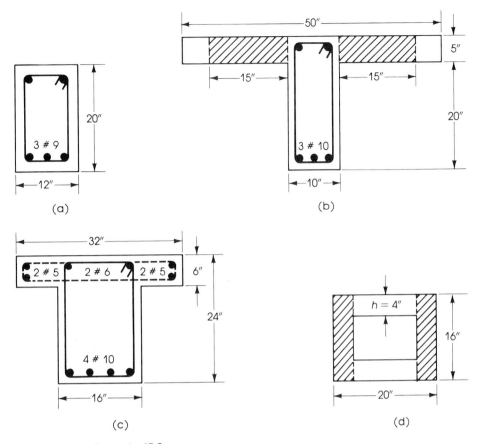

FIGURE 15.8. Example 15.2.

3. For the T-section (Figure 15.8(c)), the overhanging flange width = (32 − 16)/2 = 8 in., which is less than 3 ⋅ 6 = 18 in. Therefore use a width of 8 in.

$$\Sigma x^2 y = (16)^2 \times 24 + 2(6)^2 \times 8 = 6720 \text{ in.}^3$$

$$v_{tu} = \frac{3(36 \times 12,000)}{0.85 \times 6720} = 227 \text{ psi}$$

4. For the box section (Figure 15.8(d)), check $h \geq \dfrac{x}{4}$:

$$h = 4 \text{ in.} \geq \frac{16}{4} = 4.0 \text{ in.}$$

Therefore use equation 15.10.

$$\Sigma x^2 y = 2(4)^2 \times 16 + 2(4)^2 \times 12 = 896 \text{ in.}^3$$

$$v_{tu} = \frac{3(36 \times 12,000)}{0.85 \times 896} = 1701 \text{ psi}$$

(which is a very high stress). ∎

15.6   Torsion Theories for Concrete Members

Various methods are available for the analysis of reinforced concrete members subjected to torsion or simultaneous torsion, bending, and shear. The design methods rely generally on two basic theories: the skew bending theory and the space truss analogy.

15.6.1   Skew Bending Theory

The skew bending concept was first presented by Lessig in 1959[2] and was further developed by Goode and Helmy,[3] Collins et al. in 1968,[4] and Below, Rangan, and Hall in 1975.[5] The concept was applied to reinforced concrete beams subjected to torsion and bending. Expressions for evaluating the torsional capacity of rectangular sections were presented by Hsu in 1968[6,7] and were adopted by the ACI Code of 1971. Torsion theories for concrete members were discussed by Zia.[8] Empirical design formulas were also presented by Victor et al. in 1976.[9]

   The basic approach of the skew bending theory, as presented by Hsu, is that failure of a rectangular section in torsion occurs by bending about an axis parallel to the wider face of the section y, and inclined at about 45° to the longitudinal axis of the beam (Figure 15.9). Based on this approach, the minimum torsional moment T_n can be evaluated as follows:

$$T_n = \frac{x^2 y}{3} \times f_r \qquad\qquad (15.12)$$

where f_r is the modulus of rupture of concrete; f_r is assumed to be $5\sqrt{f_c'}$ in this case, as compared to $7.5\sqrt{f_c'}$ adopted by the ACI Code for the computation of deflection in beams.

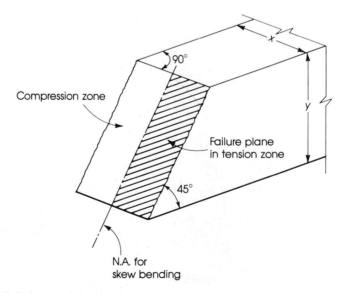

FIGURE 15.9. Failure surface due to skew bending.

The torque resisted by concrete is expressed as follows:

$$T_c = \left(\frac{2.4}{\sqrt{x}}\right) x^2 y \sqrt{f_c'} \qquad (15.13)$$

and the torque resisted by torsional reinforcement is

$$T_s = \alpha_t (x_1 y_1 A_t f_y)/s \qquad (15.14)$$

Thus, $T_n = T_c + T_s$. (These terms are explained in section 15.8.)

15.6.2   Space Truss Analogy

The space truss analogy was first presented by Rausch in 1929 and was further developed by Lampert,[10,11] who supported his theoretical approach with extensive experimental work. The Canadian Code provisions for the design of reinforced concrete beams in torsion and bending are based on the space truss analogy. Mitchell and Collins[12] presented a theoretical model for structural concrete in pure torsion. McMullen and Rangan[13] discussed the design concepts of rectangular sections subjected to pure torsion. In 1983, Solanki[14] presented a simplified design approach based on the theory presented by Mitchell and Collins.

The concept of the space truss analogy is based on the assumption that the torsional capacity of a reinforced concrete rectangular section is derived from the reinforcement and the concrete surrounding the steel only. In this case, a thin-walled section is assumed to act as a space truss (Figure 15.10). The inclined spiral concrete strips between cracks resist the compressive forces, while the longitudinal bars at the corners resist the tensile forces produced by the torsional moment.

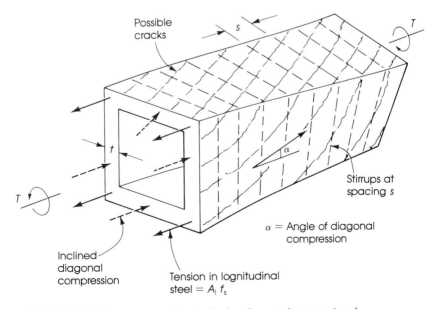

FIGURE 15.10. Forces on section in torsion (space truss analogy).

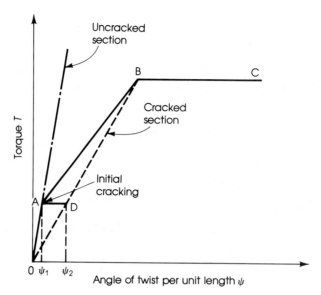

FIGURE 15.11. Idealized torque vs. twist relationship.

The behavior of a reinforced concrete beam subjected to pure torsion can be represented by an idealized graph relating the torque to the angle of twist, as shown in Figure 15.11. It can be seen that prior to cracking the concrete resists the torsional stresses and the steel is virtually unstressed.[12] After cracking, the elastic behavior of the beam is not applicable, and hence a sudden change in the angle of twist occurs, which continues to increase until the maximum torsional capacity is reached. An approximate evaluation of the torsional capacity of a cracked section may be expressed as follows:[10]

$$T_n = 2\left(\frac{A_t f_y}{s}\right) x_1 y_1 \qquad (15.15)$$

where

$$A_t = \text{area of one leg of stirrups}$$
$$s = \text{spacing of stirrups}$$
$$x_1 \text{ and } y_1 = \text{short and long distances, center to center of closed rectangular stirrups or corner bars}$$

The above expression neglects the torsional capacity due to concrete. Mitchell and Collins[12] presented the following expression to evaluate the angle of twist per unit length ψ:

$$\psi = \frac{P_o}{2A_o}\left[\frac{\varepsilon_l}{\tan\alpha} + \frac{P_h}{P_o}(\varepsilon_h \tan\alpha) + \frac{2\varepsilon_d}{\sin 2\alpha}\right] \qquad (15.16)$$

where

ε_l = strain in the longitudinal reinforcing steel

ε_h = strain in the hoop steel (stirrups)

ε_d = concrete diagonal strain at the position of the resultant shear flow

P_h = hoop centerline perimeter

α = angle of diagonal compression = $\dfrac{(\varepsilon_d + \varepsilon_l)}{[\varepsilon_d + \varepsilon_h(P_h/P_o)]}$

A_o = area enclosed by shear flow, or

= torque/2q (where q = shear flow)

P_o = perimeter of the shear flow path (perimeter of A_o)

The above twist expression is analogous to the curvature expression in flexure (Figure 15.12):

$$\phi = \text{curvature} = \frac{\varepsilon_c + \varepsilon_s}{d} \qquad (15.17)$$

where ε_c and ε_s = strains in concrete and steel, respectively. A simple equation is presented by Solanki[14] to determine the torsional capacity of a reinforced concrete beam in pure torsion, based on the space truss analogy, as follows:

$$T_u = 2A_o\left[\frac{\Sigma A_s f_{sy}}{P_o} \times \frac{A_h f_{hy}}{s}\right]^{1/2} \qquad (15.18)$$

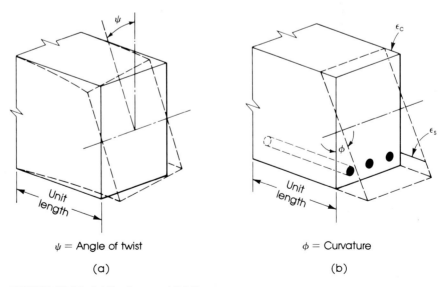

ψ = Angle of twist

(a)

ϕ = Curvature

(b)

FIGURE 15.12: (a) Torsion and (b) flexure.

where A_o, P_o, and s are as explained above,

$$\Sigma A_s f_{sy} = \text{yield force of all the longitudinal steel bars}$$
$$A_h f_{hy} = \text{yield force of the stirrups}$$

15.7   Torsional Strength of Plain Concrete Members

Concrete structural members subjected to torsion will normally be reinforced with special torsional reinforcement. In case that the torsional stresses are relatively low and need to be calculated for plain concrete members, the shear stress v_{tc} can be estimated using equation 15.10:

$$v_{tc} = \frac{3T}{\phi \Sigma x^2 y} \leq 6\sqrt{f'_c}$$

and the angle of twist $\theta = 3TL/x^3 yG$, where T is the torque applied on the section (less than the cracking torsional moment), and G is the shear modulus and can be assumed equal to 0.45 times the modulus of elasticity of concrete E_c; that is, $G = 25{,}700\sqrt{f'_c}$. The torsional cracking stress v_{tc} in plain concrete may be assumed equal to $6\sqrt{f'_c}$. Therefore, for plain concrete rectangular sections,

$$T_c = 2\phi x^2 y \sqrt{f'_c} \tag{15.19}$$

and for compound rectangular sections,

$$T_c = 2\phi \sqrt{f'_c} \, \Sigma x^2 y$$

15.8   Torsion in Reinforced Concrete Members (ACI Code Procedure)

The design procedure for torsion is similar to that for flexural shear. When the ultimate torsional moment applied on a section exceeds that which the concrete can resist, torsional cracks develop, and consequently torsional reinforcement in the form of closed stirrups or hoop reinforcement must be provided. In addition to the closed stirrups, longitudinal steel bars are provided in the corners of the stirrups and are well distributed around the section. Both types of reinforcement, closed stirrups and longitudinal bars, are essential to resist the diagonal tension forces caused by torsion; one type will not be effective without the other. The stirrups must be closed because torsional stresses occur on all faces of the section.

The reinforcement required for torsion must be added to that required for shear, bending moment, and axial forces.

The nominal ultimate torsional moment

$$T_n = T_c + T_s \quad \textbf{(ACI Code, section 11.6.5)} \tag{15.20}$$

where T_c = ultimate torsional moment resisted by concrete and T_s = ultimate torsional moment resisted by hoop reinforcement.

Introducing the factor ϕ ($\phi = 0.85$), then

$$T_u = \phi T_c + \phi T_s \tag{15.21}$$

The design procedure for pure torsion can be summarized as follows:

1. Check the torsional stress in the section using equation 15.10. If the stress v_{tu} is greater than $1.5\sqrt{f_c'}$ or the torsional moment

$$T_u > \phi(0.5\sqrt{f_c'})\Sigma x^2 y \quad \textbf{(ACI Code, section 11.6.1)} \qquad (15.22)$$

then the torsion effect must be included in the design of the member. If stress and torsional moment are both below these respective limits, then torsion may be neglected.

2. Calculate the torsional moment resisted by the concrete T_c. If the torsional moment to be contributed by the concrete in the compression zone is taken as 40 percent of the plain concrete torque capacity of equation 15.19, then

$$T_c = 0.8\sqrt{f_c'}\,(x^2 y) \qquad (15.23)$$

or

$$T_c = 0.8\sqrt{f_c'}(\Sigma x^2 y) \text{ for a compound section} \qquad (15.24)$$

This corresponds to a nominal torsional stress carried by the concrete of $2.4\sqrt{f_c'}$. This reduced value is used when a bending moment is simultaneously applied on the section.

3. Calculate the nominal torsional moment strength provided by torsion reinforcement T_s:

$$T_s = \alpha_t \left(\frac{x_1 y_1 A_t f_y}{s}\right) \qquad (15.14)$$

where

A_t = area of one leg of closed stirrup

x_1 and y_1 = shorter and longer center to center dimension of closed rectangular stirrup

s = spacing of stirrups

α_t = coefficient which is a function of y_1/x_1

$\alpha_t = [0.66 + 0.33(y_1/x_1)] < 1.5 \qquad (15.25)$

f_y = yield strength of stirrups ≤ 60 ksi

Since $T_s = T_n - T_c$ by equation 15.20, or

$$\phi T_s = T_u - \phi T_c \quad \text{and} \quad T_s = \frac{1}{\phi}(T_u - \phi T_c)$$

and

$$T_s = \alpha_t \frac{A_t}{s} x_1 y_1 f_y$$

then

$$\frac{A_t}{s} = \frac{T_s}{\alpha_t x_1 y_1 f_y} \qquad (15.26)$$

T_u and T_c are defined by equations 15.21 and 15.23. The stirrups must be of the closed type and their spacing shall not exceed 12 in. (300 mm) or $\frac{1}{4}(x_1 + y_1)$, whichever is smaller **(ACI Code, section 11.6.8)**.

4. Calculate the longitudinal reinforcement A_l: The required area of the longitudinal bars A_l, according to the **ACI Code, section 11.6.9**, is the larger value computed by the following two equations:

$$A_l = 2\left(\frac{A_t}{s}\right)(x_1 + y_1) \tag{15.27}$$

or

$$A_l = \left[\frac{400xs}{f_y}\left(\frac{T_u}{T_u + \dfrac{V_u}{3C_t}}\right) - 2A_t\right]\left(\frac{x_1 + y_1}{s}\right) \tag{15.28}$$

where V_u = factored shear force and C_t = factor relating shear and torsional stress properties = $(b_w d)/\Sigma x^2 y$. The value of A_l in equation 15.28 shall not exceed that obtained by substituting $50b_w s/f_y$ for $2A_t$ in the equation. In the case of pure torsion, $V_u = 0$ and only equation 15.27 need be used.

5. The limitations of the torsional reinforcement are as follows:

- Minimum diameter of the longitudinal bars is No. 3 (10 mm).
- The longitudinal bars must be distributed around the perimeter of the stirrups, and their spacings shall not exceed 12 in. (300 mm) **(ACI Code, section 11.6.8)**.
- At least one longitudinal bar shall be placed in each corner of the stirrups.
- Torsional reinforcement must be provided for a distance $(d + b)$ beyond the point theoretically required, where d is the effective depth and b is the width of the section **(ACI Code, section 11.6.7)**.
- The critical section is at a distance d from the face of the support, with the same reinforcement continuing to the face of the support.
- To avoid brittle failure of the concrete, the amount of reinforcement must be provided according to the following limitation **(ACI Code, section 11.6.9.4)**:

$$T_s \le 4T_c \tag{15.29}$$

- For members subjected to significant axial tension, the shear or torsional moment carried by the concrete is neglected **(ACI Code, section 11.3.1)**.

15.9   Combined Torsion and Shear for Members Without Web Reinforcement

When a concrete member is subjected simultaneously to shear and torsional forces, and web reinforcement is not provided, then the sum of square ratios of the following shear and torsional forces must not exceed unity:

$$\left(\frac{T_u}{T_o}\right)^2 + \left(\frac{V_u}{V_o}\right)^2 = 1 \tag{15.30}$$

or, in terms of stresses:

$$\left(\frac{v_{tu}}{v_{tc}}\right)^2 + \left(\frac{v_u}{v_c}\right)^2 = 1 \tag{15.31}$$

where

$$T_u = \text{ultimate torque acting with } V_u$$
$$V_u = \text{ultimate shear acting with } T_u$$
$$T_o = \text{ultimate strength under torsion alone}$$
$$V_o = \text{ultimate strength under shear alone}$$
$$v_{tu} \text{ and } v_u = \text{ultimate applied torsional and shear stresses}$$
$$v_{tc} \text{ and } v_c = \text{ultimate torsional and shear stresses carried by concrete}$$

Both equations represent a quarter circle interaction relationship between shear and torsion. After some transformation, equations 15.30 and 15.31 become:

$$T_u = \frac{T_o}{\sqrt{1 + \left(\frac{T_o}{V_o}\right)^2 \left(\frac{V_u}{T_u}\right)^2}} \qquad (15.32)$$

and

$$v_{tu} = \frac{v_{tc}}{\sqrt{1 + \left(\frac{v_{tc}}{v_c}\right)^2 \left(\frac{v_u}{v_{tu}}\right)^2}} \qquad (15.33)$$

In terms of shear and torque:

$$V_u = \frac{V_o}{\sqrt{1 + \left(\frac{V_o}{T_o}\right)^2 \left(\frac{T_u}{V_u}\right)^2}} \qquad (15.34)$$

and

$$v_u = \frac{v_c}{\sqrt{1 + \left(\frac{v_c}{v_{tc}}\right)^2 \left(\frac{v_{tu}}{v_u}\right)^2}} \qquad (15.35)$$

The equations are also applicable to reinforced concrete members with web reinforcement.

15.10 Combined Shear and Torsion (ACI Code Method)

A structural member may be subjected to bending moments and shear and torsional forces simultaneously. The effect of the three types of forces must be considered in the design of the member. The beam, for instance, can be designed for the applied bending moments and the steel reinforcement required is calculated as was explained earlier in the design of beams due to flexure. The next step is to calculate the web and longitudinal reinforcement required to resist shear and torsion. The total reinforcement is obtained by adding the steel required for torsion and shear to that required for flexure.

The **ACI Code, Chapter 11,** presents an approximate method to calculate the strength of reinforced concrete members, subjected to this combination. The code expressions provide the necessary web reinforcement due to shear and the web and the longitudinal reinforcement required for torsion. The separate web reinforcement for shear and torsion are combined for choosing the size and spacing of stirrups, and the longitudinal reinforcement for torsion and bending moment are combined for choosing the size and number of longitudinal bars.

The torsional moment strength provided by concrete when web reinforcement is provided is given by the **ACI Code equation 11.22**, as follows:

$$T_c = \frac{0.8\sqrt{f_c'}\,\Sigma x^2 y}{\sqrt{1 + \left(\frac{0.4V_u}{C_t T_u}\right)^2}} \qquad (15.36)$$

where V_u and T_u = factored shear and torsional moments (at section) and C_t = factor relating shear and torsional stress = $b_w d/\Sigma x^2 y$.

Equation 15.36 for T_c is obtained from equation 15.32 by replacing T_u with T_c, taking $T_o = 0.80\sqrt{f_c'}\Sigma x^2 y$, $V_o = 2\sqrt{f_c'}b_w d$ and

$$\frac{T_o}{V_o} = \frac{0.8\sqrt{f_c'}\Sigma x^2 y}{2\sqrt{f_c'}b_w d} = \frac{0.4\Sigma x^2 y}{b_w d} = \frac{0.4}{C_t}$$

The nominal shear strength resisted by concrete V_c, when web reinforcement is provided, is given by **ACI Code equation 11.5**, as follows:

$$V_c = \frac{2\sqrt{f_c'}b_w d}{\sqrt{1 + \left(2.5C_t\frac{T_u}{V_u}\right)^2}} \qquad (15.37)$$

This equation is obtained from equation 15.34 by replacing V_u with V_c, taking $V_o = 2\sqrt{f_c'}b_w d$ (the basic shear strength of concrete), $T_o = 0.8\sqrt{f_c'}\Sigma x^2 y$, and

$$\frac{V_o}{T_o} = \frac{2\sqrt{f_c'}b_w d}{0.8\sqrt{f_c'}\Sigma x^2 y} = \frac{2.5b_w d}{\Sigma x^2 y} = 2.5C_t$$

The design procedure for combined shear and torsion can be summarized as follows:

1. Calculate the ultimate shearing force V_u and the ultimate torsional moment T_u from the applied forces on the structural member. Critical values for shear are at a section distance d from the face of the support.
2. (a) Shear reinforcement is needed when $V_u > \phi V_c/2$, where $V_c = 2\sqrt{f_c'}b_w d$.
   (b) Torsional reinforcement is needed when $T_u > \phi T_c$, where $T_c = 0.5\sqrt{f_c'}\Sigma x^2 y$.

   If web reinforcement is needed, proceed as follows:
3. Design for shear:
   - Calculate the nominal shearing strength provided by the concrete V_c according to equation 15.37. Determine the shear to be carried by web reinforcement,

$$V_u = \phi V_c + \phi V_s \quad \text{or} \quad V_s = \frac{1}{\phi}(V_u - \phi V_c)$$

- Compare the calculated V_s with maximum permissible value of $(8\sqrt{f'_c}b_w d)$ according to the ACI Code. If calculated V_s is less, proceed in the design; if not, increase the dimensions of the concrete section.
- The shear web reinforcement is calculated as follows:

$$A_v = \frac{V_s s}{f_y d}$$

where A_v = area of two legs of the stirrup and s = spacing of stirrups. The shear reinforcement per unit length of beam is

$$\frac{A_v}{s} = \frac{V_s}{f_y d}$$

- Check A_v/s calculated with the minimum A_v/s:

$$\frac{A_v}{s}\ (\text{minimum}) = \frac{50\,b_w}{f_y}$$

The minimum $A_v = 50 b_w s/f_y$, specified by the code under the combined action of shear and torsion, is given below under step 5.

4. Design for torsion:
   - Calculate the torsional moment strength T_c provided by the concrete according to equation 15.36. The nominal torsional moment to be resisted by torsional reinforcement T_s is calculated as follows:

$$T_u = \phi T_c + \phi T_s \quad \text{or} \quad T_s = \frac{1}{\phi}(T_u - \phi T_c)$$

   - Determine the torsional reinforcement from equation 15.14:

$$T_s = \alpha_t \frac{A_t f_y x_1 y_1}{s}$$

   or

$$\frac{A_t}{s} = \frac{T_s}{\alpha_t f_y x_1 y_1} = \text{torsional reinforcement per unit length}$$

   where A_t = area of one leg of the closed stirrup and $\alpha_t = [0.66 + 0.33(y_1/x_1)] \le 1.5$.
   - Calculate the area of the longitudinal bars required according to equations 15.27 and 15.28.

5. Design closed stirrups:
   - Calculate the total area of stirrups required for both shear and torsion.

$$\text{Total area of one leg of stirrup per unit length} = \frac{A_t}{s} + \frac{A_v}{2s}$$

(A_v is the area of two legs.) Then choose stirrups and spacings.

- Check that the area obtained is equal to or greater than that computed from the following equation:

$$A_v + 2A_t \geq \frac{50 b_w s}{f_y}$$

which is the minimum web reinforcement required by the ACI Code.
- Check that the chosen spacing $s \leq (x_1 + y_1)/4 \leq 12$ in. (300 mm).

6. Longitudinal bars should be distributed around the perimeter with a spacing of no more than 12 in. (300 mm). Normally one-third is added to the flexural reinforcement, one-third at mid-height, and one-third at the compression side of the section (when spacing between top and bottom bars exceeds 12 in.).

7. Limitations are the same as given earlier in section 15.7. The following examples will explain the design procedures for torsion and for the combined torsion and shear.

15.11   Examples

EXAMPLE   15.3

Design the necessary reinforcement for an ultimate torsional moment $T_u = 220$ k·in. acting at the rectangular section shown in Figure 15.13. Use $f'_c = 3$ ksi and $f_y = 40$ ksi.

Solution

1. Check if $T_u \geq \phi(0.5\sqrt{f'_c})\Sigma x^2 y$:

    $x = 12$ in. $y = 20$ in.

    Right-hand side $= 0.85 \times 0.5 \sqrt{3000} \times (12)^2(20) = 67$ k·in.

    Since the applied $T_u > 67$ k·in., therefore torsional reinforcement must be provided.

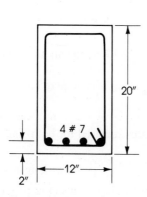
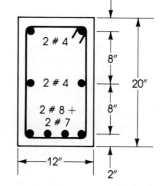

FIGURE 15.13. Example 15.3.

2. Calculate the torsional moment provided by the concrete:

$$T_c = 0.8\sqrt{f_c'}\,(x^2 y) = 0.8\sqrt{3000} \times (12)^2(20) = 126.2 \text{ k·in.}$$

$$T_s = \frac{1}{\phi}(T_u - \phi T_c) = \frac{1}{0.85}(220 - 0.85 \times 126.2) = 132.6 \text{ k·in.}$$

3. Check that $T_s \le 4T_c$:

$$4T_c = 4 \times 126.2 = 504.8 \text{ k·in.} > T_s$$

4. Determine the area of the closed stirrups required per inch spacing:

$$\frac{A_t}{s} = \frac{T_s}{\alpha_t f_y x_1 y_1} \qquad\qquad (15.26)$$

Assume No. 4 stirrups to calculate x_1 and y_1:

$$x_1 = 12 - 2 \times 1.5 - 0.5 = 8.5 \text{ in.}$$
$$y_1 = 20 - 2 \times 1.5 - 0.5 = 16.5 \text{ in.}$$
$$\alpha_t = \left(0.66 + 0.33\,\frac{y_1}{x_1}\right) = \left(0.66 + 0.33 \times \frac{16.5}{8.5}\right) = 1.30 < 1.5$$
$$\frac{A_t}{s} = \frac{132.6}{1.30 \times 40 \times 8.5 \times 16.5} = 0.0182 \text{ in.}^2/\text{in.}$$

A_t is the area of one leg of stirrup. Choose No. 3 stirrups (area of one leg = 0.11 in.^{2}).

$$\text{Spacing } s = \frac{0.11}{0.0182} = 6.04 \text{ in., say, } 6.0 \text{ in.}$$

$$\text{Minimum } \frac{A_t}{s} = \frac{50b_w}{f_y} = \frac{50 \times 12}{40,000} = 0.015 \text{ in.}^2/\text{in.} < 0.0182$$

5. Check maximum spacings:

$$s_{max} = \frac{x_1 + y_1}{4} = \frac{16.5 + 8.5}{4} = 6.25 \text{ in.} \quad \text{or} \quad s_{max} = 12 \text{ in.}$$

Therefore, use No. 3 stirrups spaced at 6.0 in.

6. Longitudinal bars A_l:

$$A_l = 2\,\frac{A_t}{s}\,(x_1 + y_1) = 2 \times 0.0182(8.5 + 16.5) = 0.91 \text{ in.}^2$$

Place one-third at top, steel area = 0.91/3 = 0.31 in.^{2}, one-third at mid-depth = 0.31 in.^{2}, and one-third at the bottom = 0.31 in.^{2}.

Total steel at tension side = $4 \times 0.6 + 0.31 = 2.71$ in.^{2}

Use 2 No. 8 and 2 No. 7 bars, $A_s = 2.77$ in.^{2}. Use 2 No. 4 bars at top and 2 No. 4 bars at mid-depth.

$$\text{Spacing between bars} = \frac{y_1}{2} = \frac{16.5}{2} = 8.25 \text{ in.}$$

which is less than 12 in. maximum spacing. The No. 4 bars used are greater than the minimum No. 3 bars required. The final section is shown in Figure 15.13. ∎

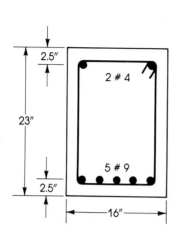
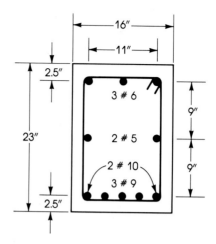

FIGURE 15.14. Example 15.4.

EXAMPLE 15.4

Determine the necessary reinforcement for combined torsion and shear on a reinforced concrete beam with the section shown in Figure 15.14. The beam is subjected to a maximum ultimate shear $V_u = 36$ k and an ultimate torsional moment $T_u = 350$ k·in. at a section located at a distance d from the face of the support. Use $f'_c = 3$ ksi and $f_y = 50$ ksi.

Solution   1. Check if torsional reinforcement is needed, that is, if

$$T_u > \phi(0.5\sqrt{f'_c})x^2y$$

Right-hand side $= 0.85 \times 0.5\sqrt{3000} \times (16)^2(23) = 137$ k·in.

T_u applied $= 350$ k·in. > 137 k·in.

therefore torsional reinforcement must be provided.

2. Calculate $\phi V_c = 2\phi\sqrt{f'_c}b_w d$ and compare with the applied shear.

$$\phi V_c = 2 \times 0.85\sqrt{3000} \times 16 \times 20.5 = 30.5 \text{ k}$$

$(\phi V_c/2)$ is less than the applied shear; therefore shear reinforcement should be provided (see Chapter 8, section 8.6).

3. Calculate shear reinforcement:

$$\bullet \; V_c = \frac{2\sqrt{f'_c}b_w d}{\sqrt{1 + \left(2.5C_t \dfrac{T_u}{V_u}\right)^2}} \tag{15.37}$$

$$C_t = \frac{b_w d}{x^2 y} = \frac{16 \times 20.5}{(16)^2(23)} = 0.056$$

$$V_c = \frac{2\sqrt{3000} \times 16 \times 20.5}{\sqrt{1 + \left(2.5 \times 0.056 \times \frac{350}{36}\right)^2}} = 21.3 \text{ k}$$

$$V_u = \phi V_c + \phi V_s \qquad V_s = \frac{1}{\phi}(V_u - \phi V_c)$$

$$V_s = \frac{1}{0.85}(36 - 0.85 \times 21.3) = 21 \text{ k}$$

- Maximum $V_s = 8\sqrt{f_c'}b_w d = 8\sqrt{3000} \times 16 \times 20.5 = 143.7 \text{ k}$

$$V_s < 143.7 \text{ k}$$

- $\dfrac{A_v}{s} = \dfrac{V_s}{f_y d} = \dfrac{21}{50 \times 20.5} = 0.02 \text{ in.}^2/\text{in.}$

    Minimum $\dfrac{A_v}{s}$ required by ACI Code $= \dfrac{50\,b_w}{f_y} = \dfrac{50 \times 16}{50,000} = 0.016 \text{ in.}^2/\text{in.}$

    $\dfrac{A_v}{s}$ for one leg $= \dfrac{0.02}{2} = 0.01 \text{ in.}^2/\text{in.}$

4. Calculate torsional reinforcement:

- $T_c = \dfrac{0.8\sqrt{f_c'}x^2 y}{\sqrt{1 + \left(\dfrac{0.4}{C_t} \times \dfrac{V_u}{T_u}\right)^2}}$ \hfill (15.36)

$$= \frac{0.8\sqrt{3000}(16)^2(23)}{\sqrt{1 + \left(\dfrac{0.4}{0.056} \times \dfrac{36}{350}\right)^2}} = 207.9 \text{ k·in.}$$

$$T_u = \phi T_c + \phi T_s \quad \text{and} \quad T_s = \frac{1}{0.85}(350 - 0.85 \times 207.9) = 203.8 \text{ k·in.}$$

- $\dfrac{A_t}{s} = \dfrac{T_s}{\alpha_t f_y x_1 y_1}$ \hfill (15.24)

Assume No. 4 stirrups:

$$x_1 = 16 - 3.5 = 12.5 \text{ in.} \qquad y_1 = 23 - 3.5 = 19.5 \text{ in.}$$

$$\alpha_t = \left(0.66 + 0.33\frac{y_1}{x_1}\right) \le 1.5$$

$$\alpha_t = \left(0.66 + 0.33 \times \frac{19.5}{12.5}\right) = 1.2 < 1.5$$

$$\frac{A_t}{s} = \frac{203.8}{1.2 \times 50 \times 12.5 \times 19.5} = 0.014 \text{ in.}^2/\text{in. (area of one leg)}$$

5. Design closed stirrups.

- Total area of stirrups required for one leg

$$\frac{A_t}{s} + \frac{A_v}{2s} = 0.014 + \frac{0.02}{2} = 0.024 \text{ in.}^2/\text{in.}$$

Choose No. 3 stirrups, area = 0.11 in.²

$$\text{Spacing } s = \frac{0.11}{0.024} = 4.6 \text{ in.}$$

If No. 4 stirrups are used, area = 0.2 in.²

$$s = \frac{0.2}{0.024} = 8.3 \text{ in.}$$

Choose No. 4 stirrups spaced at 8 in.

- Maximum spacing $= \dfrac{x_1 + y_1}{4} = \dfrac{12.5 + 19.5}{4} = 8$ in.

or $s_{max} = 12$ in. Therefore, the choice of No. 4 stirrups spaced at 8 in. is adequate.
- Check minimum area of stirrups:

$$\text{Minimum } A = \frac{50 b_w s}{f_y} \leq A_v + 2A_t$$

$$\text{Minimum } \frac{A}{s} = \frac{50 \times 16}{50,000} = 0.016 \text{ in.}^2/\text{in.}$$

$$\text{Area of stirrups provided} = \frac{A_v}{s} + \frac{2A_t}{s}$$

$$= 0.02 + 2 \times 0.014 = 0.048 \text{ in.}^2/\text{in.}$$

which is greater than the required minimum of 0.016 in.²/in.

6. Longitudinal bars:

- $A_l = 2 \dfrac{A_t}{s} (x_1 + y_1)$ \hfill (15.27)

$$= 2 \times 0.014(12.5 + 19.5) = 0.90 \text{ in.}^2$$

- Check minimum A_l according to equation 15.28:

$$\text{Minimum } A_l = \left[\left(\frac{400xs}{f_y}\right) \left(\frac{T_u}{T_u + \dfrac{V_u}{3C_t}}\right) - 2A_t \right] \left(\frac{x_1 + y_1}{s}\right)$$

or

$$\text{Minimum } A_l = \left[\left(\frac{400x}{f_y}\right) \left(\frac{T_u}{T_u + \dfrac{V_u}{3C_t}}\right) - 2\frac{A_t}{s} \right] (x_1 + y_1)$$

$$= \left[\left(\frac{400 \times 16}{50,000}\right) \left(\frac{350}{350 + \dfrac{36}{3 \times 0.056}}\right) - 2\frac{A_t}{s} \right] (12.5 + 19.5)$$

$$= 2.54 - 64\frac{A_t}{s}$$

For $A_t/s = 0.014$ in.²/in., $A_l = 1.64$ in.², but not more than that obtained by substituting $50b_w s/f_y$ for $2A_t$ ($A_t/s = 50b_w/2f_y = 0.008$). Since A_t/s used $= 0.014 >$ 0.016/2, then $A_l = 1.64$ in.² controls.

- Distribute $A_l = 1.64$ in.² as follows:
  One-third at top $= 1.64/3 = 0.55$ in.². Then total top steel $A_s' = 0.55 + 0.39$ (from 2 No. 4 bars) $= 0.94$ in.². Use 3 No. 6 bars ($A_s = 1.20$ in.²).
  One-third at mid-depth $= 0.55$ in.²; use 2 No. 6 bars ($A_s = 0.80$ in.²).
  One-third at the bottom of the section $= 0.55 + 5.0$ (from 5 No. 9 bars) $= 5.55$ in.²; use 3 No. 9 and 2 No. 10 bars ($A_s = 5.53$ in.²).
  Distances between bars are less than 12 in., as shown in Figure 15.14.   ■

EXAMPLE 15.5

A reinforced concrete spandrel beam subjected to bending, shear, and torsion has a section as shown in Figure 15.15 located at a distance d from the face of the support. Determine the necessary reinforcement if the section is subjected to an ultimate negative moment $M_u = 120$ k·ft, an ultimate shearing force $V_u = 40$ k, and an ultimate torsional moment $T_u = 20$ k·ft. Use $f_c' = 4$ ksi and $f_y = 60$ ksi.

Solution

1. Determine the necessary bending reinforcement, given $b = 13$ in. and $d = 20 - 2.5 = 17.5$ in.:

$$R_u = \frac{M_u}{bd^2} = \frac{120 \times 12,000}{13(17.5)^2} = 361 \text{ psi}$$

From Table A.2 in Appendix A, the steel percentage $\rho = 0.71$ percent:

$$A_s = 0.0071 \times 13 \times 17.5 = 1.62 \text{ in.}^2$$

Use 3 No. 7 top bars ($A_s = 1.8$ in.²). Additional steel area may be needed due to torsion.

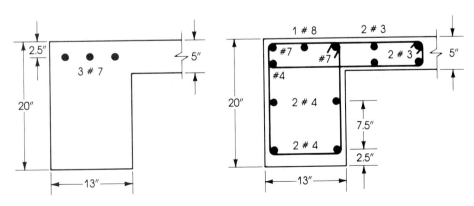

FIGURE 15.15. Example 15.5.

2. Torsional reinforcement is needed if

$$T_u > \phi(0.5\sqrt{f_c'})\Sigma x^2 y$$

Divide the given L-shape section to a main rectangle 13 by 20 in. and a flange rectangle = $5 \times (3 \times 5)$ in. The width of the flange rectangle is 3 times the slab thickness.

$$\Sigma x^2 y = (13)^2(20) + (5)^2(15) = 3755 \text{ in.}^3$$

$$\phi(0.5\sqrt{f_c'})\Sigma x^2 y = 0.85 \times 0.5\sqrt{4000} \times 3755 = 100.9 \text{ k·in.}$$

$$T_u \text{ applied} = 20 \times 12 = 240 \text{ k·in.} > 100.9 \text{ k·in.}$$

Therefore torsional reinforcement must be provided.

3. Check shear strength:

- $V_c = \dfrac{2\sqrt{f_c'}\, b_w d}{\sqrt{1 + \left(2.5 C_t \dfrac{T_u}{V_u}\right)^2}}$

$$C_t = \frac{b_w d}{\Sigma x^2 y} = \frac{13 \times 17.5}{3755} = 0.06$$

$$V_c = \frac{2\sqrt{4000} \times 13 \times 17.5}{\sqrt{1 + \left(2.5 \times 0.06 \times \dfrac{240}{40}\right)^2}} = 21.4 \text{ k} \qquad \phi V_c = 18.2 \text{ k}$$

Since $V_u > \phi V_c/2$ (9.1 k), calculate A_v according to

$$\frac{A_v}{s} = \frac{V_s}{f_y d}$$

$$V_s = \frac{1}{\phi}(V_u - \phi V_c) = \frac{1}{0.85}(40 - 0.85 \times 21.4) = 25.7 \text{ k}$$

- Maximum $V_s = 8\sqrt{f_c'}\, b_w d = 8\sqrt{4000} \times 13 \times 17.5 = 115.1 \text{ k}$

$$V_s = 25.7 \text{ k} < V_s \text{ (maximum)}$$

- $\dfrac{A_v}{s} = \dfrac{V_s}{f_y d} = \dfrac{25.7}{60 \times 17.5} = 0.024 \text{ in.}^2/\text{in.}$

    Minimum $\dfrac{A_v}{s} = \dfrac{50 b_w}{f_y} = \dfrac{50 \times 13}{60,000} = 0.011 \text{ in.}^2/\text{in.}$

which is less than 0.024. If minimum A_v/s is greater, then check that $A_v + 2A_t \leq 50 b_w s/f_y$, as will be done later.

$$\frac{A_v}{s} \text{ for one leg} = \frac{0.024}{2} = 0.012 \text{ in.}^2/\text{in.}$$

4. Calculate torsional reinforcement:

- $T_c = \dfrac{0.8\sqrt{f_c'}(\Sigma x^2 y)}{\sqrt{1 + \left(\dfrac{0.4 V_u}{C_t T_u}\right)^2}}$ \hfill (15.36)

$$= \frac{0.8\sqrt{4000} \times 3755}{\sqrt{1 + \left(\dfrac{0.4 \times 40}{0.06 \times 240}\right)^2}} = 127.1 \text{ k·in.}$$

$$T_u = \phi T_c + \phi T_s \quad \text{and} \quad T_s = \frac{1}{0.85}(240 - 0.85 \times 127.1) = 155 \text{ k·in.}$$

$$\bullet \quad \frac{A_t}{s} = \frac{T_s}{\alpha_t f_y x_1 y_1}$$

Assume No. 4 stirrups to calculate x_1 and y_1:

$$x_1 = 13 - 2 \times 1.5 - 0.5 = 9.5 \text{ in.}$$
$$y_1 = 20 - 2 \times 1.5 - 0.5 = 16.5 \text{ in.}$$
$$\alpha_t = \left(0.66 + 0.33\frac{y_1}{x_1}\right) \le 1.5$$
$$= \left(0.66 + 0.33 \times \frac{16.5}{9.5}\right) = 1.23 < 1.5 \text{ in.}$$
$$\frac{A_t}{s} = \frac{155}{1.23 \times 60 \times 9.5 \times 16.5} = 0.013 \text{ in.}^2/\text{in.}$$

5. Design closed stirrups:

   - Total area of one leg of stirrups = $A_t/s + A_v/2s = 0.013 + 0.012 = 0.025$ in.^{2}/in. of beam length. Choose No. 4 stirrups (area = 0.2 in.^{2}).

   $$\text{Spacing} = \frac{0.2}{0.025} = 8 \text{ in.}$$

   - Minimum $\left(\dfrac{A_v}{s} + \dfrac{2A_t}{s}\right) = \dfrac{50b_w}{f_y} = \dfrac{50 \times 13}{60,000} = 0.011$ in.^{2}/in.

   $\left(\dfrac{A_v}{s} + \dfrac{2A_t}{s}\right)$ provided = $0.024 + 2 \times 0.013 = 0.05$ in.^{2}/in.

   which is greater than the required minimum of 0.011 in.^{2}/in.

   - Maximum spacing $= \dfrac{x_1 + y_1}{4} = \dfrac{9.5 + 16.5}{4} = 6.5$ in.

   or $s_{max} = 12$ in. Since 6.5 in. is the maximum permissible spacing, then use No. 4 stirrups spaced at 6.5 in.
   - Use similar stirrups in slab and beam as shown in Figure 15.15.

6. Design longitudinal bars:

   $$\bullet \quad A_l = \frac{2A_t}{s}(x_1 + y_1) = 2 \times 0.013(9.5 + 16.5) = 0.68 \text{ in.}^2 \tag{15.27}$$

- Check A_l (minimum) according to equation 15.28:

$$\text{Minimum } A_l = \left[\left(\frac{400x}{f_y}\right)\left(\frac{T_u}{T_u + \frac{V_u}{3C_t}}\right) - 2\frac{A_t}{s}\right](x_1 + y_1)$$

$$= \left[\left(\frac{400 \times 13}{60,000}\right)\left(\frac{240}{240 + \frac{40}{3 \times 0.06}}\right) - 2\frac{A_t}{s}\right](9.5 + 16.5)$$

$$= 1.17 - 52.0\frac{A_t}{s}$$

For $A_t/s = 0.013$, $A_l = 0.5$ in.^{2}, but not more than that obtained by substituting $50b_w/f_y = 0.011$ for $2A_t/s$. For $2A_t/s = 0.011$, $A_l = 0.89$ in.$^2 > 0.5$ in.^{2}; thus it is not critical. Use $A_l = 0.68$ in.^{2}.

- Distribute one-third at top = $0.68/3 = 0.23$ in.^{2}. Then total top steel = $0.23 + 1.62$ (from flexure) = 1.85 in.^{2}. Use 2 No. 7 and 1 No. 8 bars ($A_s = 1.99$ in.^{2}). Use one-third at mid-depth = 0.23 in.^{2} (2 No. 4 bars, $A_s = 0.39$ in.^{2}). Use one-third at bottom = 0.23 in.^{2} (2 No. 4 bars). Details of stirrups and bars are shown in Figure 15.15.

15.12   S.I. Equations

The S.I. conversion of the equations used in this chapter are given in Table 15.2. Note that $1\sqrt{f_c'}$ in psi is equivalent to $0.08\sqrt{f_c'}$ in MPa (N/mm^2).

TABLE 15.2. Summary of Equations

ACI section (formula)	Textbook equation	U.S. customary units	S.I. units
—	15.10	$v_{tu} = 3T_u/\phi\Sigma x^2 y$	Same
—	—	$T_u = \phi\frac{v_{tu}}{3}\Sigma x^2 y$	Same
—	—	$v_{tu} = 2T_u/\phi\pi R^3$ (Circular section)	Same
—	15.19	$T_c = 2\phi\Sigma x^2 y\sqrt{f_c'}$ (Plain concrete)	$T_c = 0.17\phi\Sigma x^2 y\sqrt{f_c'}$
11.6.5 (11-21)	15.20	$T_n = T_c + T_s$	Same
11.6.5 (11-20)	—	$T_u \le \phi T_n$	Same
11.6.1	15.22	$T_u > \phi(0.5\sqrt{f_c'})\Sigma x^2 y$	$T_u > \phi(0.04\sqrt{f_c'})\Sigma x^2 y$
11.6.6	15.23	$T_c = 0.8x^2 y\sqrt{f_c'}$	$T_c = 0.07x^2 y\sqrt{f_c'}$
—	—	$v_{tc} = 2.4\sqrt{f_c'}$	$v_{tc} = 0.2\sqrt{f_c'}$
11.6.9 (11-23)	15.14	$T_s = \alpha_t x_1 y_1 A_t f_y/s$	Same

TABLE 15.2 Summary of Equations (*continued*)

ACI section (formula)	Textbook equation	U.S. customary units	S.I. units
11.6.9	15.25	$\alpha_t = \left(0.66 + 0.33\,\dfrac{y_1}{x_1}\right) \leq 1.5$	Same
11.6.9	15.26	$A_t/s = T_s/\alpha_t x_1 y_1 f_y$	Same
11.6.8	—	Min. $s \leq (x_1 + y_1)/4 \leq 12$ in.	Same ≤ 300 mm
11.6.9 (11-24)	15.27	$A_1 = \dfrac{2A_t}{s}(x_1 + y_1)$	Same
11.6.9 (11-25)	15.28	$A_1 = \left[\left(\dfrac{400xs}{f_y}\right)\left(\dfrac{T_u}{T_u + \dfrac{V_u}{3C_t}}\right) - 2A_t\right]\dfrac{(x_1 + y_1)}{s}$ *	$A_1 = \left[\left(\dfrac{2.76xs}{f_y}\right)\left(\dfrac{T_u}{T_u + \dfrac{V_u}{3C_t}}\right) - 2A_t\right]\dfrac{(x_1 + y_1)}{s}$ *
11.0	—	$C_t = \dfrac{b_w d}{\Sigma x^2 y}$	Same
11.6.9	—	Max. $T_s \leq 4T_c$	Same
11.6.6 (11-22)	15.36	$T_c = \dfrac{0.8\sqrt{f_c'}\,\Sigma x^2 y}{\sqrt{1 + \left(\dfrac{0.4V_u}{C_t T_u}\right)^2}}$	$T_c = \dfrac{0.07\sqrt{f_c'}\,\Sigma x^2 y}{\sqrt{1 + \left(\dfrac{0.4V_u}{C_t T_u}\right)^2}}$
11.3.1 (11-5)	15.37	$V_c = \dfrac{2\sqrt{f_c'}\,b_w d}{\sqrt{1 + \left(2.5C_t\dfrac{T_u}{V_u}\right)^2}}$	$V_c = \dfrac{0.17\sqrt{f_c'}\,b_w d}{\sqrt{1 + \left(2.5C_t\dfrac{T_u}{V_u}\right)^2}}$
11.5.5 (11-14)	—	Min. $\dfrac{A_v}{s} = \dfrac{50 b_w}{f_y}$	Min. $\dfrac{A_v}{s} = 0.34\dfrac{b_w}{f_y}$
11.5.5	—	For min. A_t: $\dfrac{A_v}{s} + \dfrac{2A_t}{s} \geq \dfrac{50 b_w}{f_y}$	$\dfrac{A_v}{s} + \dfrac{2A_t}{s} \geq 0.34\dfrac{b_w}{f_y}$
11.6.7	—	Extend torsion reinforcement to $(d + b)$ beyond theoretical point.	

* But not to exceed A_1 obtained by substituting $50b_w s/f_y$ ($0.35b_w s/f_y$) for $2A_t$.

REFERENCES

1. S. Timoshenko and J. N. Goodier: *Theory of Elasticity*, McGraw-Hill, New York, 1951.
2. N. N. Lessig: "The Determination of the Load-bearing Capacity of Reinforced Concrete Elements with Rectangular Sections Subjected to Flexure and Torsion," Concrete and Reinforced Concrete Institute, Moscow, Work No. 5, 1959. (Translation: Foreign Literature Study No. 371, PCA Research and Development Laboratories, Skokie, IL.)
3. C. D. Goode and M. A. Helmy: "Ultimate Strength of Reinforced Concrete Beams Subjected to Combined Torsion and Bending," in *Torsion of Structural Concrete*, Special Publication 18, American Concrete Institute, Detroit, 1968.
4. M. P. Collins, P. F. Walsh, F. E. Archer and A. S. Hall: "Ultimate Strength of Reinforced Concrete Beams Subjected to Combined Torsion and Bending," in *Torsion of Structural Concrete*, Special Publication 18, American Concrete Institute, Detroit, 1968.

5. K. D. Below, B. V. Rangan and A. S. Hall: "Theory for Concrete Beams in Torsion and Bending," *Journal of Structural Division, ASCE,* August 1975.

6. T. C. Hsu: "Ultimate Torque of Reinforced Concrete Beams," *Journal of Structural Division, ASCE,* No. 94, February 1968.

7. T. C. Hsu: "Torsion of Structural Concrete—A Summary of Pure Torsion," in *Torsion of Structural Concrete,* Special Publication 18, American Concrete Institute, Detroit, 1968.

8. P. Zia: "Torsion Theories for Concrete Members," in *Torsion of Structural Concrete,* Special Publication 18, American Concrete Institute, Detroit, 1968.

9. D. J. Victor, N. Lakshmanan and N. Rajagopalan: "Ultimate Torque of Reinforced Concrete Beams," *Journal of Structural Division, ASCE,* 102, July 1976.

10. P. Lampert: *Postcracking Stiffness of Reinforced Concrete Beams in Torsion and Bending,"* Special Publication 35, American Concrete Institute, Detroit, 1973.

11. P. Lampert and B. Thürlimann: *Ultimate Strength and Design of Reinforced Concrete Beams in Torsion and Bending,* International Association for Bridge and Structural Engineering, Publication No. 31-I, 1971.

12. D. Mitchell and M. P. Collins: "The Behavior of Structural Concrete Beams in Pure Torsion," Department of Civil Engineering, University of Toronto, Publication No. 74-06, 1974.

13. A. E. McMullen and V. S. Rangan: "Pure Torsion in Rectanguar Sections, a Re-examination," *ACI Journal,* 75, October 1978.

14. H. T. Solanki: "Behavior of Reinforced Concrete Beams in Torsion," *Proceedings of the Institution of Civil Engineers,* 75, March 1983.

15. Alan H. Mattock: "How to Design for Torsion," in *Torsion of Structural Concrete,* Special Publication 18, American Concrete Institute, Detroit, 1968.

16. T. C. Hsu and E. L. Kemp: "Background and Practical Application of Tentative Design Criteria for Torsion," *ACI Journal,* 66, January 1969.

17. D. Mitchell and M. P. Collins: "Detailing for Torsion," *ACI Journal,* 73, September 1976.

PROBLEMS

15.1. Design the necessary web reinforcement for a simply supported beam subjected to an ultimate torque T_u = 28 k·ft (38 kN·m). The section of the beam is rectangular with b =

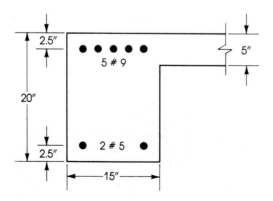

FIGURE 15.16. Problem 15.3.

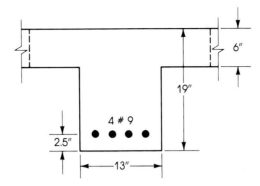

FIGURE 15.17. Problem 15.4.

12 in. (300 mm), depth h = 21 in. (525 mm), d = 18.5 in. (460 mm) and reinforced on one side with 4 No. 7 bars (4 × 22 mm). Use f'_c = 3 ksi (21 MPa) and f_y = 40 ksi (280 MPa).

15.2. Repeat problem 15.1 using f'_c = 4 ksi (28 MPa) and f_y = 60 ksi (420 MPa).

15.3. Determine the necessary web reinforcement for the member which has the section shown in Figure 15.16. The section is subjected to an ultimate torsional moment T_u = 38 k·ft (51.5 kN·m). Use f'_c = 3 ksi (21 MPa) and f_y = 40 ksi (280 MPa).

15.4. Repeat problem 15.3 for the section shown in Figure 15.17.

15.5. The section shown in Figure 15.18 is subjected to an ultimate shearing force V_u = 32 k (142 kN) and an ultimate torsional moment T_u = 18 k·ft (24.4 kN·m). Design the necessary web reinforcement. Use f'_c = 3 ksi (21 MPa) and f_y = 40 ksi (280 MPa).

15.6. The section of a spandrel beam is shown in Figure 15.19. The beam is subjected to an ultimate shearing force V_u = 60 k (267 kN) and an ultimate torsional moment T_u = 24 k·ft (33 kN·m). Determine the necessary web reinforcement. Use f'_c = 4 ksi (28 MPa) and f_y = 60 ksi (420 MPa).

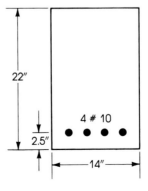

FIGURE 15.18. Problem 15.5.

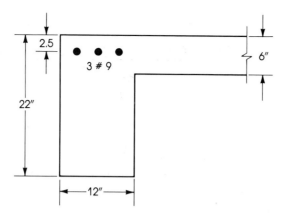

FIGURE 15.19. Problem 15.6.

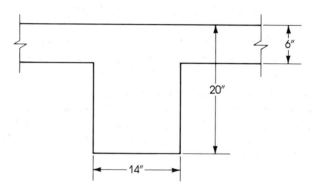

FIGURE 15.20. Problem 15.7.

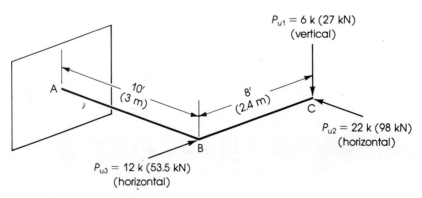

FIGURE 15.21. Problem 15.8.

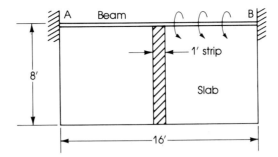

FIGURE 15.22. Problem 15.9.

15.7. The T-section shown in Figure 15.20 is subjected to an ultimate shearing force $V_u = 28$ k (125 kN), an ultimate bending moment $M_u = 124$ k·ft (168 kN·m), and an ultimate torsional moment $T_u = 25$ k·ft (34 kN·m). Design the necessary steel reinforcement for the section. Use $f'_c = 4$ ksi (28 MPa) and $f_y = 60$ ksi (420 MPa).

15.8. The cantilever beam shown in Figure 15.21 is subjected to the ultimate forces shown. (a) Draw the axial and shearing forces, the bending and torsional moment diagrams. (b) Design the beam section at A using a steel percentage $\leq \rho_{max}$ for bending moment. Use $b = 16$ in. (300 mm), $f'_c = 4$ ksi (28 MPa), and $f_y = 60$ ksi (420 MPa).

15.9. The size of the slab shown in Figure 15.22 is 16 by 8 ft; it is supported by the beam AB, which is fixed at both ends. The uniform dead load on the slab (including its own weight) equals 100 psf and the uniform live load equals 80 psf. Design the section at support A of beam AB using $f'_c = 4$ ksi, $f_y = 60$ ksi, $b_w = 14$ in., $h = 20$ in., a slab thickness = 5 in., and the ACI Code requirements.

Continuous Beams
and Frames

16.1   Introduction

Reinforced concrete buildings consist of different types of structural members, such as slabs, beams, columns, and footings. These structural members may be cast in separate units as precast concrete slabs, beams, and columns, or with the steel bars extending from one member to the other, forming a monolithic structure. Precast units are designed as structural members on simple supports unless some type of continuity is provided at their ends. In monolithic members, continuity in the different elements is provided, and the structural members are analyzed as statically indeterminate structures.

The analysis and design of continuous one-way slabs were discussed in Chapter 9 and the design coefficients and reinforcement details shown in Figures 9.8 and 9.9. In one-way floor systems, the loads from slabs are transferred to the supporting beams, as shown in Figure 16.1(a). If the factored load on the slab is w_u psf, the uniform load on beams AB and BC per unit length is $w_u s$ plus the self-weight of the beam. The uniform load on beams DE and EF is $w_u s/2$ plus the self-weight of the beam.

In two-way rectangular slabs supported by adequate beams on four sides, the floor loads are transferred to the beam from tributary areas bounded by 45° lines, as shown in Figure 16.1(b). Part of the floor loads are transferred to the long beams AB, BC, DE, and EF from trapezoidal areas, while the rest of the floor loads are transferred to the short beams AD, BE, and CF from triangular areas. In square slabs, loads are transferred to all surrounding beams from triangular floor areas. Interior beams carry loads from both sides, whereas end beams carry loads from one side only. Beams in both directions are usually cast monolithically with the slabs; therefore they should be analyzed as statically indeterminate continuous beams. The beams transfer their loads in turn to the supporting columns.

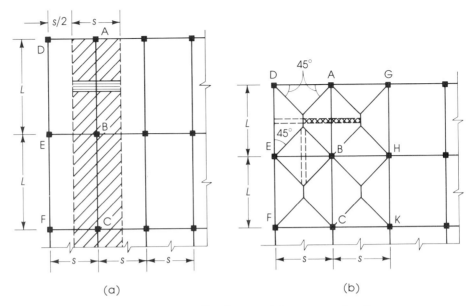

FIGURE 16.1. Slab loads on supporting beams: (a) one-way direction, $L/s > 2$; and (b) two-way direction, $L/s \leq 2$.

16.2   Maximum Moments in Continuous Beams

16.2.1   Basic Analysis

The computation of bending moments and shear forces in reinforced concrete continuous beams is generally based on the elastic theory. When reinforced concrete sections are designed using the strength design method, the results are not entirely consistent with the elastic analysis. However, the ACI Code does not include provisions for a plastic design or limit design of reinforced concrete continuous structures except in allowing moment redistribution, as explained in section 16.15.

16.2.2   Loading Application

The bending moment at any point in a continuous beam depends not only on the position of loads on the same span, but also on the loads on the other spans. In the case of dead loads, all spans must be loaded simultaneously, because the dead load is fixed in position and magnitude. In the case of moving loads or occasional live loads, the pattern of loading must be considered to determine the maximum moments at the critical sections. Influence lines are usually used to determine the position of the live load to calculate the maximum and minimum moments. However, in this chapter, simple rules based on load-deflection curves are used to determine the loading pattern that produces maximum moments.

16.2.3   Maximum and Minimum Positive Moments Within a Span

The maximum positive bending moment in a simply supported beam subjected to a uniform load w k/ft is at midspan, and $M = wl^2/8$. If one or both ends are continuous, the restraint at the continuous end will produce a negative moment at the support and shift slightly the location of the maximum positive moment from midspan. The deflected shape of the continuous beam for a single span loading is shown in Figure 16.2(a); downward deflection indicates a positive moment and upward deflection indicates a negative moment. If all spans deflected downward are loaded, each load will increase the positive moment at the considered span AB (Figure 16.2(d)). Therefore, to calculate the maximum positive moment within a span, place the live load on that span and on every alternate span on both sides. The factored live load moment, calculated as ex-

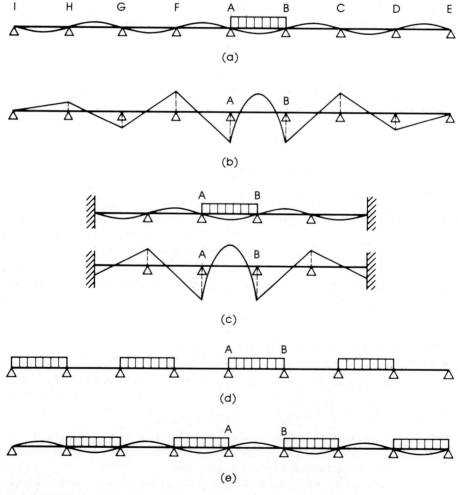

FIGURE 16.2. Loadings for maximum and minimum moment within span AB.

plained above, must be added to the factored dead load moment at the same section to obtain the maximum positive moment.

The bending moment diagram due to a uniform load on AB is shown in Figure 16.2(b). The deflections and the bending moments decrease rapidly with the distance from the loaded span AB. Therefore, to simplify the analysis of continuous beams, the moments in any span can be computed by considering the loaded span and two spans on either side of the considered span AB, assuming fixed supports at the far ends (Figure 16.2(c)).

If the spans adjacent to span AB are loaded, the deflection curve will be as shown in Figure 16.2(e). The deflection within span AB will be upward and a negative moment will be produced in span AB. This negative moment must be added to the positive moment due to dead load to obtain the final bending moment. Therefore, to calculate the minimum positive moment (or maximum negative moment) within a span AB, place the live load on the adjacent spans and on every alternate span on both sides of AB (Figure 16.2(e)).

16.2.4   Maximum Negative Moments at Supports

In this case, it is required to determine the maximum negative moment at any support, say, support A (Figure 16.3). When span AB is loaded, a negative moment is produced at support A. Similarly, the loading of span AF will produce a negative moment at A. Therefore, to calculate the maximum negative moment at any support, place the live load on the two adjacent spans and on every alternate span on both sides (Figure 16.3).

16.2.5   Moment at the Face of Support

In the structural analysis of continuous beams, the span length is taken from center to center of the supports, which are treated as knife-edge supports. In practice, the supports are always made wide enough to take the loads transmitted by the beam, usually the moments acting at the face of supports. To calculate the design moment at the face of the support, it is quite reasonable to deduct a moment $= V_u c/3$ from the factored moment at the centerline of the support, where V_u is the factored shear and c is the column width.

16.3   Moments in Continuous Beams

Continuous beams may be analyzed using approximate methods. The **ACI Code, section 8.3,** gives approximate coefficients for calculating the bending moments and shear forces in continuous beams and slabs. These coefficients were given earlier, in Chapter 9

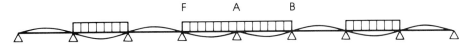

FIGURE 16.3. Loading for maximum negative moment at support A.

and Figure 9.8. The moments obtained using the ACI coefficients will be somewhat larger than those arrived at by exact analysis. The limitations stated in Figure 9.8 must be met.

EXAMPLE 16.1

The slab-beam floor system shown in Figure 16.4 carries a uniform live load of 120 psf and a dead load which consists of the slab's own weight plus 60 psf. Using the ACI moment coefficients, design a typical interior continuous beam and draw detailed sections. Use $f'_c = 3$ ksi, $f_y = 60$ ksi, beam width $b = 12$ in., columns 12 by 12 in., and a slab thickness of 5.0 in.

Solution   1. Design of slabs: The floor slabs act as one-way slabs because the ratio of the long to the short side is greater than 2. The design of a typical continuous slab was discussed in Chapter 9, example 9.4.
2. Loads on slabs:

$$\text{Dead load} = \frac{5}{12} \times 150 + 60 = 122.5 \text{ psf} \qquad \text{Live load} = 120 \text{ psf}$$

$$\text{Ultimate load } w_u = 1.4(122.5) + 1.7(120) = 375.5 \text{ psf}$$

Loads on beams: A typical interior beam ABC carries slab loads from both sides of the beam, with a total slab width = 12 ft.

$$\text{Ultimate load on beam} = 12 \times 375.5 + 1.4 \times (\text{self-weight of beam web})$$

The depth of the beam can be estimated using the coefficients of minimum thickness of beams shown in Table A.6. For $f_y = 60$ ksi, the minimum thickness of the first beam AB = $L/18.5 = (24 \times 12)/18.5 = 15.6$ in. Assume a total depth = 22 in. and a web depth = $22 - 5 = 17$ in. Therefore, ultimate load on beam ABCD

$$w_u = 12(375.5) + 1.4 \left(\frac{17 \times 12}{144} \times 150 \right) = 4804 \text{ lb/ft, say, 4.8 k/ft}$$

3. Moments in beam ABC:
Moment coefficients are shown in Figure 9.8, Chapter 9. The beam is continuous on five spans and symmetrical about the centerline at D. Therefore, it is sufficient to design half the beam ABCD as the other half will have similar dimensions and reinforcement. Since the spans AB and BC are not equal, and the ratio 26/24 < 1.2, the ACI moment coefficients can be applied to this beam. Moreover, the average of the adjacent clear spans is used to calculate the negative moments at the supports.
Moments at critical sections are calculated as follows (Figure 16.4):

$$M_u = \text{coefficient} \times w_u l_n^2$$

Location	1	2	3	4	5	6
Moment coeff.	$-\dfrac{1}{16}$	$+\dfrac{1}{14}$	$-\dfrac{1}{10}$	$+\dfrac{1}{16}$	$-\dfrac{1}{11}$	$+\dfrac{1}{16}$
M_u (k·ft)	-158.7	181.4	-276.5	187.5	-272.7	187.5

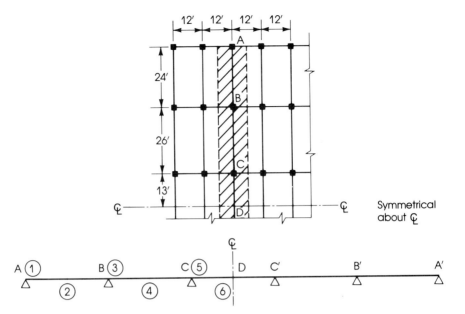

FIGURE 16.4. Example 16.1.

4. Determine beam dimensions and reinforcement.

- Maximum negative moment = -276.5 k·ft. Using $\rho_{max} = 0.0161$,

$$R_u(\text{max}) = 702 \text{ psi}$$

$$d = \sqrt{\frac{M_u}{R_u b}} = \sqrt{\frac{276.5 \times 12}{0.702 \times 12}} = 19.8 \text{ in.}$$

For one row of reinforcement, total depth = $19.8 + 2.5 = 22.3$ in., say, 23 in., and actual $d = 20.5$ in.

$$A_s = 0.0161 \times 12 \times 19.8 = 3.8 \text{ in.}^2$$

Use 4 No. 9 bars in one row.

- Note that total depth used here = 23 in. > 22 in. assumed to calculate the weight of the beam. The additional load is negligible, and there is no need to revise the calculations.
- The sections at the supports act as rectangular sections with tension reinforcement placed within the flange. The reinforcements required at the supports are as follows:

Location	1	3	5
M_u (k·ft)	-158.7	-276.5	-272.7
$R_u = \dfrac{M_u}{bd^2}$ (psi)	378	658	649
ρ (%)	0.77	1.48	1.45
A_s (in.²)	1.9	3.7	3.6
No. 9 bars	2	4	4

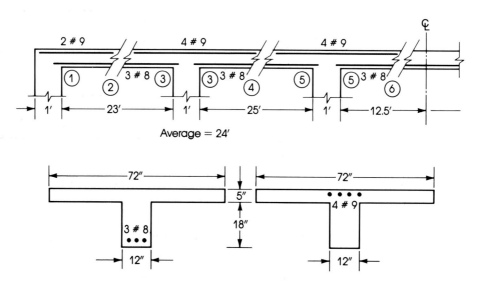

FIGURE 16.5. Reinforcement details, example 16.1.

- For the midspan T-sections, $M_u = +187.5$ k·ft. For $a = 1.0$ in., and flange width = 72 in.,

$$A_s = \frac{M_u}{\phi f_y \left(d - \dfrac{a}{2}\right)} = \frac{187.5 \times 12}{0.9 \times 60(20.5 - 1/2)} = 2.1 \text{ in.}^2$$

Check a:

$$a = \frac{A_s f_y}{0.85 f'_c b} = \frac{2.1 \times 60}{0.85 \times 3 \times 72} = 0.7 \text{ in.}$$

Revised a gives $A_s = 2.07$ in.². Therefore use 3 No. 8 bars ($A_s = 2.35$ in.²) for all midspan sections. Reinforcement details are shown in Figure 16.5.

5. Design the beam for shear as explained in Chapter 8.
6. Check deflection and cracking as explained in Chapter 6.   ■

16.4   Building Frames

A building frame is a three-dimensional structural system consisting of straight members that are built monolithically and have rigid joints. The frame may be one bay long and one story high, such as the portal frames and gable frames shown in Figure 16.6(a), or it may consist of multiple bays and stories as shown in Figure 16.6(b). All members of the frame are considered continuous in the three directions, and the columns participate with the beams in resisting external loads. Besides reducing moments due to continuity, a building frame tends to distribute the loads more uniformly on the frame. The effects of

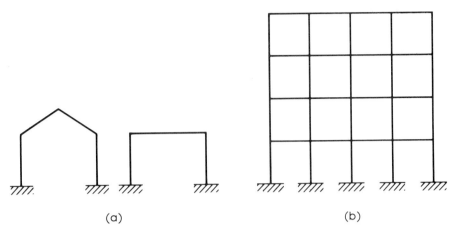

FIGURE 16.6. (a) Gable and portal frames (schematic) and (b) multi-bay, multi-story frame.

lateral loads such as wind and earthquakes are also spread over the whole frame, increasing its safety. For design purposes, approximate methods may be used by assuming a two-dimensional frame system.

A frame subjected to a system of loads may be analyzed by the equivalent frame method. In this method, the analysis of the floor under consideration is made assuming that the far ends of the columns above and below the slab level are fixed (Figure 16.7). Usually, however, the analysis is performed using the moment distribution method.

In practice, the size of panels, distance between columns, number of stories, and the height of each story are known because they are based upon architectural design and utility considerations. The sizes of beams and columns are estimated first and their relative stiffnesses based on the gross concrete sections are used. Once the moments are calculated, the sections assumed previously are checked and adjusted as necessary. More accurate analysis can be performed using computers, which is recommended in the structural analysis of statically indeterminate structures with several redundants. Methods of analysis are described in many books on structural analysis.

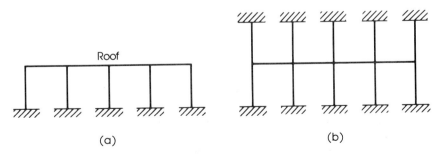

FIGURE 16.7. Assumption of fixed column ends for frame analysis.

16.5   Portal Frames

A portal frame consists of a reinforced concrete stiff girder poured monolithically with its supporting columns. The joints between the girder and the columns are considered rigidly fixed, with the sum of moments at the joint equal to zero. Portal frames are used in building large span halls, sheds, bridges, and viaducts. The top member of the frame may be horizontal (portal frame) or inclined (gable frame) (Figure 16.8). The frames may be fixed or hinged at the base.

A statically indeterminate portal frame may be analyzed by the moment distribution method or any other method used to analyze statically indeterminate structures. The frame members are designed for moments, shear, and axial forces, while the footings are designed to carry the forces acting at the column base.

Girders and columns of frames may be of uniform or variable depths, as shown in Figure 16.8. The forces in a single-bay portal frame of uniform sections may be calculated as follows.

16.5.1   Two Hinged Ends

The forces in the members of a portal frame with two hinged ends[2] can be calculated using the following expressions (Figure 16.9).

For the case of a uniform load on top member BC, let

$$K = 3 + 2 \left(\frac{I_2}{I_1} \times \frac{h}{L} \right)$$

where

I_1 and I_2 = column and beam moments of inertia

h and L = height and span of frame

The bending moments at joints B and C are

$$M_B = M_C = - \frac{wL^2}{4K}$$

Maximum positive moment at midspan BC $= \frac{wL^2}{8} + M_B$

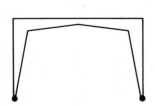 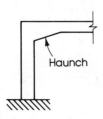 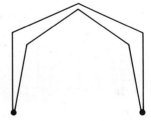

Haunch

FIGURE 16.8. Portal and gable frames.

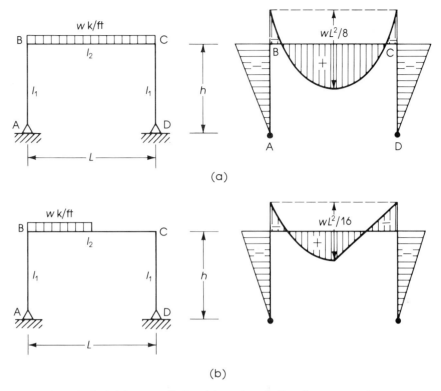

FIGURE 16.9. Portal frame with two hinged ends. Bending moments are drawn on the tension side.

The horizontal reaction at A is

$$H_A = H_D = \frac{M_B}{h}$$

The vertical reaction at A is

$$V_A = V_D = \frac{wL}{2}$$

For a uniform load on half the top member BC (Figure 16.9(b)),

$$M_B = M_C = -\frac{wL^2}{8K}$$

$$H_A = H_D = \frac{M_B}{h}$$

$$V_A = \frac{3}{8}wL \quad \text{and} \quad V_D = \frac{1}{8}wL$$

16.5.2   Two Fixed Ends

The forces in the members of a portal frame with two fixed ends[2] can be calculated as follows (Figure 16.10).

For a uniform load on top member BC, let

$$K_1 = 2 + \left(\frac{I_2}{I_1} \times \frac{h}{L}\right)$$

Then

$$M_B = M_C = -\frac{wL^2}{6K_1}$$

$$M_A = M_D = M_B/2 \qquad M(\text{midspan}) = \frac{wL^2}{8} + M_B$$

$$H_A = H_D = \frac{3M_A}{h} \quad \text{and} \quad V_A = V_D = \frac{wL}{2}$$

For a uniform load on half the top member BC, let

$$K_2 = 1 + 6\left(\frac{I_2}{I_1} \times \frac{h}{L}\right)$$

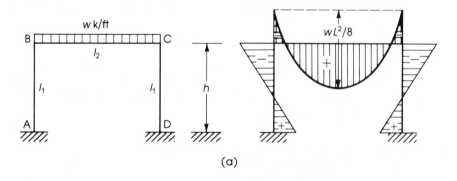

(a)

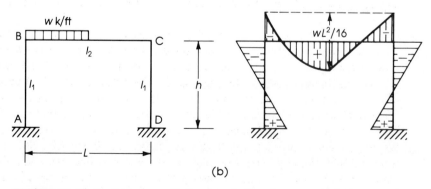

(b)

FIGURE 16.10. Portal frame with fixed ends. Bending moments are drawn on the tension side.

Then

$$M_A = \frac{wL^2}{8}\left(\frac{1}{3K_1} - \frac{1}{8K_2}\right) \qquad M_B = \frac{wL^2}{8}\left(\frac{2}{3K_1} + \frac{1}{8K_2}\right)$$

$$M_C = \frac{wL^2}{8}\left(\frac{2}{3K_1} - \frac{1}{8K_2}\right) \qquad M_D = \frac{wL^2}{8}\left(\frac{1}{3K_1} + \frac{1}{8K_2}\right)$$

$$H_A = H_D = \frac{wL^2}{8} \times \frac{1}{K_1 h}$$

$$V_A = \frac{wL}{2} - V_D \quad \text{and} \quad V_D = \frac{wL}{8}\left(1 - \frac{1}{4K_2}\right)$$

16.6   General Frames

The main feature of a frame is its rigid joints, which connect the horizontal or inclined girders of the roof to the supporting structural members. The continuity between the members tends to distribute the bending moments inherent in any loading system to the different structural elements according to their relative stiffnesses. Frames may be classified as

- statically determinate frames (Figure 16.11(a), page 498),
- statically indeterminate frames without ties (Figures 16.12 and 16.13, page 499), or
- statically indeterminate frames with ties (Figure 16.14, page 500).

Different methods for the analysis of frames and other statically indeterminate structures are described in books dealing with structural analysis. Once the bending moments and shear and axial forces are determined, the sections can be designed similar to the examples in this book.

16.7   Design of Frame Hinges

The four practical types of hinges used in concrete structures are as follows[19]:

1. Mesnager hinge. The forces that usually act on a hinge are a horizontal force H and a vertical force P. The resultant of the two forces R is transferred to the footing through the crossing bars A and B shown in Figure 16.15 (page 501). The inclination of bars A and B to the horizontal varies between 30° and 60° with a minimum distance a, measured from the lower end of the frame column, equal to $8D$, where D is the diameter of the inclined bars. The gap between the frame column and the top of the footing y varies between 1 in. and $1.3h'$, where h' is the width of the concrete section at the hinge level. A practical gap height ranges between 2 and 4 in. The rotation of the frame ends is taken by the hinges, and the gap is usually filled with bituminous cork or similar flexible material. The bitumen protects the cork in contact with the soil from deterioration. The crossing bars A and B are subjected to compressive stresses that must not exceed one-third the yield strength of the steel bars f_y under service loads or $0.55f_y$

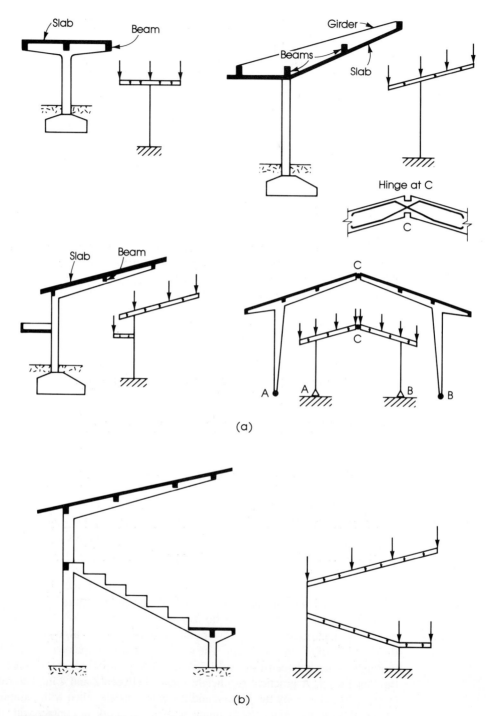

Slab Beam Girder Beams Slab Hinge at C C C A B

Slab Beam

(a)

(b)

FIGURE 16.11. (a) Statically determinate frames and (b) reinforced concrete stadium.

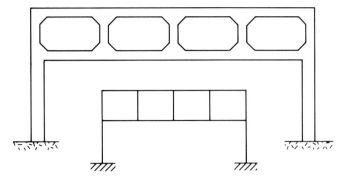

FIGURE 16.12. Vierendeel girder.

under factored loads. The low stress is assumed because any rotation at the hinge tends to bend the bars and induces secondary flexural stresses. It is generally satisfactory to keep the compression stresses low rather than to compute secondary stresses. The areas of bars A and B are calculated as follows:

$$\text{Area of bars A, } A_{s1} = \frac{R_1}{0.55 f_y} \qquad (16.1)$$

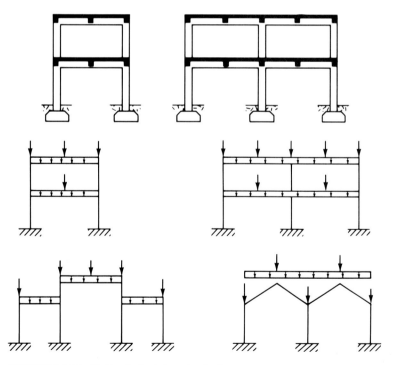

FIGURE 16.13. Statically indeterminate frames.

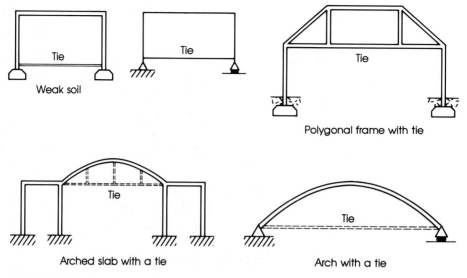

FIGURE 16.14. Structures with ties.

$$\text{Area of bars B, } A_{s2} = \frac{R_2}{0.55 f_y} \tag{16.2}$$

where R_1 and R_2 are the components of the resultant R in the direction of the inclined bars A and B using factored loads. The components R_1 and R_2 are usually obtained by a simple graphical solution as shown in Figure 16.15 or by analysis as follows:

$$H + R_2 \sin \theta = R_1 \sin \theta$$

$$R_2 = R_1 - \frac{H}{\sin \theta}$$

Also, $(R_1 + R_2) \cos \theta = P_u$.

$$R_1 = \frac{P_u}{\cos \theta} - R_2$$

$$= \frac{P_u}{\cos \theta} - \left[R_1 - \frac{H}{\sin \theta} \right]$$

$$= \frac{1}{2} \left[\frac{P_u}{\cos \theta} + \frac{H}{\sin \theta} \right]$$

The inclined hinge bars transmit their force through the bond along the embedded lengths in the frame columns and footings. Consequently, the bars exert a bursting force which must be resisted by ties. The ties should extend a distance $a = 8D$ (the larger bar diameter of bars A and B) in both columns and footings. The bursting force F can be estimated as

$$F = \frac{P_u}{2} \tan \theta + \frac{Ha}{0.85 d} \tag{16.3}$$

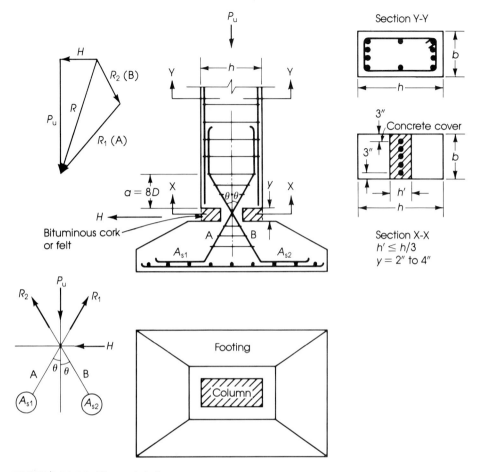

FIGURE 16.15. Hinge details.

If the contribution of concrete is neglected, then the area of tie reinforcement A_{st} required to resist F is

$$A_{st} = \frac{F}{\phi f_y} = \frac{F}{0.85 f_y} \tag{16.4}$$

The stress in the ties can also be computed as follows:

$$f_s \text{ (ties)} = \frac{\dfrac{P_u}{2} \tan \theta + \dfrac{Ha}{0.85d}}{0.005ab + A_{st} \text{ (ties)}} \le 0.85 f_y \tag{16.5}$$

where

A_{st} = area of ties within a distance $a = 8D$ (16.6)

d = effective depth of column section

b = width of column section

This type of hinge is used for moderate forces and limited by the maximum number of inclined bars that can be placed within the column width.

2. Considère hinge. The difference between this type and the Mesnager one is that the normal force P_u is assumed to be transmitted to the footing by one or more short spirally reinforced columns extending deep into the footing, while the horizontal force H is assumed to be resisted by the inclined bars A and B (Figure 16.16(a)). The load capacity of the spirally reinforced short column may be calculated using equation 10.7 of Chapter 10, neglecting the factor 0.85 for minimum eccentricity.

$$P_u = \phi P_n = 0.75[0.85 f_c'(A_g - A_{st}) + A_{st}f_y] \tag{16.7}$$

where A_g = area of concrete hinge section = bh' and A_{st} = area of longitudinal bars within the spirals. Ties should be provided in the column up to a distance equal to the long side of the column section h.

3. Lead hinges. Lead hinges are sometimes used in reinforced concrete frames. In this type of hinge a lead plate, usually 0.75 to 1.0 in. thick, is used to transmit the normal force P_u to the footing. The horizontal force H is resisted by vertical bars placed at the center of the column and extended to the footing (Figure 16.17). At the base of the column, the axial load P_u should not exceed the bearing strength specified by the **ACI Code, section 10.15,** of $\phi(0.85 f_c'A_1)$, where $\phi = 0.7$ and $A_1 = bh'$. For bearing on the footing, the bearing strength may be multiplied by $\sqrt{A_2/A_1} \leq 2$, where A_2 is the area of the portion of the footing that is geometrically similar and concentric with the column (see section 13.9, Chapter 13). The area of the vertical bars A_s can be calculated as follows:

$$A_s = \frac{H}{0.6 f_y}$$

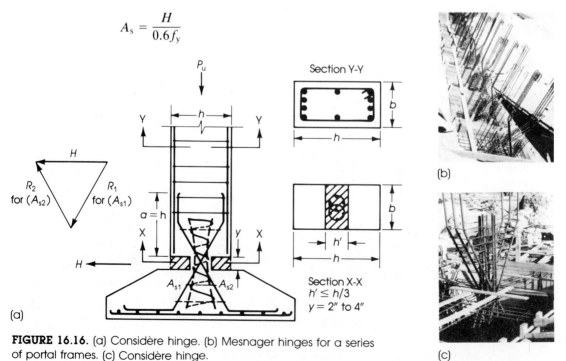

FIGURE 16.16. (a) Considère hinge. (b) Mesnager hinges for a series of portal frames. (c) Considère hinge.

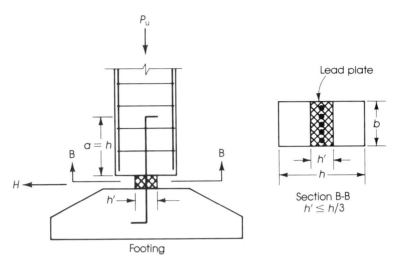

FIGURE 16.17. Lead hinge.

where H = factored horizontal force at the hinge and f_y = the yield strength of steel bars used to resist H. The transfer of load from the column to the footing through the lead plate width h' will cause a concentration of stress at the hinge and a transverse splitting force F_s (Figure 16.18). The force F_s can be estimated by assuming that the transfer of the load P_u occurs within a height $a = h$ within the column and F_s acts at $a/2$ from the hinge. Taking moments about c for half the column only, then

$$\frac{P_u}{2}\left(\frac{h}{4} - \frac{h'}{4}\right) = F_s \times \frac{a}{2} = F_s \times \frac{h}{2}$$

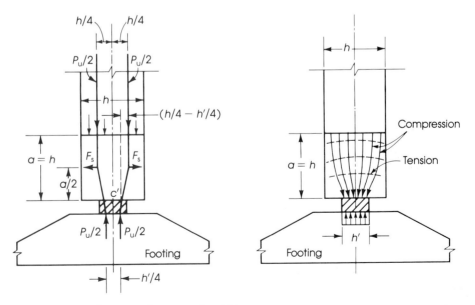

FIGURE 16.18. Forces acting on a lead hinge.

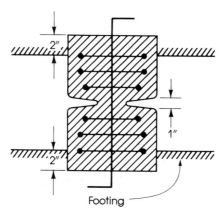

FIGURE 16.19. Precast hinge.

For $h' = 0.25h$ ($h' \leq h/3$), then $F_s = \frac{3}{16}P_u$. Ties should be provided within a height $a = h$ to resist the force F_s. The area of ties required is $F_s/0.85f_y$.

4. Concrete hinges.   Precast reinforced concrete hinges may be used when the frame is subjected to small angles of rotation. The hinge is placed in a small recess within the footing, as shown in Figure 16.19. Vertical bars are inserted within holes in the concrete hinge to tie both structural members above and below the hinge.

Steel hinges may be chosen like those commonly used in bridges, but they are expensive and impractical for reinforced concrete frames.

EXAMPLE 16.2

An 84 by 40 ft hall is to be covered by reinforced concrete slabs supported on hinged-end portal frames spaced at 12 ft on centers (Figure 16.20). The frame height is 15 ft and no columns are allowed within the hall area. The dead load on the slabs is that due to self-weight plus 60 psf from roof finish. The live load on the slab is 80 psf. Design a typical interior frame using $f'_c = 4$ ksi, $f_y = 60$ ksi for the frame, and $f_y = 40$ ksi for the slabs, and a column width $b = 16$ in.

Solution   1. The main structural design of the building will consist of

- design of one-way slabs
- analysis of a portal frame
- design of the frame girder
- design of the columns
- design of hinges
- design of footings

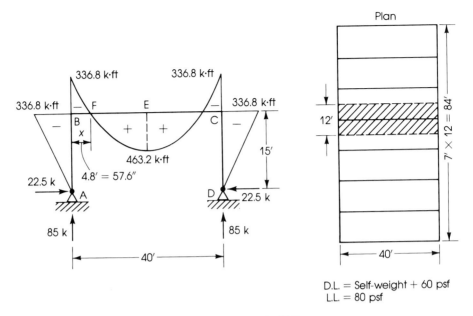

FIGURE 16.20. Design of portal frame, example 16.2.

2. One-way roof slab:
   The minimum thickness of the first slab is $L/28$ as one end is continuous and the other end is discontinuous (Table A.6 in Appendix A).

$$\text{Minimum depth} = \frac{12 \times 12}{28} = 5.1 \text{ in.}$$

Assume a slab thickness of 5.0 in. and design the slab following the steps of example 9.5, Chapter 9.

3. Analysis of an interior portal frame:
   • The loads on slabs are

$$\text{Dead load on slabs} = 60 + \left(\frac{5}{12} \times 150\right) = 122.5 \text{ psf}$$

(Reinforced concrete weighs 150 pcf.)

$$\text{Ultimate load on slabs} = 1.4 \times 122.5 + 1.7 \times 80 = 307.5 \text{ psf}$$

   • Determine loads on frames: The interior frame carries a load from a 12-ft slab in addition to its own weight. Assume that the depth of the beam is $L/24 = (40 \times 12)/24 = 20$ in. Use a projection below the slab = 16 in., giving a total beam depth of 21 in.

$$\text{Dead load from self-weight of beam} = \left(\frac{16}{12}\right)^2 \times 150 = 267 \text{ lb/ft}$$

$$\text{Total ultimate load on frame} = (307.5 \times 12) + 1.4 \times 267 = 4063 \text{ lb/ft}$$

$$w_u = 4.0 \text{ k/ft (or 4.1 k/ft)}$$

- Determine the moment of inertia of the beam and columns sections. The beam acts as a T-section. The effective width of slab acting with the beam is the smallest of span/4 = 40 × 12/4 = 120 in., $16h_s + b_w = 16 × 5 + 16 = 96$ in., or 12 ft × 12 = 144 in. Use $b = 96$ in. The centroid of the section from the top fibers

$$\bar{y} = \frac{96 × 5 × 2.5 + 16 × 16 × 13}{96 × 5 + 16 × 16} = 6.2 \text{ in.}$$

$$I_b \text{ (beam)} = \left[\frac{96}{12}(5)^3 + 96 × 5(3.7)^2\right] + \left[\frac{16}{12}(16)^3 + 16 × 16(6.8)^2\right]$$

$$= 24,870 \text{ in.}^4$$

It is a common practice to consider an approximate moment of inertia of a T-beam as equal to twice the moment of inertia of a rectangular section having the total depth of the web and slab:

$$I_b \text{ (beam)} = 2 × \frac{16}{12}(21)^3 = 24,696 \text{ in.}^4$$

(For an edge beam, approximate $I = 1.5 × bh^3/12$.) Assume a column section 16 by 20 in. (having the same width as the beam).

$$I_c \text{ (column)} = \frac{16}{12}(20)^3 = 10,667 \text{ in.}^4$$

- Let the factor

$$K = 3 + 2\left(\frac{I_b}{L} × \frac{h}{I_c}\right) = 3 + 2\left(\frac{24,870}{40} × \frac{15}{10,667}\right) = 4.75$$

Referring to Figure 16.20 and for a uniform load $w_u = 4.0$ k/ft on BC,

$$M_B = M_C = -\frac{w_u L^2}{4K} = -\frac{4.0(40)^2}{4 × 4.75} = -336.8 \text{ k·ft}$$

The maximum positive bending moment at midspan of BC equals

$$w_u \frac{L^2}{8} + M_B = \frac{4.0(40)^2}{8} - 336.8 = 463.2 \text{ k·ft}$$

Horizontal reaction at A, $H_A = H_D = \frac{M_B}{h} = \frac{336.8}{15} = 22.5$ k

Vertical reaction at A, $V_A = V_D$

$$= \frac{w_u L}{2} + \text{weight of column}$$

$$= 4.0 × \frac{40}{2} + \frac{20}{12} × \frac{16}{12} × 0.150 × 15 \text{ ft} = 85.0 \text{ k}$$

The bending moment diagram is shown in Figure 16.20.
- To consider the sidesway effect on the frame, the live load is placed on half the beam BC, and the moments are calculated at the critical sections. This case is not critical in this example.

- The maximum shear at the two ends of beam BC occurs when the beam is loaded with the factored load w_u, but the maximum shear at midspan occurs when the beam is loaded with half the live load and with the full dead load:

$$V_u \text{ at support} = 4.0 \times \frac{40}{2} = 80.0 \text{ k}$$

$$V_u \text{ at midspan} = W_L \frac{L}{8} = (1.7 \times 80 \times 12) \times \frac{40}{8} = 8160 \text{ lb} = 8.16 \text{ k}$$

- Axial force in each column, $V_A = V_D = 85.0$ k.
- Let the point of zero moment in BC be at a distance x from B; then

$$M_B = w_u L \frac{x}{2} - w_u \frac{x^2}{2}$$

$$336.8 = 4.0\left(\frac{40}{2}x - \frac{x^2}{2}\right) \quad \text{or} \quad x^2 - 40x + 168.4 = 0$$

$$x = 4.8 \text{ ft} = 57.6 \text{ in. from B}$$

4. Design of beam BC:

- Design of critical section at midspan: $M_u = 463.2$ k·ft, web width $b_w = 16$ in., flange width $b = 96$ in., and $d = 21 - 3.5 = 17.5$ in. (assuming two rows of steel bars).

   Check if the section acts as a rectangular section with effective $b = 96$ in. Assume $a = 1.0$ in.; then

$$A_s = \frac{M_u}{\phi f_y\left(d - \frac{a}{2}\right)} = \frac{463.2 \times 12}{0.9 \times 60\left(17.5 - \frac{1.0}{2}\right)} = 6.05 \text{ in.}^2$$

$$a = \frac{A_s f_y}{0.85 f'_c b} = \frac{6.05 \times 60}{0.85 \times 4 \times 96} = 1.1 \text{ in.} < 5.0 \text{ in.}$$

The assumed a equals approximately the calculated a. The section acts as a rectangular section; therefore use 6 No. 9 bars. Check b_{min} (to place bars in one row):

$$b_{min} = 11 \times \frac{9}{8} + 2 \times \frac{3}{8} + 3.0 = 16.125 \text{ in.} > 16 \text{ in.}$$

Place bars in two rows, as shown in Figure 16.21.

- Design of critical section at joint B: $M_u = 336.8$ k·ft, $b = 16$ in., and $d = 21 - 2.5 = 18.5$ in. (for one row of steel bars). The slab is under tension, and reinforcement bars are placed on top of the section.

$$R_u = \frac{M_u}{bd^2} = \frac{336.8 \times 12{,}000}{16(18.5)^2} = 738 \text{ psi}$$

From tables in Appendix A, $\rho = 0.016 < \rho_{max} = 0.0214$.

$$A_s = 0.016 \times 16 \times 18.5 = 4.73 \text{ in.}^2$$

Use 5 No. 9 bars in one row.

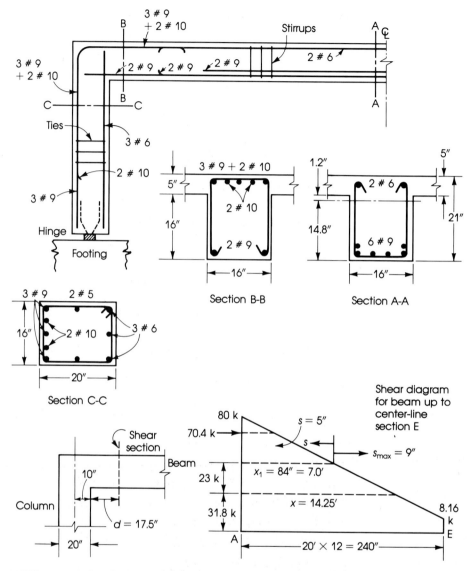

FIGURE 16.21. Reinforcement details of frame sections, example 16.2.

5. Design of beam BC due to shear:

- The critical section is at a distance d from the face of the column with a distance from the column centerline of $10 + 18.5 = 28.5$ in. $= 2.4$ ft. Thus

$$V_u \text{ (at distance } d) = 80 - 4 \times 2.4 = 70.4 \text{ k}$$

- The shear strength provided by concrete

$$\phi V_c = \phi (2 \sqrt{f'_c}) b_w d$$

$$= \frac{0.85 \times 2}{1000} \times \sqrt{4000} \times 16 \times 18.5 = 31.8 \text{ k}$$

Shear force to be provided by web reinforcement

$$\phi V_s = V_u - \phi V_c = 70.4 - 31.8 = 38.6 \text{ k} \quad \text{and} \quad V_s = \frac{38.6}{0.85} = 45.4 \text{ k}$$

- Choose No. 3 stirrups, $A_v = 2 \times 0.11 = 0.22 \text{ in.}^2$. Thus

$$s = \frac{A_v f_y d}{V_s} = \frac{0.22 \times 60 \times 18.5}{45.4} = 5.4 \text{ in., say, 5 in.}$$

or use No. 4 stirrups spaced at $\left(\frac{0.4 \times 60 \times 18.5}{45.4} \right) = 9.8$ in., say, 9.5 in.

- Maximum spacing of No. 3 stirrups

$$s_{max} = \frac{d}{2} = \frac{18.5}{2} = 9.25 \text{ in., say, 9 in.}$$

or

$$s_{max} = \frac{A_v f_y}{50 b_w} = \frac{0.22 \times 60,000}{50 \times 16} = 16.5 \text{ in.}$$

Check for maximum spacing of $d/4$, $V_s \leq 4\sqrt{f_c'}\, b_w d$ or

$$V_s \leq 4\sqrt{4000} \times \frac{16 \times 18.5}{1000} = 74.9 \text{ k}$$

V_s of 45.4 < 74.9 k, so use $s_{max} = 9$ in.

$$V_s' (\text{for } s_{max} = 9 \text{ in.}) = \frac{A_v f_y d}{s} = \frac{0.22 \times 60 \times 18.5}{9} = 27.1 \text{ k}$$

$$\phi V_s = 27.1 \times 0.85 = 23 \text{ k}$$

The distance from the centerline of the column where $s_{max} = 9$ in. can be used is equal to 84 in. = 7.0 ft (from the triangle of shear forces).
- Distribution of stirrups:

First stirrups at $s/2 = \quad 2.5$ in.

17 stirrups at 5 in. $= \quad 85.0$ in.

15 stirrups at 9 in. $= \underline{135.0 \text{ in.}}$

222.5 in.

The distance from the face of the column to the centerline of the beam is 240 − 10 = 230 in. Use the same distribution for the second half of the beam, and place one stirrup at midspan.
- Alternate solution: use No. 4 stirrups spaced at 9.5 in. for the whole beam BC.

6. Design of column section at joint B: $M_u = 336.8 \text{ k·ft}$,

$$P_u = 80 \text{ k}, \ b = 16 \text{ in., and } h = 20 \text{ in.}$$

- Assuming that the frame under the given loads will not be subjected to sidesway, then the effect of slenderness may be neglected and the column can be designed as a short column when

$$\frac{KL_u}{r} < 34 - \frac{12M_{1b}}{M_{2b}} \qquad \text{(see Chapter 12, section 12.5)}$$

$$M_{1b} = 0 \quad \text{and} \quad M_{2b} = 336.8 \text{ k·ft}$$

Let $K = 0.8$ (Figure 12.2), $L_u = 15 - 21/(2 \times 12) = 14.125$ ft, $r = 0.3h = 0.3 \times 20 = 6$ in.; then

$$\frac{KL_u}{r} = 0.8 \times \frac{14.125 \times 12}{6} = 22.6 < 34$$

If K is assumed equal to 1.0, then

$$\frac{KL_u}{r} = 28.25 < 34$$

Therefore, design the member as a short column.
• The design procedure will be similar to that of example 11.12:

$$\text{Eccentricity } e = \frac{M_u}{P_u} = \frac{336.8 \times 12}{80} = 50.5 \text{ in.}$$

This is a large eccentricity, and it will be assumed that tension controls.

$$\phi = 0.9 - \frac{2P_u}{f'_c A_g} \geq 0.7$$

$$= 0.9 - \frac{2 \times 80}{4 \times (16 \times 20)} = 0.77$$

$$M_n = \frac{336.8}{0.77} = 437.4 \text{ k·ft} \quad \text{and} \quad P_n = \frac{80}{0.77} = 103.9 \text{ k}$$

$$d = 20 - 2.5 = 17.5 \text{ in.}$$

• Let the eccentricity with respect to the steel centroid

$$e' = e + d - h/2 = 50.5 + 17.5 - \frac{20}{2} = 58 \text{ in.}$$

$$M'_n = P_n e' = 103.9 \times 58 = 6026 \text{ k·in.}$$

$$R = \frac{M'_n}{bd^2} = \frac{6026 \times 1000}{16(17.5)^2} = 1230 \text{ psi}$$

$$R_{max} \text{ for a single reinforced section} = \frac{936}{0.9} = 1040 \text{ psi}$$

$R > R_{max}$, which indicates that reinforcement in both sides is needed.
• Let $d' = 2.5$ in.

$$M_{n1} = R_{max} bd^2 = 1.040 \times 16(17.5)^2 = 5096 \text{ k·in.} = 424.7 \text{ k·ft}$$

$$A'_s = \frac{M'_n - M_{n1}}{f_y(d - d')} = \frac{6026 - 5096}{60(15)} = 1.03 \text{ in.}^2$$

Use 3 No. 6 bars ($A_s = 1.32$ in.^{2}). It is a common practice to use at least $A_s/4$ in the compression side of the column.

$$A_s = \left(\rho_{max} bd - \frac{P_n}{f_y} \right) + A'_s = \left(0.0214 \times 16 \times 17.5 - \frac{103.9}{60} \right) + 1.03$$

$A_s = 5.29$ in.², so use 3 No. 9 and 2 No. 10 bars (extend to the top of beam BC). Compression steel yields since ρ_{max} is used for $(\rho - \rho')$. The section can be checked using equation 11.24:

$$P_n = 0.85 f'_c bd \left[\rho'm' - \rho m + \left(1 - \frac{e'}{d} \right) \right.$$
$$\left. + \sqrt{ \left(1 - \frac{e'}{d} \right)^2 + 2 \left\{ \frac{e'}{d} (\rho m - \rho'm') + \rho'm' \left(1 - \frac{d'}{d} \right) \right\} } \right]$$

$$m = \frac{f_y}{0.85 f'_c} = \frac{60}{0.85 \times 4} = 17.65 \qquad m' = m - 1 = 16.65$$

$$\rho = \frac{A_s}{bd} = \frac{5.5}{16 \times 17.5} = 0.0196 \qquad \rho' = \frac{1.32}{16 \times 17.5} = 0.0047$$

$$e' = 58 \text{ in.} \qquad d' = 2.5 \text{ in.}$$

Therefore $P_n = 108.8$ k > 103.9 k required.

- Choose No. 3 ties spaced at the least of

$$b = 16 \text{ in.}, \ 16 \times 9/8 = 18 \text{ in.}, \text{ or } 48 \times 3/8 = 18 \text{ in.}$$

Use No. 3 ties spaced at 16 in.

7. Check adequacy of column section at midheight of column, 7.5 ft from A.

- $M_u = \dfrac{336.8}{2} = 168.4$ k·ft

$P_u = 80 + 2.5$ k (weight of half the column) $= 82.5$ k

Use 3 No. 9 bars ($A_s = 3.0$ in.²) on the tension side and 3 No. 6 bars ($A_s = 1.32$ in.²) on the compression side.

$$e = \frac{168.4 \times 12}{82.5} = 24.5 \text{ in.}$$

Calculate the balanced load P_b to determine tension controls (similar to example 11.2):

$$c_b = \frac{87}{87 + f_y} \times d = \frac{87}{87 + 60} (17.5) = 10.36 \text{ in.}$$

$$a_b = 0.85 \times 10.35 = 8.8 \text{ in.}$$

$$P_b = C_c + C_s - T = 0.85 f'_c a_b b + A'_s f_y - A_s f_y$$
$$= 0.85 \times 4 \times 8.8 \times 16 + 60(1.32 - 3) = 377.9 \text{ k}$$

$$\phi = 0.9 - \frac{2 P_u}{f'_c A_g} \geq 0.7 \qquad \phi = 0.9 - \frac{2 \times 82.5}{4 \times 16 \times 20} = 0.77$$

Applied $P_n = 82.5/0.77 = 107.1$ k $< P_b$, therefore tension controls.

• Check the load capacity of the section using equation 11.24:

$$m = 17.65 \quad m' = 16.65 \quad \rho = \frac{3}{16 \times 17.5} = 0.0107$$

$$\rho' = \frac{1.32}{16 \times 17.5} = 0.0047 \quad e' = 24.5 + 17.5 - \frac{20}{10} = 32 \text{ in.}$$

$$P_n = 159.7 \text{ k} > 107.1 \text{ k}$$

The section is adequate. Extend the terminated 2 No. 10 bars a development length in compression of 25 in. below midheight toward A.

8. Design of the hinge at A: $M_u = 0$, $H = 22.5$ k, $P_u = 85$ k.

• Choose a Mesnager hinge. From the force diagram shown in Figure 16.22, $R_1 = 72$ k and $R_2 = 27$ k.

$$A_{s1} = \frac{R_1}{0.55 f_y} = \frac{72}{0.55 \times 60} = 2.2 \text{ in.}^2$$

Choose 3 No. 8 bars ($A_s = 2.35$ in.^{2}).

$$A_{s2} = \frac{R_2}{0.55 \times f_y} = \frac{27}{0.55 \times 60} = 0.82 \text{ in.}^2$$

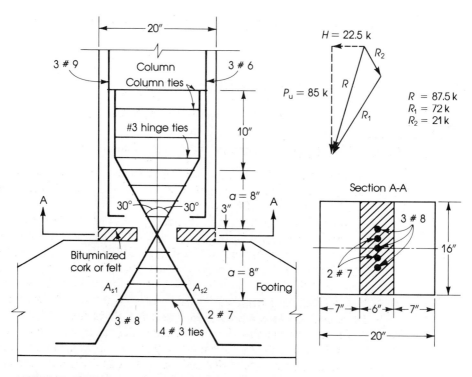

FIGURE 16.22. Hinge details, example 16.2.

Choose 2 No. 7 bars ($A_s = 1.2$ in.²). Arrange the crossing bars by placing one No. 8 bar then one No. 7 bar, as shown in Figure 16.22.

• Lateral ties should be placed along a distance $a = $ 8-bar diameter $= 8.0$ in. within the column and footing. The bursting force

$$F = \frac{P_u}{2} \tan \theta + \frac{Ha}{0.85d}$$

For $\theta = 30°$, $d = 17.5$ in., and $a = 8.0$ in.,

$$F = \frac{85}{2} \tan 30° + \frac{22.5 \times 8}{0.85 \times 17.5} = 36.6 \text{ k}$$

$$\text{Area of ties} = \frac{36.6}{0.85 \times 60} = 0.72 \text{ in.}^2$$

If No. 3 closed ties (two branches) are chosen, then the area of one tie $= 2 \times 0.11 = 0.22$ in.². The number of ties $= 0.72/0.22 = 3.27$, say, 4 ties spaced at $8/3 = 2.7$ in., as shown in Figure 16.22.

9. Design of footing: If the height of the footing is assumed to be h', then the forces acting on the footing is the axial load P and a moment $M = Hh'$. The soil pressure is calculated from equation 13.14 of Chapter 13:

$$q = +\frac{P}{A} \pm \frac{Mc}{I} \le \text{allowable soil pressure}$$

The design procedure of the footing is similar to that of example 13.7, which can be followed easily.   ■

16.8   Introduction to Limit Design

16.8.1   General

Ultimate load design of a structure falls into three distinct steps:

1. Determination of the ultimate design load or factored load, obtained by multiplying the dead and live loads by load factors. The 1983 ACI Code adopted the loads factors given in Chapter 3.
2. Analysis of the structure under ultimate or factored loads to determine the ultimate moments and forces at failure or collapse of the structure. This method of analysis has proved satisfactory for steel design; in reinforced concrete design, this type of analysis has not been fully adopted by the ACI Code because of the lack of ductility of reinforced concrete members. The code allows only partial redistribution of moments in the structure based on an empirical percentage, as will be explained later in this chapter.
3. Design of each member of the structure to fail at the ultimate moments and forces determined from structural analysis. This method is fully established now for reinforced concrete design and the ACI Code permits the use of the strength design method, as was explained in earlier chapters.

16.8.2   Limit Design Concept

Limit design in reinforced concrete refers to the redistribution of moments that occurs throughout a structure as the steel reinforcement at a critical section reaches its yield strength. The ultimate strength of the structure can be increased as more sections reach their ultimate capacity. Although the yielding of the reinforcement introduces large deflections, which should be avoided under service loads, a statically indeterminate structure does not collapse when the reinforcement of the first section yields. Furthermore, a large reserve of strength is present between the initial yielding and the collapse of the structure.

In steel design, the term plastic design is used to indicate the change in the distribution of moments in the structure as the steel fibers, at a critical section, are stressed to their yield strength. The development of stresses along the depth of a steel section under increasing load is shown in Figure 16.23. Limit analysis of reinforced concrete developed as a result of earlier research on steel structures and was based mainly on the investigations of Prager,[4] Beedle,[5] and J. F. Baker.[6] A. L. L. Baker[7] worked on the principles of limit design, while Cranston[8] tested portal frames to investigate the rotation capacity of reinforced concrete plastic hinges. However, more research work is needed before limit design can be adopted by the ACI Code.

16.8.3   Plastic Hinge Concept

The curvature ϕ of a member increases with the applied bending moment M. For an under-reinforced concrete beam, the typical moment-curvature and the load-deflection curves are shown in Figure 16.24. A balanced or an over-reinforced concrete beam is not permitted by the ACI Code, because it fails by the crushing of concrete and shows a small curvature range at ultimate moment (Figure 16.25).

The significant part of the moment-curvature curve in Figure 16.24 is that between B and C, in which M_u remains substantially constant for a wide range of ϕ values. In limit analysis, the moment-curvature curve can be assumed to be of the idealized form shown in Figure 16.26, where the curvature ϕ between B and C is assumed to be constant, forming a plastic hinge. As concrete is a brittle material, it is usually considered that there is a limit at which the member fails completely at maximum curvature at C.

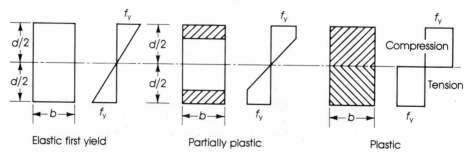

FIGURE 16.23. Distribution of yield stresses in a yielding steel rectangular section.

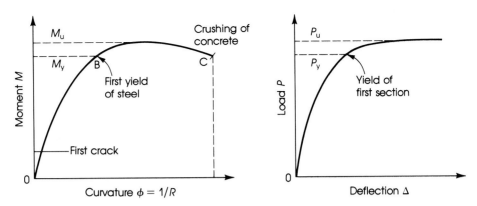

FIGURE 16.24. Yielding behavior of an under-reinforced concrete beam.

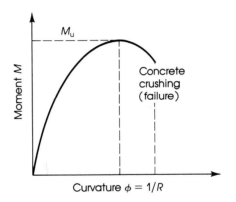

FIGURE 16.25. Yielding behavior of an over-reinforced concrete beam.

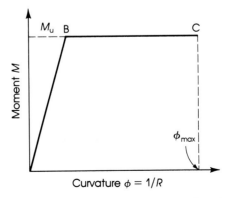

FIGURE 16.26. Idealized moment-curvature behavior of reinforced concrete beams.

Cranston[8] reported that in normally designed reinforced concrete frames, ample rotation capacity is available, and the maximum curvature at point C will not be reached until the failure or collapse of the frame. Therefore, when the member carries a moment equal to its ultimate moment M_u, the curvature continues to increase between B and C without a change in the moment, producing a plastic hinge. The increase in curvature allows other parts of the statically indeterminate structure to carry additional loading.

16.9   The Collapse Mechanism

In limit design, the ultimate strength of a reinforced concrete member is reached when it is on the verge of collapse. The member collapses when there are sufficient numbers of plastic hinges to transform it into a mechanism. The required number of plastic hinges n depends upon the degree of redundancy r of the structure. The relation between n and r to develop a mechanism is:

$$n = 1 + r \qquad (16.8)$$

For example, in a simply supported beam no redundants exist and $r = 0$. Therefore, the beam becomes unstable and collapses when one plastic hinge develops at the section of maximum moment, as shown in Figure 16.27(a). Applications to beams and frames are also shown in Figure 16.27.

16.10   Principles of Limit Design

Under working loads, the distribution of moments in a statically indeterminate structure is based on elastic theory, and the whole structure remains in the elastic range. In limit design, where factored loads are used, the distribution of moments at failure, when a mechanism is reached, is different from that distribution based on elastic theory. This change reflects moment redistribution.

For limit design to be valid, four conditions must be satisfied:

1. *Mechanism condition:* Sufficient plastic hinges must be formed to transform the whole or part of the structure into a mechanism.
2. *Equilibrium condition:* The bending moment distribution must be in equilibrium with the applied loads.
3. *Yield, condition:* The ultimate moment is not exceeded at any point in the structure.
4. *Rotation condition:* Plastic hinges must have enough rotation capacity to permit the development of a mechanism.

Only the first three conditions apply to plastic design, as sufficient rotation capacity exists in ductile materials as steel. The fourth condition puts more limitations on the limit design of reinforced concrete members as compared to plastic design.

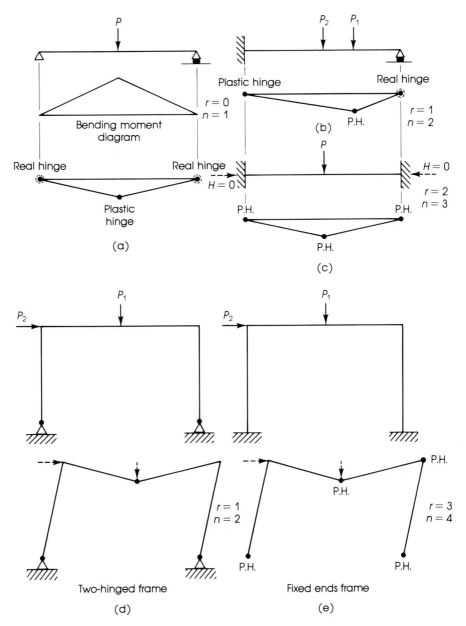

FIGURE 16.27. Development of plastic hinges (P.H.).

16.11   Upper and Lower Bounds to Load Factors

A structure on the verge of collapse must have developed the required number of plastic hinges to transform it into a mechanism. For arbitrary locations of the plastic hinges on the structure, the collapse loads can be calculated, which may be equal to or greater than the actual loads. Since the calculated loads cannot exceed the true collapse loads for the structure, then this approach indicates an upper or kinematic bound of the true collapse loads.[10] Therefore, if all possible mechanisms are investigated, the lowest M_u will be caused by the actual loads. Horne[11] explained the upper bound by assuming a mechanism, then calculating the external work W_e done by the applied loads and the internal work W_i done at the plastic hinges. If $W_e = W_i$, then the applied loads are either equal to or greater than the collapse loads.

If any arbitrary moment diagram is developed to satisfy the static equilibrium under the applied loads at failure, then the applied loads are either equal to or less than the true collapse loads. For different moment diagrams, different ultimate loads can be obtained. Higher values of the lower or static bound are obtained when the moments at several sections for the assumed moment diagram reach the collapse moment. Horne[11] explained the lower bound by assuming different moment distributions to obtain the one that is in equilibrium with the applied loads and satisfies the yield condition all over the structure. In this case, the applied loads are either equal to or less than the collapse loads.

16.12   Limit Analysis

For the analysis of structures by the limit design procedure, two methods can be used, the virtual work method and the equilibrium method.

In the virtual work method, the work done by the ultimate load P_u (or w_u) to produce a given virtual deflection Δ is equated to the work absorbed at the plastic hinges. The external work done by loads $= W_e = \Sigma(w_u \Delta)$ or $\Sigma(P_u \Delta)$. The work absorbed by the plastic hinges $=$ internal work $= W_i = \Sigma(M_u \theta)$.

EXAMPLE 16.3

The beam shown in Figure 16.28 carries a concentrated load at midspan. Calculate the collapse moment at the critical sections.

Solution

1. The beam is once statically indeterminate ($r = 1$), and the number of plastic hinges to transform the beam into a mechanism $n = 1 + 1 = 2$ plastic hinges, at A and C. The first plastic hinge develops at A and the beam acts as a simply supported member until a mechanism is reached.

2. If a rotation θ occurs at the plastic hinge at the fixed end A, the rotation at the sagging hinge C $= 2\theta$. The deflection of C under the load $= (L/2)\theta$ (Figure 16.28).

$$W_e = \text{External work} = \Sigma P_u \Delta = P_u \left(\frac{L\theta}{2}\right)$$

$$W_i = \text{Internal work} = \Sigma M_u \theta = M_{u1}(\theta) + M_{u2}(2\theta)$$

If the two sections at A and C have the same dimensions and reinforcement, then $M_{u1} = M_{u2} = M_u$, and $W_i = 3M_u \theta$. Equating W_e and W_i, then

$$M_{u1} + 2M_{u2} = P_u \frac{L}{2} = 3M_u \quad \text{and} \quad M_u = \frac{P_u L}{6}$$

    ■

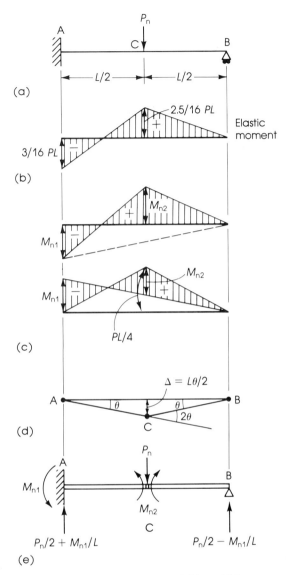

FIGURE 16.28. Example 16.3. $P_u = \phi P_n$ and $M_u = \phi M_n$.

EXAMPLE 16.4

Calculate the collapse moments at the critical sections for the beam shown in Figure 16.29 due to a uniform load w_u.

Solution   1. Number of plastic hinges = 2.

2. For a deflection at C = 1.0, the rotation at A, $\theta_A = 1/a$, $\theta_B = 1/b$, and

$$\theta_C = \theta_A + \theta_B = \frac{1}{a} + \frac{1}{b} = \frac{a+b}{ab} = \frac{L}{ab}$$

3. External work,

$$W_e = \Sigma w_u \Delta = w_u \left(\frac{1 \times L}{2} \right) = \frac{w_u L}{2}$$

Internal work,

$$W_i = \Sigma M_u \theta = M_{u1} \theta_A + M_{u2} \theta_C$$

$$= M_{u1} \left(\frac{1}{a} \right) + M_{u2} \left(\frac{1}{a} + \frac{1}{b} \right)$$

Equating W_e and W_i, then

$$w_u = \frac{2}{L} \left(\frac{M_{u1}}{a} + \frac{M_{u2}}{a} + \frac{M_{u2}}{L-a} \right) \tag{16.9}$$

If both moments are equal, then

$$w_u = \frac{2M_u}{L} \left[\frac{2}{a} + \frac{1}{(L-a)} \right] = \frac{2M_u}{L} \left[\frac{(2L-a)}{a(L-a)} \right] \tag{16.10}$$

4. To determine the position of the plastic hinge at C which produces the minimum value of the collapse load w_u, differentiate equation 16.9 with respect to a and equate to zero:

$$\frac{\delta w_u}{\delta a} = 0 \qquad -\left(\frac{M_{u1}}{a^2} + \frac{M_{u2}}{a^2} - \frac{M_{u2}}{(L-a)^2} \right) = 0$$

If $M_{u1} = M_{u2} = M_u$, then

$$\frac{2}{a^2} = \frac{1}{(L-a)^2} \quad \text{or} \quad a = L(2 - \sqrt{2}) = 0.586L$$

From equation 16.10, the collapse load $w_u = 11.66(M_u/L^2)$, and the collapse moment $M_u = 0.0858 w_u L^2$. The reaction at A $= 0.586 w_u L$, and the reaction at B $= 0.414 w_u L$. ∎

In the equilibrium method, the equilibrium of the beam or separate segments of the beam are studied under the forces present at collapse. To illustrate analysis by this method, the two previous examples will be repeated here.

EXAMPLE 16.5

For the beam shown in Figure 16.28, calculate the collapse moments using the equilibrium method.

Solution   Two plastic hinges will develop at A and C. Referring to Figure 16.28(e), the reaction at A $= (P_u/2) + (M_{u1}/L)$ and the reaction at B $= (P_u/2) - (M_{u1}/L)$. Considering the equilibrium of beam BC and taking moments about C:

$$\left(\frac{P_u}{2} - \frac{M_{u1}}{L} \right) \left(\frac{L}{2} \right) = M_{u2}$$

or

$$M_{u1} + 2M_{u2} = P_u \frac{L}{2}$$

which is the same equation obtained in example 16.3. When $M_{u1} = M_{u2} = M_u$, then

$$3M_u = P_u \frac{L}{2} \quad \text{or} \quad M_u = P_u \frac{L}{6}$$ ■

EXAMPLE 16.6

Calculate the collapse moments for the beam shown in Figure 16.29 by the equilibrium method.

Solution   1. Two plastic hinges will develop in this beam at A and C. Referring to Figure 16.29(d), the reaction at $A = w_u(L/2) + (M_{u1}/L)$ and the reaction at $B = w_u(L/2) - (M_{u1}/L)$. The load on BC $= w_u b$ acting at $b/2$ from B, and $b = (L - a)$. Considering the equilibrium of segment BC and taking moments about C:

$$\left(w_u \frac{L}{2} - \frac{M_{u1}}{L} \right) b - (w_u b) \frac{b}{2} = M_{u2}$$

If $M_{u1} = M_{u2} = M_u$, then

$$w_u \frac{b}{2} (L - b) = M_u \left(1 + \frac{b}{L} \right) = \frac{M_u}{L} (2L - a)$$

and

$$w_u = \frac{2M_u}{L} \times \frac{(2L - a)}{a(L - a)}$$

which is similar to the results obtained in example 16.4.

$$M_u = \frac{w_u L}{2} \times \frac{a(L - a)}{(2L - a)}$$

2. The position of a can be determined as before, where $a = 0.586L$, $M_u = 0.0858 w_u L_2$, and $w_u = 11.66(M_u/L^2)$. ■

16.13   Rotation of Plastic Hinges

16.13.1   Plastic Hinge Length

The assumption that the inelastic rotation of concrete occurs at the point of maximum moment while other portions of the member act elastically is a theoretical one; in fact, the plastic rotation occurs on both sides of the maximum moment section over a finite

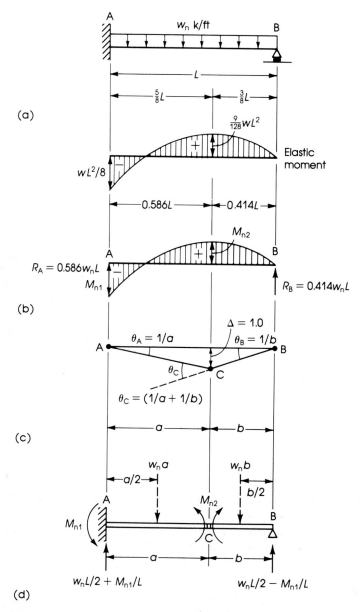

FIGURE 16.29. Example 16.4. $M_u = \phi M_n$ and $w_u = \phi w_n$.

length. This length is called the plastic hinge length l_p. The hinge length l_p is a function of the effective depth d and the distance from the section of highest moment to the point of contraflexure (zero moment).

Referring to Figure 16.30(a), the length $l_p/2$ represents the plastic hinge length on one side of the center of support. M_u and ϕ_u indicate the ultimate moment and ultimate curvature at the critical section, while M_y and ϕ_y indicate the moment and curvature at

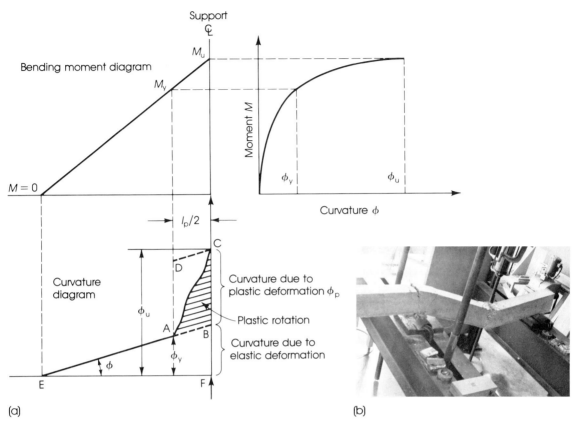

FIGURE 16.30. (a) Plastic rotation from moment-curvature and moment gradient. (b) Development of plastic hinges in a reinforced concrete continuous beam.

first yield. The plastic curvature at the critical section ϕ_p is equal to $(\phi_u - \phi_y)$ and the rotation capacity is equal to $(\phi_p l_p)$.

The estimated length of the plastic hinge was reported by many investigators. A. L. L. Baker[7] assumed that the length of the plastic hinge is approximately equal to the effective depth d. Corley[12] proposed the following expression for the equivalent length of the plastic hinge:

$$l_p = 0.5d + 0.2\sqrt{d}(z/d) \qquad (16.11)$$

where z = distance of the critical section to the point of contraflexure and d = effective depth of the section. Mattock[13] suggested a simpler form:

$$l_p = 0.5d + 0.05z \qquad (16.12)$$

Recent tests[14] on reinforced concrete continuous beams showed that l_p can be assumed equal to $1.06d$. They also showed that the length of the plastic hinge, in reinforced concrete continuous beams containing hooked-end steel fibers, increases with the increase in the amount of the steel fibers and the main reinforcing steel according to the following expression:

$$l_p = (1.06 + 0.13\rho\rho_s)d \qquad (16.13)$$

where ρ = percentage of main steel in the section and ρ_s = percentage of steel fibers by volume, $0 \le \rho_s \le 1.2$. For example, if $\rho = 1.0$ percent and $\rho_s = 0.8$ percent, then $l_p = 1.164d$.

16.13.2 Curvature Distribution Factor

Another important factor involving the calculation of plastic rotations is the curvature distribution factor β. The curvature along the plastic hinge varies significantly, and in most rotation estimations this factor is ignored, which leads to an overestimation of the plastic rotations. Referring to Figure 16.30, the shaded area ABC represents the inelastic rotation that can occur at the plastic hinge, while the unshaded area EBF represents the elastic contribution to the rotation over the length of the member. The shaded area ABC can be assumed equal to β times the total area ABCD within the plastic hinge length $l_p/2$ on one side of the critical section. The curvature distribution factor β represents the ratio of the actual plastic rotation θ_{pc} to ϕl_p, where ϕ is the curvature and l_p is the length of the plastic hinge. The value of β was reported to vary between 0.5 and 0.6. Recent tests[14] showed that β can be assumed equal to 0.56. When hooked-end steel fibers were used in concrete beams, the value of β decreased according to the following expression:

$$\beta = 0.56 - 0.16\rho_s \qquad (16.14)$$

where ρ_s = percentage of steel fibers, $0 \le \rho_s \le 1.2$ percent. The reduction of the curvature distribution factor of fibrous concrete does not imply that the rotation capacity is reduced: the plastic curvature of fibrous concrete is substantially higher than that of concrete without fibers. Figure 16.31 shows the distribution of the curvature along the plastic hinge length. The area ABC_1 represents the plastic rotation for concrete that does not contain steel fibers, $\beta = 0.56$, while the areas ABC_2 and ABC_3 represent the plastic rotation for concretes containing 0.8 percent and 1.2 percent steel fibers, respectively.

16.13.3 Ductility Index

The ratio of ultimate to first-yield curvature is called the ductility index $\mu = \phi_u/\phi_y$. The ductility index of reinforced concrete beams was reported[15] to vary between 4 and 6. If steel fibers are used in concrete beams, the ductility index increases according to the following expression:[14]

$$\mu' = (1.0 + 3.8\rho_s)\mu \qquad (16.15)$$

where

μ = the ratio of ultimate to first-yield curvature

μ' = ductility index of the fibrous concrete

ρ_s = percentage of steel fibers by volume, $0 \le \rho_s \le 1.2$ percent

16.13.4 Required Rotation

The rotation of a plastic hinge in a reinforced concrete indeterminate structure required to allow other plastic hinges to develop and the structure to reach a mechanism can be

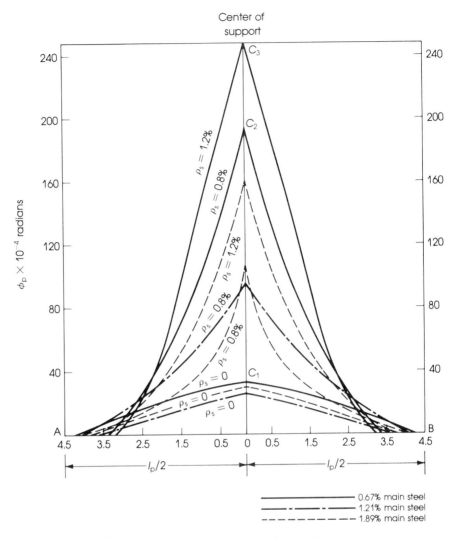

FIGURE 16.31. Curvature distribution along the plastic hinge.

determined approximately by slope deflection from the following expression.[9.20] For a segment AB between two plastic hinges, the rotation at A is

$$\theta_A = \frac{L}{6E_cI}\,[2(M_A - M_{FA}) + (M_B - M_{FB})] \tag{16.16}$$

where

M_A and M_B = ultimate moments at A and B, respectively

M_{FA} and M_{FB} = elastic fixed end moments at A and B

E_c = modulus of elasticity of concrete = $33w^{1.5}\sqrt{f_c'}$

I = moment of inertia of a cracked section (see Chapter 5)

16.13.5   Rotation Capacity Provided

Typical tensile plastic hinges at the support and midspan sections of a frame are shown in Figure 16.32. The rotation capacity depends mainly on

1. The ultimate strain capacity of concrete ε_c', which may be assumed to be 0.003, or 0.0035 as used by Baker.[7]
2. The length l_p over which yielding occurs at the plastic hinge, which can be assumed to be approximately equal to the effective depth of the section where the plastic hinge developed ($l_p = d$).
3. The depth of the compressive block c in concrete at failure, at the section of the plastic hinge.

The angle of rotation θ of a tensile plastic hinge can be determined as follows:

$$\theta = \frac{\varepsilon_p l_p}{c} \qquad\qquad (16.17)$$

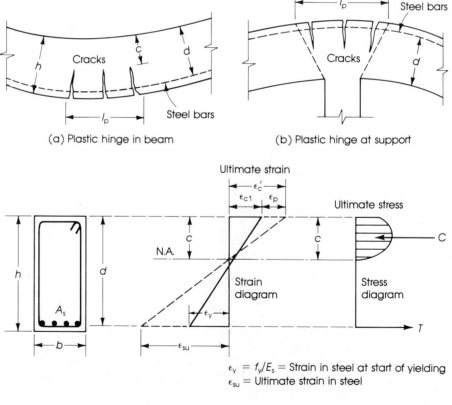

(a) Plastic hinge in beam   (b) Plastic hinge at support

$\epsilon_y = f_y/E_s =$ Strain in steel at start of yielding
$\epsilon_{su} =$ Ultimate strain in steel

(c) Stress and strain diagrams

FIGURE 16.32. Plastic hinge and typical stress and strain distribution.[2]

where ε_p = the increase in the strain in the concrete measured from the initial yielding of steel reinforcement in the section (see Figure 16.32(c)):

$$\varepsilon_p = \varepsilon_c' - \varepsilon_{cl} = 0.0035 - \varepsilon_{cl}$$

If $l_p = d$, and the ratio $c/d = \lambda \leq 0.5$,

$$\theta = \frac{(0.0035 - \varepsilon_{cl})d}{\lambda d} = \frac{0.0035 - \varepsilon_{cl}}{\lambda}$$

From strain triangles (Figure 16.32):

$$\varepsilon_{cl} = \varepsilon_y \left(\frac{c}{d-c}\right) = \frac{f_y}{E_s}\left(\frac{\lambda d}{d - \lambda d}\right) = \frac{f_y}{E_s}\left(\frac{\lambda}{1 - \lambda}\right)$$

where f_y = yield strength of steel bars and E_s = modulus of elasticity of steel = 29×10^6 psi. Therefore

$$\theta = \frac{0.0035}{\lambda} - \frac{\varepsilon_{cl}}{\lambda} = \frac{0.0035}{\lambda} - \frac{f_y}{E_s(1 - \lambda)} \qquad (16.18)$$

For grade 40 steel, f_y = 40 ksi, and using maximum value of $\lambda = 0.50$, then

$$\theta_{min} = \frac{0.0035}{0.50} - \frac{40}{29,000 \times (1 - 0.50)} = 0.00424 \text{ radian}$$

For grade 60 steel, f_y = 60 ksi, $\lambda_{max} = 0.44$:

$$\theta_{min} = \frac{0.0035}{0.44} - \frac{60}{29,000(1 - 0.44)} = 0.00426 \text{ radian}$$

The θ_{min} calculated above is from one side only, and the total permissible rotation at the plastic hinge equals 2θ or $2\theta_{min}$.

The actual λ can be calculated as follows, given

$$a = \beta_1 c \text{ and } \beta_1 = 0.85 \text{ for } f_c' \leq 4 \text{ ksi,}$$

$$c = \frac{a}{0.85} = \frac{A_s f_y}{(0.85)^2 f_c' b} \qquad (3.11)$$

and

$$\lambda = \frac{c}{d} = \frac{A_s f_y}{0.72 f_c' bd} = \frac{\rho f_y}{0.72 f_c'} \leq 0.5 \qquad (16.19)$$

where $\rho = A_s/bd$. (λ_{max} is obtained when ρ_{max} is used.)

If rotation provided is not adequate, increase the section dimensions or reduce the percentage of steel reinforcement to obtain a smaller c, a smaller λ, and greater θ. A.L.L. Baker[3] indicated that if special binding or spirals are used, the ultimate crushing strain in bound concrete may be as high as 0.012. For a compression plastic hinge (as in columns),

$$\theta = \frac{\varepsilon_p l_p}{h} \qquad (16.20)$$

where h = overall depth of the section and l_p = length over which yielding occurs. In compression hinges, l_p varies between $0.5h$ and h.

At a concrete ultimate stress of f_c', $\varepsilon_c = 0.002$; thus $\varepsilon_p = \varepsilon_c' - 0.002 = 0.0035 - 0.002 = 0.0015$ is the minimum angle of rotation on one side. Therefore

$$\theta_{min} = \frac{0.0015 \times 0.5h}{h} = 0.00075 \text{ radian}$$

With special binding or spirals, θ may be increased to

$$\theta_{max} = (0.012 - 0.002) \times \frac{0.5h}{h} = 0.005 \text{ radian}$$

The extreme value of $\varepsilon_c' = 0.012$ is quite high, and a smaller value may be used with proper spirals; otherwise a different section must be adopted.

In reinforced concrete continuous beams containing steel fibers, the plastic rotation may be estimated as follows:[14]

$$\theta_p = \gamma\beta \left(\frac{0.0035}{\lambda} - \frac{f_y}{E_s(1 - \lambda)} \right) \qquad (16.21)$$

where

$$\gamma = (4.3 + 2.24\rho_s - 0.043f_y + 4.17\rho\rho_s) \qquad (16.22)$$
$$\beta = (0.56 - 0.16\rho_s) \qquad (16.14)$$

f_y = yield strength of steel, ksi

E_s = modulus of elasticity of main steel

ρ = percentage of main steel

ρ_s = percentage of steel fibers

From equation 16.21, it is obvious that the plastic rotation of fibrous reinforced concrete is dependent upon the percentage of steel fibers and percentage of the main steel and its yield strength. Raising the yield strength of the main steel reduces the plastic rotation. Equation 16.21 also includes the effect of the plastic hinge length on rotation. A simplified form can be presented:[14]

$$\theta_p = \gamma\beta \left(\frac{0.003}{\lambda} \right) \qquad (16.23)$$

For example, if $\rho_s = 0$ and $f_y = 60$ ksi, then $\theta_{p1} = 0.00289/\lambda$, and for $\rho_s = 1.0$ percent, $\rho = 1.5$ percent, and $f_y = 60$ ksi, then $\theta_{p2} = 0.01222/\lambda$. This means that the rotation capacity of a concrete beam may be increased by about four times if 1 percent of steel fibers is used.

16.14   Summary of Limit Design Procedure

1. Compute the factored loads using the load factors given in Chapter 3:

$$w_u = 1.4D + 1.7L$$

2. Determine the mechanism, plastic hinges, and ultimate moments M_u.
3. Design the critical sections using the strength design method.
4. Determine the required rotation of plastic hinges.
5. Calculate the rotation capacity provided at the sections of plastic hinges. The rotation capacity must exceed that required.
6. Check the factor against yielding of steel and excessive cracking, that is, ϕM_n/elastic moment at service load.
7. Check deflection and cracking under service loads.
8. Check that adequate shear reinforcement is provided at all sections.

For more details, see reference 21.

EXAMPLE 16.7

The beam shown in Figure 16.33 is fixed at both ends and carries a uniform dead load = 1.5 k/ft, a uniform live load = 2.0 k/ft, a concentrated dead load = 10 k, and a concentrated live load = 20 k. Design the beam using the limit design procedure. Use b = 14 in., f'_c = 3 ksi, and f_y = 40 ksi.

Solution

1.   Ultimate uniform load $w_u = 1.4D + 1.7L = 1.4 \times 1.5 + 1.7 \times 20 = 5.5$ k/ft

Ultimate concentrated load $P_u = 1.4 \times 10 + 1.7 \times 20 = 48$ k

2. The plastic hinges will develop at A, B, and C, causing a mechanism as shown in Figure 16.33. Using the virtual work method of analysis and assuming a unit deflection at C, then the external work is equal to

$$W_e = 48 \times 1 + 5.5\left(24 \times \frac{1}{2}\right) = 114 \text{ k·ft}$$

The internal work absorbed by the plastic hinges

$$W_i = M_u\theta \text{ (at A)} + M_u\theta \text{ (at B)} + M_u(2\theta) \text{ at C}$$

$$= 4M_u\theta = 4M_u\left(\frac{1}{12}\right) = \frac{M_u}{3}$$

Equating W_e and W_i:

$$114 = \frac{M_u}{3}$$

$$M_u = 342 \text{ k·ft}$$

The general analysis gives directly:

$$M_u = \frac{w_u L^2}{16} + P_u \frac{L}{8} = \frac{5.5}{16}(24)^2 + 48\frac{(24)}{8} = 342 \text{ k·ft}$$

3. Design the critical sections at A, B, and C for M_u = 342 k·ft. From tables in Appendix A, and for f'_c = 3 ksi, f_y = 40 ksi, and a steel percentage ρ = 0.013, R_u = 420 psi.

$$M_u = R_u b d^2$$

$$342 \times 12 = 0.42 \times 14(d)^2$$

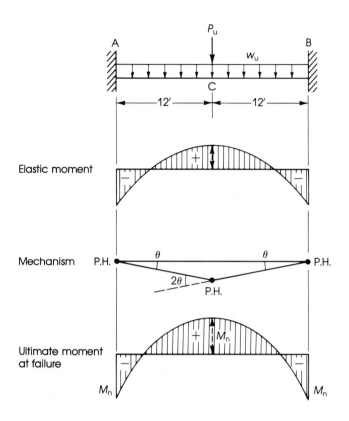

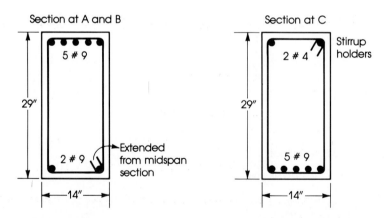

FIGURE 16.33. Example 16.7.

$d = 26.4$ in. and total depth $h = 26.4 + 2.5 = 28.9$ in., say, 29 in.

$$A_s = \rho bd = 0.013 \times 14 \times 26.4 = 4.8 \text{ in.}^2$$

Use 5 No. 9 bars in one row; A_s provided $= 5.0$ in.^{2}. $b_{min} = 9 \times \frac{9}{8} + 2 \times 1.5 + 2 \times \frac{3}{8} = 13.875$ in. < 14 in.

$$a = \frac{A_s f_y}{0.85 f'_c b} = \frac{5.0 \times 40}{0.85 \times 3 \times 14} = 5.6 \text{ in.}$$

$$c = \frac{a}{0.85} = 6.6 \text{ in.} \qquad \lambda = \frac{c}{d} = \frac{6.6}{26.4} = 0.25$$

4. Required rotation of plastic hinges:

$$\theta_A = \frac{L}{6E_c I} [2(M_A - M_{FA}) + (M_B - M_{FB})]$$

• $E_c = 57{,}400 \sqrt{f'_c} = 3.144 \times 10^6$ psi

$$E_s = 29 \times 10^6 \text{ psi} \quad \text{and} \quad n = \frac{E_s}{E_c} = 9.2$$

• Determine the fixed end moments at A and B using factored loads:

$$M_{FA} = M_{FB} = \frac{w_u L^2}{12} \text{ (uniform load)} + \frac{P_u L}{8} \text{(concentrated load)}$$

$$= 5.5\frac{(24)^2}{12} + 48 \times \frac{24}{8} = 408 \text{ k·ft}$$

Plastic M_A = plastic M_B = 342 k·ft

• The cracked moment of inertia can be calculated from

$$I_{cr} = b\frac{x^3}{3} + nA_s(d - x)^2$$

where x is the distance from compression fibers to the neutral axis (kd). To determine x (see Chapter 5),

$$\frac{bx^2}{2} - nA_s(d - x) = 0$$

$$14\frac{x^2}{2} - 9.2 \times 5(26.5 - x) = 0$$

$$x = 10.3 \text{ in.}$$

$$I_{cr} = 14\frac{(10.3)^3}{3} + 9.2 \times 5(26.5 - 10.3)^2 = 17{,}172 \text{ in.}^4$$

Required minimum rotation: Considering all moments at supports A and B are negative, then

$$\theta_A = \frac{24 \times 12}{6 \times 3.144 \times 10^6 \times 17{,}172} [2(-342 + 408) + (-342 + 408)](12{,}000)$$

$$= 0.00211 \text{ radian}$$

5. Rotation capacity provided:

$$\theta_A = \frac{0.0035}{\lambda} - \frac{f_y}{E_s(1 - \lambda)} = \frac{0.0035}{0.25} - \frac{40}{29,000(1 - 0.25)}$$

$$= 0.0122 \text{ radian} > 0.00211 \text{ required}$$

The rotation capacity provided is about 5.5 times that required, indicating that the section is adequate.

6. Check the ratio of ultimate to elastic moment at service load,

$$M_A = M_B = \frac{wL^2}{12} + \frac{PL}{8}$$

$$= 3.5 \frac{(24)^2}{12} + \frac{30 \times 24}{8} = 258 \text{ k·ft}$$

Actual $\phi M_n = \phi A_s f_y[d - (a/2)] = 0.9 \times 5 \times 40[26.5 - (5.6/2)]/12 = 356$ k·ft. Ratio = $356/258 = 1.38$, which represents the factor of safety against the yielding of steel bars at the support.

7. Check maximum deflection due to service load (at midspan): For a uniform load w,

$$\Delta_1 = \frac{wL^4}{384EI}$$

For a concentrated load at midspan,

$$\Delta_2 = \frac{PL^3}{192EI}$$

$$\text{Total deflection } \Delta = \frac{\left(\frac{3500}{12}\right)(24 \times 12)^4}{384(17,172)(3.144 \times 10^6)} + \frac{30,000(24 \times 12)^3}{192(17,172)(3.144 \times 10^6)}$$

$$= 0.166 \text{ in.}$$

$$\frac{\Delta}{L} = \frac{0.166}{24 \times 12} = \frac{1}{1735}$$

which is a very small ratio.

8. Adequate shear reinforcement must be provided to avoid any possible shear failure. ■

16.15   Moment Redistribution

The ACI Code does not cover the limit design procedure, as explained above; however, it specifies another conservative approach to take into account the redistribution of moments in indeterminate structures at failure. **Section 8.4.1 of the ACI Code** indicates that the maximum negative moments calculated by elastic theory at the supports of continuous flexural members may be increased or decreased by not more than

$$q = 20\left(1 - \frac{\rho - \rho'}{\rho_b}\right) \qquad (16.24)$$

where

$$\rho = A_s/bd$$

$$\rho' = A_s'/bd$$

$$\rho_b = 0.85\,\beta_1 \frac{f_c'}{f_y}\left(\frac{87}{87 + f_y}\right) \qquad\qquad (3.13)$$

Equation 16.24 must not be applied to the moments calculated from approximate structural analysis or to the alternate design method of reinforced concrete based on elastic theory. The ACI Code specifies a different factor for the redistribution of negative moments in continuous prestressed concrete members.

The code limits the steel ratio ρ (or $\rho - \rho'$) at the section where the moment is reduced to a maximum ratio of $0.5\rho_b$. The minimum steel ratio allowed in the section, for flexural design, $\rho_{min} = 200/f_y$. If the extreme values, $0.5\rho_b$ and ρ_{min}, are substituted in equation 16.24, the maximum and minimum redistribution percentages will be as shown in Table 16.1.

In practice, a minimum steel percentage ρ_{min} is not adopted for a good design, and therefore a maximum redistribution percentage will not practically be used. A steel percentage of $0.5\rho_b$ or $0.5\rho_{max}$ is very common in the design of sections; thus a redistribution percentage of 10 percent to 15 percent may be expected.

The ACI Code percentages for the redistribution of negative moments are more conservative relative to a maximum value of 30 percent adopted by the British Code CP110, although the British Code CP114 of 1957 recommended 15 percent.

Whatever percentage of moment redistribution is used, it is essential to ensure that no section is likely to suffer local damage or excessive cracking at service loads and that adequate rotation capacity is maintained at every critical section in the structure. The redistribution of moments in a statically indeterminate structure will result in a reduction in the negative moments at the supports and in the positive moments within the spans. This reduction will not imply that the safety of the structure has been reduced or jeopardized as compared with determinate structures. In fact, continuity in structures provides additional strength, stability, and economy in the design.

Bending moment coefficients for equal-spans continuous beams with 10 percent and 15 percent moment redistribution percentages are shown in Figures 16.34 through 16.37.

TABLE 16.1. Maximum and minimum q

f_c' (ksi)	f_y (ksi)	ρ_b	ρ_{min}	q_{max} (percent) (for ρ_{min})	q_{min} (percent) (for $0.5\rho_b$)
3	40	0.0371	0.0050	17.3	10
	60	0.0215	0.0033	16.9	10
4	40	0.0496	0.0050	18.0	10
	60	0.0285	0.0033	17.7	10
5	40	0.0581	0.0050	18.3	10
	60	0.0339	0.0033	18.0	10

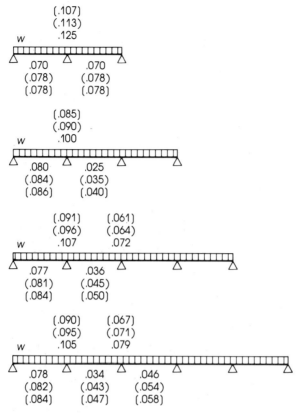

FIGURE 16.34. Bending moment coefficients K for uniform dead load on equal spans. Moment $M = KwL^2$, where w is uniform load per unit length and L is span. Values in parentheses are for moment redistribution of 10 percent; those in brackets are for 15 percent.

EXAMPLE 16.8

Determine the maximum elastic moments at the supports and midspans of the continuous beam of four equal spans shown in Figure 16.38(a) (page 538). The beam has a uniform section and carries a uniform dead load = 8 k/ft and a live load = 6 k/ft. Assume 10 percent maximum redistribution of moments and consider the following two cases: (1) When the live load is placed on alternate spans, calculate the maximum positive moments within the spans, and (2) when the live load is placed on adjacent spans, calculate the maximum negative moments at the supports.

Solution   1. The beam has a uniform moment of inertia I and has the same E; thus EI is constant. Using the three-moment equation to analyze the beam and for a constant EI, the equation is

$$M_A L_1 + 2M_B(L_1 + L_2) + M_C L_2 = -\frac{w_1 L_1^3}{4} - \frac{w_2 L_2^3}{4}$$

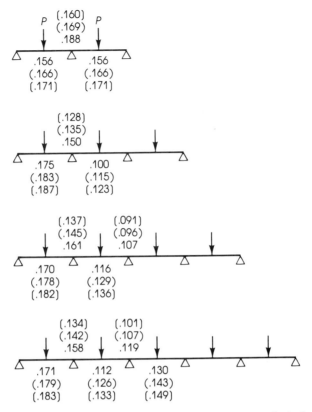

FIGURE 16.35. Bending moment coefficients K for concentrated dead load at midspan of equal spans. Moment $M = KPL$, where P is concentrated load at each midspan and L is span. Values in parentheses are for moment redistribution of 10 percent; those in brackets are for 15 percent.

Since the spans are equal, then

$$M_A + 4M_B + M_C = -\frac{L^2}{4}(w_1 + w_2) \qquad (16.25)$$

In this example $M_A = M_E = 0$. Six different cases of loading will be considered, as shown in Figure 16.38:

Case 1: Dead load is placed on the whole beam ABCDE (Figure 16.38(b)).

Case 2: Live load is placed on AB and CD for maximum positive moments within AB and CD (Figure 16.38(c)).

Case 3: Similar to case 2 for beams BC and DE (Figure 16.38(d)).

Case 4: Live load is placed on AB, BC, and DE for a maximum negative moment at B (Figure 16.38(e)).

Case 5: Live load is placed on spans CD and DE (Figure 16.38(f)).

Case 6: Live load is placed on BC and CD for a maximum negative moment at C (Figure 16.38(g)).

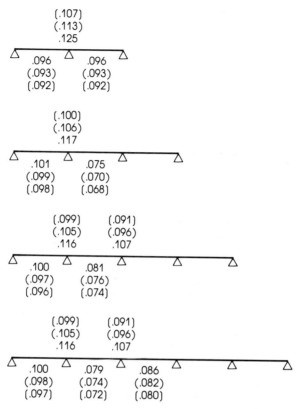

FIGURE 16.36. Bending moment coefficients K for uniform live load on equal spans. Moment $M = KwL^2$, where w is uniform load per unit length and L is span. Values in parentheses are for moment redistribution of 10 percent; those in brackets are for 15 percent. Note that values of K are for live load acting on different spans, producing maximum and minimum moments.

2. Case 1: Apply equation 16.25 to the beam segments ABC, BCD, and CDE, respectively:

$$4M_B + M_C = -\frac{(20)^2}{4}(8 + 8) = 1600 \text{ k·ft}$$

$$M_B + 4M_C + M_D = -1600 \text{ k·ft}$$

$$M_C + 4M_D = -1600 \text{ k·ft}$$

Solve the three equations to get:

$$M_B = M_D = -342.8 \text{ k·ft} \quad \text{and} \quad M_C = -228.6 \text{ k·ft}$$

For a 10 percent reduction in moments,

$$M'_B = M'_D = 0.9(-342.8) = -308.5 \text{ k·ft}$$

$$M'_C = 0.9(-228.6) = -205.7 \text{ k·ft}$$

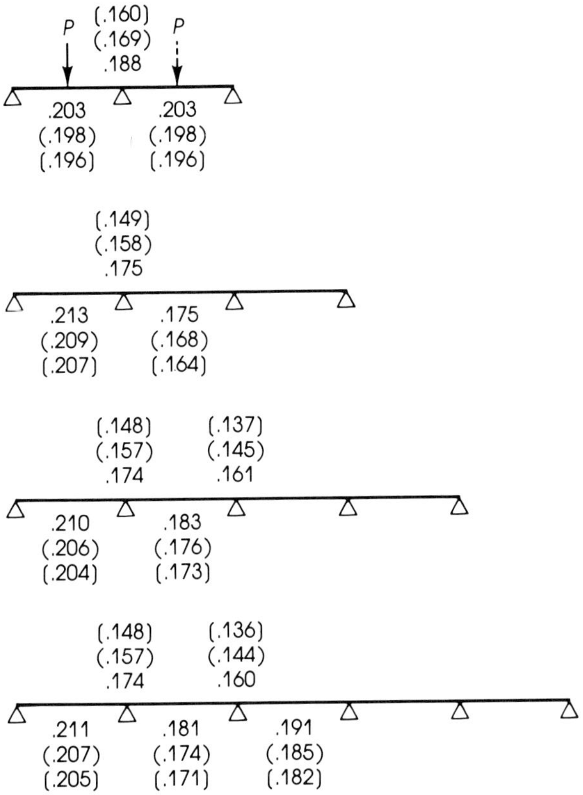

FIGURE 16.37. Bending moment coefficients K for concentrated live load at midspan of equal spans. Moment $M = KPL$, where P is concentrated load at midspan and L is span. Values in parentheses are for moment redistribution of 10 percent; those in brackets are for 15 percent. Note that values of K are for live load acting on different spans, producing maximum and minimum moments.

The corresponding midspan moments are

$$\text{Span AB} = \text{DE} = \frac{w_D L^2}{8} + \frac{1}{2} M_B = \frac{8(20)^2}{8} - \frac{1}{2} \times 308.5 = 245.8 \text{ k·ft}$$

$$\text{Span BC} = \text{CD} = \frac{w_D L^2}{8} - \frac{1}{2}(308.5 + 205.7) = \frac{8(20)^2}{8} - 257.1 = 142.9 \text{ k·ft}$$

3. Case 2: Apply equation 16.25 to ABC, BCD, and CDE, respectively:

$$4 M_B + M_C = -\frac{(20)^2}{4}(6) = -600 \text{ k·ft}$$

$$M_B + 4 M_C + M_D = -600 \text{ k·ft}$$

$$M_C + 4 M_D = -600 \text{ k·ft}$$

538

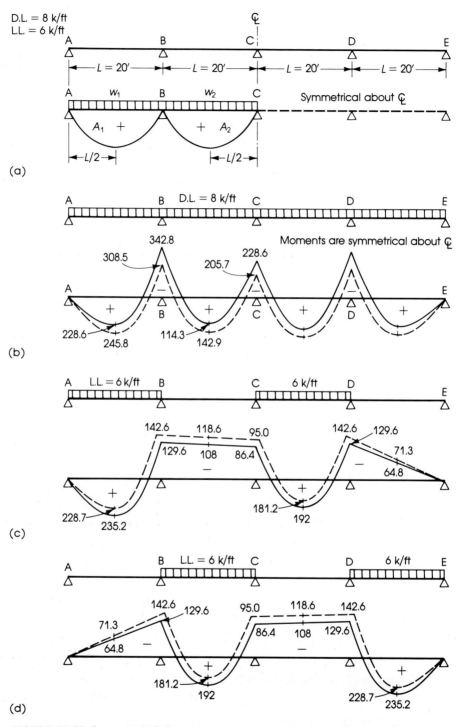

FIGURE 16.38. Example 16.8. Bending moments are drawn on the tension side.

539

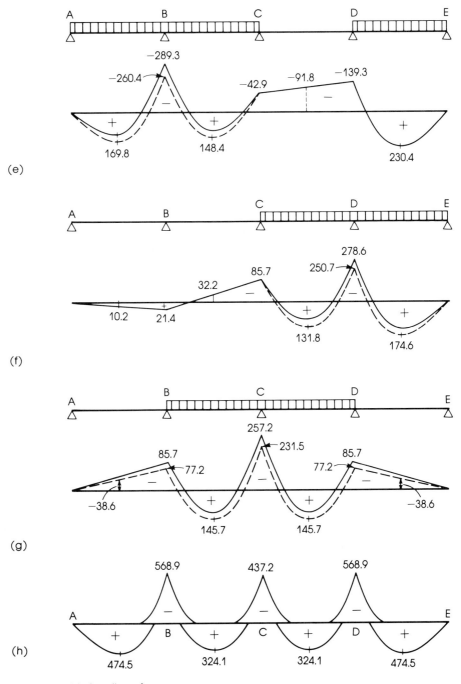

FIGURE 16.38. (continued)

Solve the three equations to get

$$M_B = M_D = -129.6 \text{ k·ft} \qquad M_C = -86.4 \text{ k·ft}$$

The corresponding elastic midspan moments are:

$$\text{Beam AB} = \frac{w_L L^2}{8} + \frac{M_B}{2} = \frac{6(20)^2}{8} - \frac{129.6}{2} = +235.2 \text{ k·ft}$$

$$BC = 0 - \frac{1}{2}(129.6 + 86.4) = -108 \text{ k·ft}$$

$$CD = \frac{w_L L^2}{8} - \frac{1}{2}(129.6 + 86.4) = \frac{6(20)^2}{8} - 108 = +192 \text{ k·ft}$$

$$DE = 0 - \frac{1}{2} \times 129.6 = -64.8 \text{ k·ft}$$

To reduce the span positive moment, increase the support moments by 10 percent and calculate the corresponding positive span moments. The resulting positive moment must be ≥90 percent of the first calculated moments given above.

$$M'_B = M'_D = 1.1(-129.6) = -142.6 \text{ k·ft}$$
$$M'_C = 1.1(-86.4) = -95.0 \text{ k·ft}$$

The corresponding midspan moments are

$$\text{Beam AB} = \frac{w_L L^2}{8} + \frac{M'_B}{2} = \frac{6(20)^2}{8} - \frac{142.6}{2} = +228.7 \text{ k·ft}$$

$$BC = -\frac{1}{2}(142.6 + 95) = -118.8 \text{ k·ft}$$

$$CD = \frac{w_L L^2}{8} + \frac{1}{2}(M'_C + M'_D) = \frac{6(20)^2}{8} - \frac{1}{2}(95 + 142.6) = +181.2 \text{ k·ft}$$

$$DE = -\frac{1}{2} \times 142.6 = -71.3 \text{ k·ft}$$

4. Case 3: This case is similar to case 2 and the moments are shown in Figure 16.38(d).
5. Case 4: Consider the spans AB, BC, and DE loaded with live load to determine the maximum negative moment at support B:

$$4M_B + M_C = -\frac{w_L L^2}{2} = -\frac{6(20)^2}{2} = -1200 \text{ k·ft}$$

$$M_B + 4M_C + M_D = -\frac{w_L L^2}{4} = -\frac{6(20)^2}{4} = -600 \text{ k·ft}$$

$$M_C + 4M_D = -\frac{6(20)^2}{4} = -600 \text{ k·ft}$$

Solve the three equations to get

$$M_C = -42.9 \text{ k·ft}$$
$$M_B = -289.3 \text{ k·ft}$$
$$M_D = -139.3 \text{ k·ft}$$

For 10 percent reduction in moment at support B,

$$M'_B = 0.9 \times (-289.3) = -260.4 \text{ k·ft}$$

The corresponding midspan moments are:

$$\text{Beam AB} = \frac{w_L L^2}{8} + \frac{M_B}{2} = \frac{6(20)^2}{8} - \frac{260.4}{2} = 169.8 \text{ k·ft}$$

$$\text{BC} = \frac{w_L L^2}{8} - \frac{1}{2}(260.4 + 42.9) = +148.4 \text{ k·ft}$$

$$\text{CD} = -\frac{1}{2}(42.9 + 139.3) = -91.1 \text{ k·ft}$$

$$\text{DE} = 300 - \frac{1}{2} \times 139.3 = +230.4 \text{ k·ft}$$

6. Case 5: This is similar to case 4 except that one end span is not loaded to produce maximum positive moment at support B (or support D for similar loading). The bending moment diagrams are shown in Figure 16.38(f).
7. Case 6: Consider the spans BC and CD loaded with live load to determine the maximum negative moment at support C:

$$4M_B + M_C = \frac{w_L L^2}{4} = -600 \text{ k·ft}$$

$$M_B + 4M_C + M_D = -\frac{w_L L^2}{2} = -1200 \text{ k·ft}$$

$$M_C + 4M_D = -\frac{w_L L^2}{4} = -600 \text{ k·ft}$$

Solve the three equations to get

$$M_C = -257.2 \text{ k·ft}$$
$$M_B = M_D = -85.7 \text{ k·ft}$$

For 10 percent reduction in support moments:

$$M'_C = 0.9 \times (-257.2) = -231.5 \text{ k·ft}$$
$$M'_B = M'_D = 0.9 \times (-85.7) = -77.2 \text{ k·ft}$$

The corresponding midspan moments are

$$\text{Beam AB} = \text{DE} = -\frac{77.2}{2} = -38.6 \text{ k·ft}$$

$$\text{BC} = \text{CD} = \frac{w_L L^2}{8} - \frac{1}{2}(231.5 + 77.2) = \frac{6(20)^2}{8} - 154.3 = 145.7 \text{ k·ft}$$

8. The final maximum and minimum moments after moment redistribution are shown in Table 16.2. The moment envelope is shown in Figure 16.38(h).
9. In this example, the midspan sections are used for simplicity; the midspan moments are not necessarily the maximum positive moments. In the case of the end spans AB and DE, the maximum moment after 10 percent moment redistribution is equal to $(w_D L^2)/12.2$ and

TABLE 16.2. Final moments of example 16.8 after moment redistribution

Case	1	2	3	4	5
				D.L. + L.L. (1) + (2)	D.L. + L.L. (1) + (3)
Section location	D.L. moments	L.L. maximum negative	L.L. maximum positive	maximum negative	maximum positive
Support					
A	0	0	0	0	0
B	−308.5	−260.4	+21.4	−568.9*	−287.1
C	−205.7	−231.5	—	−437.2*	−205.7
D	−308.5	−260.4	+21.4	−568.9*	−287.1
E	0	0	0	0	0
Midspan					
AB	245.8	− 71.3	228.7	+174.5	+474.5*
BC	142.9	−118.6	181.2	+ 24.3	+324.1*
CD	142.9	−118.6	181.2	+ 24.3	+324.1*
DE	245.8	− 71.3	228.7	+174.5	+474.5*

* Final maximum or minimum design moment.

occurs at $0.4L$ from A and D. Bending moment coefficients for design moments with 10 percent and 15 percent redistribution of moments are shown in Figures 16.34 through 16.37.

SUMMARY

SECTIONS 16.1 to 16.3

In continuous beams, the maximum and minimum moments are obtained by considering the dead load acting on all spans, while pattern loading is considered for live or moving loads, as shown in Figures 16.2 and 16.3. The ACI moment coefficients given in Chapter 9, Figure 9.8, may be used to compute approximate values for the maximum and minimum moments and shears.

SECTIONS 16.4 to 16.6

A frame subjected to a system of loads may be analyzed by the equivalent frame method. Frames may be statically determinate or indeterminate.

SECTION 16.7

There are several types of frame hinges: Mesnager, Considère, lead, and concrete hinges. The steel for a Mesnager hinge is calculated as follows:

$$A_{s1} = \frac{R_1}{0.55f_y} \quad \text{and} \quad A_{s2} = \frac{R_2}{0.55f_y} \tag{16.2}$$

$$\text{Bursting force } F = \frac{P_u}{2} \tan\theta + \frac{Ha}{0.85d} \tag{16.3}$$

$$\text{Stress in ties } f_s = \frac{F}{0.005ab + A_{st}(\text{ties})} \leq 0.85f_y \tag{16.5}$$

SECTIONS
16.8 and
16.9

Limit design in reinforced concrete refers to redistribution of moments, which occurs throughout the structure as steel reinforcement reaches its yield strength. Ultimate strength is reached when the structure is on the verge of collapse. This case occurs when a number of plastic hinges n develops in a structure with redundants r, such that $n = 1 + r$.

SECTIONS
16.10 to
16.12

For limit design to be valid, four conditions must be satisfied: mechanism, equilibrium, yield, and rotation. Two methods of analysis may be used: the virtual work method and the equilibrium method, which are both explained in examples 16.3 through 16.6.

SECTIONS
16.13 and
16.14

The plastic hinge length l_p can be considered equal to the effective depth d. In fibrous concrete,

$$l_p = (1.06 + 0.13\rho\rho_s)d \tag{16.13}$$

Ductility index $\mu = \phi_u/\phi_y$

For fibrous concrete,

$$\mu' = (1.0 + 3.8\rho_s)\mu \tag{16.15}$$

$$\text{Angle of rotation } \theta = \frac{0.0035}{\lambda} - \frac{f_y}{E_s(1 - \lambda)} \tag{16.18}$$

$$\lambda = \frac{\rho f_y}{0.72 f'_c} \le 0.5 \tag{16.19}$$

A summary of limit design procedure is given in section 16.14.

SECTION
16.15

Moment redistribution may be taken into account in the analysis of statically indeterminate structures. In this case, the maximum negative moments calculated by the elastic theory may be increased or decreased by not more than the ratio q, where

$$q = 20 \left(1 - \frac{\rho - \rho'}{\rho_b}\right) \tag{16.24}$$

Table 16.1 gives the different values of q. Moment redistribution is explained in detail in example 16.8.

REFERENCES

1. J. C. McCormac: *Structural Analysis*, Intext, New York, 1975.
2. M. N. Hassoun: *Ultimate Load Design of Reinforced Concrete*, Cement and Concrete Association, London, 1981.
3. A. L. L. Baker: *The Ultimate Load Theory Applied to the Design of Reinforced and Prestressed Concrete Frames*, Concrete Publications, London, 1956.
4. W. Prager and P. G. Hodge: *Theory of Perfectly Plastic Solids*, Wiley, New York, 1951.
5. L. S. Beedle, B. Thürlimann and R. L. Ketter: *Plastic Design in Structural Steel*, American Institute of Steel Construction, New York, 1955.
6. J. F. Baker, M. R. Horne and J. Heyman: *The Steel Skeleton*, Vol. 2, Cambridge University Press, London, 1956.

7. A. L. L. Baker: *Limit-State Design of Reinforced Concrete,* Cement and Concrete Association, London, 1970.
8. W. B. Cranston: "Tests on Reinforced Concrete Portal Frames," Cement and Concrete Association Technical Report TRA-392, London, 1965.
9. B. G. Neal: *The Plastic Methods of Structural Analysis,* Chapman and Hall, London, 1956.
10. L. S. Beedle: *Plastic Design of Steel Frames,* Wiley, New York, 1958.
11. M. R. Horne: *Plastic Theory of Structures,* The M.I.T. Press, Cambridge, MA, 1971, pp. 16–17.
12. W. G. Corley: "Rotational Capacity of Reinforced Concrete Beams," *ASCE Journal, Structural Division,* 92, October 1966.
13. A. H. Mattock, "Discussion" [of paper, ref. 12], *ASCE Journal, Structural Division,* 93, April 1967.
14. K. Sahebjam: "The Effect of Steel Fibers on the Plastic Rotation of Reinforced Concrete Continuous Beams," M.Sc. Thesis, South Dakota State University, Brookings, SD, 1984.
15. W. Chan: "The Ultimate Strength and Deformation of Plastic Hinges in Reinforced Concrete Frameworks," *Magazine of Concrete Research,* November 1955.
16. S. K. Kaushik, L. M. Ramamurthy and C. B. Kukreja: "Plasticity in Reinforced Concrete Continuous Beams, with Parabolic Soffits," *ACI Journal,* September–October 1980.
17. R. W. Furlong: "Design of Concrete Frames by Assigned Moment Limits," *ACI Journal,* 67, April 1970.
18. P. G. Hodge: *Plastic Analysis of Structures,* McGraw-Hill, New York, 1959.
19. M. Hilal, *Design of Reinforced Concrete Halls,* J. Marcou and Company, Cairo, 1971.
20. V. Ramakrishnan and P. D. Arthur: *Ultimate Strength Design for Structural Concrete,* Wheeler Publishing, India, 1977.
21. *Structural Design of Tall Concrete and Masonry Buildings,* American Society of Civil Engineers, New York, 1978.

PROBLEMS

16.1. The slab-beam floor system shown in Figure 16.39 carries a uniformly distributed dead load (excluding weight of slab and beam) of 40 psf and a live load of 100 psf. Using the ACI Code coefficients, design the interior continuous beam ABCD and draw detailed sections. Given: $f'_c = 3$ ksi, $f_y = 40$ ksi, width of beam web = 12 in., slab thickness = 4.0 in., and column dimensions = 14 by 14 in.

16.2. Repeat problem 16.1 using span lengths of the beams shown in Figure 16.39 as follows: $L_1 = 20$ ft, $L_2 = 24$ ft, $L_3 = 20$ ft, and $L_4 = 10$ ft.

16.3. For the beam shown in Figure 16.40, compute the reactions at A, B, and C using constant *EI.* Draw the shear and bending moment diagrams and design all critical sections, using $b = 14$ in., $h = 25$ in., $f'_c = 4$ ksi, $f_y = 60$ ksi, and a load factor = 1.6.

16.4. Repeat problem 16.3 using span lengths of beams as follows: span AB = 20 ft and span BC = 16 ft.

16.5. The two-hinged portal frame ABCD shown in Figure 16.41 carries a uniform dead load (excluding self-weight) = 2.0 k/ft and a uniform live load = 1.8 k/ft. Design the frame ABCD, the hinges, and footings using $f'_c = 4$ ksi, $f_y = 60$ ksi, and a beam width $b = 16$ in. The footing is placed 5 ft below ground level and the allowable bearing soil pressure = 5 ksf.

16.6. Design the portal frame ABCD of problem 16.5 if the frame ends at A and D are fixed.

16.7 and 16.8. Calculate the collapse moments at the critical sections of the beams shown in Figures 16.42 and 16.43.

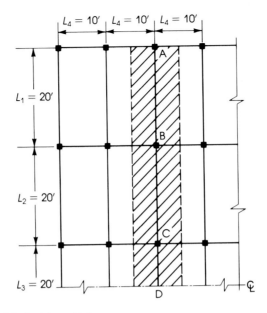

FIGURE 16.39. Problem 16.1.

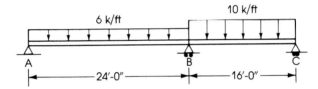

FIGURE 16.40. Problem 16.3.

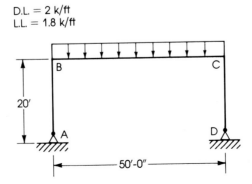

FIGURE 16.41. Problem 16.5.

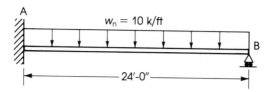

FIGURE 16.42. Problem 16.7.

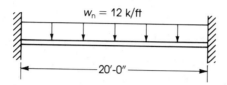

FIGURE 16.43. Problem 16.8.

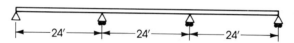

FIGURE 16.44. Problem 16.10.

16.9. If the beam shown in Figure 16.43 carries a uniform dead load of 2 k/ft and a live load of 2.4 k/ft, design the beam using the limit design procedure. Use $f'_c = 4$ ksi, $f_y = 60$ ksi, and a beam width $b = 14$ in.

16.10. Determine the maximum and minimum elastic moments at the supports and midspans of the three-span continuous beam shown in Figure 16.44. The beam has a uniform rectangular section and carries a uniform dead load of 6 k/ft and a live load of 5 k/ft. Assuming 10 percent maximum redistribution of moments, recalculate the maximum and minimum moments at the supports and midspans of the beam ABC. Note: Place the live load on alternate spans to calculate maximum positive moments and on adjacent spans to calculate the maximum negative (minimum) moments (example 16.8).

16.11. Repeat problem 16.10 if the beam consists of four equal spans of 24-ft lengths.

17

Design of Two-Way Slabs

17.1 Introduction

Slabs can be considered as structural members whose depth h is small as compared to their length L and width S. The simplest form of a slab is one supported on two opposite sides, which primarily deflects in one direction and is referred to as a one-way slab. The design of one-way slabs was discussed in Chapter 9.

When the slab is supported on all four sides, and the length L is less than twice the width S, the slab will deflect in two directions and the loads on the slab are transferred to all four supports. This slab is referred to as a two-way slab. The bending moments and deflections in such slabs are considerably less than those in one-way slabs; thus, the same slab can carry more load when supported on four sides. The load in this case is carried in two directions, and the bending moment in each direction is much less than the bending moment in the slab if the load were carried in one direction only. Typical slab-beam-girder arrangements of one-way and two-way slabs are shown in Figure 17.1.

In one-way slabs, the main reinforcement is placed along the short direction, while shrinkage and temperature reinforcement is placed along the long direction. Loads on the slab are transferred to the supporting beams on two sides only, which in turn are supported by girders or columns (Figure 17.1(a)). In two-way slabs, reinforcement in each direction is considered as main reinforcement. Loads on the slab are transferred to the supporting beams on four sides, which in turn are supported by columns (Figure 17.1(b)).

When a two-way slab is not supported by beams, the loads are transferred directly to the supporting columns, and the slab is referred to as a flat slab or a flat plate (Figure 17.2). Due to the absence of supporting beams, the columns tend to punch through the slab, causing punching shear stresses in the slab and inclined cracking around the columns. To prevent the failure of the slab due to punching shear, the thickness of the slab around the column may be increased by (1) using a drop panel, (2) enlarging the top of the column in the shape of a frustrum known as the column capital, or (3) providing both a drop panel and a column capital. This type of slab is referred to as a flat slab

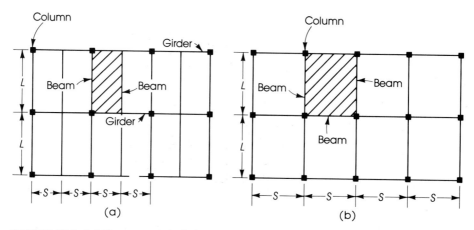

FIGURE 17.1. (a) One-way slab, $L/S > 2$, and (b) two-way slab, $L/S \leq 2$.

(Figure 17.2(b)). If the drop panel and the column capital are not used, the slab will be of uniform depth and is referred to as a flat plate (Figure 17.2(a)). In this case, the shear resistance is obtained by increasing the slab thickness or by the use of special shear reinforcement. The reinforcement may consist of steel I or channel shapes or multiple bar stirrups placed in the slab above the columns.

In the flat-slab floor system, the slab may have recesses made by using special

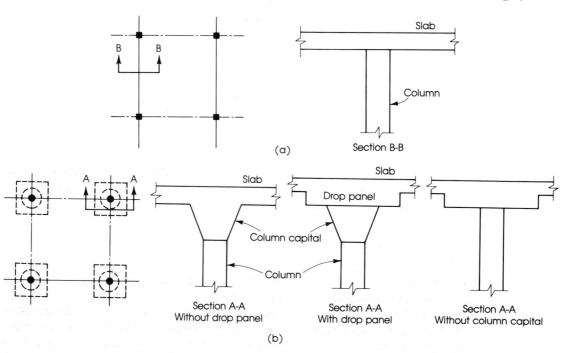

FIGURE 17.2. Two-way slabs without beams: (a) flat plate floor and section; (b) flat slab floor and sections; (c) ribbed slab and sections.

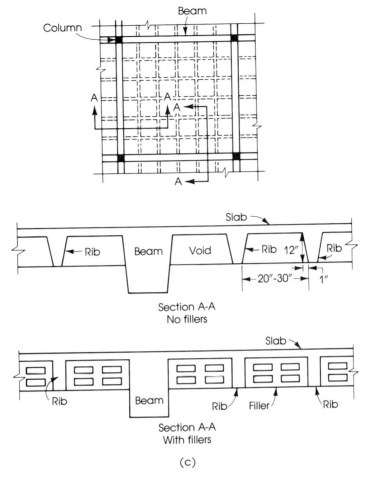

FIGURE 17.2 (continued).

removable pans to form a series of ribs or joists in two directions. This type of slab is referred to as a waffle or two-way ribbed slab (Figure 17.2(c)). The details of ribbed slabs are given in Chapter 9, section 9.2.

17.2   Design Concepts

An exact analysis of forces and displacements in a two-way slab is complex due to its highly indeterminate nature; this is true even when the effects of creep and nonlinear behavior of the concrete are neglected. Numerical methods such as finite elements can be used, but simplified methods such as those presented by the ACI Code are more suitable for practical design. The **ACI Code, Chapter 13,** assumes that the slabs behave as wide shallow beams that form, with the columns above and below them, a rigid frame. The validity of this assumption of dividing the structure into equivalent frames has been verified by analytical[1,2] and experimental[3,4] research. It is also established[3,5] that ultimate load capacity of two-way slabs with restrained boundaries is about twice that calculated by theoretical analysis, because a great deal of moment redistribution occurs in the slab before failure. At high loads, large deformations and deflections are expected; thus a minimum slab thickness is required to maintain adequate deflection and cracking conditions under service loads.

The ACI Code specifies two methods for the design of two-way slabs:

1. The direct design method **(ACI Code, section 13.6)** is an approximate procedure for the analysis and design of two-way slabs. It is limited to slab systems
   subjected to uniformly distributed loads and supported on equally or nearly
   equally spaced columns. The method uses a set of coefficients to determine the
   design moments at critical sections. Two-way slab systems that do not meet
   the limitations of the **ACI Code, section 13.6.1,** must be analyzed by more accurate procedures.
2. The equivalent frame method **(ACI Code, section 13.7)** is one in which a three-
   dimensional building is divided into a series of two-dimensional equivalent
   frames by cutting the building along lines midway between columns. The resulting frames are considered separately in the longitudinal and transverse directions of the building and treated floor by floor, as shown in Figure 17.3.

The two ACI Code procedures are based on the results of elastic analysis of the structure as a whole using factored loads.

In addition to the ACI Code procedures, a number of other alternatives are available for the analysis and design of slabs. The resulting slabs may have a greater or lesser amount of reinforcement. The analytical methods may be classified in terms of the basic relationship between load and deformation as elastic, plastic, and nonlinear.

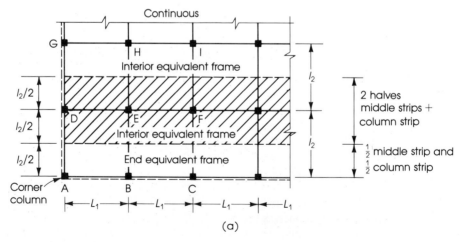

FIGURE 17.3. (a) Longitudinal and (b) transverse equivalent frames in plan view and (c) in elevation and perspective views.

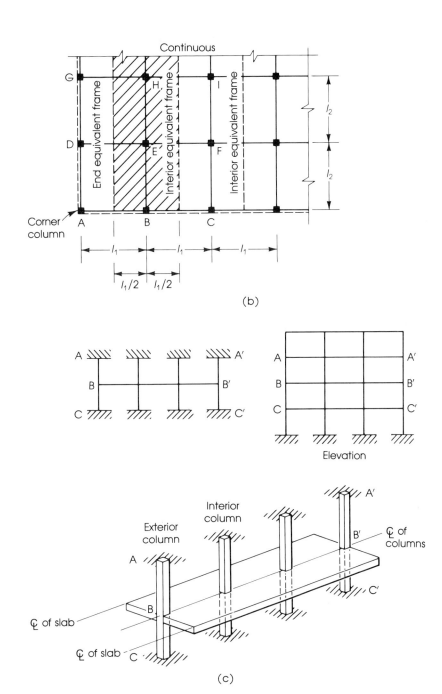

FIGURE 17.3 (continued).

1. In *elastic analysis,* a concrete slab may be treated as an elastic plate. The flexure, shear, and deflection may be calculated by the fourth differential equation relating load to deflection for thin plates with small displacements, as presented by Timoshenko.[6] Finite-difference as well as finite-element solutions have been proposed to analyze slabs and plates.[7,8] In the finite-element method, the slab is divided into a mesh of triangles or quadrilaterals. The displacement functions of the nodes (intersecting mesh points) are usually established and the stiffness matrices are developed for computer analysis.

2. For *plastic analysis,* three methods are available. The *yield line* method was developed by Johansen[9] to determine the limit state of the slab by considering the yield lines which occur in the slab as a collapse mechanism. The *strip* method was developed by Hillerborg.[10] The slab is divided into strips, and the load on the slab is distributed in two orthogonal directions. The strips are analyzed as simple beams. The third method is *optimal analysis.* There has been considerable research into optimal solutions. Rozvany and others[11] presented methods for minimizing reinforcement based on plastic analysis. Optimal solutions are complex in analysis and produce complex patterns of reinforcement.

3. *Nonlinear analysis* simulates the true load-deformation characteristics of a reinforced concrete slab when the finite-element method takes into consideration the nonlinearity of the stress-strain relationship of the individual elements.[11,12] In this case, the solution becomes complex unless simplified empirical relationships are assumed.

The above methods are presented very briefly to introduce the reader to the different methods of analysis of slabs. Experimental work on slabs has not been extensive in recent years, but more research is probably needed to simplify current design procedures with adequate safety, serviceability, and economy.[11]

17.3   Column and Middle Strips

Figure 17.4 shows an interior panel of a two-way slab supported on the columns A, B, C, and D. If the panel is loaded uniformly, the slab will deflect in both directions, with maximum deflection at the center O. The highest points will be at the columns A, B, C, and D; thus the part of the slab around the columns will have a convex shape. A gradual change in the shape of the slab occurs, from convexity at the columns to concavity at the center of the panel O, each radial line crossing a point of inflection. Sections at O, E, F, G, and H will have positive bending moments, while the periphery of the columns will have maximum negative bending moments. Considering a strip along AFB, the strip bends like a continuous beam (Figure 17.4(b)), having negative moments at A and B and positive bending moment at F. This strip extends between the two columns A and B and continues on both sides of the panel, forming a column strip.

Similarly, a strip along EOG will have negative bending moments at E and G and a positive moment at O, forming a middle strip. A third strip along DHC will behave similarly to strip AFB. Therefore, the panel can be divided into three strips, one in the middle along EOG referred to as the ''middle strip'' and one on each side, along AFB and DHC, referred to as ''column strips'' (Figure 17.5(a)). Each of the three strips

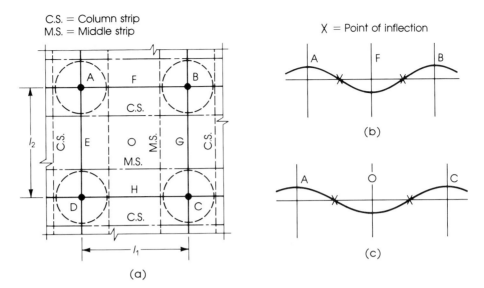

FIGURE 17.4. Plan of an interior panel.

behaves as a continuous beam. In a similar way, the panel is divided into three strips in the other direction, one middle strip along FOH and two column strips along AED and BGC, respectively (Figure 17.5(b)).

Referring to Figure 17.4(a), it can be seen that the middle strips are supported on the column strips, which in turn transfer the loads onto the columns, A, B, C, and D in this panel. Therefore, the column strips carry more load than the middle strips. Consequently, the positive bending moment in each column strip (at E, F, G, and H) is greater

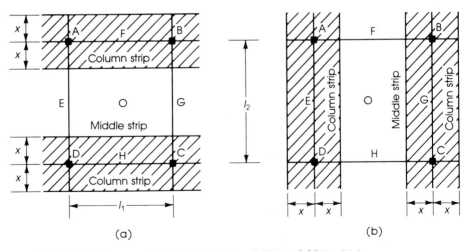

FIGURE 17.5. Column and middle strips; $x = 0.25 \, l_1$ or $0.25 \, l_2$, whichever is smaller.

than the positive bending moment at O in the middle strip. Also, the negative moments at the columns A, B, C, and D in the column strips are greater than the negative moments at E, F, G, and H in the middle strips. The portions of the design moments assigned to each critical section of the column and middle strips will be discussed in section 17.6.

The extent of each of the column and middle strips in a panel is defined by the **ACI Code, section 13.2.1**. The column strip is defined by a slab width on each side of the column centerline, x in Figure 17.5, equal to one-fourth the smaller of the panel dimensions l_1 and l_2, including beams if they are present, where

> l_1 = span length, center to center of supports, in the direction moments are being determined
>
> l_2 = span length, center to center of supports, in the direction perpendicular to l_1.

The portion of the panel between the two column strips defines the middle strip.

17.4   Minimum Slab Thickness to Control Deflection

The **ACI Code, section 9.5**, specifies a minimum slab thickness in two-way slabs to control deflection. The magnitude of a slab's deflection depends on many variables, including the flexural stiffness of the slab, which in turn is a function of the slab thickness h. By increasing the slab thickness, the flexural stiffness of the slab is increased and consequently the slab deflection is reduced.[13] Because the calculation of deflections in two-way slabs is complicated, and to avoid excessive deflections, the ACI Code limits the thickness of these slabs by adopting the following three empirical equations, which are based on experimental research. If these limitations are met, it will not be necessary to compute deflections.

$$\text{Minimum } h = \frac{l_n(800 + 0.005f_y)}{36,000 + 5000\beta[\alpha_m - 0.5(1 - \beta_s)(1 + 1/\beta)]}$$

(ACI equation 9.11)   *(17.1)*

but not less than

$$h = \frac{l_n(800 + 0.005f_y)}{36,000 + 5000\beta(1 + \beta_s)}$$ **(ACI equation 9.12)**   *(17.2)*

Also, the slab thickness need not be more than:

$$h_{max} = \frac{l_n(800 + 0.005f_y)}{36,000}$$ **(ACI equation 9.13)**   *(17.3)*

where

> l_n = clear span in the long direction measured face to face of columns (or face to face of beams for slabs with beams)
>
> β = the ratio of the long to the short clear spans
>
> β_s = the ratio of the length of continuous edges to the total perimeter of the slab panel under consideration

α_m = the average value of α for all beams on the sides of a panel

α = the ratio of flexural stiffness of a beam section $E_{cb}I_b$ to the flexural stiffness of the slab $E_{cs}I_s$, bounded laterally by the centerlines of the panels on each side of the beam

That is,

$$\alpha = \frac{E_{cb}I_b}{E_{cs}I_s} \qquad (17.4)$$

where E_{cb} and E_{cs} are the moduli of elasticity of concrete in the beam and the slab, respectively,

I_b = the gross moment of inertia of the beam section about the centroidal axis (the beam section includes a slab length on each side of the beam equal to the projection of the beam above or below the slab, whichever is greater, but not more than four times the slab thickness)

I_s = the moment of inertia of the gross section of the slab

If no beams are used, as in the case of flat plates, then $\alpha = 0$. The ACI Code equations for calculating slab thickness h take into account the effect of the span length, the panel shape, the steel reinforcement yield stress f_y, continuity at the boundaries, and the flexural stiffness of beams. When beams are not used, $\alpha_m = 0$, and the denominator in equation 17.1 becomes negative and the minimum slab thickness is determined from equation 17.3. When very stiff beams are used, equation 17.1 may give a small slab thickness, and equation 17.2 may control.

Other ACI Code limitations are summarized below:

1. For panels with discontinuous edges, end beams with a minimum α equal to 0.8 must be used, otherwise the minimum slab thickness calculated by equations 17.1 through 17.3 must be increased by at least 10 percent **(ACI Code, section 9.5.3.3)**.
2. When drop panels are used without beams, the minimum slab thickness may be reduced by 10 percent **(ACI Code, section 9.5.3.2)**. The drop panels should extend in each direction from the centerline of support a distance not less than one-sixth of the span length in that direction between center to center of supports, and also project below the slab at least $h/4$.
3. Regardless of the values obtained by equations 17.1 through 17.3, the thickness of two-way slabs shall not be less than the following: (1) for slabs without beams or drop panels: 5 in. (125 mm); (2) for slabs without beams, but with drop panels: 4 in. (100 mm); (3) for slabs with beams on all four sides with $\alpha_m \geq 2.0$: $3\frac{1}{2}$ in. (90 mm) **(ACI Code, section 9.5.3.1)**.

EXAMPLE 17.1

A flat-plate floor system with panels 24 by 20 ft is supported on 20 in. square columns. Using the ACI Code equations, determine the minimum slab thickness required for the interior and corner panels shown in Figure 17.6. Edge beams are not used. Use $f'_c = 3$ ksi and $f_y = 60$ ksi.

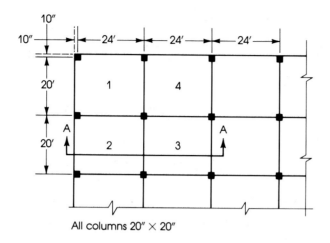

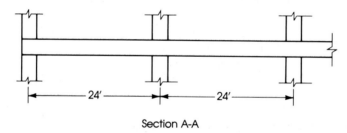

Section A-A

FIGURE 17.6. Example 17.1.

Solution　　1. For corner panel 1:
Using equations 17.1 through 17.3,

$$\text{Clear span} = l_n = 24 - \frac{20}{12} = 22.33 \text{ in.}$$

in the long direction and

$$l_n = 20 - \frac{20}{12} = 18.33 \text{ in.}$$

in the short direction.

$$\beta = \frac{22.33}{18.33} = 1.22$$

$$\alpha = \frac{I_b}{I_s} = 0 \qquad \text{then } \alpha_m = 0$$

$$\beta_s = \frac{24 + 20}{2 \times 24 + 2 \times 20} = 0.5$$

Therefore minimum thickness

$$h_{min} = \frac{(22.33 \times 12)(800 + 0.005 \times 60,000)}{36,000 + (5000 \times 1.22)[0 - 0.5(1 - 0.5)(1 + 1/1.22)]} = 8.86 \text{ in.}$$

but not less than

$$h = \frac{(22.33 \times 12)(800 + 0.005 \times 60,000)}{36,000 + (5000 \times 1.22)(1 + 0.5)} = 6.53 \text{ in.}$$

or more than

$$h_{max} = \frac{(22.33 \times 12)(800 + 0.005 \times 60,000)}{36,000} = 8.19 \text{ in.}$$

Minimum $h = 5$ in.

Therefore, $h = 8.19$ in. controls. Since no edge beams are used, h must be increased by at least 10 percent:

Minimum slab thickness = $8.19 \times 1.1 = 9.0$ in.

2. For the interior panel 3,

$\beta_s = 1.0$ (all sides are continuous)

$$h_{min} = \frac{(22.33 \times 12)(800 + 0.005 \times 60,000)}{36,000 + (5000 \times 1.22)[0 - 0.5(1 - 1)(1 + 1/1.22)]} = 8.19 \text{ in.}$$

but not less than

$$h = \frac{(22.33 \times 12)(800 + 0.005 \times 60,000)}{36,000 + (5000 \times 1.22)(1 + 1)} = 6.12 \text{ in.}$$

or more than 8.19 (as calculated above).

Minimum $h = 5.00$ in.

Therefore $h = 8.19$ in. controls. If a uniform slab thickness is adopted for all panels, then the slab thickness in panel 1 controls and a final choice of $h = 9.00$ in. will be adopted. ∎

EXAMPLE **17.2**

The floor system shown in Figure 17.7 consists of solid slabs and beams in two directions supported on 20-in. square columns. Using the ACI Code equations, determine the minimum slab thickness required for an interior panel. Use $f'_c = 3$ ksi and $f_y = 60$ ksi.

Solution
1. To use equation 17.1, α_m should be calculated first. Therefore, it is required to determine I_b, I_s, and α for the beams and slabs in the long and short directions.
2. The gross moment of inertia of the beam I_b is calculated for the section shown in Figure 17.7(b) made up of the beam and the extension of the slab on each side of the beam $x = y$, but not more than 4 times the slab thickness. Assume $h = 7$ in., to be checked later; then

$x = y = 22 - 7 = 15$ in. $< 4 \times 7 = 28$ in.

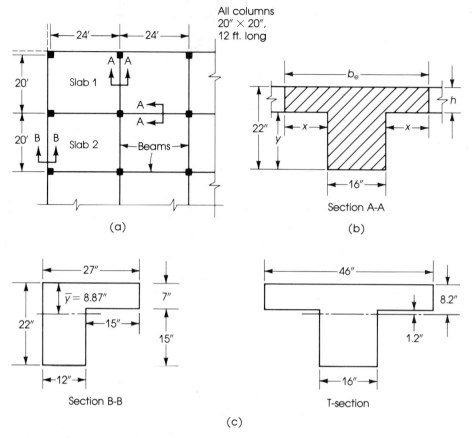

FIGURE 17.7. Example 17.2.

Therefore, $b_e = 16 + 2 \times 15 = 46$ in. and the T-section is shown in Figure 17.7(c). Determine the centroid of the section by taking moments about the top of the flange:

Area of flange $= 7 \times 46 = 322$ in.^{2} Area of web $= 16 \times 15 = 240$ in.^{2}

Total area $= 562$ in.^{2}

$(322 \times 3.5) + 240 \times (7 + 7.5) = 562\bar{y}$

$$\bar{y} = 8.20 \text{ in.}$$

$$I_b = \left[\frac{46}{12}(7)^3 + 322 \times (4.7)^2\right] + \left[\frac{16(15)^3}{12} + 240(7.5 - 1.2)^2\right] = 22{,}453 \text{ in.}^4$$

3. The moment of inertia of the slab in the long direction $I_s = (bh^3)/12$, where $b = 20$ ft and $h = 7$ in.

$$I_s = \frac{(20 \times 12)(7)^3}{12} = 6860 \text{ in.}^4$$

$$\alpha_l \text{ (in the long direction)} = \frac{EI_b}{EI_s} = \frac{22{,}453}{6860} = 3.27$$

4. The moment of inertia of the slab in the short direction

$$I_s = (bh^3)/12, \text{ where } b = 24 \text{ ft and } h = 7 \text{ in.}$$

$$I_s = \frac{(24 \times 12)(7)^3}{12} = 8232 \text{ in.}^4 \qquad \alpha_s = \frac{EI_b}{EI_s} = \frac{22{,}453}{8232} = 2.72$$

5. α_m is the average of α_l and α_s: $\alpha_m = \dfrac{3.27 + 2.72}{2} = 3.0$

6. $\beta = \dfrac{\left(24 - \dfrac{20}{12}\right)}{\left(20 - \dfrac{20}{12}\right)} = \dfrac{22.33}{18.33} = 1.22 \qquad \beta_s = \dfrac{\text{Continuous sides}}{\text{Total perimeter}} = 1.0$

7. Determine $h_{\min}$ using equation 17.1 ($l_n = 22.33$ ft):

$$h_{\min} = \frac{(22.33 \times 12)(800 + 0.005 \times 60{,}000)}{36{,}000 + (5000 \times 1.22)[3.0 - 0.5(1 - 1)(1 + 1/1.22)]} = 5.38 \text{ in.}$$

but must not be less than h given by equation 17.2:

$$h = \frac{294{,}756}{36{,}000 + (5000 \times 1.22)(1 + 1)} = 6.12 \text{ in.}$$

nor more than

$$h_{\max} = \frac{294{,}756}{36{,}000} = 8.19 \text{ in.}$$

Minimum $h = 3.5$ in.

Therefore, $h = 6.12$ in. controls > 3.5 in. A slab thickness of 6.5 in. or 7.0 in. may be adopted.

17.5   Shear Strength of Slabs

In a two-way floor system, the slab must have adequate thickness to resist both bending moments and shear forces at the critical sections. To investigate the shear capacity of two-way slabs, the following cases should be considered.

17.5.1   Two-Way Slabs Supported on Beams

The critical sections are at a distance d from the face of the supporting beams, and the shear capacity of each section $\phi V_c = \phi(2\sqrt{f_c'}\,bd)$. When the supporting beams are stiff and capable of transmitting floor loads to the columns, they are assumed to carry loads acting on floor areas bounded by 45° lines drawn from the corners, as shown in Figure 17.8. The loads on the trapezoidal areas will be carried by the long beams AB and CD while the loads on the triangular areas will be carried by the short beams AC and BD. The shear per unit width of slab is highest between E and F in both directions, and $V_u = w_u(l_2/2)$, where $w_u =$ uniform factored load per unit area.

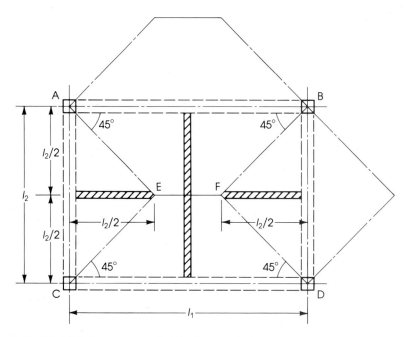

FIGURE 17.8. Areas supported by beams in two-way slab floor system.

If no shear reinforcement is provided, the shearing force at a distance d from the face of the beam V_{ud} must be

$$V_{ud} \leq \phi V_c \leq \phi(2\sqrt{f_c'}bd)$$

where $V_{ud} = w_u \left(\dfrac{l_2}{2} - d\right)$.

17.5.2   Two-Way Slabs Without Beams

In flat plats and flat slabs, beams are not provided, and the slabs are directly supported by columns. In such slabs, two types of shear stresses must be investigated; the first is one-way shear or beam shear. The critical sections are taken at a distance d from the face of the column and the slab is considered as a wide beam spanning between supports, as in the case of one-way beams. The shear capacity of the concrete section is $\phi V_c = \phi(2\sqrt{f_c'}bd)$. The second type of shear to be studied is two-way or punching shear, as was previously discussed in the design of footings. Shear failure occurs along a truncated cone or pyramid around the column. The critical section is located at a distance $d/2$ from the face of the column, column capital, or drop panel (Figure 17.9). If shear reinforcement is not provided, the shear strength

$$\phi V_c = v_c b_o d$$

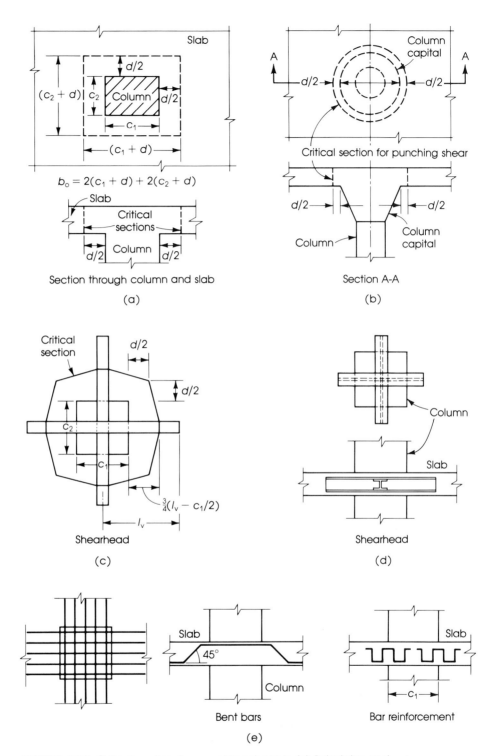

$b_o = 2(c_1 + d) + 2(c_2 + d)$

Section through column and slab

(a)

Critical section for punching shear

Section A-A

(b)

$\frac{3}{4}(l_v - c_1/2)$

Shearhead

(c)

Shearhead

(d)

Bent bars

Bar reinforcement

(e)

FIGURE 17.9. Critical section for punching shear in (a) flat plates and (b) flat slabs, and reinforcement by (c, d) shearheads and (e) anchored bars.

where

$$v_c = \phi\left(2 + \frac{4}{\beta_c}\right)\sqrt{f'_c} \leq 4\phi\sqrt{f'_c} \qquad (17.5)$$

b_o = perimeter of the critical section

β_c = ratio of the long side of column (or loaded area) to the short side

When

$\beta_c \leq 2.0,\ v_c = 4\phi\sqrt{f'_c}$ and

$$\phi V_c = \phi(4\sqrt{f'_c}b_o d) \qquad (17.6)$$

When shear reinforcement is provided, the shear strength should not exceed

$$\phi V_n \leq \phi(6\sqrt{f'_c}b_o d) \qquad (17.7)$$

17.5.3   Shear Reinforcement in Two-Way Slabs Without Beams

In flat-slab and flat-plate floor systems, the thickness of the slab selected may not be adequate to resist the applied shear stresses. In this case, either the slab thickness must be increased or shear reinforcement must be provided. The ACI Code allows the use of shear reinforcement by shearheads and anchored bars or wires.

Shearheads consist of steel I or channel shapes welded into four cross arms and placed in the slabs above the column (Figure 17.9(c,d)). Shearhead designs do not apply to exterior columns, where large torsional and bending moments must be transferred between slab and column. The **ACI Code, section 11.11.4.8**, indicates that on the critical section the nominal shear strength V_n should not exceed $4\sqrt{f'_c}b_o d$; but if shearhead reinforcement is provided, V_n should not exceed $7\sqrt{f'_c}b_o d$. To determine the size of the shearhead, the **ACI Code, section 11.11.4**, gives the following limitations:

1. The ratio α_v between the stiffness of shearhead arm $E_s I$ and that of the surrounding composite cracked section of width $(c_2 + d)$ must not be less than 0.15.
2. The compression flange of the steel shape must be located within $0.3d$ of the compression surface of the slab.
3. The depth of the steel shape must not exceed 70 times the web thickness.
4. The plastic moment capacity M_p of each arm of the shearhead is computed by

$$\phi M_p = \frac{V_u}{2\eta}\left[h_v + \alpha_v\left(l_v - \frac{c_1}{2}\right)\right] \qquad \text{(ACI Code equation 11.38)} \quad (17.8)$$

where

$\phi = 0.9$

V_u = factored shear force around the periphery of the column face

η = number of arms

h_v = depth of the shearhead

l_v = length of the shearhead measured from the centerline of the column

5. The critical slab section for shear must cross each shearhead arm at a distance equal to $\frac{3}{4}(l_v - c_1/2)$ from the column face to the end of the shearhead arm, as shown in Figure 17.9(c). The critical section must have a minimum perimeter b_0, but should not be closer than $d/2$ from the face of the column.

6. The shearhead is considered to contribute a moment resistance M_v to each slab column strip as follows:

$$M_v = \frac{\phi}{2\eta} \, \alpha_v V_u \left(l_v - \frac{c_1}{2}\right) \qquad \textbf{(ACI Code equation 11.39)} \qquad (17.9)$$

but not more than the smallest of 30 percent of the factored moment required in the column strip, or the change in the column strip moment over the length l_v, or M_p given in equation 17.8.

The use of anchored bent bars or wires is permitted by the **ACI Code, section 11.11.3**. The bars are placed on top of the column, and the possible arrangements are shown in Figure 17.9(e). When bars or wires are used as shear reinforcement, the nominal shear strength

$$V_n = V_c + V_s = (2\sqrt{f_c'})b_0 d + \frac{A_v f_y d}{s} \qquad (17.10)$$

where A_v is the total stirrup bar area and b_0 is the length of the critical section of two-way shear at a distance $d/2$ from the face of the column. The nominal shear strength V_n should not exceed $6\sqrt{f_c'}b_0 d$ **(ACI Code, section 11.11.3.2)**.

17.6   Analysis of Two-Way Slabs by the Direct Design Method

The direct design method is an approximate method established by the ACI Code to determine the design moments in uniformly loaded two-way slabs. To use this method, some limitations must be met, as indicated by the **ACI Code, section 13.6.1**.

17.6.1   Limitations

1. There must be a minimum of three continuous spans in each direction.
2. The panels must be square or rectangular; the ratio of the longer to the shorter span within a panel must not exceed 2.0.
3. Adjacent spans in each direction must not differ by more than one-third of the longer span.
4. Columns must not be offset by a maximum of 10 percent of the span length, in the direction of offset, from either axis between centerlines of successive columns.
5. All loads must be uniform, and the ratio of the live to dead load must not exceed 3.0.
6. If beams are present along all sides, the ratio of the relative stiffness of beams in two perpendicular directions $\alpha_1 l_2^2 / \alpha_2 l_1^2$ must not be less than 0.2 nor greater than 5.0.

17.6.2   Total Factored Static Moment

If a simply supported beam carries a uniformly distributed load w k/ft, then the maximum positive bending moment occurs at midspan and equals $M_o = wl_1^2/8$, where l_1 is the span length. If the beam is fixed at both ends or continuous with equal negative moments at both ends, then the total moment $M_o = M_p$ (positive moment at midspan) + M_n (negative moment at support) = $wl_1^2/8$ (Figure 17.10). Now, if the beam AB carries the load W from a slab that has a width l_2 perpendicular to l_1, then $W = w_u l_2$, and the total moment

$$M_o = \frac{(w_u l_2) l_1^2}{8}$$

where w_u = load intensity k/ft^2. In this expression, the actual moment occurs when l_1 equals the clear span between supports A and B. If the clear span is denoted by l_n, then

$$M_o = (w_u l_2) \frac{l_n^2}{8} \quad \textbf{(ACI Code equation 13.3)} \qquad (17.11)$$

The clear span l_n is measured face to face of supports in the direction in which moments are considered, but not less than 0.65 times the span length from center to center of supports. The face of the support where the negative moments should be calculated is illustrated in Figure 17.11. The length l_2 is measured in a direction perpendicular to l_n and equals the distance between center to center of supports (width of slab). The total moment M_o calculated in the long direction will be referred to here as M_{ol} and that in the short direction as M_{os}.

Once the total moment M_o is calculated in one direction, it is divided into a positive moment M_p and a negative moment M_n such that $M_o = M_p + M_n$ (Figure 17.10). Then each moment M_p and M_n is distributed across the width of the slab between the column and middle strips, as will be explained below.

17.6.3   Longitudinal Distribution of Moments in Slabs

In a typical *interior panel,* the total static moment M_o is divided into two moments, the positive moment M_p at midspan, equal to $0.35 M_o$, and the negative moment M_n at each

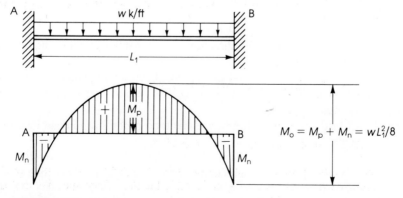

FIGURE 17.10. Bending moment in a fixed-end beam.

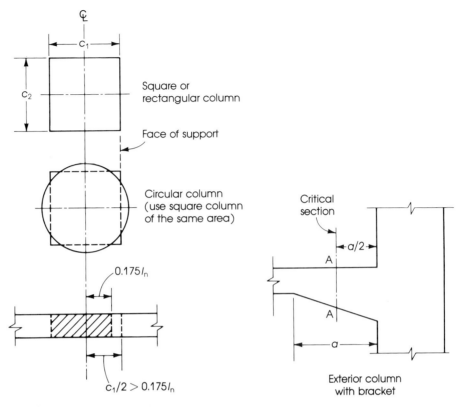

FIGURE 17.11. Critical sections for negative design moments. A-A, section for negative moment at exterior support with bracket.

support, equal to $0.65 M_o$, as shown in Figure 17.12. These values of moment are based on the assumption that the interior panel is continuous in both directions, with approximately equal spans and loads, so that the interior joints have no significant rotation. Moreover, the moment values are approximately the same as those in a fixed-end beam subjected to uniform loading where the negative moment at the support is twice the positive moment at midspan. In Figure 17.12, if $l_1 > l_2$, then the distribution of moments in the long and short directions is as follows:

$$M_{ol} = (w_u l_2) \frac{l_{nl}^2}{8}$$

$$M_{pl} = 0.35 M_{ol} \quad \text{and} \quad M_{nl} = 0.65 M_{ol}$$

$$M_{os} = (w_u l_1) \frac{l_{n2}^2}{8}$$

$$M_{ps} = 0.35 M_{os} \quad \text{and} \quad M_{ns} = 0.65 M_{os}$$

If the magnitudes of the negative moments on opposite sides of an interior support are different because of unequal span lengths, the ACI Code specifies that the larger moment should be considered to calculate the required reinforcement.

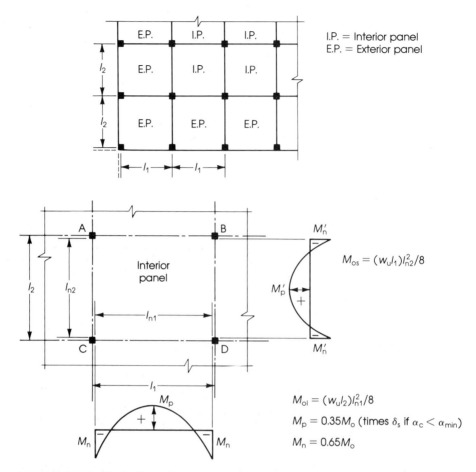

FIGURE 17.12. Distribution of moments in an interior panel.

In an *exterior panel,* the slab load is applied to the exterior column from one side only, causing an unbalanced moment and a rotation at the exterior joint. Consequently, there will be an increase in the positive moment at midspan and in the negative moment at the first interior support. The magnitude of the rotation of the exterior joint determines the increase in the moments at midspan and at the interior support. For example, if the exterior edge is a simple support, as in the case of a slab resting on a wall (Figure 17.13(a)), the slab moment at the face of the wall there is zero, the positive moment at midspan can be taken as $M_p = 0.63M_o$, and the negative moment at the interior support $M_n = 0.75M_o$. These values satisfy the static equilibrium equation

$$M_o = M_p + \frac{1}{2} M_n = 0.63M_o + \frac{1}{2}(0.75M_o)$$

In a slab-column floor system, there is some restraint at the exterior joint provided by the flexural stiffness of the slab and by the flexural stiffness of the exterior columns.

According to **section 13.6.3.3 of the ACI Code,** the total static moment M_o in an

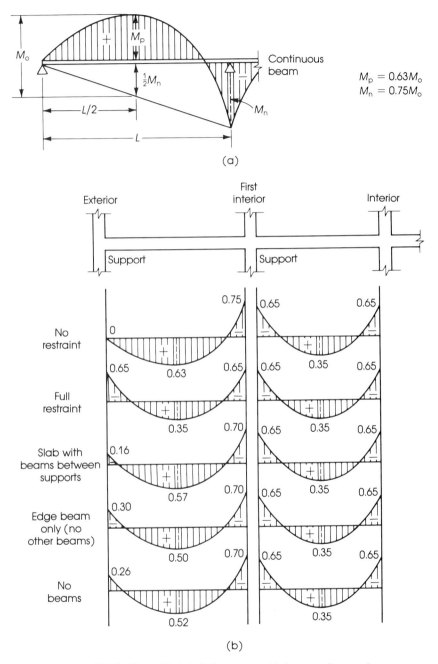

FIGURE 17.13. Distribution of total static moment into negative and positive span moments: (a) explanation of quantities and (b) moment factors for the five cases of Table 17.1.

TABLE 17.1. Distribution of moments in an end panel

	Exterior edge		Slab with beams between all supports	Slab without beams between interior supports	
	Unrestrained (1)	Fully restrained (2)	(3)	With edge beam (4)	Without edge beam (5)
Exterior negative factored moment	0	0.65	0.16	0.30	0.26
Positive factored moment	0.63	0.35	0.57	0.50	0.52
Interior negative factored moment	0.75	0.65	0.70	0.70	0.70

end span is distributed in different ratios according to Table 17.1 and Figure 17.13(b). The moment coefficients in column 1 for an unrestrained edge are based on the assumption that the ratio of the flexural stiffness of columns to the combined flexural stiffness of slabs and beams at a joint, α_{ec}, is equal to zero. The coefficients of column 2 are based on the assumption that the ratio α_{ec} is equal to infinity. The moment coefficients in columns 3, 4, and 5 have been established by analyzing the slab systems with different geometries and support conditions.

The 1977 ACI Code used distribution factors as a function of the stiffness ratio α_{ec} for proportioning the static moment M_o in an end span. The same method is also applicable for usage as indicated in the 1983 ACI Commentary[14] and denoted by the Modified Stiffness Method. In this method, the stiffnesses of the slab end beam and of the exterior column are replaced by the stiffness of an equivalent column K_{ec}. The flexural stiffness of the equivalent column K_{ec} can be calculated from the following expression:

$$\frac{1}{K_{ec}} = \frac{1}{\Sigma K_c} + \frac{1}{K_t} \quad \text{or} \quad K_{ec} = \frac{\Sigma K_c}{1 + \frac{\Sigma K_c}{K_t}} \tag{17.12}$$

where

K_{ec} = flexural stiffness of the equivalent column

K_c = flexural stiffness of the actual column

K_t = torsional stiffness of the edge beam

The sum of the flexural stiffness of the columns above and below the floor slab can be taken as follows:

$$\Sigma K_c = 4E \left(\frac{I_{c1}}{L_{c1}} + \frac{I_{c2}}{L_{c2}} \right) \tag{17.13}$$

where

I_{c1} and L_{c1} = the moment of inertia and length of column above slab level

I_{c2} and L_{c2} = the moment of inertia and length of column below slab level

The torsional stiffness of the end beam K_t may be calculated as follows:

$$K_t = \Sigma \frac{9 E_{cs} C}{l_2 \left(1 - \dfrac{c_2}{l_2} \right)^3} \qquad \textbf{(ACI Code equation 13.6)} \qquad (17.14)$$

where

> c_2 = size of the rectangular or equivalent rectangular column, capital or bracket measured on transverse spans on each side of the column
>
> E_{cs} = modulus of elasticity of the slab concrete
>
> C = torsional constant determined from the following expression:

$$C = \Sigma \left(1 - 0.63 \frac{x}{y} \right) \left(\frac{x^3 y}{3} \right) \qquad \textbf{(ACI Code equation 13.7)} \qquad (17.15)$$

where x = shorter dimension of each component rectangle and y = longer dimension of each component rectangle. In calculating C, the component rectangles of the cross-section must be taken in such a way as to produce the largest value of C.

If a panel contains a beam parallel to the direction in which moments are being determined, the torsional stiffness K_t given in equation 17.14 must be replaced by a greater value, K_{ta}, computed as follows:

$$K_{ta} = K_t \times \frac{I_{sb}}{I_s} \qquad \textbf{(ACI Code, section 13.7.5.4)}$$

where

> $I_s = l_2 h^3 / 12$ = moment of inertia of a slab that has a width equal to the full width between panel centerlines (excluding that portion of the beam stem extending above or below the slab)
>
> I_{sb} = I_s above, including the portion of the beam stem extending above or below the slab

This requirement may be waived in calculation of α_{ec} for use in the direct design method **(ACI Commentary 13.6.3.3)**.

Cross-sections of some attached torsional members are shown in Figure 17.14. Once K_{ec} is calculated, the stiffness ratio α_{ec} is obtained as follows:

$$\alpha_{ec} = \frac{K_{ec}}{\Sigma (K_s + K_b)} \qquad (17.16)$$

where

> $K_s = (4 E_{cs} I_s)/l_1$ = flexural stiffness of the slab
>
> $K_b = (4 E_{cb} I_b)/l_1$ = flexural stiffness of the beam
>
> I_b = gross moment of inertia of the longitudinal beam section

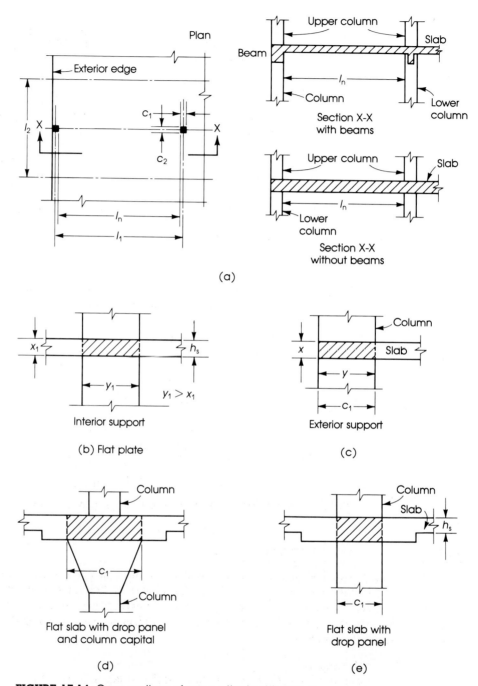

FIGURE 17.14. Cross-sections of some attached torsional members.

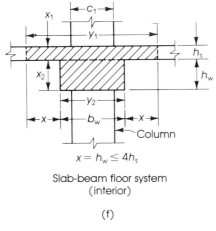
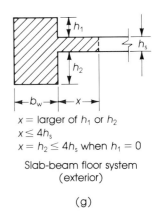

Slab-beam floor system
(interior)

Slab-beam floor system
(exterior)

(f)

(g)

FIGURE 17.14. (*continued*)

The distribution of the total static moment M_o in an exterior panel is given as a function of α_{ec} as follows **(ACI Code Commentary, section 13.6.3)**:

$$\text{Interior negative factored moment} = \left[0.75 - \frac{0.1}{\left(1 + \dfrac{1}{\alpha_{ec}}\right)}\right] M_o$$

$$\text{Positive factored moment} = \left[0.63 - \frac{0.28}{\left(1 + \dfrac{1}{\alpha_{ec}}\right)}\right] M_o$$

$$\text{Exterior negative factored moment} = \left[\frac{0.65}{\left(1 + \dfrac{1}{\alpha_{ec}}\right)}\right] M_o$$

These values are shown for a typical exterior panel in Figure 17.15.

17.6.4   Transverse Distribution of Moments

The longitudinal moment values mentioned in the previous section are for the entire width of the equivalent building frame. This frame width is the sum of the widths of two half column strips and two half middle strips of two adjacent panels, as shown in Figure 17.16. The transverse distribution of the longitudinal moments to the middle and column strips is a function of

- the ratio l_2/l_1
- the ratio $\alpha_1 = E_{cb}I_b/E_{cs}I_s$ = beam stiffness/slab stiffness   (*17.17*)
- the ratio $\beta_t = E_{cb}C/2E_{cs}I_s$ (the torsional rigidity of edge beam section divided by the flexural rigidity of a slab of width equal to beam span length)   (*17.18*)

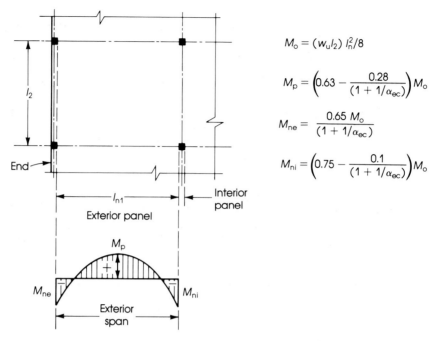

$$M_o = (w_u l_2)\, l_n^2/8$$

$$M_p = \left(0.63 - \frac{0.28}{(1 + 1/\alpha_{ec})}\right) M_o$$

$$M_{ne} = \frac{0.65\, M_o}{(1 + 1/\alpha_{ec})}$$

$$M_{ni} = \left(0.75 - \frac{0.1}{(1 + 1/\alpha_{ec})}\right) M_o$$

FIGURE 17.15. Distribution of moments in an exterior panel.

The percentages of each design moment to be distributed to column and middle strips for an interior and exterior panel are given in Tables 17.2 through 17.5.

In a typical interior panel, the portion of the design moment which is not assigned to the column strip (Table 17.2) must be resisted by the corresponding half-middle strips. Linear interpolation is permitted by the ACI Code for values of l_2/l_1 between 0.5 and 2.0 and for $\alpha_1 l_2/l_1$ between zero and one. From Table 17.2 it can be seen that when no beams are used, as in the case of flat plates or flat slabs, $\alpha_1 = 0$. The final percentage of moments in the column and middle strips as a function of M_o are given in Table 17.3.

For exterior panels, the portion of the design moment that is not assigned to the column strip (Table 17.4) must be resisted by the corresponding half-middle strips. Again, linear interpolation between values shown in Table 17.4 is permitted by the **ACI**

TABLE 17.2. Percentage of longitudinal moment in column strips, **interior** panels **(ACI Code, section 13.6.4)**

	$\alpha_1 l_2/l_1$	0.5	1.0	2.0
Negative moment	0	75	75	75
at interior support	≥ 1.0	90	75	45
Positive moment	0	60	60	60
near midspan	≥ 1.0	90	75	45

Aspect ratio l_2/l_1

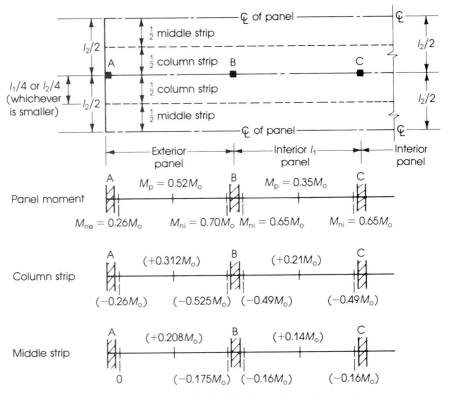

FIGURE 17.16. Width of the equivalent rigid frame (equal spans in this figure) and distribution of moments in flat plates, flat slabs, and waffle slabs with no beams.

Code, section 13.6.4.2. When no beams are used in an exterior panel, as in the case of flat slabs or flat plates with no edge (spandrel) beam, $\alpha_1 = 0$, $C = 0$, and consequently $\beta_t = 0$. This means that the end column provides the restraint to the exterior end of the slab. The applicable values of Table 17.4 for this special case are shown in Table 17.5 and Figure 17.16.

TABLE 17.3. Percentage of moments in two-way **interior** slabs without beams ($\alpha_1 = 0$). Total design moment $= M_o = (w_u l_2)(l_{nl}^2/8)$

	Negative moment	Positive moment
Longitudinal moments in one panel	$-0.65M_o$	$+0.35M_o$
Column strip	$0.75(-0.65M_o) = -0.49M_o$	$*0.60(0.35M_o) = 0.21M_o$
Middle strip	$0.25(-0.65M_o) = -0.16M_o$	$*0.40(0.35M_o) = 0.14M_o$

* When $\alpha_c < \alpha_{min}$, multiply the positive moment percentage ($0.35M_o$) by δ_s given in equation 17.19.

TABLE 17.4. Percentage of longitudinal moment in column strips, **exterior** panels (ACI Code, section 13.6.4)

	$\alpha_1 l_2/l_1$	β_t	Aspect ratio l_2/l_1		
			0.5	1.0	2.0
Negative moment at exterior support	0	0	100	100	100
		≥ 2.5	75	75	75
	≥ 1.0	0	100	100	100
		≥ 2.5	90	75	45
Positive moment near midspan	0		60	60	60
	≥ 1.0		90	75	45
Negative moment at interior support	0		75	75	75
	≥ 1.0		90	75	45

From Table 17.5 it can be seen that when no spandrel beam is used at the exterior end of the slab, $\beta_t = 0$ and 100 percent of the design moment is resisted by the column strip. The middle strip will not resist any moment; therefore, minimum steel reinforcement must be provided. The **ACI Code, section 13.6.4.3**, specifies that when the exterior support is a column or wall extending for a distance equal to or greater than 3/4 the transverse span length l_2 used to compute M_o, the exterior negative moment is to be uniformly distributed across l_2. When beams are provided along the centerlines of columns, the **ACI Code, section 13.6.5**, requires that beams must be proportioned to resist 85 percent of the moment in the column strip if $\alpha_1(l_2/l_1) \geq 1.0$. For values of $\alpha_1(l_2/l_1)$ between 1.0 and zero, the moment assigned to the beam is determined by linear interpolation. Beams must also be proportioned to resist additional moments caused by all loads applied directly to the beams, including the weight of the projecting beam stem. The portion of the moment that is not assigned to the beam must be resisted by the slab in the column strip.

TABLE 17.5. Percentage of longitudinal moment in column and middle strips, **exterior** panels (for all ratios of l_2/l_1), given $\alpha_1 = \beta_t = 0$

	%	Column strip	Middle strip	Final moment as a function of M_o and α_{ec} (column strip)
Negative moment at exterior support	100	$0.26M_o$	0	$\left[\dfrac{0.65}{\left(1 + \dfrac{1}{\alpha_{ec}}\right)}\right](M_o)$
Positive moment $(0.6 \times 0.52M_o)$	60	$0.312M_o$	$0.208M_o$	$\left[0.63 - \dfrac{0.28}{\left(1 + \dfrac{1}{\alpha_{ec}}\right)}\right](0.6M_o)$
Negative moment at interior support $(0.75 \times 0.70M_o)$	75	$0.525M_o$	$0.175M_o$	$\left[0.75 - \dfrac{0.10}{\left(1 + \dfrac{1}{\alpha_{ec}}\right)}\right](0.75M_o)$

17.6.5 ACI Provisions for Effects of Pattern Loadings

In continuous structures, the maximum and minimum bending moments at the critical sections are obtained by placing the live load in specific patterns to produce the extreme values. Placing the live load on all spans will not produce either the maximum positive or negative bending moments. The maximum and minimum moments depend mainly on:

1. The ratio of live to dead load. A high ratio will increase the effect of pattern loadings.
2. The ratio of column to beam stiffnesses. A low ratio will increase the effect of pattern loadings.
3. Maximum positive moments within the spans are less affected by pattern loadings.

To limit the increase in positive moment caused by the pattern loadings, the **ACI Code, section 13.6.10,** specifies minimum values α_{min} for the ratio of column to slab stiffnesses, as given in Table 17.6. When the ratio $\beta_a \leq 2.0$ and the ratio of the column stiffness α_c (above and below the floor slab) to slab and beam stiffness is less than α_{min} shown in Table 17.6, the positive design moments in panels supported by such columns must be multiplied by the coefficient δ_s

$$\delta_s = 1 + \frac{2 - \beta_a}{4 + \beta_a}\left(1 - \frac{\alpha_c}{\alpha_{min}}\right) \quad \text{(ACI Code equation 13.5)} \quad (17.19)$$

where

β_a = ratio of service dead to service live load per unit area (without load factors)

TABLE 17.6. Values of α_{min}, where $\alpha = E_{cb}I_b/E_{cs}I_s$

$\beta_a = \frac{D.L.}{L.L.}$	Aspect ratio l_2/l_1	Ratio of beam to slab stiffness α				
		0	0.5	1.0	2.0	4.0
2.0	0.5–2.0	0	0	0	0	0
1.0	0.50	0.6	0	0	0	0
	0.80	0.7	0	0	0	0
	1.00	0.7	0.1	0	0	0
	1.25	0.8	0.4	0	0	0
	2.00	1.2	0.5	0.2	0	0
0.5	0.50	1.3	0.3	0	0	0
	0.80	1.5	0.5	0.2	0	0
	1.00	1.6	0.6	0.2	0	0
	1.25	1.9	1.0	0.5	0	0
	2.00	4.9	1.6	0.8	0.3	0
0.33	0.50	1.8	0.5	0.1	0	0
	0.80	2.0	0.9	0.3	0	0
	1.00	2.3	0.9	0.4	0	0
	1.25	2.8	1.5	0.8	0.2	0
	2.00	13.0	2.6	1.2	0.5	0.3

α_c = ratio of column stiffness to slab and beam stiffness

$$= \frac{\Sigma K_c \text{(for upper and lower columns)}}{\Sigma(K_s + K_b)} \qquad (17.20)$$

α_{min} = values shown in Table 17.6, adopted from **ACI Code, Table 13.6.10**

Therefore, when $\alpha_c \geq \alpha_{min}$, $\delta_s = 1.0$; and when $\alpha_c < \alpha_{min}$, use δ_s.

17.6.6   Reinforcement Details

After all the percentages of the static moments in the column and middle strips are determined, the steel reinforcement can be calculated for the negative and positive moments in each strip as was done for beam sections in Chapter 4:

$$M_u = \phi A_s f_y \left(d - \frac{a}{2}\right) = R_u b d^2$$

Calculate R_u and determine the steel ratio ρ using the tables in Appendix A or use the following equation:

$$R_u = \phi \rho f_y \left(1 - \frac{\rho f_y}{1.7 f'_c}\right)$$

where $\phi = 0.9$. The steel area $A_s = \rho b d$. When the slab thickness limitations as discussed in section 17.4 are met, no compression reinforcement will be required. **Figure 13.4.8 of the ACI Code** indicates the minimum length of reinforcing bars and reinforcement details for slabs without beams; it is reproduced here as Figure 17.17. The spacing of bars in the slabs must not exceed the ACI limits of maximum spacing: 18 in. (450 mm) or twice the slab thickness, whichever is smaller.

17.6.7   Summary of the Direct Design Method

The following steps summarize the analysis and design procedure of two-way slabs by the direct design method:

1. Check the limitation requirements as explained in section 17.6.1. Estimate the slab thickness using equations 17.1 through 17.3.
2. Calculate factored loads.
3. Check the slab thickness as required by one-way and two-way shear.
4. Calculate the total static moments M_o in both directions (equation 17.11).
5. Calculate the stiffnesses and coefficients of the different structural elements as required:

   Slab stiffness: $K_s = 4EI_s/l_1$, $I_s = \Sigma(\frac{1}{2}l_2)(h_s^3/12)$
   Column stiffness: K_c (equation 17.13)
   Torsional stiffness: K_t (equation 17.14)
   Torsional constant: C (equation 17.15)
   Equivalent column stiffness: K_{ec} (equation 17.12)
   Relative stiffness: α_{ec} (equation 17.16)

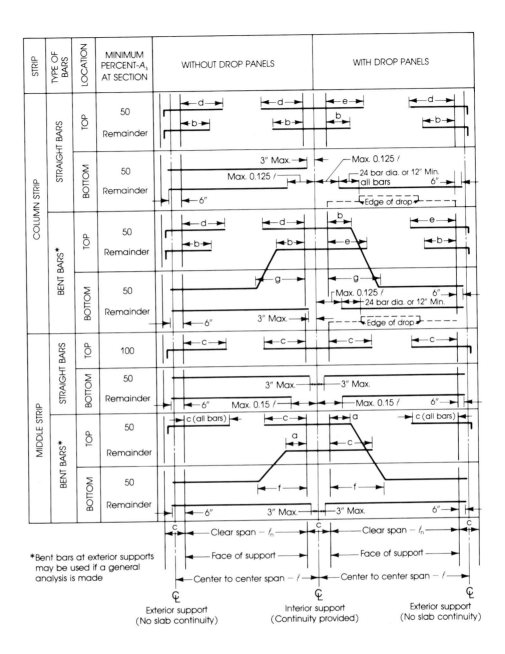

FIGURE 17.17. Minimum bend point locations and extensions for reinforcement in slabs without beams **(ACI Code Figure 13.4.8).** Courtesy American Concrete Institute.[18]

Ratio of beam to slab stiffness: α_1 (equation 17.17)
The ratio β_t (equation 17.18)
The ratio α_c column to slab and beam stiffness (equation 17.20)
α_{min} (Table 17.6)
The coefficient δ_s (equation 17.19)

6. Obtain design moments: determine the percentages of the static moments in the longitudinal and transverse directions for each column and middle strip. The different cases of interior and exterior panels can be summarized as follows:

- Interior panel without beams, $\alpha_1 = 0$ (flat plate and flat slabs): Use the moment percentages given in Table 17.3 or Figure 17.16.
- Interior panel supported on beams: For moments in the longitudinal direction, use values in Figure 17.12. For the distribution of moments in the transverse direction, use Table 17.2 for column strips. The middle strips will resist the portions of moments not assigned to column strips. Calculate α_1 from equation 17.17.
- Exterior panel without beams (flat plate and flat slabs), $\alpha_1 = \beta_t = 0$: For moments in the longitudinal direction, use values in Table 17.1 or Figure 17.13(b) (case 5) or Figure 17.15. The sections (a), (b), and (c) in Figure 17.14 are used to calculate the torsional constant C, then to calculate K_t and α_{ec}. For the distribution of moments in the transverse direction, use Table 17.5 or Figure 17.16 for column strip moment percentages. The middle strip will resist the portion of moment that is not assigned to the column strip.
- Exterior panel with edge (spandrel) beam only, no internal beams (flat plates and flat slabs with edge beams): For moments in the longitudinal direction, use values M_p, M_{ne}, and M_{ni} as shown in Figure 17.13(b) (case 4) or Figure 17.15. Section (e) or (g) in Figure 17.14 is used to calculate the torsional constant C, then to calculate K_t and α_{ec}. For the distribution of moments in the transverse direction, use Table 17.4 for the column strip. The middle strip will resist the portion of moment that is not assigned to the column strip. Calculate α_1 and β_t from equations 17.17 and 17.18, respectively.
- Exterior panel supported on beams: For moments in the longitudinal direction, use values M_p, M_{ne}, and M_{ni} as shown in Table 17.1 or Figure 17.13(b) (case 3) or Figure 17.15. Then use sections (f) and (g) in Figure 17.14 to calculate the torsional constant C, K_t, and α_{ec}. The distribution of moments in the transverse direction is similar to the previous case. Note that the beams must resist 85 percent of the moment in the column strip when $\alpha_1(l_2/l_1) \geq 1.0$. When $\alpha_1(l_2/l_1)$ varies between 1.0 and zero, the moment in the beam is determined by linear interpolation between 85 and 0 percent.

7. Calculate the steel reinforcement required for the critical sections and extend the bars throughout the slab to maintain minimum length of slab reinforcement, as detailed in Figure 17.17.

EXAMPLE **17.3**

Using the direct design method, design the typical interior flat plate panel shown in Figure 17.18. The floor system consists of four panels in each direction with a panel size of 24 by 20 ft. All panels are supported by 20 by 20 in. columns, 12 ft long. The slab carries a uniform service live load of 60 psf and a service dead load that consists of 24 psf of floor finish in addition to the slab self-weight. Use $f'_c = 3$ ksi and $f_y = 40$ ksi.

Solution

1. All the limitations stated in section 17.6.1 are met.

   • Determine the minimum slab thickness using equations 17.1 through 17.3. When no beams are used, $\alpha_m = 0$ (average ratio of beam to slab stiffness), and $\beta_s = 1.0$ (ratio of the lengths of continuous sides to total perimeter of panel).

$$\text{Clear span in the long direction} = 24 - \frac{20}{12} = 22.33 \text{ ft}$$

$$\text{Clear span in the short direction} = 20 - \frac{20}{12} = 18.33 \text{ ft}$$

   • From equation 17.1,

$$h_{min} = \frac{(22.33 \times 12)(800 + 0.005 \times 40{,}000)}{36{,}000 + 0.0} = 7.44 \text{ in.}$$

   From equation 17.2,

$$h_{min} = \frac{(22.33 \times 12)(800 + 0.005 \times 40{,}000)}{36{,}000 + (5000 \times 1.22)(1 + 1)} = 5.56 \text{ in.}$$

   but not more than h_{max} of equation 17.3 = 7.44 in.

$$h_{min} = 5.0 \text{ in.}$$

   Thus $h = 7.44$ in. controls. Therefore use a slab thickness of $h = 8.0$ in.

2. Calculate the factored loads:

$$w_D = 24 + \text{self-weight of slab} = 24 + \frac{8.0}{12} \times 150 = 124 \text{ psf}$$

$$w_u = 1.4 \times (124) + 1.7 \times (60) = 276 \text{ psf}$$

3. Check one- and two-way shears:

   • Check punching shear at a distance $d/2$ from the face of the column (two-way action): Assuming $\frac{3}{4}$ in. concrete cover and No. 4 bars, then the average $d = 8.0 - 0.75 - 0.5 = 6.75$ in. (see Figure 17.18(b)).

$$b_o = 4(20 + 6.75) = 107 \text{ in.}$$

$$V_u = \left[l_1 l_2 - \left(\frac{26.75}{12} \times \frac{26.75}{12} \right) \right] \times w_u = (24 \times 20 - 5.0) \times 0.276 = 131 \text{ k}$$

$$\phi V_c = \phi(4\sqrt{f'_c})b_o d = \frac{0.85 \times 4}{1000} \times \sqrt{3000} \times 107 \times 6.75 = 134.5 \text{ k}$$

   which is greater than V_u.

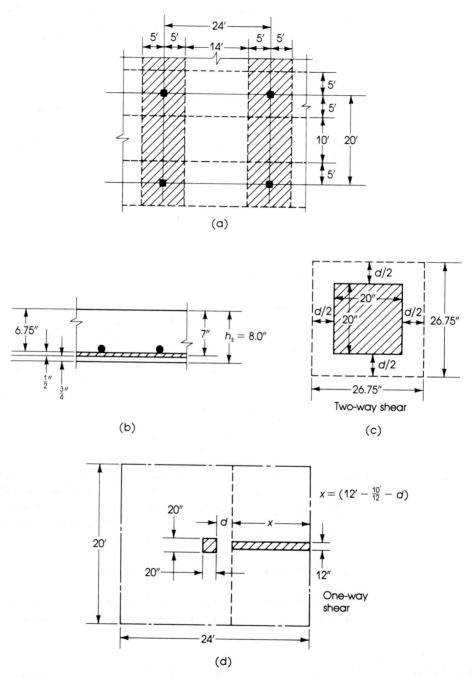

FIGURE 17.18. Example 17.3, interior flat plate.

- Check beam shear at a distance d from the face of the column, average $d = 6.75$ in. Consider a 1-ft strip (Figure 17.18(d)), the length of the strip

$$x = 12 - \frac{10}{12} - \frac{6.75}{12} = 10.6 \text{ ft}$$

$$V_u = w_u(1 \times 10.6) = 0.276 \times 1 \times 10.6 = 2.9 \text{ k}$$

$$\phi V_c = \phi(2\sqrt{f_c'})bd = \frac{0.85 \times 2}{1000} \times \sqrt{3000} \times (12 \times 6.75) = 7.5 \text{ k}$$

which is greater than $V_u = 2.9$ k. In normal loadings, one-way shear does not control.

4. Calculate the total static moments in the long and short directions. In the long direction,

$$M_{ol} = \frac{w_u l_2 l_{n1}^2}{8} = \frac{0.276}{8} \times 20 \times (22.33)^2 = 343.6 \text{ k·ft}$$

In the short direction,

$$M_{os} = \frac{w_u l_1 l_{n2}^2}{8} = \frac{0.276}{8} \times 24 \times (18.33)^2 = 277.8 \text{ k·ft}$$

Since $l_2 < l_1$, the width of half a column strip in the long direction $= 0.25 \times 20 = 5$ ft, and the width of the middle strip $= 20 - 2 \times 5 = 10$ ft. The width of half the column strip in the short direction $= 5$ ft, and the width of the middle strip $= 24 - 2 \times 5 = 14$ ft. To calculate the effective depth d in each direction, assume that steel bars in the short direction are placed on top of the bars in the long direction. Therefore

$$d \text{ (long direction)} = 8.0 - 0.75 - \frac{0.5}{2} = 7.0 \text{ in.}$$

$$d \text{ (short direction)} = 8.0 - 0.75 - 0.5 - \frac{0.5}{2} = 6.5 \text{ in.}$$

The design procedure can be conveniently arranged in a table form as in Table 17.7. The details of the bars selected for this interior slab are shown in Figure 17.19 using the bent bars system. Minimum lengths of the bars must meet those shown in Figure 17.17. Straight bars and $f_y = 60$ ksi steel bars are more preferred by contractors and may also be adopted with minimum bar lengths, as shown in the same figure.

$$\text{Maximum spacing} = \frac{\text{width of panel}}{\text{no of bars}} = \frac{168}{12} = 14 \text{ in.}$$

occurs at the middle strip in the short direction; this spacing of 14 in. is adequate because it is less than $2h_s = 16$ in. and less than 18 in. specified by the ACI Code. ∎

EXAMPLE 17.4

Using the direct design method, design an exterior flat plate panel which has the same dimensions, loads, and concrete and steel strengths given in example 17.3. No beams are used along the edges (Figure 17.20, page 584).

TABLE 17.7. Design of an interior flat plate panel

| | Long direction $M_o = 343.6$ k·ft $M_n = -0.65 M_o = -223.3$ k·ft $M_p = +0.35 M_o = 120.3$ k·ft | | | | Short direction $M_o = 277.8$ k·ft $M_n = -0.65 M_o = -180.6$ k·ft $M_p = +0.35 M_o = 97.2$ k·ft | | | |
| | Column strip | | Middle strip | | Column strip | | Middle strip | |
	Negative	Positive	Negative	Positive	Negative	Positive	Negative	Positive
Moment distribution (%)	75	60	25	40	75	60	25	40
M_u (k·ft)	$0.75 M_n$ $= -167.5$	$0.6 M_p$ $= +72.2$	$0.25 M_n$ $= -55.8$	$0.4 M_p$ $= +48.1$	$0.75 M_n$ $= -135.5$	$0.6 M_p$ $= +58.3$	$0.25 M_n$ $= -45.1$	$0.4 M_p$ $= +38.9$
Width of strip b (in.)	120	120	120	120	120	120	168	168
Effective depth d (in.)	7.0	7.0	7.0	7.0	6.5	6.5	6.5	6.5
$R_u = \dfrac{M_u}{bd^2}$ (psi)	342	147	114	98	321	138	76	66
Steel ratio ρ	0.0103	0.0043	0.0033	0.0028	0.0096	0.004	0.0022	0.002
$A_s = \rho bd$ (in.²)	8.65	3.61	2.77	2.35	7.49	3.12	2.40	2.18
Min. $A_s = 0.002 b h_s$ (in.²)	1.92	1.92	1.92	1.92	1.92	1.92	2.69	2.69
Bars selected	28 No. 5	12 No. 5	14 No. 4	12 No. 4	24 No. 5	10 No. 5	12 No. 4	12 No. 4
Straight	16	6	2	6	14	5	0	6
Bent	12	6	12	6	10	5	12	6
Spacing $\leq 2h_s$	4.3 in.	10	8.6	10	5	12	14 < 16	14 < 16

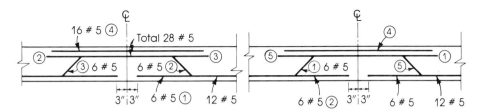

Column strip, long direction

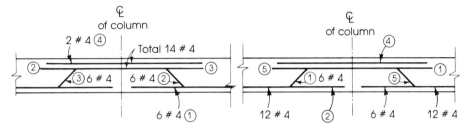

Middle strip, long direction

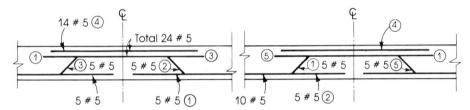

Column strip, short direction

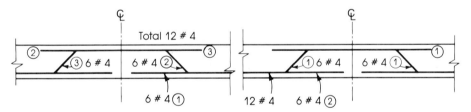

Middle strip, short direction

FIGURE 17.19. Example 17.3, reinforcement details. Circled numbers are used to trace bars across diagram.

Solution 1. Determine the slab thickness. Since no beams are used, $\alpha_m = 0$.

$$\beta_s = \frac{2 \times 24 + 20}{2 \times 24 + 2 \times 20} = 0.77$$

Clear span long direction = 22.33 ft

Clear span short direction = 18.33 ft

$$\beta = \frac{22.33}{18.33} = 1.22$$

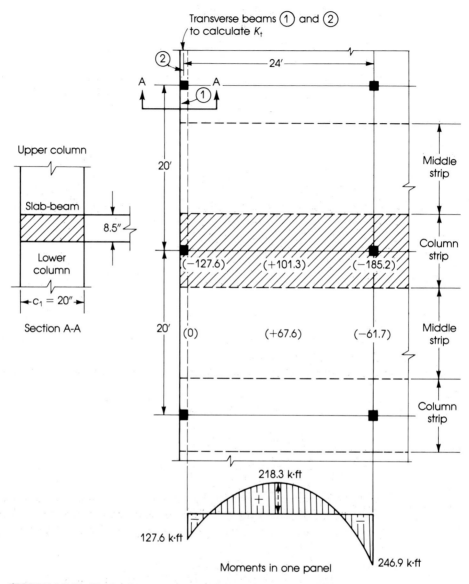

FIGURE 17.20. Distribution of bending moments, example 17.4.

From equation 17.1,

$$h_{\min} = \frac{(22.33 \times 12)(800 + 0.005 \times 40{,}000)}{36{,}000 + (5000 \times 1.22)[0 - 0.5(1 - 0.77)(1 + 1/1.22)]} = 7.72 \text{ in.}$$

From equation 17.2,

$$h_{\min} = \frac{(22.33 \times 12)(1000)}{36{,}000 + (5000 \times 1.22)(1 + 0.77)} = 5.73 \text{ in.}$$

but not more than

$$h_{max} = \frac{(22.33 \times 12)(1000)}{36,000} = 7.44 \text{ in.}$$

$h_{min} = 5.0$ in., then $h = 7.44$ in. controls. The ACI Code states that when no edge beams are used, and $\alpha < 0.8$, the calculated slab thickness must be increased by 10 percent; therefore $h = 1.1 \times 7.44 = 8.18$ in., say, 8.5 in.

2. Calculate factored loads:

$$w_D = 24 + \text{Weight of slab} = 24 + \frac{8.5}{12} \times 150 = 130.25 \text{ psf}$$

$$w_u = 1.4(130.25) + 1.7(60) = 284.35 \text{ psf, say, 285 psf}$$

3. Check punching shear at the interior support. Assuming 0.75 in. concrete cover and No. 4 bars, then the average depth $d = 8.5 - 0.75 - 0.5 = 7.25$ in.

$$b_o = 4(20 + 7.25) = 109 \text{ in.}$$

$$V_u = \left[l_1 l_2 - \left(\frac{27.25}{12} \right)^2 \right] w_u = (24 \times 20 - 5.16) \times 0.285 = 135 \text{ k}$$

$$\phi V_c = \phi(4\sqrt{f_c'} b_o d) = \frac{0.85 \times 4}{1000} \times \sqrt{3000} \times 109 \times 7.25 = 147 \text{ k}$$

which is greater than V_u. One-way or beam shear can be checked as in the previous example, and it is adequate.

4. Calculate the total static moments in the long and short directions. In the long direction,

$$M_{ol} = w_u l_2 \frac{l_{n1}^2}{8} = \frac{0.285}{8} \times 20 \times (22.33)^2 = 355.3 \text{ k·ft}$$

In the short direction,

$$M_{os} = w_u l_1 \frac{l_{n2}^2}{8} = \frac{0.285}{8} \times 24 \times (18.33)^2 = 287.3 \text{ k·ft}$$

Width of half the column strip in both directions $= 0.25 l_2 = 0.25 \times 20 = 5$ ft. The widths of the middle strips in the long and short directions are 10 ft and 14 ft, respectively. Assuming that the steel bars in the short direction will be placed on top of the bars in the long direction, then

$$d \text{ (long direction)} = 8.5 - 0.75 - \frac{0.5}{2} = 7.5 \text{ in.}$$

$$d \text{ (short direction)} = 8.5 - 0.75 - 0.5 - \frac{0.5}{2} = 7.0 \text{ in.}$$

5. Calculate the design moments in the long direction, $l_1 = 24$ ft (refer to Table 17.5 or Figure 17.16). The distribution of the total moment, M_{ol} of 355.3 k·ft in the column and middle strips, is computed as follows:

• Column strip:
Interior negative moment $= -0.525 M_o = -0.525 \times 355.3 = -186.5$ k·ft
Positive moment within span $= 0.312 M_o = 0.312 \times 355.3 = 110.9$ k·ft
Exterior negative moment $= -0.26 M_o = -0.26 \times 355.3 = -92.4$ k·ft

- Middle strip:
  Interior negative moment $= -0.175 M_o = -0.175 \times 355.3 = -62.2$ k·ft
  Positive moment within span $= 0.208 M_o = 0.208 \times 355.3 = 73.9$ k·ft
  Exterior negative moment $= 0$

6. An alternate solution uses the Modified Stiffness Method and the equivalent column stiffness K_{ec}. First calculate K_{ec}:

$$\frac{1}{K_{ec}} = \frac{1}{\Sigma K_c} + \frac{1}{K_t}$$

It can be assumed that the part of the slab strip between exterior columns acts as a beam resisting torsion. The section of the slab-beam is 20 in. (width of the column) $\times$ 8.5 in. (thickness of the slab), as shown in Figure 17.20. Determine the torsional stiffness K_t from equation 17.14:

$$C = \left(1 - 0.63 \frac{x}{y}\right) \frac{x^3 y}{3}$$

$x = 8.5$ in.   and $y = 20$ in.

$$C = \left(1 - 0.63 \times \frac{8.5}{20}\right)\left[\frac{(8.5)^3 \times 20}{3}\right] = 2998 \text{ in.}^3$$

$$K_t = \frac{9 E_c C}{l_2 \left(1 - \frac{c_2}{l_2}\right)^3} = \frac{9 E_c \times 2998}{(20 \times 12)\left(1 - \frac{20}{20 \times 12}\right)^3} = 146 E_c$$

For the two adjacent slabs (on both sides of the column) acting as transverse beams,

$$K_t = 2 \times 146 E_c = 292 E_c$$

Calculate the column stiffness K_c; the column height $L_c = 12$ ft:

$$K_c = \frac{4 E_c I_c}{L_c} = \frac{4 E_c}{(12 \times 12)} \times \frac{(20)^4}{12} = 370.4 E_c$$

For two columns above and below the floor slab,

$$K_c = 2 \times 370.4 E_c = 740.8 E_c$$

Thus

$$\frac{1}{K_{ec}} = \frac{1}{740.8 E_c} + \frac{1}{292 E_c}$$

To simplify the calculations, multiply by $1000 E_c$:

$$\frac{1000 E_c}{K_{ec}} = \frac{1000}{740.8} + \frac{1000}{292} = 4.77$$

$$K_{ec} = 209.4 E_c$$

7. Calculate slab stiffness and the ratio α_{ec}:

$$K_s = \frac{4 E_c I_s}{l_1}$$

$h_s = 8.5$ in. $l_2 = 20$ ft $I_s = \frac{l_2 h_s^3}{12}$

$$K_s = \frac{4E_c}{(24 \times 12)} \times \frac{(20 \times 12)(8.5)^3}{12} = 170.6E_c$$

$$\alpha_{ec} = \frac{K_{ec}}{\Sigma K_s + K_b} \qquad\qquad (17.16)$$

$K_b = 0$ (no beams are provided), thus

$$\alpha_{ec} = \frac{209.4E_c}{170.6E_c} = 1.23$$

$$\left(1 + \frac{1}{\alpha_{ec}}\right) = 1 + \frac{1}{1.23} = 1.81$$

8. Calculate the design moments in the long direction, $l_1 = 24$ ft. The distribution of moments in one panel is shown in Figure 17.15. The interior negative moment

$$M_{ni} = \left[0.75 - \frac{0.10}{\left(1 + \dfrac{1}{\alpha_{ec}}\right)} \right] M_{ol}$$

$$= \left(0.75 - \frac{0.10}{1.81}\right) \times 355.3 = -246.9 \text{ k·ft}$$

The positive moment

$$M_p = \left[0.63 - \frac{0.28}{\left(1 + \dfrac{1}{\alpha_{ec}}\right)} \right] M_{ol}$$

$$= \left(0.63 - \frac{0.28}{1.81}\right) \times 355.3 = +168.9 \text{ k·ft}$$

The exterior negative moment

$$M_{ne} = \frac{0.65}{\left(1 + \dfrac{1}{\alpha_{ec}}\right)} M_{ol}$$

$$= \frac{0.65}{1.81} \times 355.3 = -127.6 \text{ k·ft}$$

Note that in this example the moment multiplier $\delta_s = 1.0$. This can be checked as follows:

$$\alpha_c = \frac{\Sigma K_c}{\Sigma K_s} \text{ (interior)} = \frac{740.8E_c}{2 \times 170.6E_c} = 2.16$$

$$\frac{l_2}{l_1} = \frac{20}{24} = 0.83$$

$$\beta_a = \frac{\text{D.L.}}{\text{L.L.}} = \frac{130.25}{60} = 2.17$$

From Table 17.6, α_{min} required $= 0$ and therefore $\delta_s = 1.0$. If the ratio $\alpha_c \geq \alpha_{min}$ then $\delta_s = 1.0$; otherwise it can be calculated from equation 17.19.

9. Calculate the distribution of panel moments in the transverse direction to column and middle strips. The moments M_{ni}, M_p, and M_{ne} are distributed as follows (refer to Table 17.5):

   - The interior moment $M_{ni} = -246.9$ k·ft is distributed 75 percent for the column strip and 25 percent for the middle strip.

     Column strip = $0.75(-246.9) = -185.2$ k·ft

     Middle strip = $0.25(-246.9) = -61.7$ k·ft

   - The positive moment $M_p = 168.9$ k·ft is distributed 60 percent for the column strip and 40 percent for the middle strip.

     Column strip = $0.6(168.9) = 101.3$ k·ft

     Middle strip = $0.4(168.9) = 67.6$ k·ft

   - The exterior negative moment $M_{ne} = -127.6$ k·ft is distributed according to Table 17.4:

     $$\beta_t = \frac{E_c C}{2E_c I_s} = \frac{C}{2I_s} \text{ (concrete of slab and column is the same)}$$

     $$I_s = (20 \times 12)\frac{(8.5)^3}{12} = 12,282 \text{ in.}^4$$

     $$\beta_t = \frac{2998}{2 \times 12,282} = 0.122$$

     $$\alpha_1 = \frac{E_{cb}I_b}{E_{cs}I_s} = 0 \quad \alpha_1 \frac{l_2}{l_1} = 0 \quad \frac{l_2}{l_1} = 0.83$$

From Table 17.4 and by interpolation between $\beta_t = 0$ (percentage = 100 percent) and $\beta_t = 2.5$ (percentage = 75 percent), for $\beta_t = 0.122$, the percentage is 98.8 percent. The exterior negative moment in the column strip = $0.988 \times -127.6 = -126$ k·ft and in the middle strip = $-127.6 + 126 = -1.6$ k·ft. It is practical to consider that the column strip carries in this case 100 percent of $M_{ne} = -127.6$ k·ft.

10. Determine the reinforcement required in the long direction. Moments computed in step 5 can be used, but in this example, the moments computed in steps 6–9 will be adopted.

    Width of column strip = width of middle strip = 120 in.

    Effective depth for both strips = 7.5 in.

The steel bars are determined as in the previous example and results are shown in Table 17.8 and Figure 17.21. Maximum spacing occurs in the middle strip = 120/9 = 13.3 in., which is less than $2h_s = 17$ in., or 18 in.

11. Calculate the moments in the short direction, $l_2 = 20$ ft. The panel is continuous from both sides and it will be treated in this direction similar to example 17.3. $M_{os} = 287.3$ k·ft (calculated in step 4).

    Using coefficients from Table 17.3:

    $$M_n = -0.65 M_o = -0.65 \times 287.3 = -186.7 \text{ k·ft}$$

    $$M_p = +0.35 M_o = +0.35 \times 287.3 = +100.6 \text{ k·ft}$$

TABLE 17.8. Design of exterior flat plate panel (long direction) for example 17.4

| | Column strip | | | Middle strip | | |
	(exterior)	(near midspan)	(interior)	(exterior)	(near midspan)	(interior)
M_u (k·ft)	−127.6	+101.3	−185.2	0	+67.6	−61.7
$R_u = \dfrac{M_u}{bd^2}$ (psi)	227	180	329	0	120	110
Steel ratio, ρ	0.0067	0.0052	0.0099	0	0.0035	0.0032
$A_s = \rho bd$ (in.²)	6.03	4.68	8.91	0	3.15	2.88
Min. $A_s = 0.002 bh_s$	1.8	1.8	1.8	1.8	1.8	1.8
Bars selected	20 No. 5	16 No. 5	29 No. 5	9 No. 4	16 No. 4	15 No. 4
Straight	12	8	13	1	8	5
Bent	8	8	16	8	8	16

Column strip negative moment $= -0.49\,M_o = -140.7$ k·ft

Column strip positive moment $= +0.21\,M_o = +60.3$ k·ft

Middle strip negative moment $= -0.16\,M_o = -46.0$ k·ft

Middle strip positive moment $= +0.14\,M_o = 40.3$ k·ft

Total negative moment $= -140.7 - 46.0 = -186.7 = M_n$

Total positive moment $= +60.3 + 40.3 = 100.6 = M_p$

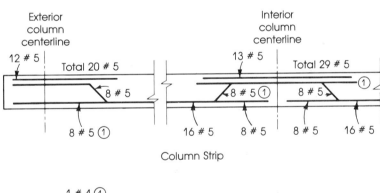

Column Strip

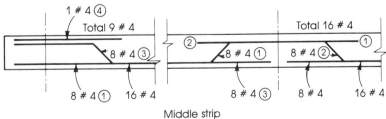

Middle strip

FIGURE 17.21. Reinforcement details (longitudinal direction), example 17.4.

TABLE 17.9. Design of flat plate panel (short direction), example 17.4

	Column strip		Middle strip	
M_u (k·ft)	−140.7	+60.3	−46.0	+40.3
Width of strip b (in.)	120	120	168	168
d (in.)	7.0	7.0	7.0	7.0
$R_u = \dfrac{M_u}{bd^2}$ (psi)	287	123	67	59
Steel ratio ρ	0.0086	0.0035	0.002	0.0017
$A_s = \rho bd$ (in.²)	7.22	2.94	2.35	2.0
$Min A_s = 0.002bh_s$	2.04	2.04	2.86	2.86
Bars selected	24 No. 5	10 No. 5	14 No. 4	14 No. 4
Straight	14	5	0	7
Bent	10	5	14	7

12. Determine the reinforcement required in the short direction.

$$\text{Width of column strip} = 10 \times 12 = 120 \text{ in.}$$
$$\text{Width of middle strip} = 14 \times 12 = 168 \text{ in.}$$

Reinforcement details are shown in Table 17.9 using a combination of bent and straight bars. Detailing of the straight bar system is simpler and can be adopted in this example. ∎

EXAMPLE **17.5**

Design an interior panel of the two-way slab floor system shown in Figure 17.7. The floor consists of six panels in each direction with a panel size of 24 by 20 ft. All panels are supported on 20 by 20 in. columns, 12 ft long. The slabs are supported by beams along the column lines with the cross-sections shown in the figure. The service live load is to be taken as 80 psf and the service dead load consists of 24 psf of floor finish in addition to the slab weight. Use $f'_c = 3$ ksi, $f_y = 60$ ksi, and the direct design method.

Solution

1. The limitations required by the ACI Code are all met. Determine the minimum slab thickness using equations 17.1 through 17.3. The slab thickness has been already calculated in example 17.2, and a 7.0-in. slab can be adopted. Generally, the slab thickness on a floor system is controlled by a corner panel, as the calculations of h_{min} for an exterior panel give greater slab thickness than for an interior panel.

2. Calculate factored loads:

$$w_D = 24 + \text{self-weight of slab} = 24 + \frac{7}{12} \times 150 = 111.5 \text{ psf}$$

$$w_u = 1.4(111.5) + 1.7(80) = 292 \text{ psf}$$

3. The shear stresses in the slab are not critical. The critical section is at a distance d from the face of the beam. For a 1-ft width:

$$V_u = w_u \left(10 \text{ ft} - \frac{1}{2} \text{ beam width} - d \right)$$

$$= 0.292 \left(10 - \frac{16}{2 \times 12} - \frac{6}{12} \right) = 2.58 \text{ k}$$

$$\phi V_c = \phi(2\sqrt{f_c'})bd = \frac{0.85 \times 2 \times \sqrt{3000} \times 12 \times 6}{1000} = 7.1 \text{ k}$$

which is much greater than V_u.

4. Calculate the total static moments in the long and short directions:

$$M_{ol} = \frac{w_u}{8} l_2(l_{n1})^2 = \frac{0.292}{8} (20)(22.33)^2 = 364.0 \text{ k·ft}$$

$$M_{os} = \frac{w_u}{8} l_1(l_{n2})^2 = \frac{0.292}{8} (24)(18.33)^2 = 294.3 \text{ k·ft}$$

5. Calculate the design moments in the long direction, $l_1 = 24$ ft.

- Distribution of moments in one panel:

    Negative moment $M_n = 0.65 M_{ol} = 0.65 \times 364 = -236.6$ k·ft

    Positive moment $M_p = 0.35 M_{ol} = 0.35 \times 364 = 127.4$ k·ft

- Distribution of panel moments in the transverse direction to the beam, column, and middle strips:

$$\frac{l_2}{l_1} = \frac{20}{24} = 0.83$$

$$\alpha_1 = \alpha_l = \frac{EI_b}{EI_s} = 3.27 \qquad \text{(from example 17.2)}$$

$$\alpha_1 \frac{l_2}{l_1} = 3.27 \times 0.83 = 2.71 > 1.0$$

- Distribute the negative moment M_n. The portion of the interior negative moment to be resisted by the column strip is obtained from Table 17.2 by interpolation and is equal to 80 percent (for $l_2/l_1 = 0.83$ and $\alpha_1(l_2/l_1) > 1.0$).

    Column strip $= 0.8 M_n = 0.8 \times 236.6 = -189.3$ k·ft

    Middle strip $= 0.2 M_n = 0.2 \times 236.6 = -47.3$ k·ft

Since $\alpha_1(l_2/l_1) > 1.0$, the **ACI Code, section 13.6.5,** indicates that 85 percent of the moment in the column strip is assigned to the beam and the balance of 15 percent is assigned to the slab in the column strip.

    Beam $= 0.85 \times 189.3 = -160.9$ k·ft

    Column strip $= 0.15 \times 189.3 = -28.4$ k·ft

    Middle strip $= -47.3$ k·ft

- Distribute the positive moment M_p. The portion of the interior positive moment to be resisted by the column strip is obtained from Table 17.2 by interpolation and is equal to 80 percent (for $l_2/l_1 = 0.83$ and $\alpha_1(l_1/l_2) > 1.0$).

Column strip $= 0.8 M_p = 0.8 \times 127.4 = +101.9$ k·ft

Middle strip $= 0.2 M_p = 0.2 \times 127.4 = +25.5$ k·ft

Since $\alpha_1(l_2/l_1) > 1.0$, 85 percent of the moment in the column strip is assigned to the beam and the balance of 15 percent is assigned to the slab in the column strip:

Beam $= 0.85 \times 101.9 = +86.6$ k·ft

Column strip $= 0.15 \times 101.9 = +15.3$ k·ft

Middle strip $= +25.5$ k·ft

Moment details are shown in Figure 17.22.

6. Calculate the design moment in the short direction, span $= 20$ ft. The procedure is similar to step 5.

Negative moment $M_n = 0.65 M_{os} = 0.65 \times 294.3 = -191.3$ k·ft

Positive moment $M_p = 0.35 M_{os} = 0.35 \times 294.3 = +103.0$ k·ft

Distribution of M_n and M_p to beam, column, and middle strips:

$$\frac{l_2}{l_1} = \frac{24}{20} = 1.2$$

$$\alpha_1 = \alpha_s = \frac{EI_b}{EI_s} = 2.72 \text{ (from example 17.2)}$$

$$\alpha_1 \frac{l_2}{l_1} = 2.72 \times 1.2 = 3.26 > 1.0$$

The percentages of the column strip negative and positive moments are obtained from Table 17.2 by interpolation. (For $l_2/l_1 = 1.2$ and $\alpha_1(l_2/l_1) > 1.0$, the percentage $= 69$ percent).

Column strip negative moment $= 0.69 M_n = 0.69 \times 191.3 = -132$ k·ft

Middle strip negative moment $= 0.31 M_n = 0.31 \times 191.3 = -59.3$ k·ft

Since $\alpha_1(l_2/l_1) > 1.0$, 85 percent of (-132 k·ft) is assigned to the beam. Therefore

Beam negative moment $= 0.85 \times 132 = -112.2$ k·ft

Column strip negative moment $= 0.15 \times 132 = -19.8$ k·ft

Beam positive moment $= (0.85)(0.69 \times 103.0) = +60.4$ k·ft

Column strip positive moment $= (0.15)(0.69 \times 103.0) = +10.7$ k·ft

Middle strip positive moment $= (1 - 0.69)(103.0) = +31.9$ k·ft

7. The steel reinforcement required and number of bars are shown in Table 17.10. ∎

EXAMPLE 17.6

Using the direct design method, determine the negative and positive moments required for the design of the exterior panel (no. 2) of the two-way slab system with beams shown in Figure 17.7. Use the loads and the data given in Example 17.5.

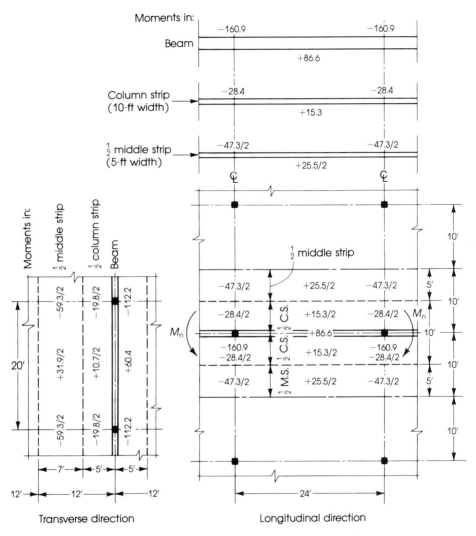

FIGURE 17.22. Example 17.5, interior slab with beams. All moments are in k·ft.

Solution   1. Limitations required by the ACI Code are satisfied in this problem. Determine the minimum slab thickness h_s using equations 17.1 through 17.3 and the following steps:

   • Assume $h_s = 7.0$ in. The sections of the interior and exterior beams are shown in Figure 17.7. Note that the extension of the slab on each side of the beam $x = y = 15$ in.

   • The moments of inertia for the interior beams and slabs were calculated earlier in example 17.2:

   I_b (in both directions) $= 22,453$ in.^{4}

   I_s (in the long direction) $= 6860$ in.^{4}

   I_s (in the short direction) $= 8232$ in.^{4}

TABLE 17.10. Design of an interior two-way slab with beams

	Long direction				Short direction			
	Column strip		Middle strip		Column strip		Middle strip	
M_u (k·ft)	−28.4	+15.3	−47.3	+25.5	−19.8	+10.7	−59.3	+31.9
Width of strip (in.)	120	120	120	120	120	120	168	168
Effective depth (in.)	6.0	6.0	6.0	6.0	5.5	5.5	5.5	5.5
$R_u = \dfrac{M_u}{bd^2}$ (psi)	79	43	132	71	65	35	196	105
Steel ratio ρ	0.0016	Low	0.0026	0.0015	Low	Low	0.0039	0.002
$A_s = \rho bd$ (in.²)	1.15	Low	1.87	1.08	Low	Low	3.60	1.85
Min. $A_s = 0.002bh_s$ (in.²)	1.68	1.68	1.68	1.68	1.68	1.68	1.85	1.85
Selected bars	9 No. 4	9 No. 4	10 No. 4	9 No. 4	9 No. 4	9 No. 4	19 No. 4	10 No. 4

- Calculate I_b and I_s for the edge beam and end slab.

$$I_b \text{ (edge beam)} = \left[\frac{27}{12} (7)^3 + (27 \times 7)(5.37)^2 \right] + \left[\frac{12}{12} (15)^3 + (12 \times 15)(5.63)^2 \right]$$

$$= 15,302 \text{ in.}^4$$

Calculate I_s for the end strip parallel to the edge beam, which has a width = 24/2 ft + $\frac{1}{2}$ column width = 12 + 10/12 = 12.83 ft.

$$I_s \text{ (end slab)} = \frac{(12.83 \times 12)}{12} (7)^3 = 4401 \text{ in.}^4$$

- Calculate α ($\alpha = EI_b/EI_s$):

$$\alpha_l \text{ (long direction)} = \frac{22,453}{6860} = 3.27$$

$$\alpha_s \text{ (short direction)} = \frac{22,453}{8232} = 2.72$$

$$\alpha \text{ (edge beam)} = \frac{15,302}{4401} = 3.48$$

$$\text{Average } \alpha = \alpha_m = \frac{3.27 + 2.72 + 3.48}{3} = 3.16$$

- $$\beta_s = \frac{\text{continuous sides}}{\text{total perimeter}} = \frac{68}{88} = 0.77$$

$$\beta = \text{ratio of long to short clear span} = \frac{22.33}{18.33} = 1.22$$

- Calculate h_s:

$$\text{Min. } h_s = \frac{(22.33 \times 12)(800 + 0.005 \times 60,000)}{36,000 + (5000 \times 1.22)[3.16 - 0.5(1 - 0.77)(1 + 1/1.22)]}$$

$$= 5.46 \text{ in.}$$

or

$$\text{Min. } h_s = \frac{294,756}{36,000 + (5000 \times 1.22)(1 + 0.77)} = 6.30 \text{ in.} \quad \text{(controls)}$$

But not more than

$$h_{max} = \frac{294,756}{36,000} = 8.19 \text{ in.}$$

Use h_s = 7 in. > 3.5 in. (minimum code limitations).

2. Calculate ultimate loads:

$$w_u = 292 \text{ psf} \quad \text{(from previous example)}$$

3. Calculate total static moments:

$$M_{ol} = 364.0 \text{ k·ft} \qquad M_{os} = 294.3 \text{ k·ft} \qquad \text{(from previous example)}$$

4. Calculate the design moments in the short direction (span = 20 ft): Since the slab is continuous in this direction, the moments are the same as those calculated in example 17.5 and shown in Figure 17.22 for an interior panel.

5. Calculate the moments in one panel using the coefficients given in Table 17.1 or Figure 17.3 (case 3):

Interior negative moment $M_{ni} = 0.7 M_o = 0.7 \times 364 = -254.8 \text{ k·ft}$

Positive moment within span $M_p = 0.57 M_o = 0.57 \times 364 = +207.5 \text{ k·ft}$

Exterior negative moment $M_{ne} = 0.16 M_o = 0.16 \times 364 = -58.2 \text{ k·ft}$

6. For the alternate solution using the Modified Stiffness Method and the equivalent column stiffness K_{ec}:

   • Calculate the design moments in the long direction (span = 24 ft), first determining K_{ec} and α_{ec}:
   Calculate the torsional stiffness K_t for the edge beam equation 17.14.

$$C = \Sigma \left(1 - 0.63 \frac{x}{y} \right) \frac{x^3 y}{3}$$

Divide the section of the edge beam into two rectangles in such a way as to obtain maximum C. Use for a beam section 12 by 22 in., $x_1 = 12$ in., $y_1 = 22$ in., and a slab section 7 by 15 in., $x_2 = 7$ in., and $y_2 = 15$ in.

$$C = \left(1 - 0.63 \times \frac{12}{22} \right) \left(\frac{12^3 \times 22}{3} \right) + \left(1 - 0.63 \times \frac{7}{15} \right) \left(\frac{7^3 \times 15}{3} \right) = 9528$$

$$K_t = \frac{9 E_c C}{l_2 \left(1 - \frac{c_2}{l_2} \right)^3} = \frac{9 E_c \times 9528}{(20 \times 12) \left(1 - \frac{20}{20 \times 12} \right)^3} = 464 E_c$$

For two adjacent slabs,

$$2K_t = 2 \times 464 E_c = 928 E_c$$

The **ACI Code, section 13.7.5.4,** specifies that where beams frame into columns in the direction of the span for which moments are being determined, K_t should be multiplied by the ratio I_{sb}/I_s in that direction. This requirement may be waived in calculation of α_{ec} for use in the direct design method (**ACI Code Commentary 13.6.3.3**).

$$K_t = 464 E_c \left(\frac{22{,}453}{6860} \right) = 1520 E_c > 928 E_c$$

Adopt $K_t = 1520 E_c$ in this example. ($928 E_c$ may also be used, producing a minimal change in final moments.)

   • Calculate the column, beam, and slab stiffnesses:

$$K_c = \frac{4 E_c I_c}{l} = \frac{4 E_c}{(12 \times 12)} \times \frac{(20)^4}{12} = 370 E_c$$

For two columns above and below the slab,

$$K_c = 2 \times 370 E_c = 740 E_c$$

$$K_b = \frac{4 E_c I_b}{l} = \frac{4 E_c}{24 \times 12} \times 22{,}453 = 312 E_c$$

$$K_s = \frac{4 E_c I_s}{l} = \frac{4 E_c}{24 \times 12} \times 6860 = 95 E_c$$

• $$\frac{1}{K_{ec}} = \frac{1}{\Sigma K_c} + \frac{1}{K_t} = \frac{1}{740 E_c} + \frac{1}{1520 E_c}$$

or

$$\frac{1000 E_c}{K_{ec}} = \frac{1000}{740} + \frac{1000}{1520}$$

$$K_{ec} = 498 E_c$$

• Calculate $\alpha_{ec} = \dfrac{K_{ec}}{\Sigma(K_s + K_b)}$

$$\alpha_{ec} = \frac{498 E_c}{(312 + 95) E_c} = 1.22 \qquad \left(1 + \frac{1}{\alpha_{ec}}\right) = 1 + \frac{1}{1.22} = 1.84$$

• Calculate the design moments:

$$\text{Interior negative moment} = \left[0.75 - \frac{0.1}{\left(1 + \dfrac{1}{\alpha_{ec}}\right)}\right] M_{ol}$$

$$M_{ni} = \left(0.75 - \frac{0.1}{1.84}\right) 364 = -253.2 \text{ k·ft}$$

$$\text{Positive moment} = \left[0.63 - \frac{0.28}{\left(1 + \dfrac{1}{\alpha_{ec}}\right)}\right] M_{ol}$$

$$M_p = \left(0.63 - \frac{0.28}{1.84}\right) 364 = +173.9 \text{ k·ft}$$

$$\text{Exterior negative moment} = \left[\frac{0.65}{1 + \dfrac{1}{\alpha_{ec}}}\right] M_{ol}$$

$$M_{ne} = \frac{0.65}{1.84} \times 364 = -128.6 \text{ k·ft}$$

This solution gives greater negative moment at the exterior support and smaller positive moment as compared to the moments computed in step 5.

7. Distribute the panel moments to beam, column, and middle strips. Here the moments computed in step 6 will be adopted.

$$\frac{l_2}{l_1} = \frac{20}{24} = 0.83 \qquad \alpha_1 = \alpha_l = 3.27$$

$$\alpha_1 \frac{l_2}{l_1} = 3.27 \times 0.83 = 2.71 > 1.0$$

$$\beta_t = \frac{E_{cb}C}{2E_{cs}I_s} = \frac{8772}{2 \times 6860} = 0.64$$

• Distribute the interior negative moment M_{ni}:
Referring to Table 17.2 and by interpolation, the percentage of moment assigned to the column strip (for $l_2/l_1 = 0.83$ and $\alpha_1 l_2/l_1 > 1.0$) = 80 percent.

> Column strip = $0.8 \times 253.2 = -202.6$ k·ft
>
> Middle strip = $0.2 \times 253.2 = -50.6$ k·ft

Since $\alpha_1 l_2/l_1 > 1.0$, 85 percent of the moment in the column strip is assigned to the beam. Therefore

> Beam = $0.85 \times 202.6 = -172.2$ k·ft
>
> Column strip = $0.15 \times 202.6 = -30.4$ k·ft
>
> Middle strip = -50.6 k·ft

• Distribute the positive moment M_p:
Referring to Table 17.2 and by interpolation, the percentage of moment assigned to the column strip = 80 percent (85 percent of this value is assigned to the beam). Therefore

> Beam = $(0.85)(0.8 \times 173.9) = +118.3$ k·ft
>
> Column strip = $(0.15)(0.8 \times 173.9) = +20.9$ k·ft
>
> Middle strip = $0.2 \times 173.9 = +34.8$ k·ft

• Distribute the exterior negative moment M_{ne}:
Referring to Table 17.4 and by interpolation, the percentage of moment assigned to the column strip (for $l_2/l_1 = 0.83$, $\alpha_1 l_2/l_1 > 1.0$, and $\beta_t = 0.64$) = 94 percent, and 85 percent of the moment is assigned to the beam. Therefore

> Beam = $(0.85)(0.94 \times 128.6) = -102.8$ k·ft
>
> Column strip = $(0.15)(0.94 \times 128.6) = -18.1$ k·ft
>
> Middle strip = $0.06 \times 128.6 = -7.7$ k·ft

Moment details are shown in Figure 17.23.

17.7   Design Moments in Columns

When the analysis of the equivalent frames is carried out by the direct design method, the moments in columns due to the unbalanced loads on adjacent panels are obtained from the following equation specified by the **ACI Code, section 13.6.9:**

$$M_u = 0.07[(w_d + 0.5w_l)l_2 l_n^2 - w_d' l_2'(l_n')^2] \qquad (17.21a)$$

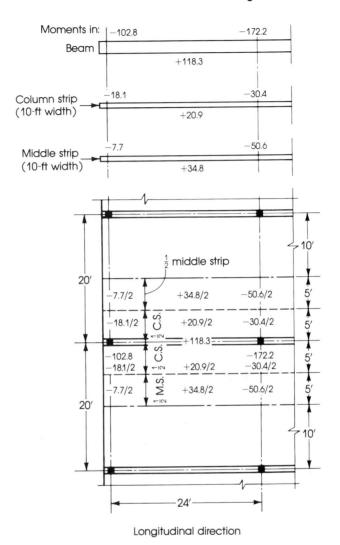

FIGURE 17.23. Example 17.6, exterior slab with beams. (All moments are in k·ft.)

If the alternate solution using K_{ec} and α_{ec} is used, then the moment M_u is computed as follows:

$$M_u = \frac{0.08[(w_d + 0.5w_l)l_2 l_n^2 - w_d' l_2' (l_n')^2]}{\left(1 + \dfrac{1}{\alpha_{ec}}\right)}$$

(ACI Code Commentary, section 13.6.3.3) $(17.21b)$

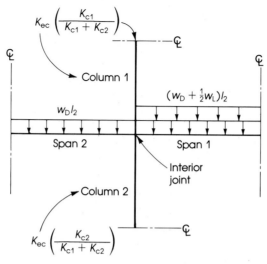

FIGURE 17.24. Interior column loading.

where

w_d and w_l = Factored dead and live loads on the longer span

w'_d = factored dead load on the shorter span

l_n and l'_n = length of the longer and shorter spans, respectively

$$\alpha_{ec} = \frac{K_{ec}}{\Sigma(K_s + K_b)} \qquad (17.16)$$

The moment in equation 17.21 should be distributed between the columns above and below the slab at the joint in proportion to their flexural stiffnesses (Figure 17.24). For equal spans, $l_2 = l'_2$ and $l_n = l'_n$,

$$M_u = 0.07(0.5w_l l_2 l_n^2) \qquad (17.22a)$$

$$M_u = \frac{0.08(0.5w_l l_2 l_n^2)}{\left(1 + \dfrac{1}{\alpha_{ec}}\right)} \qquad (17.22b)$$

The development of the above equations is based on the assumption that half the live load acts on the longer span while the dead load acts on both spans. Equation 17.21 can also be applied to an exterior column by assuming the shorter span length is zero (Figure 17.25).

17.8   Transfer of Unbalanced Moments to Columns

17.8.1   Transfer of Moments

In the analysis of an equivalent frame in a building, moments develop at the slab-column joints due to lateral loads as wind, earthquakes, or unbalanced gravity loads, causing

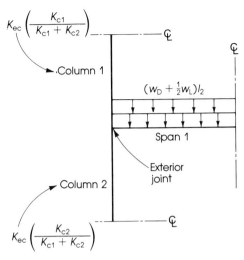

FIGURE 17.25. Exterior column loading.

unequal moments in the slab on opposite sides of columns. A fraction of the unbalanced moment in the slabs must be transferred to the columns by flexure, and the balance must be transferred by vertical shear acting on the critical sections for punching shear. Approximately 60 percent of the moment transferred to both ends of the column at a joint is transferred by flexure and the remaining 40 percent by eccentric shear (or torque) at the section located at $d/2$ from the face of the column.[14,15] The **ACI Code, section 13.3.3**, states that the fraction of the unbalanced moment transferred by flexure M_f at a slab-column connection is determined as follows:

$$M_f = \gamma_f M_u \qquad\qquad (17.23)$$

$$\gamma_f = \cfrac{1}{1 + \cfrac{2}{3}\sqrt{\cfrac{c_1 + d}{c_2 + d}}} \qquad \textbf{(ACI Code equation 13.1)} \qquad (17.24)$$

and the moment transferred by shear

$$M_v = (1 - \gamma_f)M_u = M_u - M_f \qquad\qquad (17.25)$$

where c_1 and c_2 are the lengths of the two sides of a rectangular or equivalent rectangular column. When $c_1 = c_2$, $M_f = 0.6M_u$, and $M_v = 0.4M_u$.

17.8.2   Concentration of Reinforcement Over The Column

For a direct transfer of moment to the column, it is necessary to concentrate part of the steel reinforcement in the column strip within a specified width over the column. The part of the moment transferred by flexure M_f is considered acting through a slab width equal to the transverse column width c_2 plus $1.5h_s$ on each side of the column, or to the width $(c_2 + 3h_s)$ **(ACI Code, section 13.3.3)**. Reinforcement can be concentrated over the column by closer spacing of bars or the use of additional reinforcement.

17.8.3   Shear Stresses Due to M_v

The shear stresses produced by the portion of the unbalanced moment M_v must be combined with the shear stresses produced by the shearing force V_u due to vertical loads. Both shear stresses are assumed acting around a periphery plane located at a distance $d/2$ from the face of the column[16], as shown in Figure 17.26. The equation for computing the shear stresses is

$$v_{1,2} = \frac{V_u}{A_c} \pm \frac{M_v c}{J_c} \qquad\qquad (17.26)$$

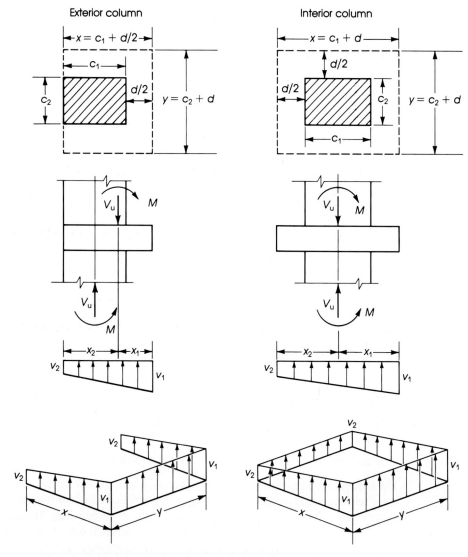

FIGURE 17.26. Shear stresses due to V_u and M.

where

    A_c = area of critical section around the column

    J_c = polar moment of inertia of the areas parallel to the applied moment in addition to that of the end area about the centroidal axis of the critical section

For an interior column,

$$A_c = 2d(x + y) \qquad (17.27)$$

and

$$J_c = \frac{d}{2}\left(\frac{x^3}{3} + x^2 y\right) + \frac{xd^3}{6} \qquad (17.28)$$

For an exterior column,

$$A_c = d(2x + y) \qquad (17.29)$$

and

$$J_c = \frac{2dx^3}{3} - (2x + y)dx_1^2 + \frac{xd^3}{6} \qquad (17.30)$$

where x, x_1, and y are as shown in Figure 17.26. The maximum shear stress

$$v_1 = \frac{V_u}{A_c} + \frac{M_v c}{J_c}$$

must be less than $\phi(4\sqrt{f_c'})$, otherwise shear reinforcement should be provided.

EXAMPLE 17.7

Determine the moments at the exterior and interior columns in the long direction of the flat plate in example 17.4.

Solution

1. Exterior column moment

$$M = \frac{0.08}{\left(1 + \dfrac{1}{\alpha_{ec}}\right)}[(w_d + 0.5w_1)l_2 l_n^2 - w_d' l_2'(l_n')^2] \qquad (17.21b)$$

From example 17.4,

$$w_d = (130.25)(1.4) = 0.18 \text{ ksf}$$

$$0.5w_1 = 0.5 \times (1.7 \times 60) = 51 \text{ psf}$$

$$l_2 = l_2' = 20 \text{ ft} \quad l_n = l_n' = 22.33 \text{ ft} \quad \left(1 + \frac{1}{\alpha_{ec}}\right) = 1.81$$

The unbalanced moment to be transferred to the exterior column

$$M_u = \frac{0.08}{1.81}[0.18 + (0.051)(20)(22.33)^2 - 0] = 101.8 \text{ k·ft}$$

If equation 17.21a is used, $M_u = 161.26$ k·ft, which is a conservative value.

2. At an interior support, the slab stiffness on both sides of the column must be used to compute α_{ec}.

$$\alpha_{ec} = \frac{K_{ec}}{\Sigma K_s + K_b} \qquad (17.16)$$

From example 17.4, $K_{ec} = 209.4E_c$, $K_s = 170.6E_c$, and $K_b = 0$. Therefore

$$\alpha_{ec} = \frac{209.4 E_c}{2 \times 170.6 E_c + 0} = 0.614$$

$$\left(1 + \frac{1}{\alpha_{ec}}\right) = 1 + \frac{1}{0.614} = 2.63$$

From equation 17.21b, the unbalanced moment at an interior support

$$M_u = \frac{0.08}{2.63}[(0.18 + 0.051)(20)(22.33)^2 - 0.18 \times 20(22.33)^2]$$

$$= 15.5 \text{ k·ft} \qquad \blacksquare$$

If equation 17.21a is used, $M_u = 35.6$ k·ft, which is a very conservative value.

EXAMPLE 17.8

For the flat plate in example 17.4, calculate the shear stresses in the slab at the critical sections due to unbalanced moments and shearing forces at an interior and exterior column. Check the concentration of reinforcement and torsional requirements at the exterior column. Use $f_c' = 4$ ksi and $f_y = 40$ ksi.

Solution

1. The unbalanced moment at the interior support, $M_u = 15.5$ k·ft (example 17.7), where

$$\gamma_f = \frac{1}{1 + \frac{2}{3}\sqrt{\frac{c_1 + d}{c_2 + d}}} = \frac{1}{1 + \frac{2}{3}\sqrt{\frac{20 + 7.5}{20 + 7.5}}} = 0.6$$

The moment to be transferred by flexure,

$$M_f = \gamma_f M_u = 0.6 \times 15.5 = 9.3 \text{ k·ft}$$

The moment to be transferred by shear,

$$M_v = 15.5 - 9.3 = 6.2 \text{ k·ft}$$

$$V_u = 0.285\left[20 \times 24 - \left(\frac{27.5}{12}\right)^2\right] = 135.3 \text{ k}$$

From Figure 17.27,

$$A_c = 4(27.5)(7.5) = 825 \text{ in.}^2$$

$$J_c = \frac{d}{2}\left(\frac{x^3}{3} + x^2 y\right) + \frac{xd^3}{6}$$

$$= \frac{7.5}{2}\left[\frac{(27.5)^3}{3} + (27.5)^2(27.5)\right] + \frac{27.5}{6}(7.5)^3 = 105,916 \text{ in.}^4$$

$$v_{max} = \frac{135,300}{825} + \frac{6.2(12,000) \times \left(\frac{27.5}{2}\right)}{105,916}$$

$$= 164.0 + 9.7 = 173.7 \text{ psi}$$

$$v_{min} = 164.0 - 9.7 = 154.3 \text{ psi}$$

Allowable $v_c = \phi 4\sqrt{f'_c} = 0.85 \times 4\sqrt{4000} = 215 \text{ psi} > 173.7 \text{ psi}$

2. For the exterior column, the unbalanced moment to be transferred by flexure M_f at a slab-column joint is equal to $\gamma_f M_u$, where $M_u = 101.8$ k·ft and

$$\gamma_f = \frac{1}{1 + \frac{2}{3}\sqrt{\frac{c_1 + d}{c_2 + d}}}$$

$c_1 = c_2 = 20$ in., $d = 7.5$ in. in the longitudinal direction, and $\gamma_f = 0.6$ for square columns.

$$M_f = 0.6 \times 101.8 = 61.1 \text{ k·ft}$$

The moment to be transferred by shear,

$$M_v = M_u - M_f = 101.8 - 61.1 = 40.7 \text{ k·ft}$$

3. For transfer by shear at exterior column, the critical section is located at a distance $d/2$ from the face of the column (Figure 17.28).

$$w_u = 285 \text{ psf}$$

$$V_u = 0.285\left(20 \times 12.83 - \frac{23.75}{12} \times \frac{27.5}{12}\right) = 71.8 \text{ k}$$

$$M_v = 40.7 \text{ k·ft}$$

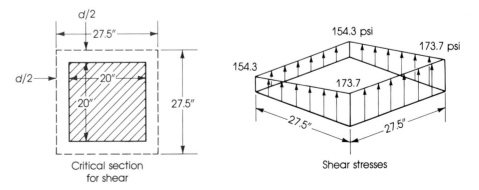

FIGURE 17.27. Example 17.8; shear stresses at interior column due to unbalanced moment.

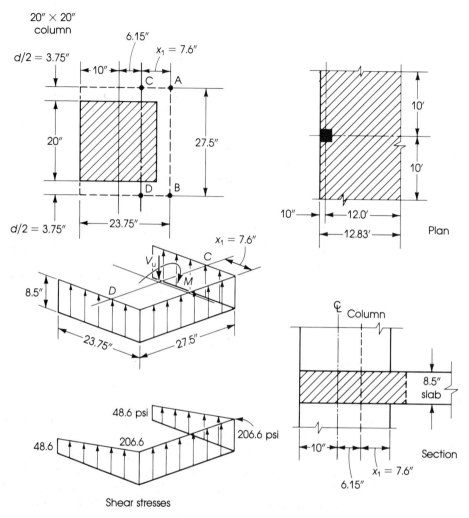

FIGURE 17.28. Example 17.8; shear stresses at exterior column due to unbalanced moment.

Locate the centroid of the critical section by taking moments about AB:

$$2\left(23.75 \times \frac{23.75}{2}\right) = (2 \times 23.75 + 27.5)x_1$$

Therefore, $x_1 = 7.6$ in. The area of the critical section A_c is

$$A_c = 2(23.75 \times 7.5) + 27.5 \times 7.5 = 562.5 \text{ in.}^2$$

Calculate $J_c = I_x + I_y$ for the two equal areas (7.5 × 23.75) with sides parallel to the direction of moment, and the area (7.5 × 27.5) perpendicular to the direction of moment, all about the axis through CD.

$$J_c = I_y + I_x = \Sigma \left(\frac{bh^3}{12} + Ax^2 \right)$$

$$= 2 \left[7.5 \frac{(23.75)^3}{12} + (7.5 \times 23.75) \left(\frac{23.75}{2} - 7.6 \right)^2 \right]$$

$$+ 2 \left[\frac{23.75}{12} (7.5)^3 \right] + [(27.5 \times 7.5)(7.6)^2] = 36{,}840 \text{ in.}^4$$

or by using equation 17.30 for an exterior column. Calculate the maximum and minimum nominal shear stresses using equation 17.26:

$$v_{max} = \frac{V_u}{A_c} + \frac{M_v c}{J_c} = \frac{71{,}800}{562.5} + \frac{40.7 \times 12{,}000 \times 7.6}{36{,}840} = 206.6 \text{ psi}$$

$$v_{min} = 48.6 \text{ psi}$$

Allowable $v_c = \phi 4 \sqrt{f_c'} = 0.85 \times 4 \sqrt{4000} = 215 \text{ psi}$

If shear stress is greater than the allowable v_c, increase the slab thickness or use shear reinforcement.

4. Check the concentration of reinforcement at exterior column; that is, check that the flexural capacity of the section is adequate to transfer the negative moment into the exterior column. The critical area of the slab extends $1.5h_s$ on either side of the column, giving an area $(20 + 3 \times 7.5) = 42.5$ in. wide and 7.5 in. deep. The total moment in the 120-in.-wide column strip is 126 k·ft, as calculated in example 17.4, step 9. Moment in a width $= d + 3h_s = 42.5$ in. is equal to

$$126 \times \frac{42.5}{120} = 44.6 \text{ k·ft}$$

If equal spacing in the column strip is used, then the additional reinforcement within 42.5-in. width will be needed for a moment $= M_f - 44.6 = 61.1 - 44.6 = 16.5$ k·ft. The required $A_s = 0.79$ in.^{2}, and 4 No. 4 bars ($A_s = 0.8$ in.^{2}) may be used. The reinforcement within the column strip may be arranged to increase the reinforcement within a width $= 42.5$ in. The amount of steel needed within this width should be enough to resist a moment of 0.6 the negative moment in the column strip $= 0.6 \times 126 = 75.6$ k·ft.

$$A_s = \frac{M_u}{\phi f_y (d - a/2)} \qquad \text{assume } a = 1.0 \text{ in.}$$

$$A_s = \frac{75.6 \times 12}{0.9 \times 40(7.5 - 0.5)} = 3.60 \text{ in.}^2$$

$$\text{Check } a = \frac{A_s f_y}{0.85 f_c' b} = \frac{3.60 \times 40}{0.85 \times 4 \times 42.5} = 1.0 \text{ in.}$$

Use 12 No. 5 bars within a width 42.5 in. divided equally at both sides from the center of the column (Figure 17.29). Additional reinforcement as indicated above provides a better solution.

5. Torque on slab: The torque from both sides of the exterior column is equal to 40 percent of the column strip moment.

$$T_u = 0.4 \times 126 = 50.4 \text{ k·ft}$$

$$\text{Torque on each side} = \frac{50.4}{2} = 25.2 \text{ k·ft} = 302.4 \text{ k·in.}$$

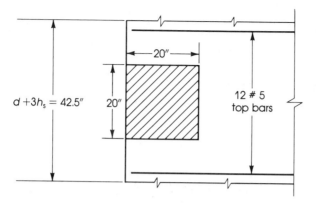

FIGURE 17.29. Concentration of reinforcement within exterior column strip, example 17.8.

A slab section of width equal to the column width will be assumed to resist the torsional stresses:

$$T_u = \frac{1}{3} v_{tu} \Sigma x^3 y$$

where $x = 8.5$ in. and $y = 20$ in. The critical section is at a distance d from the face of the column (Figure 17.30). Assuming that the torque varies in a parabolic curve to the center of the slab, then the torque at a distance d

$$T_u = 25.2 \times \left(\frac{140 - 7.5}{140}\right)^2 = 22.6 \text{ k·ft} = 271 \text{ k·in.}$$

The torsional strength of concrete

$$T_c = 0.8\sqrt{f_c'}(x^2 y) = 0.8\sqrt{4000}(8.5)^2(20) = 73.1 \text{ k·in.}$$
$$T_u = \phi T_c + \phi T_s$$

Torsional moment to be resisted by reinforcement

$$T_s = \frac{1}{0.85} (271 - 0.85 \times 73.1) = 245.7 \text{ k·in}$$

The required closed stirrups and the additional longitudinal bars are determined as was explained in Chapter 15. The final section is shown in Figure 17.30. It is advisable to provide an edge beam between the exterior columns to increase the torsional stiffness of the slab. ∎

17.9   Equivalent Frame Method

When two-way floor systems do not satisfy the limitations of the direct design method, the design moments must be computed by the equivalent frame method. In the latter method, the building is divided into equivalent frames in two directions and then analyzed elastically for all conditions of loadings. The difference between the direct design

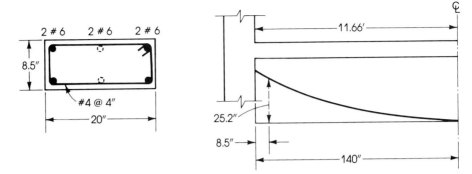

FIGURE 17.30. Reinforcement in edge of slab to resist torque, example 17.8.

and equivalent frame methods lies in the way by which the longitudinal moments along the spans of the equivalent rigid frame are determined. The design requirements can be explained as follows:

1. Description of the equivalent frame: An equivalent frame is a two-dimensional building frame obtained by cutting the three-dimensional building along lines midway between columns (Figure 17.3). The resulting equivalent frames are considered separately in the longitudinal and transverse directions of the building. For vertical loads, each floor is analyzed separately, with the far ends of the upper and lower columns assumed to be fixed. The slab-beam may be assumed to be fixed at any support two panels away from the support considered, because the vertical loads contribute very little to the moment at that support. For lateral loads, the equivalent frame consists of all the floors and extends for the full height of the building because the forces at each floor are a function of the lateral forces on all floors above the considered level.

2. Load assumptions: When the ratio of the service live load to the service dead load ≤ 0.75, the structural analysis of the frame can be made with the factored dead and live loads acting on all spans instead of a pattern loading. When the ratio of the service live load to the service dead load > 0.75, pattern loading must be used considering the following conditions:

   - Only 75 percent of the full factored live load may be used for the pattern loading analysis.
   - The maximum negative bending moment in the slab at the support is obtained by loading only the two adjacent spans.
   - The maximum positive moment near a midspan is obtained by loading only alternate spans.
   - The design moments must not be less than those occurring with full factored live load on all panels **(ACI Code, section 13.7.6).**
   - The critical negative moments are considered acting at the face of a rectangular column or at the face of the equivalent square column having the same area for nonrectangular sections.

3. Slab-beam moment of inertia: The ACI Code specifies that the variation in moment of inertia along the longitudinal axes of columns and slab-beams must be taken into account in the analysis of frames. The critical region is located between the centerline of the column and the face of the column, bracket, or capital. This region may be considered as a thickened section of the floor slab. To account for the large depth of the column and its reduced effective width in contact with the slab-beam, the **ACI Code, section 13.7.3.3.**, specifies that the moment of inertia of the slab-beam between the center of the column and the face of the support is to be assumed equal to that of the slab-beam at the face of the column divided by the quantity

$$\left(1 - \frac{c_2}{l_2}\right)^2$$

where c_2 = the column width in the transverse direction and l_2 = the width of the slab-beam. The area of the gross section can be used to calculate the moment of inertia of the slab-beam.

4. Column moment of inertia: The **ACI Code, section 13.7.4,** states that the moment of inertia of the column is to be assumed infinite from the top of the slab to the bottom of the column capital or slab-beams (Figure 17.31).

5. Column stiffness K_{ec}:

$$\frac{1}{K_{ec}} = \frac{1}{\Sigma K_c} + \frac{1}{K_t} \qquad\qquad (17.12)$$

where ΣK_c = the sum of the stiffness of the upper and lower columns at their ends,

$$K_t = \Sigma \frac{9 E_{es} C}{l_2 \left(1 - \frac{c_2}{l_2}\right)^2} \qquad \textbf{(ACI Code equation 13.6)}$$

$$C = \Sigma \left(1 - 0.63 \frac{x}{y}\right)\left(\frac{x^3 y}{3}\right) \qquad \textbf{(ACI Code equation 13.7)} \qquad (17.15)$$

6. Column moments: In frame analysis, moments determined for the equivalent columns at the upper end of the column below the slab and at the lower end of the column above the slab must be used in the design of a column.

7. Negative moments at the supports: The **ACI Code, section 13.7.7,** states that for an interior column, the factored negative moment is to be taken at the face of column or capital, but not at a distance greater than $0.175 l_1$ from the center of the column. For an exterior column, the factored negative moment is to be taken at a section located at half the distance between the face of the column and the edge of the support. Circular section columns must be treated as square columns with the same area.

8. Sum of moments: A two-way slab floor system that satisfies the limitations of the direct design method can also be analyzed by the equivalent frame method. To ensure that both methods will produce similar results, the **ACI Code, section 13.7.7,** states that the computed moments determined by the equivalent

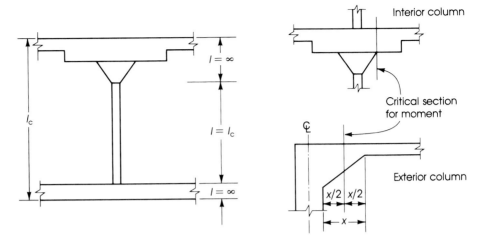

FIGURE 17.31. Critical sections for column moment, equivalent frame method.

frame method may be reduced in such proportion that the numerical sum of the positive and average negative moments used in the design must not exceed the total statical moment M_o.

EXAMPLE 17.9

By the equivalent frame method, analyze a typical interior frame of the flat-plate floor system given in example 17.3 in the longitudinal direction only. The floor system consists of four panels in each direction with a panel size of 25 by 20 ft. All panels are supported by 20 by 20 in. columns, 12 ft long. The service live load = 60 psf and the service dead load = 124 psf (including weight of slab). Use $f_c' = 3$ ksi and $f_y = 40$ ksi. Edge beams are not used. Refer to Figure 17.32.

Solution
1. A slab thickness of 8.0 in. is chosen, as explained earlier in example 17.3.
2. Ultimate load $w_u = 1.4 \times 124 + 1.7 \times 60 = 275.6$ psf. The ratio of service live load to service dead load = $60/124 = 0.48 < 0.75$; therefore the frame can be analyzed with the full factored load w_u acting on all spans instead of pattern loading.
3. Determine the slab stiffness K_s:

$$K_s = k \frac{EI_s}{l_s}$$

where k is the stiffness factor and

$$I_s = \frac{l_2 h_s^3}{12} = \frac{(20 \times 12)}{12} (8)^3 = 10{,}240 \text{ in.}^4$$

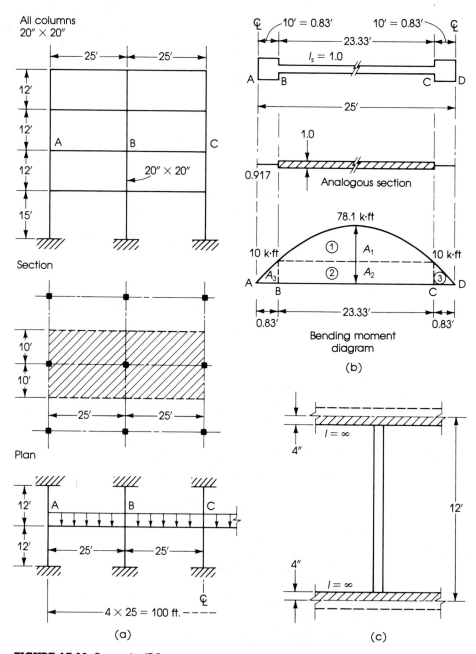

FIGURE 17.32. Example 17.9.

The stiffness factor can be determined by the column analogy method described in books on structural analysis. Considering the moment of inertia for the slab = 1.0 as a reference, the moment of inertia between the column centerline and the face of the column is

$$\frac{1.0}{\left(1 - \frac{c_2}{l_2}\right)^2} = \frac{1.0}{\left(1 - \frac{20}{20 \times 12}\right)} = 1.09$$

The width of the analogous column varies with $1/I$, as shown in Figure 17.32(b).

$$\text{Slab stiffness factor } k = l_1 \left(\frac{1}{A_a} + \frac{Mc}{I_a}\right) \qquad (17.31)$$

where

A_a = area of the analogous column section

I_a = moment of inertia of analogous column

M = moment due to a unit load at the extreme fiber of the analogous column located at the center of the slab

$$M = 1.0 \times \frac{l_1}{2}$$

$A_a = 23.33 + 2 \times (0.83 \text{ ft})(0.917) = 23.33 + 1.52 = 24.85$

$I_a = I$ (for slab portion of 23.33) + I (of end portion)

about the centerline.

$$I_a = \frac{(23.33)^3}{12} + 1.52 \left(12.5 - \frac{0.83}{2}\right)^2 = 1058 + 222 = 1280$$

neglecting the moment of inertia of the short end segments about their own centroid.

$$\text{Stiffness factor} = 25 \left[\frac{1}{24.85} + \frac{1.0 \times 12.5(12.5)}{1280}\right]$$

$$= 1.0 + 3.05 = 4.05$$

$$\text{Carry-over factor} = \frac{3.05 - 1.0}{4.05} = 0.506$$

Therefore, slab stiffness

$$K_s = \frac{4.05 E \times 10,240}{(25 \times 12)} = 138 E$$

4. Determine the column stiffness K_c:

$$K_c = K' \left(\frac{EI_c}{l_c}\right) \times 2$$

for columns above and below the slab.

k' = column stiffness factor

$$l_c = 12 \text{ ft} \qquad I_c = \frac{(20)^4}{12} = 13.333 \text{ in.}^4$$

The stiffness factor k' can be determined as follows:

$$k' = l_c \left(\frac{1}{A_a} + \frac{Mc}{I_a} \right)$$

for the column.

$$c = \frac{l_c}{2}$$

$$M = 1.0 \times \frac{l_c}{2} = \frac{l_c}{2}$$

$$A_a = l_c - h_s = 12 - \frac{8}{12} = 11.33$$

$$I_a = \frac{(l_c - h_s)^3}{12} = \frac{(11.3)^3}{12} = 121.2$$

$$k' = 12 \left[\frac{1}{11.33} + \frac{\left(1 \times \frac{12}{2}\right)\left(\frac{12}{2}\right)}{121.2} \right] = 1.06 + 3.56 = 4.62$$

$$K_c = 4.62E \times \frac{13,333}{12 \times 12} \times 2 = 856E$$

In a flat-plate floor system, the column stiffness K_c can be calculated directly as follows:

$$\frac{K_c}{E_c} = \frac{I_c}{(l_c - h_s)} + \frac{3I_c l_c^2}{(l_c - h_s)^3} \tag{17.32}$$

5. Calculate the torsional stiffness K_t of the slab at the side of the column:

$$K_t = \frac{\Sigma 9 E_{cs} C}{l_2 \left(1 - \frac{c_2}{l_2}\right)^2} \quad \text{and} \quad C = \Sigma \left(1 - 0.63 \frac{x}{y}\right) \frac{x^3 y}{3}$$

In this example, $x = 8.0$ in. (slab thickness) and $y = 20$ in. (column width) (see Figure 17.20).

$$C = \left(1 - 0.63 \times \frac{8}{20}\right) \times \frac{(8)^3 \times 20}{3} = 2553$$

$$K_t = \frac{9 E_{cs} \times 2553}{(20 \times 12)\left(1 - \frac{20}{20 \times 12}\right)^3} = 124 E_c$$

For two adjacent slabs, $K_t = 2 \times 124 E_c = 248 E_c$.

6. Calculate the equivalent column stiffness K_{ec}:

$$\frac{1}{K_{ec}} = \frac{1}{\Sigma K_c} + \frac{1}{K_t} = \frac{1}{856 E_c} + \frac{1}{248 E_c}$$

or $K_{ec} = 192 E_c$.

7. Moment distribution factors (D.F.): For the exterior joint,

$$\text{D.F. (slab)} = \frac{K_s}{K_s + K_{ec}} = \frac{138}{138 + 192} = 0.42$$

$$\text{D.F. (columns)} = \frac{K_{ec}}{\Sigma K} = 0.58$$

The columns above and below the slab have the same stiffness; therefore the distribution factor of 0.58 is divided equally between both columns, and each takes a D.F. = 0.58/2 = 0.29. For the interior joint,

$$\text{D.F. (slab)} = \frac{K_s}{2K_s + K_{ec}} = \frac{138}{2 \times 138 + 192} = 0.295$$

$$\text{D.F. (columns)} = \frac{K_{ec}}{\Sigma K} = \frac{192}{2 \times 138 + 192} = 0.41$$

Each column will have a D.F. = 0.41/2 = 0.205.
8. Fixed end moments: Since the actual L.L./D.L. < 0.75, the full factored load is assumed to act on all spans.

$$\text{Fixed end moment} = K'' w_u l_2 (l_1)^2$$

The factor K'' can be determined by the column analogy method: For a unit load $w = 1.0$ k/ft over the longitudinal span of 25 ft, the simple moment diagram is shown in Figure 17.32(b). The area of the bending moment diagram considering the variation of the moment of inertia along the span is

$$\text{Total area } A_m = A_1 + A_2 + 2A_3$$

$$= \frac{2}{3} \times 23.33(78.1 - 10) + 23.33 \times 10 + 2\left(\frac{1}{2} \times 0.83 \times 10\right)(0.917)$$

$$= 1300$$

$$\text{Fixed end moment coefficient} = \frac{A_m}{A_a l_1^2}$$

where A_a for the slab = 24.85, as calculated in step 3.

$$K'' = \frac{1300}{24.85(25)^2} = 0.0837$$

It can be seen that the fixed end moment coefficient $K'' = 0.0837$ is very close to the coefficient $1/12 = 0.0833$ usually used to calculate the fixed end moments in beams. This is expected because the part of the span that has a variable moment of inertia is very small in flat plates where no column capital or drop panels are used. In this example only parts AB and CD each = 0.83 ft have a higher moment of inertia than I_s. In flat plates where the ratio of the span to column width is high, say, ≥ 20, the coefficient 0.0833 may be used to calculate approximately the fixed end moments.

$$\text{Fixed end moment (due to } w_u = 275.6 \text{ psf)} = 0.0837(0.2756)(20)(25)^2$$

$$= 288.3 \text{ k·ft}$$

The factors K, K', and K'' can be obtained from tables prepared by Simmonds and Misic[17] to meet the ACI requirements for the equivalent frame method.
9. Moment distribution can be performed on half the frame due to symmetry. Once the end negative moments are computed, the positive moments at the center of any span can be obtained by subtracting the average value of the negative end moments from the simple beam positive moment. The moment distribution is shown in Figure 17.33. The final bending moments and shear forces are shown in Figure 17.34.
10. Slabs can be designed for the negative moments at the face of the columns as shown in Figure 17.34. ■

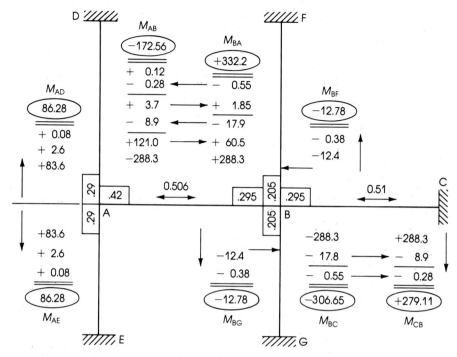

FIGURE 17.33. Analysis by moment distribution, example 17.9. All moments are in k·ft.

EXAMPLE 17.10

Design by the direct design method a typical interior flat slab with drop panels to carry a dead load of 7.86 kN/m² and a live load = 10 kN/m². The floor system consists of six panels in each direction, with a panel size of 6.0 by 5.4 m. All panels are supported by 0.4-m-diameter columns with 1.0-m-diameter column capitals. The story height is 3.0 m. Use $f'_c = 21$ MPa and $f_y = 420$ MPa.

Solution

1. All the ACI limitations to use the direct design method are met. Determine the minimum slab thickness h_s using equations 17.1 through 17.3. Diameter of the column capital equals 1.0 m. The equivalent square column section of the same area will have a side = $\sqrt{\pi R_2}$ = $\sqrt{\pi (500)^2}$ = 885 mm, say, 900 mm.

   Clear span (long direction) = 6.0 − 0.9 = 5.1 m

   Clear span (short direction) = 5.4 − 0.9 = 4.5 m

   Since no beams are used, $\alpha_m = 0$, $\beta_s = 1.0$ because all panel edges are continuous, and $\beta = 6.0$ m/5.4 m = 1.11. From equation 17.1,

$$h_{min} = \frac{l_n(800 + 0.73 f_y)}{36{,}000 + 5000\beta\left[\alpha_m - 0.5(1 - \beta_s)\left(1 + \frac{1}{\beta}\right)\right]}$$

$$= \frac{5100(800 + 0.73 \times 420)}{36{,}000 + 5000 \times 1.11(0 - 0)} = 156.7 \text{ mm}$$

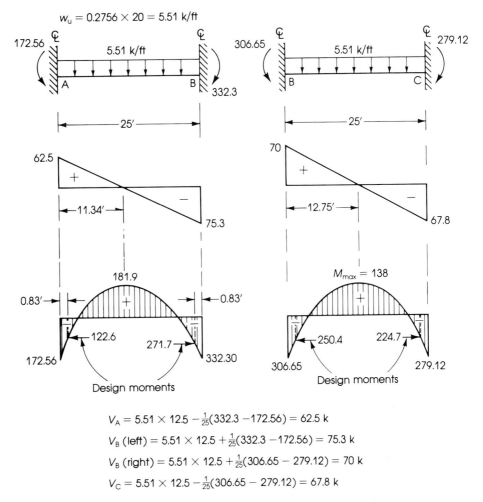

$$V_A = 5.51 \times 12.5 - \tfrac{1}{25}(332.3 - 172.56) = 62.5 \text{ k}$$

$$V_B \text{ (left)} = 5.51 \times 12.5 + \tfrac{1}{25}(332.3 - 172.56) = 75.3 \text{ k}$$

$$V_B \text{ (right)} = 5.51 \times 12.5 + \tfrac{1}{25}(306.65 - 279.12) = 70 \text{ k}$$

$$V_C = 5.51 \times 12.5 - \tfrac{1}{25}(306.65 - 279.12) = 67.8 \text{ k}$$

FIGURE 17.34. Example 17.9, equivalent frame method—final bending moments and shear forces. (Slabs can be designed for the negative moments at the face of the columns as shown.)

From equation 17.2,

$$h = \frac{l_n(800 + 0.73f_y)}{36,000 + 5000\,\beta\,(1 + \beta_s)}$$

$$= \frac{5100(800 + 0.73 \times 420)}{36,000 + 5000 \times 1.11(1 + 1)} = 120 \text{ mm}$$

but not more than

$$h_{max} = \frac{l_n(800 + 0.73f_y)}{36,000} = 156 \text{ mm}$$

$$h_{min} = 125 \text{ mm (5.0 in.)}$$

Thus $h_s = 156$ mm controls, but since a drop panel is used, the **ACI Code, section 9.5.3,** lowers h_s by 10 percent if the drop panels extend in each direction from the centerline of the support a distance $\geq l/6$ and project below the slab a distance $\geq h_s/4$. Therefore, use a slab thickness $h_s = 0.9 \times 156 = 140$ mm, and a drop panel as follows:

$$\text{Long direction } \frac{l_1}{3} = \frac{6.0}{3} = 2.0 \text{ m (both sides of support)}$$

$$\text{Short direction } \frac{l_2}{3} = \frac{5.4}{3} = 1.8 \text{ m (both sides of support)}$$

Thickness of drop panel = $1.25h_s = 1.25 \times 140 = 175$ mm. Increase drop panel thickness to 220 mm to provide adequate thickness for punching shear and to avoid the use of high percentage of steel reinforcement. Details are shown in Figure 17.35.

2. Calculate ultimate loads:

$$w_u = 1.4 \times 7.87 + 1.7 \times 10 = 28 \text{ kN/m}^2$$

3. Check two-way shear, first in drop panel: The critical section is at a distance $d/2$ around the column capital. Let $d = 220 - 30$ mm $= 190$ mm.

Diameter of shear section = 1.0 m $+ d = 1.19$ m

$$V_u = 28\left[6.0 \times 5.4 - \frac{\pi}{4}(1.19)^2\right] = 876 \text{ kN}$$

$$b_o = 2\pi\left(\frac{1.19}{2}\right) = 3.74 \text{ m}$$

$$\phi V_c = \phi(0.33\sqrt{f_c'})b_o d = \frac{0.85 \times 0.33}{1000}\sqrt{21} \times 3740 \times 190 = 913 \text{ kN}$$

which is greater than V_u of 876 kN. Then check two-way shear in slab: The critical section is at a distance $d/2$ outside the drop panel.

d (slab) $= 140 - 30 = 110$ mm

Critical area $= (2.0 + 0.11)(1.8 + 0.11) = 4.03$ m^2

$b_o = 2(2.11 + 1.91) = 8.04$ m

$V_u = 28(6 \times 5.4 - 4.03) = 794$ kN

$$\phi V_c = \frac{0.85 \times 0.33}{1000}\sqrt{21} \times 8040 \times 110 = 1137 \text{ kN} > V_u$$

One-way shear is not critical.

4. Calculate the total static moments in the long and short directions:

$$M_{ol} = \frac{w_u}{8} l_2 l_{nl}^2 = \frac{28}{8}(5.4)(5.1)^2 = 491.6 \text{ kN·m}$$

$$M_{os} = \frac{w_u}{8} l_1 l_{n2}^2 = \frac{28}{8}(6)(4.5)^2 = 425.2 \text{ kN·m}$$

Since $l_2 < l_1$, the width of the column strip in the long direction $= 2(0.25 \times 5.4) = 2.7$ m. The width of the column strip in the short direction $= 2.7$ m. Assuming that the steel bars are 12 mm in diameter and those in the short direction are placed on top of the bars in the long direction, then the effective depth in the short direction will be about 10 mm less than the effective depth in the long direction. The d values and the design proce-

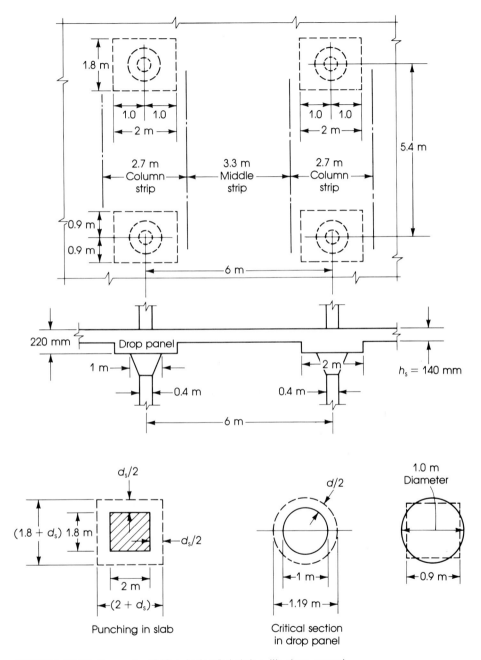

FIGURE 17.35. Example 17.10. Interior flat slab with drop panel.

TABLE 17.11. Design of an interior flat slab

	Long direction $M_o = 491.6$ kN·m $M_n = -0.65M_o = -319.5$ kN·m $M_p = +0.35M_o = +172.1$ kN·m				Short direction $M_o = 425.2$ kN·m $M_n = -0.65M_o = -276.4$ kN·m $M_p = +0.35M_o = +148.8$ kN·m			
	Column strip		Middle strip		Column strip		Middle strip	
	$0.75M_n$	$0.60M_p$	$0.25M_n$	$0.40M_p$	$0.75M_n$	$0.60M_p$	$0.25M_n$	$0.40M_p$
Moment distribution								
M_u (kN·m)	−239.6	+103.3	−79.9	+68.8	−207.3	+89.3	−69.1	+59.5
d (mm)	190	110	110	110	180	100	100	100
Strip width b (m)	2.7	2.7	2.7	2.7	2.7	2.7	3.3	3.3
$R_u = \dfrac{M_u}{bd^2}$ (MPa)	2.46	3.16	2.44	2.10	2.37	3.30	2.10	1.80
Steel ratio ρ	0.0071	0.0093	0.007	0.006	0.0069	0.010	0.006	0.005
$A_s = \rho bd$ (mm²)	3642	2762	2079	1782	3353	2700	1980	1650
Min. $A_s = 0.0018bh$ (mm²)	1070	680	680	680	1070	680	832	832
Bars selected	18 × 16 mm	14 × 16 mm	19 × 12 mm	16 × 12 mm	17 × 16 mm	14 × 16 mm	18 × 12 mm	15 × 12 mm
Straight bars	4	7	3	8	3	7	4	8
Bent bars	14	7	16	8	14	7	14	7
Spacing (mm)	150	193	142	169	159	193	183	220

dure are all arranged in Table 17.11. Minimum lengths of the selected reinforcement bars should meet the ACI Code length requirements shown in Figure 17.17.

5. Column stiffness:

$$\text{Ratio } \frac{\text{D.L.}}{\text{L.L.}} = \frac{7.86}{10} = 0.786 \quad \text{and} \quad \frac{l_2}{l_1} = 1.11$$

Determine α_{min} from Table 17.6, taking into account that the relative beam stiffness equals zero as no beams are used. By interpolation, $\alpha_{min} = 1.15$. An approximate method is used here to determine the stiffness of the column with its capital.

$$I_s \text{ (moment of inertia of slab, short direction)} = 6000 \frac{(140)^3}{12} = 1372 \times 10^6 \text{ mm}^4$$

$$K_s = \frac{4E_c I_s}{l_2} = \frac{4E_c \times 1372 \times 10^6}{5400} = 1016 \times 10^3 \, E_c$$

$$I_c \text{ (for circular column, diameter 400 mm)} = \frac{\pi D^4}{64} = \frac{\pi}{64}(400)^4 = 1257 \times 10^6 \text{ mm}^4$$

$$K_c = \frac{4E_c I_c}{l_c} = \frac{4E_c \times 1257 \times 10^6}{3000 \text{ mm}} = 1676 \times 10^3 \, E_c$$

$$\text{Ratio of column stiffness/slab stiffness} = \frac{K_c}{K_s} = \frac{1676 \times 10^3}{1016 \times 10^3} = 1.65$$

which is greater than α_{min} of 1.15. If I_s in the long direction is used, the calculated ratio of column to slab stiffness will be greater than 1.65. Therefore the column is adequate.

6. Determine the unbalanced moment in the column and check the shear stresses in the slab as explained in examples 17.7 and 17.8.

SUMMARY

SECTIONS 17.1 to 17.3

(a) A two-way slab is one that has a ratio of length to width less than 2. Two-way slabs may be classified as flat slabs, flat plates, waffle slabs, or slabs on beams.

(b) The ACI Code specifies two methods for the design of two-way slabs: the direct design method and the equivalent frame method. In the direct design method, the slab panel is divided (in each direction) into three strips, one in the middle (referred to as the "middle strip") and one on each side (referred to as "column strips").

SECTION 17.4

To control deflection, the minimum slab thickness h is limited to the values computed by equations 17.1 through 17.3 and as explained in examples 17.1 and 17.2.

SECTION 17.5

For two-way slabs without beams, the shear capacity of the concrete section in one-way shear is:

$$V_c = 2\sqrt{f_c'}bd$$

The shear capacity of the concrete section in two-way shear is:

$$V_c = \left(2 + \frac{4}{\beta_c}\right) \sqrt{f'_c} b_o d \le 4\sqrt{f'_c} b_o d$$

When shear reinforcement is provided, $V_n \le 6\sqrt{f'_c} b_o d$.

SECTION 17.6

In the direct design method, approximate coefficients are used to compute the moments in the column and middle strips of two-way slabs. The total factored moment is

$$M_o = (w_u l_2) \frac{l_n^2}{8}. \tag{17.11}$$

The distribution of M_o into negative and positive span moments is given in Figure 17.13. A summary of the direct design method is given in section 17.6.7.

SECTIONS 17.7 and 17.8

(a) Unbalanced loads on adjacent panels cause a moment in columns that can be computed by equation 17.21.

(b) Approximately 60 percent of the moment transferred to both ends of a column at a joint is transferred by flexure M_f, and 40 percent by eccentric shear M_v. The fraction of the unbalanced moment transferred by flexure $M_f = \gamma_f M_u$, where γ_f is computed from equation 17.24. The shear stresses produced by M_v must be combined with shear stresses produced by the shearing force V_u.

SECTION 17.9

In the equivalent frame method, the building is divided into equivalent frames in two directions and then analyzed for all conditions of loadings. Example 17.9 explains this procedure.

REFERENCES

1. W. G. Corley and J. O. Jirsa: "Equivalent Frame Analysis for Slab Design," *ACI Journal,* 67, November 1970.
2. W. L. Gamble: "Moments in Beam Supported Slabs," *ACI Journal,* 69, March 1972.
3. M. A. Sozen and C. P. Siess: "Investigation of Multi-panel Reinforced Concrete Floor Slabs," *ACI Journal,* 60, August 1963.
4. W. L. Gamble, M. A. Sozen and C. P. Siess: "Tests of a Two-way Reinforced Concrete Slab," *Journal of Structural Division, ASCE,* 95, June 1969.
5. R. Park and W. Gamble: *Reinforced Concrete Slabs,* Wiley, New York, 1980.
6. S. P. Timoshenko and S. W. Krieger: *Theory of Plates and Shells,* 2nd ed., McGraw-Hill, New York, 1959.
7. R. H. Wood: *Plastic and Elastic Design of Slabs and Plates,* Thames and Hudson, London, 1961.
8. O. C. Zienkiewicz: *The Finite Element Method in Engineering Science,* McGraw-Hill, New York, 1971.
9. K. W. Johansen: "Yield-Line Formulae for Slabs," Cement and Concrete Association, London, 1972.
10. A. Hillerborg: *Strip Method of Design,* Cement and Concrete Association, London, 1975.
11. R. K. Dhir and J. G. Munday: *Advances in Concrete Slab Technology,* Pergamon, New York, 1980.

12. W. C. Schnorbrich: "Finite Element Determination of Non-linear Behavior of Reinforced Concrete Plates and Shells," in *Proceedings of Symposium on Structural Analysis,* TRRL, 164VC, Crowthorne, 1975.

13. M. Fintel, ed.: *Handbook of Concrete Engineering,* Van Nostrand Reinhold, New York, 1974.

14. American Concrete Institute: *Commentary of Building Code Requirements for Reinforced Concrete,* ACI 318-83, Detroit, 1983.

15. N. W. Hanson and J. M. Hanson: "Shear and Moment Transfer Between Concrete Slabs and Columns," *PCA Journal,* 10, January 1968.

16. J. Moe: "Shear Strength of Reinforced Concrete Slabs and Footings under Concentrated Loads," *PCA Bulletin,* D47, April 1961.

17. S. H. Simmonds and J. Misic: "Design Factors for the Equivalent Frame Method," *ACI Journal,* 68, November 1971.

18. American Concrete Institute: *Building Code Requirements for Reinforced Concrete,* ACI 318-83, Detroit, 1983.

PROBLEMS

17.1. Design by the direct design method an interior flat-plate panel of a floor system that consists of five panels in each direction. All panels are square 20 by 20 ft (6 by 6 m) and supported by 16 by 16 in. (0.4 by 0.4 m) columns 10 ft (3 m) long. The slab is to carry a uniform dead load = 30 psf (1.5 kN/m^2) of floor finish in addition to its own weight and a uniform live load = 120 psf (6 kN/m^2). Use f'_c = 4 ksi (28 MPa) and f_y = 60 ksi (420 MPa). Show the reinforcement details in both directions using (1) bent bar system and (2) straight bar system.

17.2. Redesign the flat plate of problem 17.1 for the same data except with 20 by 16 ft (6 by 4.8 m) rectangular panels.

17.3. Design by the direct design method an exterior flat-plate panel which has the same dimensions as problem 17.1. No beams are used along the edges. Check the moment and shear transfer at the exterior support.

17.4. Design by the direct design method a typical interior flat slab with drop panels to carry a uniform service dead load = 120 psf (6 kN/m^2) (including self-weight) and a live load = 80 psf (4 kN/m^2). The floor system consists of four panels in each direction, with a panel size 28 by 24 ft (8.4 by 7.2 m). The floor system is supported by 18 in. (450 mm) diameter circular columns 12 ft long (3 m). Use f'_c = 3 ksi (21 MPa), f_y = 40 ksi (28 MPa), and a column capital diameter = 36 in. (900 mm).

17.5. Redesign the slab in problem 17.1 for the same data when it is supported by beams on all four sides. Each beam has a width b_w = 14 in. (350 mm) and a projection below the bottom of the slab = 18 in. (450 mm).

17.6. Redesign problem 17.5 as an exterior panel.

17.7. Redesign the flat-plate floor system of problem 17.1 using the equivalent frame method.

17.8. Redesign the slab of problem 17.5 using the equivalent frame method.

18 Introduction to Prestressed Concrete

18.1 Principles of Prestressing

To prestress a structural member is to induce internal, permanent stresses that counteract the tensile stresses in the concrete resulting from external loads; this extends the range of stress that the member can safely withstand. Prestressing force may be applied either before or at the same time as the application of the external loads. Stresses in the structural member must remain, everywhere and for all states of loading, within the limits of stress that the material can sustain indefinitely. The induced stresses, primarily compressive, are usually created by means of high-tensile steel tendons, which are tensioned and anchored to the concrete member. Stresses are transferred to the concrete either by the bond along the surface of the tendon or by anchorages at the ends of the tendon.

To explain the above discussion, consider a beam made of plain concrete which has to resist the external gravity loads shown in Figure 18.1(a). The beam section will be chosen with the tensile flexural stress as the critical criterion for design, therefore an uneconomical section will result. This is because concrete is considerably stronger in compression than in tension. The maximum flexural tensile strength of concrete, the modulus of rupture f_r, is equal to $7.5 \sqrt{f'_c}$ (Figure 18.1(a)).

In normal reinforced concrete design, the tensile strength of concrete is ignored and steel bars are placed in the tension zone of the beam to resist the tensile stresses while the concrete resists the compressive stresses (Figure 18.1(b)).

In prestressed concrete design, an initial compressive stress is introduced to the beam to offset or counteract the tensile stresses produced by the external loads (Figure 18.1(c)). If the induced compressive stress is equal to the tensile stress at the bottom fibers, then both stresses cancel themselves, while the compressive stress in the top fibers is doubled; in this case the whole section is in compression. If the induced compressive stress is less than the tensile stress at the bottom fibers, these fibers will be in tension while the top fibers are in compression.

The following numerical example illustrates some of the features of prestressed concrete.

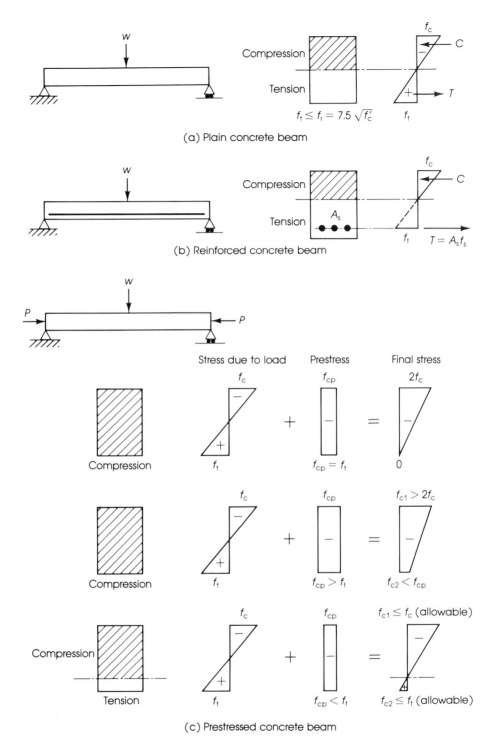

FIGURE 18.1. Effect of prestressing: (a) plain concrete, (b) reinforced concrete, and (c) prestressed concrete.

EXAMPLE 18.1

For the simply supported beam shown in Figure 18.2, determine the maximum stresses at midspan section due to its own weight and the following cases of loading and prestressing:

1. A uniform live load of 900 lb/ft.
2. A uniform live load of 900 lb/ft and an axial centroidal longitudinal compressive force $P = 259.2$ k.
3. A uniform live load of 2100 lb/ft and an eccentric longitudinal compressive force $P = 259.2$ k acting at an eccentricity $e = 4$ in.
4. A uniform live load of 2733 lb/ft and an eccentric longitudinal compressive force $P = 259.2$ k acting at the maximum practical eccentricity for this section ($e = 6$ in.).
5. The maximum live load when $P = 259.2$ k acting at $e = 6$ in.

Use $b = 12$ in., $h = 24$ in., $f'_c = 4500$ psi, and an allowable $f_c = 2050$ psi.

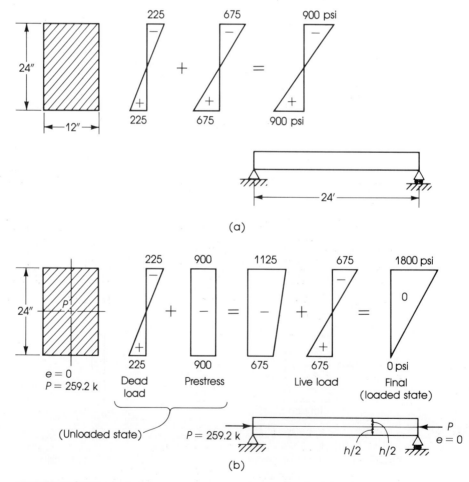

FIGURE 18.2. Example 18.1.

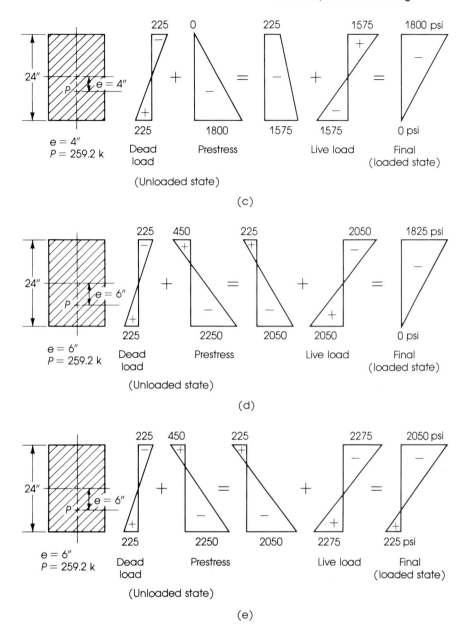

FIGURE 18.2 (continued).

Solution

1. Stresses due to dead and live loads only:

Self-weight of beam = $(1 \times 2) \times 150 = 300$ lb/ft

$$\text{Dead load moment } M_{\text{D.L.}} = \frac{wL^2}{8} = \frac{0.300(24)^2}{8} = 21.6 \text{ k·ft}$$

Stresses at the extreme fibers:

$$\sigma = \frac{Mc}{I} = \frac{M(h/2)}{bh^3/12} = \frac{6M}{bh^2}$$

$$\sigma_D = \frac{6 \times 21.6 \times 12{,}000}{12(24)^2} = \pm 225 \text{ psi}$$

Stresses due to live load $L_1 = 900$ lb/ft.

$$M_{\text{L.L.}} = \frac{0.9(24)^2}{8} = 64.8 \text{ k·ft}$$

$$\sigma_{L1} = \frac{6M}{bh^2} = \frac{6 \times 64.8 \times 12{,}000}{12(24)^2} = \pm 675 \text{ psi}$$

Adding stresses due to the dead and live loads (Figure 18.2, case 1):

Top stress $= -225 - 675 = -900$ psi (compression)

Bottom stress $= +225 + 675 = +900$ psi (tension)

The tensile stress is higher than the modulus of rupture of concrete, $f_r = 7.5\sqrt{f_c'} = 503$ psi; hence the beam will collapse.

2. Stresses due to uniform prestress:

If a compressive force $P = 259.2$ k is applied at the centroid of the section, then a uniform stress is induced on any section along the beam.

$$\sigma_p = \frac{P}{\text{area}} = \frac{259.2 \times 1000}{12 \times 24} = -900 \text{ psi (compression)}$$

Final stresses due to live and dead loads plus prestress load at the top and bottom fibers are 1800 psi and zero, respectively (Figure 18.2, case 2). In this case, the prestressing force has doubled the compressive stress at the top fibers and reduced the tensile stress at the bottom fibers to zero. The maximum compressive stress of 1800 psi is less than the allowable stress of 2050 psi.

3. Stresses due to an eccentric prestress ($e = 4$ in.):

If the prestressing force $P = 259.2$ k is placed at an eccentricity $e = 4$ in. below the centroid of the section, the stresses at the top and bottom fibers are calculated as follows.

Moment due to eccentric prestress $= Pe$

$$\sigma_p = -\frac{P}{A} \pm \frac{(Pe)c}{I} = -\frac{P}{A} \pm \frac{6(Pe)}{bh^2}$$

$$= -\frac{259.2 \times 1000}{12 \times 24} \pm \frac{6(259.2 \times 1000 \times 4)}{12(24)^2}$$

$$= -900 \pm 900$$

$$= -1800 \text{ psi at bottom fibers, zero at top fibers}$$

Consider now an increase in the live load $L_2 = 2100$ lb/ft:

$$M_{\text{L.L.}} = \frac{2.1 \times (24)^2}{8} = 151.2 \text{ k·ft}$$

$$\sigma_{L2} = \frac{6(151.2 \times 12{,}000)}{12(24)^2} = \pm 1575 \text{ psi}$$

Final stresses due to the dead, live, and prestressing loads at the top and bottom fibers are 1800 psi and zero, respectively (Figure 18.2, case 3). Note that the final stresses are exactly the same as those of the previous case when the live load was 900 lb/ft; by applying the same prestressing force but at an eccentricity of 4 in., the same beam can now support a greater live load (by 1200 lb/ft).

4. Stresses due to eccentric prestress with maximum eccentricity:

   Assuming that the maximum practical eccentricity for this section is at $e = 6$ in., leaving a 2 in. concrete cover, then the bending moment induced

   $$Pe = 259.2 \times 6 = 1555.2 \text{ k·in.} = 129.6 \text{ k·ft}$$

   Stresses due to the prestressing force:

   $$\sigma_p = -\frac{259.2 \times 1000}{12 \times 24} \pm \frac{6 \times (129.6 \times 12,000)}{12(24)^2}$$
   $$= -900 \pm 1350 \text{ psi}$$
   $$= -2250 \text{ psi and } +450 \text{ psi}$$

   Increase the live load now to $L_3 = 2733$ lb/ft. The stresses due to the live load, L_3

   $$M_{\text{L.L.}} = \frac{2.733 \times (24)^2}{8} = 196.8 \text{ k·ft}$$

   $$\sigma_{L_3} = \frac{6(196.8 \times 12,000)}{12(24)^2} = \pm 2050 \text{ psi}$$

   The final stresses at the top and bottom fibers due to the dead load, live load L_3, and the prestressing force are 1825 psi and zero, respectively (Figure 18.2, case 4). Note that the final stresses are about the same as those in the previous cases, yet the live load has been increased to 2733 lb/ft. A tensile stress of 225 psi is developed when the prestressing force is applied on the beam. This stress is less than the modulus of rupture of concrete, $f_r = 503$ psi; hence cracks will not develop in the beam.

5. Maximum live load when the eccentric force P acts at $e = 6$ in.:

   In the previous case, the final compressive stress is equal to 1825 psi, which is less than the allowable stress of 2050 psi. Therefore, the live load may be increased to $L_4 = 3033$ lb/ft.

   $$M_{\text{L.L.}} = \frac{3.033 \times (24)^2}{8} = 218.4 \text{ k·ft}$$

   $$\sigma_{L_4} = \frac{6(218.4 \times 12,000)}{12(24)^2} = \pm 2275 \text{ psi}$$

   Final stresses due to the dead load, live load L_4, and the prestressing force are -2050 psi and $+225$ psi. The compressive stress is equal to the allowable stress of 2050 psi, and the tensile stress is less than the modulus of rupture of concrete of 503 psi. In this case, the uniform live load of 3033 lb/ft has been calculated as follows:

   Add the maximum allowable compressive stress of 2050 psi to the initial tensile stress at the top fibers of 225 psi to get 2275 psi. The moment that will produce a stress at the top fibers of 2275 psi is equal to

$$M = \sigma \left(\frac{bh^2}{6} \right)$$

$$= \frac{2.275}{6} (12)(24)^2 = 2620.8 \text{ k·in.} = 218.4 \text{ k·ft}$$

$$M = \frac{W_L L^2}{8} \quad \text{and} \quad W_L = \frac{8 \times 218.4}{(24)^2} = 3.033 \text{ k/ft}$$

Notes

1. The entire concrete section is active in resisting the external loads.
2. The final tensile stress in the section is less than the modulus of rupture of concrete, which indicates that a crackless concrete section can be achieved under full load.
3. The allowable load on the beam has been increased appreciably due to the application of the prestressing force.
4. An increase in the eccentricity of the prestressing force will increase the allowable applied load, provided that the allowable stresses on the section are not exceeded. ■

EXAMPLE 18.2

For the beam and data given in the previous example, determine:

1. The area of the prestressing steel if the effective stress in the steel f_{pe} is considered equal to 160 ksi.
2. The allowable load and the area of the reinforcing steel bars required for a reinforced concrete section using the working stress design method and $f_y = 60$ ksi ($f_s = 24$ ksi).
3. The same as (2), using the strength design method.
4. Compare the results obtained from this example and example 18.1 (Figure 18.2).

Solution

1. Determine the prestressing steel A_{ps}: The prestressing force applied on the beam of the previous example, $F = 259.2$ k. The area of the prestressing steel

$$A_{ps} = \frac{F}{f_{pe}} = \frac{259.2}{160} = 1.62 \text{ in.}^2$$

2. Reinforced concrete section, working stress design method:

   • For an allowable concrete stress = 2050 psi and an allowable steel stress = 24 ksi, the modular ratio $n = E_s/E_c = 7.5$ (see Chapter 5),

$$K = \frac{2050}{2050 + \dfrac{24{,}000}{7.5}} = 0.39$$

   and

$$j = 1 - \frac{K}{3} = 0.87$$

$$\text{Balanced steel ratio} = \frac{K}{2} \times \frac{f_{ca}}{f_{sa}} = \frac{0.39}{2} \times \frac{2050}{24{,}000} = 0.0167$$

Assume two rows of bars; then $d = 24 - 3.5 = 20.5$ in.

$$A_s = \rho b d = 0.0167 \times 12 \times 20.5 = 4.1 \text{ in.}^2$$

Balanced moment $= A_s f_s j d$

$$= 4.1 \times 24 \times 0.87 \times 20.5 = 1755 \text{ k·in.} = 146.2 \text{ k·ft}$$

$$M = \frac{wL^2}{8} \quad \text{and} \quad w = \frac{8 \times 146.2}{(24)^2} = 2.03 \text{ k/ft}$$

Allowable live load $= 2.03 - 0.3 = 1.73$ k/ft

- $Kd = 0.39 \times 20.5 = 8$ in.

$$I = \frac{b(Kd)^3}{3} + nA_s(d - Kd)$$

$$= \frac{12}{3} \times (8)^3 + 7.5 \times 4.1(20.5 - 8)^2 = 6852 \text{ in.}^4$$

The compressive stress in the top fibers

$$\sigma = \frac{M(Kd)}{I}$$

$$= \frac{(146.2 \times 12,000) \times 8}{6852} = 2050 \text{ psi}$$

which is the allowable concrete stress.

3. Reinforced concrete section, strength design method:
   For $f'_c = 4500$ psi, $f_y = 60$ ksi, and $A_s = 4.1$ in.^{2} (as in part 2),

$$a = \frac{A_s f_y}{0.85 f'_c b} = \frac{4.1 \times 60}{0.85 \times 4.5 \times 12} = 5.36 \text{ in.}$$

$$M_u = \phi A_s f_y \left(d - \frac{a}{2} \right) = 0.9 \times 4.1 \times 60 \left(20.5 - \frac{5.36}{2} \right)$$

$$= 5945.4 \text{ k·in.} = 328.8 \text{ k·ft}$$

If an average load factor of 1.6 is used for dead and live loads, then the service moment will be $= 328.8/1.6 = 205.5$ k·ft.

$$M = \frac{wL^2}{8} \quad \text{and} \quad w = \frac{205.5 \times 8}{(24)^2} = 2.85 \text{ k/ft}$$

Allowable live load $= 2.85 - 0.3 = 2.55$ k/ft

If Kd is assumed $= 8$ in. (as in part 2),

$$\sigma_L = \frac{M(Kd)}{I} = \frac{(205.5 \times 12,000) \times 8}{6852} = 2880 \text{ psi}$$

The actual compressive stress in the concrete fibers under service load is greater than that allowed in the working stress design.

4. From the previous calculations, the results can be summarized as follows:

Method	Allowable live load		Area of steel, in.^{2}	
	Magnitude	Ratio	Magnitude	Ratio
Working stress	1.73 k/ft	1.0	4.1	1
Strength design	2.55 k/ft	1.47	4.1	1
Prestressed	3.03	1.75	1.62	0.4

The area of the prestressing steel is about 40 percent of that required for the reinforced concrete section designed by the strength design method and produces an increase in the allowable live load of about 19 percent (3.03/2.55 = 1.19). ∎

18.2   Prestressed versus Reinforced Concrete

The previous examples have shown that the function of the steel in the prestressed concrete beam is to provide an initial compressive stress so that the applied loads are carried safely. Virtually the only limit to the economy of the section is the value of the maximum allowable stress in the concrete. For that reason, the concrete used in prestressed concrete structures is normally of higher strength and better quality than that used for reinforced concrete.

Prestressed concrete has several advantages over reinforced concrete:

1. Prestressed concrete uses high-strength concrete and steel, and consequently higher working stresses can be allowed on the critical sections; thus the carrying capacity of the concrete member can be increased, resulting in lighter and possibly cheaper structures.
2. The entire concrete section is used to its maximum allowable stress limits, while in reinforced concrete the portion of concrete in the tension zone is neglected.
3. The absence of cracks in the concrete at working loads improves the durability of the prestressed concrete member.
4. Within certain limits, the dead load does not influence the determination of the concrete section.
5. Prestressed concrete is more resistant to shearing forces due to the slope of the tendons and the compressive stresses induced in the beam, which reduces the diagonal tension.
6. The effective deflection under sustained loads is better controlled in prestressed concrete than in reinforced concrete.
7. Prestressing makes it possible to assemble precast units in mass production under highly controlled conditions.

In general, prestressed concrete requires a high level of technology and expensive anchorages for the tendons. Therefore, several factors must be considered before deciding whether to use prestressed or reinforced concrete. Such factors include the type of structure and its serviceability, availability of materials, the structural system, maximum spans, and, most important, the final cost of the structure.

18.3   Methods of Prestressing

In practice, a concrete member may be prestressed in one of the following methods:

1. Posttensioning.   In posttensioning, the steel tendons are tensioned after the concrete has been cast and hardened. Posttensioning is performed by two main operations: tensioning the steel wires or strands by hydraulic jacks that stretch the strands while bearing against the ends of the member, then replacing the jacks by permanent anchorages that bear on the member and maintain the steel strands in tension. A tendon is generally made of wires, strands, or bars. Wires and strands can be tensioned in groups, while bars are tensioned one at a time.

In the posttensioning process, the steel tendons are placed in the framework before the concrete is cast and the tendons are prevented from bonding to the concrete by waterproof paper wrapping or a metal duct (sheath). Alternatively, ducts are formed in the concrete using inflatable rubber tubes and then threading the tendons into the duct when the concrete has hardened. When tensioning is complete, the void between the tendon and the duct is filled with a mortar grout. Grouting ensures bonding of the tendon with the concrete and reduces the risk of corrosion for the steel tendons.

Tendons bonded to the concrete as explained above are called bonded tendons. Unbonded tendons, left without grout or coated with grease, have no bond throughout the length of the tendon.

2. Pretensioning.   In pretensioning, the steel tendons are tensioned before the concrete is cast. The tendons are temporarily anchored against some abutments and then cut or released after the concrete has been placed and hardened. The prestressing force is transferred to the concrete by the bond along the length of the tendon. Pretensioning is generally done in precasting plants in permanent beds, which are used to produce pretensioned precast concrete elements for the building industry. Two standard methods are used, the unit mold method and the long-line method.

In the *unit mold method,* the tendons are placed in the mold in the specified position, and then wedged in special anchor plates bearing at the ends of rigid molds. The tendons are tensioned by forcing the anchor plates away from the ends of the mold.

The *long-line method* is the most common method used in the precast industry. In this process, a series of identical molds is arranged in line on one bed, 200 to 600 ft (60–180 m) long, and the tendons passing through all molds are anchored against rigid abutments. One stressing operation serves all the molds. Accelerated curing is normally used in the long-line method to allow for a 24-hour production cycle.

3. External Prestressing.   In this method, the prestressing force is applied by flat jacks placed between the concrete member ends and permanent rigid abutments. The member does not contain prestressing tendons as in the previous two methods (also called internal prestressing). External prestressing is not easy in practice because shrinkage and creep in concrete tend to reduce the induced compressive stresses unless the prestressing force can be adjusted.

4. Chemical Prestressing. This is another internal prestressing method. Tendons are placed untensioned in the form and concrete containing expansive cement is poured. The concrete expands during hardening and stretches the bonded tendons with it, inducing compression stresses in the concrete member.

The profile of the tendons may be straight, curved (bent), or circular, depending on the design of the structural member. Straight tendons are generally used in solid and hollow-cored slabs, while bent tendons are used in beams and most structural members. Circular tendons are used in circular structures such as tanks, silos, and pipes. The prestressing force may be applied in one or more stages, to avoid overstressing concrete, or in cases when the loads are applied in stages. In this case, part of the tendons are fully prestressed at each stage.

A considerable number of prestressing systems have been devised, among which are: Freyssinet, Magnel Blaton, B.B.R.V., Dywidag, CCL, Morandi, VSL, Western Concrete, Prescon, and INRYCO. The choice of the prestressing system for a particular job can sometimes be a problem. The engineer should consider three main factors that govern the choice of the system:

1. The magnitude of the prestressing force required.
2. The geometry of the section and the space available for the tendons.
3. Cost of the prestressing system (materials and labor). The cost of the tendon and anchors usually increases with the diameter and size of the tendon used, but this increase may be counterbalanced by the savings obtained from fixing and stressing a smaller number of tendons.

18.4   Partial Prestressing

A partially prestressed concrete member can be defined as one in which (1) there have been introduced internal stresses to counteract part of the stresses resulting from external loadings, (2) tensile stresses are developed in the concrete under working loads, and (3) nonprestressed reinforcement may be added to increase the moment capacity of the member. That definition implies that there are two cases that could be considered as partially prestressed concrete:

1. A combination of prestressed and nonprestressed steel is used in the same section. The prestressed cables induce internal stresses designed to take only part of the ultimate capacity of the concrete section. The rest of the capacity is taken by nonprestressed steel placed along the same direction as the prestressed cables. The steel used as nonprestressed steel could be any common grade of carbon steel or high-tensile-strength steel of the same kind as the prestressing cables with ultimate strength of 250 ksi (1725 N/mm^2). The choice depends on two main factors: allowable deflection and allowable crack width. As for deflection, the ACI Code specifies a maximum ratio of span to depth of reinforced concrete members.[1] With the smaller depth expected in partially

prestressed concrete,[2] and because a smaller steel percentage is used, excessive deflection under working loads must not be allowed. Cracks develop on the tension side of the concrete section or at the steel level because tensile stresses are allowed to occur under working loads. The maximum crack width allowed by the ACI Code[3] is 0.016 inch (0.41 mm) for interior members and 0.013 in. (0.33 mm) for exterior members.

2. Internal stresses act on the member from prestressed steel only, but tensioned to a lower limit. In this case cracking develops earlier than in a fully prestressed member under similar loadings.

Partially prestressed concrete can be considered an intermediate form between reinforced and fully prestressed concrete. In reinforced concrete members, cracks develop under working loads; therefore reinforcement is placed in the tension zone. In prestressed concrete members, cracks do not usually develop under working loads. The compressive stresses due to prestressing may equal or exceed the tensile stresses due to external loadings. Therefore, a partially prestressed concrete member may be considered as a reinforced concrete member in which internal stresses are introduced to counteract part of the stress from external loadings so that tensile stresses in the concrete do not exceed a limited value under working load. It reduces to reinforced concrete when no internal stresses act on the member. Full prestressing is an upper extreme of partial prestressing in which nonprestressed reinforcing steel reduces to zero.

Between a reinforced cracked member and a fully prestressed uncracked member, there exists a wide range of design in partial prestressing (Figure 18.3). A proper choice of the degree of prestressing will produce a safe and economical structure.

Figure 18.3 shows the load deflection curves of concrete beams containing different amounts and types of reinforcement. Curve a represents a reinforced concrete beam, which normally cracks at a small load W_{cr}. The cracking moment M_{cr} can be determined as follows:

$$M_{cr} = \frac{f_r I}{c}$$

where

f_r = the modulus of rupture of concrete = $7.5\sqrt{f'_c}$

I = moment of inertia of the gross concrete section

c = distance from the neutral axis to the tensile extreme fibers

The cracking load can be determined from the cracking moment when the span and the type of loading are specified. For a simply supported beam subjected to a concentrated load at midspan, $W_{cr} = (4 M_{cr})/L$.

Curves e and f represent under-reinforced and over-reinforced fully prestressed concrete beams, respectively. The over-reinforced concrete beam fails by crushing of the concrete before the steel reaches its yield strength or proof stress. The beam has small deflection and undergoes brittle failure. The under-reinforced beam fails by the steel reaching its yield or ultimate strength. It shows appreciable deflection and cracking due to elongation of the steel before the gradual crushing of the concrete and the collapse of the beam.

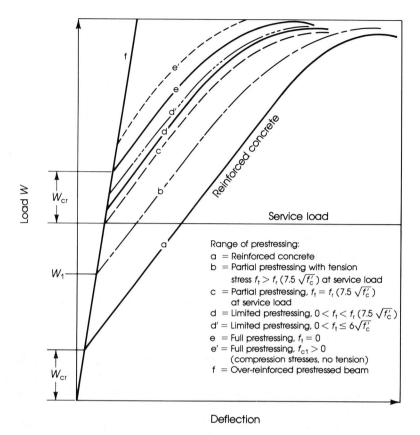

FIGURE 18.3. Load-deflection curves of concrete beams with different prestressing. The cracking load is W_{cr}.

Between curves a and e is a wide range of concrete beams with varying amounts of reinforcement and subjected to varying amounts of prestress. The beam with little prestressing is closer to curve a, while the beam with a large prestress is closer to curve e. Depending upon the allowable concrete stress, deflection, and maximum crack width, a suitable combination of prestressed and nonprestressed reinforcement may be chosen for the required design.

Curve b represents a beam which will crack under full working load. If only part of the live load L_1 occurs frequently on the structure, then W_1 represents the total dead load and that part of the live load L_1.

Curve c represents a beam that starts cracking at working load. The maximum tensile stress in the concrete $= f_r = 7.5\sqrt{f'_c}$.

Curve d represents a beam with limited prestress. The critical section of the beam will not crack under full working load, but it will have a maximum tensile stress $0 < f_t < 7.5\sqrt{f'_c}$. The maximum tensile stress in concrete allowed by the current ACI Code is $6\sqrt{f'_c}$.

Curves e and e′ represent fully prestressed concrete beams with no tensile stress under working loads (see Figure 18.4).

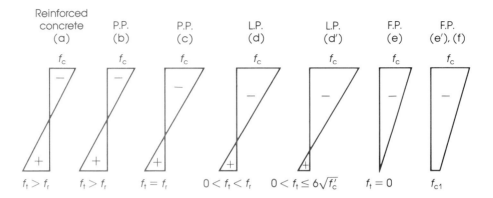

$f_r = 7.5 \sqrt{f'_c}$
P.P. = Partial prestressing
L.P. = Limited prestressing
F.P. = Full prestressing

FIGURE 18.4. Distribution of stresses in beams with varying amounts of prestressed and nonprestressed reinforcement.

The most important advantage of partial prestressing is the possibility of controlling camber. By reducing the prestressing force, the camber will be reduced and a saving in the amount of the prestressing steel, the amount of work in tensioning, and the number of end anchorages is realized.

Depending on the magnitude of the prestressing force, earlier cracking may occur in partially prestressed rather than in fully prestressed concrete members under service loads. Once cracks develop, the effective moment of inertia of the critical section is reduced and a greater deflection is expected. However, partial prestressing has been used with satisfactory results, and its practical application is increasing.

18.5 Degree of Prestressing

During the development of prestressed concrete, three schools of thought were established regarding the allowable degree of prestressing. The differences are concerned mainly with the tensile stresses that could be allowed in a prestressed structural member.

1. The French school of thought called for full prestressing. This means that tensile stresses in the concrete are not allowed and designs are based on uncracked sections. Freyssinet firmly advocated this approach.[4]
2. The German school of thought called for limited prestressing in addition to full prestressing. Tensile stresses are allowed to develop up to 80 percent of the concrete tensile stress, and designs are based on uncracked sections. Ordinary mild or medium-grade steel is allowed to meet ultimate load conditions. Finsterwalder advocated this approach, and the German Code DIN 4227 adopted it.

The 1983 ACI Code allows tensile stresses of $6\sqrt{f_c'}$, which represents the highest tensile stress that can be used for a crack-free structure. The assumed ultimate tensile strength of concrete is $7.5\sqrt{f_c'}$; thus the ACI Code allows tensile stresses equal to 80 percent of the modulus of rupture of concrete. In such a situation, fine cracks may occur under the maximum allowable live load.

3. The third school of thought is partial prestressing. Partial prestressing was applied in England in 1951, and P. W. Abeles advocated this approach. Partial prestressing was also called moderate or limited prestressing. In Europe, prestressing tendons are called active reinforcement, while untensioned steel is called passive reinforcement.

18.6   History of Prestressing

The history of reinforced concrete was presented briefly in Chapter 1. The development of full and partial prestressing went through the following stages.

Reinforced Concrete (1832–1886).   Sir March Brunel used reinforcement in concrete in an experimental arch in 1832, and Joseph Monier obtained his patent for reinforced concrete in Paris in 1867.[5]

Unidentified Partial Prestressing (1886–1922).   At this stage, partially prestressed concrete was not known by this name, as prestressing itself was not known. The aim was to improve the reinforced concrete available by developing a way to increase the strength of beams and reduce the widths of cracks in weak concrete. Precompression was introduced into structural members, but this approach failed for not knowing the effects of shrinkage and creep. Tensioning tie rods in arches was another feature of this period. In 1886, P. H. Jackson of San Francisco described methods for stretching ties in the footing of arches by using turnbuckles, screws and nuts, and wedges.[6] C. G. W. Dohring of Berlin obtained a patent in 1888 for making slabs by embedding tension wires in concrete.[7] It seems that his aim was only to reduce cracking. In 1896, J. Mandl of Austria expressed clearly the idea of counteracting the stresses due to dead loads by means of prestress, in this case, partial prestressing.[8] M. Koener of Berlin, in 1907, derived a formula for the magnitude of stretching forces in accordance with Mandl's proposal.[9] The steel was tensioned to a stress of 8500 psi (59 N/mm²). Koener realized that the initial compressive stress on the concrete was lost due to shrinkage and abandoned further tests. In 1908, E. Freyssinet built an arch bridge of 167 ft (51 m) span at Moulines, France, and joined the supports by a tie rod prestressed by 9-mm wires tensioned to 10 ksi (69 N/mm²).[10] In the same year, R. C. Steiner of California suggested tightening the steel bars against the green concrete to destroy the bond, then tensioning them to higher stresses when the concrete hardened.[8] In 1910, Zissler of Germany and Siegwart of Switzerland wrapped tensioned steel wires around pipes. K. Wettstein of Bohemia, in 1916, used highly stressed piano wires for the first time.[11] The wires were tensioned to secure them in position as conventional reinforcement.

In this stage, the development of reinforced concrete including all experiments can be considered an introduction to partial prestressing. Tensile stresses were not eliminated, and cracks existed with no scientific way to reduce them. Creep and shrinkage phenomena were not yet well understood. The induced compressive stresses were small, and consequently, tests did not give satisfactory results.

Full Prestressing (1922–1939).   Full prestressing was achieved when the effects of shrinkage and creep became known. The aim was to design a beam free from tensile stresses and cracks. Partial prestressing was ignored in this period. The first person to proceed to full prestressing was W. H. Hewett of Minneapolis. In 1922, he used tensioned wires around circular tanks and was able to eliminate tensile stresses.[12] F. von Emperger of Austria, in 1923, made reinforced concrete pipes wrapped with wires tensioned from 22 ksi (125 N/mm²) to 113 ksi (780 N/mm²). The development of prestressing led to a complete absence of cracks in concrete structural members: under working loads, tensile stresses were eliminated even after all prestressing losses occurred.

R. H. Dill of Nebraska, in 1923, made posttensioned fully prestressed beams, free from tensile stresses. Dill's process "recognizes the difference in qualities of concrete and steel and combines the two materials in a scientific manner."[13] In 1925, Professor Eduardo Torroja built the Tempul Aquaduct using tensioned high-strength wire ropes in two lever beams. He was the first to accomplish effective prestressing in a bridge structure.[14] In 1928, E. Freyssinet applied for a patent for the pretensioning of high-tensile bars with a high elastic limit. Freyssinet was the first to make clear the different functions of steel and concrete. He expressed explicitly the losses of prestress due to shrinkage and creep, which he had studied since 1911.[15] In recognition of his efforts, Freyssinet was elected the first President of the International Federation of Prestressed Concrete.

Full and Partial Prestressing (1939–1952).   In 1939, the idea and the name of partial prestressing were introduced, based on allowing concrete to crack. Partially prestressed concrete was criticized strongly, because it decreases appreciably the advantages of crackless, fully prestressed structures. In addition to the above, it was considered that fully prestressed concrete is "an entirely new material, having properties very different from those of the conventional reinforced concrete."[6] E. V. Emperger of Austria, in 1939, suggested the introduction of tensioned and nontensioned steel in the same concrete member. Emperger's purpose was to increase the permissible steel stress and to reduce cracking, although he stated that cracking was essential and beneficial.[6] This suggestion was widely criticized; W. Zerna stated that Emperger's suggestion hardly found practical application.[17]

Between 1940 and 1952, P. W. Abeles of Britain proposed "the use of high strength steel as with full prestressing, but with a reduction of the effective prestressing force to such an extent that under working loads, resultant tensile stresses occur."[6] Abeles claimed that the limiting demand of Freyssinet for only compressive stresses to occur under working load had "caused a great delay in the practical application of partially prestressed concrete." The objection to partial prestressing in Europe was strong and consistent. R. H. Evans stated in 1950 that although "the introduction of partially prestressed concrete beams has caused much controversy, I can think of no technical reason for the general condemnation of the adoption of partially prestressed concrete beams.''[18] Freyssinet in 1950 did not encourage partial prestressing: "there is not an intermediate state; a structure either is or is not prestressed.''[19]

Full Prestressing (1952–1962). In this period, engineers were able to design fully prestressed concrete members. Many patents for the different anchorage systems appeared, and fully prestressed concrete developed into a successful, profitable industry. Widespread objections were leveled at partial prestressing.

Full and Partial Prestressing (1962–). For the first time, partially prestressed concrete was discussed at the Congress of the International Federation of Prestressed Concrete (FIP) held in Rome in 1962. Peter Verna introduced the concept of partial prestressing to reduce camber resulting from fully prestressing.[20] In Japan, it was established that partial prestressing resulted in a savings of 10 percent in steel and a total savings of 3 percent in a 25-m bridge.

S. Chaikes[21] drew the distinction between passive and active steel (conventional reinforcement and tendons). In October 1965, the first symposium on partially prestressed concrete was held in Brussels. Many aspects of design, behavior, applications, and comparisons with fully prestressed and reinforced concrete were discussed.

Full prestressing in the United States greatly increased during this period. There was only one precast pretensioning factory in 1950, but 229 in 1961. In 1975 the number of precasting plants in the United States was estimated to exceed 500. The Prestressed Concrete Institute (PCI) Journal of October 1976 gave detailed information on prestressed concrete, past, present, and future.

The progress in research, development, and applications of full and partial prestressing should continue in the future with the aim of achieving optimum economy, serviceability, and speed of construction.

18.7   Materials and Allowable Stresses

18.7.1   Concrete

The physical properties of concrete were discussed in Chapter 2. While reinforced concrete members are frequently made of concrete with a compressive strength of 3 to 5 ksi (21 to 35 MPa), prestressed concrete members are made of higher-strength material, usually from 4 to 6 ksi (28 to 42 MPa). High-strength concrete may be adopted for precast prestressed concrete members where components are prepared under optimum control of mixing concrete, placing, vibrating, and curing. In Europe, the usual 28-day cube strength specified is 450 kgf/cm^2 based on 150-mm or 200-mm cubes. This is equivalent to about 5 ksi (or 35 MPa) for a standard cylinder.

The allowable stresses in concrete according to the **ACI Code, section 18.4,** are as follows:

1. Stresses after prestress transfer and before prestress losses:

   • Maximum compressive stress = $0.6f_{ci}$.
   • Maximum tensile stress except as permitted below = $3\sqrt{f_{ci}}$.

- Maximum tensile stress at the ends of simply supported members = $6\sqrt{f_{ci}}$, where f_{ci} is the ultimate strength of concrete at transfer.

If tensile stresses are exceeded, reinforcement must be provided in the tensile zone to resist the total tensile force in concrete (based on uncracked gross section).

2. Stresses at service loads after all losses:

- Maximum compressive stress = $0.45f'_c$.
- Maximum tensile stress in the precompressed zone = $6\sqrt{f'_c}$.
- Maximum tensile stress where analysis based on transformed cracked sections shows that the deflections are within the ACI Code limitations = $12\sqrt{f'_c}$.

3. The above stresses may be exceeded if shown by tests or analysis that performance is satisfactory.

18.7.2   Prestressing Steel

The most common type of steel tendons used in prestressed concrete are strands (or cables) made with several wires, usually seven or nineteen. Wires and bars are also used. The strands and wires are manufactured according to ASTM Standard A421 for uncoated stress-relieved wires and A416 for uncoated seven-wire stress-relieved strands. Properties of prestressing steel are given in Table 18.1.

Prestressing steel used in prestressed concrete must be of high-strength quality, usually of ultimate strength of 250 ksi to 270 ksi (1730–1860 MPa). High-strength steel is necessary to permit high elongation and to maintain a permanent sufficient prestress in the concrete after the inelastic shortening of the concrete.

The allowable stresses in prestressing steel according to the **ACI Code, section 18.5**, are as follows:

1. Maximum stress due to tendon jacking force must not exceed the smaller of $0.85f_{pu}$ or $0.94f_{py}$. (The smaller value must not exceed that stress recommended by the manufacturer of tendons or anchorages.)
2. Maximum stress in pretensioned tendons immediately after transfer must not exceed the smaller of $0.74f_{pu}$ or $0.82f_{py}$.
3. Maximum stress in posttensioning tendons after tendon is anchored = $0.70f_{pu}$.

18.7.3   Reinforcing Steel

Nonprestressed reinforcing steel is commonly used in prestressed concrete structural members, mainly in the prestressed precast concrete construction. The reinforcing steel is used as shear reinforcement, as supplementary reinforcement for transportation and handling the precast elements, and in combination with the prestressing steel in partially prestressed concrete members. The types and allowable stresses of reinforcing bars were discussed in Chapters 2 and 5.

TABLE 18.1. Properties of prestressing steel, nominal diameters, areas, and weights

Type	Diameter (in.)	Area (in.²)	Weight (lb/ft)	Diameter (mm)	Area (mm²)	Mass (Kg/m)
Seven-wire strand	$\frac{1}{4}$ (0.250)	0.036	0.12	6.350	23.2	0.179
(Grade 250)	$\frac{5}{16}$ (0.313)	0.058	0.20	7.950	37.4	0.298
	$\frac{3}{8}$ (0.375)	0.080	0.27	9.525	51.6	0.402
	$\frac{7}{16}$ (0.438)	0.108	0.37	11.125	69.7	0.551
	$\frac{1}{2}$ (0.500)	0.144	0.49	12.700	92.9	0.729
	(0.600)	0.216	0.74	15.240	139.4	1.101
Seven-wire strand	$\frac{3}{8}$ (0.375)	0.085	0.29	9.525	54.8	0.432
(Grade 270)	$\frac{7}{16}$ (0.438)	0.115	0.40	11.125	74.2	0.595
	$\frac{1}{2}$ (0.500)	0.153	0.53	12.700	98.7	0.789
	(0.600)	0.215	0.74	15.250	138.7	1.101
Prestressing wire	(250)	0.192	0.029	0.10	4.877	18.7	0.146
Grades	(250)	0.196	0.030	0.10	4.978	19.4	0.149
	(240)	0.250	0.049	0.17	6.350	31.6	0.253
	(235)	0.276	0.060	0.20	7.010	38.7	0.298
Prestressing bars	$\frac{3}{4}$ (0.750)	0.44	1.50	19.050	283.9	2.232
(smooth)	$\frac{7}{8}$ (0.875)	0.60	2.04	22.225	387.1	3.036
(Grade 145	1 (1.000)	0.78	2.67	25.400	503.2	3.973
or 160)	$1\frac{1}{8}$ (1.125)	0.99	3.38	28.575	638.7	5.030
	$1\frac{1}{4}$ (1.250)	1.23	4.17	31.750	793.5	6.206
	$1\frac{3}{8}$ (1.385)	1.48	5.05	34.925	954.8	7.515
Prestressing bars	$\frac{5}{8}$ (0.625)	0.28	0.98	15.875	180.6	1.458
(deformed)	$\frac{3}{4}$ (0.750)	0.42	1.49	19.050	271.0	2.218
(Grade 150–160)	1 (1.000)	0.85	3.01	25.400	548.4	4.480
	$1\frac{1}{4}$ (1.250)	1.25	4.39	31.750	806.5	6.535
	$1\frac{3}{8}$ (1.385)	1.58	5.56	34.925	1006.5	8.274

18.8   Loss of Prestress

18.8.1   Lump-Sum Losses

Following the transfer of the prestressing force from the jack to the concrete member, a continuous loss in the prestressing force occurs; the total loss of prestress is the reduction in the prestressing force during the lifespan of the structure. The amount of loss in tendon stress varies between 15 and 30 percent of the initial stress, as it depends on many factors. For most normal-weight concrete structures constructed by standard methods, the tendon stress loss due to elastic shortening, shrinkage, creep, and relaxation of steel is about 35 ksi (241 MPa) for pretensioned members and 25 ksi (172 MPa) for posttensioned members. Friction and anchorage slip are not included.

Two current recommendations to estimate the total loss in prestressed concrete members are presented by AASHTO and the Posttensioning Institute (PTI). AASHTO[23] recommends a total loss (excluding friction loss) of 45 ksi (310 MPa) for pretensioned strands and 33 ksi (228 MPa) for posttensioned strands and wires when a concrete strength $f'_c = 5$ ksi is used. The PTI[24] recommends a total lump-sum prestress loss for posttensioned members of 35 ksi (241 MPa) for beams and 30 ksi (207 MPa) for slabs (excluding friction loss). These values can be used unless a better estimate of the prestress loss by each individual source is made, as will be explained below.

In general, the sources of prestress loss are:

• elastic shortening of concrete
• shrinkage of concrete
• creep of concrete
• relaxation of steel tendons
• friction
• anchorage set

18.8.2   Loss Due to Elastic Shortening of Concrete

In pretensioned members, estimating loss proceeds as follows. Consider a pretensioned concrete member of constant section and stressed uniformly along its centroidal axis by a force F_o. After the transfer of the prestressing force, the concrete beam and the prestressing tendon shorten by an equal amount because of the bond between the two materials. Consequently, the starting prestressing force F_o drops to F_i and the loss in the prestressing force is $F_o - F_i$. Also, the strain in the concrete ε_c must be equal to the change in the tendon strain $\Delta\varepsilon_s$. Therefore, $\varepsilon_c = \Delta\varepsilon_s$ or $(f_c/E_c) = (\Delta f_s/E_s)$, and the stress loss due to the elastic shortening

$$\Delta f_s = \frac{E_s}{E_c} \times f_c = n f_c = \frac{n F_i}{A_c} \approx \frac{n F_o}{A_c} \qquad (18.1)$$

where

A_c = the area of the concrete section
$n = E_s/E_c$
f_c = the stress in the concrete at the centroid of the prestressing steel

Multiplying the stress by the area of the prestressing steel A_{sp} to get the total force, then the elastic loss

$$ES = F_o - F_i = \Delta f_s A_{sp} = (n f_c) A_{sp} \approx \left(\frac{n F_o}{A_c}\right) A_{sp} \qquad (18.2)$$

or

$$F_i = F_o - (n f_c) A_{sp} \qquad (18.3)$$

For practical design, the loss in the prestressing force Δf_s per unit A_{sp} may be taken approximately $= (n F_o/A_c)$.

If the force F_o acts at an eccentricity e, then the elastic loss due to the presence of F_o and the applied dead load at transfer

$$ES = -(nf_c)A_{sp} \text{ (due to prestress)} + (nf_c)A_{sp} \text{ (dead load)}$$

$$ES = F_o - F_i = -\left(\frac{F_i}{A} + \frac{F_i e^2}{I}\right) nA_{sp} + \left(\frac{M_D e}{I}\right) nA_{sp} \qquad (18.4)$$

An approximate value of $F_i = (0.63 f_{pu})A_{sp}$ may be used in the above equation.

$$F_o + f_c(\text{D.L.})nA_{sp} = F_i\left[1 + nA_{sp}\left(\frac{1}{A} + \frac{e^2}{I}\right)\right]$$

$$F_i = \frac{F_o + (nA_{sp})f_c(\text{D.L.})}{1 + (nA_{sp})\left(\dfrac{1}{A} + \dfrac{e^2}{I}\right)} \qquad (18.5)$$

For posttensioned members where the tendons or individual strands are not stressed simultaneously, the loss of the prestress can be taken as half the value ES for pretensioned members.

Also, it is practical to consider the elastic shortening loss in slabs equal to one-quarter of the equivalent pretensioned value because stretching of one tendon will have little effect on the stressing of the other tendons.

18.8.3   Loss Due to Shrinkage

The loss of prestress due to shrinkage is time dependent. It may be estimated as follows:

$$SH = \Delta f_s \text{ (shrinkage)} = \varepsilon_{sh}E_s \qquad (18.6)$$

where $E_s = 29 \times 10^6$ psi and ε_{sh} = shrinkage strain in concrete.

The average strain due to shrinkage may be assumed to have the following values: for pretensioned members, $\varepsilon_{sh1} = 0.0003$; for posttensioned members, $\varepsilon_{sh2} = 0.0002$. If posttensioning is carried out within 5 to 7 days of concreting, the shrinkage strain can be taken $= 0.8\varepsilon_{sh1}$. If posttensioning is carried out between 1 and 2 weeks, $\varepsilon_{sh} = 0.7\varepsilon_{sh1}$ can be used, and if more than 2 weeks, $\varepsilon_{sh} = \varepsilon_{sh2}$ can be adopted. Shrinkage loss SH can also be estimated as follows:[28]

$$SH = 8.2 \times 10^{-6}K_{sh}E_s(1 - 0.06 \, V/S)(100 - RH)$$

where V/S = volume to surface ratio, RH = average relative humidity, $K_{sh} = 1.0$ for pretensioned members and equals 0.8, 0.73, 0.64, and 0.58 for posttensioned members if posttensioning is carried out after 5, 10, 20, and 30 days, respectively.

18.8.4   Loss Due to Creep of Concrete

Creep is a time-dependent deformation that occurs in concrete under sustained loads. The developed deformation causes a loss of prestress from 5 to 7 percent of the applied force.

The creep strain varies with the magnitude of the initial stress in the concrete, the relative humidity, and time. The loss in stress due to creep can be expressed as follows:

$$CR = \Delta f_s \text{ (creep)} = C_c(nf_c) = C_c(\varepsilon_{cr}E_s) \qquad (18.7)$$

where

$$C_c = \text{creep coefficient} = \text{creep strain } \varepsilon_{cp}/\text{initial elastic strain } \varepsilon_i$$

The value of C_c may be taken as follows:[22]

Concrete strength	$f_c' \le 4\ ksi$		$f_c' > 4\ ksi$	
Relative humidity	100%	50%	100%	50%
C_c	1–2	2–4	0.7–1.5	1.5–3

Linear interpolation can be made between the above values. Considering that half the creep takes place in the first month and three-quarters in the first 6 months after transfer and under normal humidity conditions, the creep strain can be assumed for practical design as follows:

- For pretensioned members, $\varepsilon_{cr} = 48 \times 10^{-5} \times$ stress in concrete (ksi).
- For posttensioned members, $\varepsilon_{cr} = 36 \times 10^{-5} \times$ stress in concrete (ksi). This value is used when posttensioning is made within 2 to 3 weeks. For earlier posttensioning, an intermediate value may be used.

The above values apply when the strength of concrete at transfer $f_{ci} \ge 4$ ksi. When $f_{ci} < 4$ ksi, the creep strain should increase in the ratio of (4/actual strength).

$$\text{Total loss of prestress due to creep} = \varepsilon_{cr}E_s \qquad (18.8)$$

18.8.5   Loss Due to Relaxation of Steel

Relaxation of steel causes a time-dependent loss in the initial prestressing force, similar to creep in concrete. The loss due to relaxation varies for different types of steel; its magnitude is usually furnished by the steel manufacturers. The loss is generally assumed to be 3 percent of the initial steel stress for posttensioned members and 2 to 3 percent for pretensioned members. If test information is not available, the loss percentages for relaxation at 1000 hours can be assumed as follows:

- In low-relaxation strands, when the initial prestress is $0.7f_{pu}$ and $0.8f_{pu}$, relaxation RE is 2.5 percent and 3.5 percent, respectively.
- In stress-relieved strands or wire, when the initial prestress is $0.7f_{pu}$ and $0.8f_{pu}$, relaxation RE is 8 percent and 12 percent, respectively.

18.8.6   Loss Due to Friction

With pretensioned steel, friction loss occurs when wires or strands are deflected through a diaphragm. This loss is usually small and can be neglected. When the strands are deflected to follow a concordant profile, the friction loss may be considerable. In such cases, accurate load measuring devices are commonly used to determine the force in the tendon.

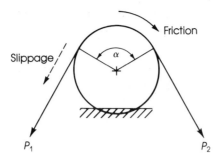

FIGURE 18.5. Belt friction.

With posttensioned steel, the effect of friction is considerable because of two main factors: the curvature of the tendon and the lack of alignment (wobble) of the duct. The curvature effect may be visualized (Figure 18.5). If a belt around a fixed cylinder is tensioned on one end with a force P_2, then the force P_1 at the other end to initiate slippage in the direction of P_1 is

$$P_1 = P_2 e^{\mu\alpha} \qquad\qquad (18.9)$$

where μ = the coefficient of static angular friction and α = the angle between P_1 and P_2. It is a general practice to treat the wobbling effect similarly:

$$P_x = P_s e^{-(\mu\alpha + Kl_x)}$$

or

$$P_s = P_x e^{(\mu\alpha + Kl_x)} \qquad \textbf{(ACI Code equation 18.1)} \qquad (18.10)$$

where

P_x = the prestressing tendon force at any point x

P_s = the prestressing tendon force at the jacking end

μ = curvature friction coefficient

α = total angular change of prestressing tendon profile, in radians, from tendon jacking end to any point x

$\quad = \dfrac{\text{length of curve}}{\text{radius of curvature}}$

K = wobble friction coefficient per foot of the prestressing tendon

As an approximation, the ACI Code gives the following expression:

$$P_s = P_x(1 + \mu\alpha + Kl_x) \qquad \textbf{(ACI Code equation 18.2)} \qquad (18.11)$$

provided that $(\mu\alpha + Kl_x) \leq 0.30$.

The frictional coefficients α and K depend on the type of prestressing strands or wires, type of duct, and the surface conditions. Some approximate values for μ and K are given in the **ACI Commentary, section 18.6,**[3] and here in Table 18.2.

TABLE 18.2. Friction coefficients for posttensioned tendons

Type of tendon	Wobble coefficient K per foot ($\times 10^{-3}$)	Curvature coefficient μ
Tendon in flexible metal sheathing (grouted)		
Wire tendons	1.0–1.5	0.15–0.25
7-wire strand	0.5–2.0	0.15–0.25
High-strength bars	0.1–0.6	0.08–0.30
Pregreased unbonded tendon		
Wire tendons and 7-wire strand	0.3–2.0	0.05–0.15
Mastic-coated unbonded tendons		
Wire tendons and 7-wire strand	1.0–2.0	0.05–0.15

Friction loss in the jack is variable and depends on many factors, including the length of travel of the arm over a given load range. The use of accurate load cells to measure directly the force in the tendon is recommended. The use of pressure gauges may lead to inaccuracies unless they are calibrated against a known force in the tendon.

The friction loss in the anchorage is mainly dependent upon the type of anchorage and the amount of deviation of the tendon as it passes through the anchorage. This loss is usually small and may be neglected. Guidance in particular cases should be obtained from the manufacturers.

18.8.7   Loss Due to Anchor Set

When the force in a tendon is transferred from the jack to the anchorage unit, a small inward movement of the tendon takes place due to the seating of the gripping device or wedges. The slippage causes a shortening of the tendon, which results in a loss in the prestressing force. The magnitude of slippage varies between 0.1 and 0.25 in. (2.5 and 6 mm) and is usually specified by the manufacturer. The loss due to the anchor set may be calculated as follows:

$$\Delta f_s = \Delta\varepsilon E_s = \frac{\Delta L}{L} \times E_s \qquad (18.12)$$

where

    $\Delta\varepsilon$ = magnitude of the anchor slippage

    $E_s = 29 \times 10^6$ psi

    L = length of the tendon

Since the loss in stress is inversely proportional to the length of the tendon (or approximately half the length of the tendon if it is stressed from both ends simultaneously), the percentage loss in steel stress decreases as the length of the tendon increases. If the tendon is elongated by $\Delta\varepsilon$ at transfer, the loss in prestress due to slippage is neglected.

EXAMPLE **18.3**

A 36-ft-span pretensioned simply supported beam has a rectangular cross-section, $b = 18$ in. and $h = 32$ in. Calculate the elastic loss and all time-dependent losses. Given: prestressing force at transfer $F_i = 435$ k, area of prestressing steel $A_{ps} = 3.0$ in.^{2}, $f_c' = 5$ ksi, $E_c = 5000$ ksi, $E_s = 29,000$ ksi; profile of tendon is parabolic, eccentricity at midspan $= 6.0$ in., and eccentricity at ends $= 0$.

Solution

1. Elastic shortening:

$$\text{Stress due to the prestressing force at transfer } \frac{F_i}{A_{ps}} = \frac{435}{3} = 145 \text{ ksi}$$

$$\text{Strain in prestressing steel } = \frac{f_s}{E_s} = \frac{145}{29,000} = 0.005$$

Using equation 18.1:

$$n = \frac{E_s}{E_c} = \frac{29,000}{5000} = 5.8 \text{ or } 6$$

$$\Delta f_s = \frac{nF_i}{A_c} = \frac{6 \times 435}{32 \times 18} = 4.5 \text{ ksi}$$

Considering the variation in the eccentricity along the beam,

$$\text{Strain at end of section } = \frac{F_i}{A_c E_c} = \frac{435}{(18 \times 32) \times 5000} = 0.151 \times 10^{-3}$$

$$\text{Strain at midspan } = \frac{F_i}{A_c E_c} + \frac{F_i e^2}{I E_c}$$

$$I = \frac{bh^3}{12} = \frac{18(32)^3}{12} = 49,152 \text{ in.}^4$$

$$\text{Strain} = 0.151 \times 10^{-3} + \frac{435(6)^2}{49,152(5000)} = 0.215 \times 10^{-3}$$

$$\text{Average strain } = \frac{1}{2} (0.151 + 0.215) \times 10^{-3} = 0.183 \times 10^{-3}$$

$$\text{Prestress loss } = \text{strain} \times E_s = 0.183 \times 10^{-3} \times 29,000 = 5.3 \text{ ksi}$$

$$\text{Percent loss } = \frac{5.3}{145} = 3.66 \text{ percent}$$

2. Loss due to shrinkage:

$$\text{Shrinkage strain } = 0.0003$$

$$\Delta f_s = \varepsilon_{sh} E_s = 0.0003 \times 29,000 = 8.7 \text{ ksi}$$

$$\text{Percent loss } = \frac{8.7}{145} = 6 \text{ percent}$$

3. Loss due to creep of concrete: assuming $C_c = 2.0$, then

$$\Delta f_s = C_c(\varepsilon_{cr} E_s)$$

$$\text{Elastic strain} = \frac{F_i}{A_c E_c} = 0.151 \times 10^{-3}$$

$$\Delta f_s = 2(0.151 \times 10^{-3} \times 29{,}000) = 8.8 \text{ ksi}$$

$$\text{Percent loss} = \frac{8.8}{145} = 6.1 \text{ percent}$$

Or approximately: $\varepsilon_{cr} = 48 \times 10^{-5} \times$ stress in the concrete (ksi):

$$\varepsilon_{cr} = 48 \times 10^{-5} \left(\frac{435}{32 \times 18}\right) = 36 \times 10^{-5}$$

$$\Delta f_s = \varepsilon_{cr} E_s = 36 \times 10^{-5} \times 29{,}000 = 10.4 \text{ ksi}$$

$$\text{Percent loss} = \frac{10.4}{145} = 7.2 \text{ percent}$$

This is a conservative value, and the same ratio is obtained if $C_c = 2.38$ is adopted in the above calculations.

4. Loss due to relaxation of steel: For low-relaxation strands, the loss is assumed 2.5 percent.

$$\Delta f_s = 0.025 \times 145 = 3.6 \text{ ksi}$$

5. Assume the losses due to bending, friction of cable spacers, and the end block of the pretensioning system are 2 percent.

$$\Delta f_s = 0.02 \times 145 = 2.9 \text{ ksi}$$

6. Loss due to friction in tendon = 0.
7. Total losses:

Elastic shortening loss	5.3 ksi	3.6%
Shrinkage loss	8.7 ksi	6.0%
Creep of concrete loss	8.8 ksi	6.1%
Relaxation of steel loss	3.6 ksi	2.5%
Other losses	2.9 ksi	2.0%
Total losses	29.3 ksi	20.2%

Effective prestress $= 145 - 24 = 121$ ksi

Effective prestressing force $F = 121 \times 3 \text{ in.}^2 = 363$ k

$$F = (1 - 0.166)F_i = 0.834 F_i$$

For $F = \eta F_i$, $\eta = 0.834$. ∎

EXAMPLE 18.4

Calculate all losses of a 120-ft-span posttensioned beam that has an I-section with the following details. Area of concrete section, $A_c = 760 \text{ in.}^2$; moment of inertia, $I_g = 1.64 \times 10^5 \text{ in.}^4$; prestressing force at transfer, $F_i = 1110$ k; area of prestressing steel, $A_{ps} = 7.5 \text{ in.}^2$; $f_c' =$

5 ksi, $E_c = 5000$ ksi, and $E_s = 29,000$ ksi; profile of tendon is parabolic, eccentricity at mid-span = 20 in., and eccentricity at ends = 0.

Solution

1. Loss due to elastic shortening:

$$\text{Steel stress at transfer} = \frac{F_i}{A_{ps}} = \frac{1110}{7.5} = 148 \text{ ksi}$$

$$\text{Stress in concrete at end section} = \frac{1110}{760} = 1.46 \text{ ksi}$$

$$\text{Stress in concrete at midspan} = \frac{F_i}{A_c} + \frac{F_i e^2}{I} - \frac{M_D e}{I}$$

$$\text{Weight of beam} = \frac{760}{144} \times 150 = 790 \text{ lb/ft}$$

$$M_D = 0.79 \frac{(120)^2}{8} = 1422 \text{ k·ft}$$

$$\text{Stress at midspan} = \frac{1110}{760} + \frac{1110(20)^2}{164,000} - \frac{(1422 \times 12)(20)}{164,000}$$

$$= 1.46 + 2.71 - 2.08 = 2.09 \text{ ksi}$$

$$\text{Average stress} = \frac{1.46 + 2.09}{2} = 1.78 \text{ ksi}$$

$$\text{Average strain } \varepsilon_c = \frac{1.78}{E_c} = \frac{1.78}{5000} = 0.356 \times 10^{-3}$$

$$\text{Elastic loss } \Delta f_s = \varepsilon_c E_s = 0.356 \times 10^{-3} \times 29,000 = 10.3 \text{ ksi}$$

assuming that the tendons are tensioned two at a time. The first pair will have the greatest loss while the last pair will have zero loss. Therefore, average $\Delta f_s = 10.3/2 = 5.15$ ksi.

$$\text{Percent loss} = \frac{5.15}{148} = 3.5 \text{ percent}$$

2. Loss due to shrinkage of concrete:

$$\Delta f_s \text{ (shrinkage)} = 0.0002 E_s = 0.0002 \times 29,000 = 5.8 \text{ ksi}$$

$$\text{Percent loss} = \frac{5.8}{148} = 3.9 \text{ percent}$$

3. Loss due to creep of concrete: Assume $C_c = 1.5$.

$$\text{Elastic strain} = \frac{F_i}{A_c E_c} = \frac{1110}{760 \times 5000} = 0.292 \times 10^{-3}$$

$$\Delta f_s \text{ (creep)} = C_c(\varepsilon_{cr} E_s)$$

$$= 1.5(0.292 \times 10^{-3} \times 29,000) = 12.7 \text{ ksi}$$

$$\text{Percent loss} = \frac{12.7}{148} = 8.6 \text{ percent}$$

4. Loss due to relaxation of steel: For low-relaxation strands, the loss is 2.5 percent.

$$\Delta f_s = 0.025 \times 148 = 3.7 \text{ ksi}$$

5. Slip in anchorage: For tensioning from one end only, assume a slippage of 0.15 in. Length of cable $= 120 \times 12 = 1440$ in.

$$\Delta f_s = \frac{\Delta L}{L} \times E_s = \frac{0.15}{1440} \times 29{,}000 = 3 \text{ ksi} \qquad (18.12)$$

To allow for anchorage slip, set the tensioned force to $148 + 3 = 151$ ksi on the pressure gauge to leave a net stress of 148 ksi in the tendons.

6. Loss due to friction: The equation of parabolic profile is

$$e_x = \frac{4e}{L^2} (Lx - x^2)$$

where $e_x =$ the eccentricity at a distance x measured from the support and $e =$ eccentricity at midspan.

$$\frac{d(e_x)}{dx} = \frac{4e}{L^2} (L - 2x)$$

is the slope of the tendon at any point.

At the support, $x = 0$ and the slope

$$\frac{d(e_x)}{dx} = \frac{4e}{L} = \frac{4 \times 20}{120 \times 12} = 0.056$$

Slope at midspan $= 0$, therefore $\alpha = 0.056$. Using flexible metallic sheath, $\mu = 0.5$ and $K = 0.001$. At midspan, $x = 60$ ft. Check if $(\mu\alpha + Kl_x) \leq 0.30$:

$$\mu\alpha + Kl_x = 0.5 \times 0.056 + 0.001 \times 60 = 0.088 < 0.3$$
$$P_s = P_x(1 + \mu\alpha + Kl_x) \qquad (18.11)$$
$$= P_x(1 + 0.088) = 1.088 P_x$$
$$= 1.088 \times 148 = 161 \text{ k (force at jacking end)}$$
$$\Delta f_s = 161 - 148 = 13 \text{ ksi}$$
$$\text{Percent loss} = \frac{13}{148} = 8.8 \text{ percent}$$

7. Total losses:

Elastic shortening loss	5.2 ksi	3.5%
Shrinkage loss	5.8 ksi	3.9%
Creep of concrete loss	12.7 ksi	8.6%
Relaxation of steel loss	3.7 ksi	2.5%
Friction loss	13.0 ksi	8.8%
Total losses	40.4 ksi	27.3%

Effective prestress $= 148 - 35.2 = 112.8$ ksi

Effective prestressing force $F = (1 - 0.238) F_i = 0.762 F_i$

$$F = 0.762 \times 1110 = 846 \text{ k}$$

For $F = \eta F_i$, $\eta = 0.762$. ∎

18.9   Concepts of Prestressed Concrete

In considering the stresses in a prestressed concrete beam, three different concepts may be applied to analyze its behavior under service load.

The Homogeneous Beam Concept.   This concept treats prestressed concrete as an elastic uncracked composite material. The stresses in the section are calculated using the flexural formula. For the combined action of an eccentric prestressing force P on a concrete section, the stresses at the extreme fibers are

$$\sigma_1 = -\frac{P}{A} - \frac{Mc_1}{I} \quad \text{and} \quad \sigma_2 = -\frac{P}{A} + \frac{Mc_2}{I}$$

where

A and I = the area and moment of inertia of the section

$M = Pe$ = the force times its eccentricity

c_1 and c_2 = the distances from the centroid of the section to its extreme fibers

The Internal Couple Concept.   Prestressed concrete is treated similar to reinforced concrete, as was discussed in Chapter 5. It is assumed that the concrete carries the compressive force C while the steel carries the tensile force T. From equilibrium, $C = T$, and the internal moment $M = C \times (\text{Arm } jd) = T \times (\text{Arm } jd)$ where jd is the distance between C and T. Once C and T are determined, the stresses in the concrete and steel can be calculated from the principles of statics.

The Load Balancing Concept.   This concept treats prestressing as a process of balancing loads on the flexural member. The prestressing force and its eccentricity are selected so as to balance the moments produced from the applied loads, resulting in zero flexural stress and zero deflection. To explain this approach, consider a simply supported beam subjected to a uniform load and a parabolic prestress steel profile, as shown in Figure 18.6. From the equilibrium of half the beam AC, the maximum prestress moment at midspan = Fe_{max}. This prestress moment is equal to the moment produced by the uniform load on the beam of $W_1L^2/8$. Therefore,

$$Fe_{max} = \frac{W_1L^2}{8}$$

or

$$e_{max} = \frac{W_1L^2}{8F} \quad \text{and} \quad F = \frac{W_1L^2}{8e_{max}}$$

Stresses at the centerline section are

$$\sigma_t \text{ (top)} = -\frac{F}{A} - \frac{M_2 y_t}{I}$$

$$\sigma_b \text{ (bottom)} = -\frac{F}{A} + \frac{M_2 y_b}{I}$$

where M_2 = moment due to external load W_2 ($W_2 = W - W_1$).

EXAMPLE 18.5

Assume that the beam shown in Figure 18.6 has a span of 40 ft and carries a uniform load $W_1 = 2.2$ k/ft. The cross-section of the beam is rectangular, with $b = 12$ in. and $h = 36$ in. Determine the least amount of the prestressing force that will balance the uniform load applied on the beam. If an additional uniform load $W_2 = 1.2$ k/ft is applied later on the beam, compute the top and bottom fiber stresses.

Solution   1. In order to balance the moments produced by the uniform load $W_1 = 2.2$ k/ft with the least amount of prestressing force, use maximum eccentricity at midspan. Assuming a concrete cover of 3 in.,

$$e_{max} \text{ at midspan} = \frac{h}{2} - 3 = \frac{36}{2} - 3 = 15 \text{ in.} = 1.25 \text{ ft}$$

$$F = \frac{W_1 L^2}{8 e_{max}} = \frac{2.2(40)^2}{8(1.25)} = 352 \text{ k}$$

A parabolic profile for the tendon is adopted; therefore the beam will be subjected to a zero deflection and to a uniform compressive stress

$$\sigma = -\frac{F}{A_c} = -\frac{352,000}{12 \times 36} = -815 \text{ psi}$$

2. For an additional uniform load $W_2 = 1.2$ k/ft on the beam, the moment at midspan is

$$M_2 = \frac{W_2 L^2}{8} = \frac{1.2(40)^2}{8} = 240 \text{ k·ft}$$

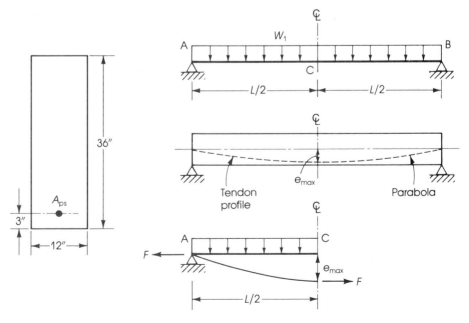

FIGURE 18.6. Load balancing, example 18.5.

The final stresses at the extreme fibers of the midspan section are

$$\sigma_t \text{ (top)} = -\frac{F}{A} - \frac{Mc}{I}$$

$$\sigma_b \text{ (bottom)} = -\frac{F}{A} + \frac{Mc}{I}$$

$$c = \frac{36}{2} = 18 \text{ in.}$$

$$I = \frac{bh^3}{12} = \frac{12(36)^3}{12} = 46{,}656 \text{ in.}^4$$

Therefore,

$$\sigma_t = -815 - \frac{(240)(12{,}000)18}{46{,}656}$$

$$= -815 - 1111 = -1926 \text{ psi (compression)}$$

$$\sigma_b = -815 + 1111 = +296 \text{ psi (tension)}$$

18.10   Elastic Analysis of Flexural Members

18.10.1   Stresses Due to Loaded and Unloaded Conditions

In the analysis of prestressed concrete beams, two extreme loadings are generally critical. The first occurs at transfer when the beam is subjected to the prestressing force F_i and the weight of the beam or the applied dead load at the time of transfer of the prestressing force. No live load or additional dead loads are considered. In this unloaded condition, the stresses at the top and bottom fibers of the critical section must not exceed the allowable stresses at transfers f_{ci} and f_{ti} for the compressive and tensile stresses in concrete, respectively.

The second case of loading occurs when the beam is subjected to the prestressing force after all losses F, and all dead and live loads. In this loaded condition, the stresses at the top and bottom fibers of the critical section must not exceed the allowable stresses f_c and f_t for the compressive and tensile stresses in concrete, respectively.

The above conditions can be expressed mathematically as follows:

1. For the unloaded condition (at transfer):

   • At top fibers,

$$\sigma_{ti} = -\frac{F_i}{A} + \frac{(F_i e)y_t}{I} - \frac{M_D y_t}{I} \leq f_{ti} \qquad (18.13)$$

   • At bottom fibers,

$$\sigma_{bi} = -\frac{F_i}{A} - \frac{(F_i e)y_b}{I} + \frac{M_D y_b}{I} \geq -f_{ci} \qquad (18.14)$$

2. For the loaded condition (all loads are applied after all losses):

• At top fibers,

$$\sigma_t = -\frac{F}{A} + \frac{(Fe)y_t}{I} - \frac{M_D y_t}{I} - \frac{M_L y_t}{I} \geq -f_c \qquad (18.15)$$

• At bottom fibers,

$$\sigma_b = -\frac{F}{A} - \frac{(Fe)y_b}{I} + \frac{M_D y_b}{I} + \frac{M_L y_b}{I} \leq f_t \qquad (18.16)$$

where

F_i and F = the prestressing force at transfer and after all losses

f_{ti} and f_t = allowable tensile stress in concrete at transfer and after all losses

f_{ci} and f_c = allowable compressive stress in concrete at transfer and after all losses

M_D and M_L = moments due to dead load and live load

y_t and y_b = distances from the neutral axis to the top and bottom fibers

In the above analysis, it is assumed that the materials behave elastically within the working range of stresses applied.

18.10.2 Kern Limits

If the prestressing force is applied at the centroid of the cross-section, uniform stresses will develop. If the prestressing force is applied at an eccentricity e below the centroid such that the stress at the top fibers is equal to zero, that prestressing force is considered acting at the lower Kern point (Figure 18.7(a)). In this case e is denoted by K_b, and the stress distribution is triangular, with maximum compressive stress at the extreme bottom fibers. The stress at the top fibers is

$$\sigma_t = -\frac{F_i}{A} + \frac{(F_i e)y_t}{I} = 0$$

or

$$e = K_b = \text{lower Kern} = \frac{I}{Ay_t} \qquad (18.17)$$

Similarly, if the prestressing force is applied at an eccentricity e' above the centroid such that the stress at the bottom fibers is equal to zero, that prestressing force is considered acting at the upper Kern point (Figure 18.7(b)). In this case the eccentricity e' is denoted by K_t, and the stress distribution is triangular, with maximum compressive stress at the extreme top fibers. The stress at the bottom fibers is

$$\sigma_b = -\frac{F_i}{A} + \frac{(F_i e')y_b}{I} = 0$$

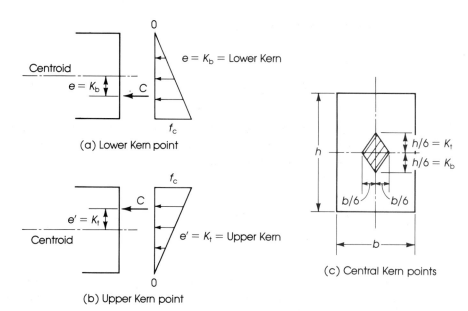

FIGURE 18.7. Kern points: (a) lower, (b) upper, and (c) central.

or

$$e' = K_t = \text{upper Kern} = \frac{I}{Ay_b} \qquad (18.18)$$

The Kern limits of a rectangular section are shown in Figure 18.7(c).

18.10.3 Limiting Values of Eccentricity

The four stress equations 18.13 through 18.16 can be written as a function of the eccentricity e for the various loading conditions. For example, equation 18.13 can be rewritten as follows:

$$\sigma_{ti} = -\frac{F_i}{A} + \frac{(F_i e)y_t}{I} - \frac{M_D y_t}{I} \le f_{ti}$$

$$\frac{(F_i e)y_t}{I} \le f_{ti} + \frac{F_i}{A} + \frac{M_D y_t}{I}$$

$$e \le \frac{I}{F_i y_t}\left(\frac{F_i}{A} + \frac{M_D y_t}{I} + f_{ti}\right) \qquad (18.19)$$

If the lower Kern limit $K_b = I/Ay_t$ is used, then

$$e \le K_b + \frac{M_D}{F_i} + \frac{f_{ti}AK_b}{F_i} \qquad (18.20)$$

This value of e represents the maximum eccentricity based on the *top fibers, unloaded condition*.

Similarly, from equation 18.14,

$$e \leq \frac{I}{F_i y_b}\left(-\frac{F_i}{A} + \frac{M_D y_b}{I} + f_{ci}\right) \qquad (18.21)$$

$$e \leq -K_t + \frac{M_D}{F_i} + \frac{f_{ci} A K_t}{F_i} \qquad (18.22)$$

This value of e represents the maximum eccentricity based on the *bottom fibers, unloaded condition*. The two maximum values of e should be calculated from the above equations and the smaller value used.

From equation 18.15,

$$e \geq \frac{I}{F y_t}\left(\frac{F}{A} + \frac{M_T y_t}{I} - f_c\right) \qquad (18.23)$$

$$e \geq K_b + \frac{M_T}{F} - \frac{f_c A K_b}{F} \qquad (18.24)$$

where M_T = moment due to dead and live loads = $(M_D + M_L)$. This value of e represents the minimum eccentricity based on the *top fibers, loaded condition*.

From equation 18.16,

$$e \geq \frac{I}{F y_b}\left(-\frac{F}{A} + \frac{M_T y_b}{I} - f_t\right) \qquad (18.25)$$

$$e \geq -K_t + \frac{M_T}{F} - \frac{f_t A K_t}{F} \qquad (18.26)$$

This value of e represents the minimum eccentricity based on the *bottom fibers, loaded condition*. The two minimum values of e should be calculated from the above equations and the larger of the two minimum eccentricities used.

18.10.4 Limiting Values of the Prestressing Force at Transfer F_i

Considering that $F = \eta F_i$, where η represents the ratio of the net prestressing force after all losses, and for the different cases of loading, equations 18.20, 18.22, 18.24, and 18.26 can be rewritten as follows:

$$(e - K_b) F_i \leq M_D + f_{ti} A K_b \qquad (18.27)$$
$$(e + K_t) F_i \leq M_D + f_{ci} A K_t \qquad (18.28)$$
$$(e - K_b) F_i \geq \frac{M_D}{\eta} + \frac{M_L}{\eta} - \frac{1}{\eta}(f_c A K_b) \qquad (18.29)$$
$$(e + K_t) F_i \geq \frac{M_D}{\eta} + \frac{M_L}{\eta} - \frac{1}{\eta}(f_t A K_t) \qquad (18.30)$$

Subtract equation 18.27 from equation 18.30 to get

$$F_i(K_b + K_t) \geq M_D\left(\frac{1}{\eta} - 1\right) + \frac{M_L}{\eta} - \frac{f_t A K_t}{\eta} - f_{ti} A K_b$$

or

$$F_i \geq \frac{1}{(K_b + K_t)}\left[\left(\frac{1}{\eta} - 1\right) M_D + \frac{M_L}{\eta} - \left(\frac{f_t AK_t}{\eta}\right) - (f_{ti} AK_b)\right] \qquad (18.31)$$

This value of F_i represents the *minimum* value of the prestressing force at transfer without exceeding the allowable stresses under the loaded and unloaded conditions.

Subtract equation 18.29 from equation 18.28 to get

$$F_i \leq \frac{1}{(K_b + K_t)}\left[\left(1 - \frac{1}{\eta}\right) M_D - \frac{M_L}{\eta} + \left(\frac{f_c AK_b}{\eta}\right) + (f_{ci} AK_t)\right] \qquad (18.32)$$

This value of F_i represents the *maximum* value of the prestressing force at transfer without exceeding the allowable stresses under the loaded and unloaded conditions.

Subtracting equation 18.31 from equation 18.32, then

$$\left(1 - \frac{1}{\eta}\right) 2M_D - \frac{2M_L}{\eta} + \left(f_{ti} + \frac{f_c}{\eta}\right) AK_b + \left(f_{ci} + \frac{f_t}{\eta}\right) AK_t \geq 0 \qquad (18.33)$$

This equation indicates that (maximum F_i) − (minimum F_i) ≥ 0. If this equation is checked for any given section and proved to be satisfactory, then the section is adequate.

EXAMPLE 18.6

A pretensioned simply supported beam of the section shown in Figure 18.8 is to be used on a span of 48 ft. The beam must carry a dead load of 900 lb/ft (excluding its own weight), which will be applied at a later stage, and a live load of 1100 lb/ft. Assuming that prestressing steel is made of 20 tendons of $\frac{7}{16}$ in. diameter, with $E_s = 29 \times 10^6$ psi, $F_o = 175$ ksi, and ultimate strength $f_{pu} = 250$ ksi, it is required to:

1. Determine the location of the upper and lower limits of the tendon profile (centroid of the prestressing steel) for the section at midspan and for three other sections between the midspan section and the beam end.
2. Locate the tendon to satisfy these limits by harping some of the tendons at one-third points of the span. Check the limiting values of the prestressing force at transfer.
3. Revise the prestress losses, taking into consideration the chosen profile of the tendons and the variation of the eccentricity e.

Use f_{ci} (at transfer) = 4 ksi, $f'_c = 5$ ksi, $E_c = 4000$ ksi, and $E_{ci} = 3600$ ksi.

Solution   1. Determine the properties of the section:

$$\text{Area} = 18 \times 6 + 24 \times 6 + 12 \times 10 = 372 \text{ in.}^2$$

Determine the centroid of the section by taking moments about the base line.

$$y_b = \frac{1}{372}(120 \times 5 + 144 \times 22 + 108 \times 37) = 20.8 \text{ in.}$$

$$y_t = 40 - 20.8 = 19.2 \text{ in.}$$

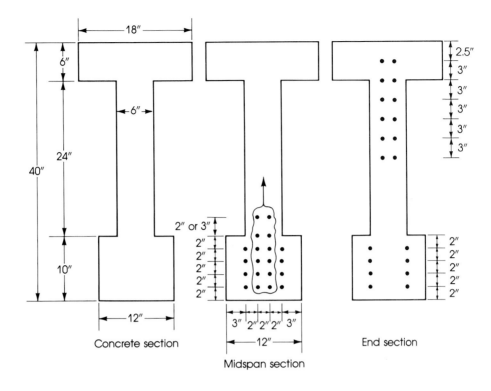

Concrete section

Midspan section

End section

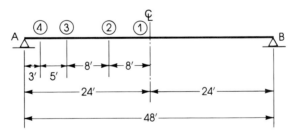

FIGURE 18.8. Example 18.6.

Calculate the gross moment of inertia I_g:

$$I_g = \left[\frac{18(6)^3}{12} + 108(16.2)^2\right] + \left[\frac{6(24)^3}{12} + 144(1.2)^2\right] + \left[\frac{12(10)^3}{12} + 120(15.8)^2\right]$$

$$= 66{,}862 \text{ in.}^4$$

$$K_b = \frac{I}{Ay_t} = \frac{66{,}862}{372 \times 19.2} = 9.4 \text{ in.}$$

$$K_t = \frac{I}{Ay_b} = \frac{66{,}862}{372 \times 20.8} = 8.6 \text{ in.}$$

2. Estimate prestress losses, given $F_o = 175$ ksi.

- Assume elastic loss = 4 percent = $0.04 \times 175 = 7$ ksi.
- Loss due to shrinkage = $0.0003 E_s = 0.0003 \times 29000 = 8.7$ ksi.
- Loss due to creep of concrete: A good first estimate of creep loss is 1.67 times the elastic loss:

$$1.67 \times 7 = 11.7 \text{ ksi}$$

- Loss due to relaxation of steel is 4 percent:

$$0.04 \times 175 = 7 \text{ ksi}$$

$$\text{Time-dependent losses} = 8.7 + 11.7 + 7 = 27.4 \text{ ksi}$$

$$\text{Percentage} = \frac{27.4}{175} = 15.7 \text{ percent}$$

- The total loss = $27.4 + 7$ (elastic loss) = 34.4 ksi. The percentage of total loss = $34.4/175 = 19.7$ percent.
- Prestress stresses:

$$F_i = 175 - 7 = 168 \text{ ksi (at transfer)}$$
$$F = 175 - 34.4 = 140.6 \text{ ksi}$$
$$F = \eta F_i$$
$$\eta = 1 - \text{time-dependent losses ratio} = 140.6/168 = 0.837$$

3. Limits of the eccentricity e at midspan section: Calculate the allowable stresses and moments. At transfer, $f'_{ci} = 4000$ psi, $f_{ci} = 0.6 \times 4000 = 2,400$ psi, and $f_{ti} = 3\sqrt{f'_{ci}} = 190$ psi. At service load, $f'_c = 5000$ psi, $f_c = 0.45 f'_c = 2250$ psi, and $f_t = 6\sqrt{f'_c} = 424$ psi.

$$\text{Self-weight of beam} = \frac{\text{Area}}{144} \times 150 = \frac{372}{144} \times 150 = 388 \text{ lb/ft}$$

$$M_D \text{ (self-weight)} = \frac{0.388}{8}(48)^2 \times 12 = 1341 \text{ k·in.}$$

$$M_a \text{ (additional load and live load)} = \frac{w_a L^2}{8}$$

$$= \frac{(0.9 + 1.1)}{8}(48)^2 \times 12 = 6912 \text{ k·in.}$$

Total moment $M_T = M_D + M_a = 8253$ k·in.

F_i = stress at transfer × area of prestressing steel

Area of 20 tendons, $\frac{7}{16}$ in. diameter = $20 \times 0.1089 = 2.178$ in.²

$F_i = 2.178 \times 168 = 365.9$ k

$F = 2.178 \times 140.6 = 306.2$ k

- Consider the section at midspan.
Top fibers, *unloaded* condition:

$$e \leq K_b + \frac{M_D}{F_i} + \frac{f_{ti} A K_b}{F_i} \qquad (18.20)$$

$$\leq 9.4 + \frac{1341}{365.9} + \frac{0.190(372)(9.4)}{365.9} \leq 14.9 \text{ in.}$$

Bottom fibers, *unloaded* condition:

$$e \leq -K_t + \frac{M_D}{F_i} + \frac{f_{ci}AK_t}{F_i} \qquad (18.22)$$

$$\leq -8.6 + \frac{1341}{365.9} + \frac{2.4(372)(8.6)}{365.9} \leq 16.1 \text{ in.}$$

Maximum $e = 14.9$ in. controls.

Top fibers, *loaded* condition:

$$e \geq K_b + \frac{M_T}{F} - \frac{f_c AK_b}{F} \qquad (18.24)$$

$$\geq 9.4 + \frac{8253}{306.2} - \frac{2.25(372)(9.4)}{306.2} \geq 10.7 \text{ in.}$$

Bottom fibers, *loaded* condition:

$$e \geq -K_t + \frac{M_T}{F} - \frac{f_t AK_t}{F} \qquad (18.26)$$

$$\geq -8.6 + \frac{8253}{306.2} - \frac{0.424(372)(8.6)}{306.2} \geq 13.9 \text{ in.}$$

Minimum $e = 13.9$ in. controls.

• Consider a section 8 ft from the midspan (section 2, Figure 18.8):

$$M_D \text{ (self-weight)} = R_A(16) - \frac{w_D}{2} \times (16)^2$$

$$= 0.388(24)(16) - \frac{0.388}{2}(16)^2 = 99.3 \text{ k·ft} = 1192 \text{ k·in.}$$

$$M_a = 2(24)(16) - \frac{2}{2}(16)^2 = 512 \text{ k·ft} = 6144 \text{ k·in.}$$

$$M_T = 6144 + 1192 = 7336 \text{ k·in.}$$

Top fibers, *unloaded* condition:

$$e \leq 9.4 + \frac{1192}{365.9} + \frac{0.190(372)(9.4)}{365.9} \leq 14.5 \text{ in.}$$

Bottom fibers, *unloaded* condition:

$$e \leq -8.6 + \frac{1192}{365.9} + \frac{2.4(372)(8.6)}{365.9} \leq 15.6 \text{ in.}$$

Maximum $e = 14.5$ in. controls.

Top fibers, *loaded* condition:

$$e \geq 9.4 + \frac{7336}{306.2} - \frac{2.25(372)(9.4)}{306.2} \geq 7.7 \text{ in.}$$

Bottom fibers, *loaded* condition:

$$e \geq -8.6 + \frac{7336}{306.2} - \frac{0.424(372)(8.6)}{306.2} \geq 11.0 \text{ in.}$$

Minimum $e = 11.0$ in. controls.

• Consider a section 16 ft from midspan (section 3, Figure 18.8):

$$M_D \text{ (self-weight)} = 745 \text{ k·in.} \quad M_a = 3840 \text{ k·in.} \quad M_T = 4585 \text{ k·in.}$$

Top fibers, unloaded condition, $e \leq 13.3$ in. (max) controls.
Bottom fibers, unloaded condition, $e \leq 14.4$ in.
Top fibers, loaded condition, $e \geq -1.3$ in.
Bottom fibers, loaded condition, $e \geq 1.9$ in. (min) controls.

• Consider a section 3 ft from the end (anchorage length):

$$M_D = 314 \text{ k·in.} \quad M_a = 1620 \text{ k·in.} \quad M_T = 1934 \text{ k·in.}$$

Top fibers, unloaded condition, $e \leq 12.1$ in. (max) controls.
Bottom fibers, unloaded condition, $e \leq 13.3$ in.
Top fibers, loaded condition, $e \geq -10$ in.
Bottom fibers, loaded condition, $e \geq -6.7$ in. (min) controls.

4. The tendon profile is shown in Figure 18.9. The eccentricity chosen at midspan $e = 14.5$ in., which is adequate for section B at 8 ft from midspan. The centroid of the prestressing steel is horizontal between A and B and then harped linearly between B and the end section at E. The eccentricities at sections C and D are calculated by establishing the slope of line BE = 14.5/16 = 0.91 in./ft. The eccentricity at C = 7.25 in. and at D = 2.72 in. The tendon profile chosen satisfies the upper and lower limits of the eccentricity at all sections.

Harping of tendons is performed as follows:

• Place the 20 tendons ($\frac{7}{16}$ in. diameter) within the middle third of the beam at spacings of 2 in. as shown in Figure 18.8. To calculate the actual eccentricity at midspan

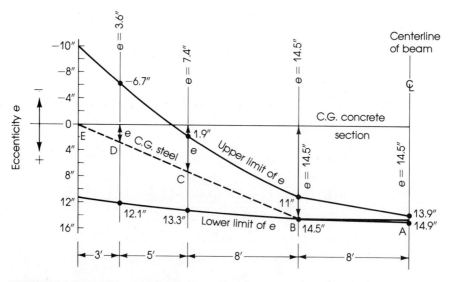

FIGURE 18.9. Tendon profile, example 18.6.

section, take moments for the number of tendons about the base line of the section:

$$\text{Distance from base} = \frac{1}{20}(16 \times 5 + 4 \times 11) = 6.2 \text{ in.}$$

$$e \text{ (midspan)} = y_b - 6.2 \text{ in.}$$
$$= 20.8 - 6.2 = 14.6 \text{ in.}$$

which is close to 14.5 in. assumed. If the top two tendons are placed at 3 in. from the row below them, then the distance from the base becomes

$$\frac{1}{20}(16 \times 5 + 2 \times 10 + 2 \times 13) = 6.3 \text{ in.}$$

The eccentricity becomes $20.8 - 6.3 = 14.5$ in., which is equal to the assumed eccentricity. Practically, all tendons may be left at 2 in. spacing by neglecting the difference of 0.1 in.

• Harp the central 12 tendons only. The distribution of tendons at the end section is shown in Figure 18.8. To check the eccentricity of tendons, take moments about the centroid of the concrete section for the 12 tendons at top and the 8 tendons left at bottom:

$$e = \frac{1}{20}(8 \times 14.5 - 12 \times 9.2) = 0.28 \text{ in.}$$

This value of e is small and adequate. The actual eccentricity at 3 ft from the end section is

$$e = \frac{3}{16}(14.5 - 0.28) + 0.28 = 2.95 \text{ in. (3 in.)}$$

The actual eccentricity at 8 ft from the end section is

$$e = \frac{1}{2}(14.5 - 0.28) + 0.28 = 7.39 \text{ in. (7.4 in.)}$$

5. Limited values of F_i:
   The value of F_i used in the above calculations is $F_i = 365.9$ k. Check minimum F_i by equation 18.31:

$$\text{Min. } F_i = \frac{1}{(K_b + K_t)}\left[\left(\frac{1}{\eta} - 1\right)M_D + \frac{M_L}{\eta} - \frac{(f_t AK_t)}{\eta} - (f_{ti}AK_b)\right]$$

$$= \frac{1}{(9.4 + 8.6)}\left[\left(\frac{1}{0.843} - 1\right)1341 + \frac{6912}{0.843}\right.$$

$$\left. - \frac{(0.424 \times 372 \times 8.6)}{0.843} - (0.19 \times 372 \times 9.4)\right]$$

$$= 343.1 \text{ k}$$

which is less than f_i used.

Check maximum F_i by equation 18.32:

$$\text{Max. } F_i = \frac{1}{(K_b + K_t)}\left[\left(1 - \frac{1}{\eta}\right)M_D - \frac{M_L}{\eta} + \frac{(f_c A K_b)}{\eta} + (f_{ci} A K_t)\right]$$

$$= \frac{1}{18}\left[\left(1 - \frac{1}{0.843}\right)1341 - \frac{6912}{0.843} + \frac{(2.25 \times 372 \times 9.4)}{0.843}\right.$$

$$\left. + (2.4 \times 372 \times 8.6)\right]$$

$$= 475.7 \text{ k}$$

which is greater than F_i used. Therefore, the critical section at midspan is adequate.

6. Check prestress losses, recalling that $F_o = 175$ ksi and $A_{ps} = 2.178$ in.^{2}

$$\text{Total } F_o = 2.178 \times 175 = 381 \text{ k}$$

$$E_c = 4000 \text{ ksi}$$

$$n = \frac{E_s}{E_c} = \frac{29}{4.0} = 7.25; \; n \text{ can be assumed} = 7$$

$$M_D \text{ at midspan} = 1341 \text{ k·in.}$$

$$F_i = \frac{F_o + nA_{ps}f_c(\text{D.L.}) \times \frac{2}{3}}{1 + (nA_{ps})\left(\frac{1}{A} + \frac{e^2}{I}\right)} \tag{18.5}$$

The value of f_c due to the distributed dead load is multiplied by $\frac{2}{3}$ to reflect the parabolic variation of the dead load stress along the span, giving a better approximation of F_i.

- Determine the average value of e^2 as adopted in the beam. The curve representing e^2 is shown in Figure 18.10.

$$\text{Average } e^2 = \frac{1}{24}\left[\left(\frac{1}{3} \times 3 \times 9\right) + (9 \times 13)\right.$$

$$\left. + \left(\frac{1}{3} \times 13 \times 201\right) + (210 \times 8)\right]$$

$$= 111.5 \text{ in.}^2$$

$$e = 10.56 \text{ in.}$$

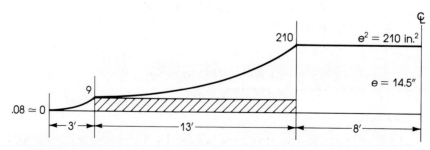

FIGURE 18.10. Average e^2, example 18.6.

The area of a parabola $= \frac{1}{3}$ area of its rectangle.

- Stress due to dead load at the level of the tendons:

$$f_c(\text{D.L.}) = \frac{1341 \times 10.56}{66,862} = 0.212 \text{ ksi}$$

Therefore

$$F_i = \frac{381 + 7(2.178) \times 0.212 \times \frac{2}{3}}{1 + (7 \times 2.178)\left(\dfrac{1}{372} + \dfrac{111.5}{66,862}\right)} = 358 \text{ k}$$

$$\text{Elastic loss} = 381 - 358 = 23 \text{ k}$$
$$= 6.1 \text{ percent}$$

This value is greater than the assumed elastic loss of 4 percent.

$$\text{Elastic loss per unit steel area} = \frac{23}{2.178} = 10.6 \text{ ksi}$$

$$F_i \text{ per unit steel area} = \frac{358}{2.178} = 164.4 \text{ ksi}$$

- Time-dependent losses:

$$\text{Loss due to shrinkage} = 8.7 \text{ ksi (as before)}$$

Loss due to creep:

$$\text{Elastic strain} = \frac{F_i}{A_c E_c} = \frac{358}{372 \times 4000} = 0.240 \times 10^{-3}$$

$$\Delta f_s = C_c(\varepsilon_{cr} E_s)$$

Let $C_c = 1.5$. Then

$$\Delta f_s = 1.5(0.24 \times 10^{-3} \times 29,000) = 10.4 \text{ ksi}$$

$$\text{Percent loss} = \frac{10.4}{164.4} = 6.3 \text{ percent}$$

Loss due to relaxation of steel $= 7$ ksi (as before). Time-dependent losses equal $8.7 + 10.4 + 7 = 26.1$ ksi, for a percentage loss of $26.1/164.4 = 15.8$ percent, which is very close to the previously estimated value of 15.7 percent.

$$F = \eta F_i = (1 - 0.158)F_i = 0.842 F_i$$
$$\eta = 0.842 \qquad \blacksquare$$

18.11   Strength Design of Flexural Members

18.11.1   General

In the previous section, it was emphasized that the stresses at the top and bottom fibers of the critical sections of a prestressed concrete member must not exceed the allowable stresses for all cases or stages of loading. In addition to these conditions, a prestressed concrete member must be designed with an adequate factor of safety against failure. The ACI Code requires that the moment due to the factored service loads M_u must not exceed ϕM_n, the flexural strength of the designed cross-section.

For the case of an under-reinforced prestressed concrete beam, failure begins when the steel stress exceeds the yield strength of steel used in the concrete section. The high-tensile prestressing steel will not exhibit a definite yield point like that of the ordinary mild steel bars used in reinforced concrete. But under additional increments of load, the strain in the steel increases at an accelerated rate, and failure occurs when the maximum compressive strain in the concrete reaches a value of 0.003 (Figure 18.11).

18.11.2 Rectangular Sections

The nominal ultimate moment capacity of a rectangular section may be determined as follows (refer to Figure 18.11):

$$M_n = C\left(d - \frac{a}{2}\right) = T\left(d - \frac{a}{2}\right) \tag{18.34}$$

where $T = A_{ps}f_{ps}$ and $C = 0.85f'_c ab$.

For $C = T$,

$$a = \frac{A_{ps}f_{ps}}{0.85f'_c b} = \frac{\rho_p f_{ps}}{0.85f'_c}d \tag{18.35}$$

where the prestressing steel ratio $\rho_p = A_{ps}/bd$, and A_{ps} and f_{ps} refer to the area and tensile stress of the prestressing steel. Let

$$\omega_p = \rho_p\left(\frac{f_{ps}}{f'_c}\right)$$

Then

$$a = \frac{\omega_p}{0.85}d \tag{18.36}$$

The quantity ω_p is a direct measure of the force in the tendon. To ensure an under-reinforced behavior, the **ACI Code, section 18.8.1,** specifies that ω_p must not exceed $0.36\beta_1$, where $\beta_1 = 0.85$ for $f'_c \leq 4$ ksi and reduces by 0.05 for each 1 ksi greater than 4 ksi **(ACI Code, section 10.2.7.3)**. M_n can also be written as follows:

$$M_n = A_{ps}f_{ps}\left(d - \frac{a}{2}\right)$$

$$M_n = A_{ps}f_{ps}d\left(1 - \frac{\rho_p f_{ps}}{1.7f'_c}\right) \tag{18.37}$$

$$M_n = A_{ps}f_{ps}d\left(1 - \frac{\omega_p}{1.7}\right) \tag{18.38}$$

and $M_u = \phi M_n$.

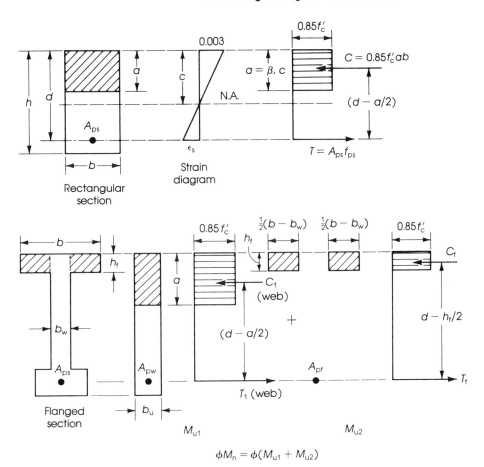

FIGURE 18.11. Ultimate moment capacity of prestressed concrete beams.

In the above equations, f_{ps} indicates the stress in the prestressing steel at failure. The actual value of f_{ps} may not be easily determined. Therefore, the **ACI Code, section 18.7.2**, permits f_{ps} to be evaluated as follows (all stresses are in psi). For *bonded* tendons,

$$f_{ps} = f_{pu} \left[1 - \frac{\gamma_p}{\beta_1} \left(\rho_p \times \frac{f_{pu}}{f'_c} \right) \right] \tag{18.39}$$

For *unbonded* tendons in members with a span-to-depth ratio ≤ 35,

$$f_{ps} = \left(f_{se} + 10{,}000 + \frac{f'_c}{100\rho_p} \right) \quad \textbf{(ACI Code equation 18.4)} \tag{18.40}$$

provided that $f_{se} \geq 0.5 f_{pu}$ and that f_{ps} for unbonded tendons must not exceed either f_{py} or $(f_{se} + 60{,}000)$ psi. For *unbonded* tendons in members with a span-to-depth ratio >35,

$$f_{ps} = (f_{se} + 10{,}000 + f'_c/300\rho_p) \tag{18.41}$$

but not greater than f_{py} or $(f_{se} + 30{,}000)$ psi

where

γ_p = factor for the type of prestressing tendon
  = 0.4 for f_{py}/f_{pu} not less than 0.85
  = 0.28 for f_{py}/f_{pu} not less than 0.9
f_{pu} = ultimate tensile strength of prestressing steel
f_{se} = effective stress in prestressing steel after all losses
f_{py} = specified yield strength of prestressing steel

In the event that $\omega_p > 0.36\beta_1$, an over-reinforced prestressed concrete beam may develop (refer to curve f, Figure 18.3). To ensure a ductile failure, the ACI Code limits ω_p to a maximum value of $0.36\beta_1$. For $\omega_p = 0.36\beta_1$, $a = 0.424\beta_1 d$ (from equation 18.36). Substituting this value of a in equation 18.38,

$$M_n = A_{ps}f_{ps}d\left(1 - \frac{0.36\beta_1}{1.7}\right)$$
$$= (\rho_p bd)f_{ps}d(1 - 0.21\beta_1)$$
$$= \omega_p f'_c(1 - 0.21\beta_1)bd^2$$
$$= (0.36\beta_1 - 0.0756\beta_1^2)f'_c bd^2 \tag{18.42}$$

For $f'_c = 5$ ksi, $\beta_1 = 0.8$. Then

$$M_n = 0.24f'_c bd^2 = 1.2bd^2$$

Similarly, for $f'_c = 4$ ksi, $M_n = bd^2$ and for $f'_c = 6$ ksi, $M_n = 1365bd^2$.

18.11.3   Flanged Sections

For flanged sections (T- or I-sections), if the neutral axis lies within the flange, it will be treated as a rectangular section. If the neutral axis lies within the web, then the web may be treated as a rectangular section using the web width b_w, and the excess flange width $(b - b_w)$ will be treated similar to that of reinforced concrete T-sections discussed in Chapters 3 and 4. The ultimate moment capacity of a flanged section can be calculated as follows (see Figure 18.11):

$M_n = M_{n1}$ (moment capacity of the web) + M_{n2} (moment capacity of excess flange)

$$M_n = A_{pw}f_{ps}\left(d - \frac{a}{2}\right) + A_{pf}f_{ps}\left(d - \frac{h_f}{2}\right) \tag{18.43}$$

$$M_u = \phi M_n \quad \text{and} \quad a = \frac{A_{pw}f_{ps}}{0.85f'_c b_w}$$

where

$A_{pw} = A_{ps} - A_{pf}$
$A_{pf} = [0.85f'_c(b - b_w)h_f]/f_{ps}$
h_f = thickness of the flange

Note that the total prestressed steel A_{ps} is divided into two parts, with A_{pw} developing the web moment capacity and A_{pf} developing the excess flange moment capacity. For flanged sections, the reinforcement index ω_{pw} must not exceed $0.36\beta_1$ where

$$\omega_{pw} = \left(\frac{A_{pw}}{b_w d}\right)\left(\frac{f_{ps}}{f'_c}\right) = \text{prestressed web steel ratio} \times \left(\frac{f_{ps}}{f'_c}\right)$$

18.11.4   Nonprestressed Reinforcement

In some cases, nonprestressed reinforcing bars (A_s) are placed in the tension zone of a prestressed concrete flexural member together with the prestressing steel (A_{ps}) to increase the ultimate moment capacity of the beam. In this case, the total steel ($A_{ps} + A_s$) is considered in the moment analysis. For rectangular sections containing prestressed and nonprestressed steel, the design moment strength ϕM_n may be computed as follows ($\phi = 0.9$):

$$M_n = A_{ps}f_{ps}\left(d_p - \frac{a}{2}\right) + A_s f_y\left(d - \frac{a}{2}\right) \tag{18.44}$$

where

$$a = \frac{A_{ps}f_{ps} + A_s f_y}{0.85 f'_c b}$$

d_p and d = distance from extreme compression fibers to the centroid of the prestressed and nonprestressed steels, respectively. For flanged sections,

$$M_n = A_{pw}f_{ps}\left(d_p - \frac{a}{2}\right) + A_s f_y\left(d - \frac{a}{2}\right) + A_{pf}f_{ps}\left(d - \frac{h_f}{2}\right) \tag{18.45}$$

where

$$A_{pw} = A_{ps} - A_{pf}$$

and

$$a = \frac{A_{pw}f_{ps} + A_s f_y}{0.85 f'_c b_w}$$

For rectangular sections with compression reinforcement,

$$M_n = A_{ps}f_{ps}\left(d_p - \frac{a}{2}\right) + A_s f_y\left(d - \frac{a}{2}\right) + A'_s f_y(d - d') \tag{18.46}$$

where

$$a = \frac{A_{ps}f_{ps} + A_s f_y - A'_s f_y}{0.85 f'_c b}$$

This equation is valid only if compression steel yields. The condition for compression steel to yield is

$$\left(\frac{A_{ps}f_{ps} + A_s f_y - A'_s f_y}{bd}\right) \geq 0.85\beta_1 \frac{f'_c}{f_y}\frac{d'}{d}\left(\frac{87}{87 - f_y}\right)$$

If this condition is not met, than compression steel does not yield. In this case, A_s' may be neglected (let $A_s' = 0$), or alternatively, the stress in A_s' may be determined by general analysis, as explained in Chapter 3.

When prestressed and nonprestressed reinforcement are used in the same section, equation 18.39 should read as follows:

$$f_{ps} = f_{pu} \left[1 - \frac{\gamma_p}{\beta_1} \left\{ \rho_p \frac{f_{pu}}{f_c'} + \frac{d}{d_p} (\omega - \omega') \right\} \right]$$

(ACI Code equation 18.3)   (18.47)

If any compression reinforcement is taken into account when calculating f_{ps}, the term

$$\rho_p \frac{f_{pu}}{f_c'} + \frac{d}{d_p} (\omega - \omega')$$

must be ≥ 0.17 and d' must be $\leq 0.15 d_p$, where d, d', and d_p = distance from the extreme compression fibers to the centroid of the nonprestressed tension steel, compression steel, and prestressed reinforcement, respectively,

γ_p = factor for type of prestressing tendon

   = 0.40 for f_{py}/f_{pu} not less than 0.85

   = 0.28 for f_{py}/f_{pu} not less than 0.90

β_1 = 0.85 for $f_c' \leq 4$ ksi less 0.05 for each 1 ksi increase in f_c' but $\beta_1 \geq 0.65$

$\omega = \rho f_y / f_c'$ and $\omega' = \rho' f_y / f_c'$

For rectangular sections, the **ACI Code, section 18.8,** limits the reinforcement ratio as follows:

$$\omega_p + \frac{d}{d_p} \omega \leq 0.36 \beta_1$$

where

$\omega_p = \rho_p (f_{ps}/f_c')$ and $\rho_p = A_{ps}/bd$ (prestressed steel)

$\omega = \rho (f_y/f_c')$ and $\rho = A_s/bd$ (non-prestressed steel)

If ordinary reinforcing bars A_s' are used in the compression zone, then the condition becomes

$$\omega_p + \frac{d}{d_p} (\omega - \omega') \leq 0.36 \beta_1$$

where $\omega' = \rho'(f_y/f_c')$ and $\rho' = A_s'/bd$. This reinforcement limitation is necessary to ensure a plastic failure of under-reinforced concrete beams.

For flanged sections, the steel area required to develop the strength of the web (A_{pw}) is used to check the reinforcement index.

$$\omega_{pw} \text{ (web)} = \rho_{pw} \left(\frac{f_{ps}}{f_c'} \right) \leq 0.36 \beta_1$$

where $\rho_{pw} = A_{pw}/b_w d$.

If nonprestressed reinforcement is used then the reinforcement limitations are

$$\omega_{pw} + \frac{d}{d_p}(\omega_w - \omega_w') \le 0.36\beta_1$$

where

$$\omega_w \text{ and } \omega_w' = \frac{A_s}{b_w d}\left(\frac{f_y}{f_c'}\right) \quad \text{and} \quad \frac{A_s'}{b_w d}\left(\frac{f_y}{f_c'}\right)$$

respectively. When compression steel A_s' is not used, then $\omega_w' = 0$. The above reinforcement conditions must be met in the analysis and design of partially prestressed concrete members.

18.12 Cracking Moment

Cracks may develop in a prestressed concrete beam when the tensile stress at the extreme fibers of the critical section equals or exceeds the modulus of rupture of concrete f_r. The value of f_r for normal weight concrete may be assumed equal to $7.5\sqrt{f_c'}$.

The stress at the bottom fibers of a simply supported beam produced by the prestressing force and the cracking moment is

$$\sigma_b = -\frac{F}{A} - \frac{(Fe)y_b}{I} + \frac{M_{cr}y_b}{I}$$

When $\sigma_b = f_r = 7.5\sqrt{f_c'}$, then the cracking moment

$$M_{cr} = \frac{I}{y_b}\left(7.5\sqrt{f_c'} + \frac{F}{A} + \frac{(Fe)y_b}{I}\right) \tag{18.48}$$

The maximum tensile stress allowed by the ACI Code after all losses is $6\sqrt{f_c'}$, which represents $0.8f_r$. In this case, prestressed concrete beams may remain uncracked at service loads. To ensure adequate strength against cracking, the **ACI Code, section 18.8.3**, requires that the ultimate moment capacity of the member ϕM_n be at least 1.2 times the cracking moment M_{cr}.

EXAMPLE 18.7

For the beam of example 18.6, check the ultimate strength and cracking moment against the ACI Code requirements.

Solution   1. Check if the neutral axis lies within the flange.

$$a = \frac{A_{ps}f_{ps}}{0.85f_c'b} \tag{18.35}$$

A_{ps} (of 20 tendons $\frac{7}{16}$ in. diameter) = 2.178 in.^{2}

Let $f_{py}/f_{pu} = 0.85$, $\gamma_p = 0.4$, and $\gamma_p/\beta_1 = 0.4/0.8 = 0.5$. For bonded tendons,

$$f_{ps} = f_{pu}\left(1 - \frac{\gamma_p}{\beta_1}\,\rho_p \times \frac{f_{pu}}{f_c'}\right) \tag{18.39}$$

$d = 40 - 6.3 = 33.7$ in.

$$\rho_p = \frac{A_{ps}}{bd} = \frac{2.178}{18 \times 33.7} = 0.00359$$

Given

$f_{pu} = 250$ ksi,

$$f_{ps} = 250\left[1 - 0.5(0.00359) \times \frac{250}{5}\right] = 228 \text{ ksi}$$

$$a = \frac{2.178 \times 228}{0.85 \times 5 \times 18} = 6.5 \text{ in.}$$

which is greater than 6 in. Therefore, the section acts as a flanged section.
2. For flanged sections,

$$M_n = A_{pw}f_{ps}\left(d - \frac{a}{2}\right) + A_{pf}f_{ps}\left(d - \frac{h_f}{2}\right) \tag{18.43}$$

where A_{pw} (web) $= A_{ps} - A_{pf}$ (flange).

$$A_{pf} = \frac{1}{f_{ps}}\,[0.85f_c'(b - b_w)h_f]$$

$$= \frac{1}{228}\,[0.85 \times 5(18 - 6)6] = 1.342 \text{ in.}^2$$

$A_{pw} = 2.178 - 1.342 = 0.836$ in.^{2}

$$a = \frac{A_{pw}f_{ps}}{0.85f_c'b_w} = \frac{0.836(228)}{0.85 \times 5 \times 6} = 7.5 \text{ in.}$$

$$M_n = 0.836(228)\left(33.7 - \frac{7.5}{2}\right) + 1.342 \times 228\left(33.7 - \frac{6}{2}\right)$$

$$= 15{,}102 \text{ k·in.} = 1258.5 \text{ k·ft}$$

$$M_u = \phi M_n = 0.9(1258.5) = 1132.7 \text{ k·ft}$$

Check the reinforcement index for the flanged section:

$$\rho_{pw} \text{ (web)} = \frac{A_{pw}}{b_w d} = \frac{0.836}{6 \times 33.7} = 0.00413$$

$$\omega_{pw} \text{ (web)} = \rho_{pw}\,\frac{f_{ps}}{f_c'} \le 0.36\beta_1 \le 0.36 \times 0.8 \le 0.288$$

($\beta_1 = 0.8$ for $f_c' = 5$ ksi)

$$\omega_{pw} = 0.00413\,\frac{(228)}{5} = 0.188 < 0.288$$

3. Calculate the external ultimate moment due to dead and live loads.

$$\text{Dead load} = \text{self-weight} + \text{additional dead load}$$
$$= 0.388 + 0.9 = 1.29 \text{ k/ft}$$
$$\text{Live load} = 1.1 \text{ k/ft}$$
$$U = 1.4D + 1.7L$$
$$M_u = \frac{(48)^2}{8}[1.4(1.29) + 1.7(1.1)] = 1058.7 \text{ k·ft}$$

This external moment is less than the ultimate moment capacity of the section of 1132.7 k·ft; therefore the section is adequate.

4. The cracking moment (equation 18.48) is

$$M_{cr} = \frac{I}{y_b}\left(7.5\sqrt{f'_c} + \frac{F}{A} + (Fe)\frac{y_b}{I}\right)$$

From the previous example, $F = 306.2$ k, $A = 372$ in.^{2}, $e = 14.5$ in., $y_b = 20.8$ in., $I = 66,862$ in.^{4}, $f'_c = 5$ ksi, and $7.5\sqrt{f'_c} = 7.5\sqrt{5000} = 530$ psi.

$$M_{cr} = \frac{66,862}{20.8}\left[0.53 + \frac{306.2}{372} + \frac{(306.2)(14.5)(20.8)}{66,862}\right]$$
$$= 8790 \text{ k·in.} = 732.5 \text{ k·ft}$$

Check that $1.2M_{cr} \le \phi M_n$:

$$1.2M_{cr} = 1.2 \times 732.5 = 879 \text{ k·ft}$$

This value is less than $\phi M_n = 1132.7$ k·ft; thus the beam is adequate against cracking. ■

18.13 Deflection

Deflection of a point in a beam is the total movement of the point, either downward or upward, due to the application of load on that beam. In a simply supported prestressed concrete beam, the prestressing force is usually applied below the centroid of the section, causing an upward deflection called camber. The self-weight of the beam and any external gravity loads acting on the beam will cause a downward deflection. The net deflection will be the algebraic sum of both deflections.

In computing deflections, it is important to consider both the short-term or immediate deflection and the long-term deflection. To ensure that the structure remains serviceable, the maximum short- and long-term deflections at all critical stages of loading must not exceed the limiting values specified by the ACI Code (see Chapter 6, section 6.3).

The deflection of a prestressed concrete member may be calculated by standard deflection equations or by the conventional methods given in books on structural analysis. For example, the midspan deflection of a simply supported beam subjected to a uniform gravity load w is equal to $(5wL^4/384EI)$. The modulus of elasticity of concrete $E_c = 33w^{1.5}\sqrt{f'_c} = 57,000\sqrt{f'_c}$ for normal-weight concrete.

The moment of inertia of the concrete section I is calculated based on the properties of the gross section for an uncracked beam. This case is appropriate when the maximum tensile stress in the concrete extreme fibers does not exceed the modulus of rupture of concrete $f_r = 7.5\sqrt{f_c'}$, or more conservatively, $6\sqrt{f_c'}$, as recommended by the ACI Code. When the maximum tensile stress based on the properties of the gross section exceeds $6\sqrt{f_c'}$, the effective moment of inertia I_e based on the cracked and uncracked sections must be used (as explained in Chapter 6, equation 6.5). Typical midspan deflections for simply supported beams due to gravity loads and prestressing forces are shown in Table 18.3.

EXAMPLE **18.8**

For the beam of example 18.6, calculate the camber at transfer and then calculate the final anticipated immediate deflection at service load.

Solution   1. Deflection at transfer:

- Calculate the downward deflection due to dead load at transfer, self weight in this case. For a simply supported beam subjected to a uniform load,

$$\Delta_D \text{ (dead load)} = \frac{5wL^4}{384EI}$$

From example 18.6, $W_D = 388$ lb/ft, $L = 48$ ft, $E_{ci} = 3600$ ksi, and $I = 66,862$ in.^{4}.

$$\Delta_D = \frac{5\left(\frac{0.388}{12}\right)(48 \times 12)^4}{384(3600)(66,862)} = +0.192 \text{ in. (downward)}$$

- Calculate the camber due to the prestressing force. For a simply supported beam harped at one-third points with the eccentricity $e_1 = 14.5$ in. at the middle third and $e_2 = 0$ at the ends,

$$\Delta_p = \frac{23(F_i e_1)L^2}{216 E_{ci} I} \qquad\qquad (Table\ 18.3)$$

$$= \frac{23(365.9 \times 14.5)(48 \times 12)^2}{216(3600)(66,862)} = -0.779 \text{ in. (upward)}$$

- Final camber at transfer $= -0.779 + 0.192 = -0.587$ in. (upward).

2. Deflection at service load:
The total uniform service load

$$W_T = 0.388 + 0.9 + 1.1 = 2.388 \text{ k/ft} \qquad E_c = 4000 \text{ ksi}$$

The downward deflection due to W_T

$$\Delta_w = \frac{5W_T L^4}{384 E_c I}$$

$$= \frac{5\left(\frac{2.388}{12}\right)(48 \times 12)^4}{384(4000)(66,862)} = +1.067 \text{ in. (downward)}$$

TABLE 18.3. Midspan deflections of simply supported beams

Schematic	Deflection equations

Camber due to prestressing force

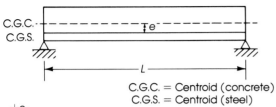

$$\Delta = \frac{(Fe)L^2}{8EI} \quad (1)$$

(Horizontal tendons)

C.G.C. = Centroid (concrete)
C.G.S. = Centroid (steel)

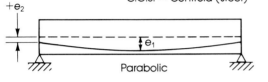

Parabolic

$$\Delta = \frac{FL^2}{8EI}\left[\frac{5}{6}e_1 + \frac{1}{6}e_2\right] \quad (2)$$

When $e_2 = 0$:

$$\Delta = \frac{5(Fe_1)L^2}{48EI} \quad (3)$$

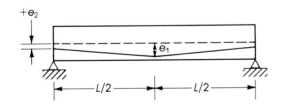

$$\Delta = \frac{FL^2}{8EI}\left[e_1 + \frac{4}{3}\left(\frac{a}{L}\right)^2(e_2 - e_1)\right] \quad (4)$$

When $a = \frac{L}{3}$:

$$\Delta = \frac{FL^2}{8EI}\left[e_1 + \frac{4}{27}(e_2 - e_1)\right] \quad (5)$$

When $a = \frac{L}{3}$ and $e_2 = 0$:

$$\Delta = \frac{23(Fe_1)L^2}{216EI} \quad (6)$$

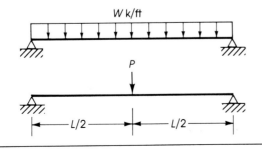

$$\Delta = \frac{FL^2}{24EI}[2e_1 + e_2] \quad (7)$$

When $e_2 = 0$:

$$\Delta = \frac{(Fe_1)L^2}{12EI} \quad (8)$$

Deflection due to gravity loads

W k/ft

$$\Delta = \frac{5wL^4}{384EI} \quad (9)$$

P

$$\Delta = \frac{PL^3}{48EI} \quad (10)$$

The camber due to prestressing force $F = 306.2$ k and $E_c = 4000$ ksi

$$\Delta_p = \frac{23(306.2 \times 14.5)(48 \times 12)^2}{216(4000)(66,862)} = -0.587 \text{ in. (upward)}$$

The final immediate deflection at service load

$$\Delta = \Delta_w - \Delta_p = 1.067 - 0.587 = +0.48 \text{ in. (downward)} \qquad \blacksquare$$

18.14   Design for Shear

The design approach to determine the shear reinforcement in a prestressed concrete beam is almost identical to that used for reinforced concrete beams. Shear cracks are assumed to develop at 45° measured from the axis of the beam. In general, two types of shear-related cracks form. One type is due to a combined effect of flexure and shear; the cracks start as flexural cracks and then deviate and propagate at an inclined direction due to the effect of diagonal tension. The second type, web-shear cracking, occurs in beams with narrow webs when the magnitude of principal tensile stress is high in comparison to flexural stress. Stirrups must be used to resist the principal tensile stresses in both cases. The ACI design criteria for shear will be adopted here.

18.14.1   Basic Approach

The ACI design approach is based on ultimate strength requirements using the load factors mentioned in Chapter 3. When the factored shear force V_u exceeds half the nominal shear strength $(\phi V_c)/2$, shear reinforcement must be provided. The required design shear force V_u at each section must not exceed the nominal design strength ϕV_n of the cross section based on the combined nominal shear capacity of concrete and web reinforcement:

$$V_u \leq \phi V_n \leq \phi(V_c + V_s) \qquad (18.49)$$

where

V_c = nominal shear capacity of concrete
V_s = nominal shear capacity of reinforcement
ϕ = capacity reduction factor = 0.85

When the factored shear force V_u is less than ϕV_c, minimum shear reinforcement is required.

18.14.2   Shear Strength Provided by Concrete

The **ACI Code, section 11.4,** presents a simple empirical expression to estimate the nominal ultimate shear capacity of a prestressed concrete member in which the tendons have an effective prestress f_{se} at least 40 percent of the ultimate tensile strength f_{pu}:

$$V_c = \left(0.6\sqrt{f_c'} + 700\frac{V_u d}{M_u}\right)b_w d \qquad \textbf{(ACI Code equation 11.10)} \qquad (18.50)$$

where

V_u and M_u = factored shear and moment at the section under consideration

b_w = width of web

d $\left(\text{in the term } \dfrac{V_u d}{M_u}\right)$ = the distance from the compression fibers to the centroid of the prestressing steel

d (in V_{ci} or V_{cw} equations) = the larger of the above d or $0.8h$ **(ACI Code, section 11.4.2)**

The use of equation 18.50 is limited to the following conditions:

- The quantity $\dfrac{V_u d}{M_u} \leq 1.0$ (to account for small values of V_u and M_u)
- $V_c \geq (2\sqrt{f'_c})b_w d$ (minimum V_c)
- $V_c \leq (5\sqrt{f'_c})b_w d$ (maximum V_c)   **(ACI Code, section 11.4.1)**

The variation of the concrete shear capacity for a simply supported prestressed concrete beam subjected to a uniform load is shown in Figure 18.12. Note that the maximum shear reinforcement may be required near the supports and near one-fourth of the span where ϕV_s reaches maximum values. In contrast, similar reinforced concrete beams require maximum shear reinforcement (or minimum spacings) only near the support where maximum ϕV_s develops.

The values of V_c calculated by equation 18.50 may be conservative sometimes, therefore the **ACI Code, section 11.4.2,** gives an alternative approach to calculate V_c that takes into consideration the additional strength of concrete in the section. In this approach, V_c is taken as the smaller of two calculated values of the concrete shear strength V_{ci} and V_{cw} (Figure 18.12). Both are explained below.

The shear strength V_{ci} is based on the assumption that flexural-shear cracking occurs near the interior extremity of a flexural crack at approximately a distance $d/2$ from the load point in the direction of decreasing moment. The ACI Code specifies that V_{ci} be computed as follows:

$$V_{ci} = (0.6\sqrt{f'_c})b_w d + \left(V_d + \frac{V_i M_{cr}}{M_{max}}\right) \quad \textbf{(ACI Code equation 11.11)} \quad (18.51)$$

but not less than $(1.7\sqrt{f'_c})b_w d$ where

V_d = shear force at section due to unfactored dead load

V_i = factored shear force at section due to externally applied loads occurring simultaneously with M_{max}

M_{max} = maximum factored moment at section due to externally applied loads

M_{cr} = cracking moment

The cracking moment can be determined from the following expression:

$$M_{cr} = \frac{I}{y_t}(6\sqrt{f'_c} + f_{pe} - f_d) \quad \textbf{(ACI Code equation 11.12)} \quad (18.52)$$

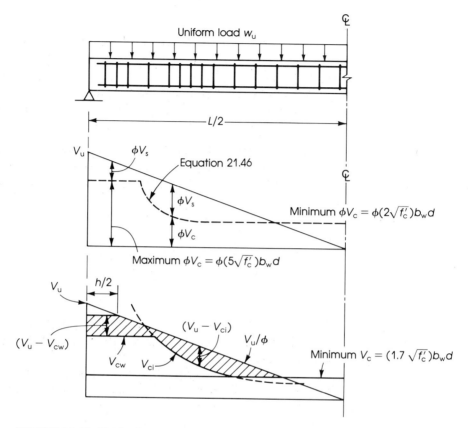

FIGURE 18.12. Distribution of shear forces along span. Middle diagram shows shear capacity of a simply supported prestressed concrete beam. Bottom diagram shows ACI analysis (stirrups are required for shaded areas).

where

I = moment of inertia of the section resisting external factored loads

y_t = distance from the centroidal axis of the gross section neglecting reinforcement to the extreme fiber in tension

f_{pe} = compressive strength at the extreme fibers of the concrete section due to the effective prestress force after all losses

f_d = stress due to the unfactored dead load at the extreme fiber where tensile stress is caused by external loads

The web-shear strength V_{cw} is based on shear cracking in a beam that has not cracked by flexure. Such cracks develop near the supports of beams with narrow webs. The **ACI Code, section 11.4.2.2,** specifies that V_{cw} be computed as follows:

$$V_{cw} = (3.5\sqrt{f'_c} + 0.3f_{pc})b_w d + V_p \quad \textbf{(ACI Code equation 11.13)} \quad (18.53)$$

where

V_p = vertical component of the effective prestress force at the section considered

f_{pc} = compressive stress (psi) in the concrete (after allowance for prestress losses) at the centroid of the section resisting the applied loads, or at the junction of the web and flange when the centroid lies within the flange

Alternatively, V_{cw} may be determined as the shear force that produces a principal tensile stress of $4\sqrt{f'_c}$ at the centroidal axis of the member, or at the intersection of the flange and web when the centroid lies within the flange. The equation for the principal stresses may be expressed as follows:

$$f_t = 4\sqrt{f'_c} = \sqrt{v_{cw}^2 + \left(\frac{f_{pc}}{2}\right)^2} - \frac{f_{pc}}{2}$$

or

$$V_{cw} = f_t \left(\sqrt{1 + \frac{f_{pc}}{f_t}}\right) b_w d \tag{18.54}$$

where $f_t = 4\sqrt{f'_c}$. When applying equations 18.51 and 18.53 or 18.54, the value of d is taken as the distance between the compression fibers and the centroid of the prestressing tendons but not less than $0.8h$.

The critical section for maximum shear is to be taken at $h/2$ from the face of the support. The same shear reinforcement must be used at sections between the support and the section at $h/2$.

18.14.3 Shear Reinforcement

The value of V_s must be calculated to determine the required area of shear reinforcement.

$$V_u = \phi(V_c + V_s) \tag{18.49}$$

or

$$V_s = \frac{1}{\phi}(V_u - \phi V_c) \tag{18.55}$$

For vertical stirrups,

$$V_s = \frac{A_v f_y d}{s} \tag{18.56}$$

and

$$A_v = \frac{V_s s}{f_y d} \quad \text{or} \quad s = \frac{A_v f_y d}{V_s} \tag{18.57}$$

where A_v = area of vertical stirrups and s = spacing of stirrups. Equations for inclined stirrups are the same as those discussed in Chapter 8.

18.14.4 Limitations

1. Maximum spacing s_{max} of stirrups must not exceed $\frac{3}{4}h$ or 24 in. If V_s exceeds $(4\sqrt{f'_c})b_w d$, the maximum spacing must be reduced to half the above values **(ACI Code, section 11.5.4)**.
2. Maximum shear V_s must not exceed $(8\sqrt{f'_c})b_w d$, otherwise increase the dimensions of the section **(ACI Code, section 11.5.6.8)**.
3. Minimum shear reinforcement A_v required by the ACI Code

$$A_v(\min) = \frac{50 b_w s}{f_y} \quad \textbf{(ACI Code equation 11.14)} \qquad (18.58)$$

When the effective prestress $f_{pe} \geq 0.4 f_{pu}$, the minimum A_v

$$A_v = \frac{A_{ps}}{80} \times \frac{f_{pu}}{f_y} \times \frac{s}{d} \times \sqrt{\frac{d}{b_w}} \quad \textbf{(ACI Code equation 11.15)} \qquad (18.59)$$

The effective depth d need not be taken less than $0.8h$. Generally, equation 18.59 requires greater minimum shear reinforcement than equation 18.58.

EXAMPLE 18.9

For the beam of example 18.6, determine the nominal shear strength and the necessary shear reinforcement. Check the sections at $h/2$ and 10 ft from the end of the beam. Use $f_y = 40$ ksi for the shear reinforcement.

Solution

1. For the section at $h/2$:

$$\frac{h}{2} = \frac{40}{2} = 20 \text{ in.} = 1.67 \text{ ft from the end}$$

2. Ultimate uniform load on beam,

$$W_u = 1.4(0.388 + 0.9) + 1.7 \times 1.1 = 3.68 \text{ k/ft}$$

$$V_u \text{ at a distance } \frac{h}{2} = 3.68(24 - 1.67) = 82.2 \text{ k}$$

Using the simplified ACI method (equation 18.50), determine M_u at section $h/2$.

$$M_u = (3.68 \times 24) \times 1.67 - 3.68 \frac{(1.67)^2}{2} = 142.4 \text{ k·ft} = 1708 \text{ k·in.}$$

d at section $h/2$ from the end (Figure 18.9) is

$$d = 33.7 \text{ (at midspan)} - \frac{(16 - 1.67)}{16} \times 14.5 = 20.7 \text{ in.}$$

$$\frac{V_u d}{M_u} = \frac{82.2 \times 20.7}{1708} = 0.996 \leq 1.0$$

as required by the ACI Code.

$$V_c = \left(0.6\sqrt{f_c'} + 700\,\frac{V_u d}{M_u}\right) b_w d$$

$$= (0.6\sqrt{5000} + 700 \times 0.996)6 \times 20.7 = 91{,}800\ \text{lb} = 91.8\ \text{k}$$

$$\text{Minimum } V_c = 2\sqrt{f_c'}\,b_w d = 2\sqrt{5000} \times 6 \times 20.7 = 17.6\ \text{k}$$

$$\text{Maximum } V_c = 5\sqrt{f_c'}\,b_w d = 43.9\ \text{k}$$

The maximum V_c of 43.9 k controls.

3. The alternative approach presented by the ACI Code is that V_c may be taken as the smaller value of V_{ci} and V_{cw}.

   • Based on the flexural-shear cracking strength,

$$V_{ci} = (0.6\sqrt{f_c'})b_w d + \left(V_d + \frac{V_i M_{cr}}{M_{max}}\right) \qquad (18.51)$$

   Calculate each item separately:

$$(0.6\sqrt{f_c'})b_w d = 0.6\sqrt{5000} \times 6 \times 20.7 = 5.3\ \text{k}$$

$$V_d = \text{unfactored dead load shear} = 1.288(24 - 1.67) = 28.8\ \text{k}$$

   M_{max} = maximum factored moment at section (except for weight of beam)

   Factored load = $1.4 \times 0.9 + 1.7 \times 1.1 = 3.13$ k/ft

$$M_{max} = 3.13\left[24 \times 1.67 - \frac{(1.67)^2}{2}\right] = 121\ \text{k·ft} = 1453\ \text{k·in.}$$

$$V_i = 3.13(24 - 1.67) = 69.9\ \text{k}$$

$$M_{cr} = \frac{I}{y_t}\left(6\sqrt{f_c'} + f_{pe} - f_d\right) \qquad (18.52)$$

$$I = 66{,}862\ \text{in.}^4 \quad y_b = 20.8\ \text{in.}$$

   f_{pe} = compressive stress due to prestressing force

$$= \frac{F}{A} + \frac{F e y_b}{I}$$

$$= \frac{306.2}{372} + \frac{306.2(1.5)(20.8)}{66{,}862} = 0.966\ \text{ksi}$$

$$f_d = \text{Dead load stress} = \frac{M_D y_b}{I}$$

$$M_D = (1.288)\left[24 \times 1.67 - \frac{(1.67)^2}{2}\right] = 49.8\ \text{k·ft} = 598\ \text{k·in.}$$

$$f_d = \frac{598 \times 20.8}{66{,}862} = 0.186\ \text{ksi}$$

$$M_{cr} = \frac{66{,}862}{20.8}(6\sqrt{5000} + 966 - 186) = 3871\ \text{k·in.}$$

Therefore,

$$V_{ci} = (5.3) + (28.8) + 69.9 \left(\frac{3871}{1453}\right) = 220.3 \text{ k}$$

V_{ci} must not be less than $(1.7\sqrt{f'_c})b_wd = (1.7\sqrt{5000}) \times 6 \times 20.7 = 14.9$ k.
- Shear strength based on web-shear cracking:

$$V_{cw} = (3.5\sqrt{f'_c} + 0.3f_{pc})b_wd + V_p \qquad (18.53)$$

$$f_{pc} = \frac{306.2}{372} = 0.823 \text{ ksi}$$

$$d = 20.7 \text{ in.} \quad \text{or} \quad 0.8h = 0.8 \times 40 = 32 \text{ in.}$$

Use $d = 32$ in.

$$V_p = 306.2 \times \frac{1}{13.2} = 23.2 \text{ k}$$

where $1/13.2$ = slope of tendon profile = 14.5 in./(16 × 12).

$$3.5\sqrt{f'_c} = 3.5\sqrt{5000} = 248 \text{ psi}$$

Therefore,

$$V_{cw} = (0.248 + 0.3 \times 0.823) \times 6 \times 32 + 23.2 = 118.2 \text{ k}$$

- Since $V_{cw} < V_{ci}$, the value $V_{cw} = 118.2$ k represents the nominal shear strength at section $h/2$ from the end of the beam. In most cases, V_{cw} controls at $h/2$ from the support.

4. Web reinforcement:

$$V_u = 82.2 \text{ k} \qquad \phi V_{cw} = 0.85 \times 118.2 = 100.5 \text{ k}$$

Since $V_u < \phi V_{cw}$, $V_s = 0$ and therefore use minimum stirrups. Use No. 3 stirrups, $A_v = 2 \times 0.11 = 0.22$ in.^{2}. Maximum spacing is the least of

$$s_1 = \frac{3}{4}h = \frac{3}{4} \times 40 = 30 \text{ in.} \qquad s_2 = 24 \text{ in.}$$

Calculate s_3 from the equation of minimum web reinforcement:

$$\text{Min. } A_v = \frac{A_{ps}}{80} \times \frac{f_{pu}}{f_y} \times \frac{s}{d} \times \sqrt{\frac{d}{b_w}} \qquad (18.59)$$

$$0.22 = \frac{2.178}{80} \times \frac{250}{40} \times \frac{s_3}{20.7} \sqrt{\frac{20.7}{6}}$$

$$s_3 = 14.4 \text{ in. (14 in.)}$$

Also,

$$\text{Min. } A_v = \frac{50b_w s}{f_y}$$

or

$$s_4 = \frac{A_v f_y}{50b_w} = \frac{0.22 \times 40,000}{50 \times 6} = 29 \text{ in.}$$

$s_{max} = s_3 = 14$ in. controls. Thus, use No. 3 stirrups spaced at 14 in.

5. For the section at 10 ft from the end, the calculation procedure is similar to that for the section at $h/2$. Using the ACI simplified method:

$$V_u = 3.68(24 - 10) = 51.5 \text{ k}$$

$$M_u = 3.68 \left[24 \times 10 - \frac{(10)^2}{2} \right] = 699.2 \text{ k·ft} = 8390 \text{ k·in.}$$

$$d = 33.7 \text{ (at midspan)} - \frac{(16 - 10)}{16} \times 14.5 = 28.3 \text{ in.}$$

$$\frac{V_u d}{M_u} = \frac{51.5 \times 28.3}{8390} = 0.174 < 1.0$$

$$V_c = (0.6\sqrt{5000} + 0.174 \times 700)6 \times 28.3 = 27{,}886 \text{ lb} = 27.9 \text{ k (controls)}$$

Minimum $V_c = 17.6$ k, maximum $V_c = 43.9$ k.

6. Using the ACI Code equations to compute V_{ci} and V_{cw}, calculate V_{ci} first (which controls at this section):

$$0.6\sqrt{f_c'} b_w d = 0.6\sqrt{5000} \times 6 \times 28.3 = 7.2 \text{ k}$$

$$V_d = 1.288(24 - 10) = 18 \text{ k}$$

$$M_{max} = 3.13 \left[24 \times 10 - \frac{(10)^2}{2} \right] = 594.7 \text{ k·ft} = 7136 \text{ k·in.}$$

$$V_i = 3.13(24 - 10) = 43.8 \text{ k}$$

$$f_{pe} = \frac{306.2}{372} + \frac{306.2(9.1)(20.8)}{66{,}862} = 1.69 \text{ ksi}$$

$$M_D = 1.288 \left[24 \times 10 - \frac{(10)^2}{2} \right] = 244.7 \text{ k·ft} = 2937 \text{ k·in.}$$

$$f_d = \frac{2937 \times 20.8}{66{,}862} = 0.914 \text{ ksi} \qquad M_{cr} = 3858 \text{ k·in.}$$

Therefore,

$$V_{ci} = 7.2 + 18 + \frac{43.8(3858)}{7136} = 48.9 \text{ k}$$

$$V_{ci}(\text{min}) = (1.7\sqrt{5000})6 \times 28.3 = 20.4 \text{ k}$$

Thus the minimum is met. Then calculate V_{cw}:

$$f_{pc} = 0.823 \text{ ksi} \qquad V_p = 23.2 \text{ k (as before)}$$
$$d = 28.3 \text{ in.} \quad \text{or} \quad 0.8h = 32 \text{ in.}$$

Use $d = 32$ in.

$$V_{cw} = (3.5\sqrt{f_c'} + 0.3f_{pc})b_w d + V_p$$
$$= (0.248 + 0.3 \times 0.823)6 \times 32 + 23.2 = 118.2 \text{ k}$$

This value of V_{cw} is not critical. At about Span/4, the critical shear value is V_{ci} (Figure 18.12).

7. Web reinforcement:

$$V_u = 51.5 \text{ k} \qquad \phi V_{ci} = 0.85 \times 48.9 = 41.5 \text{ k}$$

$$V_u = \phi(V_c + V_s)$$

$$V_s = \frac{1}{0.85}(51.5 - 41.5) = 10 \text{ k}$$

Use No. 3 stirrups, $A_v = 0.22$ in.^{2}. Check maximum spacing: $s_{max} = 14$ in. (as before).

$$\text{Required } A_v = \frac{V_s s}{f_y d} = \frac{10 \times 14}{40 \times 28.3} = 0.124 \text{ in.}^2$$

$$A_v \text{ used} = 0.22 \text{ in.}^2 > 0.124 \text{ in.}^2$$

Therefore, use No. 3 stirrups spaced at 14 in.   ■

18.15   Preliminary Design of Prestressed Concrete Flexural Members

18.15.1   Shapes and Dimensions

The detailed design of prestressed concrete members often involves a considerable amount of computation. A good guess at the dimensions of the section can result in a saving of time and effort. Hence it is important to ensure, by preliminary design, that the dimensions are reasonable before starting the detailed design.

At the preliminary design stage, some data are usually available to help choose proper dimensions. For example, the bending moments due to the applied external loads, the permissible stresses, and the data for assessing the losses are already known or calculated.

The shape of the cross-section of a prestressed concrete member may be a rectangular, T-, I-, or box section. The total depth of the section h may be limited by headroom considerations or may not be specified. Given the freedom of selection, an empirical practical choice of dimensions for a preliminary design is as follows (Figure 18.13):

1. Total depth of section $h = 1/20$ to $1/30$ of the span L: for heavy loading $h = L/20$, for light loading $h = L/30$ or $h = 2\sqrt{M_D + M_L}$, where M is in k·ft.
2. Depth of top flange $h_f = h/8$ to $h/6$.
3. Width of top flange $b \geq 2h/5$.
4. Thickness of web $b_w \geq 4$ in. Usually b_w is taken as $h/30 + 4$ in.
5. b_w and t are chosen to accommodate and uniformly distribute the prestressing tendons, keeping appropriate concrete cover protection.
6. Approximate area of the concrete section required

$$A_c \text{ (ft}^2) = \frac{M_D + M_L}{30h}$$

where $M_D + M_L$ are in k·ft and h is in feet.

In S.I. units:

$$A_c \text{ (m}^2) = \frac{M_D + M_L}{1450h} \qquad (M_D + M_L \text{ in kN·m and } h \text{ in m})$$

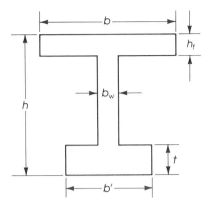

FIGURE 18.13. Proportioning prestressed concrete sections.

For practical and economical design of prestressed concrete beams and floor slabs, the precast concrete industry has introduced a large number of standardized shapes and dimensions from which the designer can choose an adequate member. Tables of standard sections are available in the PCI Design Handbook.[25] AASHTO[23] has also presented standard girders to be used in bridge construction. Tables 18.4 to 18.6 give properties of some standard shapes, while Tables 18.7 and 18.8 give properties of general types of T- and I-sections. Z in the above tables indicates the section modulus; $Z_b = I/c_b$, and $Z_t = I/c_t$.

18.15.2   Prestressing Force and Steel Area

Once the shape, depth, and other dimensions of the cross-section have been selected, approximate values of the prestressing force and the area of the prestressing steel A_{ps} can be determined. From the internal couple concept, the total moment M_T due to the service dead and live loads is equal to the tension force T times the moment arm jd.

TABLE 18.4. Properties of rectangular sections

b (in.)	h (in.)	A (in.^{2})	I (in.^{4})	y_b (in.)	Z (in.^{3})
12	16	192	4,096	8	512
12	20	240	8,000	10	800
12	24	288	13,824	12	1152
12	28	336	21,952	14	1568
12	32	384	32,768	16	2048
12	36	432	46,656	18	2592
16	24	384	18,432	12	1536
16	28	448	29,269	14	2091
16	32	512	43,691	16	2731
16	36	576	62,208	18	3456
16	40	640	85,333	20	4267

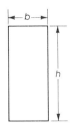

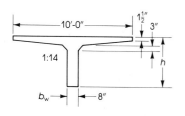

TABLE 18.5. Properties of single T-sections, normal weight concrete[25]

h (in.)	A (in.^{2})	I (in.^{4})	y_b (in.)	Z_b (in.^{3})	Z_t (in.^{3})	Weight (lb/ft)
24	590	22,914	18.73	1223	4,348	615
28	622	36,005	21.67	1662	5,688	648
32	654	53,095	24.51	2166	7,089	681
36	686	74,577	27.27	2735	8,543	715
40	718	100,819	29.97	3364	10,052	748
44	750	132,171	32.61	4053	11,604	781
48	782	168,968	35.19	4802	13,190	815

TABLE 18.6. Properties of double T-sections, normal weight concrete[25]

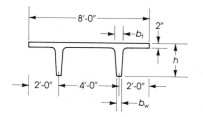

h (in.)	b_t/b_w (in.)	A (in.^{2})	I (in.^{4})	y_b (in.)	Z_b (in.^{3})	Z_t (in.^{3})	Weight (lb/ft)
16	8.00/6.00	388	8,944	11.13	804	1837	404
18	9.75/7.75	472	14,623	11.84	1235	2374	492
20	8.00/5.50	435	16,117	13.72	1175	2566	453
20	9.75/7.50	503	19,354	13.06	1482	2789	523
24	8.00/5.00	478	25,686	16.33	1573	3349	498
24	9.75/7.00	560	31,192	15.51	2011	3674	583
32	8.00/4.00	549	51,286	21.71	2362	4984	572
32	9.75/6.00	665	64,775	20.47	3164	5618	692

$$M_T = T(jd) = C(jd)$$

$$M_T = A_{ps}f_{se}(jd) \qquad A_{ps} = \frac{M_T}{f_{se}(jd)}$$

where A_{ps} is the area of the prestressing steel and f_{se} is the effective prestressing stress after all losses. The value of the moment arm jd varies from $0.4h$ to $0.8h$, with a practical range of $0.6h$ to $0.7h$. An average value of 0.65 may be used. Therefore,

$$A_{ps} = \frac{M_T}{(0.65h)f_{se}} \qquad\qquad (18.60)$$

and the prestressing force

$$F = T = A_{ps}f_{se} = \frac{M_T}{0.65h} \qquad\qquad (18.61)$$

The prestressing force at transfer

$$F_i = \frac{F}{\eta}$$

where η = factor of time-dependent losses.

TABLE 18.7. Properties of T-sections and I-sections

b_w/b	t/h	A	c_b	c_t	I	K_t	K_b
				T-sections			
0.1	0.1	0.19	0.714	0.286	0.0179	0.132	0.333
0.1	0.2	0.28	0.756	0.244	0.0192	0.0910	0.282
0.1	0.3	0.37	0.755	0.245	0.0193	0.0689	0.212
0.1	0.4	0.46	0.735	0.265	0.0202	0.0597	0.165
0.2	0.1	0.28	0.629	0.371	0.0283	0.161	0.272
0.2	0.2	0.36	0.678	0.322	0.0315	0.129	0.272
0.2	0.3	0.44	0.691	0.309	0.0319	0.105	0.234
0.2	0.4	0.52	0.684	0.316	0.0316	0.090	0.195
0.3	0.1	0.37	0.585	0.415	0.0365	0.169	0.237
0.3	0.2	0.44	0.626	0.374	0.0408	0.148	0.248
0.3	0.3	0.51	0.645	0.355	0.0417	0.127	0.231
0.3	0.4	0.58	0.645	0.355	0.0417	0.112	0.203
0.4	0.1	0.46	0.559	0.441	0.0440	0.171	0.216
0.4	0.2	0.52	0.592	0.408	0.0486	0.158	0.229
0.4	0.3	0.58	0.609	0.391	0.0499	0.141	0.220
0.4	0.4	0.64	0.612	0.388	0.0502	0.128	0.205
1.0	1.0	1.00	0.500	0.500	0.0833	0.167	0.167
				I-sections			
0.1	0.1	0.23	0.597	0.403	0.0326	0.238	0.352
0.1	0.2	0.36	0.611	0.389	0.0464	0.210	0.331
0.1	0.3	0.49	0.606	0.394	0.0535	0.180	0.274
0.2	0.1	0.31	0.572	0.428	0.0373	0.210	0.282
0.2	0.2	0.42	0.595	0.405	0.0488	0.195	0.286
0.2	0.3	0.53	0.599	0.401	0.0540	0.170	0.254
0.3	0.1	0.39	0.557	0.430	0.0443	0.198	0.250
0.3	0.2	0.48	0.582	0.418	0.0510	0.183	0.255
0.3	0.3	0.57	0.592	0.408	0.0553	0.164	0.238
Multiplier		bh	h	h	bh^3	h	h

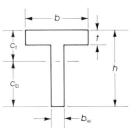

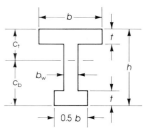

TABLE 18.8. AASHTO girders, normal weight concrete[25]

Designation	A (in.²)	I (in.⁴)	y_b (in.)	Z_b (in.³)	Z_t (in.³)	Weight (lb/ft)
Type II	369	50,979	15.83	3,220	2527	384
Type III	560	125,390	20.27	6,186	5070	593
Type IV	789	260,741	24.73	10,544	8908	822

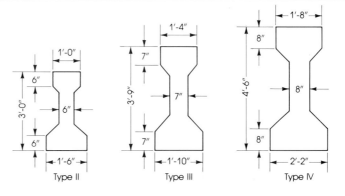

The compressive force on the section C is equal to the tension force T:

$$C = T = A_{ps}f_{se}$$

In terms of stresses,

$$\frac{C}{A_c} = \frac{A_{ps}f_{se}}{A_c} = f_{c1}$$

where f_{c1} is an assumed uniform stress on the section.

For preliminary design, a triangular stress distribution is assumed with maximum allowable compressive stress f_{ca} on one extreme fiber; therefore, the average stress $= 0.5f_{ca} = f_{c1}$. The allowable compressive stress in concrete $f_{ca} = 0.45f'_c$. Thus, the required concrete area A_c can be estimated from the force T as follows:

$$A_c = \frac{T}{f_{c1}} = \frac{A_{ps}f_{se}}{f_{c1}} = \frac{A_{ps}f_{se}}{0.5f_{ca}} = \frac{A_{ps}f_{se}}{0.225f'_c} \qquad (18.62)$$

or

$$A_c = \frac{T}{0.5f_{ca}} = \frac{M_T}{(0.65h)(0.5f_{ca})} = \frac{M_T}{0.33f_{ca}} = \frac{M_T}{0.15f'_c} \qquad (18.63)$$

The above analysis is based on the design for service loads and not for the factored loads. The eccentricity e is measured from the centroid of the section to the centroid of the prestressing steel and can be estimated approximately as follows:

$$e = K_b + \frac{M_D}{F_i} \qquad (18.64)$$

where K_b = the lower Kern limit and M_D = moment due to the service dead load.

18.16   End Block Stresses

18.16.1   Pretensioned Members

Much as a specific development length is required in every bar of a reinforced concrete beam, the prestressing force in a prestressed concrete beam must be transferred to the concrete by embedment or end anchorage or a combination thereof. In pretensioned members, the distance over which the effective prestressing force is transferred to the concrete is called the transfer length l_t. After transfer, the stress in the tendons at the extreme end of the member is equal to zero, while the stress at a distance l_t from the end is equal to the effective prestress f_{pe}. The transfer length l_t depends on the size and type of the tendon, surface condition, concrete strength f'_c, stress, and method of force transfer. A practical estimation of l_t ranges between 50 and 100 times the tendon diameter. For strands, a practical value of l_t is equal to 50 tendon diameters while for single wires, l_t is equal to 100 wire diameters.

In order that the tension in the prestressing steel develops full ultimate flexural strength, a bond length is required. The purpose is to prevent general slip before the failure of the beam at its full design strength. The development length l_d is equal to the bond length plus the transfer length l_t. Based on established tests, the **ACI Code, section 12.9.1,** gives the following expression for computing the development length of 3- or 7-wire pretensioning strands:

$$l_d \text{ (in.)} = \left(f_{ps} - \frac{2}{3} f_{se}\right) d_b \qquad (18.65)$$

where

f_{ps} = stresses in prestressed reinforcement at nominal strength (ksi)
f_{se} = effective stress in prestressed reinforcement after all losses (ksi)
d_b = nominal diameter of wire or strand (in.)

In pretensioned members, high tensile stresses exist at the end zones, for which special reinforcement must be provided. Such reinforcement in the form of vertical stirrups is uniformly distributed within a distance $h/5$ measured from the end of the beam. The first stirrup is usually placed at 1 to 3 in. from the beam end or as close as possible. It is a common practice to add nominal reinforcement for a distance d measured from the end of the beam. The area of the vertical stirrups A_v to be used at the end zone can be calculated approximately from the following expression:

$$A_v = 0.021 \frac{F_i h}{f_{sa} l_t} \qquad (18.66)$$

where f_{sa} = allowable stress in the stirrups (usually 20 ksi) and l_t = 50 tendon diameters.

EXAMPLE 18.10

Determine the necessary stirrup reinforcement required at the end zone of the beam given in example 18.6.

Solution F_i = 365.9 k, h = 40 in., f_s = 20 ksi, and $l_t = 50 \times \dfrac{7}{16}$ = 22 in. Therefore

$$A_v = 0.021 \times \frac{365.9 \times 40}{20 \times 22} = 0.7 \text{ in.}^2$$

$$\frac{h}{5} = \frac{40}{5} = 8 \text{ in.}$$

Use 4 No. 3 closed stirrups within the first 8 in. distance from the support. A_v(provided) = $4 \times 0.22 = 0.88$ in.^{2}. ∎

18.16.2 Posttensioned Members

In posttensioned concrete members, the prestressing force is transferred from the tendons to the concrete, for both bonded and unbonded tendons, at the ends of the member by special anchorage devices. Within an anchorage zone at the end of the member, very high compressive stresses and transverse tensile stresses develop, as shown in Figure 18.14. In practice, it is found that the length of the anchorage zone does not exceed the depth of the end of the member; nevertheless, the state of stress within this zone is extremely complex.

The stress distribution due to one tendon within the anchorage zone is shown in Figure 18.15. At a distance h from the end section, the stress distribution is assumed uniform all over the section. Considering the lines of force (trajectories) as individual elements acting as curved struts, the trajectories tend to deflect laterally toward the centerline of the beam in Zone A, inducing compressive stresses. In Zone B, the curvature is reversed in direction and the struts deflect outward, inducing tensile stresses. In Zone C, struts are approximately straight, inducing uniform stress distribution.

The reinforcement required for the end anchorage zones of posttensioned members generally consists of a closely spaced grid of vertical and horizontal bars throughout the length of the end block to resist the bursting and tensile stresses. It is a common practice to space the bars not more than 3 in. in each direction and to place the bars not more than 1.5 in. from the inside face of the bearing plate. Approximate design methods are presented in references 24 to 27.

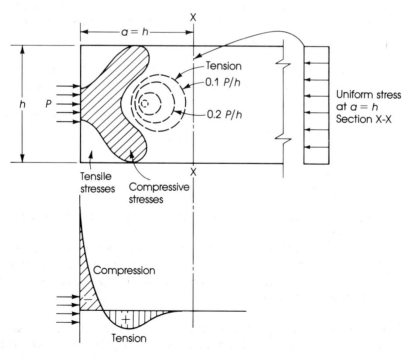

FIGURE 18.14. Tension and compression zones in a posttensioned member.

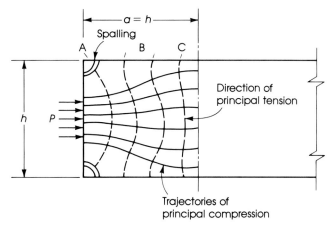

FIGURE 18.15. Tension and compression trajectories in a posttensioned member.

SUMMARY

SECTIONS 18.1 and 18.2

The main objective of prestressing is to offset or counteract all or most of the tensile stresses in a structural member produced by external loadings, thus giving some advantages over a reinforced concrete member.

SECTIONS 18.3 to 18.6

A concrete member may be pretensioned or posttensioned. Nonprestressed reinforcement may also be added to the concrete member to increase its ultimate strength. In this case, cracks may develop under working loads, and the member is referred to as partially prestressed. The range of partial prestressing is explained in Figure 18.3.

SECTION 18.7

(a) The allowable stresses in concrete at transfer are:

Maximum compressive stress $= 0.6f_{ci}$

Maximum tensile stress $= 3\sqrt{f_{ci}}$

The allowable stresses after all losses are $0.45f'_c$ for compression and $6\sqrt{f'_c}$ for tension.

(b) The allowable stress in a pretensioned tendon at transfer is the smaller of $0.74f_{pu}$ or $0.82f_{py}$. The maximum stress due to the tendon jacking force must not exceed $0.85f_{pu}$ or $0.94f_{py}$; and the maximum stress in a posttensioned tendon after the tendon is anchored is $0.70f_{pu}$.

SECTION 18.8

The sources of prestress loss are the elastic shortening, shrinkage, and creep of concrete; relaxation of steel tendons; and friction. An approximate lump sum loss is 35 ksi for pretensioned members and 25 ksi for posttensioned members (friction is not included).

Loss due to elastic shortening $= nF_i/A_c$ $\qquad$ (18.1)

Loss due to shrinkage $= \varepsilon_{sh}E_s$ $\qquad$ (18.6)

Loss due to creep $= C_c(\varepsilon_c E_s)$ $\qquad$ (18.7)

Loss due to relaxation of steel varies between 2.5 and 12 percent. Loss due to friction in posttensioned members stems from the curvature and wobbling of the tendon.

$$P_s = P_x e^{(\mu\alpha + Klx)} \qquad (18.10)$$

$$P_s = P_x(1 + \mu\alpha + Klx) \qquad (18.11)$$

SECTION 18.9

Three different concepts may be used to analyze a prestressed concrete beam: the homogeneous beam, internal couple, and balancing concepts.

SECTION 18.10

Elastic stresses in a flexural member due to loaded and unloaded conditions are given by equations 18.13 through 18.16. The limiting values of the eccentricity e are given by equations 18.20, 18.22, 18.24, and 18.26. The minimum and maximum values of F_i are given by equations 18.31 and 18.32, respectively.

SECTION 18.11

The nominal ultimate strength of a rectangular prestressed concrete member is:

$$M_n = T\left(d - \frac{a}{2}\right) = A_{ps}f_{ps}d\left(1 - \frac{\rho_p f_{ps}}{1.7f_c'}\right) \qquad (18.37)$$

The values of f_{ps} are given by equations 18.39 and 18.40. For flanged sections:

$$M_n = A_{pw}f_{ps}\left(d - \frac{a}{2}\right) + A_{pf}f_{ps}\left(d - \frac{h_f}{2}\right) \qquad (18.43)$$

If nonprestressed reinforcement is used in the flexural member, then

$$M_n = A_{ps}f_{ps}\left(d_p - \frac{a}{2}\right) + A_s f_y\left(d - \frac{a}{2}\right) \qquad (18.44)$$

where $a = (A_{ps}f_{ps} + A_s f_y)/0.85f_c'b$. For M_n of flanged and rectangular sections with compression reinforcement, refer to equations 18.46 and 18.47, respectively.

SECTIONS 18.12 and 18.13

(a) The cracking moment is:

$$M_{cr} = \frac{I}{y_b}\left[7.5\sqrt{f_c'} + \frac{F}{A} + \frac{(Fe)y_b}{I}\right] \qquad (18.48)$$

(b) Midspan deflections of simply supported beams are summarized in Table 18.3.

SECTION 18.14

Shear strength of concrete $V_c = \left(0.6\sqrt{f_c'} + 700\dfrac{V_u d}{M_u}\right)b_w d.$ $\qquad$ (18.50)

Minimum $V_c = 2\sqrt{f_c'}b_w d$

Maximum $V_c = 5\sqrt{f_c'}b_w d$

The shear strength V_{ci}, based on flexural shear, is given by equation 18.51, and the web-shear strength V_{cw} is given by equation 18.53

$$V_s = \frac{1}{\phi}(V_u - \phi V_c) \quad \text{and} \quad A_v = \frac{A_{ps}}{80}\frac{f_{pu}}{f_y}\frac{S}{d}\frac{d}{b_w}\sqrt{\frac{d}{b_w}} \qquad (18.59)$$

SECTION 18.15 Empirical practical dimensions for the preliminary design of prestressed concrete members are suggested in this section.

SECTION 18.16 The development length of 3 to 7 wire strands is:

$$l_d = \left(f_{ps} - \frac{2}{3}f_{se}\right)d_b \qquad (18.65)$$

The area of stirrups in an end block is $A_v = 0.021\left(\dfrac{F_i h}{f_{sa}l_t}\right)$ $\qquad (18.66)$

REFERENCES

1. American Concrete Institute: *Building Code Requirements for Reinforced Concrete (318-83)*, Detroit, 1983.
2. P. W. Abeles: "Design of Partially Prestressed Concrete Beams," *ACI Journal*, 64, October 1967.
3. American Concrete Institute: *Commentary on the ACI Code*, Detroit, 1983.
4. E. Freyssinet: "A Revolution in the Technique of Utilization of Concrete," *Structural Engineer*, 14, May 1936.
5. H. Straub: *History of Civil Engineering*, Leonard Hill, London, 1952.
6. P. W. Abeles and L. Czuprynski: "Partial Prestressing, Its History, Research, Application and Future Development," Convention on Partial Prestressing, Brussels, October 1965. *Annales des Travaux Publics de Belgique*, 2, April 1966.
7. P. W. Abeles: "Fully and Partly Prestressed Reinforced Concrete," *ACI Journal*, 41, January 1945.
8. F. Leonherdt: "Prestressed Concrete Design and Construction," Wilhelm Ernest and Son, Berlin, 1964.
9. P. W. Abeles: *Introduction to Prestressed Concrete*, Vols. I and II, Concrete Publications, London, 1966.
10. *S.T.U.P. Gazette:* "The Death of Mr. Freyssinet," Vol. 11, 1962.
11. P. W. Abeles: "Partial Prestressing and Its Possibilities for Its Practical Application," *PCI Journal*, 4, No. 1, June 1959.
12. W. H. Hewett: "A New Method of Constructing Reinforced Concrete Tanks," *ACI Journal*, 19, 1932.
13. R. E. Dill: "Some Experience with Prestressed Steel in Small Concrete Units," *ACI Journal*, 38, November 1941.
14. F. W. Dodge: *The Structures of Eduardo Torroja*, McGraw-Hill, New York, 1959.
15. E. Freyssinet: "The Birth of Prestressing." *Cement and Concrete Association Translation No. 59*, London, 1956.
16. G. Mangel: *Prestressed Concrete*, Concrete Publications, London, 1954.
17. W. Zerna: "Partially Prestressed Concrete" (in German), *Beton Stahlbeton* No. 12, 1956.

18. R. H. Evans: "Research and Development in Prestressing," *Journal of the Institution of Civil Engineers,* 35, February 1951.

19. E. Freyssinet: "Prestressed Concrete, Principles and Applications," *Journal of the Institution of Civil Engineers,* 4, February 1950.

20. P. J. Verna: "Economics of Prestressed Concrete in Relation to Regulations, Safety, Partial Prestressing . . . in the U.S.A.," *Fourth Congress of the FIP, Rome-Naples,* Theme III, paper 1, 1962.

21. S. Chaikes: "The Reinforced Prestressed Concrete," *Fourth Congress of the FIP, Rome-Naples,* Theme III, paper 2, 1962.

22. ACI Committee 435: "Deflection of Prestressed Concrete Members," *ACI Journal,* 60, December 1963.

23. American Association of Highway and Transportation Officials: *AASHTO Specifications for Bridges,* Washington, D.C., 1975.

24. Posttensioning Institute: *Posttensioning Manual,* Phoenix, 1976.

25. Prestressed Concrete Institute: *PCI Design Handbook,* Chicago, 1980.

26. P. Gergley and M. A. Sozen: "Design of Anchorage Zone Reinforcement in Prestressed Concrete Beams," *PCI Journal,* 12, April 1967.

27. Y. Guyon: *Prestressed Concrete,* Wiley and Sons, New York, 1960.

28. P. Zia, H. K. Peterson, N. L. Scott, and E. B. Workman: "Estimating Prestress Losses," *Concrete International,* 1, June 1979.

PROBLEMS

18.1. A 60-ft-span simply supported pretensioned beam has the section shown in Figure 18.16. The beam is prestressed by a force F_i = 395 k at transfer (after the elastic loss). The prestress force after all losses is F = 320 k, f'_{ci} (compressive stress at transfer) = 4 ksi, and f'_c (compressive stress after all losses) = 6 ksi. For the midspan section and using the ACI Code allowable stresses, (1) calculate the extreme fiber stresses due to the prestressing force plus dead load and (2) calculate the allowable uniform live load on the beam.

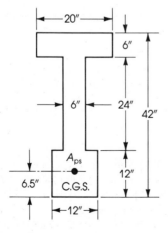

FIGURE 18.16. Problem 18.1.

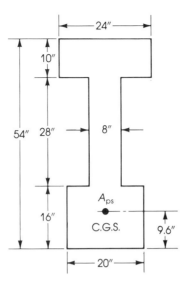

FIGURE 18.17. Problem 18.3.

18.2. For the beam of problem 18.1 (Figure 18.16), calculate the elastic loss and all time-dependent losses using the following data: $F_i = 405$ k, $A_{ps} = 2.39$ in.^{2} located at 6.5 in. from the base, $f'_{ci} = 4$ ksi, $f'_c = 6$ ksi, $E_c = 57,000 \sqrt{f'_c}$, and $E_s = 28,000$ ksi. The profile of the tendon is parabolic and the eccentricity at the supports $= 0$.

18.3. The cross-section of a 56-ft-span simply supported posttensioned girder which is prestressed by 30 cables $\frac{7}{16}$ in. in diameter (area of one cable $= 0.1089$ in.^{2}) is shown in Figure 18.17. The cables are made of 7-wire stress-relieved strands. The profile of the cables is parabolic with the centroid of the prestressing steel (C.G.S.) located at 9.6 in. from the base at the midspan section and located at the centroid of the concrete section ($e = 0$) at the ends. Calculate the elastic loss of prestress and all other losses. Given: $f'_{ci} = 4$ ksi, $f'_c = 6$ ksi, $E_c = 57,000 \sqrt{f'_c}$, $E_s = 28,000$ ksi, $f_{pu} = 250$ ksi, $F_o = 175$ ksi, D.L. $= 1.0$ k/ft (excluding self-weight), and L.L. $= 1.6$ k/ft.

18.4. For the girder of problem 18.3:
  (a) Determine the location of the upper and lower limits of the tendon profile for the section at midspan and for at least two other sections between midspan and support. (Choose sections at 12 ft, 18 ft, and 25 ft from midspan section.)
  (b) Check if the parabolic profile satisfies these limits.

18.5. For the girder of problem 18.3, check the limiting values of the prestressing force at transfer F_i.

18.6. A 64-ft-span simply supported pretensioned girder has the section shown in Figure 18.18. The loads on the girder consist of a dead load $= 1.2$ k/ft (excluding own weight) that will be applied at a later stage and a live load $= 0.6$ k/ft. The prestressing steel consists of 24 cables $\frac{1}{2}$ in. diameter (area of one cable $= 0.144$ in.^{2}) with $E_s = 28,000$ ksi, $F_o = 175$ ksi, and $f_{pu} = 250$ ksi. The strands are made of 7-wire stress-relieved steel. The concrete compressive strength at transfer $f'_{ci} = 4$ ksi, and at 28 days $f'_c = 5$ ksi. The modulus of elasticity $E_c =$

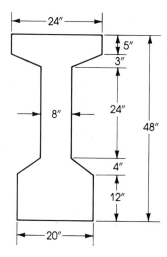

FIGURE 18.18. Problem 18.6.

$57,000\sqrt{f_c'}$. For the beam described above:

(a) Determine the upper and lower limits of the tendon profile for the section at midspan and three other sections between the midspan section and the support. (Choose sections at 3 ft, 11 ft, and 22 ft from the support.)

(b) Locate the tendons to satisfy these limits using straight horizontal tendons within the middle 20 ft. length of the span.

(c) Check the limiting values of the prestressing force at transfer.

18.7. For the girder of problem 18.6:

(a) Harp some of the tendons at one-third points and draw sections at midspan and at the end of the beam showing the distribution of tendons.

(b) Revise the prestress losses, taking into consideration the variation of the eccentricity e along the beam.

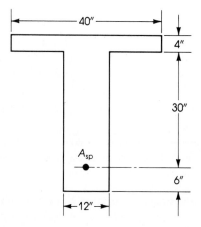

FIGURE 18.19. Problem 18.10.

(c) Check the ultimate moment capacity of the section at midspan.

(d) Determine the cracking moment.

18.8. For the girder of problem 18.6:

(a) Calculate the camber at transfer.

(b) Calculate the immediate deflection at service load.

18.9. For the girder of problem 18.6, determine the shear capacity of the section and calculate the necessary web reinforcemertt using $f_y = 60$ ksi.

18.10. Determine the nominal ultimate moment capacity M_n of a pretensioned concrete beam that has the cross-section shown in Figure 18.19. Given: $f'_c = 5$ ksi, $f_{pu} = 270$ ksi, $f_{se} = 160$ ksi, and $A_{sp} = 2.88$ in.^{2}.

APPENDIX A: Design Tables (U.S. Customary Units)

TABLE A.1. Values of R_u and a/d for $f'_c = 3$ ksi

ρ (%)	$f_y = 40$ ksi		$f_y = 50$ ksi		$f_y = 60$ ksi		$f_y = 75$ ksi	
	R_u	a/d	R_u	a/d	R_u	a/d	R_u	a/d
0.2	71	0.031	88	0.039	106	0.047	131	0.059
0.3	105	.047	131	.059	156	.071	192	.089
0.4	140	.062	173	.078	206	.094	254	.118
0.5	173	.078	214	.098	254	.118	310	.148
0.6	206	.094	254	.118	301	.141	368	.177
0.7	238	.110	293	.138	347	.165	421	.207
0.8	270	.126	332	.157	391	.189	475	.236
0.9	301	.142	369	.177	434	.213	524	.266
1.0	332	.157	406	.196	476	.238	572	.295
1.1	362	.173	441	.216	517	.260	620	.325
1.2	390	.188	476	.235	556	.282	648	.340
1.3	420	.204	510	.255	594	.306	($\rho_{max} = 1.16$)	
1.4	450	.220	543	.274	631	.330		
1.5	476	.236	575	.294	667	.353		
1.6	504	.252	607	.314	700	.376		
1.7	530	.267	637	.334	702	.378		
1.8	556	.282	667	.353	($\rho_{max} = 1.61$)			
1.9	582	.298	695	.373				
2.0	607	.314	723	.392				
2.1	630	.330						
2.2	655	.345	740	.405				
2.3	677	.360	($\rho_{max} = 2.06$)					
2.4	700	.376						
2.5	723	.392						
2.6	745	.408						
2.7	767	.424						
	783	.436						
	($\rho_{max} = 2.78$)							

Note: Last values are the maximum allowed by the ACI Code.

TABLE A.2. Values of R_u and a/d for $f'_c = 4$ ksi

ρ (%)	$f_y = 40$ ksi		$f_y = 50$ ksi		$f_y = 60$ ksi		$f_y = 75$ ksi	
	R_u	a/d	R_u	a/d	R_u	a/d	R_u	a/d
0.2	71	0.024	89	0.029	106	0.035	132	0.044
0.3	106	.036	132	.044	158	.053	194	.066
0.4	140	.047	175	.059	208	.071	257	.088
0.5	175	.059	217	.074	258	.089	317	.110
0.6	208	.071	260	.088	307	.106	378	.132
0.7	242	.083	300	.103	355	.123	434	.154
0.8	274	.094	340	.118	400	.141	490	.176
0.9	307	.106	378	.132	447	.158	545	.198
1.0	340	.118	419	.147	492	.176	600	.220
1.1	370	.130	455	.161	536	.194	650	.242
1.2	400	.141	492	.176	580	.212	702	.264
1.3	432	.153	530	.191	620	.230	752	.286
1.4	462	.165	565	.206	662	.247	801	.308
1.5	492	.177	600	.221	700	.265	846	.330
1.6	522	.188	635	.236	742	.282	867	.342
1.7	550	.200	670	.250	780	.300	($\rho_{max} = 1.55$)	
1.8	580	.212	702	.265	818	.318		
1.9	607	.224	735	.280	853	.335		
2.0	635	.236	768	.294	890	.353		
2.1	662	.248	800	.309	924	.370		
2.2	690	.260	830	.323	936	.378		
2.3	717	.271	860	.338	($\rho_{max} = 2.14$)			
2.4	742	.282	890	.352				
2.5	767	.294	918	.367				
2.6	792	.306	946	.381				
2.7	817	.318	974	.396				
2.8	842	.330	988	.405				
2.9	866	.342	($\rho_{max} = 2.75$)					
3.0	890	.354						
3.1	893	.365						
3.2	935	.376						
3.3	958	.388						
3.4	980	.400						
3.5	1000	.412						
3.6	1020	.424						
3.7	1040	.436						
	1044	.438						
	($\rho_{max} = 3.72$)							

Note: Last values are the maximum allowed by the ACI Code.

TABLE A.3. Values of R_u and a/d for $f'_c = 5$ ksi

ρ (%)	$f_y = 40$ ksi		$f_y = 50$ ksi		$f_y = 60$ ksi		$f_y = 75$ ksi	
	R_u	a/d	R_u	a/d	R_u	a/d	R_u	a/d
0.2	71	0.019	89	0.024	106	0.028	132	0.035
0.3	106	.029	133	.036	159	.042	196	.052
0.4	141	.038	176	.047	210	.056	260	.070
0.5	176	.047	218	.060	260	.070	322	.088
0.6	210	.056	260	.071	310	.085	384	.106
0.7	244	.066	302	.083	360	.100	442	.123
0.8	277	.075	343	.094	408	.113	500	.141
0.9	310	.085	383	.106	455	.127	556	.159
1.0	343	.094	424	.118	502	.141	612	.177
1.1	375	.104	463	.130	550	.155	667	.195
1.2	408	.113	500	.141	593	.169	722	.212
1.3	440	.123	540	.153	637	.183	776	.230
1.4	470	.132	578	.165	681	.198	830	.247
1.5	502	.141	615	.177	724	.212	875	.265
1.6	532	.150	652	.188	766	.226	920	.282
1.7	563	.160	688	.200	808	.240	970	.300
1.8	593	.169	724	.212	848	.254	1020	.318
1.9	623	.179	760	.224	890	.268	1033	.322
2.0	652	.188	794	.235	927	.282	($\rho_{max} = 1.83$)	
2.1	681	.198	830	.247	965	.292		
2.2	710	.207	862	.259	1003	.311		
2.3	738	.217	894	.271	1040	.325		
2.4	766	.226	927	.282	1076	.339		
2.5	794	.235	958	.294	1112	.358		
2.6	821	.244	990	.306	1120	.364		
2.7	848	.254	1021	.318	($\rho_{max} = 2.52$)			
2.8	875	.263	1052	.329				
2.9	900	.272	1082	.341				
3.0	927	.282	1111	.353				
3.1	952	.292	1140	.365				
3.2	978	.301	1168	.376				
3.3	1003	.311						
3.4	1028	.320	1179	.381				
3.5	1052	.329	($\rho_{max} = 3.24$)					
3.6	1076	.338						
3.7	1100	.348						
3.8	1122	.357						
3.9	1145	.367						
4.0	1168	.376						
4.1	1190	.385						
4.2	1212	.395						
4.3	1235	.404						
	1248	.410						
	($\rho_{max} = 4.37$)							

Note: Last values are the maximum allowed by the ACI Code.

TABLE A.4. Values of ρ_{max}, $R_u(max)$, ρ_b, and ρ_{min}

| | $f_y = 40\ ksi$ | | | | $f_y = 50\ ksi$ | | | |
| | | R_u | | | | R_u | | |
f'_c (ksi)	ρ_{max} (%)	(max) (psi)	ρ_b (%)	ρ_{min} (%)	ρ_{max} (%)	(max) (psi)	ρ_b (%)	ρ_{min} (%)
2.5	2.32	652	3.09	0.5	1.72	616	2.29	0.4
3.0	2.78	783	3.71	0.5	2.06	740	2.75	0.4
4.0	3.72	1047	4.96	0.5	2.75	987	3.67	0.4
5.0	4.36	1244	5.81	0.5	3.24	1181	4.32	0.4
6.0	4.90	1424	6.53	0.5	3.64	1364	4.85	0.4

| | $f_y = 60\ ksi$ | | | | $f_y = 75\ ksi$ | | | |
| | | R_u | | | | R_u | | |
f'_c (ksi)	ρ_{max} (%)	(max) (psi)	ρ_b (%)	ρ_{min} (%)	ρ_{max} (%)	(max) (psi)	ρ_b (%)	ρ_{min} (%)
2.5	1.34	588	1.78	0.33	0.97	542	1.29	0.27
3.0	1.61	704	2.15	0.33	1.16	646	1.55	0.27
4.0	2.14	937	2.85	0.33	1.55	870	2.07	0.27
5.0	2.54	1115	3.39	0.33	1.83	1040	2.44	0.27
6.0	2.83	1281	3.77	0.33	2.06	1180	2.75	0.27

Note: $\rho_{max} = 0.64\beta_1 \dfrac{f'_c}{f_y} \times \dfrac{87}{87 + f_y}$ $R_u = \phi\rho f_y \left(1 - \dfrac{\rho f_y}{1.7f'_c}\right)$

TABLE A.5. Suggested design steel ratios ρ_s and comparison with other steel ratios

f'_c (psi)	f_y (ksi)	ρ_b (%)	ρ_{max} (%)	ρ_s (suggested) (%)	R_u for ρ_s (psi)	ρ_b elastic (%)	ρ for $q = 1.8$ (%)	Ratio ρ_s/ρ_{max}	Weight of ρ_s (lb/ft³) of concrete
2.5	40	3.09	2.32	1.2	384	1.02	1.13	0.517	6
	50	2.29	1.72	1.2	464	0.90	0.90	0.700	6
3.0	40	3.71	2.78	1.4	450	1.29	1.35	0.500	7
	50	2.75	2.06	1.2	476	1.11	1.08	0.583	6
	60	2.15	1.61	1.2	556	0.96	0.90	0.745	6
4.0	40	4.96	3.72	1.4	462	1.88	1.80	0.376	7
	50	3.67	2.75	1.4	565	1.62	1.44	0.510	7
	60	2.85	2.14	1.4	662	1.41	1.20	0.654	7
5.0	40	5.81	4.36	1.4	470	2.50	2.25	0.321	7
	50	4.32	3.24	1.4	578	2.15	1.80	0.432	7
	60	3.39	2.54	1.4	681	1.87	1.50	0.551	7

TABLE A.6. Minimum thickness of beams and one-way slabs (L = span length)

Member	Yield strength f_y (ksi)	Simply supported	One end continuous	Both ends continuous	Cantilever
Solid one-way slabs	40	$L/25$	$L/30$	$L/35$	$L/12.5$
	50	$L/22$	$L/27$	$L/31$	$L/11$
	60	$L/20$	$L/24$	$L/28$	$L/10$
Beams or ribbed one-way slabs	40	$L/20$	$L/23$	$L/26$	$L/10$
	50	$L/18$	$L/20.5$	$L/23.5$	$L/9$
	60	$L/16$	$L/18.5$	$L/21$	$L/8$

TABLE A.7. Minimum beam width (in.), using stirrups

Size of bars	Number of bars in single layer of reinforcement							Add for each added bar (in.)
	2	3	4	5	6	7	8	
No. 4	6.1	7.6	9.1	10.6	12.1	13.6	15.1	1.50
No. 5	6.3	7.9	9.6	11.2	12.8	14.4	16.1	1.63
No. 6	6.5	8.3	10.0	11.8	13.5	15.3	17.0	1.75
No. 7	6.7	8.6	10.5	12.4	14.2	16.1	18.0	1.88
No. 8	6.9	8.9	10.9	12.9	14.9	16.9	18.9	2.00
No. 9	7.3	9.5	11.8	14.0	16.3	18.6	20.8	2.26
No. 10	7.7	10.2	12.8	15.3	17.8	20.4	22.9	2.54
No. 11	8.0	10.8	13.7	16.5	19.3	22.1	24.9	2.82
No. 14	8.9	12.3	15.6	19.0	22.4	25.8	29.2	3.39
No. 18	10.5	15.0	19.5	24.0	28.6	33.1	37.6	4.51

TABLE A.8. Values of bd^2 (in.^{3})

d (in.)	Values of b (in.)											
	6	7	8	9	10	11	12	13	14	15	16	20
4	96	112	128	144	160	176	192	208	224	240	256	320
4.5	122	142	162	182	202	223	244	264	284	305	325	405
5	150	175	200	225	250	275	300	325	350	375	400	500
5.5	182	212	242	273	303	333	364	394	424	455	485	605
6	216	252	288	324	360	396	432	468	504	540	576	720
6.5	255	297	340	382	425	467	510	552	595	637	680	850
7	294	343	392	441	490	539	588	637	686	735	784	980
8	384	448	512	576	640	704	768	832	896	960	1,024	1,280
9	486	567	648	729	810	891	972	1,053	1,134	1,215	1,296	1,620
10	600	700	800	900	1,000	1,100	1,200	1,300	1,400	1,500	1,600	2,000
11	726	847	968	1,089	1,210	1,331	1,452	1,573	1,694	1,815	1,936	2,420
12	864	1,008	1,152	1,296	1,440	1,584	1,728	1,872	2,016	2,160	2,304	2,880
13	1,014	1,183	1,352	1,521	1,690	1,859	2,028	2,197	2,366	2,535	2,704	3,380
14	1,176	1,372	1,568	1,764	1,960	2,156	2,352	2,548	2,744	2,940	3,136	3,920
15	1,350	1,575	1,800	2,025	2,250	2,475	2,700	2,925	3,150	3,375	3,600	4,500
16	1,536	1,792	2,048	2,304	2,560	2,816	3,072	3,328	3,584	3,840	4,096	5,120
17	1,734	2,023	2,312	2,601	2,890	3,179	3,468	3,757	4,046	4,335	4,624	5,780
18	1,944	2,268	2,592	2,916	3,240	3,564	3,888	4,212	4,536	4,860	5,184	6,480
19	2,166	2,527	2,888	3,249	3,610	3,971	4,332	4,693	5,054	5,415	5,776	7,220
20	2,400	2,800	3,200	3,600	4,000	4,400	4,800	5,200	5,600	6,000	6,400	8,000
21	2,646	3,087	3,528	3,969	4,410	4,851	5,292	5,733	6,174	6,615	7,056	8,820
22	2,904	3,388	3,872	4,356	4,840	5,324	5,808	6,292	6,776	7,260	7,744	9,680
23	3,174	3,703	4,232	4,761	5,290	5,819	6,348	6,877	7,406	7,935	8,464	10,580
24	3,456	4,032	4,608	5,184	5,760	6,336	6,912	7,488	8,064	8,640	9,216	11,520
28	4,704	5,488	6,272	7,056	7,840	8,624	9,408	10,192	10,976	11,760	12,544	15,680
30	5,400	6,300	7,200	8,100	9,000	9,900	10,800	11,700	12,600	13,500	14,400	18,000
34	6,936	8,092	9,248	10,404	11,560	12,716	13,872	15,028	16,184	17,340	18,496	23,120
40	9,600	11,200	12,800	14,400	16,000	17,600	19,200	20,800	22,400	24,000	25,600	32,000

Note: $bd^2 = \dfrac{M_u}{R_u}\left(\dfrac{\text{lb in.}}{\text{psi}}\right)$

TABLE A.9. Minimum steel percentage $(\rho - \rho')$ for compression steel to yield in rectangular sections

$$(\rho - \rho') \geq 0.85\beta_1\frac{f'_c}{f_y} \times \frac{d'}{d}\frac{87}{(87 - f_y)}$$

d'/d	f'_c (ksi)	β_1	$f_y =$ 40 ksi	$f_y =$ 50 ksi	$f_y =$ 60 ksi	$f_y =$ 75 ksi
0.10	2.5	0.85	0.83	0.85	0.97	1.74
	3.0	0.85	1.00	1.02	1.16	2.09
	4.0	0.85	1.33	1.35	1.55	2.78
	5.0	0.80	1.57	1.59	1.81	3.27
	6.0	0.75	1.78	1.81	2.06	3.71
0.12	2.5	0.85	1.00	1.02	1.16	2.09
	3.0	0.85	1.20	1.22	1.39	2.51
	4.0	0.85	1.60	1.62	1.86	3.34
	5.0	0.80	1.88	1.91	2.17	3.92
	6.0	0.75	2.14	2.17	2.47	4.45
0.15	2.5	0.85	1.25	1.28	1.46	2.61
	3.0	0.85	1.50	1.53	1.74	3.14
	4.0	0.85	2.00	2.03	2.33	4.17
	5.0	0.80	2.36	2.39	2.72	4.91
	6.0	0.75	2.67	2.72	3.09	5.57

Note: Minimum $(\rho - \rho')$ for any value of $d'/d = 10 \times (d'/d) \times$ value shown in table with $d'/d = 0.10$.

TABLE A.10. Values of modulus of elasticity of concrete E_c (ksi)

Concrete cylinder strength f'_c (ksi)	Unit weight of concrete (lb/ft³)				
	90	100	110	125	145
2.5	1410	1650	1900	2300	2880
3.0	1540	1800	2080	2520	3150
4.0	1780	2090	2410	2920	3640
5.0	1990	2330	2690	3260	4060
6.0	2185	2560	2950	3580	4500
7.0	2360	2760	3190	3870	4800
8.0	2520	2950	3410	4130	5200

Note: $E_c = 33w^{1.5}\sqrt{f'_c}$
$E_c = 57,400 \sqrt{f'_c}$ for $w = 145$ lb/ft³ (normal weight concrete)

TABLE A.11. Development length l_d (in.)

Bar size	f_y (ksi)	$f'_c = 2.5$ psi l_d	Top bars	$f'_c = 3.0$ psi l_d	Top bars	$f'_c = 4.0$ psi l_d	Top bars	$f'_c = 5.0$ psi l_d	Top bars	$f'_c = 6.0$ psi l_d	Top bars
No. 3	40	12	12	12	12	12	12	12	12	12	12
	50	12	12	12	12	12	12	12	12	12	12
	60	12	12	12	13	12	13	12	13	12	13
No. 4	40	12	12	12	12	12	12	12	12	12	12
	50	12	14	12	14	12	14	12	14	12	14
	60	12	17	12	17	12	17	12	17	12	17
No. 5	40	12	14	12	14	12	14	12	14	12	14
	50	13	18	13	18	13	18	13	18	13	18
	60	15	21	15	21	15	21	15	21	15	21
No. 6	40	14	20	13	18	12	17	12	17	12	17
	50	18	25	16	22	15	21	15	21	15	21
	60	21	29	19	27	18	25	18	25	18	25
No. 7	40	19	27	18	25	15	21	14	20	14	20
	50	24	34	22	31	19	27	18	25	18	25
	60	29	40	26	37	23	32	21	29	21	29
No. 8	40	26	37	23	32	20	28	18	25	16	23
	50	32	44	29	40	25	35	22	31	20	29
	60	38	53	35	48	30	42	27	38	24	34
No. 9	40	32	44	29	41	25	35	23	32	21	29
	50	40	56	37	51	32	44	28	40	26	36
	60	48	67	44	61	38	53	34	48	31	43
No. 10	40	41	57	37	52	32	45	29	40	26	37
	50	51	71	46	65	40	56	36	50	33	46
	60	61	85	56	78	48	67	43	60	39	55
No. 11	40	50	70	46	64	39	55	35	49	32	45
	50	62	87	57	80	49	69	44	62	40	56
	60	75	105	68	96	59	83	53	74	48	68
No. 14	40	72	101	62	87	54	75	48	67	44	61
	50	90	126	78	109	67	94	60	84	55	77
	60	108	151	93	130	81	113	72	101	66	72
No. 18	40	128	179	80	113	70	97	62	87	57	80
	50	160	224	100	141	87	122	78	109	71	99
	60	172	241	121	169	104	146	93	131	85	119

Note: $l_d = 0.04 A_b f_y / \sqrt{f'_c}$ but $\geq 0.0004 d_b f_y$. For No. 14 bars, $l_d = 0.085 f_y / \sqrt{f'_c}$; for No. 18 bars, $l_d = 0.11 f_y / \sqrt{f'_c}$. Minimum length is 12 in. in all cases. For top bars, use $1.4 l_d$.

TABLE A.12. Designations, areas, perimeters, and weights of standard U.S. bars

Bar no.	Diameter (in.) Nominal	Diameter (in.) Actual	Cross-sectional area (in.²)	Perimeter (in.)	Unit weight per foot (lb)	Diameter (mm)	Area (mm²)
2	$\frac{1}{4}$	0.250	0.05	0.79	0.167	6.4	32
3	$\frac{3}{8}$	0.375	0.11	1.18	0.376	9.5	71
4	$\frac{1}{2}$	0.500	0.20	1.57	0.668	12.7	129
5	$\frac{5}{8}$	0.625	0.31	1.96	1.043	15.9	200
6	$\frac{3}{4}$	0.750	0.44	2.36	1.502	19.1	284
7	$\frac{7}{8}$	0.875	0.60	2.75	2.044	22.2	387
8	1	1.000	0.79	3.14	2.670	25.4	510
9	$1\frac{1}{8}$	1.128	1.00	3.54	3.400	28.7	645
10	$1\frac{1}{4}$	1.270	1.27	3.99	4.303	32.3	820
11	$1\frac{3}{8}$	1.410	1.56	4.43	5.313	35.8	1010
14	$1\frac{3}{4}$	1.693	2.25	5.32	7.650	43.0	1450
18	$2\frac{1}{4}$	2.257	4.00	7.09	13.600	57.3	2580

TABLE A.13. Areas of groups of standard U.S. bars, in square inches

Bar no.	Number of bars													
	1	2	3	4	5	6	7	8	9	10	11	12	13	14
3	0.11	0.22	0.33	0.44	0.55	0.66	0.77	0.88	1.00	1.10	1.21	1.32	1.43	1.54
4	0.20	0.39	0.58	0.78	0.98	1.18	1.37	1.57	1.77	1.96	2.16	2.36	2.55	2.75
5	0.31	0.61	0.91	1.23	1.53	1.84	2.15	2.45	2.76	3.07	3.37	3.68	3.99	4.30
6	0.44	0.88	1.32	1.77	2.21	2.65	3.09	3.53	3.98	4.42	3.86	5.30	5.74	6.19
7	0.60	1.20	1.80	2.41	3.01	3.61	4.21	4.81	5.41	6.01	6.61	7.22	7.82	8.42
8	0.79	1.57	2.35	3.14	3.93	4.71	5.50	6.28	7.07	7.85	8.64	9.43	10.21	11.00
9	1.00	2.00	3.00	4.00	5.00	6.00	7.00	8.00	9.00	10.00	11.00	12.00	13.00	14.00
10	1.27	2.53	3.79	5.06	6.33	7.59	8.86	10.12	11.39	12.66	13.92	15.19	16.45	17.72
11	1.56	3.12	4.68	6.25	7.81	9.37	10.94	12.50	14.06	15.62	17.19	18.75	20.31	21.87
14	2.25	4.50	6.75	9.00	11.25	13.50	15.75	18.00	20.25	22.50	24.75	27.00	29.25	31.50
18	4.00	8.00	12.00	16.00	20.00	24.00	28.00	32.00	36.00	40.00	44.00	48.00	52.00	57.00

TABLE A.14. Areas of bars in slabs (square inches per foot)

Spacing (in.)	Bar no.								
	3	4	5	6	7	8	9	10	11
3	0.44	0.78	1.23	1.77	2.40	3.14	4.20	5.06	6.25
$3\frac{1}{2}$	0.38	0.67	1.05	1.51	2.06	2.69	3.43	4.34	5.36
4	0.33	0.59	0.92	1.32	1.80	2.36	3.00	3.80	4.68
$4\frac{1}{2}$	0.29	0.52	0.82	1.18	1.60	2.09	2.67	3.37	4.17
5	0.26	0.47	0.74	1.06	1.44	1.88	2.40	3.04	3.75
$5\frac{1}{2}$	0.24	0.43	0.67	0.96	1.31	1.71	2.18	2.76	3.41
6	0.22	0.39	0.61	0.88	1.20	1.57	2.00	2.53	3.12
$6\frac{1}{2}$	0.20	0.36	0.57	0.82	1.11	1.45	1.85	2.34	2.89
7	0.19	0.34	0.53	0.76	1.03	1.35	1.71	2.17	2.68
$7\frac{1}{2}$	0.18	0.31	0.49	0.71	0.96	1.26	1.60	2.02	2.50
8	0.17	0.29	0.46	0.66	0.90	1.18	1.50	1.89	2.34
9	0.15	0.26	0.41	0.59	0.80	1.05	1.33	1.69	2.08
10	0.13	0.24	0.37	0.53	0.72	0.94	1.20	1.52	1.87
12	0.11	0.20	0.31	0.44	0.60	0.78	1.00	1.27	1.56

TABLE A.15. Common styles of welded wire fabric

Style designation	Steel area (in.²/ft)		Weight (lb/100 ft²)
	Longitudinal	Transverse	
6 × 6—W1.4 × W1.4	0.03	0.03	21
6 × 6—W2 × W2	0.04	0.04	29
6 × 6—W2.9 × W2.9	0.06	0.06	42
6 × 6—W4 × W4	0.08	0.08	58
6 × 6—W5.5 × W5.5	0.11	0.11	80
4 × 4—W1.4 × W1.4	0.04	0.04	31
4 × 4—W2 × W2	0.06	0.06	43
4 × 4—W2.9 × W2.9	0.09	0.09	62
4 × 4—W4 × W4	0.12	0.12	86

TABLE A.16. Size and pitch of spirals

f_y (ksi)	Diameter of column (in.)	Out to out of spiral (in.)	f'_c		
			3 ksi	4 ksi	5 ksi
40	14, 15	11, 12	$\frac{3}{8}$–$1\frac{3}{4}$	$\frac{1}{2}$–$2\frac{1}{2}$	$\frac{1}{2}$–$1\frac{3}{4}$
	16	13	$\frac{3}{8}$–$1\frac{3}{4}$	$\frac{1}{2}$–$2\frac{1}{2}$	$\frac{1}{2}$–2
	17–19	14–16	$\frac{3}{8}$–$1\frac{3}{4}$	$\frac{1}{2}$–$2\frac{1}{2}$	$\frac{1}{2}$–2
	20–23	17–20	$\frac{3}{8}$–$1\frac{3}{4}$	$\frac{1}{2}$–$2\frac{1}{2}$	$\frac{1}{2}$–2
	24–30	21–27	$\frac{3}{8}$–2	$\frac{1}{2}$–$2\frac{1}{2}$	$\frac{1}{2}$–2
60	14, 15	11, 12	$\frac{3}{8}$–$2\frac{3}{4}$	$\frac{3}{8}$–2	$\frac{1}{2}$–$2\frac{3}{4}$
	16–23	13–20	$\frac{3}{8}$–$2\frac{3}{4}$	$\frac{3}{8}$–2	$\frac{1}{2}$–$2\frac{3}{4}$
	24–29	21–26	$\frac{3}{8}$–3	$\frac{3}{8}$–$2\frac{1}{4}$	$\frac{1}{2}$–3
	30	27	$\frac{3}{8}$–3	$\frac{3}{8}$–$2\frac{1}{4}$	$\frac{1}{2}$–$3\frac{1}{4}$

TABLE A.17. Safe loads (kips) on square, rectangular, and circular columns, given f'_c = 4 ksi and f_y = 60 ksi

A_s	A'_s	$0.56P_0$	M_u (k·ft)	P_b	M_b (k·ft)	e_b	P_u (kips) for eccentricity e of (in.)										
							2	4	6	8	12	16	20	24	28	32	40
Section 12 by 12 in., $0.1 \times f'_c \times A_g$ = 57 k																	
2 No. 6	2 No. 6	329	36	137	63	5.6	277	181	121	77	44	32	24	20	17	14	11
2 No. 7	2 No. 7	350	47	136	71	6.3	293	195	141	94	53	39	30	25	21	18	14
2 No. 8	2 No. 8	373	59	134	79	7.1	312	211	155	111	64	46	37	30	26	23	18
2 No. 9	2 No. 9	400	73	132	89	8.1	334	228	169	134	74	54	43	36	31	27	21
2 No. 10	2 No. 10	434	89	130	101	9.3	360	247	185	147	88	64	50	43	36	32	25
2 No. 11	2 No. 11	471	105	127	113	10.6	388	268	202	161	103	75	57	49	42	37	29
3 No. 5	3 No. 5	331	38	138	65	5.6	279	184	125	80	45	33	25	20	17	15	12
3 No. 6	3 No. 6	357	52	136	75	6.6	301	202	147	101	57	42	33	27	23	20	16
3 No. 7	3 No. 7	388	67	134	86	7.7	325	222	164	124	72	51	41	34	29	25	20
3 No. 8	3 No. 8	423	85	132	99	9.0	353	244	182	145	86	62	49	41	35	31	24
3 No. 9	3 No. 9	464	104	130	113	10.5	385	268	202	162	104	75	57	49	42	37	29
3 No. 10	3 No. 10	514	127	127	131	12.3	424	297	225	181	130	85	71	55	50	42	36
4 No. 5	4 No. 5	352	49	138	73	6.4	296	199	145	98	55	40	32	26	22	19	15
4 No. 6	4 No. 6	386	67	135	86	7.6	324	222	165	124	72	51	41	34	29	25	20
4 No. 7	4 No. 7	426	88	133	101	9.2	357	248	186	148	90	64	50	42	36	32	25
4 No. 8	4 No. 8	473	110	130	118	10.9	394	276	209	167	110	79	60	51	44	39	31
4 No. 9	4 No. 9	527	136	127	138	13.0	436	307	234	189	136	92	75	58	52	45	38
Section 14 by 14 in., $0.1 \times f'_c \times A_g$ = 78 k																	
2 No. 7	2 No. 7	449	59	193	104	6.5	403	282	206	142	75	53	41	33	28	24	19
2 No. 8	2 No. 8	472	75	191	115	7.3	424	300	223	165	90	64	50	41	34	30	23
2 No. 9	2 No. 9	499	92	189	128	8.1	447	320	240	191	107	74	59	48	41	36	28
2 No. 10	2 No. 10	533	113	186	143	9.2	476	343	260	209	128	88	68	57	49	43	34
2 No. 11	2 No. 11	570	135	183	159	10.4	507	368	281	227	148	103	79	66	57	50	40
2 No. 14	2 No. 14	658	185	178	196	13.2	580	424	327	266	193	134	106	84	73	63	52
3 No. 5	3 No. 5	430	47	195	96	5.9	387	268	191	122	64	44	34	27	23	19	15
3 No. 6	3 No. 6	456	65	193	109	6.8	411	290	213	152	81	57	44	36	30	26	20
3 No. 7	3 No. 7	487	85	191	123	7.7	438	313	234	182	102	70	55	45	38	33	26

3 No. 8	522	107	189	140	8.9	468	338	256	205	124	85	66	55	47	41	32
3 No. 9	563	133	186	159	10.2	503	367	281	226	149	103	78	65	56	49	39
3 No. 10	613	163	183	181	11.9	546	400	309	250	177	123	94	76	66	58	47
3 No. 11	669	197	180	206	13.7	593	437	339	276	201	144	112	90	76	67	55
4 No. 5	451	61	195	106	6.6	406	286	210	147	77	55	42	34	29	25	19
4 No. 6	485	84	192	123	7.7	437	313	234	182	101	70	55	45	38	33	26
4 No. 7	525	111	190	143	9.0	472	342	260	209	128	88	68	56	48	42	33
4 No. 8	572	140	187	164	10.6	512	375	288	233	156	108	82	68	59	51	41
4 No. 9	626	174	184	190	12.4	559	412	319	259	189	130	101	80	70	61	49
4 No. 10	693	214	180	220	14.7	616	456	355	290	212	157	122	98	82	72	59
4 No. 11	769	258	176	253	17.2	679	504	395	323	237	188	141	118	97	85	67
6 No. 6	541	122	190	151	9.6	487	355	272	219	140	96	73	61	52	46	36
6 No. 7	602	161	187	181	11.6	540	398	308	250	178	123	94	76	66	58	46
6 No. 8	671	173	166	180	13.0	574	407	309	248	178	114	97	74	68	57	49

Section 14 by 16 in., $0.1 \times f'_c \times A_g = 89\ k$

Vertical bars distributed along short sides

2 No. 7	502	70	226	133	7.1	473	347	259	192	99	67	51	41	34	29	23
2 No. 8	526	89	224	147	7.9	495	367	278	219	120	80	62	50	42	36	28
2 No. 9	553	110	222	162	8.8	520	389	298	239	143	95	73	60	51	44	35
2 No. 10	586	136	220	181	9.9	551	415	321	260	169	114	86	71	60	52	42
2 No. 11	624	163	217	201	11.1	584	442	345	281	196	134	101	82	70	61	49
2 No. 14	711	226	211	247	14.0	662	505	398	327	240	177	135	108	90	79	64
3 No. 6	510	77	227	138	7.3	481	355	267	204	108	72	55	45	37	32	25
3 No. 7	540	101	225	156	8.4	510	380	291	233	135	88	69	56	47	41	32
3 No. 8	575	129	222	176	9.5	542	408	316	255	164	109	83	68	58	50	40
3 No. 9	616	160	220	200	10.9	579	440	344	280	195	132	99	81	69	60	48
3 No. 10	666	198	217	228	12.6	624	477	375	308	226	159	120	96	81	72	57
3 No. 11	723	239	213	259	14.5	675	517	410	338	249	188	143	115	95	83	67
3 No. 14	854	331	206	328	19.1	791	609	487	403	300	239	192	155	130	111	87
4 No. 5	504	72	228	135	7.1	476	351	263	198	103	69	53	42	35	30	24
4 No. 6	538	100	226	156	8.3	508	380	291	233	135	88	69	56	47	41	32
4 No. 7	579	132	223	180	9.7	546	412	320	259	169	113	85	70	59	52	41
4 No. 8	625	168	220	207	11.3	588	448	351	287	204	139	105	84	72	63	50
4 No. 9	680	210	217	238	13.1	638	489	386	317	233	169	128	102	86	75	60

(continued)

TABLE A.17. (continued)

A_s	A'_s	$0.56P_0$	M_u (k·ft)	P_b	M_b (k·ft)	e_b	Value of P_u (kips) for eccentricity e of (in.)										
							2	4	6	8	12	16	20	24	28	32	40
Vertical bars distributed along short sides							Section 14 by 16 in., $0.1 \times f'_c \times A_g = 89 k$										
4 No. 10	4 No. 10	747	259	214	275	15.5	698	538	428	353	262	204	155	125	104	89	72
4 No. 11	4 No. 11	822	314	210	317	18.1	765	592	473	392	292	232	184	149	124	107	84
6 No. 6	6 No. 6	594	146	224	190	10.2	561	426	333	270	184	123	92	75	64	56	45
6 No. 7	6 No. 7	655	193	220	226	12.3	617	473	373	307	225	159	119	95	81	71	57
6 No. 8	6 No. 8	725	215	200	233	14.0	661	493	383	312	227	158	125	99	85	74	61
6 No. 9	6 No. 9	806	263	195	268	16.5	729	545	425	347	254	200	141	123	98	88	69
Vertical bars distributed along long sides																	
2 No. 7	2 No. 7	502	60	221	114	6.2	451	313	227	149	79	56	42	34	29	25	19
2 No. 8	2 No. 8	526	76	218	125	6.9	472	332	244	174	93	67	52	42	35	31	24
2 No. 9	2 No. 9	553	94	216	137	7.6	495	352	262	200	111	78	61	50	42	37	29
2 No. 10	2 No. 10	586	115	214	152	8.6	524	376	283	226	131	90	71	59	50	44	35
2 No. 11	2 No. 11	624	137	211	169	9.6	555	401	304	244	153	106	82	68	59	51	41
2 No. 14	2 No. 14	711	187	205	205	12.0	628	457	351	284	205	136	108	86	76	65	54
3 No. 6	3 No. 6	510	66	221	118	6.4	459	321	234	160	84	60	46	37	31	27	21
3 No. 7	3 No. 7	540	86	219	133	7.3	486	345	256	191	105	74	57	47	40	34	27
3 No. 8	3 No. 8	575	109	217	149	8.3	516	370	279	222	127	87	69	57	48	42	33
3 No. 9	3 No. 9	616	135	214	168	9.4	551	399	304	244	153	105	81	68	58	51	40
3 No. 10	3 No. 10	666	165	211	191	10.9	594	433	333	269	181	126	96	79	68	60	48
3 No. 11	3 No. 11	723	199	207	215	12.5	641	470	363	295	214	146	115	91	79	69	57
3 No. 14	3 No. 14	854	272	199	271	16.3	750	553	430	351	257	202	147	126	101	91	71
4 No. 5	4 No. 5	504	62	223	116	6.2	454	317	231	155	81	57	44	35	30	25	20
4 No. 6	4 No. 6	538	85	220	133	7.2	485	345	256	192	105	73	57	47	39	34	27
4 No. 7	4 No. 7	579	112	218	152	8.4	520	375	283	226	132	90	71	59	50	43	34
4 No. 8	4 No. 8	625	142	215	174	9.7	560	408	312	251	161	111	85	71	61	53	42
4 No. 9	4 No. 9	680	175	211	199	11.3	607	445	343	278	192	134	102	83	72	63	51
4 No. 10	4 No. 10	747	216	208	229	13.3	664	490	380	309	225	159	124	99	85	74	61
4 No. 11	4 No. 11	822	260	203	263	15.5	727	538	420	343	251	189	147	118	100	86	70
6 No. 6	6 No. 6	594	123	218	161	8.9	535	388	295	237	144	98	76	63	54	47	37
6 No. 7	6 No. 7	655	162	215	190	10.6	588	431	332	269	182	126	96	79	68	59	48

Section 14 by 20 in., $0.1 \times f'_c \times A_g = 111\ k$

Vertical bars distributed along short sides

2 No. 8	117	632	291	218	9.0	632	502	396	320	198	123	91	72	60	51	40
2 No. 9	146	659	289	239	9.9	659	527	420	342	229	148	107	86	72	62	48
2 No. 10	181	693	287	265	11.1	693	556	447	368	262	177	128	102	86	74	58
2 No. 11	220	730	284	294	12.4	730	588	476	394	292	207	152	119	99	87	68
2 No. 14	307	818	278	359	15.5	815	660	539	452	339	268	204	161	132	112	90
3 No. 6	100	616	294	206	8.4	616	488	383	307	179	109	81	64	53	45	34
3 No. 7	134	647	292	231	9.5	647	517	411	334	217	138	100	81	67	58	45
3 No. 8	171	682	289	258	10.7	682	548	440	362	253	170	122	98	82	71	55
3 No. 9	214	723	287	291	12.2	723	583	473	392	290	205	149	117	98	85	67
3 No. 10	266	773	284	329	13.9	773	626	511	426	319	242	181	142	116	101	80
3 No. 11	323	829	280	372	15.9	828	672	552	463	349	279	215	170	140	118	94
3 No. 14	452	961	273	470	20.6	954	778	644	545	415	334	280	231	191	163	124
4 No. 6	132	645	293	230	9.4	645	516	410	333	216	137	100	80	67	57	44
4 No. 7	176	685	290	263	10.9	685	552	444	366	258	175	126	100	84	73	57
4 No. 8	225	732	287	299	12.5	732	592	481	400	297	214	157	122	102	89	70
4 No. 9	282	786	284	342	14.4	786	638	523	438	328	253	192	150	124	106	84
4 No. 10	351	853	281	394	16.8	853	694	572	482	364	292	232	184	152	128	100
4 No. 11	427	929	277	452	19.6	925	756	625	529	403	325	270	221	182	155	118
6 No. 6	193	701	291	276	11.4	701	567	458	379	275	191	138	108	91	79	62
6 No. 7	258	761	287	325	13.6	761	619	506	423	316	238	179	140	114	99	79
6 No. 8	300	831	267	347	15.6	819	656	531	442	328	255	192	153	126	108	87
6 No. 9	371	913	262	399	18.3	895	718	584	488	366	292	229	185	153	131	102
6 No. 10	454	1014	256	462	21.7	987	792	647	543	409	327	273	218	186	157	122

(continued)

TABLE A.17. (continued)

Section 14 by 20 in., $0.1 \times f'_c \times A_g = 111$ k

Vertical bars distributed along long sides

A_s	A'_s	$0.56P_0$	M_u (k·ft)	P_b	M_b (k·ft)	e_b	header →										

Value of P_u (kips) for eccentricity e of (in.)

A_s	A'_s	$0.56P_0$	M_u (k·ft)	P_b	M_b (k·ft)	e_b	2	4	6	8	12	16	20	24	28	32	40
2 No. 9	2 No. 9	659	96	272	156	6.9	591	415	305	218	117	84	65	53	45	39	30
2 No. 10	2 No. 10	693	118	269	171	7.6	620	440	327	249	139	97	76	63	53	46	36
2 No. 11	2 No. 11	730	141	266	187	8.5	651	465	350	278	160	111	88	73	62	54	43
2 No. 14	2 No. 14	818	192	259	224	10.4	724	522	398	320	207	145	111	93	80	70	56
3 No. 6	3 No. 6	616	68	278	137	5.9	554	383	273	175	92	64	49	39	33	28	22
3 No. 7	3 No. 7	647	88	275	152	6.6	581	408	299	209	111	79	61	50	42	36	28
3 No. 8	3 No. 8	682	111	272	168	7.4	612	434	323	243	134	94	74	60	51	44	35
3 No. 9	3 No. 9	723	138	269	187	8.3	647	464	349	279	160	110	87	72	61	53	42
3 No. 10	3 No. 10	773	169	266	209	9.5	690	499	379	304	190	131	101	85	72	63	50
3 No. 11	3 No. 11	829	202	262	234	10.7	737	536	411	331	221	154	117	97	84	74	59
3 No. 14	3 No. 14	961	277	254	290	13.7	846	620	479	389	283	200	158	125	108	94	77
4 No. 5	4 No. 5	611	64	279	135	5.8	550	379	267	169	88	61	46	37	31	26	21
4 No. 6	4 No. 6	645	88	277	152	6.6	580	408	299	209	110	79	61	49	41	36	28
4 No. 7	4 No. 7	685	115	274	171	7.5	616	439	328	250	138	96	75	62	52	45	36
4 No. 8	4 No. 8	732	145	270	193	8.6	656	473	358	286	168	115	90	75	64	56	44
4 No. 9	4 No. 9	786	179	267	218	9.8	703	511	390	314	201	139	106	89	76	67	53
4 No. 10	4 No. 10	853	220	263	248	11.3	760	556	428	346	239	167	127	104	90	79	64
4 No. 11	4 No. 11	929	264	258	281	13.1	824	605	469	381	277	194	152	121	104	91	74

Section 16 by 16 in., $0.1 \times f'_c \times A_g = 102$ k

Value of P_u (kips) for eccentricity e of (in.)

A_s	A'_s	ρ_t	$0.56P_0$	M_u (k·ft)	P_b	e_b	2	4	6	8	12	16	20	24	28	32	40
2 No. 8	2 No. 8	1.23	586	90	257	7.4	553	407	306	234	124	84	64	52	44	37	29
2 No. 9	2 No. 9	1.56	614	112	255	8.2	578	429	327	261	148	98	76	62	52	45	35
2 No. 10	2 No. 10	1.98	647	137	252	9.2	608	456	350	282	175	116	89	73	62	54	43
2 No. 11	2 No. 11	2.44	685	165	249	10.3	642	484	375	304	202	137	102	85	73	63	50
2 No. 14	2 No. 14	3.52	772	228	243	12.8	719	546	429	351	257	180	137	109	93	82	66
3 No. 6	3 No. 6	1.03	571	78	260	7.0	538	395	294	217	111	75	57	46	38	33	25
3 No. 7	3 No. 7	1.41	601	102	257	7.9	567	421	319	253	139	93	72	58	49	42	33

716

3 No. 8	3 No. 8	1.84	636	130	255	8.9	599	449	345	278	169	112	86	71	60	52	41
3 No. 9	3 No. 9	2.34	677	162	252	10.1	637	481	374	303	202	136	101	84	72	62	49
3 No. 10	3 No. 10	2.96	727	199	249	11.6	682	518	406	332	238	163	123	98	85	74	59
3 No. 11	3 No. 11	3.66	784	241	245	13.3	732	560	441	362	266	192	145	116	98	86	69
3 No. 14	3 No. 14	5.27	915	333	238	17.2	849	652	519	429	318	252	194	158	131	113	90
2 No. 6	2 No. 6	1.04	571	101	224	8.4	535	394	296	234	138	93	72	58	49	42	33
4 No. 6	4 No. 6	1.38	599	101	259	7.8	566	420	319	252	139	92	71	58	48	42	33
2 No. 7	2 No. 7	1.41	602	134	208	10.2	561	417	318	254	169	115	88	73	62	53	42
4 No. 7	4 No. 7	1.88	640	134	256	9.0	603	454	350	282	174	115	88	72	61	53	42
2 No. 8	2 No. 8	1.84	636	170	191	12.4	590	441	340	274	196	139	105	87	75	65	52
4 No. 8	4 No. 8	2.45	686	170	253	10.4	646	490	382	310	211	142	106	87	75	65	52
2 No. 9	2 No. 9	2.34	677	210	171	15.6	624	468	364	295	213	165	126	102	88	77	62
4 No. 9	4 No. 9	3.13	740	211	250	12.0	696	531	417	342	250	172	130	103	89	78	62
2 No. 10	2 No. 10	2.96	727	258	146	20.5	665	501	391	318	231	181	149	121	102	90	73
4 No. 10	4 No. 10	3.95	808	261	246	14.0	756	580	459	378	279	208	157	126	105	92	74
2 No. 11	2 No. 11	3.66	784	310	118	28.4	710	535	419	342	250	196	162	137	120	103	84
4 No. 11	4 No. 11	4.88	883	316	242	16.3	823	634	505	417	309	246	186	151	125	108	86
2 No. 14	2 No. 14	5.27	915	395	54	76.3	813	613	481	394	289	228	188	160	139	123	100
4 No. 14	4 No. 14	7.03	1057	438	232	21.7	977	756	606	503	376	300	249	202	172	146	114
3 No. 6	3 No. 6	1.55	613	147	204	10.9	571	426	327	263	181	125	95	78	67	58	46
6 No. 6	6 No. 6	2.07	655	147	257	9.5	619	467	363	293	190	126	95	78	67	58	46
3 No. 7	3 No. 7	2.12	659	194	182	14.0	610	458	356	288	208	156	119	96	83	73	58
6 No. 7	6 No. 7	2.82	716	195	253	11.3	675	515	404	331	237	162	121	97	84	73	58
3 No. 8	3 No. 8	2.76	711	246	156	18.7	653	493	385	314	228	179	145	117	99	87	71
6 No. 8	6 No. 8	3.68	785	249	249	13.4	738	567	450	370	273	201	151	121	101	89	71
3 No. 9	3 No. 9	3.52	772	276	106	28.3	686	506	390	315	227	177	145	123	107	94	76
6 No. 9	6 No. 9	4.69	867	266	225	14.9	786	586	455	370	270	192	153	121	103	90	74
3 No. 10	3 No. 10	4.45	848	322	68	49.9	743	547	423	342	247	193	159	135	117	103	84
6 No. 10	6 No. 10	5.93	968	320	218	17.6	868	647	504	412	301	237	169	149	118	108	82
6 No. 11	6 No. 11	7.32	1081	362	211	20.7	959	713	557	456	334	263	217	154	148	113	100
4 No. 7	4 No. 7	2.82	716	233	137	19.9	645	478	368	297	213	166	136	110	94	83	67
8 No. 7	8 No. 7	3.76	792	226	232	13.0	726	543	422	343	249	170	134	105	92	80	65
4 No. 8	4 No. 8	3.68	785	292	102	30.6	699	518	401	324	234	183	150	128	111	98	80
8 No. 8	8 No. 8	4.91	885	283	226	15.6	805	603	471	385	281	206	162	130	110	95	78
4 No. 9	4 No. 9	4.69	867	343	61	58.0	761	564	437	355	257	201	166	141	122	108	87
8 No. 9	8 No. 9	6.25	994	346	220	18.6	895	672	526	431	316	250	185	160	129	115	89
8 No. 10	8 No. 10	7.91	1128	401	212	22.4	1006	753	591	486	357	283	234	175	162	128	109

(continued)

TABLE A.17. (continued)

Value of P_u (kips) for eccentricity e of (in.)

Section 20 × 20 in., $0.1 \times f'_c \times A_g = 159\ k$

A_s	A'_s	ρ_t	$0.56P_0$	M_u (k·ft)	P_b	e_b	2	4	6	8	12	16	20	24	28	32	40
2 No. 9	2 No. 9	1.00	888	150	415	8.6	888	702	550	441	260	159	119	95	78	67	51
2 No. 10	2 No. 10	1.26	921	186	412	9.4	921	732	580	470	301	190	140	112	94	80	62
2 No. 11	2 No. 11	1.56	959	225	409	10.3	959	765	611	499	341	223	160	130	109	94	73
2 No. 14	2 No. 14	2.25	1046	313	403	12.4	1045	838	678	561	414	293	215	168	141	123	97
3 No. 8	3 No. 8	1.17	910	175	416	9.2	910	724	573	463	291	182	134	107	89	76	59
3 No. 9	3 No. 9	1.50	951	219	413	10.2	951	761	608	497	337	220	158	128	107	92	72
3 No. 10	3 No. 10	1.89	1001	271	409	11.4	1001	804	648	534	385	263	191	150	127	110	87
3 No. 11	3 No. 11	2.34	1058	329	405	12.8	1058	852	691	574	425	308	226	177	147	128	102
3 No. 14	3 No. 14	3.38	1189	459	397	16.0	1184	960	786	659	495	396	304	240	198	168	133
3 No. 18	3 No. 18	6.00	1522	772	378	24.3	1499	1222	1011	857	653	528	443	381	319	273	209
4 No. 8	2 No. 8	1.18	910	229	352	11.3	907	718	570	464	328	225	164	133	112	97	76
4 No. 8	4 No. 8	1.57	960	229	414	10.4	960	770	617	506	348	230	165	133	112	96	75
4 No. 9	2 No. 9	1.50	951	286	331	13.2	945	751	601	493	359	264	198	156	133	115	91
4 No. 9	4 No. 9	2.00	1015	287	410	11.7	1015	818	661	547	400	278	201	157	133	116	91
4 No. 10	2 No. 10	1.90	1002	355	306	15.9	990	790	636	525	386	304	234	187	155	136	109
4 No. 10	4 No. 10	2.53	1082	356	406	13.4	1082	875	712	594	442	330	244	191	157	137	109
4 No. 11	2 No. 11	2.34	1058	429	278	19.4	1041	832	673	559	413	328	269	219	182	156	126
4 No. 11	4 No. 11	3.13	1157	433	402	15.2	1156	937	768	644	483	381	290	228	188	159	127
4 No. 14	2 No. 14	3.38	1189	594	214	30.8	1155	925	753	629	469	374	311	266	232	205	160
4 No. 14	4 No. 14	4.50	1332	605	392	19.6	1323	1078	891	752	571	460	383	311	256	219	167
4 No. 18	4 No. 18	8.00	1775	1020	368	31.1	1742	1426	1186	1010	776	630	530	458	403	357	273
6 No. 7	3 No. 7	1.35	933	263	343	12.3	929	738	590	483	351	249	185	147	125	108	85
6 No. 7	6 No. 7	1.80	990	263	414	11.1	990	798	644	531	380	260	187	148	125	108	85
6 No. 8	3 No. 8	1.77	985	335	317	15.0	977	780	628	518	381	294	225	179	150	131	104
6 No. 8	6 No. 8	2.35	1060	337	410	12.9	1060	859	699	582	433	317	234	183	151	132	104
6 No. 9	3 No. 9	2.25	1046	419	287	18.6	1032	826	669	556	412	326	265	215	179	154	124
6 No. 9	6 No. 9	3.00	1141	422	406	14.9	1141	928	761	638	479	375	286	225	185	157	125
6 No. 10	3 No. 10	2.85	1122	518	250	24.2	1100	882	718	599	446	355	295	252	213	183	146
6 No. 10	6 No. 10	3.79	1242	525	400	17.4	1241	1011	834	704	533	428	344	275	226	192	149

Table (continued from previous page)

A_s	$\rho\%$	$0.56P_0$	P_b	M_b (k·ft)	e_b	— P_u (kips) for eccentricity e (in.) 2	4	6	8	12	16	20	24	28	32	36
6 No. 11	3.51	1207	209	626	32.7	1174	943	770	644	482	385	320	274	240	213	168
6 No. 11	4.68	1355	394	639	20.3	1349	1103	914	774	590	476	399	328	272	231	176
6 No. 14	5.06	1403	134	1042	69.3	1369	1115	924	783	596	480	402	346	303	270	222
6 No. 14	6.75	1617	401	988	28.1	1617	1354	1142	982	763	623	527	457	403	350	275
8 No. 7	1.80	990	316	344	15.3	982	785	633	523	385	300	230	183	153	133	106
8 No. 7	2.40	1066	412	345	13.0	1066	866	706	589	439	324	241	188	155	135	107
8 No. 8	2.35	1060	282	440	19.5	1045	838	681	566	420	334	275	223	187	160	129
8 No. 8	3.14	1159	407	444	15.4	1159	945	777	653	492	389	300	236	194	164	131
8 No. 9	3.00	1141	217	510	26.8	1101	873	702	581	429	339	281	239	207	178	143
8 No. 9	4.00	1268	375	497	17.4	1244	997	810	675	505	403	309	249	206	176	139
8 No. 10	3.79	1242	167	626	39.7	1186	940	758	629	467	370	307	262	229	203	165
8 No. 10	5.06	1403	367	609	20.6	1367	1097	894	748	562	450	375	293	250	211	164
8 No. 11	4.69	1355	110	747	67.5	1281	1014	819	681	507	403	334	285	249	221	180
8 No. 11	6.25	1553	358	729	24.1	1504	1207	986	827	624	501	418	359	285	255	190

Circular section diameter 16 in., $0.1 \times f'_c \times A_g = 80\ k\ (P_u \geq 80\ k)$

A_s	$\rho\%$	$0.56P_0$	M_u (k·ft)	P_b	M_b (k·ft)	e_b	Value of P_u (kips) for eccentricity e of (in.) — 2	4	6	8	12	16	20	24	28	32
6 No. 6	1.3	466	67	174	92	6.35	406	269	184	128	0	0	0	0	0	0
6 No. 7	1.8	496	87	169	101	7.21	430	289	201	148	89	0	0	0	0	0
6 No. 8	2.4	533	108	162	112	8.30	457	310	220	168	104	0	0	0	0	0
6 No. 9	3.0	572	127	154	124	9.61	488	333	238	184	119	85	0	0	0	0
6 No. 10	3.8	624	149	144	138	11.49	526	362	260	202	137	99	0	0	0	0
6 No. 11	4.7	679	169	125	149	14.33	564	384	277	215	148	111	87	91	0	0
6 No. 14	6.7	810	218	92	181	23.69	659	447	324	252	176	135	102	0	0	0
8 No. 5	1.2	461	64	186	91	5.86	402	265	181	126	0	0	0	0	0	0
8 No. 6	1.8	494	83	183	101	6.64	428	286	201	147	91	0	0	0	0	0
8 No. 7	2.4	534	107	178	113	7.64	460	311	223	169	107	0	0	0	0	0
8 No. 8	3.1	583	132	172	127	8.90	496	338	245	190	124	91	0	0	0	0
8 No. 9	4.0	636	160	164	142	10.39	536	367	269	209	141	103	81	0	0	0
8 No. 10	5.1	704	192	154	161	12.54	588	404	298	232	161	120	95	0	0	0
8 No. 11	6.2	778	214	133	176	15.88	637	436	319	249	173	132	105	86	0	0
10 No. 5	1.5	481	76	190	97	6.13	419	278	195	140	83	0	0	0	0	0

(continued)

TABLE A.17. (continued)

A_s	$\rho\%$	$0.56P_0$	M_u (k·ft)	P_b	M_b (k·ft)	e_b	2	4	6	8	12	16	20	24	28	32
							— Value of P_u (kips) for eccentricity e of (in.)									
10 No. 6	2.2	522	102	187	110	7.05	451	303	217	163	103	0	0	0	0	0
10 No. 7	3.0	572	129	183	125	8.22	491	333	242	187	122	89	0	0	0	0
10 No. 8	3.9	633	158	176	142	9.70	537	367	269	210	141	104	82	0	0	0
10 No. 9	5.0	699	188	169	161	11.45	588	403	297	233	161	120	96	81	0	0
10 No. 10	6.3	785	226	158	184	13.98	653	448	333	262	183	139	111	92	0	0
10 No. 11	7.8	877	256	134	203	18.11	714	487	361	283	198	153	119	101	89	0
12 No. 5	1.9	500	88	192	103	6.45	435	291	206	153	93	0	0	0	0	0
12 No. 6	2.6	550	117	189	118	7.52	474	321	231	178	114	82	0	0	0	0
12 No. 7	3.6	611	149	185	137	8.87	522	356	260	203	136	100	0	0	0	0
12 No. 8	4.7	683	183	179	157	10.58	578	396	292	229	158	118	93	0	0	0
12 No. 9	6.0	763	218	171	180	12.62	639	439	325	257	179	136	109	91	0	0
12 No. 10	7.6	865	259	160	208	15.60	717	493	367	291	204	157	126	105	91	0
14 No. 5	2.2	520	101	194	109	6.79	450	304	217	164	103	0	0	0	0	0
14 No. 6	3.1	578	132	191	127	8.00	496	339	245	191	125	91	0	0	0	0
14 No. 7	4.2	649	168	186	148	9.55	551	379	278	218	149	110	86	0	0	0
14 No. 8	5.5	733	208	180	172	11.51	616	426	315	248	173	130	104	85	0	0
14 No. 9	7.0	826	247	172	198	13.87	687	476	353	280	197	149	120	101	86	0
16 No. 6	3.5	605	148	191	136	8.51	518	356	260	203	135	99	0	0	0	0
16 No. 7	4.8	687	187	187	160	10.25	581	403	296	233	161	120	93	0	0	0
16 No. 8	6.3	783	230	180	188	12.47	655	455	337	266	187	143	114	95	81	0
16 No. 9	8.0	889	275	172	217	15.17	737	513	381	302	213	165	133	110	96	82
Circular section diameter 20 in., $0.1 \times f'_c \times A_g = 125\ k\ (P_u \geq 125\ k)$																
6 No. 7	1.1	712	119	290	180	7.45	675	491	357	264	152	0	0	0	0	0
6 No. 8	1.5	748	152	285	196	8.23	706	517	382	293	175	0	0	0	0	0
6 No. 9	1.9	788	187	280	213	9.10	740	545	408	316	199	139	0	0	0	0
6 No. 10	2.4	839	228	274	234	10.27	784	580	438	344	227	160	0	0	0	0

Circular section diameter 16 in., $0.1 \times f'_c \times A_g = 80\ k$

6 No. 11	3.0	894	258	259	252	11.68	828	613	466	366	251	179	138	0	0	0
6 No. 14	4.3	1026	333	237	302	15.28	936	697	535	423	297	225	176	143	0	0
6 No. 18	7.6	1358	504	172	420	29.18	1208	899	694	553	391	303	248	209	181	158
8 No. 6	1.1	709	119	307	178	6.99	674	489	354	261	154	0	0	0	0	0
8 No. 7	1.5	750	152	304	196	7.74	710	518	382	293	181	0	0	0	0	0
8 No. 8	2.0	798	189	301	217	8.64	751	552	412	322	207	149	0	0	0	0
8 No. 9	2.5	851	228	298	239	9.64	797	588	443	350	234	170	133	0	0	0
8 No. 10	3.2	920	278	293	267	10.95	855	633	482	384	266	194	153	126	0	0
8 No. 11	4.0	993	324	279	292	12.55	914	676	516	413	291	215	170	140	126	0
8 No. 14	5.7	1168	431	259	357	16.53	1060	785	604	486	347	267	213	176	150	131
10 No. 6	1.4	737	141	313	190	7.29	701	510	374	283	172	0	0	0	0	0
10 No. 7	1.9	788	184	311	212	8.17	746	547	407	317	205	143	0	0	0	0
10 No. 8	2.5	848	233	308	237	9.21	798	589	443	349	234	170	130	0	0	0
10 No. 9	3.2	915	278	305	264	10.38	856	634	481	382	264	196	153	130	0	0
10 No. 10	4.0	1000	333	301	299	11.91	930	690	527	422	299	224	178	153	0	0
10 No. 11	5.0	1092	384	289	331	13.72	1006	743	569	457	325	248	198	178	141	0
10 No. 14	7.2	1311	510	270	411	18.26	1191	879	677	546	393	306	247	206	177	154
12 No. 5	1.2	716	124	319	181	6.82	681	496	359	268	158	0	0	0	0	0
12 No. 6	1.7	765	166	318	202	7.64	724	533	393	303	190	134	0	0	0	0
12 No. 7	2.3	826	213	316	228	8.64	777	576	431	339	225	161	130	0	0	0
12 No. 8	3.0	898	266	314	257	9.85	840	626	474	376	258	190	153	130	0	0
12 No. 9	3.8	978	323	311	290	11.19	908	680	518	414	291	219	178	141	0	0
12 No. 10	4.9	1081	391	307	331	12.94	997	748	574	460	328	250	201	166	141	0
12 No. 11	6.0	1191	446	294	369	15.06	1086	812	623	501	358	277	224	187	160	139
14 No. 5	1.4	735	141	322	190	7.07	698	511	374	283	173	0	0	0	0	0
14 No. 6	2.0	793	189	322	214	8.01	749	553	412	322	207	147	0	0	0	0
14 No. 7	2.7	864	245	320	244	9.15	811	603	456	360	244	177	138	0	0	0
14 No. 8	3.5	948	303	318	279	10.51	884	660	504	402	280	209	164	135	0	0
14 No. 9	4.5	1041	365	316	316	12.02	964	722	556	445	316	240	190	164	157	134
14 No. 10	5.7	1161	441	312	364	13.99	1067	801	620	499	357	275	222	184	157	137
14 No. 11	7.0	1290	509	297	407	16.43	1172	877	678	546	392	305	247	208	178	155

(continued)

TABLE A.17. (continued)

A_s	$\rho\%$	$0.56P_0$	M_u (k·ft)	P_b	M_b (k·ft)	e_b	Value of P_u (kips) for eccentricity e of (in.)									
							2	4	6	8	12	16	20	24	28	32
Circular section diameter 20 in., $0.1 \times f'_c \times A_g = 125\ k$																
16 No. 6	2.2	821	211	325	227	8.38	775	573	431	338	222	159	0	0	0	0
16 No. 7	3.1	902	272	324	260	9.66	846	629	479	381	262	193	151	0	0	0
16 No. 8	4.0	998	340	322	300	11.17	930	694	534	428	301	226	180	148	0	0
16 No. 9	5.1	1105	408	320	343	12.86	1022	765	592	477	340	261	208	172	147	129
16 No. 10	6.5	1242	491	316	397	15.06	1141	855	664	537	386	299	242	202	173	151
16 No. 11	7.9	1389	566	300	446	17.82	1261	941	731	591	424	331	270	227	196	171
Circular section diameter 22 in., $0.1 \times f'_c \times A_g = 152\ k$ $(P_u \geq 152\ k)$																
6 No. 8	1.2	874	173	356	249	8.39	851	643	482	373	221	0	0	0	0	0
6 No. 9	1.6	913	213	352	269	9.16	887	673	510	400	250	173	0	0	0	0
6 No. 10	2.0	965	263	346	294	10.18	932	710	544	431	283	199	0	0	0	0
6 No. 11	2.5	1020	308	333	316	11.36	978	746	575	458	313	222	169	0	0	0
6 No. 14	3.6	1151	397	315	375	14.25	1091	836	651	525	370	277	213	174	0	0
6 No. 18	6.3	1484	599	261	515	23.61	1376	1055	831	678	485	378	290	257	220	191
8 No. 8	1.7	924	219	373	272	8.74	898	680	515	405	260	185	0	0	0	0
8 No. 9	2.1	977	264	370	297	9.64	946	719	550	436	291	211	163	0	0	0
8 No. 10	2.7	1045	321	367	330	10.80	1007	768	592	474	328	239	184	154	0	0

8 No. 11			171	210	265	360	507	631	815	1069	12.12	360	357	376	1119	3.3
8 No. 14	164	182	217	261	329	426	590	729	933	1222	15.30	438	343	508	1294	4.7
10 No. 8	0	0	0	160	210	291	435	550	720	946	9.25	295	383	267	974	2.1
10 No. 9	0	0	0	187	242	326	472	592	769	1006	10.29	326	381	327	1040	2.6
10 No. 10	0	0	177	218	274	366	518	643	830	1083	11.63	366	378	391	1126	3.3
10 No. 11	0	172	201	239	302	399	557	691	889	1162	13.18	403	367	451	1218	4.1
10 No. 14	193	215	254	302	374	477	659	811	1037	1355	16.85	499	355	599	1436	5.9
12 No. 8	0	0	0	181	233	319	466	584	760	989	9.81	319	390	308	1024	2.5
12 No. 9	0	0	173	210	267	357	509	634	818	1061	11.00	356	389	374	1104	3.2
12 No. 10	0	170	200	244	305	401	561	695	891	1152	12.53	403	386	456	1206	4.0
12 No. 11	169	193	227	273	338	439	608	751	963	1247	14.32	447	375	529	1317	4.9
12 No. 14	219	248	286	337	414	530	728	895	1142	1477	18.52	561	363	696	1579	7.1
14 No. 8	0	0	164	200	256	345	495	618	795	1036	10.40	343	396	354	1074	2.9
14 No. 9	0	162	192	232	293	388	545	675	862	1119	11.74	386	395	426	1167	3.7
14 No. 10	168	188	224	269	335	436	605	745	947	1227	13.46	441	393	515	1287	4.7
14 No. 11	189	214	252	301	371	476	660	812	1030	1337	15.48	492	381	599	1416	5.7
16 No. 8	0	0	178	218	276	370	524	649	832	1082	11.00	368	401	394	1124	3.3
16 No. 9	153	179	209	251	317	416	579	712	909	1178	12.48	417	401	480	1230	4.2
16 No. 10	183	211	244	293	364	468	647	792	1005	1301	14.39	480	400	577	1367	5.3
16 No. 11	204	238	274	322	403	515	709	867	1100	1428	16.65	537	387	666	1514	6.6

(continued)

723

TABLE A.17. (continued)

A_s	$\rho\%$	$0.56P_0$	M_u (k·ft)	P_b	M_b (k·ft)	e_b	\multicolumn{10}{c}{Value of P_u (kips) for eccentricity e of (in.)}
							2	4	6	8	12	16	20	24	28	32
						Circular section diameter 24 in., $0.1 \times f'_c \times A_g = 180\ k$ ($P_u \geq 180\ k$)										
6 No. 8	1.0	1011	193	433	310	8.62	1011	783	597	464	276	185	0	0	0	0
6 No. 9	1.3	1051	239	430	333	9.31	1047	815	627	494	309	212	0	0	0	0
6 No. 10	1.7	1102	296	425	362	10.22	1093	854	663	528	348	242	182	0	0	0
6 No. 11	2.1	1158	351	413	388	11.26	1141	893	697	559	383	269	205	0	0	0
6 No. 14	3.0	1289	465	398	455	13.73	1258	988	780	633	451	333	257	208	0	0
6 No. 18	5.3	1622	700	353	618	21.02	1553	1222	976	804	583	455	371	307	260	223
8 No. 8	1.4	1061	250	452	335	8.91	1060	823	633	498	321	224	0	0	0	0
8 No. 9	1.8	1114	302	450	364	9.73	1109	864	671	534	356	257	195	0	0	0
8 No. 10	2.2	1183	367	446	401	10.78	1172	916	716	576	398	291	226	183	0	0
8 No. 11	2.8	1256	430	437	436	11.96	1237	967	759	612	435	321	350	206	0	0
8 No. 14	4.0	1431	581	427	525	14.73	1395	1093	865	705	511	394	311	257	219	192
8 No. 18	7.1	1875	910	395	740	22.45	1795	1403	1119	921	676	536	443	372	371	278
10 No. 8	1.7	1111	302	464	362	9.36	1108	866	671	533	356	252	191	0	0	0
10 No. 9	2.2	1178	370	462	397	10.31	1169	918	716	574	396	291	223	0	0	0

10 No. 10	2.8	1263	452	460	442	11.52	1248	983	772	624	442	330	260	210	0	0
10 No. 11	3.4	1355	523	451	484	12.89	1330	1046	824	669	481	364	291	240	202	0
10 No. 14	5.0	1574	693	442	593	16.10	1529	1205	956	782	569	445	357	298	259	224
12 No. 8	2.1	1161	352	473	389	9.87	1152	906	708	567	388	280	217	0	0	0
12 No. 9	2.7	1241	427	472	431	10.95	1226	967	762	614	432	321	249	204	0	0
12 No. 10	3.4	1344	520	471	484	12.34	1320	1045	829	673	483	365	290	236	201	0
12 No. 11	4.1	1454	610	461	535	13.90	1418	1122	892	726	525	404	326	269	229	198
12 No. 14	6.0	1717	811	454	664	17.55	1656	1312	1049	860	628	493	401	336	292	255
14 No. 8	2.4	1211	403	480	416	10.41	1200	944	743	600	418	306	240	194	0	0
14 No. 9	3.1	1305	490	480	465	11.62	1287	1015	804	655	466	351	276	227	191	0
14 No. 10	3.9	1424	592	480	527	13.17	1397	1104	881	722	521	401	318	262	224	196
14 No. 11	4.8	1553	689	470	586	14.94	1512	1193	954	783	568	443	357	295	253	222
14 No. 14	7.0	1859	926	464	735	19.01	1792	1414	1136	939	687	543	445	376	321	285
16 No. 8	2.8	1261	449	486	444	10.96	1248	984	776	630	447	331	260	210	0	0
16 No. 9	3.5	1368	547	488	500	12.30	1346	1064	846	690	499	379	301	248	210	0
16 No. 10	4.5	1505	665	489	571	14.02	1472	1167	931	765	559	433	348	289	245	215
16 No. 11	5.5	1652	771	479	637	15.97	1604	1269	1014	835	612	478	389	325	277	244
16 No. 14	8.0	2002	1029	473	808	20.47	1924	1521	1220	1010	747	587	481	409	357	309

APPENDIX B: Design Tables (S.I. Units)

TABLE B.1. Values of R_u and a/d for $f_c' = 21$ MPa (R_u in MPa)

ρ (%)	$f_y = 280$ MPa		$f_y = 350$ MPa		$f_y = 420$ MPa		$f_y = 520$ MPa	
	R_u	a/d	R_u	a/d	R_u	a/d	R_u	a/d
0.2	0.50	0.031	0.62	0.039	0.75	0.047	0.92	0.059
0.3	0.74	.046	0.92	.059	1.10	.071	1.35	.089
0.4	0.98	.062	1.22	.078	1.45	.094	1.79	.118
0.5	1.21	.078	1.50	.098	1.79	.118	2.18	.148
0.6	1.45	.094	1.79	.118	2.12	.141	2.59	.177
0.7	1.68	.110	2.06	.138	2.44	.165	2.96	.207
0.8	1.90	.126	2.33	.157	2.75	.189	3.34	.236
0.9	2.12	.142	2.59	.177	3.05	.213	3.68	.266
1.0	2.33	.157	2.84	.196	3.35	.238	4.02	.295
1.1	2.55	.173	3.10	.216	3.64	.260	4.36	.325
1.2	2.74	.188	3.35	.235	3.91	.280	4.56	.340
1.3	2.95	.204	3.59	.255	4.18	.306	($\rho_{max} = 1.16$)	
1.4	3.16	.220	3.82	.274	4.44	.330		
1.5	3.35	.236	4.04	.294	4.69	.353		
1.6	3.54	.252	4.27	.314	4.92	.376		
1.7	3.73	.267	4.48	.334	4.94	.378		
1.8	3.91	.282	4.69	.353	($\rho_{max} = 1.61$)			
1.9	4.09	.298	4.89	.373				
2.0	4.27	.314	5.08	.392				
2.1	4.43	.330	5.20	.405				
2.2	4.61	.345	($\rho_{max} = 2.06$)					
2.3	4.76	.360						
2.4	4.92	.376						
2.5	5.08	.392						
2.6	5.24	.408						
2.7	5.39	.424						
	5.50	.436						
	($\rho_{max} = 2.78$)							

Note: Last values are the maximum allowed by the ACI Code.

TABLE B.2. Values of R_u and a/d for $f'_c = 28$ MPa (R_u in MPa)

ρ (%)	$f_y = 280$ MPa		$f_y = 350$ MPa		$f_y = 420$ MPa		$f_y = 520$ MPa	
	R_u	a/d	R_u	a/d	R_u	a/d	R_u	a/d
0.2	0.50	0.024	0.63	0.029	0.75	0.025	0.93	0.044
0.3	0.74	.036	0.93	.044	1.11	.053	1.36	.066
0.4	0.98	.047	1.23	.059	1.46	.071	1.81	.088
0.5	1.23	.059	1.53	.074	1.81	.089	2.23	.110
0.6	1.46	.071	1.83	.088	2.16	.106	2.66	.132
0.7	1.70	.083	2.11	.103	2.50	.123	3.05	.154
0.8	1.93	.094	2.39	.118	2.81	.141	3.45	.176
0.9	2.16	.106	2.66	.132	2.14	.158	3.83	.198
1.0	2.39	.118	2.95	.147	3.46	.176	4.22	.220
1.1	2.60	.130	3.20	.161	3.77	.194	4.57	.242
1.2	2.81	.141	3.46	.176	4.08	.212	4.94	.264
1.3	3.04	.153	3.73	.191	4.36	.230	5.29	.286
1.4	3.25	.165	3.97	.206	4.65	.247	5.63	.308
1.5	3.46	.177	4.22	.221	4.92	.265	5.95	.330
1.6	3.67	.188	4.46	.236	5.22	.282	6.10	.342
1.7	3.87	.200	4.71	.250	5.48	.300	($\rho_{max} = 1.55$)	
1.8	4.08	.212	4.94	.265	5.75	.318		
1.9	4.27	.224	5.17	.280	6.00	.335		
2.0	4.46	.236	5.40	.294	6.26	.353		
2.1	4.65	.248	5.62	.309	6.50	.370		
2.2	4.85	.260	5.84	.323	6.58	.378		
2.3	5.04	.271	6.05	.338	($\rho_{max} = 2.14$)			
2.4	5.22	.282	6.26	.352				
2.5	5.39	.294	6.43	.367				
2.6	5.57	.306	6.65	.381				
2.7	5.74	.318	6.85	.396				
2.8	5.92	.330	6.95	.405				
2.9	6.09	.342	($\rho_{max} = 2.75$)					
3.0	6.26	.354						
3.1	6.42	.365						
3.2	6.57	.376						
3.3	6.74	.388						
3.4	6.89	.400						
3.5	7.03	.412						
3.6	7.17	.424						
3.7	7.31	.436						
	7.34	.438						
	($\rho_{max} = 3.72$)							

Note: Last values are the maximum allowed by the ACI code.

TABLE B.3. Values of R_u and a/d for $f'_c = 35$ MPa (R_u in MPa)

ρ (%)	$f_y = 350\ MPa$		$f_y = 420\ MPa$		$f_y = 520\ MPa$	
	R_u	a/d	R_u	a/d	R_u	a/d
0.2	0.63	0.024	0.75	0.028	0.93	0.035
0.3	0.93	.036	1.12	.042	1.38	.052
0.4	1.24	.047	1.48	.056	1.83	.070
0.5	1.53	.060	1.83	.070	2.26	.088
0.6	1.83	.071	2.18	.085	2.70	.106
0.7	2.12	.083	2.53	.100	3.11	.123
0.8	2.41	.094	2.87	.113	3.52	.141
0.9	2.69	.106	3.20	.127	3.91	.159
1.0	2.98	.118	3.53	.141	4.30	.177
1.1	3.26	.130	3.87	.155	4.69	.195
1.2	3.52	.141	4.17	.169	5.08	.212
1.3	3.80	.153	4.48	.183	5.46	.230
1.4	4.06	.165	4.79	.198	5.84	.247
1.5	4.32	.177	5.09	.212	6.15	.265
1.6	4.58	.188	5.39	.226	6.47	.282
1.7	4.84	.200	5.68	.240	6.82	.300
1.8	5.09	.212	5.96	.254	7.17	.318
1.9	5.34	.224	6.26	.268	7.26	.322
2.0	5.58	.235	6.52	.282	($\rho_{max} = 1.83$)	
2.1	5.84	.247	6.78	.296		
2.2	6.06	.259	7.05	.311		
2.3	6.29	.271	7.31	.325		
2.4	6.52	.282	7.56	.339		
2.5	6.74	.294	7.85	.352		
2.6	6.96	.306	7.87	.354		
2.7	7.18	.318	($\rho_{max} = 2.52$)			
2.8	7.40	.329				
2.9	7.61	.341				
3.0	7.81	.353				
3.1	8.01	.365				
3.2	8.21	.376				
	8.29	.381				
	($\rho_{max} = 3.24$)					

Note: Last values are the maximum allowed by the ACI Code.

TABLE B.4. Values of ρ_{max}, R_u(max), ρ_b, and ρ_{min}

$$\rho_{max} = 0.64\beta_1\frac{f'_c}{f_y} \times \frac{600}{600 + f_y} \qquad R_u = \phi\rho f_y\left(1 - \frac{\rho f_y}{1.7f'_c}\right)$$

f'_c (MPa)	$f_y = 280$ MPa				$f_y = 350$ MPa			
	ρ_{max} (%)	R_u (max) (MPa)	ρ_b (%)	ρ_{min} (%)	ρ_{max} (%)	R_u (max) (MPa)	ρ_b (%)	ρ_{min} (%)
17.5	2.32	4.55	3.09	0.5	1.72	4.31	2.29	0.4
21.0	2.78	5.50	3.71	0.5	2.06	5.20	2.75	0.4
28.0	3.72	7.34	4.96	0.5	2.75	6.95	3.67	0.4
35.0	4.37	8.77	5.81	0.5	3.24	8.29	4.32	0.4
42.0	4.9	10.0	6.53	0.5	3.64	9.46	4.85	0.4

f'_c (MPa)	$f_y = 420$ MPa				$f_y = 520$ MPa			
	ρ_{max} (%)	R_u (max) (MPa)	ρ_b (%)	ρ_{min} (%)	ρ_{max} (%)	R_u (max) (MPa)	ρ_b (%)	ρ_{min} (%)
17.5	1.34	4.07	1.79	0.33	0.97	3.78	1.29	0.267
21.0	1.61	4.94	2.15	0.33	1.16	4.56	1.55	0.267
28.0	2.14	6.58	2.85	0.33	1.55	6.10	2.07	0.267
35.0	2.52	7.87	3.39	0.33	1.83	7.26	2.44	0.267
42.0	2.83	9.00	3.77	0.33	2.06	8.3	2.75	0.267

TABLE B.5. Suggested design steel ratios ρ_s and comparison with other steel ratios

f'_c (MPa)	f_y (MPa)	ρ_b (%)	ρ_{max} (%)	ρ_s (sug-gested) (%)	R_u for ρ_s (MPa)	ρ_b elastic (%)	ρ for $q = 1.8$ (%)	Ratio ρ_s/ρ_{max}	Weight of ρ_s (kg/m³) of concrete)
17.5	280	3.09	2.32	1.2	2.69	1.02	1.13	0.52	96
	350	2.29	1.72	1.2	3.23	0.90	0.90	0.70	96
21.0	280	3.71	2.78	1.4	3.16	1.29	1.35	0.50	112
	350	2.75	2.06	1.2	3.35	1.11	1.08	0.58	96
	420	2.15	1.61	1.2	3.91	0.96	0.90	0.74	96
28.0	280	4.96	3.72	1.4	3.25	1.58	1.80	0.38	112
	350	3.67	2.75	1.4	3.97	1.62	1.44	0.51	112
	420	2.85	2.52	1.4	4.65	1.41	1.20	0.65	112
35.0	280	5.81	4.37	1.4	3.30	2.50	2.25	0.32	112
	350	4.32	3.24	1.4	4.06	2.15	1.80	0.43	112
	420	3.39	2.52	1.4	4.79	1.87	1.50	0.55	112

TABLE B.6. Minimum thickness of beams and one-way slabs

Member	Yield strength f_y (MPa)	Simply supported	One end continuous	Both ends continuous	Cantilever
Solid one-	280	$L/25$	$L/30$	$L/35$	$L/12.5$
way slabs	350	$L/22$	$L/27$	$L/31$	$L/11$
	420	$L/20$	$L/24$	$L/28$	$L/10$
Beams or	280	$L/20$	$L/23$	$L/26$	$L/10$
ribbed one-	350	$L/18$	$L/20.5$	$L/23.5$	$L/9$
way slabs	420	$L/16$	$L/18.5$	$L/21$	$L/8$

TABLE B.7. Minimum steel percentage $(\rho - \rho')$ for compression steel to yield in rectangular sections

$$(\rho - \rho') \geq 0.85\beta_1 \frac{f'_c}{f_y} \times \frac{d'}{d} \times \frac{600}{600 - f_y} \qquad (f_y, f'_c \text{ in MPa})$$

d/d'	f'_c (MPa)	β_1	$f_y =$ 280 MPa	$f_y =$ 350 MPa	$f_y =$ 420 MPa	$f_y =$ 520 MPa
0.10	17.5	0.85	0.83	0.85	0.97	1.74
	21	0.85	1.00	1.02	1.16	2.09
	28	0.85	1.33	1.35	1.55	2.78
	35	0.80	1.57	1.59	1.81	3.27
	42	0.75	1.78	1.81	2.06	3.71
0.12	17.5	0.85	1.00	1.02	1.16	2.09
	21	0.85	1.20	1.22	1.39	2.51
	28	0.85	1.60	1.62	1.86	3.34
	35	0.80	1.88	1.91	2.17	3.92
	42	0.75	2.14	2.17	2.47	4.45
0.15	17.5	0.85	1.25	1.28	1.46	2.61
	21	0.85	1.50	1.53	1.74	3.14
	28	0.85	2.00	2.03	2.33	4.17
	35	0.80	2.36	2.39	2.72	4.91
	42	0.75	2.67	2.72	3.09	5.57

Note: Minimum $(\rho - \rho')$ for any value of $d'/d = 10 \times (d'/d) \times$ value shown in table with $(d'/d) = 0.10$.

TABLE B.8. Modulus of elasticity of normal weight concrete

f'_c (MPa)	E_c kN/mm²	E_c (kgf/cm²) $\times 10^3$
17.5	20.0	200
21.0	22.5	225
28.0	25.0	250
35.0	29.0	290
42.0	32.0	320
49.0	33.5	335
56.0	36.5	365

Note: $E_c = 0.043w^{1.5}\sqrt{f'_c}$ MPa, $E_c = 0.14w^{1.5}\sqrt{f'_c}$ (kgf/cm²). For normal weight concrete, $w = 2350$ kg/m³; thus $E_c = 4730\sqrt{f'_c}$ MPa and $E_c = 15,200\sqrt{f'_c}$ (kgf/cm²).

TABLE B.9. Development length l_d (mm)

Bar diameter (mm)	f_y (MPa)	$f'_c =$ 17.5 MPa		$f'_c =$ 21 MPa		$f'_c =$ 28 MPa		$f'_c =$ 35 MPa	
		l_d	Top bar	l_d	Top bar	l_d	Top bar	l_d	Top bar
10	280	300	300	300	300	300	300	300	300
	350	300	300	300	300	300	300	300	300
	420	300	300	300	320	300	320	300	320
12	280	300	300	300	300	300	300	300	300
	350	300	350	300	350	300	350	300	350
	420	300	420	300	420	300	420	300	420
14	280	300	350	300	350	300	350	300	350
	350	300	400	300	400	300	400	300	400
	420	350	500	350	500	350	500	350	500
16	280	300	350	300	350	300	350	300	350
	350	320	450	320	450	320	450	320	450
	420	380	520	380	520	380	520	380	520
18	280	350	500	320	450	300	420	300	420
	350	450	620	400	550	380	520	380	520
	420	520	720	480	680	450	620	450	620
20	280	400	650	360	520	340	480	340	480
	350	500	700	450	650	400	560	400	560
	420	600	850	550	780	500	700	500	700
22	280	480	680	450	620	380	520	350	500
	350	600	850	550	780	480	680	450	620
	420	720	1000	650	920	580	800	520	720
25	280	650	920	580	800	500	700	450	620
	350	800	1100	720	1000	620	880	550	780
	420	950	1320	880	1200	750	1050	650	950
28	280	800	1100	720	1000	620	880	580	800
	350	1000	1400	920	1300	800	1100	700	1000
	420	1200	1680	1100	1550	950	1320	850	1200
32	280	1020	1420	920	1300	800	1100	720	1000
	350	1280	1780	1150	1620	1000	1400	900	1250
	420	1520	2120	1400	1950	1200	1680	1080	1500
36	280	1250	1750	1150	1600	1000	1400	880	1200
	350	1550	2180	1420	2000	1250	1750	1100	1550
	420	1880	2620	1700	2400	1500	2100	1320	1850

Note: For bars 36 mm or smaller, $l_d = 0.019 A_b f_y / \sqrt{f'_c}$ (mm) $\geq 0.058 d_b f_y$. Minimum length is 300 mm. For top bars, use $1.4 l_d$.

734

TABLE B.10. Designations, areas, perimeters, and weights of bars

Diameter (mm)	Area (mm²)	Weight (kg/meter)	Perimeter (mm)	Length per ton (m/1000 kg)
6	28.3	0.222	18.9	4510
8	50.3	0.395	25.1	2532
10	78.5	0.617	31.4	1621
12	113.1	0.888	37.7	1125
14	153.9	1.208	44.0	829
16	201.1	1.578	20.3	633
18	254.5	2.000	56.5	500
20	314.2	2.466	62.8	405
22	380.1	2.980	69.1	336
24	452.4	3.551	75.4	282
25	490.9	3.854	78.5	260
28	615.7	4.830	88.0	207
30	706.9	5.549	94.2	180
32	804.2	6.313	100.5	159
34	907.9	7.127	106.8	140
36	1017.9	7.990	113.1	125
40	1256.6	9.864	125.7	101
45	1590.4	12.490	141.4	80
50	1963.5	15.410	157.1	65

TABLE B.11. Areas of groups of steel bars, A_s (mm^2)

Diameter (mm)	Number of bars					
	1	2	3	4	5	6
6	28.27	57	85	113	141	170
8	50.27	101	151	201	252	302
10	78.54	157	236	314	393	471
12	113.10	226	339	452	565	678
14	153.94	308	462	616	770	924
16	201.06	402	603	804	1005	1206
18	254.47	508	762	1016	1270	1524
20	314.16	628	942	1256	1570	1884
22	380.13	760	1140	1520	1900	2280
24	452.39	904	1356	1808	2260	2712
25	490.90	982	1473	1964	2455	2945
28	615.75	1232	1848	2464	3080	3696
30	706.86	1414	2121	2828	3535	4242
32	804.25	1608	2413	3217	4021	4825
34	907.92	1816	2724	3632	4540	5448

	Number of bars					
	7	8	9	10	15	20
6	198	226	255	283	425	566
8	352	402	453	503	755	1006
10	550	628	706	785	1178	1570
12	791	904	1017	1130	1695	2260
14	1078	1232	1386	1540	2310	3080
16	1407	1608	1809	2010	3015	4020
18	1778	2032	2286	2540	3810	5080
20	2198	2512	2826	3140	4710	6280
22	2660	3040	3420	380	5700	7600
24	3164	3616	4068	4520	6780	9040
25	3436	3927	4418	4909	7364	9818
28	4312	4928	5544	6160	9249	1231
30	4949	5656	6363	7070	1060	1414
32	5630	6434	7238	8042	1206	1608
34	6355	7263	8171	9079	1362	1816

TABLE B.12. Areas of bars in slabs per meter width (mm^2)

Spacing (mm)	Bar diameter (mm)								
	6	8	10	12	14	16	18	20	25
50	565	1005	1571	2262	3079	4021	5089	6283	9817
60	471	838	1309	1885	2566	3351	4241	5236	8181
70	404	718	1122	1616	2200	2872	3635	4488	7012
80	353	628	982	1414	1924	2513	3181	3927	6136
90	314	558	873	1257	1710	2234	2828	3491	5454
100	283	503	785	1131	1539	2011	2545	3142	4909
110	257	457	714	1028	1400	1828	2313	2856	4462
120	236	419	654	942	1283	1675	2121	2618	4090
130	217	387	604	870	1184	1547	1957	2417	3776
140	202	359	561	808	1100	1436	1818	2244	3506
150	188	335	524	754	1026	1340	1696	2094	3272
160	177	314	491	707	962	1257	1590	1963	3068
170	166	296	462	665	905	1183	1497	1848	2887
180	157	279	436	628	855	1117	1444	1745	2727
190	149	265	413	595	810	1058	1340	1653	2584
200	141	251	393	565	770	1005	1272	1571	2454
210	135	239	374	539	733	957	1212	1496	2337
220	128	228	357	514	700	914	1157	1428	2231
230	123	218	341	492	669	875	1106	1366	2134
240	118	209	327	471	641	838	1060	1309	1054
250	113	201	314	452	616	804	1018	1257	1963
260	109	193	302	435	592	773	979	1208	1888
270	105	186	291	419	570	745	942	1164	1818
280	101	179	280	404	550	718	909	1122	1753
290	97	173	271	390	531	693	877	1083	1693
300	94	168	262	377	513	670	848	1047	1636

APPENDIX C: Structural Aids

Note: S.D. stands for shearing force diagram. B.D. stands for bending moment diagram. Bending moments are drawn on the tension sides of beams.

TABLE C.1. Simple beams (w = load/unit length)

1. Uniform load

W = total load = wL

$$R_A = R_B = V_A = V_B = \frac{W}{2}$$

$$M_x = \frac{Wx}{2}\left(1 - \frac{x}{L}\right)$$

$$M_{max} = \frac{WL}{8} \quad \text{(at center)}$$

$$\Delta_{max} = \frac{5}{384} \times \frac{WL^3}{EI} \quad \text{(at center)}$$

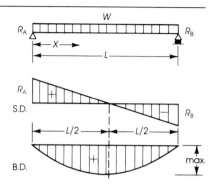

2. Uniform partial load

W = total load = wb

$$R_A = V_A = \frac{W}{L}\left(\frac{b}{2} + c\right)$$

$$R_B = V_B = \frac{W}{L}\left(\frac{b}{2} + a\right)$$

$$M_{max} = \frac{W}{2b}(x^2 - a^2) \quad \text{when } x = a + \frac{R_A b}{W}$$

$$\Delta_{max} = \frac{W}{384EI}(8L^3 - 4Lb^2 + b^3) \quad \text{when } a = c$$

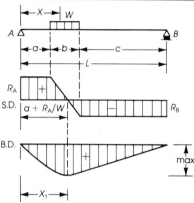

3. Uniform partial load at one end

W = total load = wa

$$R_A = V_A = W\left(1 - \frac{a}{2L}\right)$$

$$R_B = V_B = \frac{Wa}{2L}$$

$$M_{max} = \frac{Wa}{2}\left(1 - \frac{a}{2L}\right)^2 \quad \text{when } x = a\left(1 - \frac{a}{2L}\right)$$

$$\Delta = \frac{WL^4}{24aEI}\, n^2[2m^3 - 6m^2 + m(4 + n^2) - n^2]$$

when $x \geq a$

$$\Delta = \frac{WL^4 m}{24aEI}\,[n^2(2 - n)^2 - 2nm^2(2 - n) + m^3] \quad \text{when } x < a$$

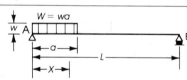
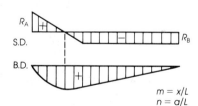

TABLE C.1. (continued)

---

4. Triangular load on span with maximum value at one end

$$W = \text{total load} = \frac{wL}{2}$$

$$R_A = V_A = \frac{W}{3}$$

$$R_B = V_B = \frac{2W}{3}$$

$$M_x = \frac{Wx}{3}\left(1 - \frac{x^2}{L^2}\right)$$

$$M_{max} = 0.128WL \quad \text{when } x = 0.5774L$$

$$\Delta_{max} = \frac{0.01304WL^3}{EI} \quad \text{when } x = 0.5193L$$

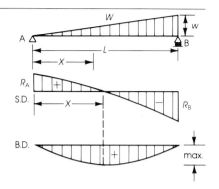

---

5. Triangular load with maximum value at midspan

$$W = \text{total load} = \frac{wL}{2}$$

$$R_A = R_B = V_A = V_B = \frac{W}{2}$$

$$M_x = Wx\left(\frac{1}{2} - \frac{2x^2}{3L^2}\right)$$

$$M_{max} = \frac{WL}{6} \quad \text{(at midspan)}$$

$$\Delta_{max} = \frac{WL^3}{60EI} \quad \text{(at midspan)}$$

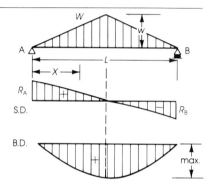

---

6. Moments at ends

$$R_A = R_B = V_A = V_B = \frac{M_A - M_B}{L}$$

$$\Delta_{max} \text{ (at midspan)} = \frac{ML^2}{8EI} \quad \text{when } M_A = M_B$$

$$\Delta \text{ (at midspan)} = \frac{M_A L^2}{16EI} \quad \text{when } M_B = 0$$

$$\Delta \text{ (at midspan)} = \frac{M_B L^2}{16EI} \quad \text{when } M_A = 0$$

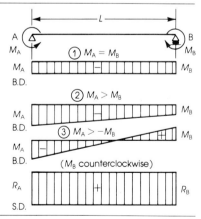

(continued)

TABLE C.1. (continued)

---

7. External moment at any point

$$R_A = -R_B = V_A = V_B = \frac{M}{L}$$

$$M_{CA} = \frac{Ma}{L}$$

$$M_{CB} = \frac{Mb}{L}$$

$$\Delta_c = \frac{-Mab}{3EIL}(a - b)$$

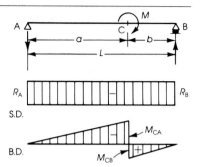

---

8. Concentrated load at midspan

$$R_A = R_B = V_A = V_B = \frac{P}{2}$$

$$M_{max} = \frac{PL}{4} \quad \text{(at midspan)}$$

$$\Delta_{max} = \frac{PL^3}{48EI} \quad \text{(at midspan)}$$

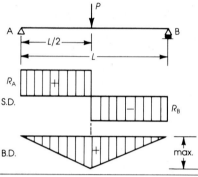

---

9. Concentrated load at any point

$$R_A = V_A = \frac{Pb}{L}$$

$$R_B = V_B = \frac{Pa}{L}$$

$$M_{max} = \frac{Pab}{L} \quad \text{(at point load)}$$

$$\Delta_C = \frac{Pa^2b^2}{3EIL} \quad \text{(at point load)}$$

$$\Delta_{max} = \frac{PL^3}{48EI}\left[\frac{3a}{L} - 4\left(\frac{a}{L}\right)^3\right] \quad \text{(when } a \geq b)$$

$$\text{at } x = \sqrt{a(b + L)/3}$$

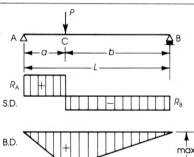

TABLE C.1. (continued)

---

10. Two symmetrical concentrated loads

$$R_A = R_B = V_A = V_B = P$$

$$M_{max} = Pa$$

$$\Delta_{max} = \frac{PL^3}{6EI}\left[\frac{3a}{4L} - \left(\frac{a}{L}\right)^3\right] \quad \text{(at midspan)}$$

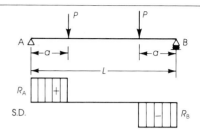

---

11. Two concentrated loads

$$R_A = V_A = \frac{P(b + 2c)}{L}$$

$$R_B = V_B = \frac{P(b + 2a)}{L}$$

$$M_C = \frac{Pa(b + 2c)}{L}$$

$$M_D = \frac{Pc(b + 2a)}{L}$$

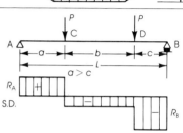

---

12. Two concentrated loads at one-third points

$$R_A = R_B = V_A = V_B = P$$

$$M_{max} = \frac{PL}{3}$$

$$\Delta_{max} = \frac{23PL^3}{648EI} \quad \text{(at midspan)}$$

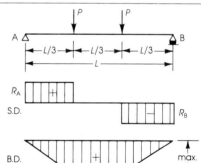

(continued)

TABLE C.1. (continued)

13. Three concentrated loads at one-fourth points

$$R_A = R_B = V_A = V_B = \frac{3P}{2}$$

$$M_C = M_E = \frac{3PL}{8}$$

$$M_D = \frac{PL}{2}$$

$$\Delta_{max} = \frac{19PL^3}{384EI} \quad \text{(at midspan)}$$

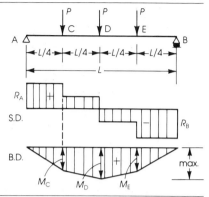

14. Three concentrated loads as shown

$$R_A = R_B = V_A = V_B = \frac{3P}{2}$$

$$M_C = M_E = \frac{PL}{4}$$

$$M_D = \frac{5PL}{12}$$

$$\Delta_{max} = \frac{53PL^3}{1296EI} \quad \text{(at midspan)}$$

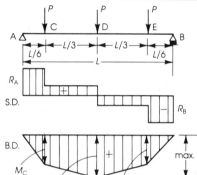

15. Uniformly distributed load and variable end moments

$$W = \text{total load} = wL$$

$$R_A = V_A = \frac{W}{2} + \frac{M_1 - M_2}{L}$$

$$R_B = V_B = \frac{W}{2} - \frac{M_1 - M_2}{L}$$

$$M_3 = \frac{WL}{8} - \frac{M_1 + M_2}{2} + \frac{(M_1 - M_2)^2}{2WL}$$

$$\text{at } x = \frac{L}{2} + \frac{M_1 - M_2}{W}$$

$$\Delta_x = \frac{Wx}{24EIL}\left[x^3 - \left(2L + \frac{4M_1}{W} - \frac{4M_2}{W}\right)x^2 + \frac{12M_1L}{W}x + L^3 - \frac{8M_1L^2}{W} - \frac{4M_2L^2}{W}\right]$$

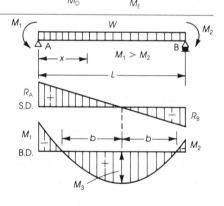

TABLE C.1. (continued)

16. Concentrated load at center and variable end moments

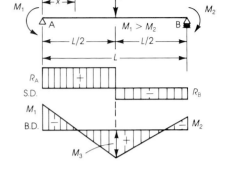

$$R_A = V_A = \frac{P}{2} + \frac{M_1 - M_2}{L}$$

$$R_B = V_B = \frac{P}{2} - \frac{M_1 - M_2}{L}$$

$$M_3 = \frac{PL}{4} - \frac{M_1 + M_2}{2} \quad \text{(at midspan)}$$

$$M_x = \left(\frac{P}{2} + \frac{M_1 - M_2}{L}\right) x - M_1 \quad \text{when } x < \frac{L}{2}$$

$$M_x = \frac{P}{2}(L - x) + \frac{(M_1 - M_2)}{L} x - M_1 \quad \text{when } x > \frac{L}{2}$$

$$\Delta_x = \frac{Px}{48EI}\left[3L^2 - 4x^2 - \frac{8(L - x)}{PL}\{M_1(2L - x) + M_2(L + x)\}\right] \quad \text{when } x < \frac{L}{2}$$

17. One concentrated moving load

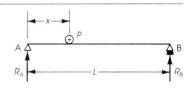

$$R_A\text{max} = V_A\text{max} = P \quad \text{at } x = 0$$

$$R_B\text{max} = V_A\text{max} = P \quad \text{at } x = L$$

$$M_{\text{max}} = \frac{PL}{4} \quad \text{at } x = \frac{L}{2}$$

$$M_x = \frac{P}{L}(L - x)x$$

18. Two equal concentrated moving loads

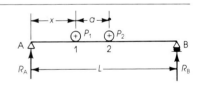

$$R_A\text{max} = V_A\text{max} = P\left(2 - \frac{a}{L}\right) \quad \text{at } x = 0$$

$$M_{\text{max}} = \frac{P}{2L}\left(L - \frac{a}{2}\right)^2$$

when $a < 0.586L$ under load 1 at $x = \frac{1}{2}\left(L - \frac{a}{2}\right)$

$$M_{\text{max}} = \frac{PL}{4} \quad \text{when } a > 0.5L \text{ with one load at midspan}$$

(continued)

TABLE C.1. (continued)

19. Two unequal concentrated moving loads

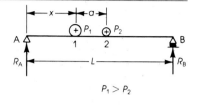

$$R_A\text{max} = V_A\text{max} = P_1 + P_2 \left(\frac{L-a}{L}\right) \quad \text{at } x = 0$$

$$M_\text{max} = (P_1 + P_2) \frac{x^2}{L}$$

under load P_1 at $x = \frac{1}{2}\left(L - \frac{P_2 a}{P_1 + P_2}\right)$

$M_\text{max} = \frac{P_1 L}{4}$ may occur with larger load at center of span and other load off span

---

20. General rules for simple beams carrying moving concentrated loads

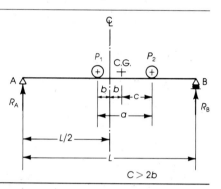

V_max occurs at one support and other loads on span (trial method)

For M_max: place center line of beam midway between center of gravity of loads and nearest concentrated load. M_max occurs under this load (here P_1)

TABLE C.2. Cantilever beams

21. Uniform load

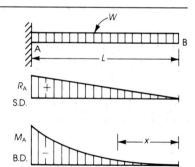

$W = \text{total load} = wL$

$R_A = V_A = W$

$M_A = \dfrac{WL}{2}$ (at support A)

$M_x = \dfrac{Wx^2}{2L}$

$\Delta_B\text{max} = \dfrac{WL^3}{8EI}$

$\Delta_x = \dfrac{W}{24EIL}(x^4 - 4L^3 x + 3L^4)$

TABLE C.2. (continued)

22. Partial uniform load starting from support

W = total load = wa

$R_A = V_A = W$

$M_A = \dfrac{Wa}{2}$ (at support A)

$M_x = \dfrac{Wx^2}{2a}$

$\Delta_C = \dfrac{Wa^3}{8EI}$

$\Delta_B\text{max} = \dfrac{Wa^3}{8EI}\left(1 + \dfrac{4b}{3a}\right)$

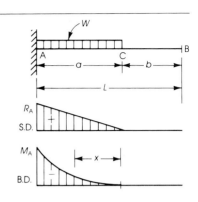

23. Concentrated load

$R_A = V_A = P$

$M_{\text{max}} = Pa$ (at support A)

$M_x = Px$

$\Delta_C = Pa^3/3EI$

$\Delta_B\text{max} = \dfrac{Pa^3}{3EI}\left(1 + \dfrac{3b}{2a}\right)$ (at free end)

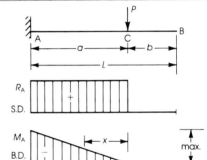

24. Concentrated load at free end

$R_A = V_A = P$

$M_{\text{max}} = PL$ (at A)

$M_x = Px$

$\Delta_B\text{max} = \dfrac{PL^3}{3EI}$

$\Delta_x = \dfrac{P}{6EI}(2L^3 - 3L^2x + x^3)$

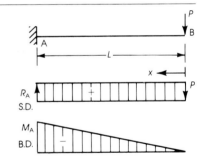

TABLE C.3. Propped beams

25. Uniform load

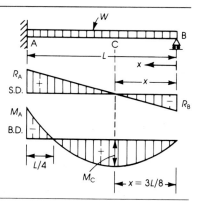

$$W = \text{total load} = wL$$

$$R_A = V_A = \frac{5W}{8} \qquad R_B = V_B = \frac{3W}{8}$$

$$M_A = -\frac{WL}{8} \qquad M_C = \frac{9WL}{128} \quad \left(\text{at } x = \frac{3}{8}L\right)$$

$$\Delta_x = \frac{WL^3}{48EI}(m - 3m^3 + 2m^4) \quad \text{where } m = \frac{x}{L}$$

$$\Delta_{max} = \frac{WL^3}{185EI} \quad \text{at a distance } x = 0.4215L \text{ (from support B)}$$

26. Partial uniform load starting from hinged support

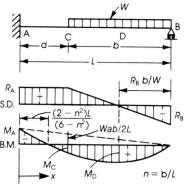

$$W = wb \qquad n = \frac{b}{L} \qquad m = \frac{x}{L}$$

$$R_A = V_A = \frac{Wn}{8}(6 - n^2)$$

$$R_B = V_B = \frac{W}{8}(n^3 - 6n + 8)$$

$$M_A = -\frac{Wb}{8}(2 - n^2) \qquad M_C = \frac{Wb}{8}(6n - n^3 - 4)$$

$$\Delta_x = \frac{WbL^2}{48EI}[(n^2 - 6)m^3 - (3n^2 - 6)m^2] \quad \text{when } x \le a$$

$$\Delta_x = \frac{WL^4}{48bEI}[2P^4 - P^3 n(n^3 - 6n + 8) + Pn^2(3n^2 - 8n + 6)] \quad \text{when } x \ge a \text{ and } P = \frac{L - x}{L}$$

27. Partial uniform load starting from fixed end

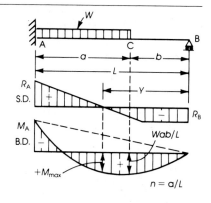

$$W = wa \qquad n = \frac{a}{L}$$

$$R_A = V_A = \frac{W}{8}[8 - n^2(4 - n)]$$

$$R_B = V_B = \frac{Wn^2}{8}(4 - n) \qquad Y = b + an^2(4 - n)$$

$$M_A = -\frac{Wa}{8}(2 - n)^2$$

$$M_{max} = \frac{Wa}{8}\left\{-\frac{[8 - n^2(4 - n)]^2}{16} + 4 - n(4 - n)\right\}$$

$$\Delta_C = \frac{Wa^3}{48EI}(6 - 12n + 7n^2 - n^3)$$

TABLE C.3. (continued)

28. Triangular load on all span L

$$W = \text{total load} = \frac{wL}{2}$$

$$R_A = V_A = \frac{4}{5} W \qquad R_B = \frac{W}{5} = V_B$$

$$M_A = -\frac{2}{15} WL$$

$$M_C = +\frac{3}{50} WL$$

$$\Delta_{max} = \frac{WL^3}{212EI} \quad (\text{at } x = 0.447L)$$

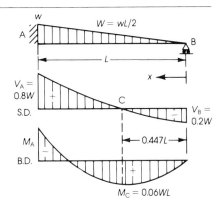

29. Triangular load on part of the span

$$W = \frac{wa}{2}$$

$$R_B = V_B = \frac{Wa^2}{20L^3} (5L - a)$$

$$R_A = W - R_B$$

$$M_A = \frac{Wa}{60L^2} (3a^2 - 15aL + 20L^2)$$

Maximum positive moment at

$$S = b + \frac{a^2}{2L} \sqrt{1 - \frac{a}{5L}}$$

M_{max}(positive) at D:

$$M_D = R_B S - \frac{WL}{3a^3} (-b + S)^3$$

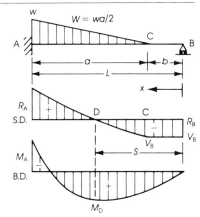

30. Concentrated load at midspan

$$R_A = V_A = \frac{11P}{16}$$

$$R_B = V_B = \frac{5P}{16}$$

$$M_A = -\frac{3PL}{16}$$

$$M_C = \frac{5PL}{32}$$

$$\Delta_C = \frac{7PL^3}{768EI}$$

$$\Delta_{max} = \frac{PL^3}{107EI} \quad (\text{at } x = 0.447L \text{ from B})$$

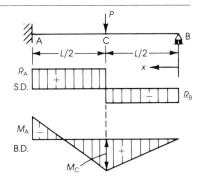

(continued)

TABLE C.3. (continued)

---

31. Concentrated load at any point

$$R_A = V_A = P - R_B \qquad R_B = V_B = \frac{Pa^2}{2L^3}(b + 2L)$$

$$M_A = -\frac{Pb(L^2 - b^2)}{2L^2}$$

$M_A$max $= 0.193PL$ when $b = 0.577L$

$$M_C = \frac{Pb}{2}\left(2 - \frac{3b}{L} + \frac{b^3}{L^3}\right)$$

$M_C$max $= 0.174PL$ when $b = 0.366L$

$$\Delta_C = \frac{Pa^3b^2}{12EIL^3}(4L - a)$$

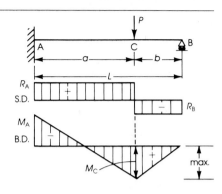

---

32. Two concentrated loads at one-third points

$$R_A = V_A = \frac{4P}{3}$$

$$R_B = V_B = \frac{2P}{3}$$

$$M_A = -\frac{PL}{3}$$

$$M_C = \frac{PL}{9} \qquad M_D = \frac{2PL}{9}$$

$$\Delta_{max} = \frac{PL^3}{65.8EI}$$

occurs at point $= 0.423L$ from support B

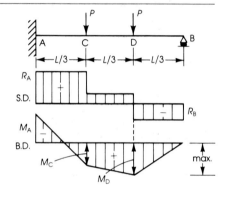

TABLE C.4. Fixed end beams

33. Uniform load

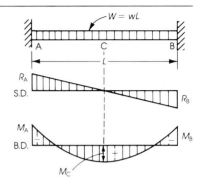

W = total load = wL

$$R_A = V_A = R_B = V_B = \frac{W}{2}$$

$$M_A = M_B = -\frac{WL}{12} \quad \text{(at support)}$$

$$M_C\text{max} = \frac{WL}{24} \quad \text{(at midspan)}$$

$$\Delta_{max} = \frac{WL^3}{384EI} \quad \text{(at midspan)}$$

$$\Delta x = \frac{Wx^2}{24EIL}(L - x)^2$$

34. Uniform partial load at one end

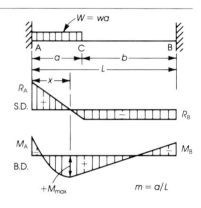

$$W \text{ = total load} = wa \qquad m = \frac{a}{L}$$

$$R_A = V_A = \frac{W(m^3 - 2m^2 + 2)}{2}$$

$$R_B = V_B = \frac{Wm^2(2 - m)}{2} = W - R_A$$

$$M_A = \frac{WLm}{12}(3m^2 - 8m + 6)$$

$$M_B = \frac{WLm^2}{12}(4 - 3m)$$

$$M_{max} = \frac{WLm^2}{12}\left(-\frac{3}{2}m^5 + 6m^4 - 6m^3 - 6m^2 + 15m - 8\right)$$

$$\text{when } x = \frac{a}{2}(m^3 - 2m^2 + 2)$$

$$\Delta_{max} = \frac{WL^3}{333EI}$$

$$\Delta_C = \frac{WL^3}{384EI}$$

(continued)

TABLE C.4. (continued)

35. Triangular load

$$W = \frac{wL}{2}$$

$$R_A = V_A = 0.7W$$

$$R_B = V_B = 0.3W$$

$$M_A = \frac{WL}{10} \qquad M_B = \frac{WL}{15}$$

$$\Delta_{max} = \frac{WL^3}{382EI} \quad \text{(at } x = 0.55L \text{ from B)}$$

$$M_C \text{ (maximum positive moment)} = +\frac{WL}{23.3} \quad \text{(at } 0.55L \text{ from B)}$$

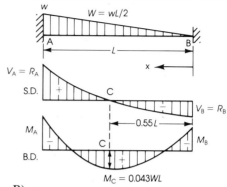

36. Triangular load on part of the span

$$W = \frac{wa}{2}$$

$$R_B = V_B = \frac{Wa^2}{10L^3}(5L - 2a)$$

$$R_A = W - R_B$$

$$M_A = \frac{Wa}{30L^2}(3a^2 + 10bL)$$

$$M_B = \frac{Wa}{30L^2}(-3a^2 + 5aL)$$

Maximum positive moment at $S = b + \dfrac{a^2}{3.16L}\sqrt{5 - \dfrac{2a}{L}}$

$$M_D = R_B S - \frac{WL}{3a^3}(a + S - L)^3 - M_B$$

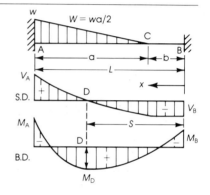

37. Triangular load, maximum intensity at midspan

$$W = \text{total load} = \frac{wL}{2}$$

$$R_A = R_B = \frac{W}{2}$$

$$M_A = M_B = -\frac{5}{48}WL$$

$$M_C \text{ (maximum positive)} = \frac{WL}{16}$$

$$\Delta_{max} = \frac{1.4WL^3}{384EI} \quad \text{(at midspan)}$$

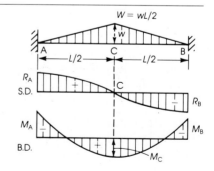

TABLE C.4. (continued)

38. Concentrated load at midspan

$$R_A = V_A = R_B = V_B = \frac{P}{2}$$

$$M_A = M_B = M_C = -\frac{PL}{8}$$

$$\Delta_{max} = \frac{PL^3}{192EI} \quad \text{(at midspan)}$$

$$\Delta x = \frac{Px^2}{48EI}(3L - 4x) \quad \left(x < \frac{L}{2}\right)$$

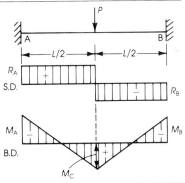

39. Two symmetrical concentrated loads

$$R_A = V_A = R_B = V_B = P$$

$$M_A = M_B = -\frac{Pa(L - a)}{L}$$

$$M_C = M_D = \frac{Pa^2}{L}$$

$$\Delta_{max} = \frac{PL^3}{6EI}\left[\frac{3a^2}{4L^2} - \left(\frac{a}{L}\right)^3\right] \quad \text{(at midspan)}$$

If $a = \dfrac{L}{3}$,

$$M_A = M_B = \frac{2}{9}PL$$

$$\Delta_{max} = \frac{5PL^3}{648EI} \quad \text{(at centerline)}$$

If $a = \dfrac{L}{4}$,

$$M_A = M_B = \frac{3}{16}PL$$

$$\Delta_{max} = \frac{PL^3}{192EI} \quad \text{(at centerline)}$$

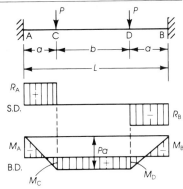

(continued)

TABLE C.4. (continued)

40. Concentrated load at any point

$$R_A = V_A = P \left(\frac{b}{L}\right)^2 \left(1 + \frac{2a}{L}\right)$$

$$R_B = V_B = P \left(\frac{a}{L}\right)^2 \left(1 + \frac{2b}{L}\right)$$

$$M_A = -\frac{Pab^2}{L^2} \qquad M_B = -\frac{Pba^2}{L^2} \qquad M_C = \frac{2Pa^2b^2}{L^3}$$

$$\Delta_C = \frac{Pa^3b^3}{3EIL^3} \quad \text{(at point C)}$$

$$\Delta_{\max} = \frac{2Pa^2b^3}{3EI(3L - 2a)^2} \quad \text{when } x = \frac{2bL}{3L - 2a} \quad \text{and} \quad b > a$$

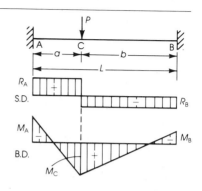

TABLE C.5. Moments in two unequal spans and values of the coefficient K (w = unit load/unit length)

1. Load on short span

$$M_B = \frac{wL_2^3}{8(L_1 + L_2)} = \frac{wL_2^2}{K}$$

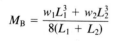

L_2/L_1	0.20	0.25	0.30	0.40	0.50	0.60	0.70	0.80	0.90	1.00
K	46.0	40.0	34.7	28.0	24.0	21.4	19.5	18.0	16.9	15.9

2. Load on long span

$$M_B = \frac{wL_1^3}{8(L_1 + L_2)} = \frac{wL_1^2}{K}$$

L_2/L_1	0.20	0.25	0.30	0.40	0.50	0.60	0.70	0.80	0.90	1.00
K	9.6	10.0	10.4	11.2	12.0	12.8	13.6	14.4	15.2	15.9

3. Both spans loaded with w_1 on L_1 and w_2 on L_2

$$M_B = \frac{w_1 L_1^3 + w_2 L_2^3}{8(L_1 + L_2)}$$

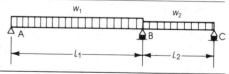

TABLE C.6. Moments in three unequal spans and values of the coefficient K (w = load/unit length)

4. Load on span CD

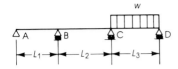

$$M_B = \frac{wL_3^2}{K} \qquad M_C = \frac{wL_3^2}{K}$$

L_2/L_3	(positive)	(negative)
0.25	100.0	9.9
0.30	90.9	10.3
0.40	76.3	11.0
0.50	70.4	11.7
0.60	65.8	12.3
0.70	62.9	13.0
0.80	61.7	13.7
1.00	59.9	14.9

5. Load on middle span

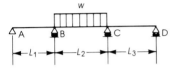

$$M_B = M_C = \frac{wL_2^2}{K}$$

L_2/L_1	(negative)
0.25	43.5
0.30	38.5
0.40	32.3
0.50	27.8
0.60	25.6
0.70	23.3
0.80	22.2
1.00	20.0

6. Load on span AB

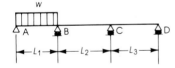

$$M_B = \frac{wL_1^2}{K} \qquad M_C = \frac{wL_1^2}{K}$$

L_2/L_1	(negative)	(positive)
0.25	9.9	100.0
0.30	10.3	90.9
0.40	11.0	76.3
0.50	11.7	70.4
0.60	12.3	65.8
0.70	13.0	62.9
0.80	13.7	61.7
1.00	14.9	59.9

TABLE C.7. Maximum and minimum moments in equal spans continuous beams

7. Uniform loads

$$M = \frac{wL^2}{K}$$

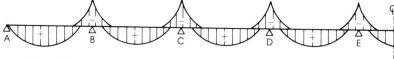

where w = (D.L. + L.L.) per unit length   D.L. = Uniform dead load   L.L. = Uniform live load

Values of coefficient K

Ratio D.L./w	First span AB (positive moment)				Second support B (negative moment)			
	Number of spans				Number of spans			
	2	3	4	5	2	3	4	5
0.0	10.5	10.0	10.2	10.1	8.0	8.6	8.3	8.3
0.1	10.8	10.2	10.4	10.3	8.0	8.7	8.4	8.5
0.2	11.1	10.4	10.6	10.6	8.0	8.8	8.5	8.6
0.3	11.4	10.6	10.9	10.8	8.0	9.0	8.6	8.7
0.4	11.8	10.9	11.1	11.0	8.0	9.1	8.6	8.8
0.5	12.1	11.1	11.4	11.3	8.0	9.2	8.8	8.9
0.6	12.5	11.4	11.7	11.6	8.0	9.4	8.9	9.0
0.7	12.9	11.6	12.0	11.9	8.0	9.5	9.0	9.1
0.8	13.3	11.9	12.3	12.2	8.0	9.7	9.1	9.2
0.9	12.8	12.2	12.6	12.5	8.0	9.8	9.2	9.4
1.0	14.3	12.5	13.0	12.8	8.0	9.9	9.3	9.5

Ratio D.L./w	Second span BC (positive moment)			Third support C (negative moment)		Third span CD (positive moment)	Interior span (positive moment)	Interior support (negative moment)
	Number of spans			Spans		Span		
	3	4	5	4	5	5		
0.0	13.4	12.4	12.7	9.3	9.0	11.7	12.0	8.8
0.1	14.3	13.2	13.5	9.7	9.3	12.3	12.6	9.1
0.2	15.4	14.0	14.3	10.0	9.6	12.9	13.3	9.8
0.3	16.7	14.9	15.3	10.4	9.9	13.6	14.1	9.5
0.4	18.2	16.0	16.5	10.8	10.2	14.3	15.0	9.9
0.5	20.0	17.2	17.9	11.5	10.5	15.2	16.0	10.1
0.6	22.2	18.7	19.5	11.7	10.9	16.2	17.2	10.5
0.7	25.0	20.4	21.4	12.2	11.3	17.3	18.4	10.8
0.8	28.6	22.4	23.8	12.7	11.7	18.5	20.0	11.2
0.9	33.3	24.9	26.6	13.3	12.2	20.0	21.8	11.6
1.0	40.0	28.3	30.0	14.0	12.7	21.7	24.0	12.0

TABLE C.7. (continued)

Example: K values

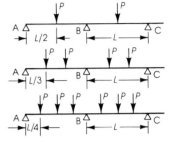

1. $\dfrac{\text{D.L.}}{w} = 0.4$

2. $\dfrac{\text{D.L.}}{w} = 1.0$

8. Concentrated loads

   P' = concentrated dead load

   P'' = concentrated live load

$$M = \left(\frac{P'}{K_1} + \frac{P''}{K_2}\right)L$$

	First span AB						Second support B					
	K_1 (D.L.)			K_2 (L.L)			K_1 (D.L.)			K_2 (L.L.)		
Number of spans	2	3	4	2	3	4	2	3	4	2	3	4
Central load	6.40	5.71	5.89	4.92	4.70	4.76	5.35	6.67	6.22	5.33	5.71	5.53
One-third- point loads	4.50	4.09	4.20	3.60	3.46	3.50	3.00	3.75	3.50	3.00	3.21	3.11
One-fourth- point loads	3.67	3.20	3.34	2.61	2.46	2.50	2.13	2.67	2.49	2.13	2.28	2.21

	Second span BC				Third supports C	
	K_1		K_2		K_1	K_2
Number of spans	3	4	3	4	4	4
Central load	10.00	8.61	5.71	5.46	9.33	6.22
One-third- point loads	15.00	9.00	5.00	4.50	5.25	3.50
One-fourth- point loads	8.00	6.05	3.20	3.01	3.73	2.49

Example: K values

K_1 (dead load) $M_{AB}(\text{max}) = \left(\dfrac{P'}{5.71} + \dfrac{P''}{4.7}\right)L$

$-M_B(\text{max}) = \left(\dfrac{P'}{6.67} + \dfrac{P''}{5.71}\right)L$

K_2 (live load) $M_{BC}(\text{max}) = \left(\dfrac{P'}{10} + \dfrac{P''}{5.71}\right)L$

TABLE C.8. Moments in unequal spans continuous beams subjected to unequal loads

9. Unequal spans and unequal loads. For approximate bending moments in continuous beams, use

$L' = 0.8L$ for spans continuous at both ends

$L' = L$ for spans continuous at only one end

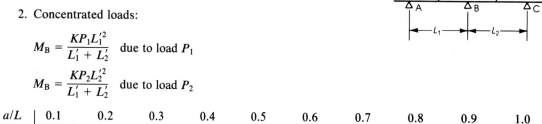

1. Uniform loads (load on two adjacent spans):

$$M_B = \frac{w_1 L_1'^3 + w_2 L_2'^3}{8.5(L_1' + L_2')}$$

2. Concentrated loads:

$$M_B = \frac{K P_1 L_1'^2}{L_1' + L_2'} \quad \text{due to load } P_1$$

$$M_B = \frac{K P_2 L_2'^2}{L_1' + L_2'} \quad \text{due to load } P_2$$

a/L	0.1	0.2	0.3	0.4	0.5	0.6	0.7	0.8	0.9	1.0
K	0.080	0.136	0.168	0.182	0.176	0.158	0.128	0.090	0.050	0.000

3. Moments within span:

- Maximum positive moment is obtained by superposing B.M. due to D.L. + L.L. and the negative moments at supports due to D.L. only.
- Maximum negative moment is obtained by superposing B.M. due to D.L. only and the negative moments at supports due to D.L. + L.L.

Example:

$$M_B = -\frac{30(0.8 \times 5)^3 + 40(0.8 \times 6)^3}{8.5(0.8 \times 5 + 0.8 \times 6)} = -84.8 \text{ kN·m}$$

$$M_C = -\frac{40(0.8 \times 6)^3 + 25(0.8 \times 4)^3}{8.5(0.8 \times 4 + 0.8 \times 6)} = -77.1 \text{ kN·m}$$

$$M_E = \text{(at centerline of BC)} = +\frac{wL^2}{8} + \frac{1}{2}(M_B + M_C) = \frac{40 \times 36}{8} + \frac{1}{2}(-84.8 - 77.1) = +99 \text{ kN·m}$$

ANSWERS TO SELECTED PROBLEMS

Chapter 3

3.1. (a) $E_c = 4027.56$ ksi $\quad f_r = 474$ psi
$E_c = 3155.93$ ksi $\quad f_r = 411$ psi
$E_c = 1940.00$ ksi $\quad f_r = 335$ psi
(b) $n = 7.2, 9.19, 15$

3.3. (a) $\rho_b = 0.0371$, $\rho_{max} = 0.0278$, $\frac{a}{d} = 0.248$, $\left(\frac{a}{d}\right)max = 0.436$
$R_u = 504$ psi, $R_u(max) = 783$ psi, $\rho = 0.0157$
(c) $\rho_b = 0.0335$, $\rho_{max} = 0.0254$, $\frac{a}{d} = 0.282$, $\left(\frac{a}{d}\right)max = 0.356$
$R_u = 927$ psi, $R_u(max) = 1110$ psi, $\rho = 0.02$

3.5. (a) over-reinforced
(b) $A_{sb} = 3.96$ in.^{2}
(c) $A_s = 4.78$ in.^{2}, $M_u = 201.2$ k·ft, ACI $M_u = 159.84$ k·ft
(d) $M_u = 117.4$ k·ft

3.7. $P_L = 13.82$ k at midspan

3.9. (a) $M_u = 118.75$ k·ft
(b) $M_u = 281.3$ k·ft

3.11. M_u (internal) $= 273.2$ k·ft
M_u (external) $= 263.1$ k·ft
The section is adequate.

3.13. $M_u = 339.6$ k·ft

3.15. $M_u = 339.6$ k·ft

Chapter 4

4.1. (a) $d = 19.4$ in., $A_s = 4.0$ in.^{2} (4 No. 9), $h = 23$ in.
(b) $d = 21.4$ in., $A_s = 3.42$ in.^{2} (6 No. 7), $h = 25$ in.
(c) $d = 24.2$ in., $A_s = 2.90$ in.^{2} (3 No. 9), $h = 27$ in.

4.3. $A_s = 2.28$ in.^{2} (3 No. 8)

4.5. Negative $M_u = 154.88$ k·ft, $h = 24$ in., $A_s = 1.74$ in.^{2}
Positive $M_u = 305.15$ k·ft, $h = 24$ in., $A_s = 4.02$ in.^{2}

4.7. At the fixed end: $d = 13.9$ in., $A_s = 4.64$ in.^{2}, $h = 18$ in.
At the free end: $h = 10$ in.

4.9. (a) $d = 21.08$ in., $A_s = 3.39$ in.^{2}

(b) $d = 18.26$ in., $A_s = 3.84$ in.^{2}, $A_s' = 0.903$ in.^{2}

(c) $d = 21.08$ in., $A_s = 3.13$ in.^{2}

Chapter 5

5.1. (a) $\rho = 0.0101$, $K = 0.36$, $j = 0.88$, $R = 178$ psi, $n = 10.0$

(b) $\rho = 0.0128$, $K = 0.378$, $j = 0.874$, $R = 223$ psi, $n = 9.13$

(c) $\rho = 0.0141$, $K = 0.375$, $j = 0.875$, $R = 295$ psi, $n = 8.0$

5.3. (a) $f_c = 1.107$ ksi, $f_s = 13.94$ ksi

(b) $f_c = 1.107$ ksi, $f_s = 13.93$ ksi

(c) $M_a = 64.6$ k·ft

5.5. $f_c = 1.948$ ksi > 1.8 ksi

$f_s = 22.14$ ksi > 22 ksi

The section is not adequate.

5.7. $M_a = 78.76$ k·ft, $W_a = 2.46$ k/ft

5.9. (a) $f_c = 476$ psi, $f_s = 10.55$ ksi, $M_a = 150.5$ k·ft

(b) Same as (a).

5.11. $M_a = 109.8$ k·ft

5.13. $M_a = 212.6$ k·ft

Chapter 6

6.1. $h = 22$ in., $A_s = 4.54$ in.^{2}, instantaneous $\Delta = 0.588$ in.

long-term $\Delta = 0.616$ in., total $\Delta = 1.204$ in.

6.3. $h = 24$ in., $A_s = 5.17$ in.^{2}, instantaneous $\Delta = 0.34$ in.

total $\Delta = 0.802$ in.

6.5. $h = 25$ in., instantaneous $\Delta = 1.046$ in., total $\Delta = 2.18$ in.

6.7. $W_{max} = 0.011$ in., $Z = 117.8$ k/in.

6.9. $W_{max} = 0.0102$ in., $Z = 111.85$ k/in.

Chapter 7

7.1. $x_1 = 42$ in., $x_2 = 18$ in., $x_3 = 18$ in., $x_4 = 48$ in.

$x_5 = 19$ in., $x_6 = 72$ in., $x_7 = 18$ in.

7.3. M_u (one bar) $= 109.8$ k·ft, $d = 35.5$, $h = 39$

Chapter 8

8.1. v(max) $= 77.4$ psi, v(average) $= 68.2$ psi

8.3. $V_s = 38.4$ k for $s = 5.5$ in.

$V_s = 17.6$ k for $s = 12$ in.

8.7. No. 3 stirrups spaced at 8 in.

8.9. 5 No. 3 stirrups spaced at 5.5 in., then 6 No. 3 stirrups spaced at 8 in.

Chapter 9

9.1. $M_u = 13.6$ k·ft
9.3. $h = 10$ in., No. 7 bars spaced at 5 in.
9.9. $h = 12$ in., 2 No. 5 bars per rib.

Chapter 10

10.1. $P_u = 601.4$ k
10.3. $P_u = 599.5$ k
10.5. 10×18 in. (6 No. 7 bars)
For ties, use No. 3 bars spaced at 10 in.
10.7. Diameter = 16 in. (6 No. 9 bars)
10.9. 14×14 in. (8 No. 7 bars)

Chapter 11

11.1. $P_b = 346.15$ k, $M_b = 443.75$ k·ft, $e_b = 15.38$ in.
11.3. $P_n = 732.8$ k, P_n (Whitney) = 717.4 k
11.5. $P_u = 126.6$ k. By analysis, $P_u = 133$ k.
11.7. 12×20 in., $A_s = 3$ No. 9 bars, $A_s' = 2$ No. 8 bars
11.9. 12×13 in.
$A_s = 2$ No. 9 bars, $A_s' = 2$ No. 9 bars
11.11. 14×23 in., $A_s = 3$ No. 9 bars
or 14×21 in., $A_s = A_s' = 4$ No. 9 bars

Chapter 12

12.1. $\delta_b = 1.166$, $P_n = 476$ k, $M_n = 545.8$ k·ft
The section is adequate.
12.3. $\delta_b = 1.2$, $M_n = 557$ k·ft
The section is adequate.
12.5. $\delta_b = 1.137$, $M_n = 255$ k·ft
The section is adequate.

Chapter 13

13.1. Width = 6.5 ft, $h = 13$ in., $A_s = $ No. 7 bars spaced at 10 in.
13.5. 7.5×7.5 ft, $h = 16.5$ in., $A_s = 9$ No. 7 bars in each direction
13.9. 12.75×9.5 ft, $h = 2$ ft, $A_{s1} = 14$ No. 9 bars, $A_{s2} = 13$ No. 8 bars

Chapter 14

14.1. $q_1 = 2.19$ ksf, $q_2 = 0.19$ ksf   **14.3.** $q_1 = 2.1$ ksf, $q_2 = 1.2$ ksf
The wall is adequate. A_s (wall) $= 0.84$ in.^{2}/ft

14.5. Sliding factor $= 1.36 < 1.5$. Increase the heel by one foot, so the base $= 12$ ft.
$q_1 = 2.75$ ksf and $q_2 = 0.95$ ksf. A_s (wall) $= 1.40$ in.^{2}/ft

14.7. $q_1 = 1.77$ ksf and $q_2 = 0.89$ ksf
A_s (wall) $= 0.56$ in.^{2}/ft

Chapter 15

15.1. For stirrups use No. 4 bars spaced at 6 in., $A_s = 3$ No. 9 bars, $A_s' = 2$ No. 5 bars, middle $= 2$ No. 5 bars.

15.3. For stirrups use No. 4 bars spaced at 5 in., $A_s = $ (3 No. 10 and 2 No. 9 bars), $A_s' = 2$ No. 8 bars, middle $= 2$ No. 6 bars.

15.5. For stirrups use No. 4 bars spaced at 7 in., $A_s = 5.61$ in.^{2}, $A_s' = 0.53$ in.^{2}, middle $= 0.53$ in.^{2}

15.7. For stirrups use No. 3 bars spaced at 6 in., $A_s = 1.97$ in.^{2}, $A_s' = 0.29$ in.^{2}, middle $= 0.29$ in.^{2}

Chapter 16

16.1. $h = 9.5$ in., $A_s = 2.12$ in.^{2}, 3.77 in.^{2}, and 3.34 in.^{2} at A, B, and C, respectively

16.3. At AB: $M_u = 416.2$ k·ft, $A_s = 5.06$ in.^{2}
At BC: $M_u = 249.1$ k·ft, $A_s = 2.83$ in.^{2}
At B: $M_u = -619.5$ k·ft, $A_s = 6.44$ in.^{2}, $A_s' = 1.34$ in.^{2}

16.5. Use 16×34 in. beam
At B and C: $M_u = -1006.6$ k·ft, $A_s = 8.31$ in.^{2}
At BC: $M_u = +1087.5$ k·ft, $A_s = 9.13$ in.^{2}
Use 16×30 in. column. $A_s = 10.46$ in.^{2}, $A_s' = 5.21$ in.^{2}

16.7. $M_p = 494.2$ k·ft   **16.8.** $M_p = 300$ k·ft   **16.9.** $h = 19$ in., A_s (use 3 No. 9 bars)

Chapter 17

17.1. $h = 9.5$ in., $M_o = 360$ k·ft   **17.3.** $h = 8$ in., $M_o = 360$ k·ft
17.5. $h = 6$ in., $M_o = 306$ k·ft

Chapter 18

18.1. At transfer: $\sigma_t = -21$ psi, $\sigma_b = -2016$ psi
After all losses: $\sigma_t = -124$ psi, $\sigma_b = -1514$ psi, $W_L = 1.35$ k/ft

18.3. $F_i = 166.5$ ksi   **18.5.** F_i(max) $= 1616.6$ kips
$F = 127$ ksi F_i(min) $= 349.5$ kips
$\eta = 0.763$

18.8. At transfer: $\Delta = 0.665$ in., upward   **18.10.** $M_n = 1950.8$ k·ft
At service load: $\Delta = 0.553$ in., downward

INDEX

Heterick Memorial Library
Ohio Northern University

DUE	RETURNED	DUE	RETURNED
1.		13.	
2.		14.	
3.		15.	
4.		16.	
5.		17.	
6.		18.	
7.		19.	
8.		20.	
9.		21.	
10.		22.	
11.		23.	
12.		24.	

CONVERSION

To convert		to	multiply by
1. LENGTH	inch	mm	25.4
	foot	mm	304.8
	yard	meter	0.9144
	meter	foot	3.281
	meter	inch	39.37
2. AREA	square inch	square mm	645
	square foot	square meter	0.0929
	square yard	square meter	0.836
	square meter	square foot	10.76
3. VOLUME	cubic inch	cubic mm	16390
	cubic foot	cubic meter	0.02832
	cubic yard	cubic meter	0.765
	cubic foot	liter	28.3
	cubic meter	cubic foot	35.31
	cubic meter	cubic yard	1.308
4. MASS	ounce (mass)	gram	28.35
	pound (lb) (mass)	kg	0.454
	pound of water	gallon	0.12
	short ton (2000 lb)	kg	907
	long ton (2240 lb)	kg	1016
	kg	pound (lb)	2.205
	slug	kg	14.59
5. DENSITY	pounds/ft^3	kg/m^3	16.02
	kg/m^3	pounds/ft^3	0.06243
6. FORCE	pound (lb)	newton (N)	4.448
	kip (1000 lb)	kilo newton (kN)	4.448
	kip (1000 lb)	kg force	453.6
	kg force	newton (N)	9.8
	kg force	pound	2.205